SOME PHYSICAL PROPERTIES

Air (dry, at 20° C and 1 atm)
    Density   1.29 kg/m^3
    Specific heat at constant pressure 1.00×10^3 J/kg·K
        0.240 cal/gm·K

    Ratio of specific heats (γ)   1.40
    Speed of sound   331 m/s
        1090 ft/s

Water (20° C and 1 atm)
    Density 1.00×10^3 kg/m^3
        1.00 gm/cm^3

    Speed of sound   1460 m/s
        4790 ft/s

    Index of refraction ($\lambda = 5890$Å)   1.33
    Specific heat at constant pressure   4180 J/kg·K
        1.00 cal/gm·K

    Heat of fusion (0° C) 3.33×10^5 J/kg
        79.7 cal/gm

    Heat of vaporization (100° C) 2.26×10^6 J/kg
        539 cal/gm

The Earth
    Mass 5.98×10^{24} kg
    Mean radius 6.37×10^6 m
        3960 mi

    Mean earth-sun distance 1.49×10^8 km
        9.29×10^7 mi

    Mean earth-moon distance 3.80×10^5 km
        2.39×10^5 mi

    Standard gravity   9.81 m/s^2
        32.2 ft/s^2

    Standard atmosphere 1.01×10^5 Pa
        14.7 lb/in^2
        760 mm-Hg
        29.9 in-Hg

physics

PART
TWO

BOOKS BY HALLIDAY (D.)
Introductory Nuclear Physics. Second Edition

BOOKS BY HALLIDAY (D.) AND RESNICK (R.)
Physics, Parts I and II Combined, Third Edition

Physics, Part II, Third Edition
Fundamentals of Physics

BOOKS BY RESNICK (R.) AND HALLIDAY (D.)
Physics, Part I, Third Edition

BOOKS BY R. RESNICK
Introduction to Special Relativity
Available in Paper Edition

Basic Concepts in Relativity and
Early Quantum Theory
Available in Cloth and Paper Editions

BY ROBERT EISBERG AND ROBERT RESNICK
Quantum Physics of Atoms, Molecules,
Solids, Nuclei, and Particles

physics

PART TWO
THIRD EDITION

PROFESSOR OF PHYSICS **DAVID HALLIDAY**
UNIVERSITY OF PITTSBURGH

PROFESSOR OF PHYSICS **ROBERT RESNICK**
RENSSELAER POLYTECHNIC INSTITUTE

JOHN WILEY & SONS
NEW YORK   CHICHESTER   BRISBANE   TORONTO

SUPPLEMENTAL MATERIAL
Student Study Aid
Available for student use with *Physics* as well as with *Fundamentals of Physics* is the *Student Study Guide* by Williams, Brownstein, and Gray.
ISBN 94801-2
This study guide is available from John Wiley and Sons.

Library of Congress Cataloging in Publication (Revised)
Resnick, Robert, 1923–
    Physics.

    Published in 1960 and 1962 under title: Physics for students of science and engineering.
    Includes bibliographical references and index.
    1. Physics. I. Halliday, David, joint author.
II. Title.
QC21.2.R47   1977   530   77-1295
ISBN 0-471-34529-6

10   9   8   7   6

preface to the third edition of part two

Physics is available in a single volume or in two separate parts; Part One includes mechanics, sound and heat, and Part Two includes electromagnetism, optics and quantum physics. The first edition was published in 1960 (*Physics for Students of Science and Engineering*) and the second in 1966 (*Physics*).

The text is intended for students studying calculus concurrently, such as students of science and engineering. The emphasis is on building a strong foundation in the principles of classical physics and on solving problems. Attention is given, however, to practical application, to the most modern theories, and to historical and philosophic issues throughout the book. This is accomplished by inclusion of special sections and thought questions, and by the entire manner of presentation of the material. There is a large set of worked-out examples, interspersed throughout the book, and an extensive collection of problems at the end of each chapter. Much care has been given to pedagogic devices that have proved effective for learning.

It has been eleven years since the publication of the second edition of *Physics*. During that time the book has continued to be well received throughout the world. We have had abundant correspondence with users over those years and concluded that a new edition is now appropriate.

In accordance with the increasing use of metric units in the United States and their general use throughout the world, we have greatly increased the emphasis on the metric system, using the Système Internationale (SI) units and nomenclature throughout. Where it seems to be sensible, in this transition period for the United States, we retain some features of the British Engineering system.

The entire book was carefully reviewed for pedagogic improvement, based chiefly on the experiences of users —, students and teachers, — and on the most recent scientific literature. As a result, we have rewritten selected areas significantly for improvements in presentation, accuracy, or physics. We have included new worked-out examples for topics or areas needing them. We have modernized all references, added new ones, and have improved many figures for greater clarity. The tables and the appendices have been expanded and updated to give newer data and more information than before. And we have added a supplementary topic on special relativity, in which the applications of this theory, scattered throughout Parts One and Two, are brought together as a cohesive whole.

Some subjects, not included in the second edition, are treated significantly in this third edition of Part Two. These include semiconductors, mutual inductance, earth magnetism, radio astronomy, virtual objects, and optical instruments. The long chapter on electromagnetic oscillations of the second edition has been divided into two chapters here, with extensive rewriting for greater clarity, and an entirely new chapter on alternating currents, for which there has been much demand, has been added.

As in Part One, we have made major improvements in the questions and the problems. In Parts One and Two combined, the number of questions has increased by 57%, from 778 in the second edition to 1219 in the present edition. For the problems the increase is 29%, from 1441 to 1864. Both problems and questions have been carefully edited, and most of the new ones have been classroom tested.

To assist students and teachers in organizing and evaluating the large number of problems, we have done several things. First, we have grouped problems within each chapter by section number, namely the first section needed to be covered in order to be able to work out the problem. Then, within each set of section problems, we have arranged the problems in the approximate order of increasing difficulty. Naturally, neither the assignment by section nor by difficulty is absolute, given different ways of solving some problems and different pedagogic values and tastes. We have coded the illustrations to the problems and have put the answers to the odd-numbered problems right at the end of these problems rather than at the end of the book. Finally, we have blended the supplementary problems, which appeared at the end of Part Two in the second edition, with the problems at the end of each chapter.

We have restyled the physical layout of the book to give it a less crowded appearance than formerly, making it easier now for the student to read the material, to make notations, and to differentiate between the various components of each chapter (text, figures, examples, tables, quotes, references, questions, problems, and so forth). We continue the practice of using somewhat reduced print for material which, in the context of a chapter, is of an advanced, specialized, or historical character.

We are grateful to John Wiley and Sons and to Donald Deneck, physics editor, for outstanding cooperation. We acknowledge the valuable assistance of Dr. Edward Derringh with the problem sets and of Mrs. Carolyn Clemente with the wide range of secretarial services required. We are indebted to the many teachers and students who have sent us constructive criticisms of the 1966 edition and particularly to Robert P. Bauman, Kenneth Brownstein, Robert Karplus, and Brian A. McInnes, who have advised or assisted us in many ways. We hope that this third edition of *Physics* will contribute to the improvement of physics education.

Hanover, New Hampshire 03755   DAVID HALLIDAY
3 Clement Road

January 1978   ROBERT RESNICK
Troy, New York 12181   Department of Physics
Rensselaer Polytechnic Institute

contents

physics
part two

26
charge and matter

The science of electricity has its roots in the observation, known to Thales of Miletus in 600 B.C., that a rubbed piece of amber will attract bits of straw. The study of magnetism goes back to the observation that naturally occurring "stones" (that is, magnetite) will attract iron. These two sciences developed quite separately until 1820, when Hans Christian Oersted (1777–1851) observed a connection between them, namely, that an *electric* current in a wire can affect a *magnetic* compass needle (Section 33-1).

The new science of electromagnetism was developed further by many workers, of whom one of the most important was Michael Faraday (1791–1867). It fell to James Clerk Maxwell (1831–1879) to put the laws of electromagnetism in essentially the form in which we know them today. These laws, called *Maxwell's equations*, are displayed in Table 40-2, which you may want to examine at this time. These laws play the same role in electromagnetism that Newton's laws of motion and of gravitation do in mechanics.

Although Maxwell's synthesis of electromagnetism rests heavily on the work of his predecessors, his own contribution was central and vital. Maxwell deduced that light is electromagnetic in nature and that its speed can be found by making purely electric and magnetic measurements. Thus the science of optics was intimately connected with those of electricity and of magnetism. The scope of Maxwell's equations is remarkable, including as it does the fundamental principles of all large-scale electromagnetic and optical devices such as motors, radio, television, microwave radar, microscopes, and telescopes.

The development of classical electromagnetism did not end with Maxwell. The English physicist Oliver Heaviside (1850–1925) and es-

26-1
ELECTROMAGNETISM— A PREVIEW

pecially the Dutch physicist H. A. Lorentz (1853–1928) contributed substantially to the clarification of Maxwell's theory. Heinrich Hertz (1857–1894)* took a great step forward when, more than twenty years after Maxwell set up his theory, he produced in the laboratory electromagnetic "Maxwellian waves" of a kind that we would now call short radio waves. It remained for Marconi and others to exploit the practical application of the electromagnetic waves of Maxwell and Hertz.

Present interest in electromagnetism takes two forms. At the level of engineering applications Maxwell's equations are used constantly and universally in the solution of a wide variety of practical problems. At the level of the foundations of the theory there is a continuing effort to extend its scope in such a way that electromagnetism is revealed as a special case of a more general theory. Such a theory would also include (say) the theories of gravitation and of quantum physics. This grand synthesis has not yet been achieved.

The rest of this chapter deals with electric charge and its relationship to matter. We can show that there are *two kinds* of charge by rubbing a glass rod with silk and hanging it from a long thread as in Fig. 26-1. If a second rod is rubbed with silk and held near the rubbed end of the first rod, the rods will repel each other. On the other hand, a rod of plastic (Lucite, say) rubbed with fur will *attract* the glass rod. Two plastic rods rubbed with fur will repel each other. We explain these facts by saying that rubbing a rod gives it an *electric charge* and that the charges on the two rods exert forces on each other. Clearly the charges on the glass and on the plastic must be different in nature.

Benjamin Franklin (1706–1790), who, among his other achievements, was the first American physicist‡, named the kind of electricity that appears on the glass *positive* and the kind that appears on the plastic (sealing wax or shell-lac in Franklin's day) *negative*; these names have remained to this day. We can sum up these experiments by saying that *like charges repel and unlike charges attract.*

Electric effects are not limited to glass rubbed with silk or to plastic rubbed with fur. Any substance rubbed with any other under suitable conditions will become charged to some extent; by comparing the unknown charge with a glass rod which had been rubbed with silk or a plastic rod which had been rubbed with fur, it can be labeled as either positive or negative.

The modern view of bulk matter is that, in its normal or neutral state, it contains equal amounts of positive and negative electricity. If two bodies like glass and silk are rubbed together, a small amount of charge is transferred from one to the other, upsetting the electric neutrality of each. In this case the glass would become positive, the silk negative.

26-2
ELECTRIC CHARGE†

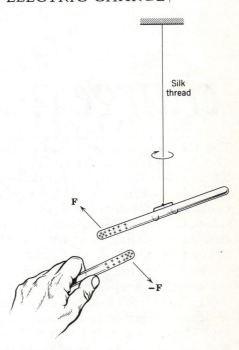

figure 26-1
Two positively charged glass rods repel each other.

---

* "Heinrich Hertz," by P. and E. Morrison, *Scientific American*, December 1957.

† To learn about practical applications of static electric charges, as in fly-ash precipitators, paint sprayers, electrostatic copying machines, etc., see "Modern Electrostatics" by A. W. Bright, *Physics Education*, **9**, 381 [1974], and "Electrostatics" by A. D. Moore, *Scientific American*, March 1972.

‡ The science historian I. Bernard Cohen of Harvard University says of Franklin in his book *Franklin and Newton*: "To say . . . that had Franklin 'Not been famous as a publisher and a statesman, he might never have been heard of as a scientist,' is absolutely wrong. Just the opposite is more nearly the case; his international fame and public renown as a scientist was in no small measure responsible for his success in international statesmanship." See also "The Lightning Discharge" by Richard E. Orville, *The Physics Teacher*, January 1976 for a description of Franklin's famous kite experiment and a review of modern concepts about the nature of lightning.

A metal rod held in the hand and rubbed with fur will not seem to develop a charge. It is possible to charge such a rod, however, if it is furnished with a glass or plastic handle and if the metal is not touched with the hands while rubbing it. The explanation is that metals, the human body, and the earth are *conductors* of electricity and that glass, plastics, etc., are *insulators* (also called *dielectrics*).

In conductors electric charges are free to move through the material, whereas in insulators they are not. Although there are no perfect insulators, the insulating ability of fused quartz is about 10^{25} times as great as that of copper, so that for many practical purposes some materials behave as if they were perfect insulators.

In metals a fairly subtle experiment called the Hall effect (see Section 33-5) shows that only negative charge is free to move. Positive charge is as immobile as it is in glass or in any other dielectric. The actual charge carriers in metals are the *free electrons*. When isolated atoms are combined to form a metallic solid, the outer electrons of the atom do not remain attached to individual atoms but become free to move throughout the volume of the solid. For some conductors, such as electrolytes, both positive and negative charges can move.

A class of materials called *semiconductors* is intermediate between conductors and insulators in its ability to conduct electricity. Among the elements, silicon and germanium are well-known examples. Semiconductors have many practical applications, among which is their use in the construction of transistors. The way a semiconductor works cannot be described adequately without some understanding of the basic principles of quantum physics. Figure 26-2, however, suggests the principal features of the distinction between conductors, semiconductors, and insulators.

In solids, electrons have energies that are restricted to certain levels, the levels being confined to certain bands. The intervals between bands are forbidden, in the sense that electrons in the solid may not possess such energies. Electrons are assigned two to a level and they may not increase their energy (which means that they may not move freely through the solid) unless there are empty levels at higher energies into which they can readily move.

Figure 26-2a shows a conductor, such as copper. Band 1 is only partially filled so that electrons can easily move to the higher empty levels and thus travel through the solid. Figure 26-2b shows a (intrinsic) semiconductor such as silicon. Here band 1 is completely filled but band 2 is so close energetically that electrons can easily "jump" (absorbing energy from, say, thermal fluctuations) into the unfilled levels of that band. Figure 22-6c shows an insulator, such as sodium chloride. Here again band 1 is filled, but band 2 is too far above band 1 energetically to permit any appreciable number of the band-1 electrons to jump the energy gap.

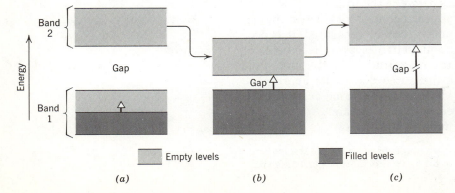

(a)   (b)   (c)

figure 26-2
Suggesting (a) a conductor, (b) an intrinsic semiconductor, and (c) an insulator. In (b) the gap is relatively small but in (c) it is relatively large. In intrinsic semiconductors the electrical conductivity can often be greatly increased by adding very small amounts of other elements such as arsenic or boron, a process called "doping."

Charles Augustin Coulomb (1736–1806) measured electrical attractions and repulsions quantitatively and deduced the law that governs them. His apparatus, shown in Fig. 26-3, resembles the hanging rod of Fig. 26-1, except that the charges in Fig. 26-3 are confined to small spheres a and b.

If a and b are charged, the electric force on a will tend to twist the suspension fiber. Coulomb canceled out this twisting effect by turning the suspension head through the angle θ needed to keep the two charges at the particular distance apart in which he was interested. The angle θ is then a relative measure of the electric force acting on charge a. The device of Fig. 26-3 is called a *torsion balance*; a similar arrangement was used later by Cavendish to measure gravitational attractions (Section 16-3).

Coulomb's first experimental results can be represented by

$$F \propto \frac{1}{r^2}.$$

Here F is the magnitude of the interaction force that acts on each of the two charges a and b; r is their distance apart. These forces, as Newton's third law requires, act along the line joining the charges but point in opposite directions. Note that the magnitude of the force on each charge is the same, even though the charges may be different.

The force between charges depends also on the magnitude of the charges. Specifically, it is proportional to their product. Although Coulomb did not prove this rigorously, he implied it and thus we arrive at

$$F \propto \frac{q_1 q_2}{r^2}, \tag{26-1}$$

where q_1 and q_2 are relative measures of the charges on spheres a and b. Equation 26-1, which is called *Coulomb's law*, holds only for charged objects whose sizes are much smaller than the distance between them. We often say that it holds only for *point charges*.

Coulomb's law resembles the inverse square law of gravitation which was already more than 100 years old at the time of Coulomb's experiments; q plays the role of m in that law. In gravity, however, the forces are always attractive; this corresponds to the fact that there are two kinds of electricity but (apparently) only one kind of mass.

Our belief in Coulomb's law does not rest quantitatively on Coulomb's experiments. Torsion balance measurements are difficult to make to an accuracy of better than a few percent. Such measurements could not, for example, convince us that the exponent in Eq. 26-1 is exactly 2 and not, say, 2.01. In Section 28-7 we show that Coulomb's law can also be deduced from an indirect experiment (1971) which shows that the exponent in Eq. 26-1 lies between the limits $2 \pm 3 \times 10^{-16}$.

Although we have established the physical concept of electric charge, we have not yet defined a unit in which it may be measured. It is possible to do so operationally by putting equal charges q on the spheres of a torsion balance and by measuring the magnitude F of the force that acts on each when the charges are a measured distance r apart. One could then define q to have a unit value if a unit force acts on each charge when the charges are separated by a unit distance and one can give a name to the unit of charge so defined.*

---

* This scheme is the basis for the definition of the unit of charge called the *statcoulomb*. However, in this book we do not use this unit or the systems of units of which it is a part; see Appendix L, however.

26-4
COULOMB'S LAW

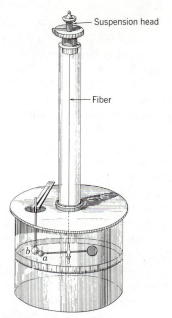

figure 26-3
Coulomb's torsion balance, from his 1785 memoir to the French Academy of Sciences.

For practical reasons having to do with the accuracy of measurements, the SI unit of charge is not defined using a torsion balance but is derived from the unit of electric current. If the ends of a long wire are connected to the terminals of a battery, it is common knowledge that an *electric current i* is set up in the wire. We visualize this current as a flow of charge. The SI unit of current is the *ampere* (abbr. A). In Section 34-4 we describe the operational procedures in terms of which the ampere is defined.

The SI unit of charge is the *coulomb* (abbr. C). *A coulomb is defined as the amount of charge that flows through any cross section of a wire in 1 second if there is a steady current of 1 ampere in the wire.* In symbols

$$q = it, \qquad (26\text{-}2)$$

where q is in coulombs if i is in amperes and t is in seconds. Thus, if a wire is connected to an insulated metal sphere, a charge of 10^{-6} C can be put on the sphere if a current of 1.0 A exists in the wire for 10^{-6} s.

EXAMPLE 1

A copper penny has a mass of 3.1 g. Being electrically neutral, it contains equal amounts of positive and negative electricity. What is the magnitude q of these charges? A copper atom has a positive nuclear charge of 4.6×10^{-18} C and a negative electronic charge of equal magnitude.

The number N of copper atoms in a penny is found from the ratio

$$\frac{N}{N_0} = \frac{m}{M},$$

where N_0 is the Avogadro number, m the mass of the coin, and M the atomic weight of copper. This yields, solving for N,

$$N = \frac{(6.0 \times 10^{23} \text{ atoms/mole})(3.1 \text{ g})}{64 \text{ g/mole}} = 2.9 \times 10^{22} \text{ atoms.}$$

The charge q is

$$q = (4.6 \times 10^{-18} \text{ C/atom})(2.9 \times 10^{22} \text{ atoms}) = 1.3 \times 10^{5} \text{ C.}$$

In a 100-watt, 110-volt light bulb the current is 0.91 ampere. Verify that it would take 40 h for a charge of this amount to pass through this bulb.

Equation 26-1 can be written as an equality by inserting a constant of proportionality. Instead of writing this simply as, say, k, it is usually written in a more complex way as $1/4\pi\epsilon_0$ or

$$F = \frac{1}{4\pi\epsilon_0} \frac{q_1 q_2}{r^2}. \qquad (26\text{-}3)$$

Certain equations that are derived from Eq. 26-3, but are used more often than it is, will be simpler in form if we do this.

In SI units we can measure q_1, q_2, r, and F in Eq. 26-3 in ways that do not depend on Coulomb's law. Numbers with units can be assigned to them. There is no choice about the so-called *permittivity constant* ϵ_0; it must have that value which makes the right-hand side of Eq. 26-3 equal to the left-hand side. This (measured) value turns out to be*

$$\epsilon_0 = 8.854187818 \times 10^{-12} \text{ C}^2/\text{N} \cdot \text{m}^2.$$

---

*For practical reasons this value is not actually measured by direct application of Eq. 26-3 but in an equivalent although more circuitous way.

In this book the value 8.9×10^{-12} C²/N · m² will be accurate enough for most problems. For direct application of Coulomb's law or in any problem in which the quantity $1/4\pi\epsilon_0$ occurs we may use, with sufficient accuracy for this book,

$$1/4\pi\epsilon_0 = 9.0 \times 10^9 \text{ N} \cdot \text{m}^2/\text{C}^2.$$

EXAMPLE 2

Let the total positive and the total negative charges in a copper penny be separated to a distance such that their force of attraction is 1.0 lb (= 4.5 N). How far apart must they be?

We have (Eq. 26-3)

$$F = \frac{1}{4\pi\epsilon_0} \frac{q_1 q_2}{r^2}.$$

Putting $q_1 q_2 = q^2$ (see Example 1) and solving for r yields

$$r = q \sqrt{\frac{1/4\pi\epsilon_0}{F}} = 1.3 \times 10^5 \text{ C} \sqrt{\frac{9.0 \times 10^9 \text{ N} \cdot \text{m}^2/\text{C}^2}{4.5 \text{ N}}}$$

$$= 5.8 \times 10^9 \text{ m} = 3.6 \times 10^6 \text{ mi}.$$

This is 910 earth radii and it suggests that it is not possible to upset the electrical neutrality of gross objects by any very large amount. What would be the force between the two charges if they were placed 1.0 m apart?

If more than two charges are present, Eq. 26-3 holds for every pair of charges. Let the charges be q_1, q_2, q_3, etc.; we calculate the force exerted on any one (say q_1) by all the others from the vector equation

$$\mathbf{F}_1 = \mathbf{F}_{12} + \mathbf{F}_{13} + \mathbf{F}_{14} + \cdots, \qquad (26\text{-}4)$$

where $\mathbf{F}_{12}$, for example, is the force exerted on q_1 by q_2.

EXAMPLE 3

Figure 26-4 shows three fixed charges q_1, q_2, and q_3. What force acts on q_1? Assume that $q_1 = -1.0 \times 10^{-6}$ C, $q_2 = +3.0 \times 10^{-6}$ C, $q_3 = -2.0 \times 10^{-6}$ C, $r_{12} = 15$ cm, $r_{13} = 10$ cm, and $\theta = 30°$.

From Eq. 26-3, ignoring the signs of the charges, since we are interested only in the magnitudes of the forces,

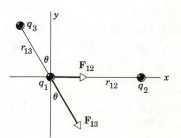

$$F_{12} = \frac{1}{4\pi\epsilon_0} \frac{q_1 q_2}{r_{12}^2}$$

$$= \frac{(9.0 \times 10^9 \text{ N} \cdot \text{m}^2/\text{C}^2)(1.0 \times 10^{-6} \text{ C})(3.0 \times 10^{-6} \text{ C})}{(1.5 \times 10^{-1} \text{ m})^2}$$

$$= 1.2 \text{ N}$$

and

$$F_{13} = \frac{(9.0 \times 10^9 \text{ N} \cdot \text{m}^2/\text{C}^2)(1.0 \times 10^{-6} \text{ C})(2.0 \times 10^{-6} \text{ C})}{(1.0 \times 10^{-1} \text{ m})^2}$$

$$= 1.8 \text{ N}.$$

figure 26-4
Example 3. Showing the forces exerted on q_1 by q_2 and q_3.

The directions of $\mathbf{F}_{12}$ and $\mathbf{F}_{13}$ are as shown in the figure.

The components of the resultant force $\mathbf{F}_1$ acting on q_1 (see Eq. 26-4) are

$$F_{1x} = F_{12x} + F_{13x} = F_{12} + F_{13} \sin \theta$$

$$= 1.2 \text{ N} + (1.8 \text{ N})(\sin 30°) = 2.1 \text{ N}$$

and

$$F_{1y} = F_{12y} + F_{13y} = 0 - F_{13} \cos \theta$$

$$= -(1.8 \text{ N})(\cos 30°) = -1.6 \text{ N}.$$

Find the magnitude of $\mathbf{F}_1$ and the angle it makes with the x-axis.

In Franklin's day electric charge was thought of as a continuous fluid, an idea that was useful for many purposes. The atomic theory of matter, however, has shown that fluids themselves, such as water and air, are not continuous but are made up of atoms. Experiment shows that the "electric fluid" is not continuous either but that it is made up of integral multiples of a certain minimum electric charge. This fundamental charge, to which we give the symbol e, has the magnitude $1.6021892 \times 10^{-19}$ C. Any physically existing charge q, no matter what its origin, can be written as ne where n is a positive or a negative integer.

When a physical property such as charge exists in discrete "packets" rather than in continuous amounts, the property is said to be *quantized*. Quantization is basic to modern quantum physics. The existence of atoms and of particles such as the electron and the proton indicates that *mass* is quantized also. Later you will learn that several other properties prove to be quantized when suitably examined on the atomic scale; among them are energy and angular momentum.

The quantum of charge e is so small that the "graininess" of electricity does not show up in large-scale experiments, just as we do not realize that the air we breathe is made up of atoms. In an ordinary 110-volt, 100-watt light bulb, for example, 6×10^{18} elementary charges enter and leave the bulb every second.

There exists today no theory that predicts the quantization of charge (or the quantization of mass, that is, the existence of fundamental particles such as protons, electrons, muons, and so ons). Even assuming quantization, the classical theory of electromagnetism and Newtonian mechanics are incomplete in that they do not correctly describe the behavior of charge and matter on the atomic scale. The classical theory of electromagnetism, for example, describes correctly what happens when a bar magnet is thrust through a closed copper loop; it fails, however, if we wish to explain the magnetism of the bar in terms of the atoms that make it up. The more detailed theories of quantum physics are needed for this and similar problems.

26-5
CHARGE IS QUANTIZED

26-6
CHARGE AND MATTER

Matter as we ordinarily experience it can be regarded as composed of three kinds of particles, the proton, the neutron, and the electron. Table 26-1 shows their masses and charges. Note that the masses of the neutron and the proton are approximately equal but that the electron is less massive by a factor of about 1840.

Table 26-1
Some properties of three particles

Particle	Symbol	Charge	Mass
Proton	p	$+e$	$1.6726485 \times 10^{-27}$ kg
Neutron	n	0	$1.6749543 \times 10^{-27}$ kg
Electron	e^-	$-e$	9.109534×10^{-31} kg

Atoms are made up of a dense, positively charged *nucleus*, surrounded by a cloud of electrons; see Fig. 26-5. The nucleus varies in radius from about 1×10^{-15} m for hydrogen to about 7×10^{-15} m for the heaviest atoms. The outer diameter of the electron cloud, that is, the diameter of the atom, lies in the range 1–3×10^{-10} m, about 10^5 times larger than the nuclear diameter.

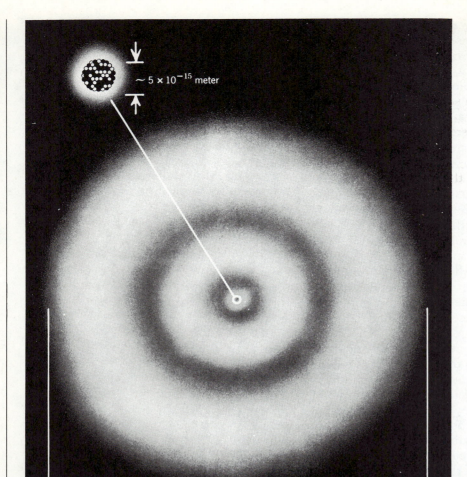

~ 5×10^{-15} meter

~ 2×10^{-10} meter

figure 26-5
An atom, suggesting the electron cloud and, above, an enlarged view of the nucleus.

EXAMPLE 4

The distance r between the electron and the proton in the hydrogen atom is about 5.3×10^{-11} m. What are the magnitudes of (a) the electrical force and (b) the gravitational force between these two particles?

From Coulomb's law,

$$F_e = \frac{1}{4\pi\epsilon_0} \frac{q_1 q_2}{r^2}$$

$$= \frac{(9.0 \times 10^9 \text{ N} \cdot \text{m}^2/\text{C}^2)(1.6 \times 10^{-19} \text{ C})^2}{(5.3 \times 10^{-11} \text{ m})^2}$$

$$= 8.1 \times 10^{-8} \text{ N}.$$

The gravitational force is given by Eq. 16-1, or

$$F_g = G \frac{m_1 m_2}{r^2}$$

$$= \frac{(6.7 \times 10^{-11} \text{ N} \cdot \text{m}^2/\text{kg}^2)(9.1 \times 10^{-31} \text{ kg})(1.7 \times 10^{-27} \text{ kg})}{(5.3 \times 10^{-11} \text{ m})^2}$$

$$= 3.7 \times 10^{-47} \text{ N}.$$

Thus the electrical force is about 10^{39} times stronger than the gravitational force.

The significance of Coulomb's law goes far beyond the description of the forces acting between charged spheres. This law, when incorporated into the structure of quantum physics, correctly describes (a) the electric forces that bind the electrons of an atom to its nucleus, (b) the forces that bind atoms together to form molecules, and (c) the forces that bind atoms or molecules together to form solids or liquids. Thus most of the forces of our daily experience that are not gravitational in nature are electrical. A force transmitted by a steel cable is basically an electrical force because, if we pass an imaginary plane through the cable at right angles to it, it is only the attractive electrical interatomic forces acting between atoms on opposite sides of the plane that keep the cable from parting. We ourselves are an assembly of nuclei and electrons bound together in a stable configuration by Coulomb forces.

In the atomic *nucleus* we encounter a new force which is neither gravitational nor electrical in nature. This strong attractive force, which binds together the protons and neutrons that make up the nucleus, is called simply *the nuclear force* or *the strong interaction.* If this force were not present, the nucleus would fly apart at once because of the strong Coulomb repulsion force that acts between its protons. The nature of the nuclear force is only partially understood today and forms the central problem of present day researches in nuclear physics.

EXAMPLE 5

What repulsive Coulomb force exists between two protons in a nucleus of iron? Assume a separation of 4.0×10^{-15} m.

From Coulomb's law,

$$F = \frac{1}{4\pi\epsilon_0} \frac{q_1 q_2}{r^2}$$

$$= \frac{(9.0 \times 10^9 \text{ N} \cdot \text{m}^2/\text{C}^2)(1.6 \times 10^{-19} \text{ C})^2}{(4.0 \times 10^{-15} \text{ m})^2}$$

$$= 14 \text{ N}.$$

This enormous repulsive force (3.2 lb acting on a single proton) must be more than compensated for by the strong attractive nuclear forces. This example, combined with Example 4, shows that nuclear binding forces are much stronger than atomic binding forces. Atomic binding forces are, in turn, much stronger than gravitational forces for the same particles separated by the same distance.

The repulsive Coulomb forces acting between the protons in a nucleus make the nucleus less stable than it otherwise would be. The spontaneous emission of alpha particles from heavy nuclei and the phenomenon of nuclear fission are evidences of this instability.

The fact that heavy nuclei contain significantly more neutrons than protons is still another effect of the Coulomb forces. Consider Fig. 26-6 in which a particular atomic species is represented by a circle, the coordinates being Z, the number of protons in the nucleus (that is, the *atomic number*), and N, the number of neutrons in the nucleus (that is, the *neutron number*). Stable nuclei are represented by filled circles and radioactive nuclei, that is, nuclei which disintegrate spontaneously, emitting electrons or α-particles, by open circles. Note that all elements (xenon, for example, for which $Z = 54$; see arrow) exist in a number of different forms, called *isotopes*.

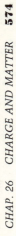

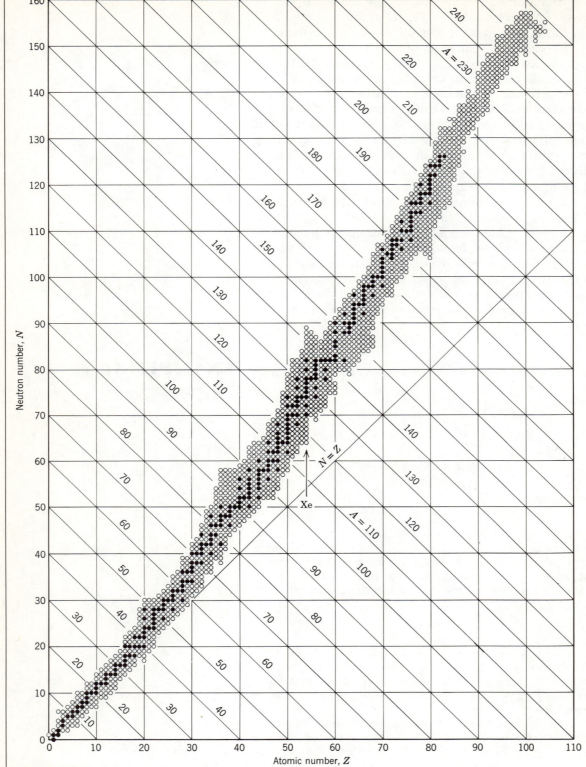

figure 26-6
The filled circles represent stable nuclei and the open ones represent radioactive nuclei. Note, for example, that xenon (Z = 54) has 26 isotopes, 9 of them stable and 17 radioactive. Each xenon isotope has 54 protons (and 54 extranuclear electrons for neutral atoms). The number of neutrons ranges from N = 64 to N = 89 and the mass number A (= N + Z) ranges from 118 to 143. No other element has as many isotopes.

Figure 26-6 shows that light nuclei, for which the Coulomb forces are relatively unimportant,* lie on or close to the line labeled "$N = Z$" and thus have about equal numbers of neutrons and protons. The heavier nuclei have a pronounced neutron excess, U^{238} having 92 protons and $238 - 92$ or 146 neutrons.† In the absence of Coulomb forces we would assume, extending the $N = Z$ rule, that the most stable nucleus with 238 particles would have 119 protons and 119 neutrons. However, such a nucleus, if assembled, would fly apart at once because of Coulomb repulsion. Relative stability is found only if 27 of the protons are replaced by neutrons, thus diluting the total Coulomb repulsion. Even in U^{238} Coulomb repulsion is still very important because (a) this nucleus is radioactive and emits α-particles and (b) it may break up into two large fragments (fission) if it absorbs a neutron; both processes result in separation of the nuclear charge and are Coulomb repulsion effects. Figure 26-6 shows that all nuclei with $Z \gtrsim 83$ are unstable.

We have pointed out that matter, as we ordinarily experience it, is made up of electrons, neutrons, and protons. Nature exhibits much more variety than this, however. There are very many more particles than these. Appendix F, which lists some properties of some of these particles, shows that, like the more familiar particles of Table 26-1, their charges are quantized, the quantum of charge again being e. An understanding of the nature of these particles and of their relationships to each other is one of the most significant research goals of modern physics.

26-7
CHARGE IS CONSERVED

When a glass rod is rubbed with silk, a positive charge appears on the rod. Measurement shows that a negative charge of equal magnitude appears on the silk. This suggests that rubbing does not create charge but merely transfers it from one object to another, disturbing slightly the electrical neutrality of each. This hypothesis of the *conservation of charge* has stood up under close experimental scrutiny both for large-scale events and at the atomic and nuclear level; no exceptions have ever been found.

An interesting example of charge conservation comes about when an electron (charge $= -e$) and a positron (charge $= +e$) are brought close to each other. The two particles may simply disappear, converting all their rest mass into energy according to the well-known $E = mc^2$ relationship; this annihilation process was described in Section 8-9. The energy appears in the form of two oppositely directed gamma rays, which are similar in character to X-rays; thus:

$$e^- + e^+ \rightarrow \gamma + \gamma. \qquad (26\text{-}5)$$

The net charge is zero both before and after the event so that charge is conserved. Rest mass is *not* conserved, being turned completely into energy.

Another example of charge conservation is found in radioactive decay, of which the following process is typical:

$$U^{238} \rightarrow Th^{234} + He^4. \qquad (26\text{-}6)$$

The radioactive "parent" nucleus, U^{238}, contains 92 protons (that is, its

---

* Coulomb forces are important in relation to the strong nuclear attractive forces only for large nuclei, because Coulomb repulsion occurs between *every pair* of protons in the nucleus but the attractive nuclear force does not. In U^{238}, for example, every proton exerts a force of repulsion on each of the other 91 protons. However, each proton (and neutron) exerts a nuclear attraction on only a small number of other neutrons and protons that happen to be near it. As we proceed to larger nuclei, the amount of energy associated with the repulsive Coulomb forces increases much faster than that associated with the attractive nuclear forces.

† The superscript in this notation is the *mass number A* $(= N + Z)$. This is the total number of particles in the nucleus. See the sloping lines in Figure 26-6.

atomic number $Z = 92$). It disintegrates spontaneously by emitting an α-particle (He⁴; $Z = 2$) transmuting itself into the nucleus Th²³⁴, with $Z = 90$. Thus the amount of charge present before disintegration $(+92e)$ is the same as that present after the disintegration.

An additional example of charge conservation is found in nuclear reactions, of which the bombardment of Ca⁴⁴ with cyclotron-accelerated protons is typical. In a particular collision a neutron may emerge from the nucleus, leaving Sc⁴⁴ as a "residual" nucleus:

$$Ca^{44} + p \rightarrow Sc^{44} + n.$$

The sum of the atomic numbers before the reaction $(20 + 1)$ is exactly equal to the sum of the atomic numbers after the reaction $(21 + 0)$. Again charge is conserved.

A final example of charge conservation is the decay of the K-meson (see Appendix F) which, in one mode, goes as

$$K^0 \rightarrow \pi^+ + \pi^-.$$

The resultant charge is zero both before and after this decay process.

questions

1. You are given two metal spheres mounted on portable insulating supports. Find a way to give them equal and opposite charges. You may use a glass rod rubbed with silk but may not touch it to the spheres. Do the spheres have to be of equal size for your method to work?

2. In Question 1, find a way to give the spheres equal charges of the same sign. Again, do the spheres need to be of equal size for your method to work?

3. A charge rod attracts bits of dry cork dust which, after touching the rod, often jump violently away from it. Explain.

4. In Section 26-2 can there not be four kinds of charge, that is, on glass, silk, plastic, and fur? What is the argument against this?

5. If you rub a coin briskly between your fingers, it will not seem to become charged by friction. Why?

6. If you walk briskly down the carpeted corridor of a hotel, you often experience a spark upon touching a door knob. (a) What causes this? (b) How might it be prevented?

7. Why do electrostatic experiments not work well on humid days?

8. An insulated rod is said to carry an electric charge. How could you verify this and determine the sign of the charge?

9. If a charged glass rod is held near one end of an insulated uncharged metal rod as in Fig. 26-7, electrons are drawn to one end, as shown. Why does the flow of electrons cease? There is an almost inexhaustible supply of them in the metal rod.

10. In Fig. 26-7 does any net electrical force act on the metal rod? Explain.

11. A person standing on an insulated stool touches a charged, insulated conductor. Is the conductor discharged completely?

12. (a) A positively charged glass rod attracts a suspended object. Can we conclude that the object is negatively charged? (b) A positively charged glass rod repels a suspended object. Can we conclude that the object is positively charged?

13. Is the Coulomb force that one charge exerts on another changed if other charges are brought nearby?

14. You are given a collection of small charged spheres, the sign and magnitude of the charge and the mass of the sphere being at your disposal. Do you think

Metal

Insulating support

Glass rod

figure 26-7
Questions 9, 10

that a *stable* equilibrium position is possible, involving only electrostatic forces? Test several arrangements. A rigorous answer is not required.

15. Suppose that someone told you that in Eq. 26-3 the product of the charges $(q_1 q_2)$ should be replaced by their algebraic sum $(q_1 + q_2)$. What experimental facts refute this statement? What if the square root of their product $\sqrt{q_1 q_2}$ were proposed?

16. The quantum of charge is 1.60×10^{-19} C. Is there a corresponding single quantum of mass?

17. A nucleus U^{238} splits into two identical parts. Are the nuclei so produced likely to be stable or radioactive?

18. In the decay mode

$$\Xi^0 \to \Lambda + \pi^0$$

what is the charge of the Λ particle? See Appendix F.

19. Verify the fact that the decay schemes for the elementary particles in Appendix F are consistent with charge conservation.

20. What does it mean to say that a physical quantity is (a) quantized or (b) conserved? Give some examples.

SECTION 26-4

problems

1. The electrostatic force between two like ions that are separated by a distance of 5.0×10^{-10} m is 3.7×10^{-9} N. (a) What is the charge on each ion? (b) How many electrons are missing from each ion?
   Answer: (a) 3.2×10^{-19} C (b) Two.

2. Two fixed charges, $+1.0 \times 10^{-6}$ C and -3.0×10^6 C, are 10 cm apart. (a) Where may a third charge be located so that no force acts on it? (b) Is the equilibrium of this third charge stable or unstable?

3. Each of two small spheres is charged positively, the combined charge being 5.0×10^{-5} C. If each sphere is repelled from the other by a force of 1.0 N when the spheres are 2.0 m apart, how is the total charge distributed between the spheres? *Answer:* 1.2×10^{-5} C and 3.8×10^{-5} C.

4. Two equal positive point charges are held a fixed distance $2a$ apart. A point test charge is located in a plane which is normal to the line joining these charges and midway between them. (a) Calculate the radius r of the circle of symmetry in this plane for which the force on the test charge has a maximum value. (b) What is the direction of this force, assuming a positive test charge.

5. A certain charge Q is to be divided into two parts, q and $Q - q$. What is the relationship of Q to q if the two parts, placed a given distance apart, are to have a maximum Coulomb repulsion? *Answer:* $q = \frac{1}{2}Q$.

6. How far apart must two protons be if the electrical repulsive force acting on either one is equal to its weight at the earth's surface. The mass of a proton is 1.7×10^{-27} kg.

7. Two equal positive charges, Q, are fixed at a distance $2a$ apart. The force on a small positive test charge q midway between them is zero. If the test charge is displaced a short distance either (a) toward one of the charges or (b) at right angles to the line joining the charges, find the direction of the force on it. Is the equilibrium stable or unstable in each case?
   Answer: (a) Toward the original position; stable (b) Away from the original position; unstable.

8. Two *free* point charges $+q$ and $+4q$ are a distance l apart. A third charge is so placed that the entire system is in equilibrium. Find the location, magnitude, and sign of the third charge. Is the equilibrium stable?

9. Two similar conducting balls of mass m are hung from silk threads of length l and carry similar charges q as in Fig. 26-8. Assume that θ is so small that $\tan\theta$ can be replaced by its approximate equal, $\sin\theta$. To this approximation (a) show that

$$x = \left(\frac{q^2 l}{2\pi\epsilon_0 mg}\right)^{1/3}$$

where x is the separation between the balls. (b) If $l = 120$ cm, $m = 10$ g, and $x = 5.0$ cm, what is q? *Answer:* (b) $\pm 2.4 \times 10^{-8}$ C.

10. Assume that each ball in Problem 9 is losing charge at the rate of 1.0×10^{-9} C/s. At what instantaneous relative speed $(= dx/dt)$ do the balls approach each other initially?

11. If the balls of Fig. 26-8 are conducting, (a) what happens to them after one is discharged? (b) Find the new equilibrium separation.
Answer: (a) They touch and repel. (b) 3.1 cm.

12. The charges and coordinates of two charged particles held fixed in the x-y plane are: $q_1 = +3.0 \times 10^{-6}$ C; $x = 3.5$ cm, $y = 0.50$ cm, and $q_2 = -4.0 \times 10^{-6}$ C; $x = -2.0$ cm, $y = 1.5$ cm. (a) Find the magnitude and direction of the force on q_2. (b) Where you locate a third charge $q_3 = +4.0 \times 10^{-6}$ C such that the total force on q_2 is zero?

13. Two identical conducting spheres, having charges of opposite sign, attract each other with a force of 0.108 N when separated by 0.500 m. The spheres are connected by a conducting wire, which is then removed, and thereafter repel each other with a force of 0.0360 N. What were the initial charges on the spheres?
Answer: $\pm 1.0 \times 10^{-6}$ C; $\mp 3.0 \times 10^{-6}$ C.

14. Two engineering students (John at 200 lb and Mary at 100 lb) are 100 ft apart. Let each have an 0.01% imbalance in their amount of positive and negative charge, one student being positive and the other negative. Estimate roughly the electrostatic force of attraction between them. (Hint: Replace the students by equivalent spheres of water.)

15. Two equally charged particles, 3.2×10^{-3} m apart, are released from rest. The acceleration of the first particle is observed to be 7.0 m/s² and that of the second to be 9.0 m/s². If the mass of the first particle is 6.3×10^{-7} kg, what are (a) the mass of the second particle and (b) the common charge?
Answer: (a) 4.9×10^{-7} kg. (b) 7.1×10^{-11} C.

16. (a) How many electrons would have to be removed from a penny to leave it with a charge of $+1.0 \times 10^{-7}$ C (b) What fraction of the electrons in the penny does this correspond to?

17. (a) What equal positive charges would have to be placed on the earth and on the moon to neutralize their gravitational attraction? (b) Do you need to know the lunar distance to solve this problem? (c) How many tons of hydrogen would be needed to provide the positive charge calculated in a?
Answer: (a) 5.7×10^{13} C. (b) No. (c) 630 tons.

18. Estimate *roughly* the number of coulombs of positive charge in a glass of water. Assume the volume of the water to be 250 cm³.

19. Protons in the cosmic rays strike the earth's upper atmosphere at a rate, averaged over the earth's surface, of 0.15 protons/cm²·s. What total current does the earth receive from beyond its atmosphere in the form of incident cosmic ray protons? The earth's radius is 6.4×10^6 m. *Answer:* 0.12 A.

20. Three charged particles lie on a straight line and are separated by a distance d as shown in Fig. 26-9. Charges q_1 and q_2 are held fixed. If q_3 is free to move but in fact remains stationary, then how are q_1 and q_2 related?

21. Three small balls, each of mass 10 g, are suspended separately from a common point by silk threads, each 1.0 m. long. The balls are identically charged and hang at the corners of an equilateral triangle 0.1 m long on a side. What is the charge on each ball? *Answer:* 6.0×10^{-8} C.

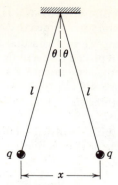

figure 26-8
Problems 9, 10, 11

figure 26-9
Problem 20

22. Three point charges of $+4.0 \times 10^{-6}$ C are fixed at the corners of an equilateral triangle of side 10 cm. What force (magnitude and direction) acts on a typical charge?

23. A charge Q is fixed at each of two opposite corners of a square. A charge q is placed at each of the other two corners. (a) If the resultant electrical force on Q is zero, how are Q and q related? (b) Could q be chosen to make the resultant force on *every* charge zero?
Answer: (a) $Q = -2\sqrt{2}\, q$. (b) No.

24. In Fig. 26-10 what is the resultant force on the charge in the lower left corner of the square? Assume that $q = 1.0 \times 10^{-7}$ C and $a = 5.0$ cm. The changes are fixed in position.

25. A cube of edge a carries a point charge q at each corner. (a) Show that the magnitude of the resultant force on any one of the charges is

$$F = \frac{0.262 q^2}{\epsilon_0 a^2}.$$

(b) What is the direction of **F** relative to the cube edges?
Answer: (b) Along a body diagonal, directed away from the cube.

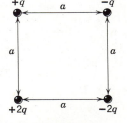

figure 26-10
Problem 24

26. Figure 26-11 shows a long insulating, massless rod of length l, pivoted at its center and balanced with a weight W at a distance x from the left end. At the left and right ends of the rod are attached positive charges q and $2q$, respectively. A distance h directly beneath each of these charges is a fixed positive charge Q. (a) Find the distance x for the position of the weight when the rod is balanced. (b) What value should h have so that the rod exerts no vertical force on the bearing when balanced? Neglect the interaction between charges at the opposite ends of the rod.

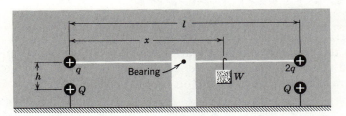

figure 26-11
Problem 26

SECTION 26-7

27. An electron is projected with an initial speed of 3.24×10^5 m/s directly toward a proton that is essentially at rest. If the electron is initially a great distance from the proton, at what distance from the proton is its speed instantaneously equal to twice its initial value? (Hint: Use the work-energy theorem.) *Answer:* 1.6×10^{-9} m.

28. In the radioactive decay of U^{238} (see Eq. 26-6) the center of the emerging α-particle is, at a certain instant, 9.0×10^{-15} m from the center of the residual nucleus Th^{234}. At this instant (a) what is the force on the α-particle and (b) what is its acceleration?

27
the electric field

With every point in space near the earth we can associate a *gravitational field* vector **g** (see Eq. 16-12). This is the gravitational acceleration that a test body, placed at that point and released, would experience. If m is the mass of the body and **F** the gravitational force acting on it, **g** is given by

$$\mathbf{g} = \mathbf{F}/m. \qquad (27\text{-}1)$$

This is an example of a *vector field*. For points near the surface of the earth the field is often taken as *uniform*; that is, **g** is the same for all points.

The flow of water in a river provides another example of a vector field, called a *flow field* (see Section 18-7). Every point in the water has associated with it a vector quantity, the velocity **v** with which the water flows past the point. If **g** and **v** do not change with time, the corresponding fields are described as *stationary*. In the case of the river note that even though the water is moving, the vector **v** at any point does not change with time for steady-flow conditions.

If we place a test charge in the space near a charged rod, an electrostatic force will act on the charge. We speak of an *electric field* in this space. In a similar way we speak of a *magnetic field* in the space around a bar magnet. In the classical theory of electromagnetism the electric and magnetic fields are central concepts. In this chapter we deal with electric fields associated with charges viewed from a reference frame in which they are at rest, that is, with *electrostatics*.

Before Faraday's time, the force acting between charged particles was thought of as a direct and instantaneous interaction between the two particles. This *action-at-a-distance* view was also held for magnetic and for gravitational forces. Today we prefer to think in terms of electric fields as follows:

1. Charge q_1 in Fig. 27-1 sets up an electric field in the space around itself. This field is suggested by the shading in the figure; later we shall show how to represent electric fields more concretely.
2. The field acts on charge q_2; this shows up in the force $\mathbf{F}$ that q_2 experiences.

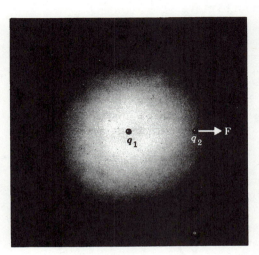

figure 27-1
Charge q_1 sets up a field that exerts a force $\mathbf{F}$ on charge q_2.

The field plays an intermediary role in the forces between charges. There are two separate problems: (a) calculating the fields that are set up by given distributions of charge and (b) calculating the forces that given fields will exert on charges placed in them. We think in terms of

$$\text{charge} \rightleftharpoons \text{field} \rightleftharpoons \text{charge}$$

and not, as in the action-at-a-distance point of view, in terms of

$$\text{charge} \rightleftharpoons \text{charge.}$$

In Fig. 27-1 we can also imagine that q_2 sets up a field and that this field acts on q_1, producing a force $-\mathbf{F}$ on it in accord with Newton's third law. The situation is completely symmetrical, each charge being immersed in a field associated with the other charge.

If the only problem in electromagnetism was that of the forces between stationary charges, the field and the action-at-a-distance points of view would be perfectly equivalent. Suppose, however, that q_1 in Fig. 27-1 suddenly accelerates to the right. How quickly does the charge q_2 learn that q_1 has moved and that the force which it (q_2) experiences must increase? Electromagnetic theory predicts that q_2 learns about q_1's motion by a *field disturbance* that emanates from q_1, traveling with the speed of light. The action-at-a-distance point of view requires that information about q_1's acceleration be communicated *instantaneously* to q_2; this is not in accord with experiment.* Accelerating electrons in the antenna of a radio transmitter influence electrons in a distant receiving antenna only after a time l/c where l is the separation of the antennas and c is the speed of light.

* By introducing other considerations it *is* possible to develop a consistent program of electromagnetism from the action-at-a-distance point of view. This is not commonly done however and we will not do so in this book.

To define the electric field operationally, we place a small test charge q_0 (assumed positive for convenience) at the point in space that is to be examined, and we measure the electrical force **F** (if any) that acts on this body. The *electric field* **E** at the point is defined as*

$$\mathbf{E} = \mathbf{F}/q_0. \qquad (27\text{-}2)$$

Here **E** is a vector because **F** is one, q_0 being a scalar. The direction of **E** is the direction of **F**, that is, it is the direction in which a resting positive charge placed at the point would tend to move.

The definition of gravitational field **g** is much like that of electric field, except that the mass of the test body rather than its charge is the property of interest. Although the units of **g** are usually written as m/s², they could also be written as N/kg (Eq. 27-1); those for **E** are N/C (Eq. 27-2). Thus both **g** and **E** are expressed as a force divided by a property (mass or charge) of the test body.

What is the magnitude of the electric field **E** such that an electron, placed in the field, would experience an electrical force equal to its weight?

EXAMPLE 1

From Eq. 27-2, replacing q_0 by e and F by mg, where m is the electron mass, we have

$$E = \frac{F}{q_0} = \frac{mg}{e}$$

$$= \frac{(9.1 \times 10^{-31} \text{ kg})(9.8 \text{ m/s}^2)}{1.6 \times 10^{-19} \text{ C}}$$

$$= 5.6 \times 10^{-11} \text{ N/C}.$$

This is a very weak electric field. Which way will **E** have to point if the electric force is to cancel the gravitational force?

In applying Eq. 27-2 we must use a test charge as small as possible. A large test charge might disturb the primary charges that are responsible for the field, thus changing the very quantity that we are trying to measure. Equation 27-2 should, strictly, be replaced by

$$\mathbf{E} = \lim_{q_0 \to 0} \frac{\mathbf{F}}{q_0}. \qquad (27\text{-}3)$$

This equation instructs us to use a smaller and smaller test charge q_0, evaluating the ratio $\mathbf{F}/q_0$ at every step. The electric field **E** is then the limit of this ratio as the size of the test charge approaches zero.†

The concept of the electric field as a vector was not appreciated by Michael Faraday, who always thought in terms of *lines of force*. The lines of force still form a convenient way of visualizing electric-field patterns. We shall use them for this purpose but we shall not employ them quantitatively.

The relationship between the (imaginary) lines of force and the electric field vector is this:

---

*This definition of **E**, though conceptually sound and quite appropriate to our present purpose, is rarely carried out in practice because of experimental difficulties. **E** is normally found by calculation from more readily measurable quantities such as the electric potential; see Section 29-7.

† Of course, q_0 can never be less than the electronic charge e

1. The tangent to a line of force at any point gives the *direction* of **E** at that point.
2. The lines of force are drawn so that the number of lines per unit cross-sectional area (perpendicular to the lines) is proportional to the *magnitude* of **E**. Where the lines are close together E is large and where they are far apart E is small.

It is not obvious that it is possible to draw a continuous set of lines to meet these requirements. Indeed, it turns out that if Coulomb's law were not true, it would *not* be possible to do so; see Problem 7.

Figure 27-2 shows the lines of force for a uniform sheet of positive charge. We assume that the sheet is infinitely large, which, for a sheet of finite dimensions, is equivalent to considering only those points whose distance from the sheet is small compared to the distance to the nearest edge of the sheet. A positive test charge, released in front of such a sheet, would move away from the sheet along a perpendicular line. Thus the electric field vector at any point near the sheet must be at right angles to the sheet. The lines of force are uniformly spaced, which means that **E** has the same magnitude for all points near the sheet.

Figure 27-3 shows the lines of force for a negatively charged sphere. From symmetry, the lines must lie along radii. They point inward because a free positive charge would be accelerated in this direction. The electric field E is not constant but decreases with increasing distance from the charge. This is evident in the lines of force, which are farther apart at greater distances. From symmetry, E is the same for all points that lie a given distance from the center of the charge.

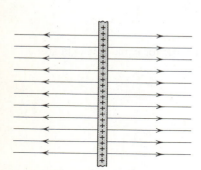

figure 27-2
Lines of force for a section of an infinitely large sheet of positive charge.

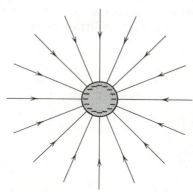

figure 27-3
Lines of force for a negatively charged sphere.

In Fig. 27-3 how does E vary with the distance r from the center of the charged sphere?

Suppose that N lines terminate on the sphere. Draw an imaginary concentric sphere of radius r; the number of lines per unit cross-sectional area at every point on the sphere is $N/4\pi r^2$. Since E is proportional to this, we can write that

$$E \propto 1/r^2.$$

We derive an exact relationship in Section 27-4. How does E vary with distance from an infinitely long uniform cylinder of charge?

EXAMPLE 2

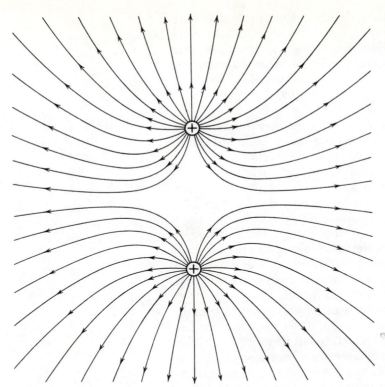

figure 27-4
Lines of force for two equal
positive charges.

Figures 27-4 and 27-5 show the lines of force for two equal like charges and for two equal unlike charges, respectively. Michael Faraday, as we have said, used lines of force a great deal in his thinking. They were more real for him than they are for most scientists and engineers today. It is possible to sympathize with Faraday's point of view. Can

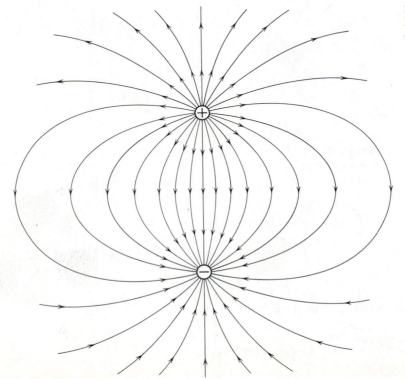

figure 27-5
Lines of force for equal but opposite charges.

we not almost "see" the charges being pushed apart in Fig. 27-4 and pulled together in Fig 27-5 by the lines of force? Compare Fig. 27-5 with Fig. 18-14, which represents a flow field. Figure 27-6 shows a representation of lines of force around charged conductors, using grass seeds suspended in an insulating liquid.

Lines of force give a vivid picture of the way **E** varies through a given region of space. However, the equations of electromagnetism (see Table 40-2) are written in terms of the electric field **E** and other field vectors and not in terms of the lines of force. The electric field **E** varies in a perfectly continuous way as any path in the field is traversed; see Fig. 27-7.

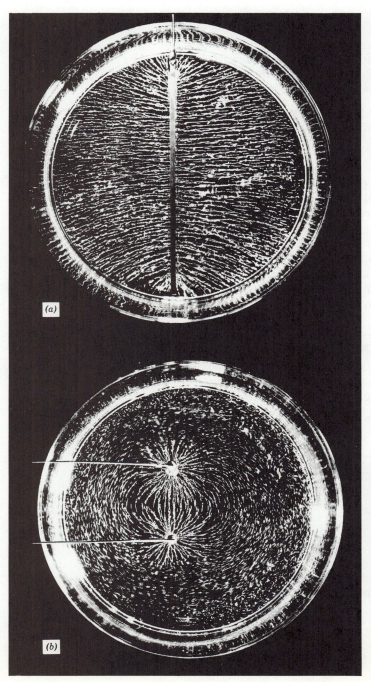

(a)

(b)

figure 27-6
Photographs of the patterns of electric lines of force around (a) a charged plate (compare Fig. 27-2), and (b) two rods with equal and opposite charges (compare Fig. 27-5). The patterns were made by suspending grass seed in an insulating liquid. (Courtesy Educational Services Incorporated, Watertown, Mass.)

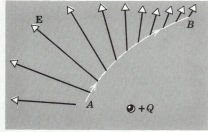

figure 27-7
E varies continuously as we move along any path AB in the field set up by point charge +Q. In general, the path AB and the field vectors **E** will not lie in the plane of the figure.

In this section we deal with the charge-field interaction by showing how we may calculate **E** for various points near given charge distributions, starting with the simple case of a point charge q.

Let a test charge q_0 be placed a distance r from a point charge q. The magnitude of the force acting on q_0 is given by Coulomb's law, or

$$F = \frac{1}{4\pi\epsilon_0} \frac{qq_0}{r^2}.$$

The electric field at the site of the test charge is given by Eq. 27-2, or

$$E = \frac{F}{q_0} = \frac{1}{4\pi\epsilon_0} \frac{q}{r^2}. \qquad (27\text{-}4)$$

The direction of **E** is on a radial line from q, pointing outward if q is positive and inward if q is negative.

To find **E** for a group of point charges: (a) Calculate $\mathbf{E}_n$ due to each charge at the given point *as if it were the only charge present*. (b) Add these separately calculated fields vectorially to find the resultant field **E** at the point. In equation form,

$$\mathbf{E} = \mathbf{E}_1 + \mathbf{E}_2 + \mathbf{E}_3 + \cdots = \Sigma\mathbf{E}_n \qquad n = 1, 2, 3, \ldots. \qquad (27\text{-}5)$$

The sum is a vector sum, taken over all the charges. Equation 27-5 (like Eq. 26-4) is an example of the *principle of superposition* which states, in this context, that at a given point the electric fields due to separate charge distributions simply add up (vectorially) or superimpose independently. The principle of superposition is very important in physics. It applies, for example, to gravitational and magnetic situations as well.*

If the charge distribution is a continuous one, the field it sets up at any point P can be computed by dividing the charge into infinitesimal elements dq. The field $d\mathbf{E}$ due to each element at the point in question is then calculated, treating the elements as point charges. The magnitude of $d\mathbf{E}$ (see Eq. 27-4) is given by

$$dE = \frac{1}{4\pi\epsilon_0} \frac{dq}{r^2}, \qquad (27\text{-}6)$$

where r is the distance from the charge element dq to the point P. The resultant field at P is then found from the superposition principle by adding (that is, integrating) the field contributions due to all the charge elements, or,

$$\mathbf{E} = \int d\mathbf{E}. \qquad (27\text{-}7)$$

The integration, like the sum in Eq. 27-5, is a vector operation; in Example 5 we will see how such an integral is handled in a simple case.

27-4
CALCULATION OF E

An electric dipole. Figure 27-8 shows a positive and a negative charge of equal magnitude q placed a distance $2a$ apart, a configuration called an electric dipole. The pattern of lines of force is that of Fig. 27-5, which also shows an electric dipole. What is the field **E** due to these charges at point P, a distance r along the perpendicular bisector of the line joining the charges? Assume $r \gg a$.

EXAMPLE 3

---

* Formally, the principle of superposition holds in physics only to the extent that the differential equation defining the situation is linear. To the extent that the amplitudes of mechanical or electromagnetic oscillations become relatively large the principle tends to fail. We do not discuss such cases in this book. In particular, the principle holds absolutely in electrostatics.

Equation 27-5 gives the vector equation

$$\mathbf{E} = \mathbf{E}_1 + \mathbf{E}_2,$$

where, from Eq. 27-4,*

$$E_1 = E_2 = \frac{1}{4\pi\epsilon_0} \frac{q}{a^2 + r^2}.$$

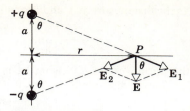

figure 27-8
Example 3.

The vector sum of $\mathbf{E}_1$ and $\mathbf{E}_2$ points vertically downward and has the magnitude

$$E = 2E_1 \cos \theta.$$

From the figure we see that

$$\cos \theta = \frac{a}{\sqrt{a^2 + r^2}}.$$

Substituting the expressions for E_1 and for $\cos \theta$ into that for E yields

$$E = \frac{2}{4\pi\epsilon_0} \frac{q}{(a^2 + r^2)} \frac{a}{\sqrt{a^2 + r^2}} = \frac{1}{4\pi\epsilon_0} \frac{2aq}{(a^2 + r^2)^{3/2}}.$$

If $r \gg a$, we can neglect a in the denominator; this equation then reduces to

$$E \cong \frac{1}{4\pi\epsilon_0} \frac{(2a)(q)}{r^3}. \qquad (27\text{-}8a)$$

The essential properties of the charge distribution in Fig. 27-8, the magnitude of the charge q and the separation $2a$ between the charges, enter Eq. 27-8a only as a product. This means that, if we measure $\mathbf{E}$ at various distances from the electric dipole (assuming $r \gg a$), we can never deduce q and $2a$ separately but only the product $2aq$; if q were doubled and a simultaneously cut in half, the electric field *at large distances from the dipole* would not change.

The product $2aq$ is called the *electric dipole moment p*. Thus we can rewrite this equation for E, *for distant points along the perpendicular bisector*, as

$$E = \frac{1}{4\pi\epsilon_0} \frac{p}{r^3}. \qquad (27\text{-}8b)$$

The result for distant points *along the dipole axis* (see Problem 25) and the general result for any distant point (see Problem 28) also contain the quantities $2a$ and q only as the product $2aq$ ($= p$). The variation of E with r in the general result for distant points is also as $1/r^3$, as in Eq. 27-8b.

The dipole of Fig. 27-8 is two equal and opposite charges placed close to each other so that their separate fields at distant points almost, but not quite, cancel. On this point of view it is easy to understand that $E(r)$ for a dipole varies at large distances as $1/r^3$ (Eq. 27-8b), whereas for a point charge $E(r)$ drops off more slowly, namely as $1/r^2$ (Eq. 27-4).

Figure 27-9 shows a charge q_1 ($= +1.0 \times 10^{-6}$ C) 10 cm from a charge q_2 ($= +2.0 \times 10^{-6}$ C). At what point on the line joining the two charges is the electric field zero?

The point must lie between the charges because only here do the forces exerted by q_1 and q_2 on a test charge (no matter whether it is positive or negative) oppose each other. If $\mathbf{E}_1$ is the electric field due to q_1 and $\mathbf{E}_2$ that due to q_2, we must have

$$E_1 = E_2$$

or (see Eq. 27-4)

$$\frac{1}{4\pi\epsilon_0} \frac{q_1}{x^2} = \frac{1}{4\pi\epsilon_0} \frac{q_2}{(l - x)^2}$$

EXAMPLE 4

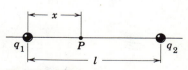

figure 27-9
Example 4.

* Note that the r's in Eq. 27-4 and in this equation have different meanings.

where x is the distance from q_1 and l equals 10 cm. Solving for x yields

$$x = \frac{l}{1 + \sqrt{q_2/q_1}} = \frac{10 \text{ cm}}{1 + \sqrt{2}} = 4.1 \text{ cm}.$$

Supply the missing steps. On what basis was the second root of the resulting quadratic equation discarded?

Ring of charge. Figure 27-10 shows a ring of charge q and radius a. Calculate $\mathbf{E}$ for points on the axis of the ring a distance x from its center.   **EXAMPLE 5**

Consider a differential element of the ring of length ds, located at the top of the ring in Fig. 27-10. It contains an element of charge given by

$$dq = q\,\frac{ds}{2\pi a}$$

where $2\pi a$ is the circumference of the ring. This element sets up a differential electric field $d\mathbf{E}$ at point P.

The resultant field $\mathbf{E}$ at P is found by integrating the effects of all the elements that make up the ring. From symmetry this resultant field must lie along the ring axis. Thus only the component of $d\mathbf{E}$ parallel to this axis contributes to the final result. The component perpendicular to the axis is canceled out by an equal but opposite component established by the charge element on the opposite side of the ring.

Thus the general vector integral (Eq. 27-7)

$$\mathbf{E} = \int d\mathbf{E}$$

becomes a scalar integral $E = \int dE \cos\theta.$

The quantity dE follows from Eq. 27-6, or

$$dE = \frac{1}{4\pi\epsilon_0}\frac{dq}{r^2} = \frac{1}{4\pi\epsilon_0}\left(\frac{q\,ds}{2\pi a}\right)\frac{1}{a^2 + x^2}.$$

From Fig. 27-10 we have $\cos\theta = \dfrac{x}{\sqrt{a^2 + x^2}}.$

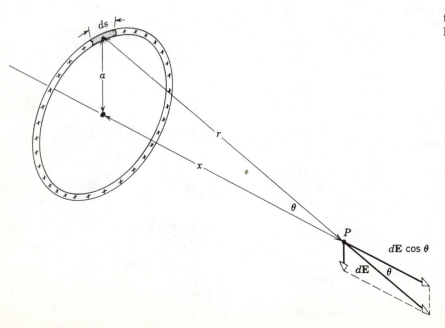

figure 27-10
Example 5

Noting that, for a given point P, x has the same value for all charge elements and is not a variable and that s is the variable of integration, we obtain

$$E = \int dE \cos \theta = \int \frac{1}{4\pi\epsilon_0} \frac{q \, ds}{(2\pi a)(a^2 + x^2)} \frac{x}{\sqrt{a^2 + x^2}}$$

$$= \frac{1}{4\pi\epsilon_0} \frac{qx}{(2\pi a)(a^2 + x^2)^{3/2}} \int ds.$$

The integral is simply the circumference of the ring $(= 2\pi a)$, so that

$$E = \frac{1}{4\pi\epsilon_0} \frac{qx}{(a^2 + x^2)^{3/2}}.$$

Does this expression for E reduce to an expected result for $x = 0$? For $x \gg a$ we can neglect a in the denominator of this equation, yielding

$$E \cong \frac{1}{4\pi\epsilon_0} \frac{q}{x^2}.$$

This is an expected result (compare Eq. 27-4) because at great enough distances the ring behaves like a point charge q.

Infinite line of charge. Figure 27-11 shows a section of an infinite line of charge whose linear charge density (that is, the charge per unit length, measured in C/m) has the constant value λ. Calculate the field $\mathbf{E}$ a distance y from the line.

The magnitude of the field contribution dE due to charge element dq $(= \lambda \, dx)$ is given, using Eq. 27-6, by

$$dE = \frac{1}{4\pi\epsilon_0} \frac{dq}{r^2} = \frac{1}{4\pi\epsilon_0} \frac{\lambda \, dx}{y^2 + x^2}.$$

EXAMPLE 6

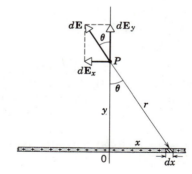

figure 27-11
Example 6. A section of an infinite line of charge.

The vector $d\mathbf{E}$, as Fig. 27-11 shows, has the components

$$dE_x = -dE \sin \theta \quad \text{and} \quad dE_y = dE \cos \theta.$$

The minus sign in front of dE_x indicates that $d\mathbf{E}_x$ points in the negative x direction. The x and y components of the resultant vector $\mathbf{E}$ at point P are given by

$$E_x = \int dE_x = -\int_{x=-\infty}^{x=+\infty} \sin \theta \, dE \quad \text{and} \quad E_y = \int dE_y = \int_{x=-\infty}^{x=+\infty} \cos \theta \, dE.$$

E_x must be zero because every charge element on the right has a corresponding element on the left such that their field contributions in the x direction cancel. Thus $\mathbf{E}$ points entirely in the y direction. Because the contributions to E_y from the right- and left-hand halves of the rod are equal, we can write

$$E = E_y = 2\int_{x=0}^{x=+\infty} \cos \theta \, dE.$$

Note that we have changed the lower limit of integration and have introduced a compensating factor of two.

Substituting the expression for dE into this equation gives

$$E = \frac{\lambda}{2\pi\epsilon_0} \int_{x=0}^{x=\infty} \cos\theta \, \frac{dx}{y^2 + x^2}.$$

From Fig. 27-11 we see that the quantities θ and x are not independent. We must eliminate one of them, say x. The relation between x and θ is (see figure)

$$x = y \tan\theta.$$

Differentiating, we obtain $dx = y \sec^2\theta \, d\theta.$

Substituting these two expressions leads finally to

$$E = \frac{\lambda}{2\pi\epsilon_0 y} \int_{\theta=0}^{\theta=\pi/2} \cos\theta \, d\theta.$$

You should check this step carefully, noting that the limits must now be on θ and not on x. For example, as $x \to +\infty$, $\theta \to \pi/2$, as Fig. 27-11 shows. This equation integrates readily to

$$E = \frac{\lambda}{2\pi\epsilon_0 y} (\sin\theta) \Big|_0^{\pi/2} = \frac{\lambda}{2\pi\epsilon_0 y}.$$

You may wonder about the usefulness of solving a problem involving an infinite rod of charge when any actual rod must have a finite length (see Problem 23). However, for points close enough to finite rods and not near their ends, the equation that we have just derived yields results that are so close to the correct values that the difference can be ignored in many practical situations. It is usually unnecessary to solve exactly every geometry encountered in practical problems. Indeed, if idealizations or approximations are not made, the vast majority of significant problems of all kinds in physics and engineering cannot be solved at all.

27-5
A POINT CHARGE IN AN ELECTRIC FIELD

Here and in the following section, in contrast with Section 27-4, we investigate the other half of the charge-field interaction, namely, if we are given a field $\mathbf{E}$, what forces and torques will act on a charge configuration placed in it? We start with the simple case of a point charge in a uniform electric field.

An electric field will exert a force on a charged particle given by (Eq. 27-2)

$$\mathbf{F} = \mathbf{E}q.$$

This force will produce an acceleration

$$\mathbf{a} = \mathbf{F}/m,$$

where m is the mass of the particle. We will consider two examples of the acceleration of a charged particle in a uniform electric field. Such a field can be produced by connecting the terminals of a battery to two parallel metal plates which are otherwise insulated from each other. If the spacing between the plates is small compared with the dimensions of the plates, the field between them will be fairly uniform except near the edges. Note that in calculating the motion of a particle in a field set up by external charges the field due to the particle itself (that is, its *self-field*) is ignored. In the same way, the earth's gravitational field can exert no net force on the earth itself but only on a second object, say a stone, placed in that field.

A particle of mass m and charge q is placed at rest in a uniform electric field (Fig. 27-12) and released. Describe its motion.

EXAMPLE 7

figure 27-12
A charge is released from rest in a uniform electric field set up between two oppositely charged metal plates P_1 and P_2.

The motion resembles that of a body falling in the earth's gravitational field. The (constant) acceleration is given by

$$a = \frac{F}{m} = \frac{qE}{m}.$$

The equations for uniformly accelerated motion (Table 3-1) then apply. With $v_0 = 0$, they are

$$v = at = \frac{qEt}{m},$$

$$y = \tfrac{1}{2}at^2 = \frac{qEt^2}{2m},$$

and

$$v^2 = 2ay = \frac{2qEy}{m}.$$

The kinetic energy attained after moving a distance y is found from

$$K = \tfrac{1}{2}mv^2 = \tfrac{1}{2}m\left(\frac{2qEy}{m}\right) = qEy.$$

This result also follows directly from the work-energy theorem because a constant force qE acts over a distance y.

Deflecting an electron beam. Figure 27-13 shows an electron of mass m and charge e projected with speed v_0 at right angles to a uniform field **E**. Describe its motion.

EXAMPLE 8

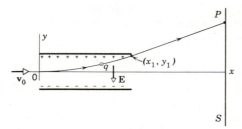

figure 27-13
Example 8. An electron is projected into a uniform electric field.

The motion is like that of a projectile fired horizontally in the earth's gravitational field. The considerations of Section 4-3 apply, the horizontal (x) and vertical (y) motions being given by

$$x = v_0 t$$

and

$$y = \tfrac{1}{2}at^2 = \frac{eE}{2m}\,t^2.$$

Eliminating t yields

$$y = \frac{eE}{2mv_0^2}\,x^2 \qquad\qquad (27\text{-}9)$$

for the equation of the trajectory.

When the electron emerges from the plates in Fig. 27-13, it travels (neglecting gravity) in a straight line tangent to the parabola of Eq. 27-9 at the exit point. We can let it fall on a fluorescent screen S placed some distance beyond the plates. Together with other electrons following the same path, it will then make itself visible as a small luminous spot; this is the principle of the electrostatic *cathode-ray oscilloscope*.

The electric field between the plates of a cathode-ray oscilloscope is 1.2×10^4 N/C. What deflection will an electron experience if it enters at right angles to the field with a kinetic energy of 2000 eV ($= 3.2 \times 10^{-16}$ J), a typical value? The deflecting assembly is 1.5 cm long.

Recalling that $K_0 = \tfrac{1}{2}mv_0^2$, we can rewrite Eq. 27-9 as

$$y = \frac{eEx^2}{4K_0}.$$

If x_1 is the horizontal position of the far edge of the plate, y_1 will be the corresponding deflection (see Fig. 27-13), or

$$y_1 = \frac{eEx_1^2}{4K_0}$$

$$= \frac{(1.6 \times 10^{-19}\ \text{C})(1.2 \times 10^4\ \text{N/C})(1.5 \times 10^{-2}\ \text{m})^2}{(4)(3.2 \times 10^{-16}\ \text{J})}$$

$$= 3.4 \times 10^{-4}\ \text{m} = 0.34\ \text{mm}.$$

The deflection measured, not at the deflecting plates but at the fluorescent screen, is much larger.

EXAMPLE 9

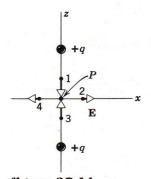

figure 27-14
Example 10. The electric field at four points near a point P which is centered between two equal positive charges.

EXAMPLE 10

A positive point test charge q_0 is placed halfway between two equal fixed positive charges q. What force acts on it at or near this point P?

From symmetry the force *at* the point is zero so that the particle is in equilibrium; the nature of the equilibrium remains to be found. Figure 27-14 (compare Fig. 27-4) shows the **E** vectors for four points near P. If the test charge is moved along the z axis, a *restoring* force is brought into play; however, the equilibrium is unstable for motion along the x (and y) axes. Thus we have the three-dimensional equivalent of *saddle point equilibrium*; see Fig. 14-8. What is the nature of the equilibrium for a negative test charge?

27-6
A DIPOLE IN AN ELECTRIC FIELD

An electric dipole moment can be regarded as a vector **p** whose magnitude p, for a dipole like that described in Example 3, is the product $2aq$ of the magnitude of either charge q and the distance $2a$ between the charges. The *direction* of **p** for such a dipole is from the *negative* to the *positive* charge. The vector nature of the electric dipole moment permits us to cast many expressions involving electric dipoles into concise form, as we shall see.

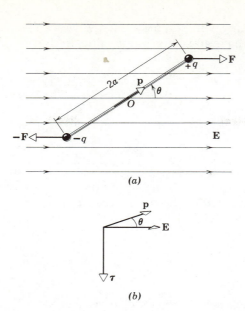

figure 27-15
(a) An electric dipole in a uniform external field. (b) Illustrating $\tau = p \times E$.

Figure 27-15a shows an electric dipole formed by placing two charges $+q$ and $-q$ a fixed distance $2a$ apart. The arrangement is placed in a uniform *external* electric field E, its dipole moment p making an angle θ with this field. Two equal and opposite forces F and $-F$ act as shown, where

$$F = qE.$$

The net force is clearly zero, but there is a net torque about an axis through O (see Eq. 12-2) given by

$$\tau = 2F(a \sin \theta) = 2aF \sin \theta.$$

Combining these two equations and recalling that $p = (2a)(q)$, we obtain

$$\tau = 2aqE \sin \theta = pE \sin \theta. \tag{27-10}$$

Thus an electric dipole placed in an external electric field E experiences a torque tending to align it with the field. Equation 27-10 can be written in vector form as

$$\tau = p \times E, \tag{27-11}$$

the appropriate vectors being shown in Fig. 27-15b.

Work (positive or negative) must be done by an external agent to change the orientation of an electric dipole in an external field. This work is stored as potential energy U in the system consisting of the dipole and the arrangement used to set up the external field. If θ in Fig. 27-15a has the initial value θ_0, the work required to turn the dipole axis to an angle θ is given (see Table 12-2) from

$$W = \int dW = \int_{\theta_0}^{\theta} \tau \, d\theta = U,$$

where τ is the torque exerted by the agent that does the work. Combining this equation with Eq. 27-10 yields

$$U = \int_{\theta_0}^{\theta} pE \sin \theta \, d\theta = pE \int_{\theta_0}^{\theta} \sin \theta \, d\theta$$

$$= pE \left(-\cos \theta \right) \Big|_{\theta_0}^{\theta}.$$

Since we are interested only in *changes* in potential energy, we can choose the reference orientation θ_0 to have any convenient value, in this case 90°. This gives

$$U = -pE \cos \theta \qquad (27\text{-}12)$$

or, in vector terms,

$$U = -\mathbf{p} \cdot \mathbf{E}. \qquad (27\text{-}13)$$

An electric dipole consists of two opposite charges of magnitude $q = 1.0 \times 10^{-6}$ C separated by $d = 2.0$ cm. The dipole is placed in an external field of 1.0×10^5 N/C.

(a) What maximum torque does the field exert on the dipole? The maximum torque is found by putting $\theta = 90°$ in Eq. 27-10 or

$$\tau = pE \sin \theta = qdE \sin \theta$$

$$= (1.0 \times 10^{-6} \text{ C})(0.020 \text{ m})(1.0 \times 10^5 \text{ N/C})(\sin 90°)$$

$$= 2.0 \times 10^{-3} \text{ N} \cdot \text{m}.$$

(b) How much work must an external agent do to turn the dipole end for end, starting from a position of alignment $(\theta = 0)$? The work is the difference in potential energy U between the positions $\theta = 180°$ and $\theta = 0$. From Eq. 27-12,

$$W = U_{180°} - U_{0°} = (-pE \cos 180°) - (-pE \cos 0)$$

$$= 2pE = 2qdE$$

$$= (2)(1.0 \times 10^{-6} \text{ C})(0.020 \text{ m})(1.0 \times 10^5 \text{ N/C})$$

$$= 4.0 \times 10^{-3} \text{ J}.$$

EXAMPLE 11

---

questions

1. Name as many scalar fields and vector fields as you can.

2. (a) In the gravitational attraction between the earth and a stone, can we say that the earth lies in the gravitational field of the stone? (b) How is the gravitational field due to the stone related to that due to the earth?

3. A positively charged ball hangs from a long silk thread. We wish to measure $\mathbf{E}$ at a point in the same horizontal plane as that of the hanging charge. To do so, we put a positive test charge q_0 at the point and measure F/q_0. Will F/q_0 be less than, equal to, or greater than E at the point in question?

4. Taking into account the quantization of electric charge (the single electron providing the basic charge unit), how can we justify the procedure suggested by Eq. 27-3?

5. In exploring electric fields with a test charge we have often assumed, for convenience, that the test charge was positive. Does this really make any difference in determining the field? Illustrate in a simple case of your own devising.

6. Electric lines of force never cross. Why?

7. In Fig. 27-4 why do the lines of force around the edge of the figure appear, when extended backward, to radiate uniformly from the center of the figure?

8. Figure 27-2 shows that $\mathbf{E}$ has the same value for all points in front of an infinite uniformly charged sheet. Is this reasonable? One might think that the field should be stronger near the sheet because the charges are so much closer.

9. If a point charge q of mass m is released from rest in a nonuniform field, will it follow a line of force?

10. A point charge is moving in an electric field at right angles to the lines of force. Does any force **F** act on it?

11. In Fig. 27-7 path *AB* is not a line of force. How can you tell?

12. In Fig. 27-6, why should grass seeds line up with electric lines of force? Grass seeds normally carry no electric charge. See "Demonstration of the Electric Fields of Current-Carrying Conductors" by O. Jefimenko, *American Journal of Physics*. January 1962.

13. Two point charges of unknown magnitude and sign are a distance *d* apart. The electric field is zero at one point between them, on the line joining them. What can you conclude about the charges?

14. Compare the way *E* varies with *r* for (*a*) a point charge (Eq. 27-4), (*b*) a dipole (Eq. 27-8*a*), and (*c*) a quadrupole (Problem 33).

15. Charges $+Q$ and $-Q$ are fixed a distance *L* apart and a long straight line is drawn through them. What is the direction of **E** on this line for points (*a*) between the charges, (*b*) outside the charges in the direction of $+Q$, and (*c*) outside the charges in the direction of $-Q$?

16. Two point charges of unknown sign and magnitude are fixed a distance *L* apart. Can we have **E** = 0 for off-axis points (excluding ∞)? Explain.

17. In what way does Eq. 27-8*b* fail to represent the lines of force of Fig. 27-5 if we relax the requirement that $r \gg a$?

18. If two dipoles of moments $\mathbf{p}_1$ and $\mathbf{p}_2$ are superimposed, is the dipole moment of the resulting configuration given by $\mathbf{p}_1 + \mathbf{p}_2$?

19. In Fig. 27-5 the force on the lower charge points up and is finite. The crowding of the lines of force, however, suggests that *E* is infinitely great at the site of this (point) charge. A charge immersed in an infinitely great field should have an infinitely great force acting on it. What is the solution to this dilemma?

20. An electric dipole is placed in a *nonuniform* electric field. Is there a net force on it?

21. An electric dipole is placed at rest in a uniform external electric field, as in Fig. 27-15*a*, and released. Discuss its motion.

22. An electric dipole has its dipole moment **p** aligned with a uniform external electric field **E**. (*a*) Is the equilibrium stable or unstable? (*b*) Discuss the nature of the equilibrium if **p** and **E** point in opposite directions.

problems

SECTION 27-2

1. What is the magnitude of a point charge chosen so that the electric field 50 cm away has the magnitude 2.0 N/C? *Answer:* 5.6×10^{-11} C.

2. What is the magnitude and direction of an electric field that will balance the weight of (*a*) an electron and (*b*) an alpha particle?

3. An electric field **E** with an average magnitude of about 150 N/C points downward in the earth's atmosphere. We wish to "float" a sulfur sphere weighing 1.0 lb in this field by charging it. (*a*) What charge (sign and magnitude) must be used? (*b*) Why is the experiment not practical? Give a qualitative reason supported by a very rough numerical calculation to prove your point.
   Answer: (*a*) -0.030 C. (*b*) Sphere would blow up because of mutual Coulomb repulsion.

4. At some instant the velocity components of an electron moving between two charged parallel plates are $v_x = 1.5 \times 10^5$ m/s and $v_y = 0.30 \times 10^4$ m/s. If the electric field between the plates is given by $\mathbf{E} = \mathbf{j} \, 1.2 \times 10^4$ N/C, (*a*) what is the acceleration of the electron? (*b*) When the *x*-coordinate of the electron has changed by 2.0 cm what will be the velocity of the electron?

5. A particle having a charge of -2.0×10^{-9} C is acted on by a downward electric force of 3.0×10^{-6} N in a uniform electric field. (a) What is the strength of the electric field? (b) What is the magnitude and direction of the electric force exerted on a proton placed in this field? (c) What is the gravitational force on the proton? (d) What is the ratio of the electric to the gravitational forces in this case?
   Answer: (a) 1.5×10^3 N/C. (b) 2.4×10^{-16} N, up. (c) 1.6×10^{-26} N. (d) 1.5×10^{10}.

6. A uniform vertical field **E** is established in the space between two large parallel plates. In this field one suspends a small conducting sphere of mass m from a string of length l. Find the period of this pendulum when the sphere is given a charge $+q$ if the lower plate (a) is charged positively; (b) is charged negatively.

SECTION 27-3

7. Assume that the exponent in Coulomb's law is not "two" but n. Show that for n $\neq$ 2 it is impossible to construct lines that will have the properties listed for lines of force in Section 27-3. For simplicity, treat an isolated point charge.

8. Sketch qualitatively the lines of force associated with a thin, circular, uniformly charged disk of radius R. (Hint: Consider as limiting cases points very close to the surface and points very far from it.) Show the lines only in a plane containing the axis of the disk.

9. Sketch qualitatively the lines of force between two concentric conducting spherical shells, charge $+ q_1$ being placed on the inner sphere and $-q_2$ on the outer. Consider the cases $q_1 > q_2$, $q_1 = q_2$, $q_1 < q_2$.

10. (a) Sketch qualitatively the lines of force associated with three long parallel lines of charge, in a perpendicular plane. Assume that the intersections of the lines of charge with such a plane form an equilateral triangle and that each line of charge has the same linear charge density λ (C/m). (b) Discuss the nature of the equilibrium of a test charge placed on the central axis of the charge assembly.

11. In Fig. 27-4 consider two neighboring lines of force leaving the upper charge at small angles with a straight line connecting the charges. If the angle between their tangents for points near the charge is θ, it becomes $\theta/\sqrt{2}$ at great distances. Verify this statement and explain it. (Hint: Consider how the lines must behave both close to either charge and far from the charges.)

SECTION 27-4

12. Three charges are arranged in an equilateral triangle as in Fig. 27-16. What is the direction of the force on $+q$?

13. Two equal and opposite charges of magnitude 2.0×10^{-7} C are 15 cm apart. (a) What are the magnitude and direction of **E** at a point midway between the charges? (b) What force (magnitude and direction) would act on an electron placed there?
    Answer: (a) 6.4×10^5 N/C, toward the negative charge. (b) 1.0×10^{-13} N, toward the positive charge.

14. Two point charges of magnitude $+2.0 \times 10^{-7}$ C and $+8.5 \times 10^{-8}$ C are 12 cm apart. (a) What electric field does each produce at the site of the other? (b) What force acts on each?

15. Two point charges of unknown magnitude and sign are placed a distance d apart. (a) If it is possible to have $\mathbf{E} = 0$ at any point *not* between the charges but on the line joining them, what are the necessary conditions and where is the point located? (b) Is it possible, for any arrangement of two point charges, to find *two* points (neither at infinity) at which $\mathbf{E} = 0$; if so, under what conditions?
    Answer: (a) Charges must be opposite in sign, the nearer charge being smaller in magnitude than the farther charge. (b) No.

16. (a) In Fig. 27-17 locate the point (or points) at which the electric field is zero. (b) Sketch qualitatively the lines of force. Take $a = 50$ cm.

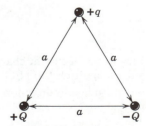

figure 27-16
Problem 12

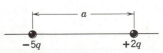

figure 27-17
Problem 16

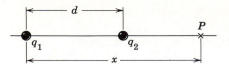

figure 27-18
Problem 17

17. Two point charges are a distance d apart (Fig. 27-18). Plot $E(x)$, assuming $x = 0$ at the left-hand charge. Consider both positive and negative values of x. Plot E as positive if **E** points to the right and negative if **E** points to the left. Assume $q_1 = +1.0 \times 10^{-6}$ C, $q_2 = +3.0 \times 10^{-6}$ C, and $d = 10$ cm.

18. What is **E** in magnitude and direction at the center of the square of Fig. 27-19? Assume that $q = 1.0 \times 10^{-8}$ C and $a = 5.0$ cm.

19. In Fig. 27-8 assume that both charges are positive. (a) Show that E at point P in that figure, assuming $r \gg a$, is given by

$$E = \frac{1}{4\pi\epsilon_0} \frac{2q}{r^2}.$$

(b) What is the direction of **E**? (c) Is it reasonable that E should vary as r^{-2} here and as r^{-3} for the dipole of Fig. 27-8?
Answer: (b) At right angles to the axis and away from it

20. Charges $+q$ and $-2q$ are fixed a distance d apart as in Fig. 27-20. (a) Find **E** at points A, B, and C. (b) Sketch roughly the lines of force.

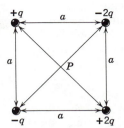

figure 27-19
Problem 18

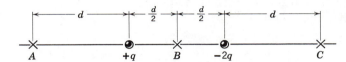

figure 27-20
Problem 20

21. Calculate **E** (direction and magnitude) at point P in Fig. 27-21.
Answer: $E = q/\pi\epsilon_0 a^2$, along bisector, away from triangle.

22. A thin glass rod is bent into a semicircle of radius R. A charge $+Q$ is uniformly distributed along the upper half and a charge $-Q$ is uniformly distributed along the lower half, as shown in Fig. 27-22. Find the electric field **E** at P, the center of the semicircle.

23. A thin nonconducting rod of finite length l carries a total charge q, spread uniformly along it. Show that E at point P on the perpendicular bisector in Fig. 27-23 is given by

$$E = \frac{q}{2\pi\epsilon_0 y} \frac{1}{\sqrt{l^2 + 4y^2}}.$$

Show that as $l \to \infty$ this result approaches that of Example 6.

24. An electron is constrained to move along the axis of the ring of charge in Example 5. Show that the electron can perform oscillations whose frequency is given by

$$\omega = \sqrt{\frac{eq}{4\pi\epsilon_0 ma^3}}.$$

This formula holds only for small oscillations, that is, for $x \ll a$ in Fig. 27-10. (Hint: Show that the motion is simple harmonic and use Eq. 15-11.)

25. *Axial field due to an electric dipole.* In Fig. 27-8, consider a point a distance r from the center of the dipole along its axis. (a) Show that, at large values of r, the electric field is

$$E = \frac{1}{2\pi\epsilon_0} \frac{p}{r^3},$$

which is twice the value given for the conditions of Example 3. (b) What is the direction of **E**?   *Answer:* (b) Parallel to **p**.

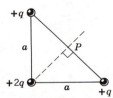

figure 27-21
Problem 21

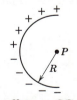

figure 27-22
Problem 22

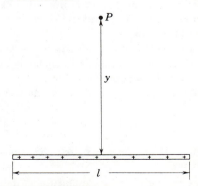

figure 27-23
Problem 23

26. For the ring of charge in Example 5, show that the maximum value of E occurs at $x = a/\sqrt{2}$.

27. Consider the ring of charge of Example 5. Suppose that the charge q is not distributed uniformly over the ring but that charge q_1 is distributed uniformly over half the circumference and charge q_2 is distributed uniformly over the other half. Let $q_1 + q_2 = q$. (a) Find the component of the electric field at any point on the axis directed *along* the axis and compare with the uniform case of Example 5. (b) Find the component of the electric field at any point on the axis *perpendicular* to the axis and compare with the uniform case of Example 5.

Answer: (a) $E = \dfrac{1}{4\pi\epsilon_0} \dfrac{qx}{(a^2 + x^2)^{3/2}}$; (b) $E = \dfrac{1}{2\pi^2\epsilon_0} \dfrac{(q_1 - q_2)a}{(a^2 + x^2)^{3/2}}$.

28. *Field due to an electric dipole.* Show that the components of **E** due to a dipole are given, at distant points, by

$$E_x = \frac{1}{4\pi\epsilon_0} \frac{3pxy}{(x^2 + y^2)^{5/2}}$$

$$E_y = \frac{1}{4\pi\epsilon_0} \frac{p(2y^2 - x^2)}{(x^2 + y^2)^{5/2}},$$

where x and y are coordinates of a point in Fig. 27-24. Show that this general result includes the special results of Eq. 27-8b and of Problem 25.

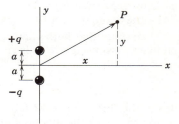

figure 27-24
Problem 28.

29. A "semi-infinite" insulating rod (Fig. 27-25) carries a constant charge per unit length of λ. Show that the electric field at the point P makes an angle of $45°$ with the rod and that this result is independent of the distance R.

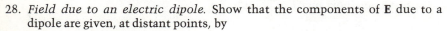

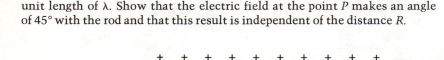

figure 27-25
Problem 29.

30. A nonconducting hemispherical cup of inner radius a has a total charge q spread uniformly over its inner surface. Find the electric field at the center of curvature.

31. A thin nonconducting rod is bent to form the arc of a circle of radius a and subtends an angle θ_0 at the center of the circle. A total charge q is spread uniformly along its length. Find the electric field at the center of the circle in terms of a, q, and θ_0.

Answer: $E = \dfrac{q}{2\pi\epsilon_0\theta_0 a^2} \sin(\theta_0/2)$.

32. A thin circular disk of radius a is charged uniformly so as to have a charge per unit area of σ. Find the electric field on the axis of the disk at a distance r from the disk.

33. *Electric quadrupole.* Figure 27-26 shows a typical electric quadrupole. It consists of two dipoles whose effects at external points do not quite cancel. Show that the value of E on the axis of the quadrupole for points distant r from its center (assume $r \gg a$) is given by

$$E = \frac{3Q}{4\pi\epsilon_0 r^4}$$

where $Q (= 2qa^2)$ is called the *quadrupole moment* of the charge distribution.

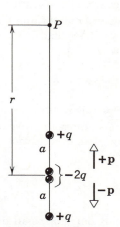

figure 27-26
Problem 33.

34. One type of "electric quadrupole" is formed by four charges located at the vertices of a square of side $2a$. Point P lies a distance R from the center of the quadrupole on a line parallel to two sides of the square as shown in

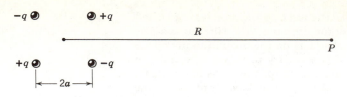

figure 27-27
Problem 34.

Fig. 27-27. For $R \gg a$, show that the electric field at P is approximately given by

$$E = \frac{3(Q)}{2\pi\epsilon_0 R^4},$$

where $Q \, (=2qa^2)$ is the quadrupole moment; see Problem 33. (Hint: Treat the quadrupole as two dipoles.)

SECTION 27-5

35. A uniform electric field exists in a region between two oppositely charged plates. An electron is released from rest at the surface of the negatively charged plate and strikes the surface of the opposite plate, 2.0 cm away, in a time 1.5×10^{-8} s. (a) What is the speed of the electron as it strikes the second plate? (b) What is the magnitude of the electric field **E**?
   Answer: (a) 2.7×10^6 m/s. (b) 1.0×10^3 N/C.

36. An electron moving with a speed of 5.0×10^8 cm/s is shot parallel to an electric field of strength 1.0×10^3 N/C arranged so as to retard its motion. (a) How far will the electron travel in the field before coming (momentarily) to rest and (b) how much time will elapse? (c) If the electric field ends abruptly after 0.8 cm, what fraction of its initial kinetic energy will the electron lose in traversing it?

37. (a) What is the acceleration of an electron in a uniform electric field of 1.0×10^6 N/C? (b) How long would it take for the electron, starting from rest, to attain one-tenth the speed of light? Assume that Newtonian mechanics holds.
   Answer: (a) 1.8×10^{17} m/s^2. (b) 1.7×10^{-10} s.

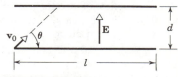

38. An electron is projected as in Fig. 27-28 at a speed of 6.0×10^6 m/s and at an angle θ of 45°; $E = 2.0 \times 10^3$ N/C (directed upward), $d = 2.0$ cm, and $l = 10.0$ cm. (a) Will the electron strike either of the plates? (b) If it strikes a plate, where does it do so?

figure 27-28
Problem 38.

39. *Millikan's oil drop experiment.* R. A. Millikan set up an apparatus (Fig. 27-29) in which a tiny, charged oil drop, placed in an electric field **E**, could be

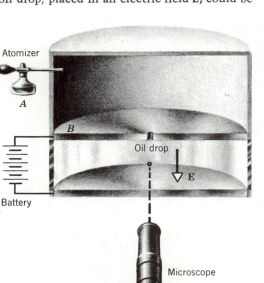

figure 27-29, Problem 39 Millikan's oil drop apparatus. Charged oil drops from the atomizer fall through the hole in the central plate.

"balanced" by adjusting E until the electric force on the drop was equal and opposite to its weight. If the radius of the drop is 1.64×10^{-4} cm and E at balance is 1.92×10^5 N/C, (a) what charge is on the drop in terms of the electronic charge e? (b) Why did Millikan not try to balance electrons in his apparatus instead of oil drops? The density of the oil is 0.851 g/cm^3. (Millikan first measured the electronic charge in this way. He measured the drop radius by observing the limiting speed that the drops attained when they fell in air with the electric field turned off. He charged the oil drops by irradiating them with bursts of X-rays.) See *The Electron* by Robert Millikan, 2d ed., University of Chicago Press, 1924.

Answer: (a) 5.0 e. (b) Cannot see electrons; also E at balance would be inconveniently small.

40. In a particular early run (1911), Millikan observed that the following measured charges, among others, appeared at different times on a single drop:

6.563×10^{-19} C $\qquad$ 13.13×10^{-19} C $\qquad$ 19.71×10^{-19} C
8.204×10^{-19} C $\qquad$ 16.48×10^{-19} C $\qquad$ 22.89×10^{-19} C
$11.50 \ \times 10^{-19}$ C $\qquad$ 18.08×10^{-19} C $\qquad$ 26.13×10^{-19} C

What value for the elementary charge e can be deduced from these data?

SECTION 27-6

41. *Dipole in a nonuniform field.* (a) Derive an expression for dE/dz at a point midway between two equal positive charges, where z is the distance from one of the charges, measured along the line joining them. (b) Would there be a force on a small dipole placed at this point, its axis being aligned with the z axis? Recall that $\mathbf{E} = 0$ at this point.

Answer: (a) $dE/dz = -8q/\pi\epsilon_0 d^3$, where d is the distance between the charges. (b) Yes.

42. Find the frequency of oscillation of an electric dipole, of moment p and rotational inertia I, for small amplitudes of oscillation about its equilibrium position in a uniform electric field E.

43. A charge $q = 3.0 \times 10^{-6}$ C is 30 cm from a small dipole along its perpendicular bisector. The magnitude of the force on the charge is 5.0×10^{-6} N. Show on a diagram (a) the direction of the force on the charge, (b) the direction of the force on the dipole, and (c) determine the magnitude of the force on the dipole.

Answer: (a) Opposite to **p**. (b, c) 5.0×10^{-6} N, parallel to **p**.

28
gauss's law

In the preceding chapter we saw how we could use Coulomb's law to calculate **E** at various points if we knew enough about the distribution of charges that set up the field. This method always works. It is straightforward but, except in the simplest cases, laborious. Given a versatile enough computor, however, we can always find the answer to any such problem, no matter how complicated.

We can express Coulomb's law in another form, called *Gauss's law*. If we use this formulation, the calculations are not laborious. However, the number of problems that we can solve by the Gauss law formulation is small. Those that we can solve we solve with grace and elegance but, by and large, the Gauss law formulation is more useful for the insights that it gives rather than for practical problem solving.

Before we discuss Gauss's law we must develop a new concept, that of the flux of a vector field.

Flux (symbol Φ) is a property of all vector fields. We are concerned in this chapter with the flux Φ_E of the electric field **E.** By way of introduction however, let us discuss semiquantitatively the more familiar flux of fluid flow Φ_v (see Chapter 18). The word *flux* is derived from the Latin word *fluere* (to flow).

Figure 28-1 shows a stationary, uniform field of fluid flow (water, say) characterized by a constant flow vector **v**, the constant velocity of the fluid at any given point. Figure 28-1a suggests, in cross section, a hypothetical plane surface, a circle of radius R and area $\mathbf{A}_a$, immersed in the flow field at right angles to **v.** The mass flux $\Phi_{v,a}$ (kg/s) through

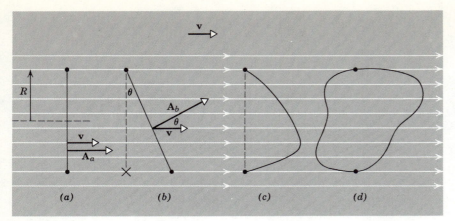

(a)   (b)   (c)   (d)

figure 28-1

figure 28-1
Showing four hypothetical surfaces immersed in a stationary, uniform flow field of an incompressible fluid (water, say) characterized by a constant field vector **v**, the velocity of the fluid at any given point. The horizontal lines are stream lines. R, in all four cases, is the radius of a circle at right angles to the stream lines.

this surface is given by

$$\Phi_{v,a} = \rho\, v\, A_a \qquad (28\text{-}1a)$$

in which ρ is the fluid density (kg/m³). Check that the dimensions are correct. We may also write this equation in vector notation as

$$\Phi_{v,a} = \rho\mathbf{v} \cdot \mathbf{A}_a. \qquad (28\text{-}1b)$$

Note that flux is a scalar.

Figure 28-1b suggests a plane surface whose projected area ($A_b \cos \theta$) is equal to A_a. It seems clear that the mass flux Φ_v (kg/s) through surface b must be the same as that through surface a. To gain some insight we can write

$$\Phi_{v,b} = \Phi_{v,a} = \rho v A_a = \rho v(A_b \cos \theta)$$

$$= \rho\mathbf{v} \cdot \mathbf{A}_b. \qquad (28\text{-}2)$$

Figure 28-1c suggests a curved hypothetical surface whose projected area is said, without proof, to equal A_a. Once more it seems clear that $\Phi_{v,c} = \Phi_{v,a}$.

Figure 28-1d suggests a *closed* surface, the preceding three having been open. We assert that the flux $\Phi_{v,d}$ for this closed surface in this flow field is zero and justify it by noting that the amount of fluid (kg/s) that enters the left portion of the surface per unit time also leaves through the right portion. In this case the fluid (assumed incompressible) neither builds up nor disappears within the surface. In the language of Chapter 18 we say that there happen to be no sources or sinks of fluid within the surface. Every stream line that enters on the left leaves on the right.

After these preliminaries we now turn our attention from Φ_v to Φ_E, the flux of the electric field. It may seem that in the latter case nothing is flowing. In a formal sense, however, Eqs. 28-1b and 28-2 do not concern themselves with flow either but deal with the (constant in this case) field vector **v**. If, in Fig. 28-1 we replace **v** by **E** and regard the stream lines as lines of force, all the discussion of this section remains true.

Finally, in what follows we deal only with closed surfaces immersed in the **E** field. This is because we are concerned here with Gauss's law, which is expressed only in terms of closed surfaces.

In the flow of incompressible fluids it is not true in general that, as in the special case of Fig. 28-1d, $\Phi_v = 0$ for *all* closed surfaces. There may be sources or sinks of fluid within the surface, as suggested in Fig. 18-14. In such cases $\Phi_v \neq 0$.

In the same way it is not true that $\Phi_E = 0$ for every closed surface. There *are* sources (positive charges; here $\Phi_E > 0$) and sinks (negative charges; here $\Phi_E < 0$) of **E** that may be located within the hypothetical closed surface immersed in the **E** field.

For closed surfaces in an electric field we shall see below that Φ_E is positive if the lines of force point outward everywhere and negative if they point inward. Figure 28-2 shows two equal and opposite charges and their lines of force. Curves S_1, S_2, S_3, and S_4 are the intersections with the plane of the figure of four hypothetical closed surfaces. From the statement just given, Φ_E is positive for surface S_1 and negative for S_2. Φ_E for surface S_3 (compare Fig. 28-1d) is zero. We shall discuss Φ_E for surface S_4 in Section 28-4. The flux of the electric field is important because Gauss's law, one of the four basic equations of electromagnetism (see Table 40-2), is expressed in terms of it.

To define Φ_E precisely, consider Fig. 28-3, which shows an arbitrary closed surface immersed in a nonuniform electric field. Let the surface be divided into elementary squares ΔS, each of which is small enough so that it may be considered to be plane. Such an element of area can be represented as a vector $\Delta \mathbf{S}$, whose magnitude is the area ΔS; the direction of $\Delta \mathbf{S}$ is taken as the *outward-drawn normal* to the surface.

At every square in Fig. 28-3 we can also construct an electric field vector **E**. Since the squares have been taken to be arbitrarily small, **E** may be taken as constant for all points in a given square.

The vectors **E** and $\Delta \mathbf{S}$ that characterize each square make an angle θ with each other. Figure 28-3b shows an enlarged view of the three

28-3
FLUX OF THE ELECTRIC FIELD

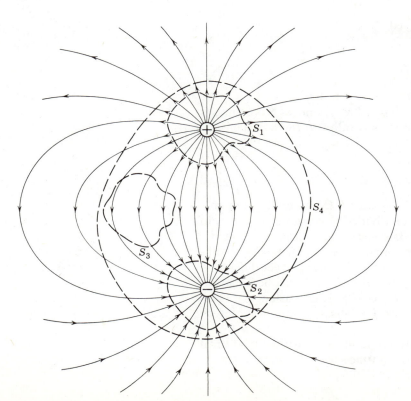

figure 28-2
Two equal and opposite charges. The dashed lines represent the intersection with the plane of the figure of hypothetical closed surfaces.

figure 28-3
(a) A hypothetical surface immersed in a nonuniform electric field.
(b) Three elements of area on this surface, shown enlarged.

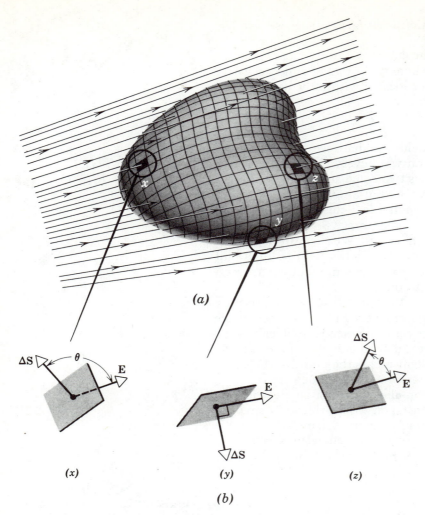

(a)

(x)   (y)   (z)

(b)

squares on the surface of Fig. 28-3a marked x, y, and z. Note that at x, $\theta > 90°$ (**E** points in); at y, $\theta = 90°$ (**E** is parallel to the surface); and at z, $\theta < 90°$ (**E** points out).

A semiquantitative definition of flux is

$$\Phi_E \cong \Sigma \mathbf{E} \cdot \Delta \mathbf{S}, \tag{28-3}$$

which instructs us to add up the scalar quantity $\mathbf{E} \cdot \Delta \mathbf{S}$ for all elements of area into which the surface has been divided. For points such as x in Fig. 28-3 the contribution to the flux is negative; at y it is zero and at z it is positive. Thus if **E** is everywhere outward, $\theta < 90°$, $\mathbf{E} \cdot \Delta \mathbf{S}$ will be positive, and Φ_E for the entire surface will be positive; see Fig. 28-2, surface S_1. If **E** is everywhere inward, $\theta > 90°$, $\mathbf{E} \cdot \Delta \mathbf{S}$ will be negative, and Φ_E for the surface will be negative; see Fig. 28-2, surface S_2. From Eq. 28-3 we see that the appropriate SI unit for Φ_E is the newton meter²/coulomb (N · m²/C).

The exact definition of electric flux is found in the differential limit of Eq. 28-3. Replacing the sum over the surface by an integral over the surface yields

$$\Phi_E = \oint \mathbf{E} \cdot d\mathbf{S}. \qquad (28\text{-}4)$$

This *surface integral* indicates that the surface in question is to be divided into infinitesimal elements of area $d\mathbf{S}$ and that the scalar quantity $\mathbf{E} \cdot d\mathbf{S}$ is to be evaluated for each element and the sum taken for the entire surface. The circle on the integral sign indicates that the surface of integration is a closed surface.*

Figure 28-4 shows a hypothetical closed cylinder of radius R immersed in a uniform electric field $\mathbf{E}$, the cylinder axis being parallel to the field. What is Φ_E for this closed surface?

EXAMPLE 1

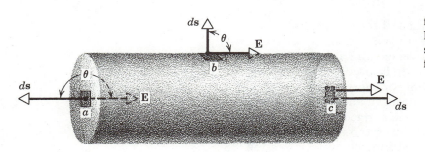

figure 28-4
Example 1. A closed cylindrical surface immersed in a uniform field $\mathbf{E}$ parallel to its axis.

The flux Φ_E can be written as the sum of three terms, an integral over (a) the left cylinder cap, (b) the cylindrical surface, and (c) the right cap. Thus

$$\Phi_E = \oint \mathbf{E} \cdot d\mathbf{S}$$
$$= \int_{(a)} \mathbf{E} \cdot d\mathbf{S} + \int_{(b)} \mathbf{E} \cdot d\mathbf{S} + \int_{(c)} \mathbf{E} \cdot d\mathbf{S}.$$

For the left cap, the angle θ for all points is $180°$, $\mathbf{E}$ has a constant value, and the vectors $d\mathbf{S}$ are all parallel. Thus

$$\int_{(a)} \mathbf{E} \cdot d\mathbf{S} = \int E \cos 180° \, dS$$
$$= -E \int dS = -ES,$$

where $S \ (= \pi R^2)$ is the cap area. Similarly, for the right cap,

$$\int_{(c)} \mathbf{E} \cdot d\mathbf{S} = +ES,$$

the angle θ for all points being zero here. Finally, for the cylinder wall,

$$\int_{(b)} \mathbf{E} \cdot d\mathbf{S} = 0,$$

because $\theta = 90°$, hence $\mathbf{E} \cdot d\mathbf{S} = 0$ for all points on the cylindrical surface. Thus

$$\Phi_E = -ES + 0 + ES = 0.$$

As we shall see in Section 28-4 we expect this because there are no sources or sinks of $\mathbf{E}$, that is, charges, within the closed surface of Fig. 28-4. Lines of (constant) $\mathbf{E}$ enter at the left and emerge at the right, just as in Fig. 28-1d.

---

* Similarly, a circle on a *line* integral sign indicates a closed *path*. It will be clear from the context and from the differential element ($d\mathbf{S}$ in this case) whether we are dealing with a surface integral or a line integral.

Gauss's law, which applies to any closed hypothetical surface (called a *Gaussian surface*), gives a connection between Φ_E for the surface and the net charge q enclosed by the surface. It is

$$\epsilon_0 \Phi_E = q \qquad (28\text{-}5)$$

or, using Eq. 28-4,
$$\epsilon_0 \oint \mathbf{E} \cdot d\mathbf{S} = q. \qquad (28\text{-}6)$$

The fact that Φ_E proves to be zero in Example 1 is predicted by Gauss's law because no charge is enclosed by the Gaussian surface in Fig. 28-4 $(q = 0)$.

Note that q in Eq. 28-5 (or in Eq. 28-6) is the *net* charge, taking its algebraic sign into account. If a surface encloses equal and opposite charges, the flux Φ_E is zero. Charge *outside* the surface makes no contribution to the value of q, nor does the exact location of the inside charges affect this value.

Gauss's law can be used to evaluate $\mathbf{E}$ if the charge distribution is so symmetric that by proper choice of a Gaussian surface we can easily evaluate the integral in Eq. 28-6. Conversely, if $\mathbf{E}$ is known for all points on a given closed surface, Gauss's law can be used to compute the charge inside. If $\mathbf{E}$ has an outward component for every point on a closed surface, Φ_E, as Eq. 28-4 shows, will be positive and, from Eq. 28-6, there must be a net positive charge within the surface (see Fig. 28-2, surface S_1). If $\mathbf{E}$ has an inward component for every point on a closed surface, there must be a net negative charge within the surface (see Fig. 28-2, surface S_2). Surface S_3 in Fig. 28-2 encloses no charge, so that Gauss's law predicts that $\Phi_E = 0$. This is consistent with the fact that lines of $\mathbf{E}$ pass directly through surface S_3, the contribution to the integral on one side canceling that on the other. For surface S_4 in Fig. 28-2, $\Phi_E = 0$ because the algebraic sum of the charges within the surface is zero. Put another way, as for surface S_3, as many lines of force leave the surface as enter it.

28-4
GAUSS'S LAW†

28-5
GAUSS'S LAW AND COULOMB'S LAW

Coulomb's law can be deduced from Gauss's law and symmetry considerations. To do so, let us apply Gauss's law to an isolated point charge q as in Fig. 28-5. Although Gauss's law holds for any surface whatever, information can most readily be extracted for a spherical surface of radius r centered on the charge. The advantage of this surface is that, from symmetry, $\mathbf{E}$ must be normal to it and must have the same (as yet unknown) magnitude for all points on the surface.

In Fig. 28-5 both $\mathbf{E}$ and $d\mathbf{S}$ at any point on the Gaussian surface are directed radially outward. The angle between them is zero and the quantity $\mathbf{E} \cdot d\mathbf{S}$ becomes simply $E\, dS$. Gauss's law (Eq. 28-6) thus reduces to

$$\epsilon_0 \oint \mathbf{E} \cdot d\mathbf{S} = \epsilon_0 \oint E\, dS = q.$$

Because E is constant for all points on the sphere, it can be factored from inside the integral sign, leaving

$$\epsilon_0 E \oint dS = q,$$

where the integral is simply the area of the sphere.* This equation gives

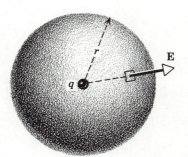

figure 28-5
A spherical Gaussian surface of radius r surrounding a point charge q.

---

* The usefulness of Gauss's law depends on our ability to find a surface over which, from symmetry, both E and θ (see Fig. 28-3) have constant values. Then $E \cos \theta$ can be factored out of the integral and E can be found simply, as in this example.

† See "Gauss," Ian Stewart, *Scientific American,* July 1977 for a fascinating account of the life of this remarkable man.

$$\epsilon_0 E(4\pi r^2) = q$$

or

$$E = \frac{1}{4\pi\epsilon_0}\frac{q}{r^2}. \qquad (28\text{-}7)$$

Equation 28-7 gives the magnitude of the electric field **E** at any point a distance r from an isolated point charge q; compare Eq. 27-4. The direction of **E** is already known from symmetry.

Let us put a second point charge q_0 at the point at which **E** is calculated. The magnitude of the force that acts on it (see Eq. 27-2) is

$$F = Eq_0.$$

Combining with Eq. 28-7 gives

$$F = \frac{1}{4\pi\epsilon_0}\frac{qq_0}{r^2},$$

which is precisely Coulomb's law. Thus we have deduced Coulomb's law from Gauss's law and considerations of symmetry.

Gauss's law is one of the fundamental equations of electromagnetic theory and is displayed in Table 40-2 as one of Maxwell's equations. Coulomb's law is not listed in that table because, as we have just proved, it can be deduced from Gauss's law and from simple assumptions about the symmetry of **E** due to a point charge.

It is interesting to note that writing the proportionality constant in Coulomb's law as $1/4\pi\epsilon_0$ (see Eq. 26-3) permits a particularly simple form for Gauss's law (Eq. 28-5). If we had written the Coulomb law constant simply as k, Gauss's law would have to be written as $(1/4\pi k)\Phi_E = q$. We prefer to leave the factor 4π in Coulomb's law so that it will not appear in Gauss's law or in other much used relations that will be derived later.

28-6 AN INSULATED CONDUCTOR

Gauss's law can be used to make an important prediction, namely: *An excess charge, placed on an insulated conductor, resides entirely on its outer surface.* This hypothesis was shown to be true by experiment (see Section 28-7) before either Gauss's law or Coulomb's law was advanced. Indeed, the experimental proof of the hypothesis is the experimental foundation upon which both laws rest: We have already pointed out that Coulomb's torsion balance experiments, although direct and convincing, are not capable of great accuracy. In showing that the italicized hypothesis is predicted by Gauss's law, we are simply reversing the historical situation.

Figure 28-6 is a cross section of an insulated conductor of arbitrary shape carrying an excess charge q. The dashed lines show a Gaussian surface that lies a small distance below the actual surface of the conductor. Although the Gaussian surface can be as close to the actual surface as we wish, it is important to keep in mind that the Gaussian surface is *inside* the conductor.

When an excess charge is placed at random on an insulated conductor, it will set up electric fields inside the conductor. These fields act on the charge carriers of the conductor (electrons) and cause them to move, that is, they set up internal currents. These currents redistribute the excess charge in such a way that the internal electric fields are automatically reduced in magnitude. Eventually the electric fields inside the conductor become zero everywhere, the currents automatically stop, and electrostatic conditions prevail. This redistribution of charge normally takes place in a time that is negligible for most purposes. What

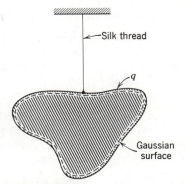

figure 28-6
An insulated metallic conductor.

can be said about the distribution of the excess charge when such electrostatic conditions have been achieved?

If, at electrostatic equilibrium, **E** is zero everywhere inside the conductor, it must be zero for every point on the Gaussian surface because this surface lies inside the conductor. This means that the flux Φ_E for this surface must be zero. Gauss's law then predicts (see Eq. 28-5) that there must be no net charge inside the Gaussian surface. If the excess charge q is not *inside* this surface, it can only be *outside* it, that is, *it must be on the actual surface of the conductor.*

28-7
EXPERIMENTAL PROOF OF GAUSS'S AND COULOMB'S LAWS

Let us turn to the experiments that prove that the hypothesis of Section 28-6 is true. For a simple test, charge a metal ball and lower it with a silk thread deep into a metal can as in Fig. 28-7. Touch the ball to the inside of the can; when the ball is removed from the can, all its charge will have vanished. When the metal ball touches the can, the ball and can together form an "insulated conductor" to which the hypothesis of Section 28-6 applies. That the charge moves entirely to the outside surface of the can can be shown by touching a small insulated metal object to the can; only on the *outside* of the can will it be possible to pick up a charge.

Benjamin Franklin seems to have been the first to notice that there can be no charge inside an insulated metal can. In 1755 he wrote to a friend:

> I electrified a silver pint cann, on an electric stand, and then lowered into it a cork-ball, of about an inch diameter, hanging by a silk string, till the cork touched the bottom of the cann. The cork was not attracted to the inside of the cann as it would have been to the outside, and though it touched the bottom, yet when drawn out, it was not found to be electrified by that touch, as it would have been by touching the outside. The fact is singular. You require the reason; I do not know it. . . .

About ten years later Franklin recommended this "singular fact" to the attention of his friend Joseph Priestley (1733–1804). In 1767 (about twenty years before Coulomb's experiments) Priestley checked Franklin's observation and, with remarkable insight, realized that the inverse square law of force followed from it. Thus the indirect approach is not only more accurate than the direct approach of Section 26-4 but was carried out earlier.

Priestley, reasoning by analogy with gravitation, said that the fact that no electric force acted on Franklin's cork ball when it was surrounded by a deep metal can is similar to the fact (see Section 16-6) that no gravitational force acts on a mass inside a spherical shell of matter; if gravitation obeys an inverse square law, perhaps the electrical force does also. In Priestley's words:

> May we not infer from this [that is, Franklin's experiment] that the attraction of electricity is subject to the same laws with that of gravitation and is therefore according to the squares of the distances; since it is easily demonstrated that were the earth in the form of a shell, a body in the inside of it would not be attracted to one side more than another?

Michael Faraday also carried out experiments designed to show that excess charge resides on the outside surface of a conductor. In particular, he built a large metal-covered box which he mounted on insulating supports and charged with a powerful electrostatic generator. In Faraday's words:

> I went into the cube and lived in it, and using lighted candles, electrometers, and all other tests of electrical states, I could not find the least influence upon them . . . though all the time the outside of the cube was very powerfully charged, and large sparks and brushes were darting off from every part of its outer surface.

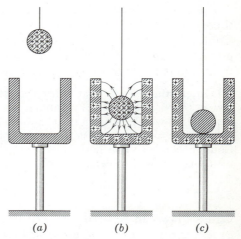

(a)   *(b)*   *(c)*

figure 28-7
The *entire* charge on the ball is transferred to the *outside* of the can. This statement and the discussion of the first paragraph of Section 28-7 are strictly correct only if the can is provided with a conducting lid which can be closed after the ball is inserted; otherwise we cannot define "outside surface."

table 28-1
Test of Coulomb's inverse square law[a]

Experimenters	Date	n
Benjamin Franklin[c]	1755	—
Joseph Priestley[c]	1767	". . . according to the squares of the distance . . ."
John Robison[b]	1769	≤ 0.06
Henry Cavendish[b,c]	1773	≤ 0.02
Charles A. Coulomb	1785	a few percent at most
James Clerk Maxwell[c]	1873	$\le 5 \times 10^{-5}$
Samuel J. Plimpton and Willard E. Lawton[c,d]	1936	$\le 2 \times 10^{-9}$
Edwin R. Williams, James E. Faller, and Henry A. Hill[c,e]	1971	$\le 2 \times 10^{-16}$

[a] Values of n (see Eq. 28-8) are subject to a probable error, not shown. All results are consistent with $n = 0$.
[b] Robison's and Cavendish's results were not made public until after Coulomb had published his results.
[c] These are "Gauss's law" experiments, in the spirit of Fig. 28-7. The others are direct tests of Coulomb's law.
[d] Work done at Worcester Polytechnic Institute.
[e] Work done at Wesleyan University.

For many reasons it is important to know whether or not the exponent in Coulomb's law is exactly "2"; experiments based on Gauss's law can help to determine this. Let us write Coulomb's law as

$$F = \frac{1}{4\pi\epsilon_0} \frac{q_1 q_2}{r^{2+n}} \qquad (28\text{-}8)$$

in which $n = 0$ yields an exact inverse square law.* Table 28-1 shows the progress made in determining how close n in Eq. 28-8 approaches zero.

Figure 28-8 is an idealized sketch of the apparatus of Plimpton and

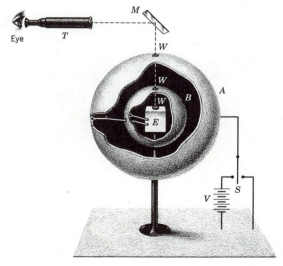

figure 28-8
A representation of the apparatus of Plimpton and Lawton.

---

* If $n \ne 0$ Maxwell's equations (Table 40-2), including Gauss's law, must be modified and the photon, the particle aspect of light, must have a nonzero mass, contrary to the usual expectation. See "The Mass of the Photon" by Alfred Scharff Goldhaber and Michael Martin Nieto, *Scientific American*, May 1976 (p. 86) for a readable account, including magnetic as well as electric field considerations.

Lawton; see Table 28-1. It consists in principle of two concentric metal shells, A and B, the former being 5 ft in diameter. The inner shell contains a sensitive electrometer E connected so that it will indicate whether any charge moves between shells A and B. If the shells are connected electrically, any charge placed on the shell assembly should reside entirely on shell A (see Section 28-6) if Gauss's law—and thus Coulomb's law—are correct as stated.

By throwing switch S to the left, a substantial charge can be placed on the sphere assembly. If any of this charge moves to shell B, it will have to pass through the electrometer and will cause a deflection, which can be observed optically using telescope T, mirror M, and windows W.

However, when the switch S is thrown alternately from left to right, thus connecting the shell assembly either to the battery or to the ground, no effect is observed on the galvanometer. Knowing the sensitivity of their electrometer, Plimpton and Lawton calculated that n in Eq. 28-8 has the value shown in Table 28-1.

Gauss's law can be used to calculate **E** if the symmetry of the charge distribution is high. One example of this, the calculation of **E** for a point charge, has already been discussed (Eq. 28-7). Here we present other examples.

28-8
GAUSS'S LAW—SOME APPLICATIONS

Spherically symmetric charge distribution. Figure 28-9 shows a spherical distribution of charge of radius R. The *charge density* ρ (that is, the charge per unit volume, measured in C/m³) at any point depends only on the distance of the point from the center and not on the direction, a condition called *spherical symmetry*. Find an expression for E for points (a) outside and (b) inside the charge distribution. Note that the object in Fig. 28-9 cannot be a conductor or, as we have seen, the excess charge would reside on its surface.

EXAMPLE 2

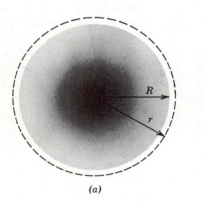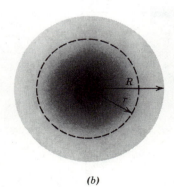

(a)   (b)

figure 28-9
Example 2. A spherically symmetric charge distribution, showing two Gaussian surfaces. The density of charge, as the shading suggests, varies with distance from the center but not with direction.

Applying Gauss's law to a spherical Gaussian surface of radius r in Fig. 28-9a (see Section 28-5) leads exactly to Eq. 28-7, or

$$E = \frac{1}{4\pi\epsilon_0}\frac{q}{r^2}, \qquad (28\text{-}7)$$

where q is the total charge. Thus for points outside a spherically symmetric distribution of charge, the electric field has the value that it would have if the charge were concentrated at its center. This reminds us that a spherically symmetric distribution of mass m behaves gravitationally, for outside points, as if the mass were concentrated at its center. At the root of this similarity lies the fact that both Coulomb's law and the law of gravitation are inverse square laws. The gravitational case was proved in detail in Section 16-6; the proof using Gauss's law in the electrostatic case is certainly much simpler.

Figure 28-9b shows a spherical Gaussian surface of radius r drawn *inside* the charge distribution. Gauss's law (Eq. 28-6) gives

$$\epsilon_0 \oint \mathbf{E} \cdot d\mathbf{S} = \epsilon_0 E(4\pi r^2) = q'$$

or

$$E = \frac{1}{4\pi\epsilon_0} \frac{q'}{r^2},$$

in which q' is that part of q contained within the sphere of radius r. The part of q that lies outside this sphere makes no contribution to $\mathbf{E}$ at radius r. This corresponds, in the gravitational case (Section 16-6), to the fact that a spherical shell of matter exerts no gravitational force on a body inside it.

A special case of a spherically symmetric charge distribution is a uniform sphere of charge. For such a sphere, which would be suggested by uniform shading in Fig. 28-9, the charge density ρ would have a constant value for all points within a sphere of radius R and would be zero for all points outside this sphere. For points inside such a uniform sphere of charge we can put

$$q' = q \frac{\frac{4}{3}\pi r^3}{\frac{4}{3}\pi R^3}$$

or

$$q' = q\left(\frac{r}{R}\right)^3$$

where $\frac{4}{3}\pi R^3$ is the volume of the spherical charge distribution. The expression for E then becomes

$$E = \frac{1}{4\pi\epsilon_0} \frac{qr}{R^3}. \tag{28-9}$$

This equation becomes zero, as it should, for $r=0$. Note that Eqs. 28-7 and 28-9 give the same result, as they must, for points on the surface of the charge distribution (that is, if $r=R$). Note that Eq. 28-9 does not apply to the charge distribution of Fig. 28-9b because the charge density, suggested by the shading, is *not* constant in that case.

EXAMPLE 3

The Thomson atom model. At one time the positive charge in the atom was thought to be distributed uniformly throughout a sphere with a radius of about 1.0×10^{-10} m, that is, throughout the entire atom. Calculate the electric field at the surface of a gold atom $(Z=79)$ on this (erroneous) assumption. Neglect the effect of the electrons.

The positive charge of the atom is Ze or $(79)(1.6 \times 10^{-19}$ C). Equation 28-7 yields, for E at the surface,

$$E = \frac{1}{4\pi\epsilon_0} \frac{q}{r^2}$$

$$= \frac{(9.0 \times 10^9 \text{ N·m}^2/\text{C}^2)(79)(1.6 \times 10^{-19} \text{ C})}{(1.0 \times 10^{-10} \text{ m})^2}$$

$$= 1.1 \times 10^{13} \text{ N/C}.$$

Figure 28-10 is a plot of E as a function of distance from the center of the atom,

$$E, 10^{13}\ N/C$$

r, 10^{-10} m

figure 28-10
Example 3. The electric field due to the positive charge in a gold atom, according to the (erroneous) Thomson model.

using Eqs. 28-7 and 28-9. We see that E has its maximum value on the surface and decreases linearly to zero at the center (see Eq. 28-9). Outside the sphere E decreases as the inverse square of the distance (see Eq. 28-7).

EXAMPLE 4

The Rutherford, or nuclear, atom model. We shall see in Section 28-9 that the positive charge of the atom is *not* spread uniformly throughout the atom (see Example 3) but is concentrated in a small region (the *nucleus*) at the center of the atom. For gold the radius of the nucleus is about 6.9×10^{-15} m. What is the electric field at the nuclear surface? Again neglect effects associated with the atomic electrons.

The problem is the same as that of Example 3, except that the radius is much smaller. This will make the electric field at the surface larger, in proportion to the ratio of the squares of the radii. Thus

$$E = (1.1 \times 10^{13}\ N/C)\frac{(1.0 \times 10^{-10}\ m)^2}{(6.9 \times 10^{-15}\ m)^2}$$

$$= 2.3 \times 10^{21}\ N/C.$$

This is an enormous electric field, much stronger than could be produced and maintained in the laboratory. It is about 10^8 times as large as the field calculated in Example 3.

EXAMPLE 5

*An infinite line of charge.** Figure 28-11 shows a section of an infinite line of charge, the *linear charge density* λ (that is, the charge per unit length, measured in C/m) being constant for all points on the line. Find an expression for E at a distance r from the line.

figure 28-11
Example 5. An infinite line of charge, showing a closed, coaxial cylindrical Gaussian surface.

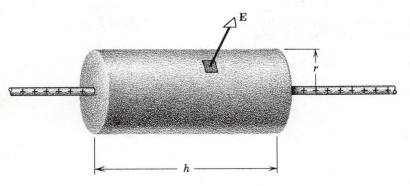

* There are in practice, of course, no such things as infinite lines or sheets of charge. It is a convenient abstraction however, as, for example, was the ideal gas concept developed in Chapter 23. In all cases discussed in this book the assumption of infinite lines (or sheets) of charge allows us to ignore end (or edge) effects.

From symmetry, **E** due to a uniform linear charge can only be radially directed. As a Gaussian surface we choose a circular cylinder of radius r and length h, closed at each end by plane caps normal to the axis. E is constant over the cylindrical surface and the flux of **E** through this surface is $E(2\pi rh)$ where $2\pi rh$ is the area of the surface. There is no flux through the circular caps because **E** here lies in the surface at every point.

The charge enclosed by the Gaussian surface of Fig. 28-11 is λh. Gauss's law (Eq. 28-6),

$$\epsilon_0 \oint \mathbf{E} \cdot d\mathbf{S} = q,$$

then becomes

$$\epsilon_0 E(2\pi rh) = \lambda h,$$

whence

$$E = \frac{\lambda}{2\pi\epsilon_0 r}. \tag{28-10}$$

The direction of **E** is radially outward for a line of positive charge.

Note how much simpler is the solution using Gauss's law than that using integration methods, as in Example 6, Chapter 27. Note too that the solution using Gauss's law is possible only if we choose our Gaussian surface to take full advantage of the cylindrical symmetry of the electric field set up by a long line of charge. We are free to choose any surface, such as a cube or a sphere, for a Gaussian surface. Even though Gauss's law holds for all such surfaces, they are not all useful for the problem at hand; only the cylindrical surface of Fig. 28-11 is appropriate in this case.

Gauss's law has the property that it provides a useful technique for calculation only in problems that have a certain degree of symmetry, but in these problems the solutions are strikingly simple.

EXAMPLE 6

An infinite sheet of charge. Figure 28-12 shows a portion of a thin, *nonconducting*, infinite sheet of charge, the *surface charge density* σ (that is, the charge per unit area, measured in C/m²) being constant. What is **E** at a distance r in front of the sheet?

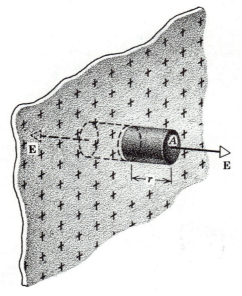

figure 28-12
Example 6. An infinite nonconducting sheet of charge pierced by a cylindrical Gaussian surface.

A convenient Gaussian surface is a closed cylinder of cross-sectional area A and height $2r$, arranged to pierce the plane as shown. From symmetry, **E** points at right angles to the end caps and away from the plane. Since **E** does not pierce the cylindrical surface, there is no contribution to the flux from this source. Thus Gauss's law,

$$\epsilon_0 \oint \mathbf{E} \cdot d\mathbf{S} = q$$

becomes
$$\epsilon_0(EA + EA) = \sigma A$$

where σA is the enclosed charge. This gives

$$E = \frac{\sigma}{2\epsilon_0}. \qquad (28\text{-}11)$$

Note that E is the same for all points on each side of the plane; compare Fig. 27-2. Although an infinite sheet of charge cannot exist physically, this derivation is still useful in that Eq. 28-11 yields substantially correct results for real (not infinite) charge sheets if we consider only points not near the edges whose distance from the sheet is small compared to the dimensions of the sheet.

EXAMPLE 7

A charged conductor. Figure 28-13 shows a *conductor* carrying on its surface a charge whose surface charge density at any point is σ; in general σ will vary from point to point. What is $\mathbf{E}$ for points a short distance above the surface?

The direction of $\mathbf{E}$ for points close to the surface is at right angles to the surface, pointing away from the surface if the charge is positive. If $\mathbf{E}$ were *not* normal to the surface, it would have a component lying in the surface. Such a component would act on the charge carriers in the conductor and set up surface currents. Since there are no such currents under the assumed electrostatic conditions, $\mathbf{E}$ must be normal to the surface.

The magnitude of $\mathbf{E}$ can be found from Gauss's law using a small flat closed cylinder of cross section A as a Gaussian surface. Since $\mathbf{E}$ equals zero everywhere inside the conductor (see Section 28-6), the only contribution to Φ_E is through the plane cap of area A that lies outside the conductor. Gauss's law

$$\epsilon_0 \oint \mathbf{E} \cdot d\mathbf{S} = q$$

becomes
$$\epsilon_0(EA) = \sigma A$$

where σA is the net charge within the Gaussian surface. This yields

$$E = \frac{\sigma}{\epsilon_0}. \qquad (28\text{-}12)$$

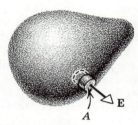

figure 28-13
Example 7. A charged insulated conductor, showing a Gaussian surface.

Comparison with Eq. 28-11 shows that the electric field is twice as great near a *conductor* carrying a charge whose surface charge density is σ as that near a *nonconducting sheet* with the same surface charge density. Compare the Gaussian surfaces in Figs. 28-12 and 28-13 carefully. In Fig. 28-12 lines of force leave the surface through *each* end cap, an electric field existing on *both* sides of the sheet. In Fig. 28-13 the lines of force leave only through the *outside* end cap, the inner end cap being inside the conductor where no electrical field exists. If we assume the same surface charge density and cross-sectional area A for the two Gaussian surfaces, the enclosed charge $(= \sigma A)$ will be the same. Since, from Gauss's law, the flux Φ_E must then be the same in each case, it follows that $E = (\Phi_E/A)$ must be twice as large in Fig. 28-13 as in Fig. 28-12. It is helpful to note that in Fig. 28-12 half the flux emerges from one side of the surface and half from the other, whereas in Fig. 28-13 all the flux emerges from the outside surface.

28-9
THE NUCLEAR MODEL OF THE ATOM

Ernest Rutherford (1871–1937) was first led, in 1911, to assume that the atomic nucleus existed when he tried to interpret some experiments carried out at the University of Manchester by his collaborators H. Geiger and E. Marsden.* The results of Examples 3 and 4 played an important part in Rutherford's analysis of these experiments.

---

* See "The Birth of the Nuclear Atom," E. N. da C. Andrade, *Scientific American*, November 1956. See also Example 6, Chapter 10.

These workers, at Rutherford's suggestion, allowed a beam of α-particles* to strike and be deflected by a thin film of a heavy element such as gold. They counted the number of particles deflected through various angles φ. Figure 28-14 shows the experimental setup schematically. Figure 28-15 shows the paths taken by typical α-particles as they scatter from atoms in a gold foil; the angles φ through which the α-particles are deflected range from 0 to 180° as the character of the collision varies from "grazing" to "head-on."

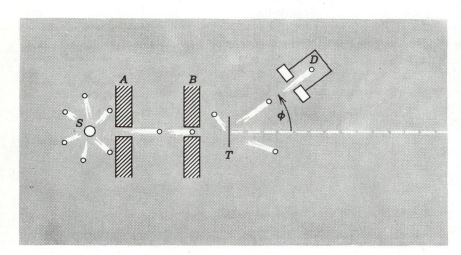

figure 28-14
Experimental arrangement for studying the scattering of α-particles. Particles from radioactive source S are allowed to fall on a thin metal "target" T; α-particles scattered by the target through an (adjustable) angle φ are counted by detector D.

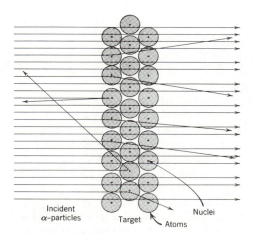

Incident α-particles Target Atoms Nuclei

figure 28-15
The deflection of the incident α-particles depends on the nature of the nuclear collision. (From "The Birth of the Nuclear Atom," E. N. Andrade, *Scientific American*, November 1956.)

The electrons in the gold atom, being so light, have almost no effect on the motion of an oncoming α-particle; the electrons are themselves strongly deflected, just as a swarm of insects would be by a stone hurled through them. Any deflection of the α-particle must be caused by the repulsive action of the positive charge of the gold atom, which was known to possess most of the mass of the atom.

At the time of these experiments most physicists believed in the so-called "plum pudding" model of the atom that had been suggested by J. J. Thomson (1856–1940). In this view (see Example 3) the positive

* α-particles are helium nuclei that are emitted spontaneously by some radioactive materials such as radium. They move with speeds of the order of one-thirtieth that of light when so emitted.

charge of the atom was thought to be spread out through the whole atom, that is, through a spherical volume of radius about 10^{-10} m. The electrons were thought to vibrate about fixed centers inside this sphere.

Rutherford showed that this model of the atom was not consistent with the α-scattering experiments and proposed instead the nuclear model of the atom that we now accept. Here the positive charge is confined to a very much smaller sphere whose radius is about 10^{-14} m (the *nucleus*). The electrons move around this nucleus and occupy a roughly spherical volume of radius about 10^{-10} m. This brilliant deduction by Rutherford laid the foundation for modern atomic and nuclear physics.

The feature of the α-scattering experiments that attracted Rutherford's attention at once was that a few α-particles are deflected through very large angles, up to 180°. To scientists accustomed to thinking in terms of the "plum pudding" model, this was a very surprising result. In Rutherford's words: "It was quite the most incredible event that ever happened to me in my life. It was almost as incredible as if you had fired a 15-inch shell at a piece of tissue paper and it came back and hit you."

The α-particle must pass through a region in which the electric field is very high indeed in order to be deflected so strongly.* Example 3 shows that, in Thomson's model, the maximum electric field is 1.1×10^{13} N/C. Compare this with the value calculated in Example 4 for a point on the surface of a gold nucleus (2.3×10^{21} N/C). Thus the deflecting force acting on an α-particle can be up to 10^8 times as great if the positive charge of the atom is compressed into a small enough region (the nucleus) at the center of the atoms. Rutherford made his hypothesis about the existence of nuclei only after a much more detailed mathematical analysis than that given here.

questions

1. By analogy with Φ_E, how would you define the flux Φ_g of a gravitational field? What is the flux of the earth's gravitational field through the boundaries of a room, assumed to contain no matter?

2. A point charge is placed at the center of a spherical Gaussian surface. Is Φ_E changed (a) if the surface is replaced by a cube of the same volume, (b) if the sphere is replaced by a cube of one-tenth the volume, (c) if the charge is moved off-center in the original sphere, still remaining inside, (d) if the charge is moved just outside the original sphere, (e) if a second charge is placed near, and outside, the original sphere, and (f) if a second charge is placed inside the Gaussian surface?

3. In Gauss's law,

$$\epsilon_0 \oint \mathbf{E} \cdot d\mathbf{S} = q,$$

is $\mathbf{E}$ the electric field attributable to the charge q?

4. Show that Eq. 18-3 illustrates what might be called *Gauss's law for incompressible fluids*, or

$$\Phi_v = \oint \mathbf{v} \cdot d\mathbf{S} = 0.$$

5. A surface encloses an electric dipole. What can you say about Φ_E for this surface?

6. Suppose that a Gaussian surface encloses no net charge. Does Gauss's law require that $\mathbf{E}$ equal zero for all points on the surface? Is the converse of this statement true, that is, if $\mathbf{E}$ equals zero everywhere on the surface, does Gauss's law require that there be no net charge inside?

* The chance that a big deflection can result from the combined effects of many small deflections can be shown to be very small.

7. Is Gauss's law useful in calculating the field due to three equal charges located at the corners of an equilateral triangle? Explain.

8. The use of line, surface, and volume densities of charge to calculate the charge contained in an element of a charged object implies a continuous distribution of charge, whereas, in fact, charge on the microscopic scale is discontinuous. How, then, is this procedure justified?

9. Is **E** necessarily zero inside a charged rubber balloon if the balloon is (a) spherical or (b) sausage-shaped? For each shape assume the charge to be distributed uniformly over the surface.

10. A spherical rubber balloon carries a charge that is uniformly distributed over its surface. How does E vary for points (a) inside the balloon, (b) at the surface of the balloon, and (c) outside the balloon, as the balloon is blown up?

11. We have seen that there are sources (positive charges) and sinks (negative charges) for the **E** field. Are there sources and/or sinks for (a) the **v** field in fluid flow and (b) for the gravitational **g** field?

12. Is it precisely true that Gauss's law states that the total number of lines of force crossing any closed surface in the outward direction is proportional to the net positive charge enclosed within the surface?

13. In Section 28-5 we have seen that Coulomb's law can be derived from Gauss's law. Does this *necessarily* mean that Gauss's law can be derived from Coulomb's law?

14. A large, insulated, hollow conductor carries a charge $+q$. A small metal ball carrying a charge $-q$ is lowered by a thread through a small opening in the top of the conductor, allowed to touch the inner surface, and then withdrawn. What is then the charge on (a) the conductor and (b) the ball?

15. Can we deduce from the argument of Section 28-6 that the electrons in the wires of a house wiring system move along the surfaces of those wires. If not, why not?

16. Does Gauss's law, as applied in Section 28-6, require that all the conduction electrons in an insulated conductor reside on the surface?

17. In Section 28-6, we assumed that **E** equals zero everywhere inside a conductor. However, there are certainly very large electric fields inside the conductor, at points close to the electrons or to the nuclei. Does this invalidate the proof of Section 28-6?

18. It is sometimes said that excess charge resides entirely on the outer surface of a conductor because like charges repel and try to get as far away as possible from one another. Comment on this plausibility argument.

19. Would Gauss's law hold if the exponent in Coulomb's law were not exactly two?

20. As you penetrate a uniform sphere of charge, E should decrease because less charge lies inside a sphere drawn through the observation point. On the other hand, E should increase because you are closer to the center of this charge. Which effect predominates and why?

21. Given a spherically symmetric charge distribution (not of uniform density of charge), is E necessarily a maximum at the surface? Comment on various possibilities.

22. Explain in your own words the factor of 2 that distinguishes Eq. 28-11 from Eq. 28-12.

23. Does Eq. 28-7 hold true for Fig. 28-9a if (a) there is a concentric spherical cavity in the body, (b) a point charge Q is at the center of this cavity, and (c) the charge Q is inside the cavity but not at its center?

24. An atom is normally *electrically neutral*. Why then should an α-particle be deflected by the atom under any circumstances?

25. If an α-particle, fired at a gold nucleus, is deflected through 135°, can you conclude (a) that any force has acted on the α-particle or (b) that any net work has been done on it?

26. Explain in your own words why the α-scattering experiments of Rutherford and his colleagues (see Example 4) render the Thomson atom model (see Example 3) untenable.

SECTION 28-1

1. Calculate Φ_E through a hemisphere of radius R. The field of **E** is uniform and is parallel to the axis of the hemisphere.   *Answer:* $\pi R^2 E$.
2. A butterfly net is in a uniform electric field as shown in Fig. 28-16. The rim, a circle of radius a, is aligned perpendicular to the field. Find the electric flux through the netting.

SECTION 28-3

3. In Example 1 compute Φ_E for the cylinder if it is turned so that its axis is perpendicular to the electric field. Calculate the flux directly without using Gauss's law.   *Answer:* Zero.
4. A point charge of 1.0×10^{-6} C is at the center of a cubical Gaussian surface 0.50 m on edge. What is Φ_E for the surface?

SECTION 28-4

5. Charge on an originally uncharged insulated conductor is separated by holding a positively charged rod nearby, as in Fig. 28-17. What can you learn from Gauss's law about the flux for the five Gaussian surfaces shown? The induced negative charge on the conductor is equal to the positive charge on the rod.
    Answer: $+q$ = charge on rod. $\Phi_{S_1} = q/\epsilon_0$. $\Phi_{S_2} = -q/\epsilon_0$. $\Phi_{S_3} = q/\epsilon_0$. $\Phi_{S_4} = 0$.
    $\Phi_{S_5} = q/\epsilon_0$.
6. A uniformly charged conducting sphere of 1.0 m diameter has a surface charge density of 8.0 C/m². What is the total electric flux leaving the surface of the sphere?
7. The intensity of the earth's electric field near its surface is $\sim$ 130 N/C, pointing down. What is the earth's charge, assuming that this field is caused by it?   *Answer:* -6×10^5 C.
8. A point charge q is placed at one corner of a cube of edge a. What is the flux through each of the cube faces? (Hint: Use Gauss's law and symmetry arguments.)
9. "Gauss's law for gravitation" is

$$\frac{1}{4\pi G}\,\Phi_g = \frac{1}{4\pi G}\oint \mathbf{g}\cdot d\mathbf{S} = -m,$$

where m is the enclosed mass and G is the universal gravitation constant (Section 16-3). Derive Newton's law of gravitation from this.
10. The electric field components in Fig. 28-18 are $E_x = bx^{1/2}$, $E_y = E_z = 0$, in which $b = 800$ N/C $\cdot$ m$^{1/2}$. Calculate (a) the flux Φ_E through the cube and (b) the charge within the cube. Assume that $a = 10$ cm.

problems

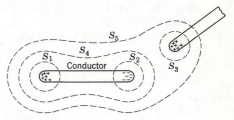

figure 28-16
Problem 2

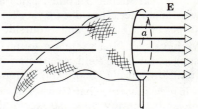

figure 28-17
Problem 5

figure 28-18
Problem 10

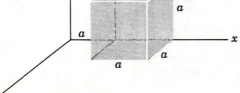

SECTION 28-6

11. Equation 28-12 ($E = \sigma/\epsilon_0$) gives the electric field at points near a charged conducting surface. Show that this equation leads to a believable result when applied to a conducting sphere of radius r, carrying a charge q.
    Answer: It leads to $E = q/4\pi\epsilon_0 r^2$.

12. An insulated conductor carries a net charge of $+10 \times 10^{-6}$ C. Inside the conductor is a hollow cavity inside of which is a point charge Q of $+3.0 \times 10^{-6}$ C. What is the charge (a) on the cavity wall and (b) on the outer surface of the conductor?

13. Figure 28-19 shows a point charge of 1.0×10^{-7} C at the center of a spherical cavity of radius 3.0 cm in a piece of metal. Use Gauss's law to find the electric field (a) at point a, halfway from the center to the surface and (b) at point b.
    Answer: (a) 4.0×10^6 N/C. (b) Zero.

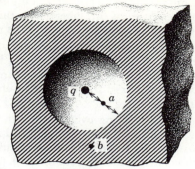

figure 28-19
Problem 13

14. Figure 28-20 shows a spherical nonconducting shell of charge of uniform density ρ (C/m³). Plot E for distances r from the center of the shell ranging from zero to 30 cm. Assume that $\rho = 1.0 \times 10^{-6}$ C/m³, $a = 10$ cm, and $b = 20$ cm.

15. An uncharged, spherical, thin, metallic shell has a point charge q at its center. Give expressions for the electric field (a) inside the shell and (b) outside the shell, using Gauss's law. (c) Has the shell any effect on the field due to q? (d) Has the presence of q any effect on the shell? (e) If a second point charge is held outside the shell, does this outside charge experience a force? (f) Does the inside charge experience a force? (g) Is there a contradiction with Newton's third law here?
    Answer: (a) $E = q/4\pi\epsilon_0 r^2$, radially outward. (b) Same as (a). (c) No. (d) Yes, charges are induced on the surfaces. (e) Yes. (f) No. (g) No.

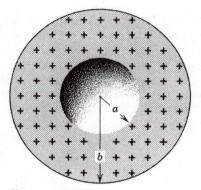

figure 28-20
Problem 14

16. A thin, metallic, spherical shell of radius a carries a charge q_a. Concentric with it is another thin, metallic, spherical shell of radius b ($b > a$) carrying a charge q_b. Use Gauss's law to find the electric field at radial points r where (a) $r < a$; (b) $a < r < b$; (c) $r > b$. (d) Discuss the criterion one would use to determine how the charges are distributed on the inner and outer surfaces of each shell.

17. A *nonconducting* sphere of radius a is placed at the center of a spherical *conducting* shell of inner radius b and outer radius c, as in Fig. 28-21. A charge $+Q$ is distributed uniformly through the inner sphere (charge density ρ, C/m³). The outer shell carries $-Q$. Find $E(r)$, (a) within the sphere ($r < a$), (b) between the sphere and the shell ($a < r < b$), (c) inside the shell ($b < r < c$), and (d) outside the shell ($r > c$). (e) What charges appear on the inner and outer surfaces of the shell?
    Answer: (a) $E = (Q/4\pi\epsilon_0 a^3)r$. (b) $E = Q/4\pi\epsilon_0 r^2$. (c) Zero. (d) Zero. (e) Inner, $-Q$; Outer, zero.

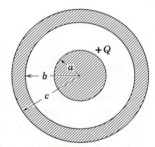

figure 28-21
Problem 17

18. An irregularly shaped conductor has an irregularly shaped cavity inside. A charge $+Q$ is placed on the conductor but there is no charge inside the cavity. Show that (a) $E = 0$ inside the cavity and (b) there is no charge on the cavity walls.

19. Two concentric conducting spherical shells have radii $R_1 = 0.145$ m and $R_2 = 0.207$ m. The inner sphere bears a charge -6.00×10^{-8} C. An electron escapes from the inner sphere with negligible speed. Assuming that the region between the spheres is a vacuum, compute the speed with which the electron strikes the outer sphere.
    Answer: 2.0×10^7 m/s.

20. The spherical region $a < r < b$ carries a charge per unit volume of $\rho = A/r$, where A is constant. At the center ($r = 0$) of the enclosed cavity is a point charge Q. What should the value of A be so that the electric field in the region $a < r < b$ has constant magnitude?

21. A solid nonconducting sphere carries a uniform charge per unit volume ρ. Let **r** be the vector from the center of the sphere to a general point P within the sphere. (a) Show that the electric field at P is given by $\mathbf{E} = \rho\mathbf{r}/3\epsilon_0$. (b) A spherical cavity is created in the above sphere, as shown in Fig. 28-22.

figure 28-22
Problem 21

Using superposition concepts show that the electric field at all points within the cavity is $\mathbf{E} = \rho\mathbf{a}/3\epsilon_0$ (uniform field), where $\mathbf{a}$ is the vector connecting the center of the sphere with the center of the cavity. Note that both these results are independent of the radii of the sphere and the cavity.

22. Figure 28-23 shows a section through a long, thin-walled metal tube of radius R, carrying a charge per unit length λ on its surface. Derive expressions for E for various distances r from the tube axis, considering both (a) $r > R$ and (b) $r < R$. Plot your results for the range $r = 0$ to $r = 5$ cm, assuming that $\lambda = 2.0 \times 10^{-8}$ C/m and $R = 3.0$ cm.

23. A long conducting cylinder (length l) carrying a total charge $+q$ is surrounded by a conducting cylindrical shell of total charge $-2q$, as shown in cross section in Fig. 28-24. Use Gauss's law to find (a) the electric field at points outside the conducting shell, (b) the distribution of the charge on the conducting shell, and (c) the electric field in the region between the cylinders.
Answer: (a) $E = q/2\pi\epsilon_0 lr$, radially inward. (b) $-q$ on both inner and outer surfaces. (c) $E = q/2\pi\epsilon_0 lr$, radially outward.

24. Two charged concentric cylinders have radii of 3.0 cm and 6.0 cm. The charge per unit length on the inner cylinder is 5.0×10^{-6} C/m and that on the outer cylinder is -7.0×10^{-6} C/m. Find the electric field at (a) $r = 4.0$ cm and (b) at $r = 8.0$ cm.

25. Figure 28-25 shows a section through two long concentric cylinders of radii a and b. The cylinders carry equal and opposite charges per unit length λ. Using Gauss's law prove (a) that $E = 0$ for $r > b$ and for $r < a$ and (b) that between the cylinders E is given by

$$E = \frac{1}{2\pi\epsilon_0}\frac{\lambda}{r}.$$

26. In Problem 25 a positive electron revolves in a circular path of radius r, between and concentric with the cylinders. What must be its kinetic energy K? Assume $a = 2.0$ cm, $b = 3.0$ cm, and $\lambda = 3.0 \times 10^{-8}$ C/m.

27. Charge is distributed uniformly throughout an infinitely long cylinder of radius R. (a) Show that E at a distance r from the cylinder axis $(r < R)$ is given by

$$E = \frac{\rho r}{2\epsilon_0},$$

where ρ is the density of charge (C/m³). (b) What result do you expect for $r > R$?   Answer: (b) $\rho R^2/2\epsilon_0 r$.

28. A metal plate 8.0 cm on a side carries a total charge of 6.0×10^{-6} C. (a) Estimate the electric field 0.50 cm above the surface of the plate near the plate's center. (b) Estimate the field at a distance of 3.0 m.

29. Two large metal plates face each other as in Fig. 28-26 and carry charges with surface charge density $+\sigma$ and $-\sigma$, respectively, on their inner surfaces. What is $\mathbf{E}$ at points (a) to the left of the sheets, (b) between them, and (c) to the right of the sheets. Consider only points not near the edges whose distance from the sheets is small compared to the dimensions of the sheet.
Answer: (a) Zero. (b) $E = \sigma/\epsilon_0$, to the left. (c) Zero.

30. Two large metal plates of area 1.0 m² face each other. They are 5.0 cm apart and carry equal and opposite charges on their inner surfaces. If E between the plates is 55 N/C, what is the charge on the plates? Neglect edge effects.

31. Two large nonconducting sheets of positive charge face each other as in Fig. 28-27. What is $\mathbf{E}$ at points (a) to the left of the sheets, (b) between them, and (c) to the right of the sheets? Assume the same surface charge density σ for each sheet. Consider only points not near the edges whose distance from the sheets is small compared to the dimensions of the sheet. (Hint: $\mathbf{E}$ at any point is the vector sum of the separate electric fields set up by each sheet.)
Answer: (a) $E = \sigma/\epsilon_0$, to the left. (b) $E = 0$. (c) $E = \sigma/\epsilon_0$, to the right.

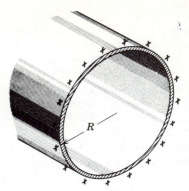

figure 28-23
Problem 22

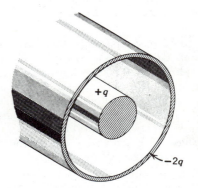

figure 28-24
Problem 23

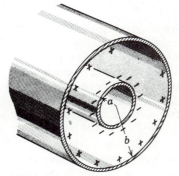

figure 28-25
Problem 25

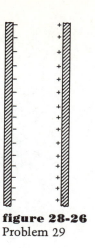

figure 28-26
Problem 29

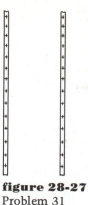

figure 28-27
Problem 31

32. A nonconducting plane slab of thickness d has a uniform volume charge density ρ. Find the magnitude of the electric field at all points in space both (a) inside and (b) outside the slab.

33. A 100-eV electron is fired directly toward a large metal plate that has a surface charge density of -2.0×10^{-6} C/m². From what distance must the electron be fired if it is to just fail to strike the plate?   *Answer:* 0.44 mm.

34. A small sphere whose mass m is 1.0×10^{-3} g carries a charge q of 2.0×10^{-8} C. It hangs from a silk thread which makes an angle of 30° with a large, charged nonconducting sheet as in Fig. 28-28. Calculate the surface charge density σ for the sheet.

35. Show that stable equilibrium under the action of electrostatic forces alone is impossible. (Hint: Assume that at a certain point P in an **E** field a charge $+q$ would be in stable equilibrium if it were placed there — which it is not. Draw a spherical Gaussian surface about P, imagine how **E** must point on this surface, and apply Gauss's law.)

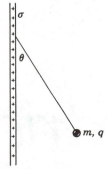

figure 28-28
Problem 34

SECTION 28-9

36. A gold foil used in a Rutherford scattering experiment is 3×10^{-5} cm thick. (a) What fraction of its surface area is "blocked out" by gold nuclei, assuming a nuclear radius of 6.9×10^{-15} m? Assume that no nucleus is screened by any other. (b) What fraction of the volume of the foil is occupied by the nuclei? (c) What fills all the rest of the space in the foil?

37. An α-particle, approaching the surface of a nucleus of gold, is a distance equal to one nuclear radius (6.9×10^{-15} m) away from that surface. What are (a) the force on the α-particle and (b) its acceleration at that point? The mass of the α-particle, which may be treated as a point, is 6.7×10^{-27} kg.
       Answer: (a) 190 N. (b) 2.9×10^{28} m/s².

29
electric potential

The electric field around a charged rod can be described not only by a (vector) electric field **E** but also by a scalar quantity, the *electric potential V*. These quantities are intimately related, and often it is only a matter of convenience which is used in a given problem.

To find the *electric potential difference* between two points A and B in an electric field, we move a test charge q_0 from A to B, always keeping it in equilibrium, and we measure the work W_{AB} that must be done by the agent moving the charge. The electric potential difference* is defined from

$$V_B - V_A = \frac{W_{AB}}{q_0}.$$
(29-1)

The work W_{AB} may be (*a*) positive, (*b*) negative, or (*c*) zero. In these cases the electric potential at B will be (*a*) higher, (*b*) lower, or (*c*) the same as the electric potential at A.

The SI unit of potential difference that follows from Eq. 29-1 is the joule/coulomb. This combination occurs so often that a special unit, the *volt* (abbr. V), is used to represent it; that is,

1 volt = 1 joule/coulomb.

Usually point A is chosen to be at a large (strictly an infinite) distance from all charges, and the electric potential V_A at this infinite distance is arbitrarily taken as zero. This allows us to define the *electric potential*

---

*This definition of potential difference, though conceptually sound and suitable for our present purpose, is rarely carried out in practice because of technical difficulties. Equivalent and more technically feasible methods are usually adopted.

at a point. Putting $V_A = 0$ in Eq. 29-1 and dropping the subscripts leads to

$$V = \frac{W}{q_0},\qquad(29\text{-}2)$$

where W is the work that an external agent must do to move the test charge q_0 from infinity to the point in question. Keep in mind that *potential differences* are of fundamental concern and that Eq. 29-2 depends on the arbitrary assignment of the value zero to the potential V_A at the reference position (infinity); this reference potential could equally well have been chosen as any other value, say -100 V. Similarly, any other agreed-upon point could be chosen as a reference position. In many problems the earth is taken as a reference of potential and assigned the value zero.

Bearing in mind the assumptions made about the reference position, we see from Eq. 29-2 that V near an isolated positive charge is positive because positive work must be done by an outside agent to push a (positive) test charge in from infinity. Similarly, the potential near an isolated negative charge is negative because an outside agent must exert a restraining force on (that is, must do negative work on) a (positive) test charge as it comes in from infinity. Electric potential as defined in Eq. 29-2 is a scalar because W and q_0 in that equation are scalars.

Both W_{AB} and $V_B - V_A$ in Eq. 29-1 are independent of the path followed in moving the test charge from point A to point B. If this were not so, point B would not have a unique electric potential (with respect to point A as a defined reference position) and the concept of potential would not be useful.

Let us now show that the potential difference is path independent for the field due to the point charge of Fig. 29-1. Then we shall show that path independence holds in *all* electrostatic situations. Figure 29-1 shows two points A and B set up in the field of a point charge q, assumed positive.

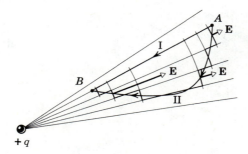

figure 29-1
A test charge q_0 is moved from A to B in the field of charge q along either of two paths. The open arrows show E at three points on path II.

Let us imagine a positive test charge q_0 being moved by an external agent from point A to point B by two different paths. Path I is a simple radial line. Path II is a completely arbitrary path between the two points.

We may approximate path II by a broken path made up of alternating elements of circular arc and of radius. Since these elements can be arbitrarily small, the broken path can be made arbitrarily close to the actual path. On path II the external agent does work *only along the radial segments* because along the arcs the force **F** and the displacement $d\mathbf{l}$ are at right angles, $\mathbf{F} \cdot d\mathbf{l}$ being zero in such cases. The sum of the work done on the radial segments that make up path II is the same as the work done on path I because each path has the same array of radial segments. Since path II is arbitrary, we have proved that the work done is the same for *all* paths connecting A and B.

This proof holds only for the field due to a point charge. However, any charge distribution (discrete or continuous) can be considered as made up of an assembly of point charges or differential charge elements. Hence, from the principle of superposition, we conclude that path independence for the electrostatic potential holds for *all* electrostatic charge configurations.

The locus of points, all of which have the same electric potential, is called an *equipotential surface*. A family of equipotential surfaces, each surface corresponding to a different value of the potential, can be used to give a general description of the electric field in a certain region of space. We have seen earlier (Section 27-3) that electric lines of force can also be used for this purpose; in later sections (see, for example, Fig. 29-15) we explore the intimate connection between these two ways of describing the electric field.

No work is required to move a test charge between any two points on an equipotential surface. This follows from Eq. 29-1,

$$V_B - V_A = \frac{W_{AB}}{q_0},$$

because W_{AB} must be zero if $V_A = V_B$. This is true, because of the path independence of potential difference, even if the path connecting A and B does not lie entirely in the equipotential surface.

Figure 29-2 shows an arbitrary family of equipotential surfaces. The work to move a charge along paths I and II is zero because both these paths begin and end on the same equipotential surface. The work to move a charge along paths I′ and II′ is not zero but is the same for each path because the initial and the final potentials are identical; paths I′ and II′ connect the same pair of equipotential surfaces.

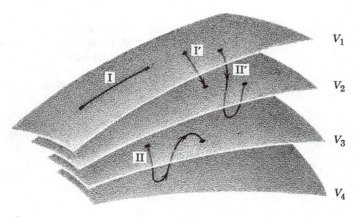

figure 29-2
Portions of four equipotential surfaces. The heavy lines show four paths along which a test charge is moved.

From symmetry, the equipotential surfaces for a spherical charge are a family of concentric spheres. For a uniform field they are a family of planes at right angles to the field. In all cases (including these two examples) the equipotential surfaces are at right angles to the lines of force and thus to **E** (see Fig. 29-15). If **E** were *not* at right angles to the equipotential surface, it would have a component lying in that surface. Then work would have to be done in moving a test charge about on the surface. Work cannot be done if the surface is an equipotential, so **E** must be at right angles to the surface.

There is a strong analogy between electrostatic forces and gravitational forces, based on the fact that their fundamental laws are inverse square laws (see Eqs. 26-3 and 16-1):

$$F_E = \frac{1}{4\pi\epsilon_0} \frac{q_1 q_2}{r^2} \quad \text{and} \quad F_g = G \frac{m_1 m_2}{r^2},$$

and also on the fact that the forces are proportional to the magnitude of the test body (charge in one case, mass in the other). Thus we can define the *gravitational potential* V_g (compare Eq. 29-2) from

$$V_g = \frac{W}{m},$$

where W is the work required to move a test body of mass m from infinity to the point in question. Gravitational equipotential surfaces can also be constructed; they prove to be everywhere at right angles to the gravitational field vector **g**. For a uniform gravitational field, such as that near the surface of the earth, these surfaces are horizontal planes. This correlates with the facts that (a) no net work is required to move a stone of mass m between two points with the same elevation and (b) the same net work is required to move a stone along any path starting on a given horizontal surface and ending on another. Gravitational forces, like coulombic forces, are conservative; see Section 8-2.

Let A and B in Fig. 29-3 be two points in a uniform electric field **E**, set up by an arrangement of charges not shown, and let A be a distance d in the field direction from B. Assume that a positive test charge q_0 is moved, by an external agent and without acceleration, from A to B along the straight line connecting them.

The electric force on the charge is $q_0\mathbf{E}$ and points down. To move the charge in the way we have described we must counteract this force by applying an external force **F** of the same magnitude but directed up. The work W done by the agent that supplies this force is

$$W_{AB} = Fd = q_0 Ed. \tag{29-3}$$

Substituting this into Eq. 29-1 yields

$$V_B - V_A = \frac{W_{AB}}{q_0} = Ed. \tag{29-4}$$

This equation shows the connection between potential difference and field strength for a simple special case. Note from this equation that another SI unit for **E** is the volt/meter (V/m). You may wish to prove that a volt/meter is identical with a newton/coulomb (N/C); this latter unit was the one first presented for **E** in Section 27-2.

In Fig. 29-3 B has a higher potential than A. This is reasonable because an external agent would have to do positive work to push a positive test charge from A to B. Figure 29-3 could be used as it stands to illustrate the act of lifting a stone from A to B in the uniform gravitational field near the earth's surface.

What is the connection between V and **E** in the more general case in which the field is *not* uniform and in which the test body is moved along a path that is *not* straight, as in Fig. 29-4? The electric field exerts a force $q_0\mathbf{E}$ on the test charge, as shown. To keep the test charge from accelerating, an external agent must apply a force **F** chosen to be exactly equal to $-q_0\mathbf{E}$ for all positions of the test body.

If the external agent causes the test body to move through a displacement $d\mathbf{l}$ along the path from A to B, the element of work done by the

29-2
POTENTIAL AND THE ELECTRIC FIELD

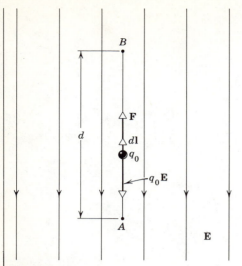

figure 29-3
A positive test charge q_0 is moved from A to B in a uniform electric field $\mathbf{E}$ by an external agent that exerts a force $\mathbf{F}$ on it.

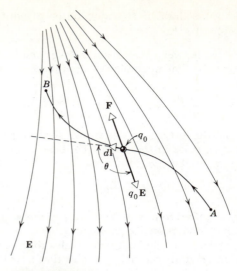

figure 29-4
A positive test charge q_0 is moved from A to B in a nonuniform electric field by an external agent that exerts a force $\mathbf{F}$ on it.

external agent is $\mathbf{F} \cdot d\mathbf{l}$. To find the total work W_{AB} done by the external agent in moving the test charge from A to B, we add up (that is, integrate) the work contributions for all the infinitesimal segments into which the path is divided. This leads to

$$W_{AB} = \int_A^B \mathbf{F} \cdot d\mathbf{l} = -q_0 \int_A^B \mathbf{E} \cdot d\mathbf{l}.$$

Such an integral is called a *line integral.* Note that we have substituted $-q_0\mathbf{E}$ for its equal, $\mathbf{F}$.

Substituting this expression for W_{AB} into Eq. 29-1 leads to

$$V_B - V_A = \frac{W_{AB}}{q_0} = -\int_A^B \mathbf{E} \cdot d\mathbf{l}. \qquad (29\text{-}5)$$

If point A is taken to be infinitely distant and the potential V_A at infinity is taken to be zero, this equation gives the potential V at point B, or, dropping the subscript B,

$$V = -\int_\infty^B \mathbf{E} \cdot d\mathbf{l}. \qquad (29\text{-}6)$$

These two equations allow us to calculate the potential difference between any two points (or the potential at any point) if $\mathbf{E}$ is known at various points in the field.

EXAMPLE 1

In Fig. 29-3 calculate $V_B - V_A$ using Eq. 29-5. Compare the result with that obtained by direct analysis of this special case (Eq. 29-4).

In moving the test charge the element of path $d\mathbf{l}$ always points in the direction of motion; this is upward in Fig. 29-3. The electric field $\mathbf{E}$ in this figure points down so that the angle θ between $\mathbf{E}$ and $d\mathbf{l}$ is 180°.

Equation 29-5 then becomes

$$V_B - V_A = -\int_A^B \mathbf{E} \cdot d\mathbf{l} = -\int_A^B E \cos 180° \, dl = \int_A^B E \, dl.$$

E is constant for all parts of the path in this problem and can thus be taken outside the integral sign, giving

$$V_B - V_A = E \int_A^B dl = Ed,$$

which agrees with Eq. 29-4, as it must.

In Fig. 29-5 let a test charge q_0 be moved without acceleration from A to B over the path shown. Compute the potential difference between A and B.

For the path AC we have $\theta = 135°$ and, from Eq. 29-5,

$$V_C - V_A = -\int_A^C \mathbf{E} \cdot d\mathbf{l} = -\int_A^C E \cos 135° \, dl = \frac{E}{\sqrt{2}} \int_A^C dl.$$

EXAMPLE 2

The integral is the length of the line AC which is $\sqrt{2}d$. Thus

$$V_C - V_A = \frac{E}{\sqrt{2}} (\sqrt{2}d) = Ed.$$

Points B and C have the same potential because no work is done in moving a charge between them, $\mathbf{E}$ and $d\mathbf{l}$ being at right angles for all points on the line CB. In other words, B and C lie on the same equipotential surface at right angles to the lines of force. Thus

$$V_B - V_A = V_C - V_A = Ed.$$

This is the same value derived for a direct path connecting A and B, a result to be expected because the potential difference between two points is path independent.

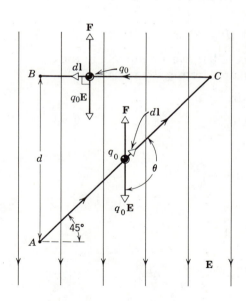

figure 29-5
Example 2. A test charge q_0 is moved along path ACB in a uniform electric field by an external agent.

Figure 29-6 shows two points A and B near an isolated positive point charge q. For simplicity we assume that A, B, and q lie on a straight line. Let us compute the potential difference between points A and B, assuming that a test charge q_0 is moved without acceleration along a radial line from A to B.

29-3
POTENTIAL DUE TO A POINT CHARGE

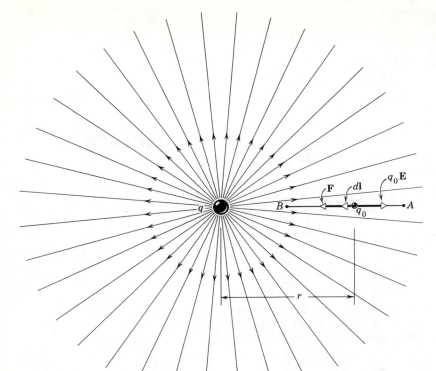

In Fig. 29-6 **E** points to the right and $d\mathbf{l}$, which is always in the direction of motion, points to the left. Therefore, in Eq. 29-5,

$$\mathbf{E} \cdot d\mathbf{l} = E \cos 180° \, dl = -E \, dl.$$

However, as we move a distance dl to the left, we are moving in the direction of decreasing r because r is measured from q as an origin. Thus

$$dl = -dr.$$

Combining yields $\mathbf{E} \cdot d\mathbf{l} = E \, dr.$

Substituting this into Eq. 29-5 gives

$$V_B - V_A = -\int_A^B \mathbf{E} \cdot d\mathbf{l} = -\int_{r_A}^{r_B} E \, dr.$$

Combining with Eq. 27-4,

$$E = \frac{1}{4\pi\epsilon_0} \frac{q}{r^2}$$

leads to $V_B - V_A = -\frac{q}{4\pi\epsilon_0} \int_{r_A}^{r_B} \frac{dr}{r^2} = \frac{q}{4\pi\epsilon_0} \left(\frac{1}{r_B} - \frac{1}{r_A} \right).$ (29-7)

Choosing reference position A to be at infinity (that is, letting $r_A \rightarrow \infty$), choosing $V_A = 0$ at this position, and dropping the subscript B leads to

$$V = \frac{1}{4\pi\epsilon_0} \frac{q}{r}. \qquad (29\text{-}8)$$

This equation shows clearly that equipotential surfaces for an isolated point charge are spheres concentric with the point charge (see Fig. 29-15a). A study of the derivation will show that this relation also holds for points external to spherically symmetric charge distributions.

EXAMPLE 3

What must the magnitude of an isolated positive point charge be for the electric potential at 10 cm from the charge to be $+100$ V?

Solving Eq. 29-8 for q yields

$$q = V4\pi\epsilon_0 r = (100 \text{ V})(4\pi)(8.9 \times 10^{-12} \text{ C}^2/\text{N} \cdot \text{m}^2)(0.10 \text{ m})$$
$$= 1.1 \times 10^{-9} \text{ C}.$$

This charge is comparable to charges that can be produced by friction.

EXAMPLE 4

What is the electric potential at the surface of a gold nucleus? The radius is 6.6×10^{-15} m and the atomic number $Z = 79$.

The nucleus, assumed spherically symmetrical, behaves electrically for external points as if it were a point charge. Thus we can use Eq. 29-8, or, recalling that the proton charge is 1.6×10^{-19} C,

$$V = \frac{1}{4\pi\epsilon_0}\frac{q}{r} = \frac{(9.0 \times 10^9 \text{ N} \cdot \text{m}^2/\text{C}^2)(79)(1.6 \times 10^{-19} \text{ C})}{6.6 \times 10^{-15} \text{ m}}$$

$$= 1.7 \times 10^7 \text{ V}.$$

29-4
A GROUP OF POINT CHARGES

The potential at any point due to a group of point charges is found by (a) calculating the potential V_n due to each charge, as if the other charges were not present and (b) adding the quantities so obtained, or (see Eq. 29-8)

$$V = \sum_n V_n = \frac{1}{4\pi\epsilon_0} \sum_n \frac{q_n}{r_n}, \tag{29-9}$$

where q_n is the value of the nth charge and r_n is the distance of this charge from the point in question. The sum used to calculate V is an *algebraic sum* and not a vector sum like the one used to calculate **E** for a group of point charges (see Eq. 27-5). Herein lies an important computational advantage of potential over electric field.

If the charge distribution is continuous, rather than being a collection of points, the sum in Eq. 29-9 must be replaced by an integral, or

$$V = \int dV = \frac{1}{4\pi\epsilon_0} \int \frac{dq}{r}, \tag{29-10}$$

where dq is a differential element of the charge distribution, r is its distance from the point at which V is to be calculated, and dV is the potential it establishes at that point.

EXAMPLE 5

What is the potential at the center of the square of Fig. 29-7? Assume that $q_1 = +1.0 \times 10^{-8}$ C, $q_2 = -2.0 \times 10^{-8}$ C, $q_3 = +3.0 \times 10^{-8}$ C, $q_4 = +2.0 \times 10^{-8}$ C, and $a = 1.0$ m.

The distance r of each charge from P is $a/\sqrt{2}$ or 0.71 m. From Eq. 29-9

$$V = \sum_n V_n = \frac{1}{4\pi\epsilon_0} \frac{q_1 + q_2 + q_3 + q_4}{r}$$

$$= \frac{(9.0 \times 10^9 \text{ N} \cdot \text{m}^2/\text{C}^2)(1.0 - 2.0 + 3.0 + 2.0) \times 10^{-8} \text{ C}}{0.71 \text{ m}}$$

$$= 500 \text{ V}.$$

Is the potential constant within the square? Does any point inside have a negative potential? Sketch roughly the intersection of the plane of Fig. 29-7 with the equipotential surface corresponding to zero volts.

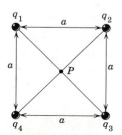

figure 29-7
Example 5

A charged disk. Find the electric potential for points on the axis of a uniformly charged circular disk whose surface charge density is σ (see Fig. 29-8).

Consider a charge element dq consisting of a flat circular strip of radius y and width dy. We have

$$dq = \sigma(2\pi y)(dy),$$

where $(2\pi y)(dy)$ is the area of the strip. All parts of this charge element are the same distance $r'\,(= \sqrt{y^2 + r^2})$ from axial point P so that their contribution dV to the electric potential at P is given by Eq. 29-8, or

$$dV = \frac{1}{4\pi\epsilon_0}\frac{dq}{r'} = \frac{1}{4\pi\epsilon_0}\frac{\sigma 2\pi y\, dy}{\sqrt{y^2 + r^2}}.$$

The potential V is found by integrating over all the strips into which the disk can be divided (Eq. 29-10), or

$$V = \int dV = \frac{\sigma}{2\epsilon_0}\int_0^a (y^2 + r^2)^{-1/2}y\, dy$$

$$= \frac{\sigma}{2\epsilon_0}\,(\sqrt{a^2 + r^2} - r).$$

This general result is valid for all values of r. In the special case of $r \gg a$ the quantity $\sqrt{a^2 + r^2}$ can be approximated as

$$\sqrt{a^2 + r^2} = r\left(1 + \frac{a^2}{r^2}\right)^{1/2} = r\left(1 + \frac{1}{2}\frac{a^2}{r^2} + \cdots\right) \cong r + \frac{a^2}{2r},$$

in which the quantity in parentheses in the second member of this equation has been expanded by the binomial theorem (see Appendix I). This equation means that V becomes

$$V \cong \frac{\sigma}{2\epsilon_0}\left(r + \frac{a^2}{2r} - r\right) = \frac{\sigma \pi a^2}{4\pi\epsilon_0 r} = \frac{1}{4\pi\epsilon_0}\frac{q}{r},$$

where $q\,(= \sigma\pi a^2)$ is the total charge on the disk. This limiting result is expected because the disk behaves like a point charge for $r \gg a$.

EXAMPLE 6

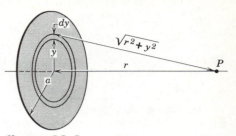

figure 29-8
Example 6. A point P on the axis of a uniformly charged circular disk of radius a.

Two equal charges of opposite sign, $\pm q$, separated by a distance $2a$, constitute an electric dipole; see Example 3, Chapter 27. The electric dipole moment $\mathbf{p}$ has the magnitude $2aq$ and points from the negative charge to the positive charge. Here we derive an expression for the electric potential V at any point of space due to a dipole, provided only that the point is not too close to the dipole ($r \gg a$).

A point P is specified by giving the quantities r and θ in Fig. 29-9. From symmetry, it is clear that the potential will not change as point P rotates about the z axis, r and θ being fixed. Thus we need only find $V(r,\theta)$ for any plane containing this axis; the plane of Fig. 29-9 is such a plane. Applying Eq. 29-9 gives,

$$V = \sum_n V_n = V_1 + V_2 = \frac{1}{4\pi\epsilon_0}\left(\frac{q}{r_1} - \frac{q}{r_2}\right) = \frac{q}{4\pi\epsilon_0}\frac{r_2 - r_1}{r_1 r_2},$$

which is an exact relationship.

We now limit consideration to points such that $r \gg 2a$. These approximate relations then follow from Fig. 29-9:

$$r_2 - r_1 \cong 2a\cos\theta \quad \text{and} \quad r_1 r_2 \cong r^2,$$

and the potential reduces to

$$V = \frac{q}{4\pi\epsilon_0}\frac{2a\cos\theta}{r^2} = \frac{1}{4\pi\epsilon_0}\frac{p\cos\theta}{r^2}, \tag{29-11}$$

29-5
POTENTIAL DUE TO A DIPOLE

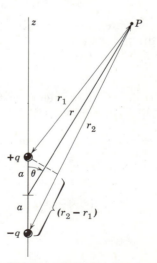

figure 29-9
A point P in the field of an electric dipole.

in which p (= $2aq$) is the dipole moment. Note that V vanishes every-where in the equatorial plane ($\theta = 90°$). This reflects the fact that it takes no work to bring a test charge in from infinity along the per-pendicular bisector of the dipole. For a given radius, V has its greatest positive value for $\theta = 0$ and its greatest negative value for $\theta = 180°$. Note that the potential does not depend separately on q and $2a$ but only on their product p.

It is convenient to call *any* assembly of charges, for which V at distant points is given by Eq. 29-11, an *electric dipole*. Two point charges separated by a small distance behave this way, as we have just proved. However, other charge con-figurations also obey Eq. 29-11. Suppose that by measurement at points outside an imaginary box (Fig. 29-10) we find a pattern of lines of force that can be de-scribed quantitatively by Eq. 29-11. We then declare that the object inside the box is an *electric dipole*, that its axis is the line zz', and that its dipole moment **p** points vertically upward.

Many molecules have electric dipole moments. That for H_2O in its vapor state is 6.1×10^{-30} C·m. Figure 29-11 is a representation of this molecule, show-ing the three nuclei and the surrounding electron cloud. The dipole moment **p** is represented by the arrow on the axis of symmetry of the molecule. In this molecule the effective center of positive charge does not coincide with the effec-tive center of negative charge. It is precisely because of this separation that the dipole moment exists.

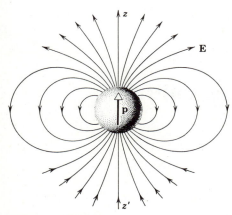

figure 29-10
If an object inside the spherical box sets up the electric field shown (described quantitatively by Eq. 29-11), it is an electric dipole.

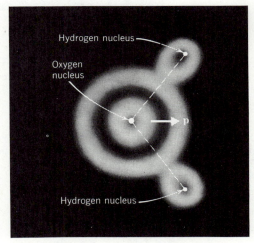

figure 29-11
A schematic representation of a water molecule, showing the three nuclei, the electron cloud, and the orientation of the dipole moment.

Atoms, and many molecules, do not have permanent dipole moments. How-ever, dipole moments may be induced by placing any atom or molecule in an external electric field. The action of the field (Fig. 29-12) is to separate the centers of positive and of negative charge. We say that the atom becomes *polarized* and acquires an *induced electric dipole moment*. Induced dipole moments disappear when the electric field is removed.

Electric dipoles are important in situations other than atomic and molecular ones. Radio and radar antennas are often in the form of a metal wire or rod in which electrons surge back and forth periodically. At a certain time one end of the wire or rod will be negative and the other end positive. Half a cycle later the

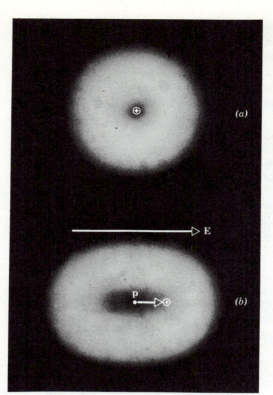

figure 29-12
(a) An atom, showing the nucleus and the electron cloud. The center of negative charge coincides with the center of positive charge, that is, with the nucleus. (b) If an external field **E** is applied, the electron cloud is distorted so that the center of negative charge, marked by the dot, and the center of positive charge no longer coincide. An electric dipole appears. The distortion is greatly exaggerated.

polarity of the ends is exactly reversed. This is an *oscillating* electric dipole. It is so named because its dipole moment changes in a periodic way with time.

EXAMPLE 7

An electric quadrupole. An electric quadrupole, of which Fig. 27-26 is an example, consists of two electric dipoles so arranged that they almost, but not quite, cancel each other in their electric effects at distant points. Calculate $V(r)$ for points on the axis of this quadrupole.

Applying Eq. 29-9 to Fig. 27-26 yields,

$$V = \sum_n V_n = \frac{1}{4\pi\epsilon_0}\left(\frac{q}{r-a} - \frac{2q}{r} + \frac{q}{r+a}\right)$$

$$= \frac{q}{4\pi\epsilon_0}\frac{2a^2}{(r-a)(r)(r+a)}.$$

Assuming that $r \gg a$ allows us to put $a = 0$ in the sum and difference terms in the denominator, yielding

$$V = \frac{1}{4\pi\epsilon_0}\frac{Q}{r^3},$$

where $Q \; (= 2qa^2)$ is the *electric quadrupole moment* of the charge assembly of Fig. 27-26. Note that V varies (a) as $1/r$ for a point charge (see Eq. 29-8), (b) as $1/r^2$ for a dipole (see Eq. 29-11), and (c) as $1/r^3$ for a quadrupole.

Note too that (a) a dipole is two equal and opposite charges that do not quite coincide in space so that their electric effects at distant points do not quite cancel, and (b) a quadrupole is two equal and opposite dipoles that do not quite coincide in space so that their electric effects at distant points again do not quite cancel. This pattern can be extended to define higher orders of charge distribution such as *octupoles*.

The potential at points at distances from an arbitrary charge distribution (continuous or discrete) that are large compared with the size of the distribution can always be written as the sum of separate potential distributions due to (a) a single charge—sometimes, in this context, called a *monopole*—(b) a dipole, (c) a quadrupole, etc. This process is called an *expansion in multipoles* and is a very useful technique in problem solving.

29-6
*ELECTRIC POTENTIAL ENERGY**

If we raise a stone from the earth's surface, the work that we do against the earth's gravitational attraction is stored as *potential energy* in the system earth + stone. If we release the stone, the stored potential energy changes steadily into kinetic energy as the stone drops. After the stone comes to rest on the earth, this kinetic energy, equal in magnitude just before the time of contact to the originally stored potential energy, is transformed into thermal energy in the system earth + stone.

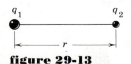

figure 29-13

A similar situation exists in electrostatics. Consider two charges q_1 and q_2 a distance r apart, as in Fig. 29-13. If we increase the separation between them, an external agent must do work that will be positive if the charges are opposite in sign and negative otherwise. The energy represented by this work can be thought of as stored in the system $q_1 + q_2$ as *electric potential energy*. This energy, like all varieties of potential energy, can be transformed into other forms. If q_1 and q_2, for example, are charges of opposite sign and we release them, they will accelerate toward each other, transforming the stored potential energy into kinetic energy of the accelerating masses. The analogy to the earth + stone system is exact, save for the fact that electric forces may be either attractive or repulsive whereas gravitational forces are always attractive.

We define the electric potential energy of a system of point charges as the work required to assemble this system of charges by bringing them in from an infinite distance. We assume that the charges are all at rest when they are infinitely separated, that is, they have no initial kinetic energy.

In Fig. 29-13 let us imagine q_2 removed to infinity and at rest. The *electric potential* at the original site of q_2, caused by q_1, is given by Eq. 29-8, or

$$V = \frac{1}{4\pi\epsilon_0} \frac{q_1}{r}.$$

If q_2 is moved in from infinity to the original distance r, the work required is, from the definition of electric potential, that is, from Eq. 29-2,

$$W = Vq_2. \tag{29-12}$$

Combining these two equations and recalling that this work W is pre-

*In mechanics the concept of *potential energy* (of compressed springs, bodies in the earth's gravitational field, etc.) is perhaps more commonly used than that of *potential* (gravitational, say). In electricity and magnetism, on the other hand, *potential* is perhaps more commonly used than *potential energy*. A shorthand distinction is that potential is potential energy *per unit charge*. With potential in units of volts and potential energy in units of joules, the volt is equivalent, therefore, to a joule per coulomb.

cisely the *electric potential energy* U of the system $q_1 + q_2$ yields

$$U \ (= W) = \frac{1}{4\pi\epsilon_0} \frac{q_1 q_2}{r_{12}}. \qquad (29\text{-}13)$$

The subscript of r emphasizes that the distance involved is that between the point charges q_1 and q_2.

For systems containing more than two charges the procedure is to compute the potential energy for every pair of charges separately and to add the results algebraically. This procedure rests on a physical picture in which (a) charge q_1 is brought into position, (b) q_2 is brought from infinity to its position near q_1, (c) q_3 is brought from infinity to its position near q_1 and q_2, etc.

The potential energy of continuous charge distributions (an ellipsoid of charge, for example) can be found by dividing the distribution into differential elements dq, treating each such element as a point charge, and using the procedures of the preceding paragraph, with the summation process replaced by an integration. We have not considered such problems in this text.

EXAMPLE 8

Two protons in a nucleus of U^{238} are 6.0×10^{-15} m apart. What is their mutual electric potential energy?

From Eq. 29-13

$$U = \frac{1}{4\pi\epsilon_0} \frac{q_1 q_2}{r} = \frac{(9.0 \times 10^9 \ \text{N} \cdot \text{m}^2/\text{C}^2)(1.6 \times 10^{-19} \ \text{C})^2}{6.0 \times 10^{-15} \ \text{m}}$$

$$= 3.8 \times 10^{-14} \ \text{J} = 2.4 \times 10^5 \ \text{eV}.$$

EXAMPLE 9

Three charges are held fixed as in Fig. 29-14. What is their mutual electric potential energy? Assume that $q = 1.0 \times 10^{-7}$ C and $a = 10$ cm.

The total energy of the configuration is the sum of the energies of each pair of particles. From Eq. 29-13,

$$U = U_{12} + U_{13} + U_{23}$$

$$= \frac{1}{4\pi\epsilon_0} \left[\frac{(+q)(-4q)}{a} + \frac{(+q)(+2q)}{a} + \frac{(-4q)(+2q)}{a} \right]$$

$$= -\frac{10}{4\pi\epsilon_0} \frac{q^2}{a}$$

$$= -\frac{(9.0 + 10^9 \ \text{N} \cdot \text{m}^2/\text{C}^2)(10)(1.0 \times 10^{-7} \ \text{C})^2}{0.10 \ \text{m}} = -9.0 \times 10^{-3} \ \text{J}.$$

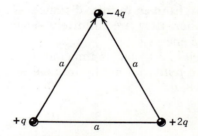

figure 29-14
Example 9. Three charges are fixed, as shown, by external forces.

The fact that the total energy is negative means that negative work would have to be done to assemble this structure, starting with the three charges separated and at rest at infinity. Expressed otherwise, 9.0×10^{-3} J of work must be done to dismantle this structure, removing the charges to an infinite separation from one another.

When, as is common practice, infinity is taken as the zero of electric potential, a positive potential energy, in simple problems like these, (as in Example 8) corresponds to repulsive electric forces and a negative potential energy (as in this example) to attractive electric forces. If the protons in Example 8 were not held in place by attractive (nonelectrical) nuclear forces, they would move away from each other. If the three particles in this example were released from their fixed positions, in which they are held by external forces, they would move toward each other.

We have stated that V and $\mathbf{E}$ are equivalent descriptions of electric fields, and have determined (Eq. 29-6) how to calculate V from $\mathbf{E}$. Let us now consider how to calculate $\mathbf{E}$ if we know V throughout a certain region.

This problem has already been solved graphically. If $\mathbf{E}$ is known at every point in space, the lines of force can be drawn; then a family of equipotentials can be sketched in by drawing surfaces at right angles. These equipotentials describe the behavior of V. Conversely, if V is given as a function of position, a set of equipotential surfaces can be drawn. The lines of force can then be found by drawing lines at right angles, thus describing the behavior of $\mathbf{E}$. It is the mathematical equivalent of this second graphical process that we seek here. Figure 29-15 shows three examples of lines of force and of the corresponding equipotential surfaces.

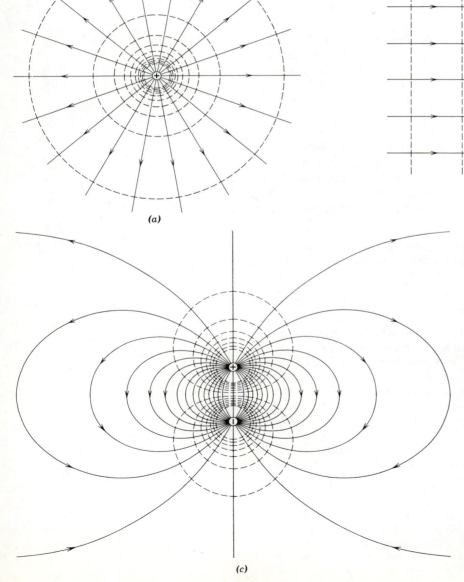

(a)

(b)

(c)

figure 29-15
Equipotential surfaces (dashed lines) and lines of force (solid lines) for (a) a point charge, (b) a uniform electric field set up by charges not shown, and (c) an electric dipole. In (a) and (c) the dashed lines represent intersections with the plane of the figure of closed surfaces; in (b) the dashed lines represent infinite sheets. In all figures there is a constant difference of potential ΔV between adjacent equipotential surfaces. Thus from Eq. 29-14, written for the case of $\theta = 180°$ as $\Delta l = -\Delta V/E$, the surfaces will be relatively close together where E is relatively large and relatively far apart where E is small. Similarly (see Section 27-3) the lines of force are relatively close together where E is large and far apart where E is small. See discussion and figures of Section 18-7 for other examples.

Figure 29-16 shows the intersection with the plane of the figure of a family of equipotential surfaces. The figure shows that **E** at a typical point P is at right angles to the equipotential surface through P, as it must be.

Let us move a test charge q_0 from P along the path marked Δl to the equipotential surface marked $V + \Delta V$. The work that must be done by the agent exerting the force **F** (see Eq. 29-1) is $q_0\Delta V$.

From another point of view we can calculate the work from*

$$\Delta W = \mathbf{F} \cdot \mathbf{\Delta l},$$

where **F** is the force that must be exerted on the charge to overcome exactly the electrical force $q_0\mathbf{E}$. Since **F** and $q_0\mathbf{E}$ have opposite signs and are equal in magnitude,

$$\Delta W = -q_0\mathbf{E} \cdot \mathbf{\Delta l} = -q_0 E \cos (\pi - \theta) \, \Delta l = q_0 E \cos \theta \, \Delta l.$$

These two expressions for the work must be equal, which gives

$$q_0 \, \Delta V = q_0 E \cos \theta \, \Delta l$$

or $$E \cos \theta = \frac{\Delta V}{\Delta l}. \qquad (29\text{-}14)$$

Now $E \cos \theta$ is the component of **E** in the direction $-l$ in Fig. 29-16; the quantity $-E \cos \theta$, which we call E_l, would then be the component of **E** in the $+l$ direction. In the differential limit Eq. 29-14 can then be written as

$$E_l = -\frac{dV}{dl}. \qquad (29\text{-}15)$$

In words, this equation says: If we travel through an electric field along a straight line and measure V as we go, the rate of change of V with distance that we observe, when changed in sign, is the component of **E** in that direction. The minus sign implies that **E** points in the direction of decreasing V, as in Fig. 29-16. It is clear from Eq. 29-15 that appropriate units for **E** are volt/meter (V/m).

There will be one direction l for which the quantity $-dV/dl$ is a maximum. From Eq. 29-15, E_l will also be a maximum for this direction and will in fact be E itself. Thus

$$E = -\left(\frac{dV}{dl}\right)_{max}. \qquad (29\text{-}16)$$

The maximum value of dV/dl at a given point is called the *potential gradient* at that point. The direction l for which dV/dl has its maximum value is always at right angles to the equipotential surface, corresponding to $\theta = 0$ in Fig. 29-16.

If we take the direction l to be, in turn, the directions of the x, y, and z axes, we can find the three components of **E** at any point, from Eq. 29-15.

$$E_x = -\frac{\partial V}{\partial x}; \qquad E_y = -\frac{\partial V}{\partial y}; \qquad E_z = -\frac{\partial V}{\partial z}. \qquad (29\text{-}17)$$

Thus if V is known for all points of space, that is, if the function $V(x, y, z)$ is known, the components of **E**, and thus **E** itself, can be found by taking derivatives.†

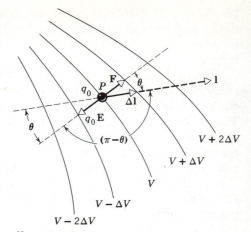

figure 29-16
A test charge q_0 is moved from one equipotential surface to another along an arbitrarily selected direction marked **l**.

---

* We assume that the equipotentials are so close together that **F** is constant for all parts of the path Δl. In the limit of a differential path (dl) there will be no difficulty.

† The symbol $\partial V/\partial x$ denotes a *partial derivative*. In taking this derivative of the function $V(x, y, z)$ the quantity x is to be viewed as a variable and y and z are to be regarded as constants. Similar considerations hold for $\partial V/\partial y$ and $\partial V/\partial z$.

Calculate $E(r)$ for a point charge q, using Eq. 29-16 and assuming that $V(r)$ is given as (see Eq. 29-8) **EXAMPLE 10**

$$V = \frac{1}{4\pi\epsilon_0}\frac{q}{r}.$$

From symmetry, **E** must be directed radially outward for a (positive) point charge. Consider a point P in the field a distance r from the charge. It is clear that $-dV/dl$ at P has its greatest value if the direction l is identified with that of r. Thus, from Eq. 29-16,

$$E = -\frac{dV}{dr} = -\frac{d}{dr}\left(\frac{1}{4\pi\epsilon_0}\frac{q}{r}\right)$$

$$= -\frac{q}{4\pi\epsilon_0}\frac{d}{dr}\left(\frac{1}{r}\right) = \frac{1}{4\pi\epsilon_0}\frac{q}{r^2}.$$

This result agrees exactly with Eq. 27-4, as it must.

E *for a dipole.* Figure 29-17 shows a (distant) point P in the field of a dipole located at the origin of an xy-axis system. V is given by Eq. 29-11, or **EXAMPLE 11**

$$V = \frac{1}{4\pi\epsilon_0}\frac{p\cos\theta}{r^2}.$$

Calculate **E** as a function of position.

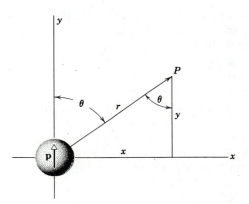

figure 29-17
Showing a point P in the field of an electric dipole **p**.

From symmetry, **E**, for points in the plane of Fig. 29-17, lies in this plane. Thus it can be expressed in terms of its components E_x and E_y, E_z being zero. Let us first express the potential function in rectangular coordinates rather than polar coordinates, making use of

$$r = (x^2 + y^2)^{1/2} \quad \text{and} \quad \cos\theta = \frac{y}{(x^2 + y^2)^{1/2}}.$$

The result is
$$V = \frac{p}{4\pi\epsilon_0}\frac{y}{(x^2 + y^2)^{3/2}}.$$

We find E_y from Eq. 29-17, recalling that x is to be treated as a constant in this calculation:

$$E_y = -\frac{\partial V}{\partial y} = -\frac{p}{4\pi\epsilon_0}\frac{(x^2 + y^2)^{3/2} - y\frac{3}{2}(x^2 + y^2)^{1/2}(2y)}{(x^2 + y^2)^3}$$

$$= -\frac{p}{4\pi\epsilon_0}\frac{x^2 - 2y^2}{(x^2 + y^2)^{5/2}}.$$

Note that putting $x = 0$ describes distant points along the dipole axis (that is, the *y* axis), and the expression for E_y reduces to

$$E_y = \frac{2p}{4\pi\epsilon_0} \frac{1}{y^3}.$$

This result agrees exactly with that found in Chapter 27 (see Problem 25), for, from symmetry, E_x equals zero on the dipole axis.

Putting $y = 0$ in the expression for E_y describes distant points in the median plane of the dipole and yields

$$E_y = -\frac{p}{4\pi\epsilon_0} \frac{1}{x^3},$$

which agrees exactly with the result found in Chapter 27 (see Example 3), for, again from symmetry, E_x equals zero in the median plane. The minus sign in this equation indicates that **E** points in the negative *y* direction (see Fig. 29-10).

The component E_x is also found from Eq. 29-17, recalling that *y* is to be taken as a constant during this calculation:

$$E_x = -\frac{\partial V}{\partial x} = -\frac{py}{4\pi\epsilon_0} (-\tfrac{3}{2})(x^2 + y^2)^{-5/2}(2x)$$

$$= \frac{3p}{4\pi\epsilon_0} \frac{xy}{(x^2 + y^2)^{5/2}}.$$

As expected, E_x vanishes both on the dipole axis $(x = 0)$ and in the median plane $(y = 0)$; see Fig. 29-10.

29-8
AN INSULATED CONDUCTOR

We proved in Section 28-6, using Gauss's law, that after a steady state is reached an excess charge *q* placed on an insulated conductor will move to its outer surface. We now assert that this charge *q* will distribute itself on this surface so that all points of the conductor, including those on the surface and those inside, have the same potential.

Consider any two points *A* and *B* in or on the conductor. If they were not at the same potential, the charge carriers in the conductor near the point of lower potential would tend to move toward the point of higher potential. We have assumed, however, that a steady-state situation, in which such currents do not exist, has been reached; thus all points, both on the surface and inside it, must have the same potential. Since the surface of the conductor is an equipotential surface, **E** for points on the surface must be at right angles to the surface.

We saw in Section 28-6 that a charge placed on an insulated conductor will spread over the surface until **E** equals zero for all points inside. We now have an alternative way of saying the same thing; the charge will move until all points of the conductor (surface points and interior points) are brought to the same potential, for if *V* is constant in the conductor, then **E** is zero everywhere in the conductor $(E_l = -dV/dl)$.

Figure 29-18*a* is a plot of potential against radial distance for an isolated spherical conducting shell of 1.0-m radius carrying a positive charge of 1.0×10^{-6} C. For points outside the shell $V(r)$ can be calculated from Eq. 29-8 because the charge *q* behaves, for such points, as if it were concentrated at the center of the sphere. Equation 29-8 is correct right up to the surface of the shell. Now let us push the test charge through the surface, assuming that there is a small hole, and into the interior. No extra work is needed because no electrical forces act on the test charge once it is inside the shell. Thus the potential everywhere inside is the same as that on the surface, as Fig. 29-18*a* shows.

Figure 29-18*b* shows the electric field for this same sphere. Note that *E* equals zero inside. The lower of these curves can be derived from the

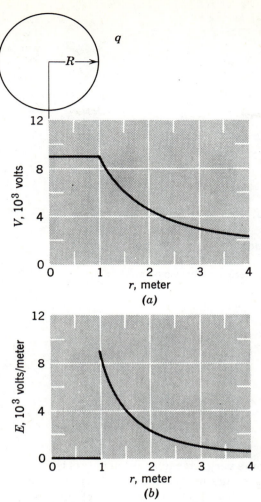

figure 29-18
(a) The potential and (b) the electric field, for points near a conducting spherical shell of 1.0 m radius carrying a charge of $+1.0 \times 10^{-6}$ C.

upper by differentiation with respect to r, using Eq. 29-16; the upper can be derived from the lower by integration with respect to r, using Eq. 29-6.

Figure 29-18 holds without change if the conductor is a solid sphere rather than a spherical shell as we have assumed. It is constructive to compare Fig. 29-18b (conducting shell or sphere) with Fig. 28-10, which holds for a *nonconducting* sphere. Try to understand the difference between these two figures, bearing in mind that in the first the charge lies on the surface, whereas in the second it was assumed to have been spread uniformly throughout the volume of the sphere.

Finally we note that, as a general rule, the charge density tends to be high on isolated conducting surfaces whose radii of curvature are small, and conversely. For example, the charge density tends to be relatively high on sharp points and relatively low on plane regions on a conducting surface. The electric field E at points immediately above a charged surface is proportional to the charge density σ so that E may also reach very high values near sharp points. Glow discharges from sharp points during thunderstorms are a familiar example. The lightning rod acts in this way to neutralize charged clouds and thus prevent lightning strokes.

We can examine the qualitative relationship between σ and the curvature of the surface in a particular case by considering two spheres of different radii con-

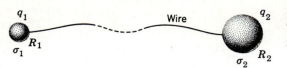

figure 29-19
Two spheres connected by a long fine wire.

nected by a very long fine wire; see Fig. 29-19. Suppose that this entire assembly is raised to some arbitrary potential V. The (equal) potentials of the two spheres are, from Eq. 29-8,*

$$V = \frac{1}{4\pi\epsilon_0} \frac{q_1}{R_1} = \frac{1}{4\pi\epsilon_0} \frac{q_2}{R_2},$$

which yields

$$\frac{q_1}{q_2} = \frac{R_1}{R_2}, \qquad (29\text{-}18)$$

where q_1 is the charge on the sphere of radius R_1 and q_2 is the charge on the sphere of radius R_2.

The *surface charge densities* for each sphere are given by

$$\sigma_1 = \frac{q_1}{4\pi R_1^2} \quad \text{and} \quad \sigma_2 = \frac{q_2}{4\pi R_2^2}.$$

Dividing gives

$$\frac{\sigma_1}{\sigma_2} = \frac{q_1}{q_2} \frac{R_2^2}{R_1^2}.$$

Combining with Eq. 29-18 yields

$$\frac{\sigma_1}{\sigma_2} = \frac{R_2}{R_1},$$

which is consistent with our qualitative statement above. Note that the larger sphere has the larger total charge but the smaller charge density.

The fact that σ, and thus **E**, can become very large near sharp points is important in the design of high-voltage equipment. *Corona discharge* can result from such points if the conducting object is raised to high potential and surrounded by air. Normally air is thought of as a nonconductor. However, it contains a small number of ions produced, for example, by the cosmic rays. A positively charged conductor will attract negative ions from the surrounding air and thus will slowly neutralize itself.

If the charged conductor has sharp points, the value of E in the air near the points can be very high. If the value is high enough, the ions, as they are drawn toward the conductor, will receive such large accelerations that, by collision with air molecules, they will produce vast additional numbers of ions. The air is thus made much more conducting, and the discharge of the conductor by this corona discharge may be very rapid indeed. The air surrounding sharp conducting points may even glow visibly because of light emitted from the air molecules during these collisions.

The electrostatic generator was conceived by Lord Kelvin in 1890 and put into useful practice in essentially its modern form by R. J. Van de Graaff in 1931. It is a device for producing electric potential differences of the order of several millions of volts. Its chief application in physics is the use of this potential difference to accelerate charged particles to

29-9
THE ELECTROSTATIC GENERATOR

---

* Equation 29-8 holds only for an isolated point charge or spherically symmetric charge distribution. The spheres must be assumed to be so far apart that the charge on either one has a negligible effect on the distribution of charge on the other.

high energies. Beams of energetic particles made in this way can be used in many different "atom-smashing" experiments. The technique is to let a charged particle "fall" through a potential difference V, gaining kinetic energy as it does so.

Let a particle of (positive) charge q move in a vacuum under the influence of an electric field from one position A to another position B whose electric potential is lower by V. The electric potential energy of the system is reduced by qV because this is the work that an external agent would have to do to restore the system to its original condition. This decrease in potential energy appears as kinetic energy of the particle, or

$$K = qV. \qquad (29\text{-}19)$$

K is in joules if q is in coulombs and V in volts. If the particle is an electron or a proton, q will be the quantum of charge e.

If we adopt the quantum of charge e as a unit in place of the coulomb, we arrive at another unit for energy, the *electron volt* (abbr. eV), which we have used before and is used extensively in atomic and nuclear physics. By substituting into Eq. 29-19,

1 electron volt = (1 quantum of charge)(1 volt)

$\qquad = (1.60 \times 10^{-19} \text{ coulomb})(1.00 \text{ volt})$

$\qquad = 1.60 \times 10^{-19} \text{ joule}.$

The electron volt can be used interchangeably with any other energy unit. Thus a 10-g object moving at 1000 cm/s can be said to have a kinetic energy of 3.1×10^{18} eV. Most physicists would prefer to express this result as 0.50 J, the electron volt being inconveniently small. In atomic, nuclear, and elementary particle physics, however, the electron volt (eV) and its multiples the keV (= 10^3 eV), MeV (= 10^6 eV), and the GeV (= 10^9 eV) are the usual units of choice.

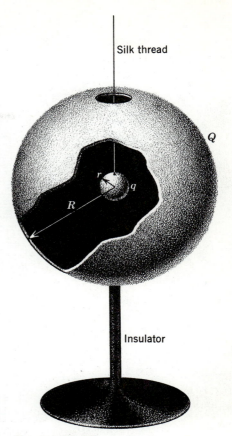

figure 29-20
Example 12. A small charged sphere of radius r is suspended inside a charged spherical shell whose outer surface has radius R.

EXAMPLE 12

The electrostatic or Van de Graaff generator. Figure 29-20, which illustrates the basic operating principle of the electrostatic generator, shows a small sphere of radius r placed inside a large spherical shell of radius R. The two spheres carry charges q and Q, respectively. Calculate their potential difference.

The potential of the large sphere is caused in part by its own charge and in part because it lies in the field set up by the charge q on the small sphere. From Eq. 29-8,

$$V_R = \frac{1}{4\pi\epsilon_0}\left(\frac{Q}{R} + \frac{q}{R}\right).$$

The potential of the small sphere is caused in part by its own charge and in part because it is inside the large sphere; see Fig. 29-18a. From Eq. 29-8,

$$V_r = \frac{1}{4\pi\epsilon_0}\left(\frac{q}{r} + \frac{Q}{R}\right).$$

The potential difference is

$$V_r - V_R = \frac{q}{4\pi\epsilon_0}\left(\frac{1}{r} - \frac{1}{R}\right).$$

Thus, assuming q is positive, the inner sphere will always be higher in potential than the outer sphere. If the spheres are connected by a fine wire, the charge q will flow *entirely* to the outer sphere, regardless of the charge Q that may already be present.

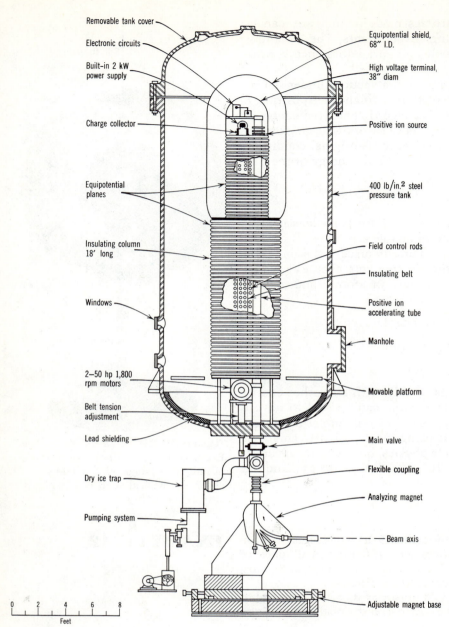

Removable tank cover

Electronic circuits

Built-in 2 kW power supply

Charge collector

Equipotential planes

Insulating column 18' long

Windows

2–50 hp 1,800 rpm motors

Belt tension adjustment

Lead shielding

Dry ice trap

Pumping system

Equipotential shield, 68" I.D.

High voltage terminal, 38" diam

Positive ion source

400 lb/in.² steel pressure tank

Field control rods

Insulating belt

Positive ion accelerating tube

Manhole

Movable platform

Main valve

Flexible coupling

Analyzing magnet

Beam axis

Adjustable magnet base

0 2 4 6 8
Feet

figure 29-21
An electrostatic generator at MIT capable of producing 9-MeV protons. The proton beam is accelerated vertically downward, being deflected into a horizontal plane by the analyzing magnet shown at the bottom. (Courtesy of J. G. Trump.)

From another point of view, we note that since the spheres when electrically connected form a single conductor at electrostatic equilibrium there can be only a single potential. This means that $V_r - V_R = 0$, which can occur only if $q = 0$.

In actual electrostatic generators charge is carried into the shell on rapidly moving belts made of insulating material. Charge is "sprayed" onto the belts outside the shell by corona discharge from a series of sharp metallic points connected to a source of moderately high potential difference. Charge is removed from the belts inside the shell by a similar series of points connected to the shell. Electrostatic generators can be built commercially to accelerate protons to energies up to 10 MeV, using a single acceleration. Figure 29-21 shows a schematic diagram of an electrostatic generator at MIT that can produce 9-MeV protons.

Generators can be built in which the accelerated particles are subject to two or three successive accelerations.

questions

1. Are we free to call the potential of the earth +100 volts instead of zero? What effect would such an assumption have on measured values of (a) potentials and (b) potential differences?

2. What would happen to a person on an insulated stand if his potential was increased by 10,000 volts?

3. Do electrons tend to go to regions of high potential or of low potential?

4. Suppose that the earth has a net charge that is *not* zero. Is it still possible to adopt the earth as a standard reference point of potential and to assign the potential $V=0$ to it?

5. Does the potential of a positively charged insulated conductor have to be positive? Give an example to prove your point.

6. Can two different equipotential surfaces intersect?

7. An electrical worker is accidentally electrocuted and a newspaper account reported: "He accidentally touched a high voltage cable and 20,000 volts of electricity surged through his body." Criticize this statement.

8. Does the amount of work per unit charge required to transfer electric charge from one point to another in an electrostatic field depend on the amount of charge transferred?

9. Advice to mountaineers caught in lightning and thunderstorms is (a) get rapidly off peaks and ridges and (b) put both feet together and crouch in the open, only the feet touching the ground. What is the basis for this good advice?

10. If **E** equals zero at a given point, must V equal zero for that point? Give some examples to prove your answer.

11. If you know **E** at a given point, can you calculate V at that point? If not, what further information do you need?

12. In Fig. 29-2 is the electric field **E** greater at the left or the right side of the figure?

13. In Fig. 29-6 is it necessary to assume that A, B, and q lie on a straight line in order to prove that Eq. 29-8 is true?

14. Is the uniformly charged, nonconducting disk of Example 6 a surface of constant potential? Explain.

15. Why can an isolated atom not have a permanent electric dipole moment?

16. If V equals a constant throughout a given region of space, what can you say about **E** in that region?

17. In Section 16-6 we saw that the gravitational field strength is zero inside a spherical shell of matter. The electrical field strength is zero not only inside an isolated charged spherical conductor but inside an isolated conductor of *any* shape. Is the gravitational field strength inside, say, a cubical shell of matter zero? If not, in what respect is the analogy not complete?

18. How can you insure that the electric potential in a given region of space will have a constant value?

19. A charge is placed on an insulated conductor in the form of a perfect cube. What will be the relative charge density at various points on the cube (surfaces, edges, corners); what will happen to the charge if the cube is in air?

20. We have seen (Fig. 29-18a) that the potential inside a thin conducting spherical shell is the same as that on its surface. (a) What if the sphere is solid? (b) What if it is solid and irregularly shaped? (c) What if it is solid, is irregularly shaped, and has an irregularly shaped cavity inside? In particular, what is V within the hollow cavity? (d) Same as (c) but suppose a point charge is suspended within the cavity?

21. A closed metal box in the form of a pyramid is placed on an insulating support and charged to a potential $+V$. Is the average potential inside the pyramid (a) greater than, (b) equal to, or (c) less than V?

22. An isolated conducting spherical shell carries a negative charge. What will happen if a positively charged metal object is placed in contact with the shell interior? Assume that the positive charge is (a) less than, (b) equal to, and (c) greater than the negative charge in magnitude.

23. An uncharged metal sphere suspended by a silk thread is placed in a uniform external electric field **E**. What is the magnitude of the electric field for points inside the sphere? Is your answer changed if the sphere carries a charge?

problems

SECTION 29-1

1. In a typical lightning flash the potential difference between discharge points is about 10^9 V and the quantity of charge transferred is about 30 C. How much ice would it melt at 0°C if all the energy released could be used for this purpose?   *Answer:* 99 tons.

2. A charge q is distributed uniformly throughout a nonconducting spherical volume of radius R. (a) Show that the potential a distance r from the center, where $r < R$, is given by

$$V = \frac{q(3R^2 - r^2)}{8\pi\epsilon_0 R^3}.$$

(b) Is it reasonable that, according to this expression, V is not zero at the center of the sphere?

3. A gold nucleus contains a positive charge equal to that of 79 protons. An α-particle ($Z = 2$) has a kinetic energy K at points far from this charge and is traveling directly toward the charge. The particle just touches the surface of the charge (assumed spherical) and is reversed in direction. (a) Calculate K, assuming a nuclear radius of 5.0×10^{-15} m. (b) The actual α-particle energy used in the experiments of Rutherford and his collaborators (see Section 28-9) was 5.0 MeV. What do you conclude?
Answer: (a) 45 MeV. (b) In these important experiments the α-particles approached, but did not "touch" the gold nuclei.

4. (a) Through what potential difference must an electron fall, according to Newtonian mechanics, to acquire a speed v equal to the speed c of light? (b) Newtonian mechanics fails as $v \rightarrow c$. Therefore, using the correct relativistic expression for the kinetic energy

$$K = mc^2 \left[\frac{1}{\sqrt{1 - (v/c)^2}} - 1 \right]$$

in place of the Newtonian expression $K = \frac{1}{2}mv^2$, determine the actual electron speed acquired in falling through the potential difference computed in (a). Express this speed as an appropriate fraction of the speed of light.

SECTION 29-2

5. An infinite charged sheet has a surface charge density σ of 1.0×10^{-7} C/m². How far apart are the equipotential surfaces whose potentials differ by 5.0 V?
Answer: 0.89 mm.

6. Two large parallel conducting plates are 10 cm apart and carry equal but opposite charges on their facing surfaces. An electron placed midway between the two plates experiences a force of 1.6×10^{-15} N. What is the potential difference between the plates?

7. In the Millikan oil drop experiment (see Fig. 27-29) an electric field of 1.92×10^5 N/C is maintained at balance across two plates separated by 1.50 cm. Find the potential difference between the plates.   *Answer:* 2900 V.

SECTION 29-3

8. Consider a point charge with $q = 1.5 \times 10^{-8}$ C. (a) What is the radius of an equipotential surface having a potential of 30 V? (b) Are surfaces whose potentials differ by a constant amount (1.0 V, say) evenly spaced in radius?

SECTION 29-4

9. A point charge has $q = +1.0 \times 10^{-6}$ C. Consider point A which is 2.0 m distant and point B which is 1.0 m distant in a direction diametrically opposite, as in Fig. 29-22a. (a) What is the potential difference $V_A - V_B$? (b) Repeat if points A and B are located as in Fig. 29-22b.
Answer: (a) −4500 V. (b) Same as (a) because potential is a scalar quantity.

10. In Fig. 29-23, locate the points (a) where $V = 0$ and (b) where $\mathbf{E} = 0$. Consider only points on the axis and choose $d = 1.0$ m.

11. In Fig. 29-23 (see Problem 10) sketch qualitatively (a) the lines of force and (b) the intersections of the equipotential surfaces with the plane of the figure. (Hint: Consider the behavior close to each point charge and at considerable distances from the pair of charges.)

12. (a) In Fig. 29-24 derive an expression for $V_A - V_B$. (b) Does your result reduce to the expected answer when $d = 0$? When $q = 0$?

13. For the charge configuration of Fig. 29-25, show that $V(r)$ for points on the vertical axis, assuming $r \gg a$, is given by

$$V = \frac{1}{4\pi\epsilon_0}\left(\frac{q}{r} + \frac{2qa}{r^2}\right).$$

Is this an expected result? (Hint: The charge configuration can be viewed as the sum of an isolated charge and a dipole.)

SECTION 29-5

14. Calculate the dipole moment of a water molecule under the assumption that all ten electrons in the molecule circulate symmetrically about the oxygen atom, that the OH distance is 0.96×10^{-8} cm, and that the angle between the two OH bonds is 104°. Compare with the value quoted on p. 631; see Fig. 29-11.

SECTION 29-6

15. A particle of mass m, charge $q > 0$, and initial kinetic energy K is projected (from "infinity") toward a heavy nucleus of charge Q, assumed to have a fixed position in our reference frame. (a) If the aim is "perfect," how close to the center of the nucleus is the particle when it comes instantaneously to rest? (b) With a particular imperfect aim the particle's closest approach to the nucleus is twice the distance determined in part (a). Determine the speed of the particle at this closest distance of approach.
Answer: (a) $qQ/4\pi\epsilon_0 K$. (b) $\sqrt{K/m}$.

16. The charges and coordinates of two charges located on the x-y plane are: $q_1 = +3.0 \times 10^{-6}$ C; $x = +3.5$ cm, $y = +0.50$ cm, and $q_2 = -4.0 \times 10^{-6}$ C; $x = -2.0$ cm, $y = +1.5$ cm. (a) Find the electric potential at the origin. (b) How much work must be done to locate these charges at their given positions, starting from infinity?

17. Derive an expression for the work required to put the four charges together as indicated in Fig. 29-26.   *Answer:* $-0.21q^2/\epsilon_0 a$.

18. A particle of charge Q is kept in a fixed position at a point P and a second particle of mass m, having the same charge Q, is initially held at rest a distance r_1 from P. The second particle is then released and is repelled from the first one. Determine its speed at the instant it is a distance r_2 from P. Let $Q = 3.1 \times 10^{-6}$ C, $m = 2.0 \times 10^{-5}$ kg, $r_1 = 9.0 \times 10^{-4}$ m, and $r_2 = 25 \times 10^{-4}$ m.

19. What is the electric potential energy of the charge configuration of Fig. 29-6? Use the numerical values of Example 5.   *Answer:* -6.4×10^{-7} J.

20. In the rectangle shown in Fig. 29-27, the sides have lengths 5.0 cm and 15.0 cm, $q_1 = -5.0 \times 10^{-6}$ C and $q_2 = +2.0 \times 10^{-6}$ C. (a) What is the electric potential at corner B? At corner A? (b) How much work is involved in moving

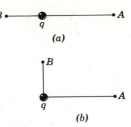

(a)

(b)

figure 29-22
Problem 9

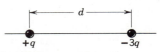

figure 29-23
Problems 10, 11

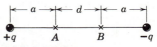

figure 29-24
Problem 12

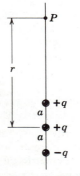

figure 29-25
Problem 13

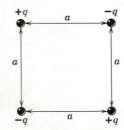

figure 29-26
Problem 17

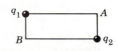

figure 29-27
Problem 20

a third charge $q_3 = +3.0 \times 10^{-6}$ C *from B to A* along a diagonal of the rectangle? (c) In this process, is external work converted into electrostatic potential energy or vice versa? Explain.

21. Two charges q ($=+2.0 \times 10^{-6}$ C) are fixed in space a distance d (2.0 cm) apart, as shown in Fig. 29-28. (a) What is the electric potential at point C? (b) You bring a third charge q ($+2.0 \times 10^{-6}$ C) very slowly from infinity to C. How much work must you do? (c) What is the potential energy U of the configuration when the third charge is in place?
    Answer: (a) 2.5×10^6 V. (b) 5.1 J. (c) 6.9 J.

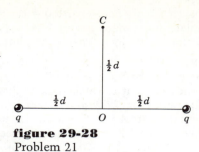

figure 29-28
Problem 21

22. Three charges of $+0.1$ C each are placed on the corner of an equilateral triangle, 1.0 m on a side. If energy is supplied at the rate of 1.0 kW, how many days would be required to move one of the charges onto the midpoint of the line joining the other two?

23. Two electrons are 2.0 m apart. Another electron is shot from infinity and comes to rest midway between the two. What must its initial velocity be?
    Answer: 32 m/s.

24. Calculate (a) the electric potential established by the nucleus of a hydrogen atom at the mean distance of the circulating electron ($r = 5.3 \times 10^{-11}$ m), (b) the electric potential energy of the atom when the electron is at this radius, and (c) the kinetic energy of the electron, assuming it to be moving in a circular orbit of this radius centered on the nucleus. (d) How much energy is required to ionize the hydrogen atom? Express all energies in electron volts.

25. A particle of (positive) charge Q is assumed to have a fixed position at P. A second particle of mass m and (negative) charge $-q$ moves at constant speed in a circle of radius r_1, centered at P. Derive an expression for the work W that must be done by an external agent on the second particle in order to increase the radius of the circle of motion, centered at P, to r_2.
    Answer: $W = \dfrac{qQ}{8\pi\epsilon_0} \left[\dfrac{1}{r_1} - \dfrac{1}{r_2} \right]$.

26. Devise an arrangement of three point charges, separated by finite distances, that has zero electric potential energy.

27. Figure 29-29 shows an idealized representation of a U^{238} nucleus ($Z=92$) on the verge of fission. Calculate (a) the repulsive force acting on each fragment and (b) the mutual electric potential energy of the two fragments. Assume that the fragments are equal in size and charge, spherical, and just touching. The radius of the initially spherical U^{238} nucleus is 8.0×10^{-15} m. Assume that the material out of which nuclei are made has a constant density.   *Answer:* (a) 3.0×10^3 N. (b) 3.8×10^{-11} J, or 240 MeV.

figure 29-29
Problem 27

SECTION 29-7

28. The electric potential varies along the x-axis as shown in the graph of Fig. 29-30. For each of the intervals shown (ignore the behavior at the end points of the intervals) determine the x-component of the electric field and plot E_x vs. x.

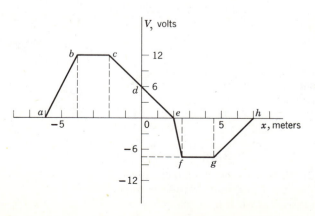

figure 29-30
Problem 28

29. (a) Show that the electric potential at a point on the axis of a ring of charge of radius a, computed directly from Eq. 29-10, is given by

$$V = \frac{1}{4\pi\epsilon_0} \frac{q}{\sqrt{x^2 + a^2}}.$$

(b) From this result derive an expression for E at axial points; compare with the direct calculation of E in Example 5, Chapter 27.

30. What is the potential gradient, in V/m, at the surface of a gold nucleus? See Problem 3.

31. In Example 6 the potential at an axial point for a charged disk was shown to be

$$V = \frac{\sigma}{2\epsilon_0} (\sqrt{a^2 + r^2} - r).$$

(a) From this result show that E for axial points is given by

$$E = \frac{\sigma}{2\epsilon_0} \left(1 - \frac{r}{\sqrt{a^2 + r^2}}\right).$$

(b) Does this expression for E reduce to an expected result for $r \gg a$ and for $r = 0$?

32. (a) Starting from Eq. 29-11, find the magnitude E_r of the radial component of the electric field due to a dipole. (b) For what value of θ is E_r zero?

33. A charge per unit length λ is distributed uniformly along a straight-line segment of length L. (a) Determine the electrostatic potential (chosen to be zero at infinity) at a point P a distance y from one end of the charged segment and in line with it (see Fig. 29-31). (b) Use the result of (a) to compute the component of the electric field at P in the y-direction (along the line). (c) Determine the component of the electric field at P in a direction perpendicular to the straight line.

Answer: (a) $\dfrac{\lambda}{4\pi\epsilon_0} \ln \dfrac{L+y}{y}.$

(b) $\dfrac{\lambda}{4\pi\epsilon_0} \dfrac{L}{y(L+y)}.$

(c) Zero.

figure 29-31
Problem 33

34. On a thin rod of length L lying along the x-axis with one end at the origin $(x = 0)$, as in Fig. 29-32, there is distributed a charge per unit length given by $\lambda = kx$, where k is a constant. (a) Taking the electrostatic potential at infinity to be zero, find V at the point P on the y-axis. (b) Determine the vertical component, E_y, of the electric field intensity at P from the result of part (a) and also by direct calculation. (c) Why cannot E_x, the horizontal component of the electric field at P, be found using the result of part (a)?

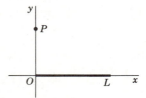

figure 29-32
Problem 34

SECTION 29-8
35. What is the charge density σ on the surface of a conducting sphere of radius 0.15 m whose potential is 200 V?   Answer: 1.2×10^{-8} C/m^2.

36. A charged sphere of radius 1.5 m contains a total charge of 3.0×10^{-6} C. (a) What is the electric field at the sphere's surface? (b) What is the electric potential at the sphere's surface? (c) At what distance from the sphere's surface has the electric potential decreased by 5000 V?

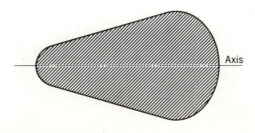

figure 29-33
Problem 37

37. The metal object in Fig. 29-33 is a figure of revolution about the horizontal axis. If it is charged negatively, sketch roughly a few equipotentials and lines of force. Use physical reasoning rather than mathematical analysis.

38. If the earth had a net charge equivalent to 1 electron/m² of surface area (a very artificial assumption), (a) what would the earth's potential be? (b) What would the electric field due to the earth be just outside its surface?

39. A charge of 10^{-8} C can be produced by simple rubbing. To what potential would such a charge raise an insulated conducting sphere of 10-cm radius? *Answer:* 900 V.

40. For the spheres of Fig. 29-19, what is the ratio of electric fields at the surface?

41. Consider a thin, isolated, conducting, spherical shell which is uniformly charged to a constant charge density σ (C/m²). How much work does it take to move a small positive test charge q_0 (a) from the surface of the shell to the interior, through a small hole, (b) from one point on the surface to another, regardless of path, (c) from point to point inside the shell, and (d) from any point P outside the shell over any path, whether or not it pierces the shell, back to P? (e) For the conditions given, does it matter whether or not the shell is conducting?   *Answer:* (a) Zero. (b) Zero. (c) Zero. (d) Zero. (e) No.

42. A thin, spherical, conducting shell of radius R is mounted on an insulated support and charged to a potential $-V$. An electron is fired from point P a distance r from the center of the shell $(r \gg R)$ with an initial speed v_0, directed radially inward. What is the value of v_0 chosen so that the electron will just reach the shell?

43. Can a conducting sphere 10 cm in radius hold a charge of 4×10^{-6} C in air without breakdown? The dielectric strength (minimum field required to produce breakdown) of air at 1 atm is 3×10^6 V/m.   *Answer:* No.

44. Two thin, insulated, concentric conducting spheres of radii R_1 and R_2 carry charges q_1 and q_2. Derive expressions for $E(r)$ and $V(r)$, where r is the distance from the center of the spheres. Plot $E(r)$ and $V(r)$ from $r = 0$ to $r = 4.0$ m for $R_1 = 0.50$ m, $R_2 = 1.0$ m, $q_1 = +2.0 \times 10^{-6}$ C, and $q_2 = +1.0 \times 10^{-6}$ C. Compare with Fig. 29-18.

45. The space between two concentric spheres of radii r_1 and r_2 is filled with a nonconducting material having a uniform charge density ρ. Find the electric potential V as a function of the distance r from the center of the spheres, considering the regions (a) $r > r_2$, (b) $r_2 > r > r_1$, and (c) $r < r_1$. (d) Do these solutions agree at $r = r_2$ and at $r = r_1$?

   Answer: (a) $\dfrac{\rho}{3\epsilon_0} \dfrac{(r_2{}^3 - r_1{}^3)}{r}$. (b) $\dfrac{\rho}{3\epsilon_0}\left(\dfrac{3}{2}r_2{}^2 - \dfrac{r^2}{2} - \dfrac{r_1{}^3}{r}\right)$. (c) $\dfrac{\rho}{2\epsilon_0}(r_2{}^2 - r_1{}^2)$. (d) Yes.

46. Two metal spheres are 3.0 cm in radius and carry charges of $+1.0 \times 10^{-8}$ C and -3.0×10^{-8} C, respectively, assumed to be uniformly distributed. If their centers are 2.0 m apart, calculate (a) the potential of the point halfway between their centers and (b) the potential of each sphere.

47. Two identical conducting spheres of radius $r = 0.15$ m are separated by a distance $a = 10.0$ m. What is the charge on each sphere if the potential of one is $+1500$ V and if the other is -1500 V?   *Answer:* $\pm 2.5 \times 10^{-8}$ C.

48. Two conducting spheres, one of radius 6.0 cm and the other of radius 12.0 cm, each have a charge of 3×10^{-8} C and are very far apart. If the spheres are connected by a conducting wire, find (a) the direction of motion and the magnitude of the charge transferred and (b) the final charge on and potential of each sphere.

49. In Fig. 29-19 let $R_1 = 1.0$ cm and $R_2 = 2.0$ cm. Before the spheres are connected by the fine wire, a charge of 2.0×10^{-7} C is placed on the smaller sphere, the larger sphere being uncharged. Calculate (a) the charge, (b) the surface charge density, and (c) the potential for each sphere after they are connected.

   Answer: (a) $q_1 = 0.67 \times 10^{-7}$ C, $q_2 = 1.33 \times 10^{-7}$ C. (b) $\sigma_1 = 5.33 \times 10^{-5}$ C/m², $\sigma_2 = 2.65 \times 10^{-5}$ C/m². (c) $V_1 = V_2 = 6.0 \times 10^4$ V.

50. A spherical drop of water carrying a charge of 3×10^{-11} C has a potential of 500 V at its surface. (a) What is the radius of the drop? (b) If two such drops of the same charge and radius combine to form a single spherical drop, what is the potential at the surface of the new drop so formed?

SECTION 29-9

51. (a) How much charge is required to raise an isolated metallic sphere of 1.0-m radius to a potential of 1.0×10^6 V? Repeat for a sphere of 1.0-cm radius. (b) Why use a large sphere in an electrostatic generator since the same potential can be achieved for a smaller charge with a small sphere?
    Answer: (a) 1.1×10^{-4} C; 1.1×10^{-6} C. (b) Because of the larger E field at the surface of the smaller sphere (verify), charge will leak off more rapidly.

52. An alpha particle is accelerated through a potential difference of one million volts in an electrostatic generator. (a) What kinetic energy does it acquire? (b) What kinetic energy would a proton acquire under these same circumstances? (c) Which particle would acquire the greater speed, starting from rest?

53. Let the potential difference between the shell of an electrostatic generator and the point at which charges are sprayed onto the moving belt be 3.0×10^6 V. If the belt transfers charge to the shell at the rate of 3.0×10^{-3} C/s, what power must be provided to drive the belt, considering only electrical forces?
    Answer: 9.0 kW.

54. The high-voltage electrode of an electrostatic generator is a charged spherical metal shell having a potential V ($+9.0 \times 10^6$ V). (a) Electrical breakdown occurs in the gas in this machine at a field E (1.0×10^8 V/m). In order to prevent such breakdown, what restriction must be made on the radius r of the shell? (b) A long moving rubber belt transfers charge to the shell at 3.0×10^{-4} C/s, the potential of the shell remaining constant because of leakage. What minimum power is required to transfer the charge? (c) The belt is of width w (0.50 m) and travels at speed v (30 m/s). What is the surface charge density on the belt?

30
capacitors
and dielectrics

Figure 30-1 shows a generalized *capacitor*, consisting of two insulated conductors, *a* and *b*, of arbitrary shape (later, regardless of their geometry, we will call them *plates*). We assume that they are totally isolated from objects in their surroundings and carry equal and opposite charges +*q* and −*q*, respectively. Every line of force that originates on *a* terminates on *b*. We further assume, for the time being, that the conductors *a* and *b* exist in a vacuum.

The capacitor of Fig. 30-1 is characterized by *q*, the magnitude of the *charge on either conductor*, and by *V*, the *potential difference between the conductors*. Note (*a*) that *q* is *not* the net charge on the capacitor, which is zero, and (*b*) that *V* is *not* the potential of either conductor, referred perhaps to $V \to 0$ at ∞, but the potential difference between them.

It is not hard to put equal and opposite charges on conductors such as those of Fig. 30-1. We need not charge them separately. All that we have to do is to connect the conductors momentarily to opposite poles of a battery. Equal and opposite charges (±*q*) will automatically appear.

For the moment we state without proof that *q* and *V* for a capacitor are proportional to each other, or

$$q = CV \qquad (30\text{-}1)$$

in which *C*, the constant of proportionality, is called the *capacitance* of the capacitor. We state, again for the moment without proof, that *C* depends on the shapes and relative positions of the conductors. We will show in Section 30-2, for three important special cases, that *C* does indeed depend on these variables. *C* also depends on the medium in

650

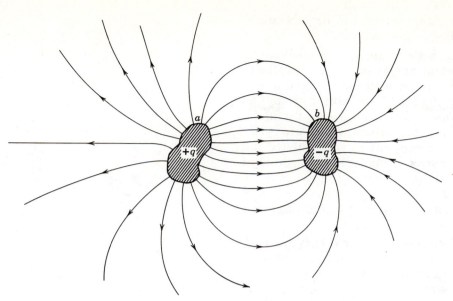

figure 30-1
Two insulated conductors, totally isolated from their surroundings and carrying equal and opposite charges, form a capacitor.

which the conductors are immersed but for the present (see Section 30-4 however) we assume this to be a vacuum.

The SI unit of capacitance that follows from Eq. 30-1 is the coulomb/volt. A special unit, the *farad* (abbr. F), is used to represent it. It is named in honor of Michael Faraday who, among other contributions, developed the concept of capacitance. Thus

$$1 \text{ farad} = 1 \text{ coulomb/volt.}$$

The submultiples of the farad, the *microfarad* (1 μF = 10^{-6} F) and the *picofarad* (1 pF = 10^{-12} F), are more convenient units in practice.

An analogy can be made between a capacitor carrying a charge q and a rigid container of volume V containing n moles of an ideal gas. The gas pressure p is directly proportional to n for a fixed temperature, according to the ideal gas law (Eq. 23-2)

$$n = \left(\frac{V}{RT}\right) p.$$

For the capacitor (Eq. 30-1)

$$q = (C)V.$$

Comparison shows that the capacitance C of the capacitor, assuming a fixed temperature for the gas, is analogous to the volume V of the container.

Note that any amount of charge can be put on the capacitor, and any mass of gas can be put in the container, up to certain limits. These correspond to electrical breakdown ("arcing over") for the capacitor and to rupture of the walls for the container.

Capacitors are very useful devices, of great interest to engineers and physicists. For example:

1. In this book we stress the importance of *fields* to the understanding of natural phenomena. A capacitor can be used to establish desired electric field configurations for various purposes. In Section 27-5 we described the deflection of an electron beam in a uniform field set up by a capacitor, although we did not use this term in that section. In later sections we discuss the behavior of dielectric materials when placed in an electric field (provided conveniently by a capacitor) and

we shall see how the laws of electromagnetism can be generalized to take the presence of dielectric bodies into account.

2. A second concept stressed in this book is *energy*. By analyzing a charged capacitor we show that electric energy may be considered to be stored in the electric field between the plates and indeed in any electric field, however generated. Because capacitors can confine strong electric fields to small volumes, they can serve as useful devices for storing energy. In electron synchrotrons, which are cyclotron-like devices for accelerating electrons, energy accumulated and stored in a large bank of capacitors over a relatively long period of time is made available intermittently to accelerate the electrons by discharging the capacitor in a much shorter time. Many researches and devices in plasma physics also make use of bursts of energy stored in this way.

3. The electronic age could not exist without capacitors. They are used, in conjunction with other devices, to reduce voltage fluctuations in electronic power supplies, to transmit pulsed signals, to generate or detect electromagnetic oscillations at radio frequencies, to provide electronic time delays, and in many other ways. In most of these applications the potential difference between the plates will not be constant, as we assume in this chapter, but will vary with time, often in a sinusoidal or a pulsed fashion. In later chapters we consider some aspects of the capacitor used as a circuit element.

Figure 30-2 shows a *parallel-plate* capacitor in which the conductors of Fig. 30-1 take the form of two parallel plates of area A separated by a distance d. If we connect each plate momentarily to the terminals of a battery, a charge $+q$ will automatically appear on one plate and a charge $-q$ on the other. If d is small compared with the plate dimensions, the electric field $\mathbf{E}$ between the plates will be uniform, which means that the lines of force will be parallel and evenly spaced. The laws of electromagnetism (see Problem 14, Chapter 35) require that there be some "fringing" of the lines at the edges of the plates, but for small enough d we can neglect it for our present purpose.

We can calculate the capacitance of this device using Gauss's law, another illustration of the usefulness of this law in situations of simple geometry. Figure 30-2 shows (dashed lines) a Gaussian surface of height h closed by plane caps of area A that are the shape and size of the capacitor plates. The flux of $\mathbf{E}$ is zero for the part of the Gaussian surface that lies inside the top capacitor plate because the electric field inside a conductor carrying a static charge is zero. The flux of $\mathbf{E}$ through the vertical wall of the Gaussian surface is zero because, to the extent that we can neglect the fringing of the lines of force, $\mathbf{E}$ lies in the wall.

This leaves only the face of the Gaussian surface that lies between

30-2
CALCULATING CAPACITANCE

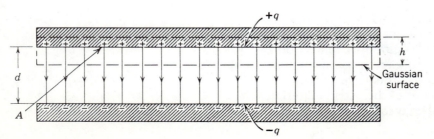

+q

d

h

Gaussian surface

A

−q

figure 30-2
A parallel-plate capacitor with conductors (plates) of area A. The dashed line represents a Gaussian surface whose height is h and whose top and bottom caps are the same shape and size as the capacitor plates.

the plates. Here $\mathbf{E}$ is constant and the flux Φ_E is simply EA. Gauss's law (Eq. 28-5) gives

$$\epsilon_0 \Phi_E = \epsilon_0 EA = q. \qquad (30\text{-}2)$$

The work required to carry a test charge q_0 from one plate to the other can be expressed either as $q_0 V$ (see Eq. 29-1) or as the product of a force $q_0 E$ times a distance d or $q_0 Ed$. These expressions must be equal, or

$$V = Ed. \qquad (30\text{-}3)$$

Formally, Eq. 30-3 is a special case of the general relation (Eq. 29-5; see also Example 1, Chapter 29)

$$V = -\int \mathbf{E} \cdot d\mathbf{l},$$

where V is the difference in potential between the plates. The integral may be taken over any path that starts on one plate and ends on the other because each plate is an equipotential surface and the electrostatic force is path independent. Although the simplest path between the plates is a perpendicular straight line, Eq. 30-3 follows no matter what path of integration we choose.

If we substitute Eqs. 30-2 and 30-3 into the relation $C = q/V$, we obtain

$$C = \frac{q}{V} = \frac{\epsilon_0 EA}{Ed} = \frac{\epsilon_0 A}{d}. \qquad (30\text{-}4)$$

Equation 30-4 holds only for capacitors of the parallel-plate type; different formulas hold for capacitors of different geometry. This equation shows, for a particular case, that C does indeed depend on the geometry of the conductors (plates) as we pointed out in Section 30-1. Both A and d are geometrical factors.

In Section 26-4 we stated that ϵ_0, which we first met in connection with Coulomb's law, was not measured in terms of that law because of experimental difficulties. Equation 30-4 suggests that we might determine ϵ_0 by building a capacitor of accurately known plate area and plate spacing and measuring its capacitance experimentally by measuring q and V in the relation $C = q/V$. Thus we can solve Eq. 30-4 for ϵ_0 and find a numerical value in terms of the measured quantities A, d, and C; ϵ_0 has indeed been measured in this way.

EXAMPLE 1

The parallel plates of an air-filled capacitor are everywhere 1.0 mm apart. What must the plate area be if the capacitance is to be 1.0 F?

From Eq. 30-4

$$A = \frac{dC}{\epsilon_0} = \frac{(1.0 \times 10^{-3} \text{ m})(1.0 \text{ F})}{8.9 \times 10^{-12} \text{ C}^2/\text{N} \cdot \text{m}^2} = 1.1 \times 10^8 \text{ m}^2.$$

This is the area of a square sheet more than 6 miles on edge; the farad is indeed a large unit.

EXAMPLE 2

A cylindrical capacitor. A cylindrical capacitor consists of two coaxial cylinders (Fig. 30-3) of radii a and b and length l. What is the capacitance of this device? Assume that the capacitor is very long (that is, that $l \gg b$) so that we can ignore the fringing of the lines of force at the ends for the purpose of calculating the capacitance.

As a Gaussian surface construct a coaxial cylinder of radius r and length l, closed by plane caps. Gauss's law (Eq. 28-6)

$$\epsilon_0 \oint \mathbf{E} \cdot d\mathbf{S} = q$$

gives

$$\epsilon_0 E(2\pi r)(l) = q,$$

the flux being entirely through the cylindrical surface and not through the end caps. Solving for E yields

$$E = \frac{q}{2\pi\epsilon_0 rl}.$$

The potential difference between the plates is given by Eq. 29-5 [note that $\mathbf{E}$ and $d\mathbf{l}$ $(= d\mathbf{r})$ point in opposite directions] or

$$V = -\int_a^b \mathbf{E} \cdot d\mathbf{l} = \int_a^b E \, dr = \int_a^b \frac{q}{2\pi\epsilon_0 l} \frac{dr}{r} = \frac{q}{2\pi\epsilon_0 l} \ln \frac{b}{a}.$$

Finally, the capacitance is given by

$$C = \frac{q}{V} = \frac{2\pi\epsilon_0 l}{\ln(b/a)}.$$

Like the relation for the parallel-plate capacitor (Eq. 30-4), this relation also depends only on geometrical factors, l, b, and a.

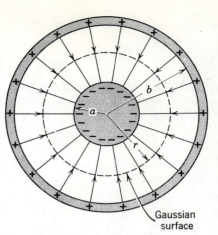

figure 30-3
Example 2. A cross section of a cylindrical capacitor. The dashed circle is a cross section of a closed cylindrical Gaussian surface of radius r and length l.

EXAMPLE 3

The Capacitance of an isolated sphere. In Section 29-8 we showed that the potential of an isolated conducting sphere of radius R carrying a charge q is given by

$$V = \frac{1}{4\pi\epsilon_0} \frac{q}{R}. \qquad (30\text{-}5)$$

We can regard this sphere as one plate of a capacitor, the other plate being a conducting sphere of infinite radius, with V chosen to be zero on the sphere at infinity.

The capacitance of the sphere of radius R is then given, from Eq. 30-5, by

$$C = \frac{q}{V} = 4\pi\epsilon_0 R. \qquad (30\text{-}6)$$

Again, the only relevant geometric factor, the sphere radius R, appears.

EXAMPLE 4

Capacitors in parallel. Figure 30-4 shows three capacitors connected in parallel. What single capacitance C is equivalent to this combination? "Equivalent" means that if the parallel combination and the single capacitor were each in a box with wires a and b connected to terminals, it would not be possible to distinguish the two by electrical measurements external to the box.

The potential difference across each capacitor in a parallel arrangement (Fig. 30-4) *is the same.* This follows because all of the upper plates are connected together and to terminal a, whereas all of the lower plates are connected together and to terminal b. Applying the relation $q = CV$ to each capacitor yields

$$q_1 = C_1 V; \qquad q_2 = C_2 V; \qquad \text{and} \qquad q_3 = C_3 V.$$

The total charge q on the combination is

$$q = q_1 + q_2 + q_3$$
$$= (C_1 + C_2 + C_3)V.$$

The equivalent capacitance C is

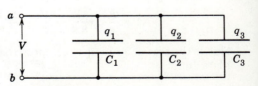

figure 30-4
Example 4. Three capacitors in parallel. The symbol for a capacitor (⊣⊢) is chosen for its simplicity. Although it suggests a parallel-plate capacitor, it is meant to suggest a capacitor of *any* geometry.

$$C = \frac{q}{V} = C_1 + C_2 + C_3.$$

We can easily extend this result to any number of parallel-connected capacitors.

EXAMPLE 5

Capacitors in series. Figure 30-5 shows three capacitors connected in series. What single capacitance C is "equivalent" (see Example 4) to this combination?

For capacitors connected in a series arrangement (Fig. 30-5) *the magnitude q of the charge on each plate must be the same.* This is true because the net charge on the part of the circuit enclosed by the dashed line in Fig. 30-5 must be zero; that is, the charge present on these plates initially is zero and connecting a battery between a and b will only produce a charge separation, the *net* charge on these plates still being zero. Assuming that neither C_1 nor C_2 "sparks over," there is no way for charge to enter or leave the volume suggested by the dashed line.

Applying the relation $q = CV$ to each capacitor yields

$$V_1 = q/C_1; \qquad V_2 = q/C_2; \qquad \text{and} \qquad V_3 = q/C_3.$$

The potential difference for the series combination is

$$V = V_1 + V_2 + V_3$$

$$= q \left(\frac{1}{C_1} + \frac{1}{C_2} + \frac{1}{C_3} \right).$$

The equivalent capacitance

$$C = \frac{q}{V} = \frac{1}{\dfrac{1}{C_1} + \dfrac{1}{C_2} + \dfrac{1}{C_3}},$$

or

$$\frac{1}{C} = \frac{1}{C_1} + \frac{1}{C_2} + \frac{1}{C_3}.$$

The equivalent series capacitance is always less than the smallest capacitance in the chain.

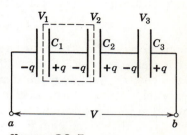

figure 30-5
Example 5. Three capacitors in series.

30-3
ENERGY STORAGE IN AN ELECTRIC FIELD

In Section 29-6 we saw that all charge configurations have a certain *electric potential energy U*, equal to the work W (which may be positive or negative) that must be done to assemble them from their individual components, originally assumed to be infinitely far apart and at rest. This potential energy reminds us of the potential energy stored in a compressed spring or the gravitational potential energy stored in, say, the earth-moon system.

For a simple example, work must be done to separate two equal and opposite charges. This energy is stored in the system and can be recovered if the charges are allowed to come together again. Similarly, a charged capacitor has stored in it an electrical potential energy U equal to the work W required to charge it. This energy can be recovered if the capacitor is allowed to discharge. We can visualize the work of charging by imagining that an external agent pulls electrons from the positive plate and pushes them onto the negative plate, thus bringing about the charge separation; normally the work of charging is done by a battery, at the expense of its store of chemical energy.

Suppose that at a time t a charge $q'(t)$ has been transferred from one plate to the other. The potential difference $V(t)$ between the plates at that moment will be $q'(t)/C$. If an extra increment of charge dq' is transferred, the small amount of additional work needed will be

$$dW = V \, dq = \left(\frac{q'}{C}\right) dq'.$$

If this process is continued until a total charge q has been transferred, the total work will be found from

$$W = \int dW = \int_0^q \frac{q'}{C} \, dq' = \frac{1}{2}\frac{q^2}{C}. \qquad (30\text{-}7)$$

From the relation $q = CV$ we can also write this as

$$W \; (= U) = \tfrac{1}{2}CV^2. \qquad (30\text{-}8)$$

It is reasonable to suppose that the energy stored in a capacitor resides in the electric field. As q or V in Eqs. 30-7 and 30-8 increase, for example, so does the electric field E; when q and V are zero, so is E.

In a parallel-plate capacitor, neglecting fringing, the electric field has the same value for all points between the plates. Thus the *energy density u*, which is the stored energy per unit volume, should also be uniform; u (see Eq. 30-8) is given by

$$u = \frac{U}{Ad} = \frac{\frac{1}{2}CV^2}{Ad},$$

where Ad is the volume between the plates. Substituting the relation $C = \epsilon_0 A/d$ (Eq. 30-4) leads to

$$u = \frac{\epsilon_0}{2}\left(\frac{V}{d}\right)^2.$$

However, V/d is the electric field E, so that

$$u = \tfrac{1}{2}\epsilon_0 E^2. \qquad (30\text{-}9)$$

Although we derived this equation for the special case of a parallel-plate capacitor, it is true in general. *If an electric field* **E** *exists at any point in space (a vacuum), we can think of that point as the site of stored energy in amount, per unit volume, of* $\tfrac{1}{2}\epsilon_0 E^2$.

A capacitor C_1 is charged to a potential difference V_0. This charging battery is then removed and the capacitor is connected as in Fig. 30-6 to an uncharged capacitor C_2.

(*a*) What is the final potential difference V across the combination?
The original charge q_0 is now shared by the two capacitors. Thus

$$q_0 = q_1 + q_2.$$

Applying the relation $q = CV$ to each of these terms yields

$$C_1 V_0 = C_1 V + C_2 V$$

or

$$V = V_0 \frac{C_1}{C_1 + C_2}.$$

This suggests a way to measure an unknown capacitance (C_2, say) in terms of a known one.

(*b*) What is the stored energy before and after the switch in Fig. 30-6 is thrown?
The initial stored energy is

$$U_0 = \tfrac{1}{2}C_1 V_0{}^2.$$

The final stored energy is

EXAMPLE 6

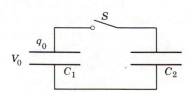

figure 30-6
Example 6. C_1 is charged to a potential difference V_0 and then connected to C_2 by closing switch S.

$$U = \tfrac{1}{2}C_1 V^2 + \tfrac{1}{2}C_2 V^2 = \tfrac{1}{2}(C_1 + C_2)\left(\frac{V_0 C_1}{C_1 + C_2}\right)^2 = \left(\frac{C_1}{C_1 + C_2}\right) U_0.$$

Thus U is less than U_0! For $C_1 = C_2$, in fact, $U = \tfrac{1}{2}U_0$.

This is *not* a violation of the principle of the conservation of energy. The example tacitly assumes an ideal (rather than an actual laboratory) circuit in that the resistance (Chapter 31) and the inductance (Chapter 36) of the connecting wires have both been assumed to be zero. In an actual laboratory circuit the "missing" energy would appear as thermal energy in the wires and/or as energy radiated away from the circuit as electromagnetic radiation (Chapter 41). For a good discussion see "On Conservation of Energy in Electric Circuits" by Camillo Cuvaj, *American Journal of Physics*, 1968.

EXAMPLE 7

An isolated conducting sphere of radius R, in a vacuum, carries a charge q. (a) Compute the total electrostatic energy stored in the surrounding space. At any radius r from the center of the sphere (assuming $r > R$) E is given by

$$E = \frac{1}{4\pi\epsilon_0}\frac{q}{r^2}.$$

The energy density at any radius r is found from Eq. 30-9, or

$$u = \tfrac{1}{2}\epsilon_0 E^2 = \frac{q^2}{32\pi^2\epsilon_0 r^4}.$$

The energy dU that lies in a spherical shell between the radii r and $r + dr$ is

$$dU = (4\pi r^2)(dr)u = \frac{q^2}{8\pi\epsilon_0}\frac{dr}{r^2},$$

where $(4\pi r^2)(dr)$ is the volume of the spherical shell. The total energy U is found by integration, or

$$U = \int dU = \frac{q^2}{8\pi\epsilon_0}\int_R^\infty \frac{dr}{r^2} = \frac{q^2}{8\pi\epsilon_0 R}.$$

Note that this relation follows at once from Eq. 30-7 $(U = q^2/2C)$, where C (see Example 3) is the capacitance $(= 4\pi\epsilon_0 R)$ of an isolated sphere of radius R.

(b) What is the radius R_0 of a spherical surface such that half the stored energy lies within it?

In the equation just given we put

$$\tfrac{1}{2}U = \frac{q^2}{8\pi\epsilon_0}\int_R^{R_0}\frac{dr}{r^2}$$

or

$$\frac{q^2}{16\pi\epsilon_0 R} = \frac{q^2}{8\pi\epsilon_0}\left(\frac{1}{R} - \frac{1}{R_0}\right),$$

which yields, after some rearrangement,

$$R_0 = 2R.$$

Thus, most of the energy is stored in space rather close to the sphere.

Equation 30-4 holds only for a parallel-plate capacitor with its plates in a vacuum. Michael Faraday, in 1837, first investigated the effect of filling the space between the plates with a dielectric (see Table 30-1 for a sampling of dielectrics used today). In Faraday's words:

The question may be stated thus: suppose A an electrified plate of metal suspended in air, and B and C two exactly similar plates, placed parallel to and on each side of A at equal distances and insulated; A will then induce equally to-

30-4
PARALLEL-PLATE CAPACITOR WITH DIELECTRIC

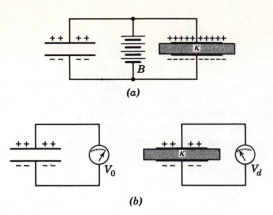

figure 30-7
(a) Battery B supplies the same potential difference to each capacitor; the one on the right has the higher charge. (b) Both capacitors carry the same charge; the one on the right has the lower potential difference, as indicated by the meter readings.

ward B and C [that is, equal charges will appear on these plates]. If in this position of the plates some other dielectric than air, as shell-lac, be introduced between A and C, will the induction between them remain the same? Will the relation of C and B to A be unaltered, notwithstanding the difference of the dielectrics interposed between them?

Faraday answered this question by constructing two identical capacitors, in one of which he placed a dielectric, the other containing air at normal pressure. When both capacitors were charged to the *same potential difference*, Faraday found by experiment that *the charge on the one containing the dielectric was greater than that on the other*; see Fig. 30-7a.

Since q is larger, for the same V, if a dielectric is present, it follows from the relation $C = q/V$ that the capacitance of a capacitor increases if a dielectric is placed between the plates. The ratio of the capacitance with the dielectric* to that without is called the *dielectric constant κ* of the material; see Table 30-1.

Instead of maintaining the two capacitors at the same potential difference, we can place the *same charge* on them, as in Fig. 30-7b. Experiment then shows that the potential difference V_d between the plates of the right-hand capacitor is smaller than that for the left-hand capacitor by the factor $1/\kappa$, or

$$V_d = V_0/\kappa.$$

We are led once again to conclude, from the relation $C = q/V$, that the effect of the dielectric is to increase the capacitance by a factor κ.

For a parallel-plate capacitor we can write, as an experimental result,

$$C = \frac{\kappa\epsilon_0 A}{d}. \qquad (30\text{-}10)$$

Equation 30-4 is a special case of this relation found by putting $\kappa = 1$, corresponding to a vacuum between the plates. Experiment shows that the capacitance of *all* types of capacitor is increased by the factor κ if the space between the plates is filled with a dielectric. Thus the capacitance of any capacitor can be written as

$$C = \kappa\epsilon_0 L,$$

where L depends on the geometry and has the dimensions of a length. For a parallel-plate capacitor (see Eq. 30-4) L is A/d; for a cylindrical capacitor (see Example 2) it is $2\pi l/\ln(b/a)$.

---

* Assumed to fill completely the space between the plates.

Table 30-1
Properties of some dielectrics*

Material	Dielectric Constant	Dielectric Strength** (kV/mm)
Vacuum	1.00000	∞
Air	1.00054	0.8
Water	78	—
Paper	3.5	14
Ruby mica	5.4	160
Porcelain	6.5	4
Fused quartz	3.8	8
Pyrex glass	4.5	13
Bakelite	4.8	12
Polyethylene	2.3	50
Amber	2.7	90
Polystyrene	2.6	25
Teflon	2.1	60
Neoprene	6.9	12
Transformer oil	4.5	12
Titanium dioxide	100	6

* These properties are at approximately room temperature and for conditions such that the electric field **E** in the dielectric does not vary with time.
** This is the maximum potential gradient that may exist in the dielectric without the occurrence of electrical breakdown. Dielectrics are often placed between conducting plates to permit a higher potential difference to be applied between them than would be possible with air as the dielectric.

EXAMPLE 8

A parallel-plate capacitor has plates with area A and separation d. A battery charges the plates to a potential difference V_0. The battery is then disconnected, and a dielectric slab of thickness d is introduced. Calculate the stored energy both before and after the slab is introduced and account for any difference.

The energy U_0 before introducing the slab is

$$U_0 = \tfrac{1}{2} C_0 V_0^2.$$

After the slab is in place, we have

$$C = \kappa C_0 \qquad \text{and} \qquad V = V_0/\kappa$$

and thus

$$U = \tfrac{1}{2} C V^2 = \tfrac{1}{2} \kappa C_0 \left(\frac{V_0}{\kappa}\right)^2 = \frac{1}{\kappa} U_0.$$

The energy after the slab is introduced is *less* by a factor $1/\kappa$. The "missing" energy would be apparent to the person who inserted the slab. He would feel a "tug" on the slab and would have to restrain it if he wished to insert the slab without acceleration. This means that he would have to do negative work on it, or, alternatively, that the capacitor + slab system would do positive work on him. This positive work is

$$W = U_0 - U = \tfrac{1}{2} C_0 V_0^2 \left(1 - \frac{1}{\kappa}\right).$$

As expected, $W = 0$ for the case of $\kappa = 1$.

The following section will give some detailed insight into how the "tug" referred to above arises, in terms of the attraction between what we will call "free" charges on the capacitor plates and "induced" charges on the dielectric.

Note from the relation $U = \frac{1}{2}CV^2$ (see Eq. 30-8) that we can derive the *energy density u*, for a parallel-plate capacitor in which a dielectric slab is present, from

$$u = \frac{U}{(Ad)} = \left(\frac{1}{Ad}\right)(\tfrac{1}{2}CV^2) = \left(\frac{1}{Ad}\right)(\tfrac{1}{2})\left(\frac{\epsilon_0 \kappa A}{d}\right)(V^2).$$

But $E = V/d$ so that we have

$$u = \tfrac{1}{2}\epsilon_0 \kappa E^2.$$

As for Eq. 30-9, this relation, although derived for a parallel-plate capacitor, holds in general; that is, at any point P in a dielectric of constant κ. As we expect, for $\kappa = 1$, this new relation reduces to Eq. 30-9.

30-5
DIELECTRICS— AN ATOMIC VIEW

We now seek to understand, in atomic terms, what happens when we place a dielectric in an electric field. There are two possibilities. The molecules of some dielectrics, like water (see Fig. 29-11), have permanent electric dipole moments. In such materials (called *polar*) the electric dipole moments **p** tend to align themselves with an external electric field, as in Fig. 30-8*b*; see also section 27-6. Because the molecules are in constant thermal agitation, the degree of alignment will not be complete but will increase as the applied electric field is increased or as the temperature is decreased.

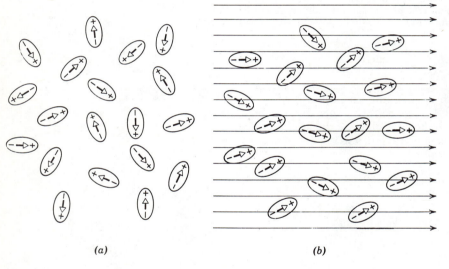

(a) (b)

figure 30-8
(*a*) Molecules with a permanent electric dipole moment, showing their random orientation in the absence of an external electric field. (*b*) An electric field is applied, producing partial alignment of the dipoles. Thermal agitation prevents complete alignment.

Whether or not the molecules have permanent electric dipole moments, they acquire them by *induction* when placed in an electric field. In Section 29-5 we saw that the external electric field tends to separate the negative and the positive charge in the atom or molecule. This *induced electric dipole moment* is present only when the electric field is present. It is proportional to the electric field (for normal field strengths) and is created already lined up with the electric field as Fig. 29-12 suggests.

Let us use a parallel-plate capacitor, carrying a fixed charge q and not connected to a battery (see Fig. 30-7*b*), to provide a uniform external electric field $\mathbf{E}_0$ into which we place a dielectric slab. The over-all effect of alignment and induction is to separate the center of positive charge of the entire slab slightly from the center of negative charge. The slab, as

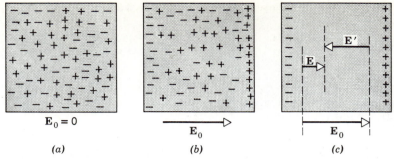

figure 30-9
(a) A dielectric slab, showing the random distribution of plus and minus charges. (b) An external field $\mathbf{E}_0$, established by putting the slab between the plates of a parallel-plate capacitor (not shown), separates the center of plus charge in the slab slightly from the center of minus charge, resulting in the appearance of surface charges. No *net* charge exists in any volume element located in the *interior* of the slab. (c) The surface charges set up a field $\mathbf{E}'$ which opposes the external field $\mathbf{E}_0$ associated with the charges on the capacitor plates. The resultant field $\mathbf{E}$ $(= \mathbf{E}_0 + \mathbf{E}')$ in the dielectric is thus less than $\mathbf{E}_0$.

a whole, although remaining electrically neutral, becomes *polarized*, as Fig. 30-9b suggests. The net effect is a pile-up of positive charge on the right face of the slab and of negative charge on the left face; within the slab no excess charge appears in any given volume element. Since the slab as a whole remains neutral, the positive *induced surface charge* must be equal in magnitude to the negative induced surface charge. Note that in this process electrons in the dielectric are displaced from their equilibrium positions by distances that are considerably less than an atomic diameter. There is no transfer of charge over macroscopic distances such as occurs when a current is set up in a conductor.

Figure 30-9c shows that the induced surface charges will always appear in such a way that the electric field set up *by them* ($\mathbf{E}'$) opposes the external electric field $\mathbf{E}_0$. The *resultant* field in the dielectric $\mathbf{E}$ is the vector sum of $\mathbf{E}_0$ and $\mathbf{E}'$. It points in the same direction as $\mathbf{E}_0$ but is smaller. *If we place a dielectric in an electric field, induced surface charges appear which tend to weaken the original field within the dielectric.*

This weakening of the electric field reveals itself in Fig. 30-7b as a reduction in potential difference between the plates of a charged isolated capacitor when a dielectric is introduced between the plates. The relation $V = Ed$ for a parallel-plate capacitor (see Eq. 30-3) holds whether or not dielectric is present and shows that the reduction in V described in Fig. 30-7b is directly connected to the reduction in E described in Fig. 30-9. More specifically, if a dielectric slab is introduced into an isolated charged parallel-plate capacitor, then

$$\frac{E_0}{E} = \frac{V_0}{V_d} = \kappa \qquad (30\text{-}11)$$

where the symbols on the left refer to Fig. 30-9 and the symbols V_0 and V_d refer to Fig. 30-7b.*

---

* Equation 30-11 does not hold if the battery remains connected while the dielectric slab is introduced. In this case V (hence E) could not change. Instead, the charge q on the capacitor plates would increase by a factor κ, as Fig. 30-7a suggests.

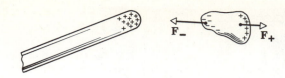

figure 30-10
A charged rod attracts an uncharged piece of paper because unbalanced forces act on the induced surface charges.

Induced surface charge is the explanation of the most elementary fact of static electricity, namely, that a charged rod will attract uncharged bits of paper, etc. Figure 30-10 shows a bit of paper in the field of a charged rod. Surface charges appear on the paper as shown. The negatively charged end of the paper will be pulled toward the rod and the positively charged end will be repelled. These two forces do not have the same magnitude because the negative end, being closer to the rod, is in a stronger field and experiences a stronger force. The net effect is an attraction. A dielectric body in a *uniform* electric field will not experience a net force.

In Example 8 we pointed out that, if we insert a dielectric slab into a parallel-plate capacitor carrying a fixed charge q, a force will act on the slab drawing it into the capacitor. This force is provided by the electrostatic attraction between the charges $\pm q$ on the capacitor plates and the induced surfaces charges $\mp q'$ on the dielectric slab. When the slab is only part way into the capacitor neither q nor q' will be uniformly distributed. Sketch qualitatively a possible distribution for q and q' when the slab is, say, halfway into the capacitor.

30-6 DIELECTRICS AND GAUSS'S LAW

So far our use of Gauss's law has been confined to situations in which no dielectric was present. Now let us apply this law to a parallel-plate capacitor filled with a dielectric of dielectric constant κ.

Figure 30-11 shows the capacitor both with and without the dielectric. We assume that the charge q on the plates is the same in each case. Gaussian surfaces have been drawn after the fashion of Fig. 30-2.

If no dielectric is present (Fig. 30-11a), Gauss's law (see Eq. 30-2) gives

$$\epsilon_0 \oint \mathbf{E} \cdot d\mathbf{S} = \epsilon_0 E_0 A = q$$

or

$$E_0 = \frac{q}{\epsilon_0 A}. \tag{30-12}$$

If the dielectric is present (Fig. 30-11b), Gauss's law gives

$$\epsilon_0 \oint \mathbf{E} \cdot d\mathbf{S} = \epsilon_0 E A = q - q'$$

or

$$E = \frac{q}{\epsilon_0 A} - \frac{q'}{\epsilon_0 A}, \tag{30-13}$$

in which $-q'$, the *induced surface charge*, must be distinguished from q, the so-called *free charge* on the plates. These two charges, both of which lie within the Gaussian surface, are opposite in sign; $q - q'$ is the *net* charge within the Gaussian surface.

Equation 30-11 shows that in Fig. 30-11

$$E = \frac{E_0}{\kappa}.$$

Combining this with Eq. 39-12, we have

$$E = \frac{E_0}{\kappa} = \frac{q}{\kappa \epsilon_0 A}. \tag{30-14}$$

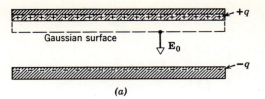

figure 30-11
A parallel-plate capacitor (a) without and (b) with a dielectric. The charge q on the plates is assumed to be the same in each case.

(a)

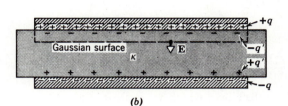

(b)

Inserting this in Eq. 30-13 yields

$$\frac{q}{\kappa\epsilon_0 A} = \frac{q}{\epsilon_0 A} - \frac{q'}{\epsilon_0 A} \qquad (30\text{-}15a)$$

or

$$q' = q\left(1 - \frac{1}{\kappa}\right). \qquad (30\text{-}15b)$$

This shows correctly that the induced surface charge q' is always less in magnitude than the free charge q and is equal to zero if no dielectric is present, that is, if $\kappa = 1$.

Now we write Gauss's law for the case of Fig. 30-11b in the form

$$\epsilon_0 \oint \mathbf{E} \cdot d\mathbf{S} = q - q', \qquad (30\text{-}16)$$

$q - q'$ again being the net charge within the Gaussian surface. Substituting from Eq. 30-15b for q' leads, after some rearrangement, to

$$\epsilon_0 \oint \kappa \mathbf{E} \cdot d\mathbf{S} = q. \qquad (30\text{-}17)$$

This important relation, although derived for a parallel-plate capacitor, is true generally and is the form in which Gauss's law is usually written when dielectrics are present. Note the following:

1. The flux integral now contains a factor κ.

2. The charge q contained within the Gaussian surface is taken to be the *free charge only*. Induced surface charge is deliberately ignored on the right side of this equation, having been taken into account by the introduction of κ on the left side. Equations 30-16 and 30-17 are completely equivalent formulations.

EXAMPLE 9

Figure 30-12 shows a dielectric slab of thickness b and dielectric constant κ placed between the plates of a parallel-plate capacitor of plate area A and separation d. A potential difference V_0 is applied with no dielectric present. The battery is then disconnected and the dielectric slab inserted. Assume that $A = 100$ cm^2, $d = 1.0$ cm, $b = 0.50$ cm, $\kappa = 7.0$, and $V_0 = 100$ V and (a) calculate the capacitance C_0 before the slab is inserted.

From Eq. 30-4, C_0 is:

$$C_0 = \frac{\epsilon_0 A}{d} = \frac{(8.9 \times 10^{-12} \text{ C}^2/\text{N} \cdot \text{m}^2)(10^{-2} \text{ m}^2)}{10^{-2} \text{ m}} = 8.9 \times 10^{-12} \text{ F} = 8.9 \text{ pF}.$$

(b) Calculate the free charge q.

From Eq. 30-1,

$$q = C_0V_0 = (8.9 \times 10^{-12} \text{ F})(100 \text{ V}) = 8.9 \times 10^{-10} \text{ C}.$$

Because of the technique used to charge the capacitor, the free charge remains unchanged as the slab is introduced. If the charging battery had *not* been disconnected, this would not be the case.

(c) Calculate the electric field in the gap.

Applying Gauss's law in the form given in Eq. 30-17 to the Gaussian surface of Fig. 30-12 (upper plate) yields

$$\epsilon_0 \oint \kappa \mathbf{E} \cdot d\mathbf{S} = \epsilon_0 E_0 A = q,$$

or

$$E_0 = \frac{q}{\epsilon_0 A} = \frac{8.9 \times 10^{-10} \text{ C}}{(8.9 \times 10^{-12} \text{ C}^2/\text{N} \cdot \text{m}^2)(10^{-2} \text{ m}^2)} = 1.0 \times 10^4 \text{ V/m}.$$

Note that we put $\kappa = 1$ here because the surface over which we evaluate the flux integral does not pass through any dielectric. Note too that E_0 remains unchanged when the slab is introduced; this derivation takes no specific account of the presence of the dielectric.

(d) Calculate the electric field in the dielectric.

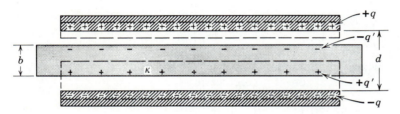

figure 30-12
Example 9. A parallel-plate capacitor containing a dielectric slab.

Applying Eq. 30-17 to the Gaussian surface of Fig. 30-12 (lower plate) yields

$$\epsilon_0 \oint \kappa \mathbf{E} \cdot d\mathbf{S} = \epsilon_0 \kappa E A = q.$$

Note that κ appears here because the surface cuts through the dielectric and that only the free charge q appears on the right. Thus we have

$$E = \frac{q}{\kappa \epsilon_0 A} = \frac{E_0}{\kappa} = \frac{1.0 \times 10^4 \text{ V/m}}{7.0} = 0.14 \times 10^4 \text{ V/m}.$$

(e) Calculate the potential difference between the plates.

Applying Eq. 29-5 to a straight perpendicular path from the lower plate (L) to the upper one (U) yields

$$V = -\int_L^U \mathbf{E} \cdot d\mathbf{l} = -\int_L^U E \cos 180° \, dl = \int_L^U E \, dl = E_0(d - b) + Eb.$$

Numerically

$$V = (1.0 \times 10^4 \text{ V/m})(5 \times 10^{-3} \text{ m}) + (0.14 \times 10^4 \text{ V/m})(5 \times 10^{-3} \text{ m}) = 57 \text{ V}.$$

This contrasts with the original applied potential difference of 100 V; compare Fig. 30-7b.

(f) Calculate the capacitance with the slab in place.

From Eq. 30-1,

$$C = \frac{q}{V} = \frac{8.9 \times 10^{-10} \text{ C}}{57 \text{ V}} = 16 \text{ pF}.$$

When the dielectric slab is introduced, the potential difference drops from 100 to 57 V and the capacitance rises from 8.9 to 16 pF, a factor of 1.8. If the dielectric slab had filled the capacitor, the capacitance would have risen by a factor of κ (= 7.0) to 62 pF.

For all situations that we encounter in this book our discussion of the behavior of dielectrics in an electric field is adequate. However, the problems that we treat are simple ones, such as that of a rectangular slab placed at right angles to a uniform external electric field. For more difficult problems, such as that of finding **E** at the center of a dielectric ellipsoid placed in a (possibly nonuniform) external electric field, it greatly simplifies the labor and leads to deeper insight if we introduce a new formalism. We do so largely so that students who take a second course in electromagnetism will have some familiarity with the concepts.

Let us rewrite Eq. 30-15a, which applies to a parallel-plate capacitor containing a dielectric, as

$$\frac{q}{A} = \epsilon_0 \left(\frac{q}{\kappa \epsilon_0 A} \right) + \frac{q'}{A}. \tag{30-18}$$

The quantity in parentheses (see Eq. 30-14) is simply the electric field E in the dielectric. The last term in Eq. 30-18 is the *induced surface charge per unit area*. We call it the *electric polarization P*, or

$$P = \frac{q'}{A}. \tag{30-19}$$

The name is suitable because the induced surface charge q' (also called the polarization charge) appears when the dielectric is polarized.

The electric polarization P can be defined in an equivalent way by multiplying the numerator and denominator in Eq. 30-19 by d, the thickness of the dielectric slab in Fig. 30-11,

$$P = \frac{q'd}{Ad}. \tag{30-20}$$

The numerator is the product $q'd$ of the magnitude of the (equal and opposite) polarization charges by their separation. It is thus the induced electric dipole moment of the dielectric slab. Since the denominator Ad is the volume of the slab, we see that the electric polarization can also be defined as the induced electric dipole moment per unit volume in the dielectric. This definition suggests that since the electric dipole moment is a vector the electric polarization is also a vector, its magnitude being P. The direction of **P** is from the negative induced charge to the positive induced charge, as for any dipole. In Fig. 30-13, which shows a capacitor with a dielectric slab filling half the space between the plates, **P** points down.

We can now rewrite Eq. 30-18 as

$$\frac{q}{A} = \epsilon_0 E + P. \tag{30-21}$$

The quantity on the right occurs so often in electrostatic problems that we give it the special name *electric displacement D*, or

$$D = \epsilon_0 E + P \tag{30-22a}$$

in which $$D = \frac{q}{A}. \tag{30-22b}$$

The name has historical significance only.

Since **E** and **P** are vectors, **D** must also be one, so that in the more general case we have

$$\mathbf{D} = \epsilon_0 \mathbf{E} + \mathbf{P}. \tag{30-23}$$

In Fig. 30-13 all three vectors point down and each has a constant magnitude for every point in the dielectric (and also at every point in the air gap) so that the vector nature of Eq. 30-23 is not very important in this case. In more compli-

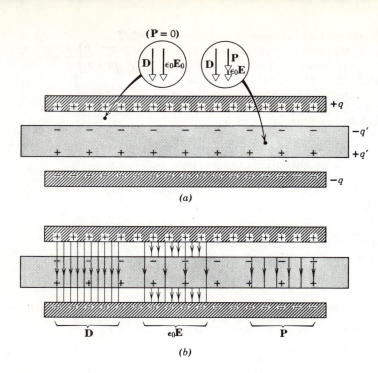

figure 30-13
(*a*) Showing **D**, ϵ_0**E**, and **P** in the dielectric (*upper right*) and in the gap (*upper left*) for a parallel-plate capacitor. (*b*) Showing samples of the lines associated with **D** (free charge), ϵ_0**E** (all charges), and **P** (polarization charge).

cated problems, however, **E**, **P**, and **D** may vary in magnitude and direction from point to point.

From their definitions we see the following:

1. **D** (see Eq. 30-22*b*) is connected with the *free charge* only. We can represent the vector field of **D** by *lines of* **D**, just as we represent the field of **E** by lines of force. Figure 30-13*b* shows that the lines of **D** begin and end on the free charges.
2. **P** (see Eq. 30-19) is connected with the *polarization charge* only. It is also possible to represent this vector field by lines. Figure 30-13*b* shows that the lines of **P** begin and end on the polarization charges.
3. **E** is connected with *all* charges that are actually present, whether free or polarization. The lines of **E** reflect the presence of both kinds of charge, as Fig. 30-13*b* shows. Note (Eqs. 30-19 and 30-22*b*) that the units for **P** and **D** (C/m²) differ from those of **E** (N/C).

The electric field vector **E**, which is what determines the force that acts on a suitably placed test charge, remains of fundamental interest. **D** and **P** are auxiliary vectors useful as aids in the solution of problems more complex than that of Fig. 30-13.

The vectors **D** and **P** can both be expressed in terms of **E** alone. A convenient starting point is the identity

$$\frac{q}{A} = \kappa \epsilon_0 \left(\frac{q}{\kappa \epsilon_0 A} \right).$$

Comparison with Eqs. 30-14 and 30-22*b* shows that this, extended to vector form, can be written as

$$\mathbf{D} = \kappa \epsilon_0 \mathbf{E}. \tag{30-24}$$

We can also write the polarization (see Eqs. 30-19 and 30-15*b*) as

$$P = \frac{q'}{A} = \frac{q}{A} \left(1 - \frac{1}{\kappa} \right).$$

Since $q/A = D$, we can rewrite this, using Eq. 30-24 and casting the result into vector form, as

$$\mathbf{P} = \epsilon_0(\kappa - 1)\mathbf{E}. \qquad (30\text{-}25)$$

This shows clearly that in a vacuum ($\kappa = 1$) the polarization vector $\mathbf{P}$ is zero.* Equations 30-24 and 30-25 show that for isotropic materials, to which a single dielectric constant κ can be assigned, $\mathbf{D}$ and $\mathbf{P}$ both point in the direction of $\mathbf{E}$ at any given point.

The definition of $\mathbf{D}$ given by Eq. 30-24 allows us to write Eq. 30-17, that is, Gauss's law in the presence of a dielectric, simply as

$$\oint \mathbf{D} \cdot d\mathbf{S} = q, \qquad (30\text{-}26)$$

where, as before, q represents the free charge only, the induced surface charges being excluded.

In Figure 30-13, using data from Example 9, calculate E, D, and P: (a) in the dielectric and (b) in the air gap.

EXAMPLE 10

(a) The electric field in the dielectric is calculated in Example 9 to be 1.43×10^3 V/m. From Eq. 30-24,

$$D = \kappa\epsilon_0 E$$
$$= (7.0)(8.9 \times 10^{-12} \text{ C}^2/\text{N}\cdot\text{m}^2)(1.43 \times 10^3 \text{ V/m})$$
$$= 8.9 \times 10^{-8} \text{ C/m}^2$$

and, from Eq. 30-25,

$$P = \epsilon_0(\kappa - 1)E$$
$$= (8.9 \times 10^{-12} \text{ C}^2/\text{N}\cdot\text{m}^2)(7.0 - 1)(1.43 \times 10^3 \text{ V/m})$$
$$= 7.5 \times 10^{-8} \text{ C/m}^2.$$

(b) The electric field E_0 in the air gap is calculated in Example 9 to be 1.00×10^4 V/m. From Eq. 30-24,

$$D_0 = \kappa\epsilon_0 E_0$$
$$= (1)(8.9 \times 10^{-12} \text{ C}^2/\text{N}\cdot\text{m}^2)(1.00 \times 10^4 \text{ V/m})$$
$$= 8.9 \times 10^{-8} \text{ C/m}^2$$

and, from Eq. 30-25, recalling, as above, that $\kappa = 1$ in the air gap,

$$P_0 = \epsilon_0(\kappa - 1)E_0 = 0.$$

Note that $\mathbf{P}$ vanishes outside the dielectric, $\mathbf{D}$ has the same value in the dielectric and in the gap, and $\mathbf{E}$ has different values in the dielectric and in the gap. Verify that Eq. 30-23 ($\mathbf{D} = \epsilon_0\mathbf{E} + \mathbf{P}$) is correct both in the gap and in the dielectric.

It can be shown from Maxwell's equations that no matter how complex the problem the component of $\mathbf{D}$ *normal* to the surface of the dielectric has the same value on each side of the surface. In this problem $\mathbf{D}$ itself is normal to the surface, there being no component but the normal one. It can also be shown that the component of $\mathbf{E}$ *tangential* to the dielectric surface has the same value on each side of the surface. This *boundary condition*, like the one for $\mathbf{D}$, is trivial

---

* Certain waxes, when polarized in their molten state, retain a permanent polarization after solidifying, even though the external polarizing field is removed. *Electrets*, manufactured in this way, are the electrostatic analog of permanent magnets in that they possess a gross permanent electric dipole moment. Materials from which electrets can be constructed are called *ferroelectric*. Electrets do *not* obey Eq. 30-25 because they have a nonvanishing value of $\mathbf{P}$ even though $\mathbf{E} = 0$.

in this problem, both tangential components being zero. In more complex problems these boundary conditions on **D** and **E** are very important. Table 30-2 summarizes the properties of the electric vectors, **E**, **D**, and **P**.

Table 30-2
Three electric vectors

Name	Symbol	Associated with	Boundary Condition
Electric field	**E**	All charges	Tangential component continuous
Electric displacement	**D**	Free charges only	Normal component continuous
Polarization (electric dipole moment per unit volume)	**P**	Polarization charges only	Vanishes in a vacuum

Defining equation for **E**	$\mathbf{F} = q\mathbf{E}$	Eq. 27-2
General relation among the three vectors	$\mathbf{D} = \epsilon_0\mathbf{E} + \mathbf{P}$	Eq. 30-23
Gauss's law when dielectric media are present	$\oint \mathbf{D} \cdot d\mathbf{S} = q$	Eq. 30-26
	(q = free charge only)	
Empirical relations for certain dielectric materials*	$\mathbf{D} = \kappa\epsilon_0\mathbf{E}$	Eq. 30-24
	$\mathbf{P} = (\kappa - 1)\epsilon_0\mathbf{E}$	Eq. 30-25

* Generally true, with κ independent of **E**, except for certain materials called *ferroelectrics*; see footnote on page 667.

questions

1. A capacitor is connected across a battery. (a) Why does each plate receive a charge of exactly the same magnitude? (b) Is this true even if the plates are of different sizes?

2. Can there be a potential difference between two adjacent conductors that carry the same positive charge?

3. The relation $\sigma \propto 1/R$, in which R is the radius of curvature (see Section 29-8), suggests that the charge placed on an isolated conductor concentrates on points and avoids flat surfaces, where $R = \infty$. How do we reconcile this with Fig. 30-2 in which the charge is definitely on the flat surface of either plate?

4. A sheet of aluminum foil of negligible thickness is placed between the plates of a capacitor as in Fig. 30-14. What effect has it on the capacitance if (a) the foil is electrically insulated and (b) the foil is connected to the upper plate?

5. You are given two capacitors, C_1 and C_2, in which $C_1 \gg C_2$. Can C_1 always hold more charge than C_2? Explain.

6. In Fig. 30-1 suppose that a and b are nonconductors, the charge being distributed arbitrarily over their surfaces. (a) Would Eq. 30-1 ($q = CV$) hold, with C independent of the charge arrangements? (b) How would you define V in this case?

7. In connection with Eq. 30-1 ($q = CV$) we said that C is a constant. Yet we pointed out (see Eq. 30-4) that it depends on the geometry (and also, as we

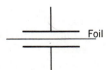
Foil

figure 30-14
Question 4.

saw later, on the medium). If C is indeed a constant, with respect to what variables does it remain constant?

8. For a finite A, does Eq. 30-4 ($C = \epsilon_0 A/d$) hold as $d \to \infty$? If not, why not?

9. Suppose that in Example 4 the three capacitors shown are identical parallel-plate capacitors with the same (square) plates of area A and the same plate separation d. Develop an argument, based only on Eq. 30-4 ($C = \epsilon_0 A/d$) that the equivalent capacitance is three times the individual capacitance, as Example 4 predicts.

10. You are given a parallel-plate capacitor with square plates of area A and separation d, in a vacuum. What is the qualitative effect of each of the following on its capacitance? (a) Reduce d. (b) Put a slab of copper between the plates, touching neither plate. (c) Double the area of both plates. (d) Double the area of one plate only. (e) Slide the plates parallel to each other so that the area of overlap is, say, 50%. (f) Double the potential difference between the plates. (g) Tilt one plate so that the separation remains d at one end but is $\frac{1}{2}d$ at the other.

11. Discuss similarities and differences when (a) a dielectric slab and (b) a conducting slab are inserted between the plates of a parallel-plate capacitor. Assume the slab thicknesses to be one-half the plate separation.

12. An oil-filled parallel-plate capacitor has been designed to have a capacitance C and to operate safely at or below a certain maximum potential difference V_m without arcing over. However, the designer did not do a good job and the capacitor occasionally arcs over. What can be done to redesign the capacitor, keeping C and V_m unchanged and using the same dielectric?

13. Would you expect the dielectric constant, for substances containing permanent molecular electric dipoles, to vary with temperature?

14. An isolated conducting sphere is given a positive charge. Does its mass increase, decrease, or remain the same?

15. A dielectric slab is inserted in one end of a charged parallel-plate capacitor (the plates being horizontal and the charging battery having been disconnected) and then released. Describe what happens. Neglect friction.

16. A capacitor is charged by using a battery, which is then disconnected. A dielectric slab is then slipped between the plates. Describe qualitatively what happens to the charge, the capacitance, the potential difference, the electric field, and the stored energy.

17. While a capacitor remains connected to a battery, a dielectric slab is slipped between the plates. Describe qualitatively what happens to the charge, the capacitance, the potential difference, the electric field, and the stored energy. Is work required to insert the slab?

18. Two identical capacitors are connected as shown in Fig. 30-15. A dielectric slab is slipped between the plates of one capacitor, the battery remaining connected. Describe qualitatively what happens to the charge, the capacitance, the potential difference, the electric field, and the stored energy for each capacitor.

19. Show that the dielectric constant of a conductor can be taken to be infinitely great.

20. For a given potential difference does a capacitor store more or less charge with a dielectric than it does without a dielectric (vacuum)? Explain in terms of the microscopic picture of the situation.

21. In this chapter we have assumed electrostatic conditions, that is, the potential difference V between the capacitor plates remains constant. Suppose however that, as it often does in practice, V varies sinusoidally with time with an angular frequency ω. Would you expect the dielectric constant κ to vary with ω?

22. In connection with Section 30-7 describe in your own words the differences between **D**, **E**, and **P** in Eq. 30-23.

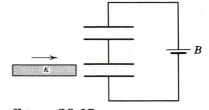

figure 30-15
Question 18.

SECTION 30-2

1. A 100-pF capacitor is charged to a potential difference of 50 V, the charging battery then being disconnected. The capacitor is then connected in parallel with a second (initially uncharged) capacitor. If the measured potential difference drops to 35 V, what is the capacitance of this second capacitor? *Answer:* 43 pF.

2. A potential difference of 300 V is applied to a 2.0-μF capacitor and an 8.0-μF capacitor connected in series. (a) What are the charge and the potential difference for each capacitor? (b) The charged capacitors are reconnected with their positive plates together and their negative plates together, no external voltage being applied. What are the charge and the potential difference for each? (c) The charged capacitors in (a) are reconnected with plates of *opposite* sign together. What are the charge and the potential difference for each?

3. If we solve Eq. 30-4 for ϵ_0, we see that its SI units are farad/meter. Show that these units are equivalent to those obtained earlier for ϵ_0, namely coulomb²/newton·meter².

4. A parallel-plate capacitor has circular plates of 8.0-cm radius and 1.0-mm separation. What charge will appear on the plates if a potential difference of 100 V is applied?

5. Figure 30-16 shows two capacitors in series, the rigid center section of length b being movable vertically. Show that the equivalent capacitance of the series combination is independent of the position of the center section and is given by

$$C = \frac{\epsilon_0 A}{a - b}.$$

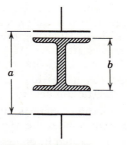

figure 30-16
Problem 5

6. In Fig. 30-17 a variable air capacitor of the type used in tuning radios is shown. Alternate plates are connected together, one group being fixed in position, the other group being capable of rotation. Consider a pile of n plates of alternate polarity, each having an area A and separated from adjacent plates by a distance d. Show that this capacitor has a maximum capacitance of

$$C = \frac{(n - 1)\epsilon_0 A}{d}.$$

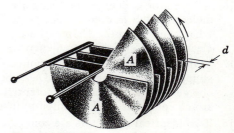

figure 30-17
Problem 6

7. A 6.0-μF capacitor is connected in series with a 4.0-μF capacitor and a potential difference of 200 V is applied across the pair. (a) What is the charge on each capacitor? (b) What is the potential difference across each capacitor? *Answer:* (a) 4.8×10^{-4} C. (b) $V_4 = 120$ V; $V_6 = 80$ V.

8. Repeat the previous problem for the same two capacitors connected in parallel.

9. How many 1.0-μF capacitors would need to be connected in parallel in order to store a charge of 1.0 C with a potential of 300 V across the capacitors? *Answer:* 3300.

figure 30-18
Problem 10

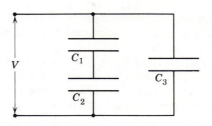

10. In Fig. 30-18 find the equivalent capacitance of the combination. Assume that $C_1 = 10$ μF, $C_2 = 5$ μF, and $C_3 = 4$ μF.

11. In Fig. 30-19 find the equivalent capacitance of the combination. Assume that $C_1 = 10$ μF, $C_2 = 5$ μF, and $C_3 = 4$ μF. *Answer:* 3.2 μF.

7.33uf

12. In Fig. 30-19 suppose that capacitor C_3 breaks down electrically, becoming equivalent to a conducting path. What *changes* in (a) the charge and (b) the potential difference occur for capacitor C_1? Assume $V = 100$ V.

13. A slab of copper of thickness b is thrust into a parallel-plate capacitor as shown in Fig. 30-20; it is exactly halfway between the plates. What is the capacitance (a) before and (b) after the slab is introduced?
Answer: (a) $\epsilon_0 A/d$. (b) $\epsilon_0 A/(d - b)$.

14. When switch S is thrown to the left in Fig. 30-21, the plates of the capacitor C_1 acquire a potential difference V_0. C_2 and C_3 are initially uncharged. The switch is now thrown to the right. What are the final charges q_1, q_2, q_3 on the corresponding capacitors?

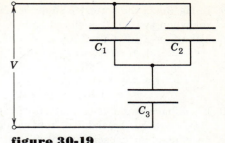
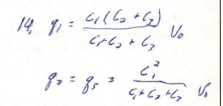

figure 30-19
Problems 11, 12, 31

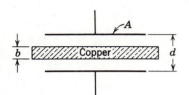

figure 30-20
Problem 13

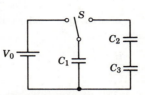

figure 30-21
Problem 14

$14, \quad q_1 = \dfrac{C_1(C_2 + C_3)}{C_1 + C_2 + C_3} V_0$

$q_2 = q_3 = \dfrac{C_1^2}{C_1 + C_2 + C_3} V_0$

15. Charges q_1, q_2, q_3 are placed on capacitors of capacitance C_1, C_2, C_3, respectively, arranged in series as shown in Fig. 30-22. Switch S is then closed. What are the final charges q_1', q_2', q_3' on the capacitors?
Answer: (a) $q_1' = \dfrac{-(C_1C_2 + C_1C_3)q_1 + C_1C_3q_2 + C_1C_2q_3}{C_1C_2 + C_1C_3 + C_2C_3}$.

(b) $q_2' = \dfrac{(C_1C_2 + C_2C_3)q_2 - C_1C_2q_3 - C_2C_3q_1}{C_1C_2 + C_1C_3 + C_2C_3}$.

(c) $q_3' = \dfrac{(C_1C_3 + C_2C_3)q_3 - C_1C_3q_2 - C_2C_3q_1}{C_1C_2 + C_1C_3 + C_2C_3}$.

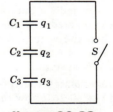

figure 30-22
Problem 15

16. You have material available to construct two parallel-plate capacitors having a combined plate area A. How would you distribute this area between the two capacitors to obtain maximum total capacitance if you intend to connect them (a) in parallel and (b) in series?

17. Capacitors C_1 (1.0 μF) and C_2 (3.0 μF) are each charged to a potential V (100 V) but with opposite polarity, so that points a and c are on the side of the respective positive plates of C_1 and C_2, and points b and d are on the side of the respective negative plates (see Fig. 30-23). Switches S_1 and S_2 are now closed. (a) What is the potential difference between points e and f? (b) What is the charge on C_1? (c) What is the charge on C_2?
Answer: (a) 50 V. (b) 0.50×10^{-4} C. (c) 1.5×10^{-4} C.

18. Find the equivalent capacitance between points x and y in Fig. 30-24. Assume that $C_2 = 10$ μF and that the other capacitors are all 4.0 μF. (Hint: Apply a potential difference V between x and y and write down all the relationships that involve the charges and potential differences for the separate capacitors.)

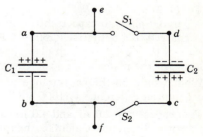

figure 30-23
Problem 17

figure 30-24
Problem 18

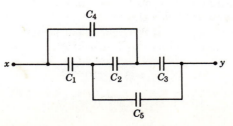

19. In Fig. 30-25 the battery B supplies 12 V. (a) Find the charge on each capacitor when switch S_1 is closed and (b) when (later) switch S_2 is also closed. Take $C_1 = 1 \ \mu F$, $C_2 = 2 \ \mu F$, $C_3 = 3 \ \mu F$, and $C_4 = 4 \ \mu F$.

Answer: (a) $q_1 = 9.0 \mu C$; $q_2 = 16 \mu C$;
$q_3 = 9.0 \mu C$; $q_4 = 16 \mu C$.
(b) $q_1 = 8.4 \mu C$; $q_2 = 17 \mu C$;
$q_3 = 11 \mu C$; $q_4 = 14 \mu C$.

20. If you have available several 2.0-μF capacitors, each capable of withstanding 200 V without breakdown, how would you assemble a combination having an equivalent capacitance of (a) 0.40 μF or of (b) 1.2 μF, each capable of withstanding 1000 V?

21. Calculate the capacitance of the earth, viewed as a spherical conductor of radius 6400 km.   Answer: 710 μF.

22. A spherical capacitor consists of two concentric spherical shells of radii a and b, with $b > a$. (a) Show that its capacitance is

$$C = 4\pi\epsilon_0 \frac{ab}{b-a}.$$

(b) Does this reduce (with $a = R$) to the result of Example 3 as $b \rightarrow \infty$?

23. Suppose that the two spherical shells of a spherical capacitor have their radii approximately equal. Under these conditions the device approximates a parallel-plate capacitor with $b - a = d$. Show that the formula in Problem 22 does indeed reduce to Eq. 30-4 in this case.

24. Two metallic spheres, radii a and b, are connected by a thin wire. Their separation is large compared with their dimensions. A charge Q is put onto this system and then the wire is disconnected. (a) How much charge resides on each sphere? (b) Apply the definition of capacitance to show that the capacitance of this system is $C = 4\pi\epsilon_0 (a + b)$.

25. A capacitor has square plates, each of side a, making an angle of θ with each other as shown in Fig. 30-26. Show that for small θ the capacitance is given by

$$C = \frac{\epsilon_0 a^2}{d} \left(1 - \frac{a\theta}{2d}\right).$$

(Hint: The capacitor may be divided into differential strips which are effectively in parallel.)

SECTION 30-3

26. A parallel-plate air capacitor having area A (40 cm²) and spacing d (1.0 mm) is charged to a potential V (600 V). Find (a) the capacitance, (b) the magnitude of the charge on each plate, (c) the stored energy, (d) the electric field between the plates and (e) the energy density between the plates.

27. What would be the capacitance required to store energy U (10 kW·h) at a potential difference V (1000 V)?   Answer: 72F.

28. Two capacitors (2.0 μF and 4.0 μF) are connected in parallel across a 300-V potential difference. Calculate the total stored energy in the system.

29. A parallel-connected bank of 2000 5.0-μF capacitors is used to store electric energy. What does it cost to charge this bank to 50,000 V, assuming a rate of 10¢/kW·h?   Answer: 35¢.

30. A parallel-plate air capacitor has a capacitance of 100 pF. (a) What is the stored energy if the applied potential difference is 50 V? (b) Can you calculate the energy density for points between the plates?

31. In Fig. 30-19 find (a) the charge, (b) the potential difference, and (c) the stored energy for each capacitor. Assume the numerical values of Problem 11, with $V = 100$ V.

Answer: (a) $q_1 = 2.1 \times 10^{-4}$ C; $q_2 = 1.1 \times 10^{-4}$ C; $q_3 = 3.2 \times 10^{-4}$ C.
(b) $V_1 = V_2 = 21$ V; $V_3 = 79$ V.
(c) $U_1 = 2.2 \times 10^{-3}$ J; $U_2 = 1.1 \times 10^{-3}$ J; $U_3 = 1.3 \times 10^{-2}$ J.

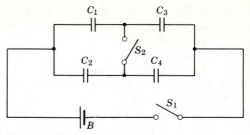

figure 30-25
Problem 19

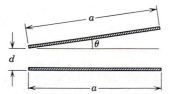

figure 30-26
Problem 25

32. For the capacitors of Problem 2, compute the energy stored for the three different connections of parts (a), (b), and (c). Compare your answers and explain any differences.

33. A parallel-plate capacitor has plates of area A and separation d, and is charged to a potential difference V. The charging battery is then disconnected and the plates are pulled apart until their separation is $2d$. Derive expressions in terms of A, d, and V for (a) the new potential difference, (b) the initial and the final stored energy, and (c) the work required to separate the plates.

Answer: (a) $V_f = 2V$. (b) $U_i = \frac{1}{2}\frac{\epsilon_0 A V^2}{d}$; $U_f = 2U_i$. (c) $W = \frac{1}{2}\frac{\epsilon_0 A V^2}{d}$.

34. In terms of the original capacitance C, find the work done in inserting a copper slab of thickness $d/2$ in Problem 13 if (a) the potential difference is held constant and (b) if the charge is held constant.

35. An isolated metal sphere whose diameter is 10 cm has a potential of 8000 V. What is the energy density at the surface of the sphere? *Answer:* 0.11 J/m³.

36. (a) If the potential difference across a cylindrical capacitor is doubled, the energy stored in the capacitor is changed by what factor? (b) If the radii of the inner and outer cylinders are each doubled, keeping the charge constant, how does the stored energy change?

37. A cylindrical capacitor has radii a and b as in Fig. 30-3. Show that half the stored electric potential energy lies within a cylinder whose radius is

$$r = \sqrt{ab}.$$

38. Show that the plates of a parallel-plate capacitor attract each other with a force given by

$$F = \frac{q^2}{2\epsilon_0 A}.$$

Prove this by calculating the work necessary to increase the plate separation from x to $x + dx$.

39. Using the result of Problem 38 show that the force per unit area (the electrostatic stress) acting on either capacitor plate is given by $\frac{1}{2}\epsilon_0 E^2$. Actually, this result is true in general, for a conductor of *any* shape with an electric field E at its surface.

40. A soap bubble of radius R_0 is slowly given a charge q. Because of mutual repulsion of the surface charges, the radius increases slightly to R. The air pressure inside the bubble drops, because of the expansion, to $p(V_0/V)$ where p is the atmospheric pressure, V_0 is the initial volume, and V the final volume. Show that

$$q^2 = 32\ \pi^2\epsilon_0\ p\ R\ (R^3 - R_0^3).$$

(Hint: Imagine the bubble to expand a further amount dR. Consider the energy changes associated with (a) the decrease in the stored electric field energy, (b) work ($= p\ dV$) done in pushing back the atmosphere, and (c) work done by the gas in the bubble. Apply conservation of energy. Neglect surface tension.)

SECTION 4

41. A certain substance has a dielectric constant of 2.8 and a dielectric strength of 18×10^6 V/m. If it is used as the dielectric material in a parallel-plate capacitor, what minimum area may the plates of the capacitor have in order that the capacitance be 7.0×10^{-2} μF and that the capacitor be able to withstand a potential difference of 4000 V? *Answer:* 0.63 m².

42. A parallel-plate capacitor is filled with two dielectrics as in Fig. 30-27. Show that the capacitance is given by

$$C = \frac{\epsilon_0 A}{d}\left(\frac{\kappa_1 + \kappa_2}{2}\right).$$

Check this formula for all the limiting cases that you can think of. (Hint: Can you justify regarding this arrangement as two capacitors in parallel?)

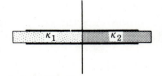

figure 30-27
Problem 42

43. A parallel-plate capacitor is filled with two dielectrics as in Fig. 30-28. Show that the capacitance is given by

$$C = \frac{2\epsilon_0 A}{d} \left(\frac{\kappa_1 \kappa_2}{\kappa_1 + \kappa_2} \right).$$

Check this formula for all the limiting cases that you can think of. (Hint: Can you justify regarding this arrangement as two capacitors in series?)

44. What is capacitance of the capacitor in Fig. 30-29? The plate area is A.

45. For making a capacitor you have available two plates of copper, a sheet of mica (thickness = 0.10 mm, $\kappa = 6$), a sheet of glass (thickness = 2.0 mm, $\kappa = 7$), and a slab of paraffin (thickness = 1.0 cm, $\kappa = 2$). To obtain the largest capacitance, which sheet (or sheets) should you place between the copper plates?   *Answer:* The mica sheet.

SECTION 30-6

46. A parallel-plate capacitor has a capacitance of 100 pF, a plate area of 100 cm², and a mica dielectric ($\kappa = 5.4$). At 50 V potential difference, calculate (a) E in the mica, (b) the magnitude of the free charge on the plates, and (c) the magnitude of the induced surface charge.

47. Two parallel plates of area 100 cm² are each given equal but opposite charges of 8.9×10^{-7} C. Within the dielectric material filling the space between the plates the electric field is 1.4×10^6 V/m. (a) Find the dielectric constant of the material. (b) Determine the magnitude of the charge induced on each dielectric surface.   *Answer:* (a) 7.1. (b) 7.7×10^{-7} C.

48. In Example 9, suppose that the 100-V battery remains connected during the time that the dielectric slab is being introduced. Calculate (a) the charge on the capacitor plates, (b) the electric field in the gap, (c) the electric field in the slab, and (d) the capacitance. For all of these quantities give the numerical values before and after the slab is introduced. Contrast your results with those of Example 9 by constructing a tabular listing.

49. A parallel-plate capacitor has plates of area 0.12 m² and a separation of 1.2 cm. A battery charges the plates to a potential difference of 120 V and is then disconnected. A dielectric slab of thickness 0.4 cm and dielectric constant 4.8 is then placed symmetrically between the plates. (a) Find the capacitance before the slab is inserted. (b) What is the capacitance with the slab in place? (c) What is the free charge q before and after the slab is inserted? (d) Determine the electric field in the space between the plates and dielectric. (e) What is the electric field in the dielectric? (f) With the slab in place what is the potential difference across the plates? (g) How much external work is involved in the process of inserting the slab?
Answer: (a) 89 pF. (b) 120 pF. (c) 1.1×10^{-8} C; 1.1×10^{-8} C. (d) 10^4 V/m. (e) 2.1×10^3 V/m. (f) 88 V. (g) 1.7×10^{-7} J.

50. In the capacitor of Example 9 the dielectric slab fills half the space between the plates. (a) What percent of the energy is stored in the air gaps? (b) What percent is stored in the slab?

51. A dielectric slab of thickness b is inserted between the plates of a parallel-plate capacitor of plate separation d. Show that the capacitance is given by

$$C = \frac{\kappa \epsilon_0 A}{\kappa d - b(\kappa - 1)}.$$

(Hint: Derive the formula following the pattern of Example 9.) Does this formula predict the correct numerical result of Example 9? Does the formula seem reasonable for the special cases of $b = 0$, $\kappa = 1$, and $b = d$?

52. In Example 8, how does the *energy density u* between the plates compare, before and after the dielectric slab is introduced?

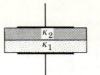

figure 30-28
Problem 43

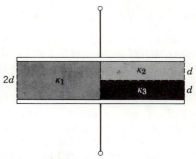

figure 30-29
Problem 44

31
current
and resistance

The free electrons in an isolated metallic conductor, such as a length of copper wire, are in random motion like the molecules of a gas confined to a container. They have no net directed motion along the wire. If we pass a hypothetical plane through the wire, the rate at which electrons pass through it from right to left is the same as the rate at which they pass through from left to right; the *net* rate is zero.*

If the ends of the wire are connected to a battery, an electric field will be set up at every point within the wire. If the potential difference maintained by the battery is 10 V and if the wire (assumed uniform) is 5 m long, the strength of this field at every point will be 2 V/m. This field **E** will act on the electrons and will give them a resultant motion in the direction of $-\mathbf{E}$. We say that an *electric current i* is established; if a net charge q passes through any cross section of the conductor in time t, the current (assumed constant) is

$$i = q/t. \tag{31-1}$$

The appropriate SI units are amperes (abbr. A) for i, coulombs for q, and seconds for t. Recall (Section 26-4) that Eq. 31-1 is the defining equation for the coulomb and that we have not yet given an operational definition of the ampere; we do so in Section 34-4.

---

*Actually, because the number of electrons is finite, there will be small statistical fluctuations in these rates and a conductor will contain a small, rapidly fluctuating current, even though, averaged over a long enough period of time, the net current i is zero. This is one aspect of the readily measurable *electrical noise* which is so familiar to those who know something about electronics.

If the rate of flow of charge with time is not constant, the current varies with time and is given by the differential limit of Eq. 31-1, or

$$i = dq/dt. \qquad (31-2)$$

In the rest of this chapter we consider only constant currents.*

The current i is the same for all cross sections of a conductor, even though the cross-sectional area may be different at different points. In the same way the rate at which water (assumed incompressible) flows past any cross section of a pipe is the same even if the cross section varies. The water flows faster where the pipe is smaller and slower where it is larger, so that the volume rate, measured perhaps in liters/second, remains unchanged. This constancy of the electric current follows because charge must be conserved; it does not pile up steadily or drain away steadily from any point in the conductor under the assumed steady-state conditions. In the language of Section 18-3 there are no "sources" or "sinks" of charge.

The existence of an electric field inside a conductor does not contradict Section 28-6, in which we asserted that **E** equals zero inside a conductor. In that section, which dealt with a state in which all net motion of charge had stopped (electrostatics), we assumed that the conductor was insulated and that no potential difference was deliberately maintained between any two points on it, as by a battery. In this chapter, which deals with charges in motion, we relax this restriction.

The electric field exerts a force $(= -e\mathbf{E})$ on the electrons in a conductor but this force does not produce a *net* acceleration because the electrons keep colliding with the atoms (strictly, ions, Cu^+ in copper) that make up the conductor. This array of ions, coupled together by strong spring-like forces of electromagnetic origin, is called the *lattice* (see Fig. 21-5). The over-all effect of these collisions is to transfer kinetic energy from the accelerating electrons into vibrational energy of the lattice. The electrons acquire a constant average *drift speed* v_d in the direction $-\mathbf{E}$. There is a close analogy to a ball bearing falling in a uniform gravitational field **g** at a constant terminal speed through a viscous oil. The gravitational force (mg) acting on the ball as it falls does not go into increasing the ball's kinetic energy (which is constant) but is transferred to the fluid by molecular collisions, producing a small rise in temperature.

Although in metals the charge carriers are electrons, in electrolytes or in gaseous conductors (plasmas) they may also be positive or negative ions, or both. We need a convention for labeling the directions of currents because charges of opposite sign move in opposite directions in a given field. A positive charge moving in one direction is equivalent in nearly all external effects to a negative charge moving in the opposite direction. Hence, for simplicity and algebraic consistency, *we assume that all charge carriers are positive and we draw the current arrows in the direction that such charges would move.* If the charge carriers are negative, they simply move opposite to the direction of the current arrow (see Fig. 31-1). When we encounter a case (as in the *Hall effect;* see Section 33-5) in which the sign of the charge carriers makes a difference in the external effects, we will disregard the convention and take the actual situation into account.

Current i is a characteristic of a particular conductor. It is a macroscopic quantity, like the mass of an object, the volume of an object, or the length of a rod. A related microscopic quantity is the current density

* A much less formal example of current flow is lightning. See "The Lightning Discharge" by Richard E. Orville, *The Physics Teacher*, January 1976.

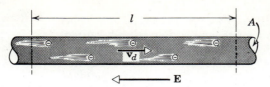

figure 31-1
Electrons drift in a direction opposite to the electric field in a conductor.

677 *CURRENT AND CURRENT DENSITY   SEC. 31-1*

j. It is a vector and is characteristic of a point inside a conductor rather than of the conductor as a whole. If the current is distributed uniformly across a conductor of cross-sectional area A, as in Fig. 31-1, the magnitude of the current density for all points on that cross section is

$$j = i/A. \tag{31-3}$$

The vector **j** at any point is oriented in the direction that a positive charge carrier would move at that point. An electron at that point moves in the direction $-$**j**. In Fig. 31-1 **j** is a constant vector and points to the left; the electrons drift to the right.

The general relationship between **j** and i is that, for a particular surface (which need not be plane) in a conductor, i is the flux of the vector **j** over that surface, or

$$i = \int \mathbf{j} \cdot d\mathbf{S}, \tag{31-4}$$

where $d\mathbf{S}$ is an element of surface area and the integral is taken over the surface in question. Equation 31-3 (written as $i = jA$) is a special case of this relationship in which the surface of integration is a plane cross section of the conductor and in which **j** is constant over this surface and at right angles to it. However, we may apply Eq. 31-4 to any surface through which we wish to know the current. Equation 31-4 shows clearly that i is a scalar because the integrand $\mathbf{j} \cdot d\mathbf{S}$ is a scalar.

The arrow often associated with the current in a wire does not indicate that current is a vector but merely shows the *sense* of charge flow. Positive charge carriers either move in a certain direction along the wire or in the opposite direction, these two possibilities being represented by $+$ or $-$ in algebraic equations. Note that (*a*) the current in a wire remains unchanged if the wire is bent, tied into a knot, or otherwise distorted, and (*b*) the arrows representing the sense of currents do not in any way obey the laws of vector addition. Thus currents can not be vectors.

We can compute the drift speed v_d of charge carriers in a conductor from the current density j. Figure 31-1 shows the conduction electrons in a wire moving to the right at an assumed constant drift speed v_d. The number of conduction electrons in the wire is nAl where n is the number of conduction electrons per unit volume and Al is the volume of the wire. A charge of magnitude

$$q = (nAl)e$$

passes out of the wire, through its right end, in a time t given by

$$t = \frac{l}{v_d}.$$

The current i is given by

$$i = \frac{q}{t} = \frac{nAle}{l/v_d} = nAev_d.$$

Solving for v_d and recalling that $j = i/A$ (Eq. 31-3) yields

$$v_d = \frac{i}{nAe} = \frac{j}{ne}. \tag{31-5}$$

An aluminum wire whose diameter is 0.10 in. is welded end to end to a copper
wire with a diameter of 0.064 in. The composite wire carries a steady current
of 10 A. What is the current density in each wire?

EXAMPLE 1

The current is distributed uniformly over the cross section of each conductor,
except near the junction, which means that the current density may be taken
as constant for all points within each wire. The cross-sectional area of the aluminum wire is 0.0079 in.². Thus, from Eq. 31-3,

$$j_{Al} = \frac{i}{A} = \frac{10 \text{ A}}{0.0079 \text{ in.}^2} = 1300 \text{ A/in.}^2.$$

The cross-sectional area of the copper wire is 0.0032 in.². Thus

$$j_{Cu} = \frac{i}{A} = \frac{10 \text{ A}}{0.0032 \text{ in.}^2} = 3100 \text{ A/in.}^2.$$

The fact that the wires are of different materials does not enter into consideration here.

What is v_d for the copper wire in Example 1?

EXAMPLE 2

We can write the current density for the copper wire (3100 A/in.²) as 480
A/cm². To compute n we start from the fact that there is one free electron per
atom in copper. The number of atoms per unit volume is dN_0/M where d is the
density, N_0 is the Avogadro number, and M is the atomic weight. The number
of free electrons per unit volume is then

$$n = \frac{dN_0}{M} = \frac{(9.0 \text{ g/cm}^3)(6.0 \times 10^{23} \text{ atoms/mol})(1 \text{ electron/atom})}{64 \text{ g/mol}}$$

$$= 8.4 \times 10^{22} \text{ electrons/cm}^3.$$

Finally, v_d is, from Eq. 31-5,

$$v_d = \frac{j}{ne} = \frac{480 \text{ A/cm}^2}{(8.4 \times 10^{22} \text{ electrons/cm}^3)(1.6 \times 10^{-19} \text{ C/electron})}$$

$$= 3.6 \times 10^{-2} \text{ cm/s.}$$

It takes 28 seconds for the electrons in this wire to drift 1.0 cm. Would you have
guessed that v_d was so low? The drift speed of electrons must not be confused
with the speed at which changes in the electric field configuration travel along
wires, a speed which approaches that of light. When we apply a pressure to one
end of a long water-filled garden hose, a *pressure wave* travels rapidly along the
hose. The speed at which *water* moves through the hose is much lower, however.

If we apply the same potential difference between the ends of geometrically similar rods of copper and of wood, very different currents result.
The characteristic of the conductor that enters here is its *resistance*.
We define the resistance of a conductor (often called a *resistor*; symbol
⎍⎍⎍-) between two points by applying a potential difference V between those points, measuring the current i, and dividing:

31-2
*RESISTANCE,
RESISTIVITY, AND
CONDUCTIVITY*

$$R = V/i. \tag{31-6}$$

If V is in volts and i in amperes, the resistance R will be in *ohms* (abbr.
Ω).

The flow of charge through a conductor is often compared with the

Table 31-1
Properties of metals as conductors

Metal	Resistivity (at 20°C) 10^{-8} $\Omega \cdot m$	Temperature Coefficient of Resistivity, α, per C° $(\times 10^{-5})$†
Silver	1.6	380
Copper	1.7	390
Aluminum	2.8	390
Tungsten	5.6	450
Nickel	6.8	600
Iron	10	500
Steel	18	300
Manganin	44	1.0
Carbon*	3500	−50

* Carbon, not strictly a metal, is included for comparison.
† This quantity, defined from

$$\alpha = \frac{1}{\rho}\frac{d\rho}{dT} \qquad (31\text{-}7)$$

is the fractional change in resistivity $(d\rho/\rho)$ per unit change in temperature. It varies with temperature, the values here referring to 20°C. For copper $(\alpha = 3.9 \times 10^{-3}/\text{C}°)$ the resistivity increases by 0.39 percent for a temperature increase of 1°C near 20°C. Note that α for carbon is negative, which means that the resistivity *decreases* with increasing temperature.

flow of water through a pipe, which occurs because there is a difference in pressure between the ends of the pipe, established perhaps by a pump. This pressure difference can be compared with the potential difference established between the ends of a resistor by a battery. The flow of water (liters/second, say) is compared with the current (coulombs/second, or amperes). The rate of flow of water for a given pressure difference is determined by the nature of the pipe. Is it long or short? Is it narrow or wide? Is it empty or filled, perhaps with gravel? These characteristics of the pipe are analogous to the resistance of a conductor.

Primary standards of resistance, kept at the National Bureau of Standards, are spools of wire whose resistances have been accurately measured. Because resistance varies with temperature, these standards, when used, are placed in an oil bath at a controlled temperature. They are made of a special alloy, called *manganin*, for which the change of resistance with temperature is very small. They are carefully annealed to eliminate strains, which also affect the resistance. These primary standard resistors are used chiefly to calibrate secondary standards for other laboratories.

Operationally, the primary resistance standards are not measured by using Eq. 31-6 but are measured in an indirect way which involves magnetic fields. Equation 31-6 is, in fact, used to measure V by setting up an accurately known current i (using a *current balance*; see Section 34-4) in an accurately known resistance R. This operational procedure for potential difference is the one normally used in place of the conceptual definition introduced in Section 29-1, in which one measures the work per unit charge required to move a test charge between two points.

Related to resistance is the *resistivity* ρ, which is a characteristic of a material rather than of a particular specimen of a material; it is defined, for isotropic materials,* from

$$\rho = \frac{E}{j}. \qquad (31\text{-}8a)$$

The resistivity of copper is 1.7×10^{-8} $\Omega \cdot$m; that of fused quartz is about 10^{16} $\Omega \cdot$m. Few physical properties are measurable over such a range of values; Table 31-1 lists some electrical properties for common metals.

Often we prefer to speak of the *conductivity* (σ) of a material rather than its resistivity. These are reciprocal quantities, related by

$$\sigma = 1/\rho. \qquad (31\text{-}8b)$$

The SI units of σ are $(\Omega \cdot \text{m})^{-1}$.

Consider a cylindrical conductor, of cross-sectional area A and length l, carrying a steady current i. Let us apply a potential difference V between its ends. If the cylinder cross sections at each end are equipotential surfaces, the electric field and the current density will be constant for all points in the cylinder and will have the values

$$E = \frac{V}{l} \quad \text{and} \quad j = \frac{i}{A}.$$

We may then write the resistivity ρ as

$$\rho = \frac{E}{j} = \frac{V/l}{i/A}.$$

But V/i is the resistance R which leads to

$$R = \rho \frac{l}{A}. \qquad (31\text{-}9)$$

V, i, and R are *macroscopic* quantities, applying to a particular body or extended region. The corresponding *microscopic* quantities are $\mathbf{E}$, $\mathbf{j}$, and ρ; they have values at every point in a body. The macroscopic quantities are related to each other by Eq. 31-6 ($V = iR$) and the microscopic quantities by Eq. 31-8a, which can be written in vector form as $\mathbf{E} = \mathbf{j}\rho$.

The macroscopic quantities can be found by integrating over the microscopic quantities, using relations already given, namely

$$i = \int \mathbf{j} \cdot d\mathbf{S} \qquad (31\text{-}4)$$

and

$$V_{ab} = -\int_a^b \mathbf{E} \cdot d\mathbf{l}. \qquad (29\text{-}5)$$

The integral in Eq. 31-4 is a surface integral, carried out over any cross section of the conductor. The integral in Eq. 29-5 is a line integral carried out along an arbitrary line drawn along the conductor, connecting any two equipotential surfaces, identified by a and b. For a long wire connected to a battery equipotential surface a might be chosen as a cross section of the wire near the positive battery terminal and b might be a cross section near the negative terminal.

We can express the resistance of a conductor between a and b in microscopic terms by dividing the two equations, or

$$R = \frac{V_{ab}}{i} = \frac{-\int_a^b \mathbf{E} \cdot d\mathbf{l}}{\int \mathbf{j} \cdot d\mathbf{S}}.$$

---

* These are materials whose properties (electrical in this case) do not vary with direction in the material.

If the conductor is a long cylinder of cross section A and length l, and if points a and b are its ends, the foregoing equation for R (see Eq. 31-8a) reduces to

$$R = \frac{El}{jA} = \rho \frac{l}{A},$$

which is Eq. 31-9.

The macroscopic quantities V, i, and R are of primary interest when we are making electrical measurements on real conducting objects. They are the quantities that one reads on meters. The microscopic quantities $\mathbf{E}$, $\mathbf{j}$, and ρ are of primary importance when we are concerned with the fundamental behavior of matter (rather than of specimens of matter), as we usually are in the research area of *solid state physics*. Thus Section 31-4 deals appropriately with an atomic view of the *resistivity* of a metal and not of the *resistance* of a metallic specimen. The microscopic quantities are also important when we are interested in the interior behavior of irregularly shaped conducting objects.

EXAMPLE 3

A rectangular carbon block has dimensions 1.0 cm × 1.0 cm × 50 cm. (*a*) What is the resistance measured between the two square ends and (*b*) between two opposing rectangular faces? The resistivity of carbon at 20°C is 3.5×10^{-5} Ω·m.

(*a*) The area of a square end is 1.0 cm² or 1.0×10^{-4} m². Equation 31-9 gives for the resistance between the square ends

$$R = \rho \frac{l}{A} = \frac{(3.5 \times 10^{-5} \ \Omega \cdot \text{m})(0.50 \ \text{m})}{1.0 \times 10^{-4} \ \text{m}^2} = 0.18 \ \Omega.$$

(*b*) For the resistance between opposing rectangular faces (area = 5.0×10^{-3} m²), we have

$$R = \rho \frac{l}{A} = \frac{(3.5 \times 10^{-5} \ \Omega \cdot \text{m})(10^{-2} \ \text{m})}{5.0 \times 10^{-3} \ \text{m}^2} = 7.0 \times 10^{-5} \ \Omega.$$

Thus a given conductor can have any number of resistances, depending on how the potential difference is applied to it. The ratio of resistances for these two cases is 2600. We assume in each that the potential difference is applied to the block in such a way that the surfaces between which the resistance is desired are equipotential. Otherwise Eq. 31-9 would not be valid.

Figure 31-2 shows (solid curve) how the resistivity of copper varies with temperature. Sometimes, for practical use, such data are expressed in equation form. If we are interested in only a limited range of temperatures extending, say, from 0 to 500°C, we can fit a straight line to the curve of Fig. 31-2, making it pass through two arbitrarily selected points; see the dashed line. We choose the point labeled T_0, ρ_0 in the figure as a reference point, T_0 being 0° C in this case and ρ_0 being 1.56×10^{-8} Ω·m. The resistivity ρ at any temperature T can be found from the empirical equation of the dashed straight line in Fig. 31-2, which is

$$\rho = \rho_0 \left[1 + \bar{\alpha}(T - T_0) \right]. \tag{31-10}$$

This relation shows correctly that $\rho \to \rho_0$ as $T \to T_0$.

If we solve Eq. 31-10 for $\bar{\alpha}$, we obtain

$$\bar{\alpha} = \frac{1}{\rho_0} \frac{\rho - \rho_0}{T - T_0}.$$

Comparison with Eq. 31-7 shows that $\bar{\alpha}$ is a *mean temperature coefficient of resistivity* for a selected pair of temperatures rather than the temperature coefficient of resistivity at a particular temperature, which is the definition of α. For most practical purposes Eq. 31-10 gives results that are within the acceptable range of accuracy.

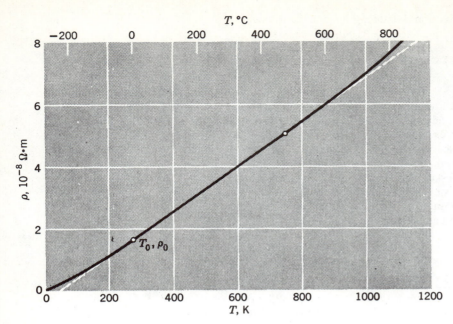

figure 31-2
The resistivity of copper as a function of temperature. The dashed line is an approximation chosen to fit the curve at the two circled points. The point marked T_0, ρ_0 is chosen as a reference point.

   The curve of Fig. 31-2 does not go to zero at the absolute zero of temperature, even though it appears to do so, the residual resistivity at this temperature being $0.02 \times 10^{-8}\ \Omega \cdot m$. For many substances the resistance *does* become zero at some low temperature. Figure 31-3 shows the resistance of a specimen of mercury for temperatures below 6 K. In the space of about 0.05 K the resistance drops abruptly to an immeasurably low value. This phenomenon, called *superconductivity,** was discovered by Kamerlingh Onnes in the Netherlands in 1911. The resistance of materials in the superconducting state seems to be truly zero; currents, once established in closed superconducting circuits, persist for weeks without diminution, even though there is no battery in the circuit. If the temperature is raised slightly above the superconducting point, or if a large enough magnetic field is applied, such currents drop rapidly to zero.

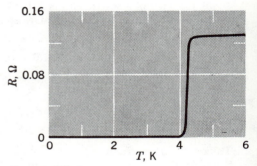

figure 31-3
The resistivity of mercury disappears below about 4 K.

31-3
OHM'S LAW

Let us apply a variable potential difference V between the ends of a 100-foot coil of #18 copper wire. For each applied potential difference, let us measure the current i and plot it against V as in Fig. 31-4. The straight line that results means that *the resistance of this conductor is the same no matter what applied voltage we use to measure it.* This important result, which holds for metallic conductors, is known as *Ohm's law.* We assume that the temperature of the conductor is essentially constant throughout the measurements.

   Many conductors do not obey Ohm's law. Figure 31-5, for example, shows a *V-i* plot for a type 2A3 vacuum tube. The plot is not straight and the resistance depends on the voltage used to measure it. Also, the current for this device is almost vanishingly small if the polarity of the applied potential difference is reversed. For metallic conductors the

* See (*a*) "Superconductivity" by B. T. Matthias, *Scientific American*, November 1957, (*b*) "The Search for High-Temperature Superconductors" by B. T. Matthias, *Physics Today*, August 1971; (*c*) "Large-Scale Applications of Superconductivity" by Brian B. Schwartz and Simon Foner, *Physics Today*, July 1977.

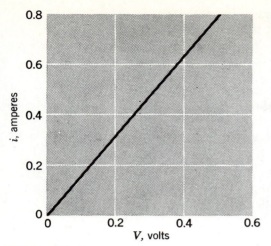

figure 31-4
The current in a particular copper
conductor as a function of potential
difference. This conductor obeys
Ohm's law.

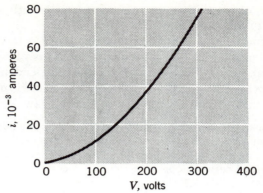

figure 31-5
The current in a type 2A3 vacuum
tube as a function of potential
difference. This conductor does *not*
obey Ohm's law.

current reverses direction when the potential difference is reversed, but
its magnitude does not change.

Figure 31-6 shows a typical *V-i* plot for another nonohmic device, a
thermistor. This is a semiconductor (see Section 26-3) with a large and
negative temperature coefficient of resistivity α (see Table 31-1) that
varies greatly with temperature. We note that two different currents
through the thermistor can correspond to the same potential difference
between its ends. Thermistors are often used to measure the rate of energy flow in microwave beams by allowing the microwave beam to fall
on the thermistor and heat it. The relatively small temperature rise so
produced results in a relatively large change in resistance, which serves
as a measure of the microwave power. Modern electronics, and therefore
much of the character of our present technological civilization, depends
in a fundamental way on the fact that many conductors, such as
transistors and vacuum tubes, do *not* obey Ohm's law.

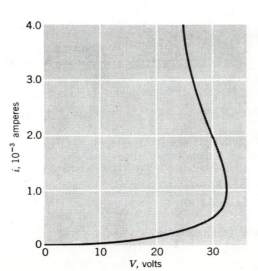

figure 31-6
A plot of current as a function of
potential difference in a Western
Electric 1-B thermistor. Again, this
conductor does not obey Ohm's law.
The shape of the curve can be
accounted for in terms of the large
negative temperature coefficient of
resistivity of the material of which
the device is made.

We stress that the relationship $V = iR$ is *not* a statement of Ohm's law. A conductor obeys this law only if its V-i curve is linear, that is, if R is independent of V and i. The relationship $R = V/i$ remains as the general definition of the resistance of a conductor whether or not the conductor obeys Ohm's law.

The microscopic equivalent of the relationship $V = iR$ is Eq. 31-8a, or $\mathbf{E} = \mathbf{j}\rho$. A conducting *material* is said to obey Ohm's law if a plot of E versus j is linear, that is, if the resistivity ρ is independent of E and j. Ohm's law is a specific property of certain materials and is not a general law of electromagnetism, for example, like Gauss's law.

A close analogy exists between the flow of charge because of a potential difference and the flow of heat because of a temperature difference. Consider a thin electrically conducting slab of thickness Δx and area A. Let a potential difference ΔV be maintained between opposing faces. The current i is given by Eqs. 31-6 ($i = V/R$) and 31-9 ($R = \rho l/A$), or

$$i = \frac{V_a - V_b}{R} = \frac{(V_a - V_b)A}{\rho l} = -\frac{(V_b - V_a)A}{\rho \Delta x}.$$

In the limiting case of a slab of thickness dx this becomes

$$i = -\frac{1}{\rho} A \frac{dV}{dx}$$

or

$$\frac{dq}{dt} = -\sigma A \frac{dV}{dx}, \tag{31-11}$$

where $\sigma\ (= 1/\rho)$, as we have seen (Eq. 31-8b), is the *conductivity* of the material. Since positive charge flows in the direction of decreasing V, we introduce a minus sign into Eq. 31-11, that is, dq/dt is positive when dV/dx is negative.

The analogous heat flow equation (see Section 22-4) is

$$\frac{dQ}{dt} = -kA \frac{dT}{dx}, \tag{31-12}$$

which shows that k, the thermal conductivity, corresponds to σ and dT/dx, the temperature gradient, corresponds to dV/dx, the potential gradient. For pure metals there is more than a formal mathematical analogy between Eqs. 31-11 and 31-12. Both heat energy and charge are carried by the free electrons in such metals; empirically, a good electrical conductor (silver, say) is also a good heat conductor.

31-4
OHM'S LAW—A MICROSCOPIC VIEW

As we have said earlier, Ohm's law is not a fundamental law of electromagnetism because it depends on the properties of the conducting medium. The law is very simple in form, and it is curious that many conductors obey it so well, whereas other conductors do not obey it at all (see Figs. 31-4, 31-5, and 31-6). Let us see if we can understand why metals obey Ohm's law, which we shall write (see Eq. 31-8a) in the microscopic form $\mathbf{E} = \rho \mathbf{j}$.

In a metal the valence electrons are not attached to individual atoms but are free to move about within the lattice and are called *conduction electrons*. In copper there is one such electron per atom, the other 28 remaining bound to the copper nuclei to form ionic cores.

Although the speed distribution of conduction electrons can be described correctly only in terms of quantum physics, the classical *free-electron model* will suit our purpose. It suffices to consider only a suitably defined average speed $\bar{v}$; for copper $\bar{v} = 1.6 \times 10^8$ cm/s. In the absence of an electric field, the directions in which the free or conduction) electrons move are completely random, like those of the molecules of a gas confined to a container.

The electrons collide constantly with the ionic cores of the conductor, that is, they interact with the lattice, often suffering sudden changes in speed and direction. These collisions remind us of the collisions of gas molecules confined to a container. As in the case of molecular collisions, we can describe electron-lattice collisions by a *mean free path* λ, where λ is the average distance that an electron travels between collisions.*

In an ideal metallic crystal at 0 K electron-lattice collisions would not occur, according to the predictions of quantum physics, that is, $\lambda \rightarrow \infty$ as $T \rightarrow 0$ K for ideal crystals. Collisions take place in actual crystals because (a) the ionic cores at any temperature T are vibrating about their equilibrium positions in a random way, (b) impurities, that is, foreign atoms, may be present, and (c) the crystal may contain lattice imperfections, such as missing atoms and displaced atoms. On this view it is not surprising that the resistivity of a metal can be increased by (a) raising its temperature, (b) adding small amounts of impurities, and (c) straining it severely, as by drawing it through a die, to increase the number of lattice imperfections.

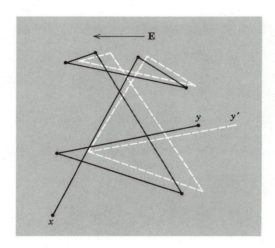

figure 31-7
The solid lines show an electron moving from x to y, making six collisions. The dashed curves show what the electron path *might* have been in the presence of an electric field **E**. Note the steady drift in the direction of $-$**E**.

When we apply an electric field to a metal, the electrons modify their random motion in such a way that they drift slowly, in the opposite direction to that of the field, with an average drift speed v_d. This drift speed is very much less (by a factor of something like 10^{10}; see Example 2) than the effective average speed $\bar{v}$ mentioned above. Figure 31-7 suggests the relationship between these two speeds. The solid lines suggest a possible random path followed by an electron in the absence of an applied field; the electron proceeds from x to y, making six collisions on the way. The dashed curves show how this same event *might* have occurred if an electric field **E** had been applied. Note that the electron drifts steadily to the right, ending at y' rather than at y. In preparing Fig. 31-7, it has been assumed that the drift speed v_d is $0.02\bar{v}$; actually, it is more like $10^{-10}\bar{v}$, so that the "drift" exhibited in the figure is greatly exaggerated.

We can calculate the drift speed v_d in terms of the applied electric field E and of $\bar{v}$ and λ. When a field is applied to an electron in the metal

---

* It can be shown that collisions between electrons occur only rarely and have little effect on the resistivity.

it will experience a force eE which will impart to it an acceleration a given by Newton's second law,

$$a = \frac{eE}{m}.$$

Consider an electron that has just collided with an ion core. The collision, in general, will momentarily destroy the tendency to drift and the electron will have a truly random direction after the collision. During the time interval to the next collision the electron's velocity will have changed, on the average, by $a(\lambda/\bar{v})$ or $a\tau$ where τ is the mean time between collisions. We call this the drift speed v_d, or

$$v_d = a\tau = \frac{eE\tau}{m} \qquad (31\text{-}13)$$

The electron's motion through the conductor is analogous to the constant terminal rate of fall of a stone in water. The gravitational force F_g on the stone is opposed by a viscous resisting force that is proportional to the velocity, or

$$F_g = mg = bv,$$

where b is a viscous coefficient (see Section 15-9). Thus the constant terminal speed of the stone is

$$v = \left(\frac{1}{b}\right) F_g.$$

We can rewrite Eq. 31-13 as

$$v_d = \left(\frac{\tau}{m}\right) F_E,$$

where F_E ($= eE$) is the electrical force. Comparison of these equations shows that the equivalent "viscous coefficient" for the motion of an electron in a particular conductor is m/τ. If τ is short, the conductor exhibits a greater "viscous effect" on the electron motion, and the drift speed v_d is proportionally lower.

We may express v_d in terms of the current density (Eq. 31-5) and combine with Eq. 31-13 to obtain

$$v_d = \frac{j}{ne} = \frac{eE\tau}{m}.$$

Combining this with Eq. 31-8a ($\rho = E/j$) leads finally to*

$$\rho = \frac{m}{ne^2\tau}. \qquad (31\text{-}14)$$

Equation 31-14 can be taken as a statement that metals obey Ohm's law *if* we can show that τ does not depend on the applied electric field E. In this case ρ will not depend on E, which (see Section 31-3) is the criterion that a material obey Ohm's law. The quantity τ depends on the speed distribution of the conduction electrons. We have seen that this distribution is affected only very slightly by the application of even a relatively large electric field, since $\bar{v}$ is of the order of 10^8 cm/s and v_d (see Example 2) only of the order of 10^{-2} cm/s, a ratio of 10^{10}. We may be sure that whatever the value of τ is (for copper at 20°C, say) in

---

* See "Drift Speed and Collision Time" by Donald E. Tilley, *American Journal of Physics,* June 1976, for a full discussion.

the absence of a field it remains essentially unchanged when the field is applied. Thus the right side of Eq. 31-14 is independent of E (which means that ρ is independent of E) and the material obeys Ohm's law.

EXAMPLE 4

What are (a) the mean time τ between collisions and (b) the mean free path for free electrons in copper?

(a) From Eq. 31-14 (see also Example 2), we have

$$\tau = \frac{m}{ne^2\rho} = \frac{(9.1 \times 10^{-31}\ \text{kg})}{(8.4 \times 10^{28}/\text{m}^3)(1.6 \times 10^{-19}\ \text{C})^2(1.7 \times 10^{-8}\ \Omega\cdot\text{m})}$$

$$= 2.5 \times 10^{-14}\ \text{s}.$$

(b) The mean free path is

$$\lambda = \tau\bar{v} = (2.5 \times 10^{-14}\ \text{s})(1.6 \times 10^8\ \text{cm/s}) = 4.0 \times 10^{-6}\ \text{cm}.$$

This is about 220 ionic diameters.*

31-5
ENERGY TRANSFERS IN AN ELECTRIC CIRCUIT

Figure 31-8 shows a circuit consisting of a battery B connected to a "black box". A steady current i exists in the connecting wires and a steady potential difference V_{ab} exists between the terminals a and b. The box might contain a resistor, a motor, or a storage battery, among other things.

Terminal a, connected to the positive battery terminal, is at a higher potential than terminal b. If a charge dq moves through the box from a to b, this charge will decrease its electric potential energy by $dq\ V_{ab}$ (see Section 29-6). The conservation-of-energy principle tells us that this energy is transferred in the box from electric potential energy to some other form. What that other form will be depends on what is in the box. In a time dt the energy dU transferred inside the box is then

$$dU = dq\ V_{ab} = i\ dt\ V_{ab}.$$

We find the *rate* of energy transfer P by dividing by the time, or

$$P = \frac{dU}{dt} = iV_{ab}. \tag{31-15}$$

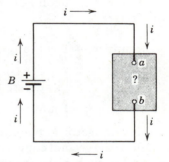

figure 31-8
A battery B sets up a current in a circuit containing a "black box", that is, a box whose contents are not known to us.

If the device in the box is a motor, the energy appears largely as mechanical work done by the motor; if the device is a storage battery that is being charged, the energy appears largely as stored chemical energy in this second battery.

If the device is a resistor, we assert that the energy appears as thermal energy in the resistor. To see this, consider a stone of mass m that falls through a height h. It decreases its gravitational potential energy by mgh. If the stone falls in a vacuum or—for practical purposes—in air, this energy is transformed into kinetic energy of the stone. If the stone falls into the depths of the ocean, however, its speed eventually becomes constant, which means that the kinetic energy no longer increases. The potential energy that is steadily being made available as the stone falls then appears as thermal energy in the stone and the surrounding water. It is the viscous, friction-like drag of the water on the surface of the stone that stops the stone from accelerating, and it is at this surface that thermal energy appears.

* See footnote on p. 686.

The course of the electrons through the resistor is much like that of the stone through water. The electrons travel with a constant drift speed v_d and thus do not gain kinetic energy. The electric potential energy that they lose is transferred to the resistor as thermal energy. On a microscopic scale we can understand this in that collisions between the electrons and the lattice (see Fig. 21-5) increase the amplitude of the thermal vibrations of the lattice; on a macroscopic scale this corresponds to a temperature increase. There can be a flow of heat out of the resistor subsequently, if the environment is at a lower temperature than the resistor.

For a resistor we can combine Eqs. 31-6 ($R = V/i$) and 31-15 and obtain either

$$P = i^2 R \qquad (31\text{-}16)$$

or

$$P = \frac{V^2}{R}. \qquad (31\text{-}17)$$

Note that Eq. 31-15 applies to electrical energy transfer of *all* kinds; Eqs. 31-16 and 31-17 apply only to the transfer of electrical energy to thermal energy in a resistor. Equations 31-16 and 31-17 are known as *Joule's law*. This law is a particular way of writing the conservation-of-energy principle for the special case in which electrical energy is transferred into thermal energy (Joule energy).

The unit of power that follows from Eq. 31-15 is the volt-ampere. We can write it as

$$1 \text{ volt·ampere} = 1 \text{ volt·ampere} \left(\frac{1 \text{ joule}}{1 \text{ volt} \times 1 \text{ coulomb}} \right)$$
$$\left(\frac{1 \text{ coulomb}}{1 \text{ ampere} \times 1 \text{ second}} \right)$$
$$= 1 \text{ joule/second}.$$

The first conversion factor in parentheses comes from the definition of the volt (Eq. 29-1); the second comes from the definition of the coulomb. The joule/second is such a common unit that it is given a special name of its own, the *watt* (abbr. W); see Section 7-7. Power is not an exclusively electrical concept, of course, and we can express in watts the power ($= \mathbf{F} \cdot \mathbf{v}$) expended by an agent that exerts a force $\mathbf{F}$ while it moves with a velocity $\mathbf{v}$.

EXAMPLE 5

You are given a 20-ft length of heating wire made of the special alloy Nichrome; it has a resistance of 24 Ω. Can you obtain more heat by winding one coil or by cutting the wire in two and winding two separate coils? In each case the coils are to be connected individually across a 110-V line.

The power P for the single coil is given by Eq. 31-17:

$$P = \frac{V^2}{R} = \frac{(110 \text{ V})^2}{24 \text{ } \Omega} = 500 \text{ W}.$$

The power for a coil of half the length is given by

$$P' = \frac{(110 \text{ V})^2}{12 \text{ } \Omega} = 1000 \text{ W}.$$

There are two "half-coils," so that the total power obtained by cutting the wire in half is 2000 W, or four times that for the single coil. This would seem to suggest that we could buy a 500-W heating coil, cut it in half, and rewind it to obtain 2000 W. Why is this not a practical idea?

1. Name other physical quantities that, like current, are scalars having a sense represented by an arrow in a diagram.

2. What conclusions can you draw by applying Eq. 31-4 to a closed surface through which a number of wires pass in random directions, carrying steady currents of different magnitudes? Does Gauss's law hold?

3. A potential difference V is applied to a copper wire of diameter d and length l. What is the effect on the electron drift speed of (a) doubling V, (b) doubling l, and (c) doubling d?

4. If the drift speeds of the electrons in a conductor under ordinary circumstances are so slow (see Example 2), why do the lights in a room turn on so quickly after the switch is closed?

5. Can you think of a way to measure the drift speed for electrons by timing their travel along a conductor?

6. Let a battery be connected to a copper cube at two corners defining a body diagonal. Pass a hypothetical plane completely through the cube, tilted at any angle. (a) Is the current i through the plane independent of the position and orientation of the plane? (b) Is there any position and orientation of the plane for which $\mathbf{j}$ is a constant in magnitude, direction, or both? (c) Does Eq. 31-4 hold for all orientations of the plane? (d) Does Eq. 31-4 hold for a closed surface of arbitrary shape, which may or may not lie entirely within the cube? If not, why; if so, what does it predict?

7. In our convention for the direction of current arrows (a) would it have been more convenient, or even possible, to have assumed all charge carriers to be negative? (b) Would it have been more convenient, or even possible, to have labeled the electron as positive, the proton as negative, etc.?

8. Explain in your own words why we can have $\mathbf{E} \neq 0$ inside a conductor in this chapter but that we took $\mathbf{E} = 0$ for granted in Chapter 28 (see, for example, Section 28-6).

9. A potential difference V is applied to a circular cylinder of carbon by clamping it between circular copper electrodes, as in Fig. 31-9. Discuss the difficulty of calculating the resistance of the carbon cylinder, using the relation $R = \rho L/A$.

10. How would you measure the resistance of a pretzel-shaped conductor? Give specific details to clarify the concept.

11. Discuss the difficulties of testing whether the filament of a light bulb obeys Ohm's law.

12. You are given a circular cylinder of aluminum, 1.00 cm in radius and 2.00 cm high. What practical laboratory arrangements would you make if you wanted to measure its resistance between parallel faces (assumed equipotentials) using Eq. 31-9 ($R = \rho l/A$).

13. You are given a cube of aluminum and access to two battery terminals. How would you connect the terminals to the cube to insure (a) a maximum and (b) a minimum resistance?

14. Does the relation $V = iR$ apply to resistors that do not obey Ohm's law?

15. The temperature coefficient of resistance of a thermistor is negative and varies greatly with temperature. Account qualitatively for the shape of the curve of i versus V for the thermistor of Fig. 31-6.

16. Why are the dashed white lines in Fig. 31-7 curved slightly?

17. A current i enters the top of a copper sphere of radius R and leaves at a diametrically opposite point. Are all parts of the sphere equally effective in dissipating thermal energy?

18. What special characteristics must (a) heating wire and (b) fuse wire have?

19. Equation 31-16 ($P = i^2R$) seems to suggest that the rate of increase of thermal energy in a resistor is reduced if the resistance is made less; Eq. 31-17 ($P = V^2/R$) seems to suggest just the opposite. How do you reconcile this apparent paradox?

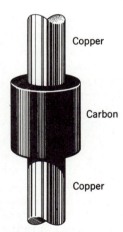

Copper

Carbon

Copper

figure 31-9
Question 9

20. Is the filament resistance lower or higher in a 500-W light bulb than in a 100-W bulb? Both bulbs are designed to operate on 100 V.

21. Five wires of the same length and diameter are connected in turn between two points maintained at constant potential difference. Will Joule energy be developed at the faster rate in the wire of (a) the smallest or (b) the largest resistance?

22. The windings of a motor (connected to a load) have a resistance of 1.0 Ω. If we apply a potential difference of 100 V to the motor, does it follow that the current through the motor will be 110 V/1.0 Ω = 110 A?

23. A cow and a man are standing in a meadow when lightning strikes the ground nearby. Why is the cow more likely to be killed than the man? The responsible phenomenon is called "step voltage".

problems

SECTION 31-1

1. A current of 5 A exists in a 10-Ω resistor for 4 min. (a) How many coulombs and (b) how many electrons pass through any cross section of the resistor in this time?   *Answer:* (a) 1.2×10^3 C. (b) 7.5×10^{21} electrons.

2. A current is established in a gas discharge tube when a sufficiently high potential difference is applied across the two electrodes in the tube. The gas ionizes; electrons move toward the positive terminal and positive ions toward the negative terminal. What are the magnitude and sense of the current in a hydrogen discharge tube in which 3.1×10^{18} electrons and 1.1×10^{18} protons move past a cross-sectional area of the tube each second?

3. A steady beam of alpha particles ($q = 2e$) traveling with constant kinetic energy 20 MeV carries a current 0.25×10^{-6} A. (a) If the beam is directed perpendicular to a plane surface, how many alpha particles strike the surface in 3.0 s? (b) At any instant, how many alpha particles are there in a given 20-cm length of the beam? (c) Through what potential difference was it necessary to accelerate each alpha particle from rest to bring it to an energy of 20 MeV?   *Answer:* (a) 2.3×10^{12}. (b) 5.0×10^3. (c) 10^7 V.

4. We have 2.0×10^8 doubly charged positive ions per cubic centimeter, all moving north with a speed of 1.0×10^7 cm/s. (a) What is the current density $\mathbf{j}$, in magnitude and direction? (b) Can you calculate the total current i in this ionic beam? If not, why?

5. A small but measurable current of 1.0×10^{-10} A exists in a copper wire whose diameter is 0.10 in. Calculate the electron drift speed. *Answer:* 1.5×10^{-15} m/s.

6. The belt of an electrostatic generator is 50 cm wide and travels at 30 m/s. The belt carries charge into the sphere at a rate corresponding to 1.0×10^{-4} A. Compute the surface charge density on the belt.

7. A current i enters one corner of a square sheet of copper and leaves at the opposite corner. Sketch arrows for various points within the square to represent the relative values of $\mathbf{j}$. Intuitive guesses rather than detailed mathematical analyses are called for.

8. You are given a conducting sphere of 10-cm radius. One wire carries a current of 1.0000020 A into it. Another wire carries a current of 1.0000000 A out of it. How long would it take for the sphere to increase in potential by 1000 V?

SECTION 31-2

9. A steel trolley-car rail has a cross-sectional area of 7.1 in.² What is the resistance of 10 miles of single track? The resistivity of the steel is 6.0×10^{-7} Ω·m.   *Answer:* 2.1 Ω.

10. A square aluminum rod is 1.0 m long and 5.0 mm on edge. (a) What is the resistance between its ends? (b) What must be the diameter of a circular 1.0 m copper rod if its resistance is to be the same?

11. A copper wire and an iron wire of the same length have the same potential

difference applied to them. (a) What must be the ratio of their radii if the current is to be the same? (b) Can the current density be made the same by suitable choices of the radii? *Answer:* (a) 2.4, iron being larger. (b) No.

12. A wire with a resistance of 6.0 Ω is drawn out through a die so that its new length is three times its original length. Find the resistance of the longer wire, assuming that the resistivity and density of the material are not changed during the drawing process.

13. A wire of Nichrome (a nickel-chromium alloy commonly used in heating elements) is 1.0 m long and 1.0 mm² in cross-sectional area. It carries a current of 4.0 A when a 2.0-V potential difference is applied between its ends. What is the conductivity σ of Nichrome?   *Answer:* 2.0×10^6 S.

14. A rod of a certain metal is 1.00 m long and 0.550 cm in diameter. The resistance between its ends (at 20°C) is 2.87×10^{-3} Ω. A round disk is formed of this same material, 2.00 cm in diameter and 1.00 mm thick. (a) What is the resistance between the opposing round faces, assuming equipotential surfaces? (b) What is the material?

15. Two conductors are made of the same material and have the same length. Conductor A is a solid wire of diameter 1.0 mm. Conductor B is a hollow tube of outside diameter 2.0 mm and inside diameter 1.0 mm. What is the resistance ratio, R_A/R_B, measured between their ends?   *Answer:* 3.0

16. Nine copper wires of length *l* and diameter *d* are connected in parallel to form a single composite conductor of resistance *R*. What must be the diameter *D* of a single copper wire of length *l* if it is to have the same resistance?

17. A copper wire and an iron wire of equal length *l* and diameter *d* are joined and a potential difference *V* is applied between the ends of the composite wire. Calculate (a) the potential difference across each wire. Assume that $l = 10$ m, $d = 2.0$ mm, and $V = 100$ V. (b) Also calculate the current density in each wire, and (c) the electric field in each wire.
Answer: (a) 15 V (copper); 85 V (iron). (b) 8.5×10^7 A/m². (c) 1.5 V/m (copper); 8.5 V/m (iron).

18. The resistance of an iron wire is 5.9 times that of a copper wire of the same dimensions. What must be the diameter of an iron wire if it is to have the same resistance as a copper wire 0.12 cm in diameter, both wires being the same length?

19. Circularly cylindrical aluminum and copper rods have the same length and are designed to have the same resistance. The resistivity of copper is 0.61 times that of aluminum but its density is 3.3 times that of aluminum. What is the ratio of the mass of the aluminum rod to that of the copper rod? *Answer:* 0.50.

20. Conductors A and B, having equal lengths of 40 m and a cross-sectional area of 0.10 m², are connected in series. A potential of 60 V is applied across the terminal points of the connected wires. The resistances of the wires are 40 and 20 Ω respectively. Determine: (a) the resistivities of the two wires; (b) the magnitude of the electric field in each wire; (c) the current density in each wire; (d) the potential difference applied to each conductor.

21. A resistor is in the shape of a truncated right circular cone (Fig. 31-10). The end radii are *a* and *b*, the altitude is *l*. If the taper is small, we may assume that the current density is uniform across any cross section. (a) Calculate the resistance of this object. (b) Show that your answer reduces to $\rho(l/A)$ for the special case of zero taper (a = b).

Answer: (a) $R = \rho \dfrac{l}{\pi ab}$.

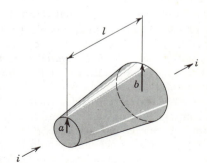

figure 31-10
Problem 21

22. (a) At what temperature would the resistance of a copper conductor be double its resistance at 0°C? (b) Does this same temperature hold for all copper conductors, regardless of shape or size?

23. The copper windings of a motor have a resistance of 50 Ω at 20°C when the motor is idle. After running for several hours the resistance rises to 58 Ω. What is the temperature of the windings?   *Answer:* 61°C.

24. When a metal rod is heated, not only its resistance but also its length and its cross-sectional area change. The relation $R = \rho l/A$ suggests that all three factors should be taken into account in measuring ρ at various temperatures. (a) If the temperature changes by 1.0°C, what percent changes in R, l, and A occur for a copper conductor? (b) What conclusion do you draw? The coefficient of linear expansion is 1.7×10^{-5}/C°.

SECTION 31-3

25. List in tabular form similarities and differences between the flow of charge along a conductor, the flow of water through a horizontal pipe, and the conduction of heat through a slab. Consider such quantities as what causes the flow, what opposes it, what particles (if any) participate, and the units in which the flow may be measured.

26. (a) Using data from Fig. 31-5, plot the resistance of the vacuum tube as a function of applied potential difference. (b) Repeat for the thermistor of Fig. 31-6.

SECTION 31-4

27. Explain why the momentum which conduction electrons transfer to the ions in a metal conductor does not give rise to a resultant force on the conductor.
    Answer: Because of Newton's third law there is no resultant force on the conductor (electrons + ions).

SECTION 31-5

28. Thermal energy is developed in a resistor at a rate of 100 W when the current is 3.0 A. What is the resistance in ohms?

29. An x-ray tube takes a current of 7.0 mA and operates at a potential difference of 80 kV. What power in watts is dissipated?   *Answer:* 560 W.

30. A potential difference of 1.0 V is applied to a 100-ft length of #18 copper wire (diameter = 0.040 in.). Calculate (a) the current, (b) the current density, (c) the electric field, and (d) the rate at which thermal energy is developed in the wire.

31. The National Board of Fire Underwriters has fixed safe current-carrying capacities for various sizes and types of wire. For #10 rubber-coated copper wire (wire diameter = 0.10 in.) the maximum safe current is 25 A. At this current, find (a) the current density, (b) the electric field, (c) the potential difference for 1000 ft of wire, and (d) the rate at which thermal energy ($= i^2R$) is developed for 1000 ft of wire.
    Answer: (a) 4.9×10^6 A/m². (b) 8.3×10^{-2} V/m. (c) 25 V. (d) 630 W.

32. A beam of 16-MeV deuterons from a cyclotron falls on a copper block. The beam is equivalent to a current of 15×10^{-6} A. (a) At what rate do deuterons strike the block? (b) At what rate is thermal energy produced in the block?

33. A 500-W immersion heater is placed in a pot containing 2.0 liters of water at 20°C. (a) How long will it take to bring the water to boiling temperature, assuming that 80% of the available energy is absorbed by the water? (b) How much longer will it take to boil half the water away?
    Answer: (a) 28 min. (b) 1.6 h.

34. A 500-W heating unit is designed to operate from a 115-V line. (a) By what percentage will its heat output drop if the line voltage drops to 110 V? Assume no change in resistance. (b) Taking the variation of resistance with temperature into account, would the actual heat output drop be larger or smaller than that calculated in (a)?

35. A 1250-W radiant heater is constructed to operate at 115 V. (a) What will be the current in the heater? (b) What is the resistance of the heating coil? (c) How many kilocalories are generated in one hour by the heater?
    Answer: (a) 11 A. (b) 11 Ω. (c) 1100 kcal.

36. A Nichrome heater dissipates 500 W when the applied potential difference is 110 V and the wire temperature is 800°C. How much power would it dissipate if the wire temperature were held to 200°C by immersion in a bath of cooling oil? The applied potential difference remains the same; $\bar{\alpha}$ for Nichrome is about 4×10^{-4}/C°.

37. (a) Derive the formulas $P = j^2\rho$ and $P = E^2/\rho$ where P = power per unit volume in a resistor. (b) A cylindrical resistor of radius 0.50 cm and length 2.0 cm has a resistivity of 3.5×10^{-5} $\Omega \cdot$m. What are the current density and potential difference when the power dissipation is 1.0 W?
    Answer: (b) $j = 1.3 \times 10^5$ A/m²; $V = 0.094$ V.

38. It is desired to make a long cylindrical conductor whose temperature coefficient of resistivity at 20°C will be close to zero. (a) If such a conductor is made by assembling alternate disks of iron and carbon, what is the ratio of the thickness of a carbon disk to that of an iron disk? Assume that the temperature remains essentially the same in each disk. (b) What is the ratio of thermal energy generation in a carbon disk to that in an iron disk? See Table 31-1.

39. An electron linear accelerator produces a pulsed beam of electrons. The pulse current is 0.50 A and the pulse duration 0.10 μs. (a) How many electrons are accelerated per pulse? (b) What is the average current for a machine operating at 500 pulses/s? (c) If the electrons are accelerated to an energy of 50 MeV, what are the average and peak power outputs of the accelerator?
    Answer: (a) 3.1×10^{11}. (b) 25 μA. (c) 1200 W (average); 2.5×10^7 W (peak).

32
electromotive force and circuits

There exist in nature certain devices such as batteries and electric generators which are able to maintain a potential difference between two points to which they are attached. We call such devices seats of *electromotive force* (symbol ε; abbr. emf). In this chapter we do not discuss their internal construction or detailed mode of action but confine ourselves to describing their gross electrical characteristics and to exploring their usefulness in electric circuits.

Figure 32-1a shows a seat of emf B, represented by a battery*, connected to a resistor R. The seat of emf maintains its upper terminal positive and its lower terminal negative, as shown by the + and − signs. In the circuit external to B positive charge carriers would be driven in the direction shown by the arrows marked *i*. In other words, a clockwise current would be set up.

An emf is represented by an arrow which is placed next to the seat and points in the direction in which the seat, acting alone, would cause a positive charge carrier to move in the external circuit. We draw a small circle on the tail of an emf arrow so that we will not confuse it with a current arrow.

A seat of emf must be able to do work on charge carriers that enter it. In the circuit of Fig. 32-1a, for example, the seat acts to move positive charges from a point of low potential (the negative terminal) through the seat to a point of high potential (the positive terminal). This reminds us of a pump, which can cause water to move from a place of low gravitational potential to a place of high potential.

In Fig. 32-1a a charge *dq* passes through *any* cross section of the circuit in time *dt*. In particular, this charge enters the seat of emf ε at its low-potential end and leaves at its high-potential end. The seat must do

*Batteries are far from being the only source of emf. Among others are generators (see Chapter 39); devices activated by temperature differences (thermocouples, etc.); devices activated by light (see "The Photovoltaic Generation of Electricity" by Bruce Chalmers, *Scientific American*, October 1976); the human heart (see "The Electrocardiograph— Teaching Physics to Premeds" by Pierre Lafrance, *The Physics Teacher*, November 1972); certain fish (see "Seeing the World Through a New Sense: Electroreception in Fish" by T. H. Bullock, *American Scientist*, May–June 1973).

an amount of work dW on the (positive) charge carriers to force them to go to the point of higher potential. The emf ε of the seat is defined from

$$\varepsilon = dW/dq. \qquad (32\text{-}1)$$

The unit of emf is the joule/coulomb (see Eq. 29-1) which is the *volt*. We might be inclined to say that a battery has an emf of 1 volt if it maintains a difference of potential of 1 volt between its terminals. This is true only under certain conditions, which we describe in Section 32-4. Note also from Eq. 32-1 that the electromotive force is not actually a force, that is, we cannot measure it in newtons. The name is involved with the early history of the subject.

If a seat of emf does work on a charge carrier, energy must be transferred within the seat. In a battery, for example, chemical energy is transferred into electrical energy. Thus we can describe a seat of emf as a device in which chemical, mechanical, or some other form of energy is changed (reversibly) into electrical energy. The chemical energy provided by the battery in Fig. 32-1a is stored in the electric and the magnetic* fields that surround the circuit. This stored energy does not increase because it is being drained away, by transfer to Joule (thermal) energy in the resistor, at the same rate at which it is supplied. The electric and magnetic fields play an intermediary role in the energy transfer process, acting as a storage reservoir.

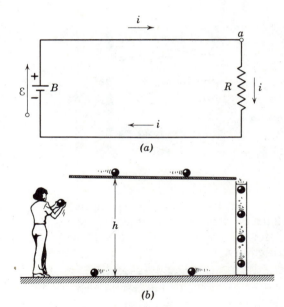

(a)

(b)

figure 32-1
(*a*) A simple electric circuit and (*b*) its gravitational analog.

Figure 32-1b shows a gravitational analog of Fig. 32-1a. In the top figure the seat of emf B does work on the charge carriers. This energy, stored in passage as electromagnetic field energy, appears eventually as thermal energy in resistor R. In the lower figure the person, in lifting the bowling balls from the floor to the shelf, does work on them. This energy is stored, in passage, as gravitational field energy. The balls roll slowly and uniformly along the shelf, dropping from the right end into a cylinder full of viscous oil. They sink to the bottom at an essentially

---

* A current in a wire is surrounded by a magnetic field, and this field, like the electric field, can also be viewed as a site of stored energy (see Section 36-4).

constant speed, are removed by a trapdoor mechanism not shown, and roll back along the floor to the left. The energy put into the system by the person appears eventually as thermal energy in the viscous fluid, resulting in a temperature rise. The energy supplied by the person comes from internal (chemical) energy. The circulation of charges in Fig. 32-1a will stop eventually if battery B is not recharged; the circulation of bowling balls in Fig. 32-1b will stop eventually if the person does not replenish the store of internal energy by eating.

Figure 32-2 shows a circuit containing two (ideal) batteries, A and B, a resistor R, and an (ideal) electric motor employed in lifting a weight. The batteries are connected so that they tend to send charges around the circuit in opposite directions; the actual direction of the current is determined by B, which supplies the larger potential difference. The energy transfers in this circuit are

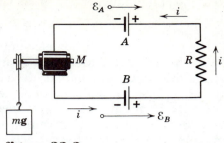

figure 32-2
Two batteries, a resistor, and a motor, connected in a single-loop circuit. It is given that $\varepsilon_B > \varepsilon_A$.

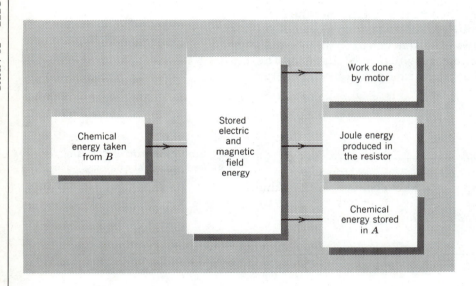

The chemical energy in B is steadily depleted, the energy appearing in the three forms shown on the right. Battery A is being charged while battery B is being discharged. Again, the electric and magnetic fields that surround the circuit act as an intermediary.

It is part of the definition of an emf that the energy transfer process be *reversible,* at least in principle. Recall that a reversible process is one that passes through equilibrium states; its course can be reversed by making an infinitesimal change in the environment of the system; see Section 25-2. A battery, for example, can either be on charge or discharge; a generator can be driven mechanically, producing electrical energy, or it can be operated backward as a motor. The (reversible) energy transfers here are

$$\text{electrical} \rightleftharpoons \text{chemical}$$

and

$$\text{electrical} \rightleftharpoons \text{mechanical}.$$

The energy transfer from electrical energy to thermal energy is not reversible. We can easily raise the temperature of a conductor by supplying electric energy to it, but it is *not* possible to set up a current in a closed copper loop by raising its temperature uniformly. Because of this lack of reversibility, we do not associate an emf with the Joule effect, that is, with energy transfers associated with Eqs. 31-16 and 31-17.

In a time dt an amount of energy given by $i^2R\,dt$ will appear in the resistor of Fig. 32-1a as Joule thermal energy. During this same time a charge $dq\,(= i\,dt)$ will have moved through the seat of emf, and the seat will have done work on this charge (see Eq. 32-1) given by

$$dW = \varepsilon\,dq = \varepsilon i\,dt.$$

From the conservation of energy principle, the work done by the seat must equal the thermal energy, or

$$\varepsilon i\,dt = i^2R\,dt.$$

Solving for i, we obtain

$$i = \varepsilon/R. \qquad (32\text{-}2)$$

We can also derive Eq. 32-2 by considering that, if electric potential is to have any meaning, a given point can have only one value of potential at any given time. If we start at any point in the circuit of Fig. 32-1a and, in imagination, go around the circuit in either direction, adding up algebraically the changes in potential that we encounter, we must arrive at the same potential when we return to our starting point. In other words, *the algebraic sum of the changes in potential encountered in a complete traversal of the circuit must be zero.*

In Fig. 32-1a let us start at point a, whose potential is V_a,* and traverse the circuit clockwise. In going through the resistor, there is a change in potential of $-iR$. The minus sign shows that the top of the resistor is higher in potential than the bottom, which must be true, because positive charge carriers move of their own accord from high to low potential. As we traverse the battery from bottom to top, there is an *increase* of potential $+\varepsilon$ because the battery does (positive) work on the charge carriers, that is, it moves them from a point of low potential to one of high potential. Adding the algebraic sum of the changes in potential to the initial potential V_a must yield the identical value V_a, or

$$V_a - iR + \varepsilon = V_a.$$

We write this as

$$-iR + \varepsilon = 0,$$

which is independent of the value of V_a and which asserts explicitly that the algebraic sum of the potential changes for a complete circuit traversal is zero. This relation leads directly to Eq. 32-2.

These two ways to find the current in single-loop circuits, based on the conservation of energy and on the concept of potential, are completely equivalent because potential differences are defined in terms of work and energy (see Section 29-1). The statement that the sum of the changes in potential encountered in making a complete loop is zero is called *Kirchhoff's second rule*; for brevity we call it the *loop theorem.* Always bear in mind that this theorem is simply a particular way of stating the law of conservation of energy for electric circuits.

To prepare for the study of more complex circuits, let us examine the rules for finding potential differences; these rules follow from the previous discussion. They are not meant to be memorized but rather to be

---

* The actual value of V_a depends on assumptions made in the definition of potential (as described in Section 29-1). The numerical value of V_a is not important because, as in most electric circuit situations, we are concerned here with *differences* of potential. Point a in Fig. 32-1a (or any other single point in that figure) could be connected to ground (symbol $\doteq$) and assigned the potential $V_a = 0$, following a common practice.

CHAP. 32   ELECTROMOTIVE FORCE AND CIRCUITS

so thoroughly understood that it becomes trivial to re-derive them on each application.

1. If a resistor is traversed in the direction of the current, the change in potential is $-iR$; in the opposite direction it is $+iR$.
2. If a seat of emf is traversed in the direction of the emf, the change in potential is $+\varepsilon$; in the opposite direction it is $-\varepsilon$.

Figure 32-3a shows a circuit which emphasizes that all seats of emf have an intrinsic internal resistance r. This resistance cannot be removed—although we would usually like to do so—because it is an inherent part of the device. The figure shows the internal resistance r and the emf separately, although, actually, they occupy the same region of space.

If we apply the loop theorem, starting at b and going around clockwise, we obtain

$$V_b + \varepsilon - ir - iR = V_b$$

or

$$+\varepsilon - ir - iR = 0.$$

Compare these equations with Fig. 32-3b, which shows the changes in potential graphically. In writing these equations, note that we traversed r and R in the direction of the current and ε in the direction of the emf. The same equation follows if we start at any other point in the circuit or if we traverse the circuit in a counterclockwise direction. Solving for i gives

$$i = \frac{\varepsilon}{R + r}. \tag{32-3}$$

32-3
OTHER SINGLE-LOOP CIRCUITS

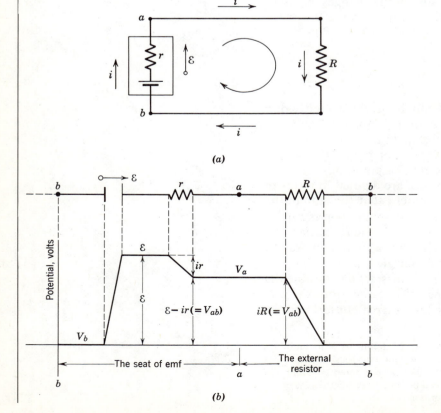

(a)

(b)

(a) A single-loop circuit. The rectangular block is a seat of emf with internal resistance r. (b) The same circuit is drawn for convenience as a straight line. Directly below are shown the changes in potential that one encounters in traversing the circuit clockwise from point b. Note that the two points marked "b" at the top of this figure are the same point.

Resistors in series. Resistors in series are connected so that there is only one conducting path through them, as in Fig. 32-4. What is the equivalent resistance R of this series combination? The equivalent resistance is the single resistance R which, substituted for the series combination between the terminals ab, will leave the current i unchanged.

Applying the loop theorem (going clockwise from a) yields

$$-iR_1 - iR_2 - iR_3 + \varepsilon = 0$$

or

$$i = \frac{\varepsilon}{R_1 + R_2 + R_3}.$$

For the equivalent resistance R

$$i = \frac{\varepsilon}{R}$$

or

$$R = R_1 + R_2 + R_3. \tag{32-4}$$

The extension to more than three resistors is clear.

EXAMPLE 1

699   POTENTIAL DIFFERENCES   SEC. 32-4

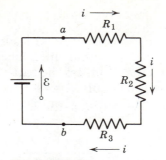

figure 32-4
Example 1. Three resistors are connected in series between terminals a and b.

32-4
POTENTIAL DIFFERENCES

We often want to compute the potential difference between two points in a circuit. In Fig. 32-3a, for example, what is the relationship between the potential difference V_{ab} $(= V_a - V_b)$ between points b and a and the fixed circuit parameters ε, r, and R? To find this relationship, let us start from point b and traverse the circuit to point a, passing through resistor R against the current. If V_b and V_a are the potentials at b and a, respectively, we have

$$V_b + iR = V_a$$

because we experience an increase in potential in traversing a resistor against the current arrow. We rewrite this relation as

$$V_{ab} = V_a - V_b = +iR,$$

which tells us that V_{ab}, the potential difference between points a and b, has the magnitude iR and that point a is more positive than point b. Combining this last equation with Eq. 32-3 yields

$$V_{ab} = \varepsilon \frac{R}{R + r}. \tag{32-5}$$

To sum up: To find the potential difference between any two points in a circuit start at one point and traverse the circuit to the other, following any path,* and add up algebraically the potential changes encountered. This algebraic sum will be the potential difference. This procedure is similar to that for finding the current in a closed loop, except that here the potential differences are added up over part of a loop and not over the whole loop.

The potential difference between any two points can have only one value; thus we must obtain the same answer for all paths that connect these points. If we consider two points on the side of a hill, the measured difference in gravitational potential (that is, in altitude) between them is the same no matter what path is followed in going from one to the other. In Fig. 32-3a let us also calculate V_{ab}, using a path passing through the seat of emf. We have

---

* Recall (see Section 29-1) that path independence is a central feature of the potential concept.

$$V_b + \varepsilon - ir = V_a$$

or (see also Fig. 32-3b)

$$V_{ab} = V_a - V_b = +\varepsilon - ir.$$

Again, combining with Eq. 32-3 leads to Eq. 32-5.

The terminal potential difference of the battery V_{ab}, as Eq. 32-5 shows, is less than ε unless the battery has no internal resistance $(r = 0)$ or if it is on open circuit $(R = \infty)$; then V_{ab} is equal to ε. Thus the emf of a device is equal to its terminal potential difference *when on open circuit*.

EXAMPLE 2

In Fig. 32-5a let ε_1 and ε_2 be 2.0 V and 4.0 V, respectively; let the resistances r_1, r_2, and R be 1.0 Ω, 2.0 Ω, and 5.0 Ω, respectively. What is the current?

Emfs ε_1 and ε_2 oppose each other, but because ε_2 is larger it controls the direction of the current. Thus i will be counterclockwise. The loop theorem, going clockwise from a, yields

$$-\varepsilon_2 + ir_2 + iR + ir_1 + \varepsilon_1 = 0.$$

Check that the same result is obtained by going around counterclockwise. Also, compare this equation carefully with Fig. 32-5b, which shows the potential changes graphically.

Solving for i yields

$$i = \frac{\varepsilon_2 - \varepsilon_1}{R + r_1 + r_2} = \frac{4.0 \text{ V} - 2.0 \text{ V}}{5.0 \ \Omega + 1.0 \ \Omega + 2.0 \ \Omega}$$

$$= 0.25 \text{ A}.$$

It is not necessary to know in advance what the actual direction of the current is. To show this, let us assume that the current in Fig. 32-5a is clockwise, an assumption that we know is incorrect. The loop theorem then yields (going clockwise from a)

$$-\varepsilon_2 - ir_2 - iR - ir_1 + \varepsilon_1 = 0$$

or

$$i = \frac{\varepsilon_1 - \varepsilon_2}{R + r_1 + r_2}.$$

Substituting numerical values (see above) yields -0.25 A for the current. The minus sign tells us that the current is in the opposite direction from the one we have assumed.

In more complex circuit problems involving many loops and branches it is often impossible to know in advance the correct directions for the currents in all parts of the circuit. We can assume directions for the currents at random. Those currents for which positive numerical values are obtained will have the correct directions; those for which negative values are obtained will be exactly opposite to the assumed directions. In all cases the numerical values will be correct.

EXAMPLE 3

What is the potential difference (a) between points b and a in Fig. 32-5a, and (b) between points a and c?

(a) For points a and b we start at b and traverse the circuit to a, obtaining

$$V_{ab} (= V_a - V_b) = -ir_2 + \varepsilon_2 = -(0.25 \text{ A})(2.0 \ \Omega) + 4.0 \text{ V}$$

$$= +3.5 \text{ V}.$$

Thus a is more positive than b and the potential difference (3.5 V) is *less than* the emf (4.0 V); see Fig. 32-5b.

(b) For points c and a, we start at c and traverse the circuit to a, obtaining

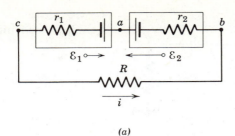

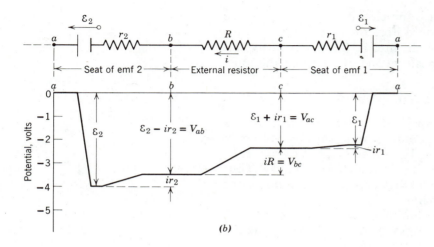

(a)

(b)

figure 32-5
Examples 2, 3. (*a*) A single-loop circuit. (*b*) The same circuit is shown schematically as a straight line, the potential differences encountered in traversing the circuit clockwise from point *a* being displayed directly below. In the lower figure the potential of point *a* was assumed to be zero for convenience. Note that the two points marked "*a*" in this figure are the same point.

$$V_{ac} \ (= V_a - V_c) = +\varepsilon_1 + ir_1 = +2.0 \text{ V} + (0.25 \text{ A})(1.0 \ \Omega)$$

$$= +2.25 \text{ V}.$$

This tells us that *a* is at a higher potential than *c*. The terminal potential difference of ε_1 (2.25 V) is *larger than* the emf (2.0 V); see Fig. 32-5*b*. Charge is being forced through ε_1 in a direction opposite to the one in which it would send charge if it were acting by itself; if ε_1 is a storage battery, it is being charged at the expense of ε_2.

Let us test the first result by proceeding from *b* to *a* along a different path, namely, through R, r_1, and ε_1. We have

$$V_{ab} = iR + ir_1 + \varepsilon_1 = (0.25 \text{ A})(5.0 \ \Omega) + (0.25 \text{ A})(1.0 \ \Omega) + 2.0 \text{ V} = +3.5 \text{ V},$$

which is the same as the earlier result.

32-5
MULTILOOP CIRCUITS

Figure 32-6 shows a circuit containing two loops. For simplicity, we have neglected the internal resistances of the batteries. There are two *junctions*, *b* and *d*, and three *branches* connecting these junctions. The branches are the left branch *bad*, the right branch *bcd*, and the central branch *bd*. If the emfs and the resistances are given, what are the currents in the various branches?

We label the currents in the branches as i_1, i_2, and i_3, as shown. Current i_1 has the same value for any cross section of the left branch from *b* to *d*. Similarly, i_2 has the same value everywhere in the right branch and i_3 in the central branch. The directions of the currents have been chosen arbitrarily. The careful reader will note that i_3 must point in a direction opposite to the one we have shown. We have deliberately drawn it in wrong to show how the formal mathematical procedures will always indicate this to us.

The three currents i_1, i_2, and i_3 carry charge either toward junction *d* or away from it. Charge does not accumulate at junction *d*, nor does it

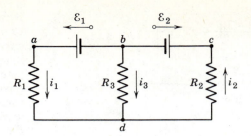

figure 32-6
A multiloop circuit.

drain away from this junction because the circuit is in a steady-state condition. Thus charge must be removed from the junction by the currents at the same rate that it is brought into it. If we arbitrarily call a current approaching the junction positive and one leaving the junction negative, then

$$i_1 + i_3 - i_2 = 0.$$

This equation suggests a general principle for the solution of multi-loop circuits: *At any junction the algebraic sum of the currents must be zero.* This *junction theorem* is also known as *Kirchhoff's first rule.* Note that it is simply a statement of the conservation of charge. Thus our basic tools for solving circuits are (*a*) the conservation of energy (see p. 697) and (*b*) the conservation of charge.

For the circuit of Fig. 32-6, the junction theorem yields only one relationship among the three unknowns. Applying the theorem at junction *b* leads to exactly the same equation, as you can easily verify. To solve for the three unknowns, we need two more independent equations; they can be found from the loop theorem.

In single-loop circuits there is only one conducting loop around which to apply the loop theorem, and the current is the same in all parts of this loop. In multiloop circuits there is more than one loop, and the current in general will not be the same in all parts of any given loop.

If we traverse the left loop of Fig. 32-6 in a counterclockwise direction, the loop theorem gives

$$\mathcal{E}_1 - i_1 R_1 + i_3 R_3 = 0. \tag{32-6}$$

The right loop gives

$$-i_3 R_3 - i_2 R_2 - \mathcal{E}_2 = 0. \tag{32-7}$$

These two equations, together with the relation derived earlier with the junction theorem, are the three simultaneous equations needed to solve for the unknowns i_1, i_2, and i_3. Doing so yields

$$i_1 = \frac{\mathcal{E}_1(R_2 + R_3) - \mathcal{E}_2 R_3}{R_1 R_2 + R_2 R_3 + R_1 R_3}, \tag{32-8a}$$

$$i_2 = \frac{\mathcal{E}_1 R_3 - \mathcal{E}_2(R_1 + R_3)}{R_1 R_2 + R_2 R_3 + R_1 R_3}, \tag{32-8b}$$

and

$$i_3 = \frac{-\mathcal{E}_1 R_2 - \mathcal{E}_2 R_1}{R_1 R_2 + R_2 R_3 + R_1 R_3}. \tag{32-8c}$$

Supply the missing steps. Equation 32-8*c* shows that no matter what numerical values are given to the emfs and to the resistances, the current i_3 will always have a negative value. This means that it will always point up in Fig. 32-6 rather than down, as we deliberately assumed. The currents i_1 and i_2 may be in either direction, depending on the particular numerical values given.

Verify that Eqs. 32-8 reduce to sensible conclusions in special cases. For $R_3 = \infty$, for example, we find

$$i_1 = i_2 = \frac{\mathcal{E}_1 - \mathcal{E}_2}{R_1 + R_2} \quad \text{and} \quad i_3 = 0.$$

What do these equations reduce to for $R_2 = \infty$?

The loop theorem can be applied to a large loop consisting of the entire circuit *abcda* of Fig. 32-6. This fact might suggest that there are more equations than we need, for there are only three unknowns and we already have three equations written in terms of them. However, the loop theorem yields for this loop

$$-i_1 R_1 - i_2 R_2 - \mathcal{E}_2 + \mathcal{E}_1 = 0,$$

which is nothing more than the sum of Eqs. 32-6 and 32-7. Thus this large loop does not yield another *independent* equation. It will never be found in solving multiloop circuits that there are more independent equations than variables.

Resistors in parallel. Figure 32-7 shows three resistors connected across the same seat of emf. Resistances across which the identical potential difference is applied are said to be in parallel. What is the equivalent resistance R of this parallel combination? The equivalent resistance is that single resistance which, substituted for the parallel combination between terminals *ab*, would leave the current i unchanged.

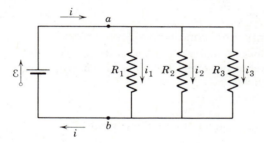

The currents in the three branches are

$$i_1 = \frac{V}{R_1}, \quad i_2 = \frac{V}{R_2}, \quad \text{and} \quad i_3 = \frac{V}{R_3},$$

where V is the potential difference that appears between points a and b. The total current i is found by applying the junction theorem to junction a, or

$$i = i_1 + i_2 + i_3 = V \left(\frac{1}{R_1} + \frac{1}{R_2} + \frac{1}{R_3} \right).$$

If the equivalent resistance is used instead of the parallel combination, we have

$$i = \frac{V}{R}.$$

Combining these two equations gives

$$\frac{1}{R} = \frac{1}{R_1} + \frac{1}{R_2} + \frac{1}{R_3}. \tag{32-9}$$

This formula can easily be extended to more than three resistances. Note that the equivalent resistance of a parallel combination is less than any of the resistances that make it up.

EXAMPLE 4

figure 32-7
Example 4. Three resistors are connected in parallel between terminals a and b.

A meter to measure currents is called an *ammeter* (or a *milliammeter, micro-ammeter*, etc., depending on the size of the current to be measured). To determine the current in a wire, it is necessary to break or cut the wire and to insert the ammeter, so that the current to be measured passes through the meter (see Fig. 32-8).*

It is essential that the resistance R_A of the ammeter be *small* compared to other resistances in the circuit. Otherwise the presence of the meter will in itself change the current to be measured. An ideal ammeter would have zero resistance. In the circuit of Fig. 32-8 the required condition, assuming that the voltmeter is not connected, is

$$R_A \ll r + R_1 + R_2.$$

A meter to measure potential differences is called a *voltmeter* (or a *milli-voltmeter* or *microvoltmeter*, etc.). To find the potential difference between two points in a circuit, it is necessary to connect one of the voltmeter terminals to each of the circuit points, without breaking the circuit (see Fig. 32-8).†

It is essential that the resistance of the voltmeter R_V be *large* compared to any circuit resistance across which the voltmeter is connected. Otherwise the meter will itself constitute an important circuit element and will alter the circuit current and the potential difference to be measured. An ideal voltmeter would have an infinite resistance. In Fig. 32-8 the required condition is

$$R_V \gg R_1.$$

In measuring potential difference in electronic circuits, where the effective circuit resistance may be of the order of 10^6 ohms or higher, it becomes necessary to use an electronic voltmeter, which is an electron tube or transistor device designed specifically to have an extremely high effective resistance between its input terminals.

Figure 32-9 shows the rudiments of a *potentiometer*, which is a device for measuring an unknown emf $\mathcal{E}_x$. The currents and emfs are marked as shown. Applying the loop theorem to loop *abcd* yields

$$-\mathcal{E}_x - ir + (i_0 - i)R = 0,$$

where $i_0 - i$, by application of the junction theorem at *a*, is the current in resistor *R*. Solving for *i* yields

$$i = \frac{i_0 R - \mathcal{E}_x}{R + r},$$

in which *R* is a variable resistor. This relation shows that if *R* is adjusted to have the value R_x where

$$i_0 R_x = \mathcal{E}_x, \tag{32-10}$$

the current *i* in the branch *abcd* becomes zero. To *balance* the potentiometer in this way, *R* must be adjusted manually until the sensitive current meter *G* reads zero.

The emf can be obtained from Eq. 32-10 if the current i_0 is known. However, it is standard practice to replace $\mathcal{E}_x$ by a known standard emf $\mathcal{E}_s$, and once again to adjust *R* to the zero-current condition. This yields, assuming the current i_0 remains unchanged,

$$i_0 R_s = \mathcal{E}_s.$$

32-6
MEASURING CURRENTS AND POTENTIAL DIFFERENCES

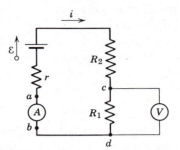

figure 32-8
An ammeter (*A*) is connected to read the current in a circuit, and a voltmeter (*V*) is connected to read the potential difference across resistor R_1.

32-7
THE POTENTIOMETER

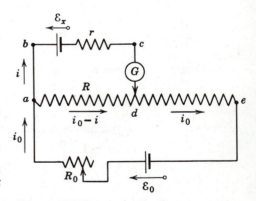

figure 32-9
Elements of a potentiometer. Point *d* represents a sliding contact.

---

* The meter must be connected so that the direction of current through it (assuming positive charge carriers) is *into* the meter terminal marked +. Otherwise the meter will deflect in a direction opposite to that intended.
† The voltmeter terminal marked + must be connected to the point of higher potential. Otherwise the meter will deflect in a direction opposite to that intended.

Combining the last two equations yields

$$\varepsilon_x = \varepsilon_s \frac{R_x}{R_s},$$ (32-11)

which allows us to compare emfs with precision. Note that the internal resistance r of the emf plays no role. In practice, potentiometers are conveniently packaged units, containing a *standard cell* which, after calibration at the National Bureau of Standards or elsewhere, serves as a convenient known standard seat of emf ε_s. Switching arrangements for replacing the unknown emf by the standard and arrangements for ascertaining that the current i_0 remains constant are also incorporated.

The preceding sections dealt with circuits in which the circuit elements were resistors and in which the currents did not vary with time. Here we introduce the capacitor as a circuit element, which will lead us to the concept of time-varying currents. In Fig. 32-10 let switch S be thrown to position a. What current is set up in the single-loop circuit so formed? Let us apply conservation of energy principles.

In time dt a charge $dq \, (= i \, dt)$ moves through any cross section of the circuit. The work done by the seat of emf $(= \varepsilon \, dq$; see Eq. 32-1) must equal the energy that appears as thermal energy in the resistor during time $dt \, (= i^2R \, dt)$ plus the increase in the amount of energy U that is stored in the capacitor $[= dU = d(q^2/2C)$; see Eq. 30-7]. In equation form

$$\varepsilon \, dq = i^2R \, dt + d\left(\frac{q^2}{2C}\right)$$

or

$$\varepsilon \, dq = i^2R \, dt + \frac{q}{C} \, dq.$$

Dividing by dt yields

$$\varepsilon \frac{dq}{dt} = i^2R + \frac{q}{C}\frac{dq}{dt}.$$

But dq/dt is simply i, so that this equation becomes

$$\varepsilon = iR + \frac{q}{C}.$$ (32-12)

This equation also follows from the loop theorem, as it must, since the loop theorem was derived from the conservation of energy principle. Starting from point x and traversing the circuit clockwise, we experience an increase in potential in going through the seat of emf and decreases in potential in traversing the resistor and the capacitor, or

$$\varepsilon - iR - \frac{q}{C} = 0,$$

which is identical to Eq. 32-12.

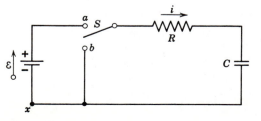

figure 32-10
An *RC* circuit.

We cannot immediately solve Eq. 32-12 because it contains two variables, q and i, which, however, are related by

$$i = \frac{dq}{dt}. \tag{32-13}$$

Substituting for i into Eq. 32-12 gives

$$\varepsilon = R\frac{dq}{dt} + \frac{q}{C}. \tag{32-14}$$

Our task now is to find the function $q(t)$ that satisfies this *differential equation*. Although this particular equation is not hard to solve, we choose to avoid mathematical complexity by simply presenting the solution, which is

$$q = C\varepsilon(1 - e^{-t/RC}). \tag{32-15}$$

We can easily test whether this function $q(t)$ is really a solution of Eq. 32-14 by substituting it into that equation and seeing whether an identity results. Differentiating Eq. 32-15 with respect to time yields

$$\frac{dq}{dt}(= i) = \frac{\varepsilon}{R}e^{-t/RC}. \tag{32-16}$$

Substituting q (Eq. 32-15) and dq/dt (Eq. 32-16) into Eq. 32-14 yields an identity, as you should verify. Thus Eq. 32-15 is a solution of Eq. 32-14.

Figure 32-11 shows some plots of Eqs. 32-15 and 32-16 for a particular case. Study of these plots and of the corresponding equations shows that (a) at $t = 0$, $q = 0$, and $i = \varepsilon/R$, and (b) as $t \to \infty$, $q \to C\varepsilon$, and $i \to 0$; that is, the current is initially ε/R and finally zero, and the charge on the capacitor plates is initially zero and finally $C\varepsilon$.

The quantity RC in Eqs. 32-15 and 32-16 has the dimensions of time (since the exponent must be dimensionless) and is called the *capacitive time constant* of the circuit. It is the time at which the charge on the capacitor has increased to within a factor of $(1 - e^{-1})$ ($\cong 63\%$) of its equilibrium value. To show this, we put $t = RC$ in Eq. 32-15 to obtain

$$q = C\varepsilon(1 - e^{-1}) = 0.63 C\varepsilon.$$

Since $C\varepsilon$ is the equilibrium charge on the capacitor, corresponding to $t \to \infty$, the foregoing statement follows.

EXAMPLE 5

After how many time constants will the energy stored in the capacitor in Fig. 32-10 reach one-half its equilibrium value?

The energy is given by Eq. 30-7, or

$$U = \frac{1}{2C}q^2,$$

the equilibrium energy U_∞ being $(1/2C)(C\varepsilon)^2$. From Eq. 32-15, we can write the energy as

$$U = \frac{1}{2C}(C\varepsilon)^2(1 - e^{-t/RC})^2$$

or

$$U = U_\infty(1 - e^{-t/RC})^2.$$

Putting $U = \frac{1}{2}U_\infty$ yields

$$\tfrac{1}{2} = (1 - e^{-t/RC})^2$$

and solving this relation for t yields finally

$$t = 1.22\,RC = 1.22 \text{ time constants.}$$

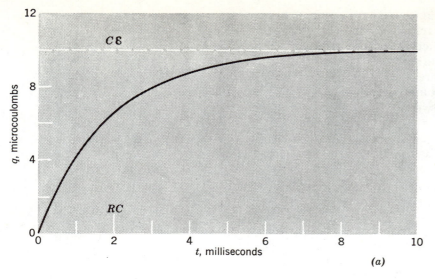

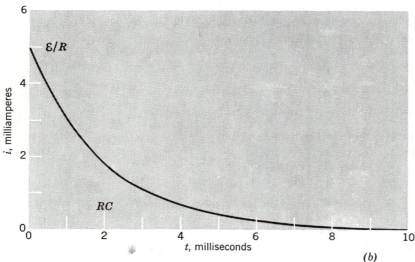

figure 32-11
If, in Fig. 32-10, we assume that $R = 2000\ \Omega$, $C = 1.0\ \mu F$, and $\varepsilon = 10$ V, then (a) shows the variation of q with t during the charging process and (b) the variations of i with t. The time constant is $RC = 2.0 \times 10^{-3}$ s.

Figure 32-11 shows that if a resistance is included in the circuit, the rate of increase of the charge of a capacitor toward its final equilibrium value is *delayed* in a way measured by the time constant RC. With no resistor present $(RC = 0)$, the charge would rise immediately to its equilibrium value. Although we have shown that this time delay follows from an application of the loop theorem to RC circuits, it is important to develop a physical understanding of the causes of the delay.

When switch S in Fig. 32-10 is closed on a, the resistor experiences instantaneously an applied potential difference of ε, and an initial current of ε/R is set up. Initially, the capacitor experiences no potential difference because its initial charge is zero, the potential difference always being given by q/C. The flow of charge through the resistor starts to charge the capacitor, which has several effects. First, the existence of a capacitor charge means that there must now be a potential difference $(= q/C)$ across the capacitor; this, in turn, means that the potential difference across the resistor must decrease by this amount, since the sum of the two potential differences must always equal ε. This decrease in the potential difference across R means that the charging current is reduced. Thus the charge of the capacitor builds up and the charging current decreases until the capacitor is fully charged. At this point the full emf ε is applied to the capacitor, there being no potential drop $(iR = 0)$ across the resistor. This is precisely the reverse of the initial situation. Review the derivations of Eqs. 32-15 and 32-16 and study Fig. 32-11 with the qualitative arguments of this paragraph in mind.

Assume now that the switch S in Fig. 32-10 has been in position a for a time t such that $t \gg RC$. The capacitor is then fully charged for all practical purposes. The switch S is then thrown to position b. How do the charge of the capacitor and the current vary with time?

With the switch S closed on b, there is no emf in the circuit and Eq. 32-12 for the circuit, with $\varepsilon = 0$, becomes simply

$$iR + \frac{q}{C} = 0. \qquad (32\text{-}17)$$

Putting $i = dq/dt$ allows us to write, as the differential equation of the circuit (compare Eq. 32-14),

$$R\frac{dq}{dt} + \frac{q}{C} = 0. \qquad (32\text{-}18a)$$

The solution is

$$q = q_0 e^{-t/RC}, \qquad (32\text{-}18b)$$

as you may readily verify by substitution, q_0 being the initial charge on the capacitor. The capacitive time constant RC appears in this expression for capacitor discharge as well as in that for the charging process (Eq. 32-15). We see that at a time such that $t = RC$ the capacitor charge is reduced to $q_0 e^{-1}$, which is about 37% of the initial charge q_0.

The current during discharge follows from differentiating Eq. 32-18b, or

$$i = \frac{dq}{dt} = -\frac{q_0}{RC}e^{-t/RC}. \qquad (32\text{-}19)$$

The negative sign shows that the current is in the direction opposite to that shown in Fig. 32-10. This is as it should be, since the capacitor is discharging rather than charging. Since $q_0 = C\varepsilon$, we can write Eq. 32-19 as

$$i = -\frac{\varepsilon}{R}e^{-t/RC},$$

in which ε/R appears as the initial current, corresponding to $t = 0$. This is reasonable because the initial potential difference for the fully charged capacitor is ε.

The behavior of the RC circuit in Fig. 32-10 during charge and discharge can be studied with a cathode-ray oscilloscope. This familiar laboratory device can display on its fluorescent screen plots of the variation of potential with time. Figure 32-12 shows the circuit of Fig. 32-10 with connections made to display (a) the potential difference V_C across the capacitor and (b) the potential difference V_R across the resistor as functions of time. V_C and V_R are given by

$$V_C = \left(\frac{1}{C}\right)q$$

and

$$V_R = (R)i,$$

the former being proportional to the charge and the latter to the current.

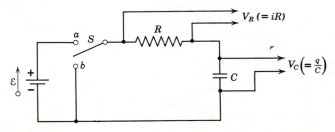

figure 32-12
The circuit of Fig. 32-10 with connections made to display the potential variations across the resistor and the capacitor on a cathode-ray oscilloscope.

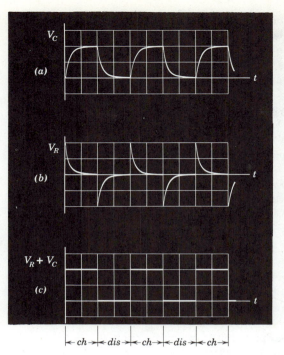

figure 32-13
In Fig. 32-10 switch S is thrown periodically, by electronic means, between positions a and b. The variations with time of the potential differences across (a) the capacitor and (b) the resistor are shown, as displayed on a cathode-ray oscilloscope. (c) The appearance of the screen when the oscilloscope is connected to display the sum of V_R and V_C. (Courtesy E. K. Hege, Rensselaer Polytechnic Institute.)

Figure 32-13 shows oscillograph plots of V_C and V_R that result when, in effect, switch S in Fig. 32-10 is thrown regularly back and forth between positions a and b, being left in each position for a time equal to several time constants. Intervals during which the charge is building up are labeled ch and those during which it is decaying are labeled dis.

The "charge" intervals in plot a (see Eq. 32-15) are represented by

$$V = \left(\frac{1}{C}\right) q = \varepsilon(1 - e^{-t/RC})$$

and the "discharge" intervals (see Eq. 32-18b) by

$$V = \left(\frac{1}{C}\right) q = \varepsilon e^{-t/RC}.$$

Note that the current, as indicated by plot b, is in opposite directions during the charge and discharge intervals, in agreement with Eqs. 32-16 and 32-19.

In plot c in Fig. 32-13 the oscillograph has been connected to show the algebraic sum of plots a and b. According to the loop theorem this sum should equal ε during the charge intervals and should be zero during the discharge intervals, when the battery is no longer in the circuit; that is,

$$V_R + V_C = \varepsilon \text{ during charge (see Eq. 32-12),}$$

$$V_R + V_C = 0 \text{ during discharge (see Eq. 32-17).}$$

Plot c is in exact agreement with this expectation.

questions

1. Does the direction of the emf provided by a battery depend on the direction of current flow through the battery?

2. In Fig. 32-2 discuss what changes would occur if we increased the mass m by such an amount that the "motor" reversed direction and became a "generator", that is, a seat of emf.

3. Discuss in detail the statement that the energy method and the loop theorem method for solving circuits are perfectly equivalent.

4. It is possible to generate a 10,000-volt potential difference by rubbing a pocket comb with wool. Why is this large voltage not dangerous when the much lower voltage provided by an ordinary electric outlet is very dangerous?

5. Devise a method for measuring the emf and the internal resistance of a battery.

6. The current passing through a battery of emf ε and internal resistance r is decreased by some external means. Does the potential difference between the terminals of the battery necessarily decrease or increase? Explain.

7. In calculating V_{ab} in Fig. 32-3a, is it permissible to follow a path from a to b that does not lie in the conducting circuit?

8. A 25-watt, 110-volt bulb glows at normal brightness when connected across a bank of batteries. A 500-watt, 110-volt bulb glows only dimly when connected across the same bank. Explain.

9. Under what circumstances can the terminal potential difference of a battery exceed its emf?

10. What is the difference between an emf and a potential difference?

11. Compare and contrast the formulae for the equivalent values of (a) capacitors and (b) resistors, in series and in parallel.

12. On what physical laws do (a) the loop theorem and (b) the junction theorem depend?

13. In Fig. 32-6 convince yourself that i_3 is drawn in the wrong direction. It was drawn so deliberately to make a point. What point?

14. Explain in your own words why the resistance of an ammeter should be very small whereas that of a voltmeter should be very large.

15. In a potentiometer, the internal resistance r of the emf plays no role. Why? See Fig. 32-9.

16. Does the time required for the charge on a capacitor in an RC circuit to build up to a given fraction of its equilibrium value depend on the value of the applied emf?

17. Devise a method whereby an RC circuit can be used to measure very high resistances.

18. In Fig. 32-10 suppose that switch S is closed on a. Explain why, in view of the fact that the negative terminal of the battery is not connected to resistance R, the initial current in R, at $t = 0$, should be ε/R, as Eq. 32-16 predicts.

19. In Fig. 32-10 suppose that switch S is closed on a. Why does the charge on capacitor C not rise instantaneously to $q = C\varepsilon$? After all, the positive battery terminal is connected to one plate of C and the negative terminal to the other.

20. Show that the product RC in Eqs. 32-15 and 32-16 has the dimensions of time, that is, that 1 second = 1 ohm × 1 farad.

21. Can you construct a water-flow analogy to capacitor discharge, using, for example, two burettes to simulate the capacitor plates and a long capillary tube to simulate the resistor? See "A Water Flow Analogy to Capacitor Discharge" by Ernest L. Madsen, *The Physics Teacher*, April 1976, for a full discussion.

problems

SECTION 32-1

1. A 5.0-A current is set up in an external circuit by a 6.0-V storage battery for 6.0 min. By how much is the chemical energy of the battery reduced?
Answer: 1.1×10^4 J.

2. A certain car battery (12 V) carries an initial charge Q (120 A·h). Assuming that the potential across the terminals stays constant until the battery is completely discharged, for how many hours can it deliver power P (100 W)?

SECTION 32-2

3. The current in a simple series circuit is 5.0 A. When an additional resistance of 2.0 Ω is inserted, the current drops to 4.0 A. What was the resistance of the original circuit? *Answer: 8.0 Ω.*

4. A battery of emf ε (2.0 V) and internal resistance r (1.0 Ω) is driving a motor. The motor is lifting a weight W (2.0 N) at constant speed v (0.50 m/s). Assuming no power losses, find (a) the current i in the circuit and (b) the potential difference V across the terminals of the motor.

SECTION 32-3

5. Thermal energy is to be generated in a 0.10-Ω resistor at the rate of 10 W by connecting it to a battery whose emf is 1.5 V. (a) What is the internal resistance of the battery? (b) What potential difference exists across the resistor? *Answer: (a) 0.050 Ω. (b) 1.0 V.*

6. A wire of resistance 5.0 Ω is connected to a battery whose emf ε is 2.0 V and whose internal resistance is 1.0 Ω. In 2.0 min (a) how much energy is transferred from chemical to electric form? (b) How much energy appears in the wire as thermal energy? (c) Account for the difference between (a) and (b).

7. (a) In the circuit of Fig. 32-3a show that the power delivered to R as thermal energy is a maximum when R is equal to the internal resistance r of the battery. (b) Show that this maximum power is $P = \varepsilon^2/4r$.

8. (a) In Fig. 32-14 what value must R have if the current in the circuit is to be 0.0010 A? Take $\varepsilon_1 = 2.0$ V, $\varepsilon_2 = 3.0$ V, and $r_1 = r_2 = 3.0$ Ω. (b) What is the rate of thermal energy transfer in R?

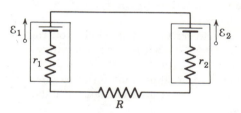

figure 32-14
Problem 8

9. In Fig. 32-3a put $\varepsilon = 2.0$ V and $r = 100$ Ω. Plot (a) the current and (b) the potential difference across R, as functions of R over the range 0 to 500 Ω. Make both plots on the same graph. (c) Make a third plot by multiplying together, for each value of R, the two curves plotted. What is the physical significance of this plot?

10. You are given a number of 10-Ω resistors, each capable of dissipating only 1.0 W. What is the minimum number of such resistors that you need to combine in series or parallel combinations to make a 10-Ω resistor capable of dissipating 5.0 W?

SECTION 32-4

11. In Fig. 32-5(a) calculate the potential difference between a and c by considering a path that contains R and ε_2.
Answer: 2.25 V, as expected. See Example 3.

12. In Example 2 an ammeter whose resistance is 0.050 Ω is inserted in the circuit. What percent change in the current results because of the presence of the meter?

13. The section of circuit AB (see Fig. 32-15) absorbs power $P = 50.0$ W and a current $i = 1.0$ A passes through it in the indicated direction. (a) What is the potential difference between A and B? (b) If the element C does not have internal resistance, what is its emf? (c) What is its polarity?
Answer: (a) 50 V. (b) 48 V. (c) B is the negative terminal.

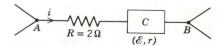

figure 32-15
Problem 13

14. Two batteries having the same emf ε but different internal resistances r_1 and r_2 are connected in series to an external resistance R. Find the value of R that makes the potential difference zero between the terminals of the first battery.

SECTION 32-5

15. Two light bulbs, one of resistance R_1 and the other of resistance R_2 $(<R_1)$ are connected (a) in parallel and (b) in series. Which bulb is brighter?
    Answer: (a) R_2. (b) R_1.

16. Two batteries of emf ε and internal resistance r are connected in parallel across a resistor R, as in Fig. 32-18b. (a) For what value of R is the thermal energy delivered to the resistor a maximum? (b) What is the maximum energy dissipation rate?

17. In Fig. 32-6 calculate the potential difference between points c and d by as many paths as possible. Assume that $\varepsilon_1 = 4.0$ V, $\varepsilon_2 = 1.0$ V, $R_1 = R_2 = 10\ \Omega$, and $R_3 = 5\ \Omega$.   *Answer:* $V_d - V_c = +0.25$ V, by all paths.

18. Two resistors, R_1 and R_2, may be connected either in series or parallel across a (resistanceless) battery with emf ε. We desire the thermal energy transfer rate for the parallel combination to be five times that for the series combination. If $R_1 = 100\ \Omega$, what is R_2?

19. (a) In Fig. 32-16 what is the equivalent resistance of the network shown? (b) What are the currents in each resistor? Put $R_1 = 100\ \Omega$, $R_2 = R_3 = 50\ \Omega$, $R_4 = 75\ \Omega$, and $\varepsilon = 6.0$ V.
    Answer: (a) $120\ \Omega$; note that R_2, R_3, and R_4 are in parallel. (b) $i_1 = 50$ mA; $i_2 = i_3 = 20$ mA; $i_4 = 13$ mA.

20. A copper wire of radius a (0.25 mm) has an aluminum jacket of outside radius b (0.38 mm). (a) If there is a current i (2.0 A) in the wire, find the current in each material. (b) What is the wire length if potential difference V (12 V) maintains the current?

21. By using only two resistance coils — singly, in series, or in parallel — you are able to obtain resistances of 3, 4, 12, and 16 Ω. What are the separate resistances of the coils?   *Answer:* $4.0\ \Omega$ and $12\ \Omega$.

22. Four 100-W heating coils are to be connected in all possible series-parallel combinations and plugged into a 100-V line. What different rates of Joule (thermal) energy dissipation are possible?

23. (a) In Fig. 32-17 what power appears as thermal energy in R_1? In R_2? In R_3? (b) What power is supplied by ε_1? By ε_2? (c) Discuss the energy balance in this circuit. Assume that $\varepsilon_1 = 3.0$ V, $\varepsilon_2 = 1.0$ V, $R_1 = 5.0\ \Omega$, $R_2 = 2.0\ \Omega$, and $R_3 = 4.0\ \Omega$.   *Answer:* (a) 0.35 W; 0.050 W; 0.71 W. (b) 1.3 W; −0.16 W.

24. You are given two batteries of emf ε and internal resistance r. They may be connected either in series or in parallel and are used to establish a current in a resistor R, as in Fig. 32-18. (a) Derive expressions for the current in R for both methods of connection. (b) Which connection yields the larger current if $R > r$ and if $R < r$?

25. In Fig. 32-19 find the current in each resistor and the potential difference between a and b. Put $\varepsilon_1 = 6.0$ V, $\varepsilon_2 = 5.0$ V, $\varepsilon_3 = 4.0$ V, $R_1 = 100\ \Omega$, and $R_2 = 50\ \Omega$.   *Answer:* $i_1 = 50$ mA, to the right; $i_2 = 60$ mA, down; $V_{ab} = 9.0$ V.

26. What is the equivalent resistance between the terminal points x and y of the circuits shown in (a) Fig. 32-20a, (b) Fig. 32-20b, and (c) Fig. 32-20c? Assume that the resistance of each resistor is 10 Ω. Do you detect any similarities between Figs. 32-20a and c?

27. (a) Find the three currents in Fig. 32-21. (b) Find V_{ab}. Assume that $R_1 = 1.0\ \Omega$, $R_2 = 2.0\ \Omega$, $\varepsilon_1 = 2.0$ V, and $\varepsilon_2 = \varepsilon_3 = 4.0$ V.
    Answer: (a) Left branch, 0.67 A, down; center branch, 0.33 A, up; right branch, 0.33 A, up. (b) 3.3 V.

28. What current, in terms of ε and R, does the meter M in Fig. 32-22 read? Assume that M has zero resistance.

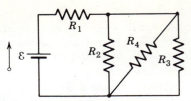

figure 32-16
Problem 19

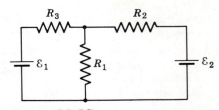

figure 32-17
Problem 23

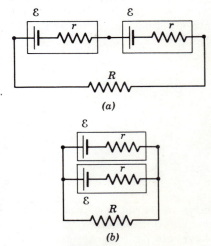

(a)

(b)

figure 32-18
Problems 16, 24

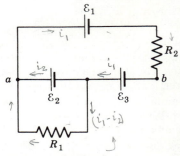

figure 32-19
Problem 25

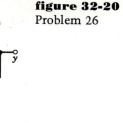

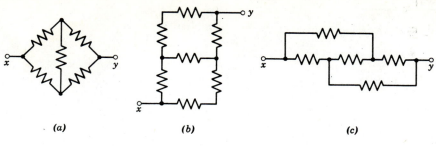

(a)   (b)   (c)

figure 32-20
Problem 26

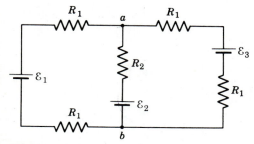

figure 32-21
Problem 27

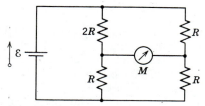

figure 32-22
Problem 28

29. (a) Two batteries of emf and internal resistance ε_1, r_1 and ε_2, r_2, are connected in parallel. Show that the effective emf of this parallel combination is

$$\varepsilon = r\left(\frac{\varepsilon_1}{r_1} + \frac{\varepsilon_2}{r_2}\right)$$

where r is defined from

$$\frac{1}{r} = \frac{1}{r_1} + \frac{1}{r_2}.$$

See "Batteries Connected in Parallel" by J. S. Wallingford and H. W. Jones, *American Journal of Physics*, **36**, 639, (1968) for an extension of this problem to an indefinite number of batteries.

30. N identical batteries of emf ε and internal resistance r may be connected all in series or all in parallel. Show that each arrangement will give the same current in an external resistor R, if $R = r$.

31. *The Wheatstone bridge.* In Fig. 32-23 R_s is to be adjusted in value until points a and b are brought to exactly the same potential. (One tests for this condition by momentarily connecting a sensitive meter between a and b; if these points are at the same potential, the meter will not deflect.) Show that when this adjustment is made, the following relation holds:

$$R_x = R_s \frac{R_2}{R_1}.$$

Unknown resistors (R_x) can be measured in terms of standards (R_s) using this device, which is called a Wheatstone bridge.

32. If points a and b in Fig. 32-23 are connected by a wire of resistance r, show that the current in the wire is

$$i = \frac{\varepsilon(R_s - R_x)}{(R + 2r)(R_s + R_x) + 2R_sR_x},$$

where ε is the emf of the battery. Assume that R_1 and R_2 are equal ($R_1 = R_2 = R$) and that R_0 equals zero. Is this formula consistent with the result of Problem 31?

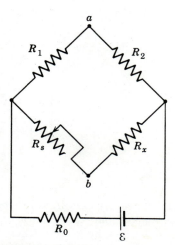

figure 32-23
Problem 31

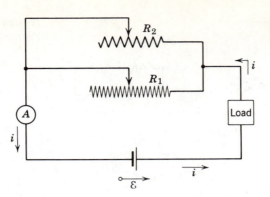

figure 32-24
Problem 33

33. For manual control of the current in a circuit, you can use a parallel combination of variable resistors of the sliding contact type, as in Fig. 32-24,
with $R_1 = 20R_2$. (a) What procedure is used to adjust the current to the desired value? (b) Why is the parallel combination better than a single-variable
resistor? (c) Can you extend this technique to three resistors in parallel?
Answer: (a) Put R_1 roughly in the middle of its range; adjust current roughly
with R_2; make fine adjustment with R_1. (b) Relatively large percentage changes in R_1 cause only small percentage changes in the
resistance of the parallel combination, thus permitting fine adjustment. The ratio is 1:21.

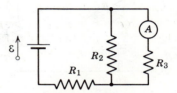

figure 32-25
Problem 34

34. In Fig. 32-25 imagine an ammeter inserted in the branch containing R_3.
(a) What will it read, assuming $\varepsilon = 5.0$ V, $R_1 = 2.0$ Ω, $R_2 = 4.0$ Ω, and $R_3 = 6.0$ Ω? (b) The ammeter and the source of emf are now physically interchanged. Show that the ammeter reading remains unchanged.

35. Twelve resistors, each of resistance R ohms, form a cube (see Fig. 32-26).
(a) Find R_{AB}, the equivalent resistance of an edge. (b) Find R_{BC}, the equivalent resistance of a face diagonal. (c) Find R_{AC}, the equivalent resistance of
a body diagonal.   *Answer:* (a) $\frac{7}{12}$ R. (b) $\frac{3}{4}$ R. (c) $\frac{5}{6}$ R.

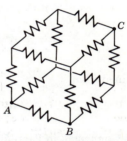

figure 32-26
Problem 35

SECTION 32-6

36. A voltmeter and an ammeter are used to determine two unknown resistances R_1 and R_2, one determination by each of the two methods shown in
Fig. 32-27. The voltmeter resistance is 307 Ω and the ammeter resistance
is 3.62 Ω; in method (a) the ammeter reads 0.317 A and the voltmeter reads
28.1 V, whereas in method (b) the ammeter reads 0.356 A and the voltmeter
23.7 V. Compute R_1 and R_2.

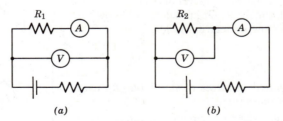

(a)   (b)

figure 32-27
Problem 36

37. *Resistance measurement.* A voltmeter (resistance R_V) and an ammeter (resistance R_A) are connected to measure a resistance R, as in Fig. 32-28a. The
resistance is given by $R = V/i$, where V is the voltmeter reading and i is the
current *in the resistor R*. Some of the current registered by the ammeter
(i') goes through the voltmeter so that the ratio of the meter readings (= V/i')

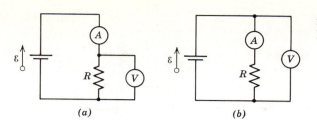

figure 32-28
Problems 37, 38

(a)   (b)

gives only an *apparent* resistance reading R'. Show that R and R' are related by

$$\frac{1}{R} = \frac{1}{R'} - \frac{1}{R_V}.$$

Note that if $R_V \gg R$, then $R \cong R'$.

38. *Resistance measurement.* If meters are used to measure resistance, they may also be connected as they are in Fig. 32-28b. Again the ratio of the meter readings gives only an apparent resistance R'. Show that R' is related to R by

$$R = R' - R_A,$$

in which R_A is the ammeter resistance. Note that if $R_A \ll R$, then $R \cong R'$.

39. In Fig. 32-8 assume that $\varepsilon = 5.0$ V, $r = 2.0$ Ω, $R_1 = 5.0$ Ω, and $R_2 = 4.0$ Ω. If $R_A = 0.10$ Ω, what percent error is made in reading the current? Assume that the voltmeter is not present.   *Answer: 0.9%.*

40. In Fig. 32-8 assume that $\varepsilon = 5.0$ V, $r = 20$ Ω, $R_1 = 50$ Ω, and $R_2 = 40$ Ω. If $R_V = 1000$ Ω, what percent error is made in reading the potential differences across R_1? Ignore the presence of the ammeter.

SECTION 32-8

41. How many time constants must elapse before a capacitor in an RC circuit is charged to within 1.0 percent of its equilibrium charge?   *Answer: 4.6.*

42. A 10,000-Ω resistor and a capacitor are connected in series and a 10-V potential is suddenly applied. If the potential across the capacitor rises to 5.0 V in 1.0 μs, what is the capacitance of the capacitor?

43. The potential difference between the plates of a leaky capacitor C (2.0 μF) drops from V_0 to V ($\frac{1}{4}V_0$) in time t (2.0 s). What is the equivalent resistance between the capacitor plates?   *Answer: 7.2×10^5 Ω.*

44. An RC circuit is discharged by closing a switch at time $t = 0$. The initial potential difference across the capacitor is 100 V. If the potential difference has decreased to 1.0 V after 10 s, (a) what will the potential difference be 20 s after $t = 0$? (b) What is the time constant of the circuit?

45. Prove that when switch S in Fig. 32-10 is thrown from a to b all the energy stored in the capacitor is transformed into thermal energy in the resistor. Assume that the capacitor is fully charged before the switch is thrown.

46. A capacitor with capacitance $C = 1.0$ μF and initial stored energy $U_0 = 0.50$ J is discharged through a resistance $R = 1.0 \times 10^6$ Ω. (a) What is the initial charge on the capacitor? (b) What is the current through the resistor when the discharge starts? (c) Determine V_C, the voltage across the capacitor, and V_R, the voltage across the resistor, as a function of time. (d) Express the rate of generation of thermal energy in the resistor as a function of time.

47. A 3.0×10^6-Ω resistor and a 1.0-μF capacitor are connected in a single-loop circuit with a seat of emf with $\varepsilon = 4.0$ V. At 1.0 s after the connection is made, what are the rates at which (a) the charge of the capacitor is increasing, (b) energy is being stored in the capacitor, (c) thermal energy is appearing in the resistor, and (d) energy is being delivered by the seat of emf? *Answer: (a) 9.6×10^{-7} C/s. (b) 1.1×10^{-6} W. (c) 2.8×10^{-6} W. (d) 3.8×10^{-6} W.*

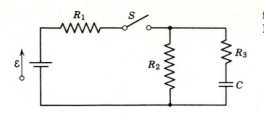

figure 32-29
Problem 48

48. In the circuit of Fig. 32-29 let i_1, i_2, and i_3 be the currents through resistors R_1, R_2, and R_3, respectively, and let V_1, V_2, V_3, and V_C be the corresponding potential differences across the resistors and across the capacitor C. (a) Plot qualitatively, as a function of time after switch S is closed, the currents and voltages listed above. (b) After being closed for a large number of time constants, the switch S is now opened. Plot qualitatively, as a function of time after the switch is opened, the currents and voltages listed above.

49. In the circuit of Fig. 32-30, $\varepsilon = 1200$ V, $C = 6.50$ μF, $R_1 = R_2 = R_3 = 7.30 \times 10^5$ Ω. With C completely uncharged, the switch S is suddenly closed ($t = 0$). (a) Determine the currents through each resistor for $t = 0$ and $t = \infty$. (b) Draw qualitatively a graph of the potential drop V_2 across R_2 from $t = 0$ to $t = \infty$. (c) What are the numerical values of V_2 at $t = 0$ and $t = \infty$? (d) Give the physical meaning of "$t = \infty$" and state a rough, but significant, numerical lower bound, in seconds, for "$t = \infty$" in this case.

    Answer: (a) At $t = 0$, $i_1 = 1.1$ mA, $i_2 = i_3 = 0.55$ mA; at $t = \infty$, $i_1 = i_2 = 0.82$ mA, $i_3 = 0$. (c) At $t = 0$, $V_2 = 400$ V; at $t = \infty$, $V_2 = 600$ V. (d) The time constant is 7.1 s.

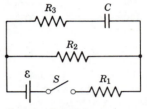

figure 32-30
Problem 49

33
the magnetic field

The science of magnetism grew from the observation that certain "stones" (magnetite) would attract bits of iron. The word *magnetism* comes from the district of Magnesia in Asia Minor, which is one of the places at which the stones were found. Figure 33-1 shows a modern permanent magnet, the lineal descendant of these natural magnets. Another "natural magnet" is the earth itself, whose orienting action on a magnetic compass needle has been known since ancient times.

In 1820 Oersted discovered that a current in a wire can also produce magnetic effects, namely, that it can change the orientation of a compass needle.* We pointed out in Section 26-1 how this important discovery linked the then separate sciences of magnetism and electricity. We can intensify the magnetic effect of a current in a wire by forming the wire into a coil of many turns and by providing an iron core. Figure 33-2 shows how this is done in a large electromagnet of a type commonly used for research involving magnetism.

We define the space around a magnet or a current-carrying conductor as the site of a *magnetic field*, just as we defined the space near a charged rod as the site of an electric field. The basic magnetic field vector **B**, which we define in the following section, is called the *magnetic induc-*

33-1
THE MAGNETIC FIELD

---

*In 1878 H. A. Rowland, at the Johns Hopkins University, discovered that a moving charged object (in his case, a rapidly rotating charged disk) also causes magnetic effects. Actually, it is far from obvious that a moving charge is equivalent to a current in a wire. See. "Rowland's Physics" by John D. Miller, *Physics Today*, July 1976 for a scientific biography of this outstanding American physicist, including a detailed discussion of this experiment.

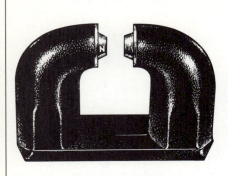

figure 33-1
A permanent magnet. Lines of magnetic induction leave the north pole face, marked N, and enter the south pole face on the other side of the air gap.

figure 33-2
A research-type electromagnet showing iron frame F, pole faces P, and coils C. The pole faces are 12 in. in diameter. (Courtesy Varian Associates.)

tion.* We can represent it by *lines of induction*, just as we represent the electric field by lines of force. As for the electric field (see Section 27-3), the magnetic field vector is related to its lines of induction in this way:

1. The tangent to a line of induction at any point gives the *direction* of **B** at that point.

2. The lines of induction are drawn so that the number of lines per unit cross-sectional area (perpendicular to the lines) is proportional to the *magnitude* of **B.** Where the lines are close together B is large and where they are far apart B is small.

As for the electric field, the magnetic field vector **B** is of fundamental importance, the lines of induction simply giving a graphic representation of the way **B** varies throughout a certain region of space. The *flux* Φ_B for a magnetic field can be defined in exact analogy with the flux Φ_E for the electric field, namely $\Phi_B = \int \mathbf{B} \cdot d\mathbf{S}$, in which the integral is taken over the surface (closed or open) for which Φ_B is defined.

In this chapter we are not concerned with the *causes* of the magnetic field; we will explore them in the next chapter. Our related concerns here are (*a*) to *define* operationally the magnitude and direction of **B** at any point P near, say, a magnet, a current-carrying conductor, or a moving charge and (*b*) to determine the *effect* of a magnetic field on objects, such as moving charges, subject to its influence.

As we did for the electric field, let us consider a charge q_0 as a test body. Let us place the test body *at rest* at a point P near, say, a perma-

33-2
THE DEFINITION† OF **B**

* *Magnetic field strength* would be a more suitable name for **B,** but it has been usurped for historical reasons by another vector connected with the magnetic field (see Section 37-8). Often in this book we shall call **B,** for brevity, the "magnetic field" in analogy to calling **E** the "electric field".

† The definition and method of measurement of **B** given in this section, though conceptually sound and suitable for our present purpose, are not carried out in practice because of experimental difficulties. The following section will show how **B** may be more conveniently measured in the laboratory.

nent magnet such as that of Fig. 33-1. We would then find by experiment that *no force* (attributable *only* to the presence or absence of the magnet) acts on q_0. However, if we fire test body q_0 through point P with a velocity **v**, we find that a *sideways force* **F** acts on it if the magnet is in place; by a sideways force we mean a force at right angles to **v**. We shall define **B** at point P in terms of **F**, **v**, and q_0.

If we vary the direction of **v** through point P, keeping the magnitude of **v** unchanged, we find that although **F** will always remain at right angles to **v**, its magnitude F will change. For a particular orientation of **v** (and also for the opposite orientation $-$**v**) the force **F** becomes zero. We define this orientation as the *direction* of **B**, the specification of the *sense* of **B** (that is, the way it points along this line) being left to the more complete definition of **B** that we give below.

Having found the direction of **B**, we are now able to orient **v** so that the test charge moves at right angles to **B**. We will find that the force **F** is now a maximum, and we define the *magnitude* of **B** from the measured magnitude of this maximum force $F_\perp$, or

$$B = \frac{F_\perp}{q_0 v}. \tag{33-1}$$

Let us regard this definition of **B** (in which we have specified its magnitude and direction, but not its sense) as preliminary to the complete vector definition that we now give: *If we fire a positive test charge q_0 with velocity **v** through a point P and if a force **F** acts on the moving charge, a magnetic field **B** is present at point P, where **B** is the vector that satisfies the relation*

$$\mathbf{F} = q_0 \mathbf{v} \times \mathbf{B}, \tag{33-2}$$

v, q_0, and **F** being measured quantities. The magnitude of the magnetic deflecting force **F**, according to the rules for vector products, is given by *

$$F = q_0 v B \sin\theta, \tag{33-3}$$

in which θ is the angle between **v** and **B**.

Figure 33-3 shows the relations among the vectors. We see from Eq. 33-2 that **F**, being at right angles to the plane formed by **v** and **B**, will always be at right angles to **v** (and also to **B**) and thus will always be a sideways force. Equation 33-2 is consistent with the observed facts that (*a*) the magnetic force vanishes as $v \to 0$, (*b*) the magnetic force vanishes if **v** is either parallel or antiparallel to the direction of **B** (in these cases $\theta = 0$ or $180°$ and $\mathbf{v} \times \mathbf{B} = 0$), and (*c*) if **v** is at right angles to **B** ($\theta = 90°$), the deflecting force has its maximum value, given by Eq. 33-1, that is, $q_0 v B$.

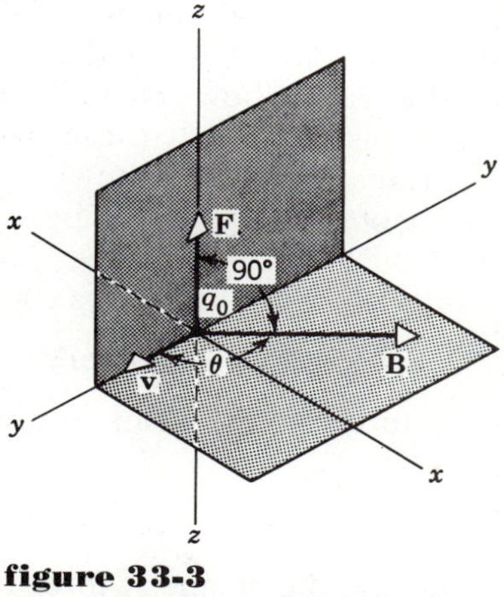

This definition of **B** is similar in spirit, although more complex, than the definition of the electric field **E**, which we can cast into this form: *If we place a positive test charge q_0 at point P and if an (electric) force **F** acts on the stationary charge, an electric field **E** is present at P, where **E** is the vector satisfying the relation*

$$\mathbf{F} = q_0 \mathbf{E},$$

q_0 and **F** being measured quantities. In defining **E**, the only characteristic direction to appear is that of the electric force $\mathbf{F}_E$ which acts on the positive test body; the direction of **E** is taken to be that of $\mathbf{F}_E$. In defining **B**, *two* characteristic directions appear, those of **v** and of the magnetic force $\mathbf{F}_B$; they prove always to be at right angles. Although we can easily solve the above equation for **E**, we cannot solve Eq. 33-2 for **B**. Why not?

* You may wish to review Section 2-4, which deals with vector products.

figure 33-3
Illustrating $\mathbf{F} = q_0 \mathbf{v} \times \mathbf{B}$ (Eq. 33-2). Test charge q_0 is fired through the origin with velocity **v**.

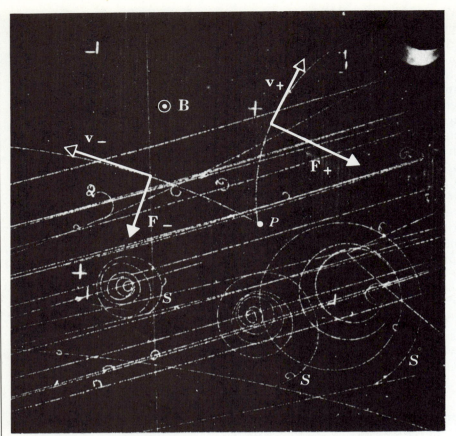

figure 33-4

A *bubble chamber* is a device for rendering visible, by means of small bubbles, the tracks of charged particles that pass through the chamber. The figure is a photograph taken with such a chamber immersed in a magnetic field **B** and exposed to radiations from a large cyclotron-like accelerator. The curved *V* at point *P* is formed by a positive and a negative electron, which deflect in opposite directions in the magnetic field. The spirals *S* are the tracks of three low-energy electrons. (Courtesy E. O. Lawrence Radiation Laboratory, University of California.)

In Fig. 33-4 a positive and a negative electron are created at point P in a bubble chamber. A magnetic field is perpendicular to the chamber, pointing out of the plane of the figure (symbol $\odot$).* The relation $\mathbf{F} = q_0 \mathbf{v} \times \mathbf{B}$ (Eq. 33-2) shows that the deflecting forces acting on the two particles are as indicated in the figure. These deflecting forces would make the tracks deflect as shown.

The unit of **B** that follows from Eq. 33-3 is the newton/(coulomb)(meter/second). This is given the SI name *tesla* (abbr. T) or weber/meter² (abbr. Wb/m²).** Recalling that a coulomb/second is an ampere, we have

$$1 \text{ tesla} = 1 \text{ weber/meter}^2 = 1 \text{ newton/(ampere} \cdot \text{meter)}.$$

An earlier unit for **B**, still in common use, is the *gauss;* the relationship is

$$1 \text{ tesla} = 1 \text{ weber/meter}^2 = 10^4 \text{ gauss.}\dagger$$

The fact that the magnetic force is always at right angles to the direction of motion means that (for steady magnetic fields at least) the work done by this force on the particle is zero. For an element of the path of the particle of length $d\mathbf{l}$, this work dW is $\mathbf{F}_B \cdot d\mathbf{l}$; dW is zero because $\mathbf{F}_B$ and $d\mathbf{l}$ are always at right angles. Thus a static magnetic field cannot

---

* The symbol $\otimes$ indicates a vector into the page, the $\times$ being thought of as the tail of an arrow; the symbol $\odot$ indicates a vector out of the page, the dot being thought of as the tip of an arrow.
** The weber is used to measure the magnetic flux, a concept we discuss in succeeding chapters. See also Sec. 33-1.
† See "Megagauss Physics" by C. M. Fowler, *Science,* April 1973 for a fascinating account of the properties of such enormous fields.

change the kinetic energy of a moving charge; it can only deflect it sideways.*

If a charged particle moves through a region in which both an electric field and a magnetic field are present, the resultant force is found by combining Eqs. 27-2 and 33-2, or

$$\mathbf{F} = q_0\mathbf{E} + q_0\mathbf{v} \times \mathbf{B}. \qquad (33\text{-}4)$$

This is sometimes called the *Lorentz equation* in tribute to H. A. Lorentz who did so much to develop and clarify the concepts of the electric and magnetic fields.

EXAMPLE 1

A uniform magnetic field **B** points horizontally from south to north; its magnitude is 1.5 T. If a 5.0-MeV proton moves vertically downward through this field, what force will act on it?

The kinetic energy of the proton is

$$K = (5.0 \times 10^6 \text{ eV})(1.6 \times 10^{-19} \text{ J/eV}) = 8.0 \times 10^{-13} \text{ J}.$$

We can find its speed from the relation $K = \frac{1}{2}mv^2$, or

$$v = \sqrt{\frac{2K}{m}} = \sqrt{\frac{(2)(8.0 \times 10^{-13} \text{ J})}{1.7 \times 10^{-27} \text{ kg}}} = 3.1 \times 10^7 \text{ m/s}.$$

Equation 33-3 gives

$$F = qvB \sin \theta = (1.6 \times 10^{-19} \text{ C})(3.1 \times 10^7 \text{ m/s})(1.5 \text{ T})(\sin 90°)$$

$$= 7.4 \times 10^{-12} \text{ N}.$$

You can show that this force is about 4×10^{14} times greater than the weight of the proton.

The relation $\mathbf{F} = q\mathbf{v} \times \mathbf{B}$ shows that the *direction* of the deflecting force is to the east. If the particle had been negatively charged, the deflection would have been to the west. This is predicted automatically by Eq. 33-2 if we substitute $-e$ for q_0.

A current is an assembly of moving charges. Because a magnetic field exerts a sideways force on a moving charge, we expect that it will also exert a sideways force on a wire carrying a current. Figure 33-5 shows a length *l* of wire carrying a current *i* and placed in a magnetic field **B**. For simplicity we have oriented the wire so that the current density vector **j** is at right angles to **B**.

33-3
MAGNETIC FORCE ON A CURRENT

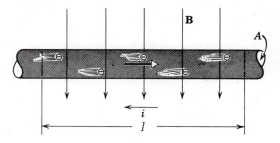

figure 33-5
A wire carrying a current *i* is placed at right angles to a magnetic field **B**. Only the drift velocity of the electrons, not their random motion, is suggested.

---

* Actually, $\mathbf{F}_B$ is at right angles to **v**, *even if* **B** varies with time. In this case, however, the theory of special relativity predicts that an *electric* field **E** will appear and act on the charged particle in such a way as to do work on it. See *Introduction to Special Relativity* by Robert Resnick, John Wiley & Sons, 1968 (Section 4.2) for details.

The current i in a metal wire is carried by the free (or conduction) electrons, n being the number of such electrons per unit volume of the wire. The magnitude of the average force on one such electron is given by Eq. 33-3, or, since $\theta = 90°$,

$$F' = q_0 vB \sin \theta = ev_d B$$

where v_d is the drift speed. From the relation $v_d = j/ne$ (Eq. 31-5),

$$F' = e\left(\frac{j}{ne}\right) B = \frac{jB}{n}.$$

The length l of the wire contains nAl free electrons, Al being the volume of the section of wire of cross section A that we are considering. The total force on the free electrons in the wire, and thus on the wire itself, is

$$F = (nAl)F' = nAl\,\frac{jB}{n}.$$

Since jA is the current i in the wire, we have

$$F = ilB. \tag{33-5}$$

The negative charges, which move to the right in the wire of Fig. 33-5, are equivalent to positive charges moving to the left, that is, in the direction of the current arrow. For such a positive charge the velocity $\mathbf{v}$ would point to the left and the force on the wire, given by Eq. 33-2 ($\mathbf{F} = q_0\mathbf{v} \times \mathbf{B}$) points up, out of the page. This same conclusion follows if we consider the actual negative charge carriers for which $\mathbf{v}$ points to the right but q_0 has a negative sign. Thus by measuring the sideways magnetic force on a wire carrying a current and placed in a magnetic field, we cannot tell whether the current carriers are negative charges moving in a given direction or positive charges moving in the opposite direction.

Equation 33-5 holds only if the wire is at right angles to $\mathbf{B}$. We can express the more general situation in vector form as

$$\mathbf{F} = i\mathbf{l} \times \mathbf{B}, \tag{33-6a}$$

where $\mathbf{l}$ is a vector whose magnitude is the length of the wire and which points along the (straight) wire in the direction of the current. Equation 33-6a is equivalent to the relation $\mathbf{F} = q_0\mathbf{v} \times \mathbf{B}$ (Eq. 33-2); either can be taken as a defining equation for $\mathbf{B}$. Note that the vector $\mathbf{l}$ in Fig. 33-5 points to the left and that the magnetic force $\mathbf{F} (= i\mathbf{l} \times \mathbf{B})$ points up, out of the page. This agrees with the conclusion obtained by analyzing the forces that act on the individual charge carriers.

If we consider a differential element of a conductor of length $d\mathbf{l}$, we can find the force $d\mathbf{F}$ acting on it, by analogy with Eq. 33-6a, from

$$d\mathbf{F} = i\,d\mathbf{l} \times \mathbf{B}. \tag{33-6b}$$

By integrating this formula in an appropriate way we can find the force $\mathbf{F}$ on a conductor which is not straight.

EXAMPLE 2

A wire bent as shown in Fig. 33-6 carries a current i and is placed in a uniform magnetic field $\mathbf{B}$ that emerges from the plane of the figure. Calculate the force acting on the wire. The magnetic field is represented by lines of induction, shown emerging from the page. The dots show that the sense of $\mathbf{B}$ is up, out of the page.

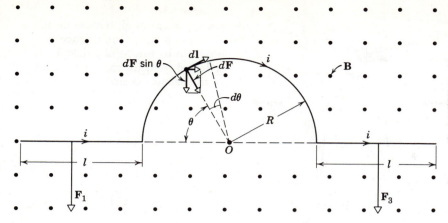

figure 33-6
Example 2

The force on each straight section, from Eq. 33-6a, has the magnitude

$$F_1 = F_3 = ilB$$

and points down as shown by the arrows in the figure.

A segment of wire of length dl on the arc has a force $d\mathbf{F}$ on it whose magnitude is

$$dF = iB \; dl = iB(R \; d\theta)$$

and whose direction is radially toward O, the center of the arc. Only the downward component of this force is effective, the horizontal component being canceled by an oppositely directed component associated with the corresponding arc segment on the other side of O. Thus the total force on the semicircle of wire about O points down and is

$$F_2 = \int_0^\pi dF \sin \theta = \int_0^\pi (iBR \; d\theta) \sin \theta = iBR \int_0^\pi \sin \theta \; d\theta = 2iBR.$$

The resultant force on the whole wire is

$$F = F_1 + F_2 + F_3 = 2ilB + 2iBR = 2iB(l + R).$$

Notice that this force is the same as that acting on a straight wire of length $2l + 2R$.

Figure 33-7 shows the arrangement used by Thomas, Driscoll, and Hipple in 1949 at the National Bureau of Standards to measure the magnetic field provided by a laboratory magnet such as that of Fig. 33-2. The rectangle is a coil of nine turns whose width a and length b are about 10 cm and 70 cm, respectively. The lower end of the coil is placed in the magnetic field $\mathbf{B}$ and the upper end is hung from the arm of a balance; $\mathbf{B}$ enters the plane of the figure at right angles.

An accurately known current i of about 0.10 A is set up in the coil in the direction shown, and weights are placed in the right-hand balance pan until the system is balanced. The magnetic force $\mathbf{F}$ ($= i\mathbf{l} \times \mathbf{B}$; see Eq. 33-6a) on the bottom leg of the coil points upward, as shown in the figure. Equation 33-5 also shows that the force on each wire at the bottom of the coil is iaB. Since there are nine wires, the total force on the bottom leg of the coil is $9iaB$. The forces on the vertical sides of the coil ($= i\mathbf{l} \times \mathbf{B}$) are sideways; because they are equal and opposite, they cancel and produce no effect.

After balancing the system, the experimenters reversed the direction of the current, which changed the sign of all the magnetic forces acting on the coil. In particular, $\mathbf{F}$ then pointed downward, which caused the rest point of the bal-

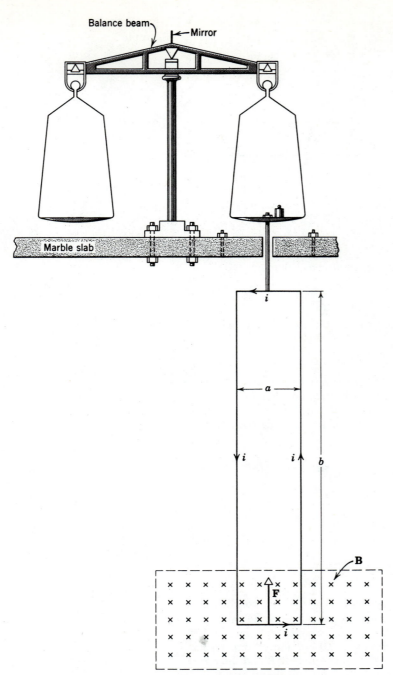

figure 33-7
An apparatus used to measure **B**. The zero point of the balance is observed by means of a light beam reflected from the mirror attached to the balance beam.

ance to move. A mass m of about 8.78 g had to be added to the left balance pan to restore the original rest point. The *change* in force when the current is reversed is $2F$, and this must equal the weight added to the left balance pan, or

$$mg = 2(9iaB) = 18iaB.$$

This gives

$$B = \frac{mg}{18ai} = \frac{(8.78 \times 10^{-3}\ \text{kg})(9.80\ \text{m/s}^2)}{(18)(0.10\ \text{m})(0.10\ \text{A})} = 0.48\ \text{T} = 4800\ \text{gauss}.$$

The Bureau of Standards' workers made this measurement with much more precision than these approximate numbers suggest. In one series of measurements, for example, they found a magnetic field of 4697.55 gauss; even greater precision is possible today, by this and other methods.

Figure 33-8 shows a rectangular loop of wire of length a and width b placed in a uniform magnetic field **B**, with sides 1 and 3 always normal to the field direction. The normal nn' to the plane of the loop makes an angle θ with the direction of **B**.

33-4
TORQUE ON A CURRENT LOOP

Assume the current to be as shown in the figure. Wires must be provided to lead the current into the loop and out of it. If these wires are twisted tightly together, there will be no net magnetic force on the twisted pair because the currents in the two wires are in opposite directions. Thus the lead wires may be ignored. Also, some way of supporting the loop must be provided. Let us imagine it to be suspended from a long string attached to the loop at its center of mass. In this way the loop will be free to turn, through a small angle at least, about any axis through the center of mass. Alternatively, we can imagine the experiment to be carried out in an orbiting earth satellite (Skylab, say) in which the effective value of **g** is zero.

The net force on the loop is the resultant of the forces on the four sides of the loop. On side 2 the vector **l** points in the direction of the current and has the magnitude b. The angle between **l** and **B** for side 2 (see Fig. 33-8b) is $90° − \theta$. Thus the magnitude of the force on this side is

$$F_2 = ibB \sin (90° − \theta) = ibB \cos \theta.$$

From the relation $\mathbf{F} = i\mathbf{l} \times \mathbf{B}$ (Eq. 33-6a), we find the direction of $\mathbf{F}_2$ to be out of the plane of Fig. 33-8b. You can show that the force $\mathbf{F}_4$ on side 4 has the same magnitude as $\mathbf{F}_2$ but points in the opposite direction. Thus $\mathbf{F}_2$ and $\mathbf{F}_4$, taken together, have no effect on the motion of the loop. The net force they provide is zero, and, since they have the same line of action, the net torque due to these forces is also zero.

The common magnitude of $\mathbf{F}_1$ and $\mathbf{F}_3$ is iaB. These forces, too, are oppositely directed so that they do not tend to move the coil bodily. As Fig. 33-8b shows, however, they do *not* have the same line of action if the coil is in the position shown; there is a net torque, which tends to rotate the coil clockwise about the line xx'. The coil can be supported on a rigid axis that lies along xx', with no loss of its freedom of motion. This torque can be represented in Fig. 33-8b by a vector pointing into the figure at point x' or in Fig. 33-8a by a vector pointing along the xx' axis from right to left.

The magnitude of the torque τ' is found by calculating the torque caused by $\mathbf{F}_1$ about axis xx' and doubling it, for $\mathbf{F}_3$ exerts the same torque about this axis that $\mathbf{F}_1$ does. Thus

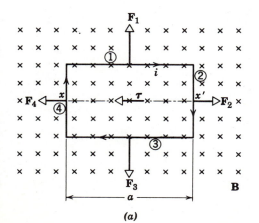

(a)

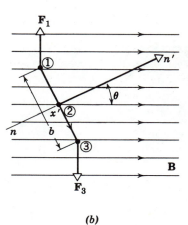

(b)

figure 33-8
A rectangular coil carrying a current i is placed in a uniform external magnetic field.

$$\tau' = 2(iaB)\left(\frac{b}{2}\right)(\sin\theta) = iabB\sin\theta.$$

This torque acts on every turn of the coil. If there are N turns, the torque on the entire coil is

$$\tau = N\tau' = NiabB\sin\theta = NiAB\sin\theta, \qquad (33\text{-}7)$$

in which A, the area of the coil, is substituted for ab.

This equation can be shown to hold for *all plane loops of area A, whether they are rectangular or not.* A torque on a current loop is the basic operating principle of the electric motor and of most electric meters used for measuring current or potential difference.

EXAMPLE 3

A galvanometer. Figure 33-9 shows the rudiments of a galvanometer, which is the basic operating mechanism of ammeters and voltmeters. The coil is 2.0 cm high and 1.0 cm wide; it has 250 turns and is mounted so that it can rotate about a vertical axis in a uniform *radial* magnetic field with $B = 2000$ gauss. A spring Sp provides a countertorque that cancels out the magnetic torque, resulting in a steady angular deflection ϕ corresponding to a given current i in the coil. If a current of 1.0×10^{-4} A produces an angular deflection of 30°, what is the *torsional constant* κ of the spring (see Eq. 15-21)?

Equating the magnetic torque to the torque caused by the spring (see Eq. 33-7) yields

$$\tau = NiAB\sin\theta = \kappa\phi$$

or

$$\kappa = \frac{NiAB\sin\theta}{\phi}$$

$$= \frac{(250)(1.0 \times 10^{-4}\ \text{A})(2.0 \times 10^{-4}\ \text{m}^2)(0.20\ \text{T})(\sin 90°)}{30°}$$

$$= 3.3 \times 10^{-8}\ \text{N}\cdot\text{m/degree.}$$

Note that the normal to the plane of the coil (that is, the pointer P) is always at right angles to the (radial) magnetic field so that $\theta = 90°$.

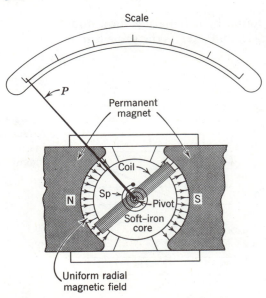

figure 33-9
Example 3. The elements of a galvanometer, showing the coil, the helical spring Sp, and pointer P.

A current loop orienting itself in an external magnetic field reminds us of the action of a compass needle in such a field. One face of the loop behaves like the north pole of the needle;* the other face behaves like the south pole. Compass needles, bar magnets, and current loops can all be regarded as *magnetic dipoles.* We show this here for the current loop, reasoning entirely by analogy with *electric* dipoles.

A structure is called an electric dipole if (a) when placed in an *external* electric field it experiences a torque given by Eq. 27-11,

$$\tau = \mathbf{p} \times \mathbf{E}, \tag{33-8}$$

where $\mathbf{p}$ is the electric dipole moment, and (b) it sets up a field of its own at distant points, described qualitatively by the lines of force of Fig. 29-10 and quantitatively by Eq. 29-11. These two requirements are not independent; if one is fulfilled, the other follows automatically.

The magnitude of the torque described by Eq. 33-8 is

$$\tau = pE \sin \theta, \tag{33-9}$$

where θ is the angle between $\mathbf{p}$ and $\mathbf{E}$. Let us compare this with Eq. 33-7, the expression for the torque on a current loop:

$$\tau = (NiA)B \sin \theta. \tag{33-7}$$

In each case the appropriate field (E or B) appears, as does a term $\sin \theta$. Comparison suggests that NiA in Eq. 33-7 can be taken as the *magnetic dipole moment* μ, corresponding to p in Eq. 33-9, or

$$\mu = NiA. \tag{33-10}$$

Equation 33-7 suggests that we write the torque on a current loop in vector form, in analogy with Eq. 33-8, or

$$\tau = \boldsymbol{\mu} \times \mathbf{B}. \tag{33-11}$$

The magnetic dipole moment of the loop $\boldsymbol{\mu}$ must be taken to lie along an axis perpendicular to the plane of the loop; its direction is given by the following rule: Let the fingers of the right hand curl around the loop in the direction of the current; the extended right thumb will then point in the direction of $\boldsymbol{\mu}$. If $\boldsymbol{\mu}$ is defined by this rule and Eq. 33-10, check carefully that Eq. 33-11 correctly describes in every detail the torque acting on a current loop in an external field (see Fig. 33-8).

Since a torque acts on a current loop, or other magnetic dipole, when it is placed in an external magnetic field, it follows that work (positive or negative) must be done by an external agent to change the orientation of such a dipole. Thus a magnetic dipole has *potential energy* associated with its orientation in an external magnetic field. This energy may be taken to be zero for any arbitrary position of the dipole. By analogy with the assumption made for electric dipoles in Section 27-6, we assume that the magnetic energy U is zero when $\boldsymbol{\mu}$ and $\mathbf{B}$ are at right angles, that is, when $\theta = 90°$. This choice of a zero-energy configuration for U is arbitrary because we are interested only in the *changes* in energy that occur when the dipole is rotated.

The magnetic potential energy in any position θ is defined as the work that an external agent must do to turn the dipole from its zero-energy position ($\theta = 90°$) to the given position θ. Thus

* The north pole of a compass needle is the end that points toward the geographic north.

$$U = \int_{90°}^{\theta} \tau \, d\theta = \int_{90°}^{\theta} NiAB \sin \theta \, d\theta = \mu B \int_{90°}^{\theta} \sin \theta \, d\theta = -\mu B \cos \theta,$$

in which Eq. 33-7 is used to substitute for τ. In vector symbolism we can write this relation as

$$U = -\boldsymbol{\mu} \cdot \mathbf{B}, \qquad (33\text{-}12)$$

which is in perfect correspondence with Eq. 27-13, the expression for the energy of an *electric* dipole in an external *electric* field,

$$U = -\mathbf{p} \cdot \mathbf{E}.$$

EXAMPLE 4

A circular coil of N turns has an effective radius a and carries a current i. How much work is required to turn it in an external magnetic field **B** from a position in which θ equals zero to one in which θ equals 180°? Assume that $N = 100$, $a = 5.0$ cm, $i = 0.10$ A, and $B = 1.5$ T.

The work required is the difference in energy between the two positions, or, from Eq. 33-12,

$$W = U_{\theta = 180°} - U_{\theta = 0} = (-\mu B \cos 180°) - (-\mu B \cos 0) = 2\mu B.$$

But $\mu = NiA$, so that

$$W = 2NiAB = 2Ni(\pi a^2) B$$

$$= (2)(100)(0.10 \text{ A})(\pi)(5 \times 10^{-2} \text{ m})^2(1.5 \text{ T}) = 0.24 \text{ J}.$$

33-5
THE HALL EFFECT

In 1879 E. H. Hall, at Harvard University, reported an experiment that gives the sign of the charge carriers in a conductor; see p. 676. Figure 33-10 shows a flat strip of copper carrying a current i in the direction shown. As usual, the direction of the current arrow, labeled i, is the direction in which the charge carriers would move *if* they were positive. The current arrow can represent either positive charges moving down (as in Fig. 33-10a) or negative charges moving up (as in Fig. 33-10b). The Hall effect can be used to decide between these two possibilities.[*]

A magnetic field **B** is set up at right angles to the strip by placing the strip between the polefaces of an electromagnet. This field exerts a deflecting force **F** on the strip (given by $i\mathbf{l} \times \mathbf{B}$), which points to the right in the figure. Since the sideways force on the strip is due to the sideways forces on the charge carriers (given by $q\mathbf{v} \times \mathbf{B}$), it follows that these carriers, whether they are positive or negative, will tend to drift toward the right in Fig. 33-10 as they drift along the strip, producing a *transverse Hall potential difference* V_{xy} between points such as x and y. The sign of the charge carriers is determined by the sign of this Hall poten-

---

[*] The connection between the eminent but self-effacing physicist H. A. Rowland and the Hall effect has often been glossed over. From the reference cited in the footnote on p. 717 we learn that Hall (Rowland's student at the Johns Hopkins University) used "... an experimental configuration devised by Rowland," and also, in Rowland's words (in 1894), "... I had already obtained the Hall effect on a small scale. ..." Rowland's colleague, Joseph Ames, wrote "... There have been several striking cases where it might have seemed to an impartial observer that Rowland's name should have appeared on the title page.

The question of priority of discovery is complex. Before drawing conclusions, you should read carefully (a) the reference in the footnote on p. 717 and (b) Hall's paper, which can be found in "Source Book in Physics" by William Francis Magie, ed., McGraw-Hill Book Co., 1935, p. 541.

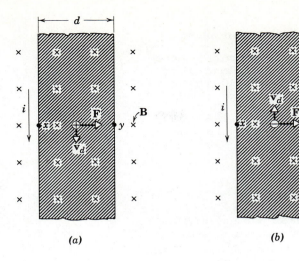

figure 33-10
A current i is set up in a copper strip placed in a magnetic field **B**, assuming (a) positive carriers and (b) negative carriers.

(a)   (b)

tial difference. If the carriers are positive, y will be at a higher potential than x; if they are negative, y will be at a lower potential than x. Experiment shows that in metals the charge carriers are negative.*

To analyze the Hall effect quantitatively, let us use the free-electron model of a metal, the same model used in Section 31-4 to help us understand why metals obey Ohm's law. The charge carriers can be assumed to move along the conductor with a certain constant drift speed v_d. The magnetic deflecting force that causes the moving charge carriers to drift toward the right edge of the strip is given by $q\mathbf{v}_d \times \mathbf{B}$ (see Eq. 33-2).

The charge carriers do not build up without limit on the right edge of the strip because the displacement of charge gives rise to a transverse *Hall electric field* $\mathbf{E}_H$, which acts, inside the conductor, to oppose the sideways drift of the carriers. This Hall electric field is another manifestation of the Hall potential difference and is related to it by

$$E_H = V_{xy}/d.$$

Eventually an equilibrium is reached in which the sideways magnetic deflecting force on the charge carriers is just canceled by the oppositely directed electric force $q\mathbf{E}_H$ caused by the Hall electric field, or

$$q\mathbf{E}_H + q\mathbf{v}_d \times \mathbf{B} = 0,$$

which we can write as

$$\mathbf{E}_H = -\mathbf{v}_d \times \mathbf{B}. \tag{33-13}$$

This equation shows explicitly that if we measure $\mathbf{E}_H$ and **B**, we can find $\mathbf{v}_d$ both in magnitude and direction; given the direction of $\mathbf{v}_d$, the sign of the charge carriers follows at once, as Fig. 33-10 shows.

The number of charge carriers per unit volume (n) can also be found from Hall effect measurements. If we write Eq. 33-13 in terms of magnitudes, for the case in which $\mathbf{v}_d$ and **B** are at right angles, we obtain $E_H = v_d B$. Combining this with Eq. 31-5 $(v_d = j/ne)$ leads to

$$E_H = \frac{j}{ne} B \quad \text{or} \quad n = \frac{jB}{eE_H}. \tag{33-14}$$

The agreement between experiment and Eq. 33-14 is rather good for monovalent metals, as Table 33-1 shows.

---

* At the time of the Hall-Rowland experiments the electron was not yet discovered (see Section 33-8). Hall's analysis was based on a fluid model of electricity but the general conclusions remain unchanged.

Table 33-1
Number of conduction electrons per unit volume

Metal	Based on Hall Effect Data, $10^{22}/cm^3$	Calculated, Assuming One Electron/Atom, $10^{22}/cm^3$
Li	3.7	4.8
Na	2.5	2.6
K	1.5	1.3
Cs	0.80	0.85
Cu	11	8.4
Ag	7.4	6.0
Au	8.7	5.9

For nonmonovalent metals, for iron and similar magnetic materials, and for semiconductors such as germanium, the simple interpretation of the Hall effect in terms of the free-electron model is not valid. A theoretical interpretation of the Hall effect based on quantum physics gives a reasonable agreement with experiment in all cases.

EXAMPLE 5

A copper strip 2.0 cm wide and 1.0 mm thick is placed in a magnetic field with $B = 1.5$ T, as in Fig. 33-10. If a current of 200 A is set up in the strip, what Hall potential difference appears across the strip?

From Eq. 33-14,

$$E_H = \frac{jB}{ne};$$

but

$$E_H = \frac{V_{xy}}{d} \quad \text{and} \quad j = \frac{i}{A} = \frac{i}{dh},$$

where h is the thickness of the strip. Combining these equations gives

$$V_{xy} = \frac{iB}{neh} = \frac{(200 \text{ A})(1.5 \text{ T})}{(8.4 \times 10^{28}/\text{m}^3)(1.6 \times 10^{-19} \text{ C})(1.0 \times 10^{-3} \text{ m})}$$

$$= 2.2 \times 10^{-5} \text{ V} = 22 \ \mu\text{V}.$$

These potential differences, though quite measurable, are not large. See p. 678 for the calculation of n.

33-6
CIRCULATING CHARGES

Figure 33-11 shows a negatively charged particle introduced with velocity $\mathbf{v}$ into a uniform magnetic field $\mathbf{B}$. We assume that $\mathbf{v}$ is at right angles to $\mathbf{B}$ and thus lies entirely in the plane of the figure. The relation $\mathbf{F} = q\mathbf{v} \times \mathbf{B}$ (Eq. 33-2) shows that the particle will experience a sideways deflecting force of magnitude qvB. This force will lie in the plane of the figure, which means that the particle cannot leave this plane.

This reminds us of a stone held by a rope and whirled in a horizontal circle on a smooth surface. Here, too, a force of constant magnitude, the tension in the rope, acts in a plane and at right angles to the velocity. The charged particle, like the stone, also moves with constant speed in a circular path. From Newton's second law we have

$$qvB = \frac{mv^2}{r} \quad \text{or} \quad r = \frac{mv}{qB}, \tag{33-15}$$

which gives the radius of the path. The three spirals in Fig. 33-4 show relatively low-energy electrons in a bubble chamber. The paths are not

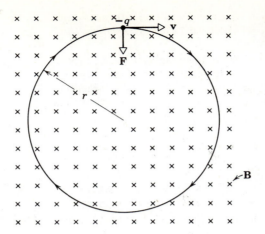

figure 33-11
A charge $-q$ circulates at right angles to a uniform magnetic field.

731   CIRCULATING CHARGES   SEC. 33-6

circles because the electrons lose energy by collisions in the chamber as they move.

The angular velocity ω is given by v/r or, from Eq. 33-15,

$$\omega = \frac{v}{r} = \frac{qB}{m}.$$

The frequency ν is given by

$$\nu = \frac{\omega}{2\pi} = \frac{qB}{2\pi m}. \qquad (33\text{-}16)$$

Note that ν does not depend on the speed of the particle. Fast particles move in large circles (Eq. 33-15) and slow ones in small circles, but all require the same time T (the *period*) to complete one revolution in the field.

The frequency ν is a characteristic frequency for the charged particle in the field and may be compared to the characteristic frequency of a swinging pendulum in the earth's gravitational field or to the characteristic frequency of an oscillating mass-spring system. It is sometimes called the *cyclotron frequency* of the particle in the field because particles circulate at this frequency in a cyclotron.

A 10-eV electron is circulating in a plane at right angles to a uniform magnetic field of 1.0×10^{-4} T ($= 1.0$ gauss).

EXAMPLE 6

(*a*) What is its orbit radius?

The velocity of an electron whose kinetic energy is K can be found from

$$v = \sqrt{\frac{2K}{m}}.$$

Verify that this yields 1.9×10^6 m/s for v. Then, from Eq. 33-15,

$$r = \frac{mv}{qB} = \frac{(9.1 \times 10^{-31} \text{ kg})(1.9 \times 10^6 \text{ m/s})}{(1.6 \times 10^{-19} \text{ C})(1.0 \times 10^{-4} \text{ T})} = 0.11 \text{ m} = 11 \text{ cm}.$$

(*b*) What is the cyclotron frequency? From Eq. 33-16,

$$\nu = \frac{qB}{2\pi m} = \frac{(1.6 \times 10^{-19} \text{ C})(1.0 \times 10^{-4} \text{ T})}{(2\pi)(9.1 \times 10^{-31} \text{ kg})} = 2.8 \times 10^6 \text{ Hz}.$$

(*c*) What is the period of revolution T?

$$T = \frac{1}{\nu} = \frac{1}{2.8 \times 10^6 \text{ Hz}} = 3.6 \times 10^{-7} \text{ s}.$$

Thus an electron requires 0.36 μs to make one revolution in a 1.0-gauss field. (d) What is the direction of circulation as viewed by an observer sighting along the field?

In Fig. 33-11 the magnetic force $q\mathbf{v} \times \mathbf{B}$ must point radially inward, since it provides the centripetal force. Since $\mathbf{B}$ points into the plane of the paper, $\mathbf{v}$ would have to point to the left at the position shown in the figure if the charge q were positive. However, the charge is an electron, with $q = -e$, which means that $\mathbf{v}$ must point to the right. Thus the charge circulates clockwise as viewed by an observer sighting in the direction of $\mathbf{B}$.

The cyclotron, first made operational in 1932 by Ernest O. Lawrence at the University of California at Berkeley, accelerates charged particles such as hydrogen nuclei (protons) and heavy hydrogen nuclei (deuterons) to high energies so that they can be used in atom-smashing experiments.* Figure 33-12 shows a cyclotron formerly operated at the University of Pittsburgh. Although conventional cyclotrons of this type are no longer used, we discuss them for two reasons: (a) they provide an excellent framework within which to discuss the action of magnetic and electric fields on charged particles and (b) the cyclotron has led to the production of several generations of improved accelerators, the proton synchrotron being one. These later devices provide even more opportunities for studying the interactions of charged particles with magnetic and electric fields although that, of course, is not their primary purpose.

33-7
CYCLOTRONS AND SYNCHROTRONS

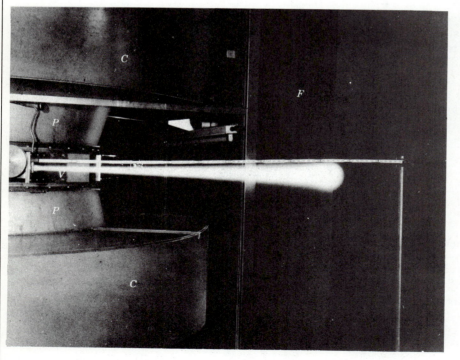

figure 33-12
The former University of Pittsburgh cyclotron. Note vacuum chamber V, magnet frame F, magnetic pole faces P, magnet coils C, and the deuteron beam emerging into the air of the laboratory through an aluminum foil "window." The rule is 6 ft. long. (Courtesy A. J. Allen.)

In an *ion source* at the center of the cyclotron molecules of deuterium (heavy hydrogen) are bombarded with electrons whose energy is high enough (say 100 eV) so that many positive ions are formed during the collisions. Many of these ions are free deuterons, which enter the cyclo-

* Lawrence received a Nobel prize in 1939 for this work.

tron proper through a small hole in the wall of the ion source and are available to be accelerated.

The cyclotron uses a modest potential difference for accelerating (say 10^5 V), but it requires the ion to pass through this potential difference a number of times. To reach 10 MeV with 10^5 V accelerating potential requires 100 passages. A magnetic field is used to bend the ions around so that they may pass again and again through the same accelerating potential.

Figure 33-13 is a top view of the part of the cyclotron that is inside the vacuum tank marked V in Fig. 33-12. The two D-shaped objects, called *dees*, are made of copper sheet and form part of an electric oscillator which establishes an accelerating potential difference across the gap between the dees. The direction of this potential difference is made to change sign some millions of times per second.

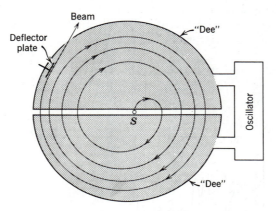

figure 33-13
The elements of a cyclotron showing the ion source S and the dees. The deflector plate, held at a suitable negative potential, deflects the particles out of the dee system.

The dees are immersed in a magnetic field ($B \cong 1.6$ T) whose direction is out of the plane of Fig. 33-13. The field is set up by a large electromagnet, part of which is shown by F in Fig. 33-12. Finally, the space in which the ions move is evacuated to a pressure of about 10^{-6} mm-Hg. If this were not done, the ions would continually collide with air molecules.

Suppose that a deuteron, emerging from the ion source, finds the dee that it is facing to be negative; it will accelerate toward this dee and will enter it. Once inside, it is screened from electric fields by the metal walls of the dees. The magnetic field is not screened by the dees so that the ion bends in a circular path whose radius, which depends on the velocity, is given by Eq. 33-15, or

$$r = \frac{mv}{qB}.$$

After a time t_0 the ion emerges from the dee on the other side of the ion source. Let us assume that the accelerating potential has now changed sign. Thus the ion *again* faces a negative dee, is further accelerated, and again describes a semicircle, of somewhat larger radius (see Eq. 33-15), in the dee. *The time of passage through this dee, however, is still t_0.* This follows because the period of revolution T of an ion circulating in a magnetic field does not depend on the speed of the ion; see Eq. 33-16. This process goes on until the ion reaches the outer edge of one dee where it is pulled out of the system by a negatively charged deflector plate.

The key to the operation of the cyclotron is that the characteristic frequency ν at which the ion circulates in the field must be equal to the fixed frequency ν_0 of the electric oscillator, or

$$\nu = \nu_0.$$

The *resonance condition* says that if the energy of the circulating ion is to increase, energy must be fed to it at a frequency ν_0 that is equal to the natural frequency ν at which the ion circulates in the field. In the same way we feed energy to a swing by pushing it at a frequency equal to the natural frequency of oscillation of the swing.

From Eq. 33-16 ($\nu = qB/2\pi m$), we can rewrite the resonance equation as

$$\frac{qB}{2\pi m} = \nu_0. \tag{33-17}$$

Once we have selected an ion to be accelerated, q/m is fixed; usually the oscillator is designed to work at a single frequency ν_0. We then "tune" the cyclotron by varying B until Eq. 33-17 is satisfied and an accelerated beam appears.

The *energy* of the particles produced in the cyclotron depends on the radius R of the dees. From Eq. 33-15 ($r = mv/qB$), the velocity of a particle circulating at this radius is given by

$$v = \frac{qBR}{m}.$$

The kinetic energy is then

$$K = \tfrac{1}{2}mv^2 = \frac{q^2B^2R^2}{2m}. \tag{33-18}$$

EXAMPLE 7

The University of Pittsburgh cyclotron had an oscillator frequency of 12×10^6 Hz and a dee radius of 53 cm ($= 21$ in.). (*a*) What value of B is needed to accelerate deuterons?

From Eq. 33-17, $\nu_0 = qB/2\pi m$, so that

$$B = \frac{2\pi\nu_0 m}{q} = \frac{(2\pi)(12 \times 10^6 \text{ Hz})(3.3 \times 10^{-27} \text{ kg})}{1.6 \times 10^{-19} \text{ C}} = 1.6 \text{ T}.$$

Note that the deuteron has the same charge as the proton but (very closely) twice the mass. (*b*) What deuteron energy results?

From Eq. 33-18,

$$K = \frac{q^2B^2R^2}{2m} = \frac{(1.6 \times 10^{-19} \text{ C})^2(1.6 \text{ T})^2(0.53 \text{ m})^2}{(2)(3.3 \times 10^{-27} \text{ kg})}$$

$$= (2.8 \times 10^{-12} \text{ J})\left(\frac{1 \text{ eV}}{1.6 \times 10^{-19} \text{ J}}\right) = 17 \text{ MeV}.$$

There are two reasons why the conventional cyclotron that we have described runs into difficulties at high energies. One deals with physics, the other with cost. Let us discuss each in turn.

(*a*) The cyclotron fails to operate at high energies because one of its assumptions, that the frequency of rotation of an ion circulating in a magnetic field is independent of its speed, is true only for speeds much less than that of light. As the particle speed increases, we must use the *relativistic mass m* in Eq. 33-16.

The relativistic mass increases with velocity (Eq. 8-20) so that at high enough speeds ν decreases with velocity. Thus the ions get out of step with the electric oscillator, and eventually the energy of the circulating ion stops increasing.

(b) The second difficulty associated with the acceleration of particles to high energies is that the size of the magnet that would be required to guide such particles in a circular path is very large. For a 30-GeV proton, for example, in a field of 1.5 T (= 15,000 gauss) the radius of curvature is 65 m. A magnet of the cyclotron type of this size (about 430 ft in diameter) would be prohibitively expensive. Incidentally, a 30-GeV proton has a speed equal to 0.99998 that of light.

Both the relativistic and the economic limitations have been removed by techniques that can be understood in terms of Eq. 33-17 ($2\pi\nu_0 m = qB$) in which m is now taken to be the relativistic mass, given by Eq. 8-20, or

$$m = \frac{m_0}{\sqrt{1 - (v/c)^2}},$$

v being the speed of the particle and c being that of light.

As the particle speed increases, the relativistic mass m also increases. To maintain the equality in Eq. 33-17, and thus insure resonance, one may decrease the oscillator frequency ν_0 as the particle (assumed to be a proton) accelerates in such a way that the product $\nu_0 m$ remains constant. Accelerators that use this technique have been called *synchrocyclotrons*.

To ameliorate the magnet cost limitation one can vary *both* B and ν_0 in a cyclic fashion in such a way that not only is Eq. 33-17 satisfied at all times but the particle orbit radius remains essentially constant during the acceleration process. This permits the use of an annular (or ring-shaped) magnet, rather than the conventional cyclotron type, at great saving in cost. With the *two* variables B and ν_0 at our disposal, it is possible to preserve *two* equalities during the acceleration process, one being Eq. 33-17 and the other being the relation

$$v = \omega_0 R_0 = (2\pi\nu_0)R_0$$

in which R_0 is the desired (fixed) orbit radius. Accelerators that use this technique are called *synchrotrons*. Table 33-2 shows some characteristics of the accelerator built at the Brookhaven National Laboratory, embodying these principles.

Table 33-2
The Brookhaven Proton Synchrotron

Maximum proton energy	33 GeV
Mean orbit radius	128 m
Maximum orbit field	1.3 T
Injection energy	50 MeV
Pulse repetition rate	2.4 Hz
Beam aperture	18 cm × 8 cm
Total weight of magnets	4000 tons

Note that even the energy at which the protons are *injected* into this accelerator (50 MeV) far exceeds the capabilities of a conventional cyclotron.*

As of this writing (1977) the protron synchrotron producing the most energetic protons (500 GeV with 1000 GeV as a goal) is located in Batavia, Illinois.† The ring, of 954 separate magnets, is 6.3 km (= 4.1 miles) in circumference!

The system for injecting protons into this ring is impressive in itself. The

* See, for example, "Introduction to Nuclear Physics" by Harald Enge, John Wiley & Sons, 1966, for a readable account with much more information about particle accelerators than we can present here.
† See "The Batavia Accelerator" by R. R. Wilson, *Scientific American*, February 1974.

protons are first accelerated to 750 keV by a transformer-rectifier arrangement. They are then directed into a 145 m-long linear accelerator from which they emerge with an energy of 200 MeV. The protons are then led into a "medium size" proton synchrotron (the "booster") from which they emerge with an energy of 80 GeV. Only then are they injected into the main accelerator ring.

In all of these processes magnetic and electric fields not only accelerate the protons, they direct them in desired directions so that they may be used for experiments; above all, they focus them so that a well-defined proton beam emerges after as much as 10^6 miles of travel through the accelerator complex. Although this is not its purpose, no better "laboratory" for demonstrating the action of magnetic and electric fields on charged particles has yet been devised. As an indication of its size and scope, we point out that its yearly budget for electric energy alone is several million (1976) dollars.

33-8

THE DISCOVERY OF THE ELECTRON

This crucial experiment, performed in 1897 by J. J. Thomson in the Cavendish Laboratory in Cambridge, England, was a measurement of the ratio of charge e to mass m of the electron by observing its deflection in combined magnetic and electric fields. It amounted to the discovery of the electron as a fundamental particle and we discuss it here as another practical example of the action of magnetic and electric fields on charged particles.

In Fig. 33-14, which is a modernized version of Thomson's apparatus, electrons are emitted from hot filament F and accelerated by an applied potential difference V. They then enter a region in which they move at right angles to an electric field $\mathbf{E}$ and a magnetic field $\mathbf{B}$; $\mathbf{E}$ and $\mathbf{B}$ are right angles to each other. The beam is made visible as a spot of light when it strikes fluorescent screen S. The entire region in which the electrons move is highly evacuated so that collisions with air molecules will not occur.

The resultant force on a charged particle moving through an electric and a magnetic field is given by Eq. 33-4, or

$$\mathbf{F} = q_0\mathbf{E} + q_0\mathbf{v} \times \mathbf{B}.$$

Study of Fig. 33-14 shows that the electric field deflects the particle upward and the magnetic field deflects it downward. If these deflecting forces are to cancel (that is, if $\mathbf{F} = 0$), this equation, for this problem, reduces to

$$eE = evB$$

or
$$E = vB. \tag{33-19}$$

Thus for a given electron speed v the condition for zero deflection can be satisfied by adjusting E or B.

Thomson's procedure was (a) to note the position of the undeflected beam spot, with $\mathbf{E}$ and $\mathbf{B}$ both equal to zero; (b) to apply a fixed electric field $\mathbf{E}$, measuring on the fluorescent screen the deflection so caused; and (c) to apply a magnetic field and adjust its value until the beam deflection is restored to zero.

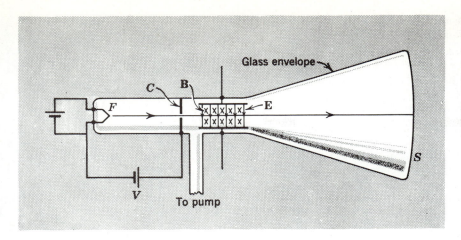

figure 33-14
Electrons from the heated filament F are accelerated by a potential difference V and pass through a hole in the screen C. After passing through a region in which perpendicular electric and magnetic fields are present, they strike the fluorescent screen S.

In Section 27-5 we saw that the deflection y of an electron in a purely electric field (step b), measured at the far edge of the deflecting plates, is given by Eq. 27-9, or, with small changes in notation,

$$y = \frac{eEl^2}{2mv^2},$$

where v is the electron speed and l is the length of the deflecting plates; y is not measurable directly, but it may be calculated from the measured displacement of the spot on the screen if the geometry of the apparatus is known. Thus y, E, and l are known; the ratio e/m and the velocity v are unknown. We cannot calculate e/m until we have found the velocity, which is the purpose of step c above.

If (step c) the electric force is set equal and opposite to the magnetic force, the net force is zero and we can write (Eq. 33-19)

$$v = \frac{E}{B}.$$

Substituting this equation into the equation for y and solving for the ratio e/m leads to

$$\frac{e}{m} = \frac{2yE}{B^2l^2}, \tag{33-20}$$

in which all the quantities on the right can be measured. Thomson's value for e/m was 1.7×10^{11} C/kg, in full agreement with the 1977 value of 1.758805×10^{11} C/kg.

questions

1. In a letter to the editor of *Sky and Telescope*, August 1976, Cicely M. Botley says, in part, "The [geomagnetic] pole in northern Canada is more correctly the *dip pole*, where a freely suspended magnetic needle is vertical. The *geomagnetic pole*, from which geomagnetic latitudes are measured, is in northwest Greenland." Discuss these two concepts.
2. Of the three vectors in the equation $\mathbf{F} = q\mathbf{v} \times \mathbf{B}$, which pairs are always at right angles? Which may have any angle between them?
3. Why do we not simply define the magnetic field $\mathbf{B}$ to point in the direction of the magnetic force that acts on the moving charge?
4. Imagine that you are sitting in a room with your back to one wall and that an electron beam, traveling horizontally from the back wall toward the front wall, is deflected to your right. What is the direction of the magnetic field that exists in the room?

5. If an electron is not deflected in passing through a certain region of space, can we be sure that there is no magnetic field in that region?

6. If a moving electron is deflected sideways in passing through a certain region of space, can we be sure that a magnetic field exists in that region?

7. A beam of protons is deflected sideways. Could this deflection be caused (a) by an electric field? (b) By a magnetic field? (c) If either could be responsible, how would you be able to tell which was present?

8. A conductor, even though it is carrying a current, has zero net charge. Why, then, does a magnetic field exert a force on it?

9. In Example 2 (see Fig. 33-6) we saw that the magnetic force was the same as if the semicircular arc had been replaced by a straight wire of length $2R$. Would this same conclusion hold if we replaced the semicircular arc by a curve of irregular shape? Give a specific example to prove your point, one way or the other.

10. Does Eq. 33-6a ($\mathbf{F} = i\mathbf{l} \times \mathbf{B}$) hold for a straight wire whose cross section varies irregularly along its length (a "lumpy" wire)?

11. A straight copper wire carrying a current i is immersed in a magnetic field $\mathbf{B}$, at right angles to it. We know that $\mathbf{B}$ exerts a sideways force on the free (or conduction) electrons. Does it do so on the bound electrons? After all, they are not at rest. Discuss.

12. In Section 33-3 we state that a magnetic field $\mathbf{B}$ exerts a sideways force on the conduction electrons in, say, a copper wire carrying a current i. We have tacitly assumed that this same force acts on the conductor itself. Are there some missing steps in this argument?

13. Equation 33-11 ($\boldsymbol{\tau} = \boldsymbol{\mu} \times \mathbf{B}$) shows that there is no torque on a current loop in an external magnetic field if the angle between the axis of the loop and the field is (a) 0° or (b) 180°. Discuss the nature of the equilibrium (that is, is it stable, neutral, or unstable?) for these two positions.

14. In Example 4 we showed that the work required to turn a current loop end-for-end in an external magnetic field is $2\mu B$. Does this hold no matter what the original orientation of the loop was?

15. Imagine that the room in which you are seated is filled with a uniform magnetic field with $\mathbf{B}$ pointing vertically upward. A circular loop of wire has its plane horizontal. For what direction of current in the loop, as viewed from above, will the loop be in stable equilibrium with respect to forces and torques of magnetic origin?

16. A rectangular current loop is in an arbitrary orientation in an external magnetic field. Is any work required to rotate the loop about an axis perpendicular to its plane?

17. You wish to modify a galvanometer (see Example 3) to make it into (a) an ammeter and (b) a voltmeter. What do you need to do in each case?

18. (a) In measuring Hall potential differences, why must we be careful that points x and y in Fig. 33-10 are exactly opposite each other? (b) If one of the contacts is movable, what procedure might we follow in adjusting it to make sure that the two points are properly located?

19. A uniform magnetic field fills a certain cubical region of space. Can an electron be fired into this cube from the outside in such a way that it will travel in a closed circular path inside the cube?

20. Imagine the room in which you are seated to be filled with a uniform magnetic field with $\mathbf{B}$ pointing vertically downward. At the center of the room two electrons are suddenly projected horizontally with the same speed but in opposite directions. (a) Discuss their motions. (b) Discuss their motions if one particle is an electron and one a positron, that is, a positively charged electron.

21. In Fig. 33-4 why are the low-energy electron tracks spirals? That is, why does the radius of curvature change in the constant magnetic field in which the chamber is immersed?

22. What are the primary functions of (*a*) the electric field and (*b*) the magnetic field in the cyclotron?

23. What central fact makes the operation of a conventional cyclotron possible? Ignore relativistic considerations.

24. For Thomson's *e/m* experiment to work properly (Section 33-8), is it essential that the electrons have a fairly constant speed?

25. The arrangement of crossed electric and magnetic fields shown in the central part of Fig. 33-14 is sometimes called a *velocity filter*. How can this name be justified?

SECTION 33-2

1. Particles 1, 2, and 3 follow the paths shown in Fig. 33-15 as they pass through the magnetic field there. What can one conclude about each particle?
   Answer: Particle 1 is positive, particle 2 is neutral, particle 3 is negative.

2. The electrons in the beam of a television tube have an energy of 12 keV. The tube is oriented so that the electrons move horizontally from south to north. The earth's magnetic field points down and has $B = 5.5 \times 10^{-5}$ T. (*a*) In what direction will the beam deflect? (*b*) What is the acceleration of a given electron? (*c*) How far will the beam deflect in moving 20 cm through the television tube?

3. An electron has a velocity given in m/s by $\mathbf{v} = 2.0 \times 10^6 \mathbf{i} + 3.0 \times 10^6 \mathbf{j}$. It enters a magnetic field given in T by $\mathbf{B} = 0.03\mathbf{i} - 0.15\mathbf{j}$. (*a*) Find the magnitude and direction of the force on the electron. (*b*) Repeat your calculation for a deuteron having the same velocity.
   Answer: (*a*) 6.2×10^{-14} **k**, N. (*b*) -6.2×10^{-14} **k**, N.

4. A beam of electrons whose kinetic energy is K emerges from a "window" at the end of an accelerator tube. There is a metal plate a distance d from this window and at right angles to the direction of the emerging beam. Show that we can stop the beam from hitting the plate if we apply a magnetic field B such that

$$B \geq \left(\frac{2mK}{e^2 d^2}\right)^{1/2},$$

in which m and e are the electron mass and charge. How should **B** be oriented?

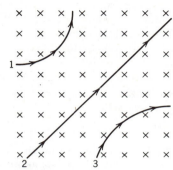

figure 33-15
Problem 1

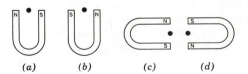

(*a*)   (*b*)   (*c*)   (*d*)

figure 33-16
Problem 5

SECTION 33-3

5. Figure 33-16 shows a magnet and a straight wire in which a current of electrons is flowing out of the page at right angles to it. In which case will there be a force on the wire that points toward the top of the page?   *Answer:* (*b*).

6. A metal wire of mass m slides without friction on two horizontal rails spaced a distance d apart, as in Fig. 33-17. The track lies in a vertical uniform magnetic field **B**. A *constant current i* flows from generator G along one rail, across the wire, and back down the other rail. Find the velocity (speed and direction) of the wire as a function of time, assuming it to be at rest at $t = 0$.

7. A wire 1.0 m long carries a current of 10 A and makes an angle of 30° with a uniform magnetic field with $B = 1.5$ T. Calculate the magnitude and direction of the force on the wire.
   Answer: 7.5 N, perpendicular to both the wire and to **B**.

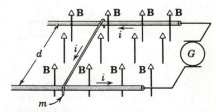

figure 33-17
Problem 6

problems

8. A wire 50 cm long lying along the *x*-axis carries a current of 0.50 A in the positive *x*-direction. A magnetic field is present that is given in T by $\mathbf{B} = 0.0030\mathbf{j} + 0.010\mathbf{k}$. Find the components of the force on the wire.

9. A wire of 60 cm length and mass 10 g is suspended by a pair of flexible leads in a magnetic field of 0.40 T. What are the magnitude and direction of the current required to remove the tension in the supporting leads? See Fig. 33-18. *Answer:* 0.41 A, from left to right.

10. An electron in a uniform magnetic field **B** has a velocity $\mathbf{v} = 4.0 \times 10^5\mathbf{i} + 7.1 \times 10^5\mathbf{j}$, in m/s. It experiences a force $\mathbf{F} = -2.7 \times 10^{-13}\mathbf{i} + 1.5 \times 10^{-13}\mathbf{j}$, in N. If $B_x = 0$, find the magnetic field.

11. Consider the possibility of a new design for an electric train. The engine is driven by the force due to the vertical component of the earth's magnetic field on a conducting axle. Current is passed down one rail, into a conducting wheel, through the axle, through another conducting wheel, and then back to the source via the other rail. (*a*) What current is needed to provide a modest 10,000 N force? Take the vertical component of **B** to be 10^{-5} T and the length of the axle to be 3.0 m. (*b*) How much power would be lost for each ohm of resistance in the rails? (*c*) Is such a train totally unrealistic or marginally unrealistic?
Answer: (*a*) 3.3×10^9 A. (*b*) 1.0×10^{17} W. (*c*) Totally unrealistic.

12. A U-shaped wire of mass *m* and length *l* is immersed with its two ends in mercury (Fig. 33-19). The wire is in a homogeneous magnetic field **B**. If a charge, that is, a current pulse $q = \int i \, dt$, is sent through the wire, the wire will jump up. Calculate, from the height *h* that the wire reaches, the size of the charge or current pulse, assuming that the time of the current pulse is very small in comparison with the time of flight. Make use of the fact that impulse of force equals $\int F \, dt$, which equals *mv*. (Hint: Try to relate $\int i \, dt$ to $\int F \, dt$.) Evaluate *q* for $B = 0.10$ T, $m = 10$ g, $l = 20$ cm, and $h = 3$ m.

13. Figure 33-20 shows a wire of arbitrary shape carrying a current *i* between points *a* and *b*. The wire lies in a plane at right angles to a uniform magnetic field **B**. Prove that the force on the wire is the same as that on a straight wire carrying a current *i* directly from *a* to *b*. (Hint: Replace the wire by a series of "steps" parallel and perpendicular to the straight line joining *a* and *b*.)

14. Figure 33-21 shows a wire ring of radius *a* at right angles to the general direction of a radially symmetric diverging magnetic field. The magnetic field at the ring is everywhere of the same magnitude *B*, and its direction at the ring is everywhere at an angle θ with a normal to the plane of the ring. The twisted lead wires have no effect on the problem. Find the magnitude and direction of the force the field exerts on the ring if the ring carries a current *i* as shown in the figure.

SECTION 33-4

15. An *N*-turn circular coil of radius *R* is suspended in a uniform magnetic field **B** that points vertically downward. The coil can rotate about a horizontal axis through its center. A mass *m* hangs by a string from the bottom of the coil. When a current *i* is put through the coil it eventually assumes an equilibrium position, in which the perpendicular to the plane of the coil makes an angle ϕ with the direction of **B**. Find ϕ and draw a sketch of this equilibrium position. Take $B = 0.50$ T, $R = 10$ cm, $N = 10$, $m = 500$ g, and $i = 1.0$ A. *Answer:* $\phi = 72°$.

16. Figure 33-22 shows a wooden cylinder with a mass *m* of 0.25 kg, a radius *R*, and a length *l* of 0.1 m with *N* equal to ten turns of wire wrapped around it longitudinally, so that the plane of the wire loop contains the axis of the cylinder. What is the least current through the loop that will prevent the cylinder from rolling down an inclined plane whose surface is inclined at

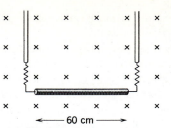

figure 33-18
Problem 9

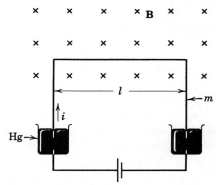

figure 33-19
Problem 12

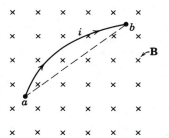

figure 33-20
Problem 13

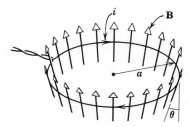

figure 33-21
Problem 14

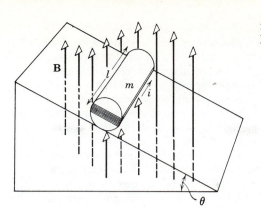

an angle θ to the horizontal, in the presence of a vertical field of magnetic induction 0.5 T, if the plane of the windings is parallel to the inclined plane?

17. A certain galvanometer has a resistance of 75.3 Ω; its needle experiences a full-scale deflection when a current 1.62×10^{-3} A passes through its coil. (*a*) Determine the value of the auxiliary resistance required to convert the galvanometer into a voltmeter that reads 1.000 V at full-scale deflection. How is it to be connected? (*b*) Determine the value of the auxiliary resistance required to convert the galvanometer into an ammeter that reads 0.0500 A at full-scale deflection. How is it to be connected?
Answer: (*a*) 540 Ω, connected in series. (*b*) 2.52 Ω, connected in parallel.

18. A circular loop of wire having a radius of 8.0 cm carries a current of 0.20 A. A unit vector parallel to the dipole moment $\boldsymbol{\mu}$ of the loop is given in A · m² by $0.60\mathbf{i} - 0.80\mathbf{j}$. If the loop is located in a magnetic field given in T by $\mathbf{B} = 0.25\mathbf{i} + 0.30\mathbf{k}$, find (*a*) the magnitude and direction of the torque on the loop and (*b*) the magnetic potential energy of the loop. Assume the same zero-energy configuration that we assumed in Section 33-4.

19. Figure 33-23 shows a rectangular twenty-turn loop of wire, 10 cm by 5.0 cm. It carries a current of 0.10 A and is hinged at one side. What torque (direction and magnitude) acts on the loop if it is mounted with its plane at an angle of 30° to the direction of a uniform magnetic field of 0.50 T?
Answer: 4.3×10^{-3} N·m. The torque vector is parallel to the long side of the coil and points down.

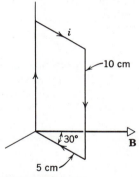

figure 33-23
Problem 19

20. Prove that the relation $\tau = NAiB \sin \theta$ holds for closed loops of arbitrary shape and not only for rectangular loops as in Fig. 33-8. (Hint: Replace the loop of arbitrary shape by an assembly of adjacent long, thin — approximately rectangular — loops which are equivalent to it as far as the distribution of current is concerned.)

21. A length l of wire carries a current i. Show that if the wire is formed into a circular coil, the maximum torque in a given magnetic field is developed when the coil has *one* turn only and the maximum torque has the value

$$\tau = \frac{1}{4\pi} \, l^2 iB.$$

SECTION 33-5
22. In a Hall effect experiment a current of 3.0 A lengthwise in a conductor 1.0 cm wide, 4.0 cm long, and 10^{-3} cm thick produced a transverse Hall voltage (across the width) of 1.0×10^{-5} V when a magnetic field of 1.5 T passed perpendicularly through the thin conductor. From these data, find (*a*) the drift velocity of the charge carriers and (*b*) the number of carriers per cubic centimeter. (*c*) Show on a diagram the polarity of the Hall voltage with a given current and magnetic field direction, assuming the charge carriers are (negative) electrons.

23. A current i, indicated by the crosses in Fig. 33-24, is established in a strip of copper of height h and width w. A uniform magnetic field **B** is applied at right angles to the strip. (a) Calculate the drift speed v_d for the electrons. (b) What are the magnitude and direction of the magnetic force **F** acting on the electrons? (c) What would the magnitude and direction of a homogeneous electric field **E** have to be in order to counter-balance the effect of the magnetic field? (d) What is the voltage V necessary between two sides of the conductor in order to create this field **E**? Between which sides of the conductor would this voltage have to be applied? (e) If no electric field is applied *from the outside*, the electrons will be pushed somewhat to one side and therefore will give rise to a uniform Hall electric field E_H across the conductor until the forces of this electrostatic field E_H balance the magnetic forces encountered in part (b). What will be the magnitude and direction of the field E_H? Assume that n, the number of conduction electrons per unit volume, is $1.1 \times 10^{29}/m^3$ and that $h = 0.020$ m, $w = 0.10$ cm, $i = 50$ A, and $B = 2.0$ T.

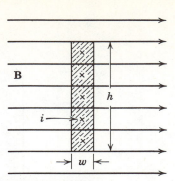

figure 33-24
Problem 23

Answer: (a) 1.4×10^{-4} m/s. (b) 4.5×10^{-23} N; down. (c) 2.8×10^{-4} V/m; down. (d) 5.7×10^{-6} V; top +, bottom −. (e) Same as (c).

24. (a) Show that the ratio of the Hall electric field E_H to the electric field E responsible for the current is

$$\frac{E_H}{E} = \frac{B}{ne\rho},$$

where ρ is the resistivity of the material.

(b) What is the angle between E_H and E? (c) Evaluate this ratio for the conditions of Example 5.

SECTION 33-6

25. A proton, a deuteron, and an α-particle, accelerated through the same potential difference, enter a region of uniform magnetic field, moving at right angles to **B**. (a) Compare their kinetic energies. (b) If the radius of the proton's circular path is 10 cm, what are the radii of the deuteron and the α-particle paths?   *Answer:* (a) $K_p = K_d = \frac{1}{2}K_\alpha$. (b) $R_d = R_\alpha = 14$ cm.

26. A proton, a deuteron, and an α-particle with the same kinetic energies enter a region of uniform magnetic field, moving at right angles to **B**. Compare the radii of their circular paths.

27. An electron is accelerated through 15,000 V and is then allowed to circulate at right angles to a uniform magnetic field with $B = 250$ gauss; 10^4 gauss $=$ 1 T. What is its path radius?   *Answer:* 1.7 cm.

28. In a nuclear experiment a 1.0-MeV proton moves in a uniform magnetic field in a circular path. What energy must (a) an alpha particle and (b) a deuteron have if they are to circulate in the same orbit?

29. (a) In a magnetic field with $B = 0.50$ T, for what path radius will an electron circulate at 0.10 the speed of light? (b) What will its kinetic energy be? *Answer:* (a) 0.34 mm. (b) 2.6 keV.

30. (a) What speed would a proton need to circle the earth at the equator, if the earth's magnetic field is everywhere horizontal there and directed along longitudinal lines? Take the magnitude of the earth's magnetic field to be 0.41×10^{-4} T at the equator. (b) Draw the velocity and magnetic field vectors corresponding to this situation.

31. What uniform magnetic field can be set up in space to permit a proton of speed 1.0×10^7 m/s to move in a circle the size of the earth's equator? *Answer:* 1.6×10^{-8} T.

32. An α-particle travels in a circular path of radius 0.45 m in a magnetic field with $B = 1.2$ T. Calculate (a) its speed, (b) its period of revolution, (c) its kinetic energy, and (d) the potential difference through which it would have to be accelerated to achieve this energy.

33. A neutral particle, viewed in a reference frame in which it is at rest, lies in a homogeneous magnetic field of magnitude B. At time $t = 0$ it decays into two charged particles each of mass m. (a) If the charge of one of the particles is $+q$, what is the charge of the other? The two particles move off in separate paths both of which lie in the plane perpendicular to **B**. (b) At a later time the particles collide. Express the time from decay until collision, t, in terms of m, B, and q.   *Answer:* (a) $-q$. (b) $t = \pi m/qB$.

34. Singly ionized chlorine atoms of 35 u (= unified atomic mass units) and 37 u, traveling with speed 2.0×10^5 m/s, enter perpendicularly a uniform magnetic field of 0.50 T. After bending through 180° the atoms strike a photographic film. What is the separation distance between the two spots on the film? $(1.00 \text{ u} = 1.67 \times 10^{-27} \text{ kg.})$

35. Show that the radius of curvature of a charged particle moving at right angles to a magnetic field is proportional to its momentum.

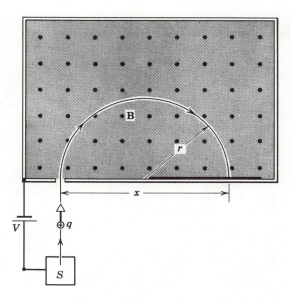

figure 33-25
Problem 36.

36. *Mass spectrometer.* Figure 33-25 shows an arrangement used by Dempster to measure the masses of ions. An ion of mass M and charge $+q$ is produced essentially at rest in source S, a chamber in which a gas discharge is taking place. The ion is accelerated by potential difference V and allowed to enter a magnetic field **B**. In the field it moves in a semicircle, striking a photographic plate at distance x from the entry slit is recorded. Show that the mass M is given by

$$M = \frac{B^2 q}{8V} x^2.$$

37. Two types of singly ionized atoms having the same charge q, and mass differing by a small amount ΔM are introduced into the mass spectrometer described in Problem 36. (a) Calculate the difference in mass in terms of V, q, M (of either), B, and the distance Δx between the spots on the photographic plate. (b) Calculate Δx for a beam of singly ionized chlorine atoms of masses 35 and 37 u if $V = 7.3 \times 10^3$ V and $B = 0.50$ T $(1.00 \text{ u} = 1.67 \times 10^{-27} \text{ kg})$.

Answer: (a) $B \left(\dfrac{mq}{2V} \right)^{1/2} \Delta x.$ (b) 8.2 mm.

figure 33-26
Problem 38.

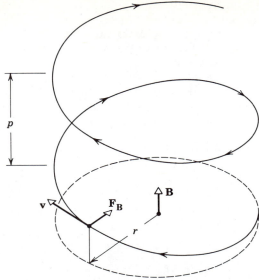

38. A 2-keV positron is projected into a uniform magnetic field **B** of 0.10 T with its velocity vector making an angle of 89° with **B**. (a) Convince yourself that the path will be a helix, its axis being the direction of **B**. Find (b) the period, (c) the pitch p, and (d) the radius r of the helix; see Fig. 33-26.

39. *Time-of-flight spectrometer.* S. A. Goudsmit has devised a method for measuring accurately the masses of heavy ions by timing their period of circulation in a known magnetic field. A singly charged ion of iodine makes 7.00 rev in a field of 4.50×10^{-2} T in about 1.29×10^{-3} s. What (approximately) is its mass in kilograms? Actually, the mass measurements are carried out to much greater accuracy than these approximate data suggest.
Answer: 2.11×10^{-25} kg, or about 127 proton masses.

40. *Zeeman effect.* In Bohr's theory of the hydrogen atom the electron can be thought of as moving in a circular orbit of radius r about the proton. Suppose that such an atom is placed in a magnetic field, with the plane of the orbit at right angles to **B**. (a) If the electron is circulating clockwise, as viewed by an observer sighting along **B**, will the angular frequency increase or decrease? (b) What if the electron is circulating counterclockwise? Assume that the orbit radius does not change. [Hint: The centripetal force is now partially electric ($\mathbf{F}_E$) and partially magnetic ($\mathbf{F}_B$) in origin.]

41. In Problem 40 show that the change in frequency of rotation caused by the magnetic field is given approximately by

$$\Delta \nu = \pm \frac{Be}{4\pi m}.$$

Such frequency shifts were observed by Zeeman in 1896. (Hint: Calculate the frequency of rotation without the magnetic field and also with it. Subtract, bearing in mind that because the effect of the magnetic field is very small, some—but not all—terms containing B can be set equal to zero with little error.)

42. (a) What is the cyclotron frequency of an electron with an energy of 100 eV in the earth's magnetic field of 1.0×10^{-4} T? (b) What is the radius of curvature of the path of this electron if its velocity is perpendicular to the magnetic field?

43. A 10-KeV electron is circulating in a plane at right angles to a uniform magnetic field. The orbit radius is 25 cm. Find (a) the magnetic field, (b) the cyclotron frequency, and (c) the period of the motion.
Answer: (a) 1.4×10^{-3} T. (b) 3.8×10^7 Hz. (c) 2.6×10^{-8} s.

SECTION 33-7

44. The cyclotron shown in Fig. 33-12 was normally adjusted to accelerate deuterons. (a) What energy of protons could it produce, using the same oscillator frequency as that used for deuterons? (b) What magnetic field would be required? (c) What energy of protons could be produced if the magnetic field was left at the value used for deuterons? (d) What oscillator frequency would then be required? (e) Answer the same questions for α-particles.

45. A deuteron in a large cyclotron is moving in a magnetic field with $B = 1.5$ T and an orbit radius of 2.0 m. Because of a grazing collision with a target, the deuteron breaks up, with a negligible loss of kinetic energy, into a proton and a neutron. Discuss the subsequent motions of each. Assume that the deuteron energy is shared equally by the proton and neutron at breakup.
    Answer: Neutron moves tangent to the original path. Proton moves in a circular orbit of 1.0-m radius.

46. In a certain cyclotron a proton moves in a circle of radius $r = 0.50$ m. The magnitude of the **B** field is 1.2 T. (a) What is the cyclotron frequency? (b) What is the kinetic energy of the proton?

47. Estimate the total path length traversed by a deuteron in the cyclotron shown in Fig. 33-12 during the acceleration process. Assume an accelerating potential between the dees of 80,000 V.   *Answer:* About 240 m.

SECTION 33-8

48. An electric field of 1500 V/m and a magnetic field of 0.40 T act on a moving electron to produce no force. (a) Calculate the minimum electron speed v. (b) Draw the vectors **E**, **B**, and **v**.

49. An electron is accelerated through a potential difference of 1000 V and directed into a region between two parallel plates separated by 0.02 m with a potential difference of 100 V between them. If the electron enters moving perpendicular to the electric field between the plates, what magnetic field is necessary perpendicular to both the electron path and the electric field so that the electron travels in a straight line?
    Answer: 2.7×10^{-4} T.

50. A 10-KeV electron moving horizontally enters a region of space in which there is a downward-directed electric field of magnitude 100 V/cm. (a) What are the magnitude and direction of the (smallest) magnetic field that will allow the electron to continue to move horizontally? Ignore the gravitational force, which is rather small. (b) Is it possible for a proton to pass through this combination of fields undeflected? If so, under what circumstances?

51. A positive point charge Q travels in a straight line with constant speed through an evacuated region in which there is a uniform electric field **E** and a uniform magnetic field **B**. (a) If **E** is directed vertically up and the charge travels horizontally from north to south with speed v, determine the least value of the magnitude of **B** and the corresponding direction of **B**. (b) Explain why **B** is not uniquely determined when **E** and **v** alone are given. (c) Suppose the charge is a proton which enters the region after having been accelerated through a potential difference of 3.10×10^5 V. If $E = 1.90 \times 10^5$ V/m, compute the value of B in part (a). (d) If in part (c) the electric field **E** is turned off, determine the radius r of the circle in which the proton now moves.
    Answer: (a) $B = E/v$; from east to west. (c) 2.47×10^{-2} T. (d) 3.26 m.

34
ampère's law

One class of problems involving magnetic fields, dealt with in Chapter 33, concerns the forces *exerted by* a magnetic field on a moving charge or on a current-carrying conductor and the torque exerted on a magnetic dipole (a bar magnet or a current loop, say). A second class of problems concerns the *production* of a magnetic field by a magnet, a current-carrying conductor, or by moving charges. This chapter deals with problems of this second class.

The discovery that currents produce magnetic effects was made by Hans Christian Oersted in 1820. Oersted made his discovery in connection with a classroom demonstration. In his paper entitled *The Action of Currents on Magnets*, Oersted wrote, translated from the Latin,

The first experiments on the subject which I undertook to illustrate were set on foot in the classes for electricity, galvanism, and magnetism, which were held by me in the winter just past.

Because of the importance of Oersted's discovery (a fundamental connection between electricity and magnetism) and especially because of its context in a teaching situation, the medal awarded annually by the American Association of Physics Teachers to a physics teacher especially noted for his or her impact on the teaching of physics is named after Oersted.

If we may deal for the moment with current-carrying wires as typical sources of magnetic fields and as typical objects on which magnetic fields may act, we may write in analogy with the argument of Section 27-1 for electric fields.

$$\text{current} \rightleftharpoons \text{field } (\mathbf{B}) \rightleftharpoons \text{current},$$

which suggests (a) that currents generate magnetic fields and (b) that magnetic fields exert forces on currents. We dealt with (b) in Section 33-3; we deal with (a) in this chapter.

Figure 34-1, which shows a wire surrounded by a number of small magnets, shows a modification of Oersted's experiment. If there is no current in the wire, all the magnets are aligned with the horizontal component of the earth's magnetic field. When a strong current is present, the magnets point so as to suggest that the magnetic field lines form closed circles around the wire. This view is strengthened by the experiment in Fig. 34-2, which shows iron filings on a horizontal glass plate, through the center of which a current-carrying conductor passes.

Today we write the quantitative relationship between current i and the magnetic field **B** as

$$\oint \mathbf{B} \cdot d\mathbf{l} = \mu_0 i, \tag{34-1}$$

which is known as *Ampère's law*. Ampère, being an advocate of the action-at-a-distance point of view, did not formulate his results in terms of fields; this was first done by Maxwell. Ampère's law, including an important extension of it made later by Maxwell, is one of the basic equations of electromagnetism (see Table 40-2).

We can gain an appreciation of the way Ampère's law developed historically by considering a hypothetical experiment which has, in fact, much in common with experiments that were actually carried out. The experiment consists of measuring **B** at various distances r from a long

figure 34-1
An array of compass needles near a central wire carrying a strong current. The black ends of the compass needles are their north poles. The central dot shows the current emerging from the page. As usual, the direction of a current is taken as the direction of flow of positive charge.

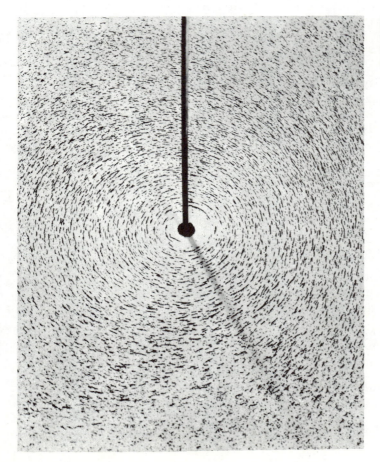

figure 34-2
Iron filings around a wire carrying a strong current. (Courtesy Physical Science Study Committee.)

straight wire of circular cross section and carrying a current i. This can be done by making quantitative the qualitative observation of Fig. 34-1.

Let us put a small compass needle a distance r from the wire. Such a needle, a small magnetic dipole, tends to line up with an external magnetic field, with its north pole pointing in the direction of **B**. Figure 34-1 makes it clear that **B** at the site of the dipole is tangent to a circle of radius r centered on the wire.

If the current in the wire of Fig. 34-1 is reversed in direction, all the compass needles would reverse end-for-end. This experimental result leads to the "right-hand rule" for finding the direction of **B** near a wire carrying a current i: *Grasp the wire with the right hand, the thumb pointing in the direction of the current. The fingers will curl around the wire in the direction of* **B**.

Let us now turn the dipole through an angle θ from its equilibrium position. To do this, we must exert an external torque just large enough to overcome the restoring torque τ that will act on the dipole. τ, θ, and **B** are related by Eq. 33-11 ($\tau = \mu \times \mathbf{B}$), which we can write in terms of magnitude as

$$\tau = \mu B \sin \theta \tag{34-2}$$

in which μ is the magnitude of the magnetic moment of the dipole and θ is the angle between the vectors μ and **B**. Even though we may not know the value of μ for the compass needle, we may take it to be a constant, independent of the position or orientation of the needle. Thus by measuring τ and θ in Eq. 34-2 we can obtain a *relative* measure of B for various distances r and for various currents i in the wire. We can describe the experimental results by the proportionality

$$B \propto \frac{i}{r}. \tag{34-3}$$

We can convert this proportionality into an equality by inserting a proportionality constant. As in the case of Coulomb's law, and for similar reasons (see Section 26-4), we do not write this constant simply as, say, k but in a more complex form, namely $\mu_0/2\pi$, in which μ_0 is called the *permeability constant*.* Equation 34-3 then becomes

$$B = \frac{\mu_0 i}{2\pi r}, \tag{34-4}$$

which we choose to write in the form

$$(B)(2\pi r) = \mu_0 i. \tag{34-5}$$

The left side of Eq. 34-5 is $\oint \mathbf{B} \cdot d\mathbf{l}$ for a path consisting of a circle of radius r centered on the wire. For all points on this circle **B** has the same (constant) magnitude B and $d\mathbf{l}$, which is always tangent to the path of integration, points in the same direction as **B**, as Fig. 34-3 shows. Thus

$$\oint \mathbf{B} \cdot d\mathbf{l} = \oint B \, dl = B \oint dl = (B)(2\pi r),$$

$\oint dl$ being simply the circumference of the circle. In this special case, therefore, we can write the experimentally observed connection between the field and the current as

$$\oint \mathbf{B} \cdot d\mathbf{l} = \mu_0 i, \tag{34-1}$$

---

* μ_0 has no connection with the dipole moment μ that appears in Eq. 34-2.

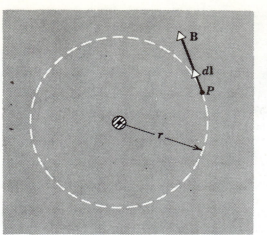

figure 34-3
A circular path of integration surrounding a wire. The central dot suggests a current i in the wire emerging from the page. Note that the angle between **B** and $d\mathbf{l}$ is zero so that $\mathbf{B} \cdot d\mathbf{l} = B \, dl$.

which is Ampère's law. A host of other experiments suggests that Eq. 34-1 is true in general for *any* magnetic field configuration, for *any* assembly of currents, and for *any* path of integration.*

In applying Ampère's law in the general case, we construct a *closed linear path* in the magnetic field as shown in Fig. 34-4. This path is divided into elements of length $d\mathbf{l}$, and for each element the quantity $\mathbf{B} \cdot d\mathbf{l}$ is evaluated. Recall that $\mathbf{B} \cdot d\mathbf{l}$ has the magnitude $B \, dl \cos \theta$ and can be interpreted as the product of dl and the component of $\mathbf{B}$ ($= B \cos \theta$) parallel to $d\mathbf{l}$. The integral is the sum of the quantities $\mathbf{B} \cdot d\mathbf{l}$ for all path elements in the complete loop; it is a *line integral* around a closed path. The term i on the right of Eq. 34-1 is the *net* current that passes through the area bounded by the closed path.

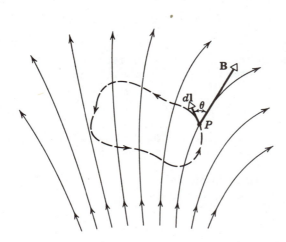

figure 34-4
A path of integration in a magnetic field.

The permeability constant in Ampère's law has an assigned value of

$$\mu_0 = 4\pi \times 10^{-7} \text{ tesla} \cdot \text{meter/ampere}.$$

Both this and the permittivity constant (ϵ_0) occur in electromagnetic formulas when SI units are used.

* We must modify Eq. 34-1 if a time-varying electric field or magnetic material are present within the path of integration. In this chapter we assume that if electric fields are present, they are constant in magnitude and direction and that no magnetic material is present.

You may wonder why ϵ_0 in Coulomb's law is a measured quantity, whereas μ_0 in Ampère's law is an assigned quantity. The answer is that the ampere, which is the SI unit for the current i in Ampère's law, is defined by a laboratory technique (the *current balance*) that involves forces exerted by magnetic fields and in which this same constant μ_0 appears. In effect, as we show in detail in Section 34-4, the size of the current that we agree to define as one ampere is adjusted so that μ_0 may have exactly the value assigned to it above. In Coulomb's law, on the other hand, the quantities **F**, q, and r are measured in ways in which the constant ϵ_0 plays no role. This constant must then take on the particular value that makes the left side of Coulomb's law equal to the right side; no arbitrary assignment is possible.

34-2
B NEAR A LONG WIRE

We have seen that the lines of **B** for a long straight cylindrical wire carrying a current i are concentric circles centered on the axis of the wire and that B at a distance r from this axis is given by Eq. 34-4:

$$B = \frac{\mu_0 i}{2\pi r}. \qquad (34\text{-}4)$$

We may regard this as an experimental result consistent with, and readily derivable from, Ampère's law.

It is interesting to compare Eq. 34-4 with the expression for the electric field near a long line of charge, or

$$E = \frac{1}{2\pi\epsilon_0}\frac{\lambda}{r}. \qquad (28\text{-}10)$$

In each case there are multiplying constants, namely $\mu_0/2\pi$ and $1/2\pi\epsilon_0$, and factors describing the device responsible for the field, namely i and λ. Finally, each field varies as $1/r$.

Equation 28-10 may be derived from Gauss's law by relating the electric field at a Gaussian surface to the net charge within this surface. The (surface) integral in Gauss's law is evaluated for a closed cylindrical surface to which the lines of **E** are everywhere perpendicular.

Equation 34-4 may be derived from Ampère's law by relating the magnetic field at a path of integration to the net current that pierces this path. The (line) integral in Ampère's law is evaluated for a closed circular path to which the lines of **B** are everywhere tangent.

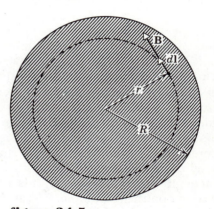

figure 34-5
Example 1. A circular path of integration inside a wire. A current i_0, distributed uniformly over the cross section of the wire, emerges from the page.

EXAMPLE 1

Derive an expression for **B** at a distance r from the center of a long cylindrical wire of radius R, where $r < R$. The wire carries a current i_0, distributed uniformly over the cross section of the wire.

Figure 34-5 shows a circular path of integration inside the wire. Symmetry suggests that **B** is tangent to the path as shown. Ampère's law,

$$\oint \mathbf{B} \cdot d\mathbf{l} = \mu_0 i,$$

gives

$$(B)(2\pi r) = \mu_0 i_0 \frac{\pi r^2}{\pi R^2},$$

since only the fraction of the current that passes through the path of integration is included in the factor i on the right. Solving for B and dropping the subscript on the current yields

$$B = \frac{\mu_0 i r}{2\pi R^2}.$$

At the surface of the wire $(r = R)$ this equation reduces to the same expression as that found by putting $r = R$ in Eq. 34-4 $(B = \mu_0 i/2\pi R)$.

Figure 34-6 shows a flat strip of copper of width a and negligible thickness carrying a current i. Find the magnetic field **B** at point P, at a distance R from the center of the strip, at right angles to the strip.

EXAMPLE 2

751   B NEAR A LONG WIRE   SEC. 34-2

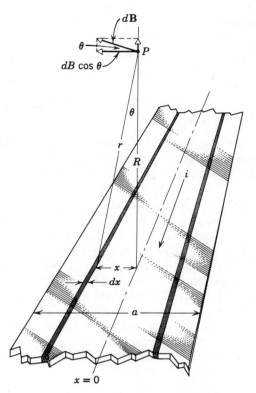

Let us subdivide the strip into long infinitesimal filaments of width dx, each of which may be treated as a wire carrying a current di given by $i(dx/a)$. The field contribution dB at point P in Fig. 34-6 is given, for the element shown, by the differential form of Eq. 34-4, or

$$dB = \frac{\mu_0}{2\pi} \frac{di}{r} = \frac{\mu_0}{2\pi} \frac{i(dx/a)}{R \sec \theta}.$$

Note that the vector $d\mathbf{B}$ is at right angles to the line marked r.

Only the horizontal component of $d\mathbf{B}$, namely $dB \cos \theta$, is effective, the vertical component being canceled by the contribution associated with a symmetrically located filament on the other side of the origin. Thus B at point P is given by the (scalar) integral

$$B = \int dB \cos \theta = \int \frac{\mu_0 i(dx/a)}{2\pi R \sec \theta} \cos \theta$$

$$= \frac{\mu_0 i}{2\pi a R} \int \frac{dx}{\sec^2 \theta}.$$

The variables x and θ are not independent, being related by

$$x = R \tan \theta$$

or

$$dx = R \sec^2 \theta \, d\theta.$$

Bearing in mind that the limits on θ are $\pm\tan^{-1}(a/2R)$ and eliminating dx from this expression for B, we find

$$B = \frac{\mu_0 i}{2\pi aR} \int \frac{R \sec^2 \theta \, d\theta}{\sec^2 \theta}$$

$$= \frac{\mu_0 i}{2\pi a} \int_{-\tan^{-1}a/2R}^{+\tan^{-1}a/2R} d\theta = \frac{\mu_0 i}{\pi a} \tan^{-1} \frac{a}{2R}.$$

At points far from the strip, $a/2R$ is a small angle, for which $\tan^{-1} \alpha \cong \alpha$. Thus we have, as an approximate result,

$$B \cong \frac{\mu_0 i}{\pi a} \left(\frac{a}{2R} \right) = \frac{\mu_0}{2\pi} \frac{i}{R}.$$

This result is expected because at distant points the strip cannot be distinguished from a cylindrical wire (see Eq. 34-4).

Figure 34-7 shows the lines of **B** representing the field of **B** near a long straight wire. Note the increase in the spacing of the lines with increasing distance from the wire. This represents the $1/r$ decrease in B predicted by Eq. 34-4.

Figure 34-8 shows the resultant magnetic lines associated with a current in a wire that is oriented at right angles to a uniform *external* field $\mathbf{B}_e$ directed to the right. At any point the resultant magnetic field **B** will be the vector sum of $\mathbf{B}_e$ and $\mathbf{B}_i$, where $\mathbf{B}_i$ is the magnetic field set up by the current in the wire. The fields $\mathbf{B}_e$ and $\mathbf{B}_i$ tend to cancel above the wire and to reenforce each other below the wire. At point P in Fig. 34-8 $\mathbf{B}_e$ and $\mathbf{B}_i$ cancel exactly. Very near the wire the field is represented by circular lines and is essentially $\mathbf{B}_i$.

Michael Faraday, who originated the concept of magnetic lines, endowed them with more reality than they are currently given. He imagined that, like stretched rubber bands, they represent the site of mechanical forces. On this picture can we not visualize that the wire in Fig. 34-8 will be pushed up? Today we use lines of **B** largely for purposes of visualization. For quantitative calculations we use the field vectors, describing the force on the wire in Fig. 34-8, for example, from the relation $\mathbf{F} = i\mathbf{l} \times \mathbf{B}$.

In applying this relation to Fig. 34-8, we recall that **B** is always the *external field* in which the wire is immersed; that is, it is $\mathbf{B}_e$ and thus points to the right. Since $\mathbf{l}$ points out of the page, the magnetic force on the wire $(= i\mathbf{l} \times \mathbf{B}_e)$ does indeed point up. It is necessary to use only the external field in such calculations because the field set up by the current in the wire cannot exert a force on the wire, just as the gravitational

34-3
LINES OF **B**

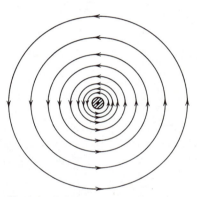

figure 34-7
Lines of **B** near a long, circularly cylindrical wire. A current i, suggested by the central dot, emerges from the page.

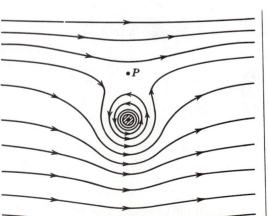

figure 34-8
Lines of **B** near a long current-carrying wire immersed in a uniform external field $\mathbf{B}_e$ that points to the right. The current i is emerging from the page. At point P, $\mathbf{B} = 0$.

field of the earth cannot exert a force on the earth itself but only on another body..In Fig. 34-7, for example, there is no magnetic force on the wire because no external magnetic field is present.

Faraday's idea of lines of **B** was instrumental in overthrowing the older action-at-a-distance theory of magnetic (and electric) attraction. Like many new ideas, it was not immediately accepted. In 1851, for example, Faraday wrote:

> I cannot refrain from again expressing my conviction of the truthfulness of the representation, which the idea of lines of force affords in regard to magnetic action. All the points that are experimentally established in regard to that action—i.e., all that is not hypothetical—appear to be well and truly represented by it.

On the other hand, four years later another well-known British scientist, Sir George Airy, wrote:

> I declare that I can hardly imagine anyone who practically and numerically knows this agreement [with the action-at-a-distance theory] to hesitate an instant in the choice between this simple and precise action, on the one hand, and anything so vague as lines of force, on the other hand.

Students who imagine that scientific pronouncements are absolute can do well to compare these two statements, each made by a distinguished contemporary. Modern examples are not lacking.

34-4
TWO PARALLEL CONDUCTORS

Figure 34-9 shows two long parallel wires separated by a distance d and carrying currents i_a and i_b. It is an experimental fact, noted by Ampère only one week after word of Oersted's experiments reached Paris, that two such conductors attract each other.

Some of Ampère's colleagues thought that in view of Oersted's experiment this attraction between two conductors was an obvious result and did not need to be proved. They reasoned that if wire a and wire b each exert forces on a compass needle they should exert forces on each other. This conclusion is wrong. When he heard it, Arago, a contemporary of Ampère, drew two iron keys from his pocket and replied, "Each of these keys attracts a magnet. Do you believe that they therefore also attract each other?"

Wire a in Fig. 34-9 will produce a magnetic field $\mathbf{B}_a$ at all nearby points. The magnitude of $\mathbf{B}_a$, due to the current i_a, at the site of the second wire is, from Eq. 34-4,

$$B_a = \frac{\mu_0 i_a}{2\pi d}.$$

The right-hand rule shows that the direction of $\mathbf{B}_a$ at wire b is down, as shown in the figure.

Wire b, which carries a current i_b, finds itself immersed in an *external* magnetic field $\mathbf{B}_a$. A length l of this wire will experience a sideways magnetic force ($= i\mathbf{l} \times \mathbf{B}$) whose magnitude is

$$F_b = i_b l B_a = \frac{\mu_0 l i_b i_a}{2\pi d}. \tag{34-6}$$

The vector rule of signs tells us that $\mathbf{F}_b$ lies in the plane of the wires and points to the left in Fig. 34-9.

We could have started with wire b, computed the magnetic field which it produces at the site of wire a, and then computed the force on wire a. The force on wire a would, for parallel currents, point to the right. The forces that the two wires exert on each other are equal and

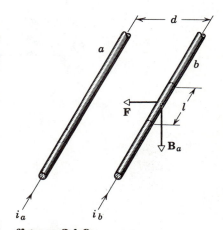

figure 34-9
Two parallel wires that carry parallel currents attract each other.

opposite, as they must be according to Newton's law of action and reaction. For antiparallel currents the two wires repel each other.

This discussion reminds us of our discussion of the electric field between two point charges in Section 27-1. There we saw that the charges act on each other through the intermediary of the electric field. The conductors in Fig. 34-9 act on each other, as we have said earlier, through the intermediary of the magnetic field **B.** We think in terms of

$$\text{current} \rightleftharpoons \text{field } (\mathbf{B}) \rightleftharpoons \text{current}$$

and not, as in the action-at-a-distance point of view, in terms of

$$\text{current} \rightleftharpoons \text{current.}$$

The attraction between long parallel wires is used to define the ampere. Suppose that the wires are one meter apart ($d = 1.0$ m, exactly) and that the two currents are equal ($i_a = i_b = i$). If this common current is adjusted until, by measurement, the force of attraction per unit length between the wires is 2×10^{-7} N/m exactly, the current is defined to be one ampere. From Eq. 34-6,

$$\frac{F}{l} = \frac{\mu_0 i^2}{2\pi d} = \frac{(4\pi \times 10^{-7} \text{ T}\cdot\text{m/A})(1 \text{ A})^2}{(2\pi)(1 \text{ m})}$$

$$= 2 \times 10^{-7} \text{ N/m}$$

as expected.*

At the National Bureau of Standards primary measurements of current are made with a *current balance*. This consists of a carefully wound coil placed between two other coils, as in Fig. 34-10. The outer pair of coils is fastened to

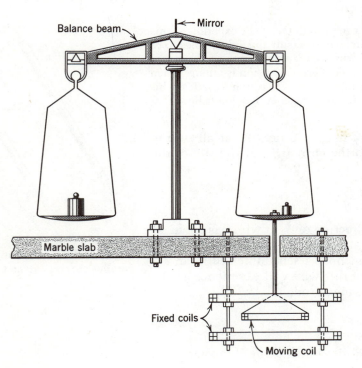

Balance beam — Mirror

Marble slab

Fixed coils

Moving coil

figure 34-10
A current balance.

---

* Note that μ_0 appears in this relation used to define the ampere. As stated on page 749, μ_0 is assigned the (arbitrary) value of $4\pi \times 10^{-7}$ tesla·meter/ampere, and the size of the current that we define as one ampere is adjusted to give the required force of attraction per unit length.

the table, and the inner one is hung from the arm of a balance. The coils are so connected that the current to be measured exists, as a common current, in all three of them.

The coils exert forces on one another—just as the parallel wires of Fig. 34-9 do—which can be measured by loading weights on the balance pan. The current is defined in terms of this measured force and the dimensions of the coils. The current balance is perfectly equivalent to the long parallel wires of Fig. 34-9 but is a much more practical arrangement. Current balance measurements are used primarily to standardize other, more convenient, secondary methods of measuring currents.

EXAMPLE 3

A long horizontal rigidly supported wire carries a current i_a of 100 A. Directly above it and parallel to it is a fine wire that carries a current i_b of 20 A and weighs 0.0050 lb/ft (= 0.073 N/m). How far above the lower wire should this second wire be strung if we hope to support it by magnetic repulsion?

To provide a repulsion, the two currents must point in opposite directions. For equilibrium, the magnetic force per unit length must equal the weight per unit length and must be oppositely directed. Solving Eq. 34-6 for d yields

$$d = \frac{\mu_0 i_a i_b}{2\pi (F/l)} = \frac{(4\pi \times 10^{-7} \text{ T·m/A})(100 \text{ A})(20 \text{ A})}{(2\pi)(0.073 \text{ N/m})}$$

$$= 5.5 \times 10^{-3} \text{ m} = 5.5 \text{ mm}.$$

We assume that the wire diameters are much smaller than their separation. This assumption is necessary because in deriving Eq. 34-6 we tacitly assumed that the magnetic field produced by one wire is uniform for all points within the second wire.

Is the equilibrium of the suspended wire stable or unstable against vertical displacements? This can be tested by displacing the wire vertically and examining how the forces on the wire change.

Suppose that the fine wire is suspended *below* the rigidly supported wire. How may it be made to "float"? Is the equilibrium against vertical displacements stable or unstable?

EXAMPLE 4

Two parallel wires a distance d apart carry equal currents i in opposite directions. Find the magnetic field for points between the wires and at a distance x from one wire.

Study of Fig. 34-11 shows that $\mathbf{B}_a$ due to the current i_a and $\mathbf{B}_b$ due to the current i_b point in the same direction at P. Each is given by Eq. 34-4 $(B = \mu_0 i/2\pi r)$ so that

$$B = B_a + B_b = \frac{\mu_0 i}{2\pi} \left(\frac{1}{x} + \frac{1}{d-x} \right).$$

This relationship does not hold for points inside the wires because Eq. 34-4 is not valid there.

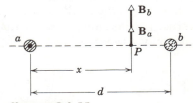

figure 34-11
Example 4

34-5
B *FOR A SOLENOID*

A *solenoid* is a long wire wound in a close-packed helix and carrying a current i. We assume that the helix is very long compared with its diameter. What is the nature of the field of **B** that is set up?

For points close to a single turn of the solenoid, the observer is not aware that the wire is bent in an arc. The wire behaves magnetically almost like a long straight wire, and the lines of **B** due to this single turn are almost concentric circles.

The solenoid field is the vector sum of the fields set up by all the

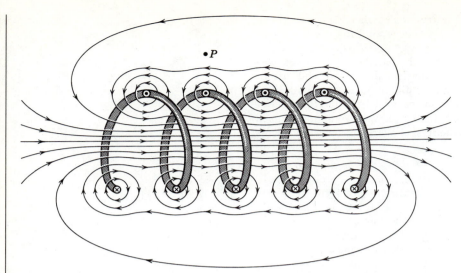

figure 34-12
A loosely wound solenoid.

turns that make up the solenoid. Figure 34-12, which shows a "solenoid" with widely spaced turns, suggests that the fields tend to cancel between the wires. It also suggests that, at points inside the solenoid and reasonably far from the wires, **B** is parallel to the solenoid axis. In the limiting case of adjacent square tightly packed wires, the solenoid becomes essentially a cylindrical current sheet and the requirements of symmetry then make the statement just given necessarily true. We assume that it is true in what follows.

For points such as P in Fig. 34-12 the field set up by the upper part of the solenoid turns (marked $\odot$) points to the left and tends to cancel the field set up by the lower part of the solenoid turns (marked $\otimes$), which points to the right. As the solenoid becomes more and more ideal, that is, as it approaches the configuration of an infinitely long cylindrical current sheet, the field **B** at outside points approaches zero. Taking the external field to be zero is not a bad assumption for a practical solenoid if its length is much greater than its diameter and if we consider only external points near the central region of the solenoid, that is, away from the ends. Figure 34-13 shows the lines of **B** for a real solenoid, which is far from ideal in that the length is not much greater than the diameter. Even here the spacing of the lines of **B** in the central plane shows that the external field is much weaker than the internal field.

Let us apply Ampère's law,

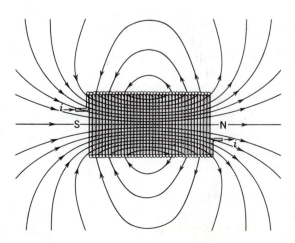

figure 34-13
A solenoid of finite length. The right end, from which lines of **B** emerge, behaves like the north pole of a compass needle. The left end behaves like the south pole.

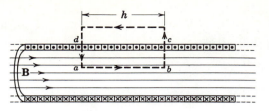

figure 34-14
A section of an ideal solenoid, made of adjacent square turns, equivalent to an infinitely long cylindrical current sheet.

$$\oint \mathbf{B} \cdot d\mathbf{l} = \mu_0 i,$$

to the rectangular path *abcd* in the ideal solenoid of Fig. 34-14. We write the integral $\oint \mathbf{B} \cdot d\mathbf{l}$ as the sum of four integrals, one for each path segment:

$$\oint \mathbf{B} \cdot d\mathbf{l} = \int_a^b \mathbf{B} \cdot d\mathbf{l} + \int_b^c \mathbf{B} \cdot d\mathbf{l} + \int_c^d \mathbf{B} \cdot d\mathbf{l} + \int_d^a \mathbf{B} \cdot d\mathbf{l}.$$

The first integral on the right is Bh, where B is the magnitude of $\mathbf{B}$ inside the solenoid and h is the arbitrary length of the path from a to b. Note that path ab, though parallel to the solenoid axis, need not coincide with it.

The second and fourth integrals are zero because for every element of these paths $\mathbf{B}$ is at right angles to the path. This makes $\mathbf{B} \cdot d\mathbf{l}$ zero and thus the integrals are zero. The third integral, which includes the part of the rectangle that lies outside the solenoid, is zero because we have taken $\mathbf{B}$ as zero for all external points for an ideal solenoid.

Thus $\oint \mathbf{B} \cdot d\mathbf{l}$ for the entire rectangular path has the value Bh. The net current i that passes through the area bounded by the path of integration is not the same as the current i_0 in the solenoid because the path of integration encloses more than one turn. Let n be the number of turns per unit length; then

$$i = i_0(nh).$$

Ampère's law then becomes

$$Bh = \mu_0 i_0 n h$$

or

$$B = \mu_0 i_0 n. \tag{34-7}$$

Although we derived Eq. 34-7 for an infinitely long ideal solenoid, it holds quite well for actual solenoids for internal points near the center of the solenoid. It shows that B does not depend on the diameter or the length of the solenoid and that B is constant over the solenoid cross section. A solenoid is a practical way to set up a known uniform magnetic field for experimentation, just as a parallel-plate capacitor is a practical way to set up a known uniform electric field.

The solenoid provides a good context in which to discuss Φ_B, the *flux* of the magnetic field $\mathbf{B}$. We discussed the flux Φ_E of the electric field $\mathbf{E}$ in Section 28-3, restricting ourselves largely, for reasons having to do with Gauss's law, to the flux for closed surfaces. We did, however, discuss the flux Φ_E for open surfaces; see Fig. 28-1.

We can, in a similar way, define the flux Φ_B for a magnetic field $\mathbf{B}$, for either a closed or an open surface. In either case it is given by

$$\Phi_B = \int \mathbf{B} \cdot d\mathbf{S},$$

in strict analogy to the discussion of Φ_E in Section 28-3. The SI unit of $\mathbf{B}$, as we have seen, is the *tesla* (1 tesla = 1 weber/meter^2) while that of Φ_B is simply the *weber* (abbr. Wb).

EXAMPLE 5

A solenoid is 1.0 m long and 3.0 cm in inner diameter. It has five layers of windings of 850 turns each and carries a current of 5.0 A.

(a) What is B at its center? From Eq. 34-7,

$$B = \mu_0 i_0 n = (4\pi \times 10^{-7} \text{ T·m/A})(5.0 \text{ A})(5 \times 850 \text{ turns/m})$$

$$= 2.7 \times 10^{-2} \text{ T} = 2.7 \times 10^{-2} \text{ Wb/m}^2.$$

We can use Eq. 34-7 even if the solenoid has more than one layer of windings because the diameter of the windings does not enter.

(b) What is the magnetic flux Φ_B for a cross section of the solenoid at its center? To the extent that $\mathbf{B}$ is constant, we can calculate the flux from

$$\Phi_B = \int \mathbf{B} \cdot d\mathbf{S} = BA,$$

where A is the effective cross-sectional area. Let us take A as the area of a circular disk whose diameter is the inner diameter of the windings (3.0 cm). The effective area can then be shown to be 7.1×10^{-4} m^2, and

$$\Phi_B = BA = (2.7 \times 10^{-2} \text{ Wb/m}^2)(7.1 \times 10^{-4} \text{ m}^2)$$

$$= 1.9 \times 10^{-5} \text{ Wb}.$$

EXAMPLE 6

A toroid. Figure 34-15 shows a toroid, which we may describe as a solenoid bent into the shape of a doughnut. Calculate $\mathbf{B}$ at interior points.

From symmetry the lines of $\mathbf{B}$ form concentric circles inside the toroid, as shown in the figure. Let us apply Ampère's law to the circular path of integration of radius r:

$$\oint \mathbf{B} \cdot d\mathbf{l} = \mu_0 i$$

or

$$(B)(2\pi r) = \mu_0 i_0 N,$$

where i_0 is the current in the toroid windings and N is the total number of turns. This gives

$$B = \frac{\mu_0}{2\pi} \frac{i_0 N}{r}.$$

In contrast to the solenoid, B is not constant over the cross section of a toroid. Show from Ampère's law that B equals zero for points outside an ideal toroid.

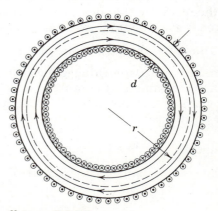

figure 34-15
Example 6. A toroid.

34-6
*THE BIOT-SAVART LAW**

We can use Ampère's law to calculate magnetic fields only if the symmetry of the current distribution is high enough to permit the easy evaluation of the line integral $\oint \mathbf{B} \cdot d\mathbf{l}$. This requirement limits the usefulness of the law in practical problems. The law does not fail; it simply becomes difficult to apply in a useful way.

Similarly, in electrostatics, we use Gauss's law to calculate electric fields only if the symmetry of the charge distribution is high enough to permit the easy evaluation of the surface integral $\oint \mathbf{E} \cdot d\mathbf{S}$. We can, for example, use Gauss's law to find the electric field due to a long uni-

* See "Ampère as a Contemporary Physicist" by R. A. Tricker, *Contemporary Physics,* August 1962; and "Electromagnetism as a Second Order Effect. III: The Biot-Savart Law" by W. G. V. Rosser, *Contemporary Physics,* October 1961, for more information about Ampère's law and the law of Biot and Savart.

formly charged rod but we cannot apply it usefully to an electric dipole, for the symmetry is not high enough in this case.

To compute **E** at a given point for an *arbitrary* charge distribution, we divided the distribution into *charge elements dq* and (see Section 27-4) we used Coulomb's law to calculate the field contribution $d\mathbf{E}$ due to each element at the point in question. We found the field **E** at that point by adding, that is, by integrating, the field contributions $d\mathbf{E}$ for the entire distribution.

We now describe a similar procedure for computing **B** at any point due to an arbitrary current distribution. We divide the current distribution into *current elements* and, using the law of Biot and Savart (which we describe below), we calculate the field contribution $d\mathbf{B}$ due to each current element at the point in question. We find the field **B** at that point by integrating the field contributions for the entire distribution.

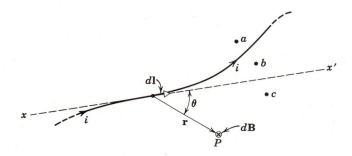

figure 34-16
The current element $d\mathbf{l}$ establishes a magnetic field contribution $d\mathbf{B}$ at point P.

Figure 34-16 shows an arbitrary current distribution consisting of a current i in a curved wire. The figure also shows a typical current element; it is a length $d\mathbf{l}$ of the conductor carrying a current i. Its direction is that of the tangent to the conductor (dashed line). A current element cannot exist as an isolated entity because a way must be provided to lead the current into the element at one end and out of it at the other. Nevertheless, we can think of an actual circuit as made up of a large number of current elements placed end to end.

Let P be the point at which we want to know the magnetic field $d\mathbf{B}$ associated with the current element. According to the Biot-Savart law, $d\mathbf{B}$ is given in magnitude by

$$dB = \frac{\mu_0 i}{4\pi} \frac{dl \, \sin \theta}{r^2}, \qquad (34\text{-}8)$$

where **r** is a displacement vector from the element to P and θ is the angle between this vector and $d\mathbf{l}$. The direction of $d\mathbf{B}$ is that of the vector $d\mathbf{l} \times \mathbf{r}$. In Fig. 34-16, for example, $d\mathbf{B}$ at point P for the current element shown is directed into the page at right angles to the plane of the figure. Note that Eq. 34-8, being an inverse square law that describes the magnetic field due to a current element, may be viewed as the magnetic equivalent of Coulomb's law, which is an inverse square law that describes the electric field due to a charge element.

We may write the law of Biot and Savart in vector form as

$$d\mathbf{B} = \frac{\mu_0 i}{4\pi} \frac{d\mathbf{l} \times \mathbf{r}}{r^3}. \qquad (34\text{-}9)$$

This formulation reduces to that of Eq. 34-8 when we express it in terms of magnitudes; it also gives complete information about the direction of $d\mathbf{B}$, namely, that it is the same as the direction of the vector $d\mathbf{l} \times \mathbf{r}$.

The resultant field at P is found by integrating Eq. 34-9, or

$$\mathbf{B} = \int d\mathbf{B}, \qquad (34\text{-}10)$$

where the integral is a vector integral.

EXAMPLE 7

A long straight wire. We illustrate the law of Biot and Savart by applying it to find $\mathbf{B}$ due to a current i in a long straight wire. We discussed this problem at length in connection with Ampère's law in Section 34-1.

Figure 34-17, a side view of the wire, shows a typical current element $d\mathbf{x}$. The magnitude of the contribution $d\mathbf{B}$ of this element to the magnetic field at P is found from Eq. 34-8, or

$$dB = \frac{\mu_0 i}{4\pi} \frac{dx \sin \theta}{r^2}.$$

The directions of the contributions $d\mathbf{B}$ at point P for all elements are the same, namely, into the plane of the figure at right angles to the page. Thus the vector integral of Eq. 34-10 reduces to a scalar integral, or

$$B = \int dB = \frac{\mu_0 i}{4\pi} \int_{x=-\infty}^{x=+\infty} \frac{\sin \theta \, dx}{r^2}.$$

Now, x, θ, and r are not independent, being related (see Fig. 34-17) by

$$r = \sqrt{x^2 + R^2}$$

and

$$\sin \theta \, [= \sin (\pi - \theta)] = \frac{R}{\sqrt{x^2 + R^2}},$$

so that the expression for B becomes

$$B = \frac{\mu_0 i}{4\pi} \int_{-\infty}^{+\infty} \frac{R \, dx}{(x^2 + R^2)^{3/2}}$$

$$= \frac{\mu_0 i}{4\pi R} \left. \frac{x}{(x^2 + R^2)^{1/2}} \right|_{x=-\infty}^{x=+\infty}$$

$$= \frac{\mu_0}{2\pi} \frac{i}{R}.$$

This is the result that we arrived at earlier for this problem (see Eq. 34-4). The law of Biot and Savart will always yield results that are consistent with Ampère's law and with experiment.

This problem reminds us of its electrostatic equivalent. We derived an expression for $\mathbf{E}$ due to a long charged rod, using Gauss's law (Section 28-8); we also solved this problem by integration methods, using Coulomb's law (Section 27-4).

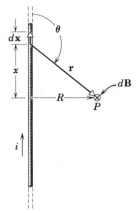

figure 34-17
Example 7

EXAMPLE 8

A circular current loop. Figure 34-18 shows a circular loop of radius R carrying a current i. Calculate $\mathbf{B}$ for points on the axis.

The vector $d\mathbf{l}$ for a current element at the top of the loop points perpendicularly out of the page. The angle θ between $d\mathbf{l}$ and $\mathbf{r}$ is 90°, and the plane formed by $d\mathbf{l}$ and $\mathbf{r}$ is normal to the page. The vector $d\mathbf{B}$ for this element is at right angles to this plane and thus lies in the plane of the figure and at right angles to $\mathbf{r}$, as the figure shows.

Let us resolve $d\mathbf{B}$ into two components, one, $d\mathbf{B}_\parallel$, along the axis of the loop and another, $d\mathbf{B}_\perp$, at right angles to the axis. Only $d\mathbf{B}_\parallel$ contributes to the total magnetic field $\mathbf{B}$ at point P. This follows because the components $d\mathbf{B}_\parallel$ for all current elements lie on the axis and add directly; however, the components $d\mathbf{B}_\perp$ point in different directions perpendicular to the axis, and their result for the complete loop is zero, from symmetry. Thus

$$B = \int dB_{\parallel},$$

where the integral is a simple scalar integration over the current elements.

For the current element shown in Fig. 34-18 we have, from the Biot-Savart law (Eq. 34-8),

$$dB = \frac{\mu_0 i}{4\pi} \frac{dl \sin 90°}{r^2}.$$

We also have
$$dB_{\parallel} = dB \cos \alpha.$$

Combining gives
$$dB_{\parallel} = \frac{\mu_0 i \cos \alpha \, dl}{4\pi r^2}.$$

Figure 34-18 shows that r and α are not independent of each other. Let us express each in terms of a new variable x, the distance from the center of the loop to the point P. The relationships are

$$r = \sqrt{R^2 + x^2}$$

and
$$\cos \alpha = \frac{R}{r} = \frac{R}{\sqrt{R^2 + x^2}}.$$

Substituting these values into the expression for $dB_{\parallel}$ gives

$$dB_{\parallel} = \frac{\mu_0 i R}{4\pi(R^2 + x^2)^{3/2}} \, dl.$$

Note that i, R, and x have the same values for all current elements. Integrating this equation, noting that $\int dl$ is simply the circumference of the loop $(= 2\pi R)$, yields

$$B = \int dB_{\parallel} = \frac{\mu_0 i R}{4\pi(R^2 + x^2)^{3/2}} \int dl$$

$$= \frac{\mu_0 i R^2}{2(R^2 + x^2)^{3/2}}. \qquad (34\text{-}11)$$

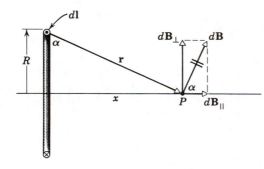

figure 34-18
Example 8. A ring of radius R carrying a current i.

If we put $x \gg R$ in Example 8 so that points close to the loop are not considered, Eq. 34-11 reduces to

$$B = \frac{\mu_0 i R^2}{2x^3}.$$

Recalling that πR^2 is the area A of the loop and considering loops with N turns, we can write this equation as

$$B = \frac{\mu_0}{2\pi} \frac{(NiA)}{x^3} = \frac{\mu_0}{2\pi} \frac{\mu}{x^3},$$

where μ is the *magnetic dipole moment* of the current loop. This reminds us of the result derived in Problem 25, Chapter 27 [$E = (1/2\pi\epsilon_0)(p/x^3)$], which is the formula for the *electric* field on the axis of an *electric* dipole.

Thus we have shown in two ways that we can regard a current loop as a magnetic dipole: It experiences a torque given by $\tau = \mu \times B$ when we place it in an *external* magnetic field (Eq. 33-11); it generates its own magnetic field given, for points on the axis, by the equation just developed.

Table 34-1 is a summary of the properties of electric and magnetic dipoles.

Table 34-1
Some dipole equations

Property	Dipole Type	Equation
Torque in an external field	electric	$\tau = p \times E$
	magnetic	$\tau = \mu \times B$
Energy in an external field	electric	$U = -p \cdot E$
	magnetic	$U = -\mu \cdot B$
Field at distant points along axis	electric	$E = \dfrac{1}{2\pi\epsilon_0}\dfrac{p}{x^3}$
	magnetic	$B = \dfrac{\mu_0}{2\pi}\dfrac{\mu}{x^3}$
Field at distant points along perpendicular bisector	electric	$E = \dfrac{1}{4\pi\epsilon_0}\dfrac{p}{x^3}$
	magnetic	$B = \dfrac{\mu_0}{4\pi}\dfrac{\mu}{x^3}$

EXAMPLE 9

In the Bohr model of the hydrogen atom the electron circulates around the nucleus in a path of radius 5.3×10^{-11} m at a frequency ν of 6.5×10^{15} Hz (= rev/s). (*a*) What value of B is set up at the center of the orbit?

The current is the rate at which charge passes any point on the orbit and is given by

$$i = e\nu = (1.6 \times 10^{-19}\text{ C})(6.5 \times 10^{15}\text{ Hz}) = 1.0 \times 10^{-3}\text{ A}.$$

B at the center of the orbit is given by Eq. 34-11 with $x = 0$, or

$$B = \frac{\mu_0 i R^2}{2(R^2 + x^2)^{3/2}} = \frac{\mu_0 i}{2R}$$

$$= \frac{(4\pi \times 10^{-7}\text{ T}\cdot\text{m/A})(1.0 \times 10^{-3}\text{ A})}{(2)(5.3 \times 10^{-11}\text{ m})}$$

$$= 12\text{ T}$$

(*b*) What is the equivalent magnetic dipole moment? From Eq. 33-10,

$$\mu = NiA = (1)(1.0 \times 10^{-3}\text{ A})(\pi)(5.3 \times 10^{-11}\text{ m})^2$$

$$= 8.8 \times 10^{-24}\text{ A}\cdot\text{m}^2.$$

1. Can the path of integration around which we apply Ampère's law pass *through* a conductor?

2. Suppose we set up a path of integration around a cable that contains twelve wires with different currents (some in opposite directions) in each wire. How do we calculate i in Ampère's law in such a case?

3. Apply Ampère's law qualitatively to the three paths shown in Fig. 34-19.

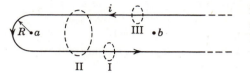

figure 34-19
Question 3,
Problem 27

4. Discuss and compare Gauss's law and Ampère's law.

5. (a) Must Ampère's law (Eq. 34-1) be applied always around a *closed* loop? Can we, for example, apply it to a semicircular arc? (b) Can we apply Ampère's law to a closed *surface?*

6. Give details of three ways in which you can measure the magnetic field **B** at a point P, a perpendicular distance r from a long straight wire carrying a constant current i. Base them on: (a) firing a particle of charge q through point P with velocity **v**, parallel, say, to the wire; (b) measuring the force per unit length exerted on a second wire, parallel to the first wire and carrying a current i'; (c) measuring the torque exerted on a small magnetic dipole located a perpendicular distance r from the wire.

7. A current is set up in a long copper pipe. Is there a magnetic field (a) inside and (b) outside the pipe?

8. Equation 34-4 ($B = \mu_0 i / 2\pi r$) suggests that a strong magnetic field is set up at points near a long wire carrying a current. Since there is a current i and a magnetic field **B**, why is there not a force on the wire in accord with the equation $\mathbf{F} = i\mathbf{l} \times \mathbf{B}$?

9. A beam of 20-MeV protons emerges from a cyclotron-like device. Is a magnetic field associated with these particles?

10. Does it necessarily follow from symmetry arguments alone that the lines of **B** around a long straight wire carrying a current i must be concentric circles?

11. A long straight wire of radius R carries a steady current i. In what sense, if any, does the magnetic field generated by this current depend on R?

12. A long straight wire carries a constant current i. Does Ampère's law (Eq. 34-1) hold for (a) a path of integration that encloses the wire but is not circular, (b) a path of integration that does not enclose the wire, and (c) a path that encloses the wire but does not all lie in one plane? Discuss.

13. Two long straight wires pass near one another at right angles. If the wires are free to move, describe what happens when currents are sent through them.

14. Is **B** constant in magnitude for points that lie on a given magnetic field line?

15. In electronics, wires that carry equal but opposite currents are often twisted together to reduce their magnetic effect at distant points. Why is this effective?

16. Two long parallel conductors carry equal currents i in the same direction. Sketch roughly the resultant lines of **B** due to the action of *both* currents. Does your figure suggest an attraction between the wires (in the same sense that Fig. 34-8 suggests an upward force on the wire in that figure)?

17. In Fig. 34-8 explain the relation between the figure and Eq. 33-6a ($\mathbf{F} = i\mathbf{l} \times \mathbf{B}$).

18. Test the "floating" wire of Example 3 for equilibrium under *horizontal* displacements. Consider that the wire floats above the rigidly supported wire and also below it. Summarize the equilibrium situation for both wire positions and for both vertical and horizontal displacements.

19. A current is sent through a vertical spring from whose lower end a weight is hanging; what will happen?

20. Comment on this statement: "The magnetic field **B** outside a long solenoid cannot be zero, if only for the reason that the helical nature of the windings produces a field for external points like that of a straight wire along the solenoid axis."

21. Does Eq. 34-7 $(B = \mu_0 in)$ hold for a solenoid of square cross section?

22. In your own words convince yourself that $B = 0$ for points outside an ideal solenoid, such as that of Fig. 34-14.

23. What is the direction of the magnetic fields at points a, b, and c in Fig. 34-16 set up by the particular current element shown?

24. In a circular loop of wire carrying a current i, is **B** uniform for all points within the loop?

25. Discuss analogies and differences between Coulomb's law and the Biot-Savart law.

26. Equation 34-9 gives the law of Biot and Savart in vector form. Write its electrostatic equivalent [that is, Eq. 27-6, or $dE = dq/(4\pi\epsilon_0 r^2)$] in vector form.

27. How might you measure the magnetic dipole moment of a compass needle?

28. What is the basis for saying that a current loop is a magnetic dipole?

29. A circular loop of wire lies on the floor of the room in which you are sitting and it carries a constant current i in a clockwise direction, as viewed from above. What is the direction of the magnetic dipole moment of this current loop?

30. As an exercise in vector representation, contrast and compare Fig. 18-12 which deals with fluid flow, with Fig. 34-7, which deals with the magnetic field. How strong an analogy can you make?

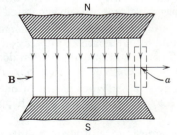

figure 34-20
Problem 1

problems

SECTION 34-1

1. Eight wires cut the page perpendicularly at the points shown in Fig. 34-20. A wire labeled with the integer k $(k = 1, 2, \ldots 8)$ bears the current ki_0. For those with odd k, the current flows up out of the page; for those with even k it flows down into the page. Evaluate $\oint \mathbf{B} \cdot d\mathbf{l}$ along the closed path shown in the direction indicated by the arrowhead.
Answer: $-10\mu_0 i_0$ (Why the minus sign?).

2. Show that it is impossible for a uniform magnetic field **B** to drop abruptly to zero as one moves at right angles to it, as suggested by the horizontal arrow in Fig. 34-21 (see point a). In actual magnets fringing of the lines of **B** always occurs, which means that **B** approaches zero in a continuous and gradual way. (Hint: Apply Ampère's law to the rectangular path shown by the dashed lines.)

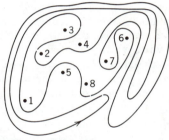

figure 34-21
Problem 2

SECTION 34-2

3. A #10 bare copper wire (0.10 in. in diameter) can carry a current of 50 A without overheating. For this current, what is B at the surface of the wire?
Answer: 7.9×10^{-3} T.

4. A surveyor is using a compass 20 ft below a power line in which there is a steady current of 100 A. Will this interfere seriously with the compass reading? The horizontal component of the earth's magnetic field at the site is 0.20 gauss.

5. If a point charge of magnitude $+q$ and speed v is a distance d from the axis of a long straight wire carrying a current i and is traveling perpendicular to

the axis of the wire, what are the direction and magnitude of the force acting on it if the charge is moving (a) toward, or (b) away from the wire?

Answer: (a) $\dfrac{\mu_0}{2\pi}\dfrac{qvi}{r}$, antiparallel to *i*. (b) Same magnitude, parallel to *i*.

6. A long straight wire carries a current of 50 A. An electron, traveling at 1.0×10^7 m/s, is 5.0 cm from the wire. What force acts on the electron if the electron velocity is directed (a) toward the wire, (b) parallel to the wire, and (c) at right angles to the directions defined by (a) and (b)?

7. A long solid cylindrical copper wire of radius R carries a current i distributed uniformly over the cross section of the wire. Sketch roughly the magnitude of the magnetic field **B** as a function of the distance r from the axis of the wire for (a) r < R and (b) r > R.

8. Four long copper wires are parallel to each other, their cross section forming a square 20 cm on edge. A 20-A current is set up in each wire in the direction shown in Fig. 34-22. What are the magnitude and direction of **B** at the center of the square?

9. A long coaxial cable consists of two concentric conductors with the dimensions shown in Fig. 34-23. There are equal and opposite currents i in the conductors. (a) Find the magnetic field B at r within the inner conductor (r < a). (b) Find B between the two conductors (a < r < b). (c) Find B within the outer conductor (b < r < c). (d) Find B outside the cable (r > c).

Answer: (a) $\dfrac{\mu_0 ir}{2\pi a^2}$. (b) $\dfrac{\mu_0 i}{2\pi r}$. (c) $\dfrac{\mu_0 i}{2\pi r}\left(\dfrac{c^2 - r^2}{c^2 - b^2}\right)$. (d) Zero.

10. Two long wires a distance d apart carry equal antiparallel currents i, as in Fig. 34-24. (a) Show that B at point P, which is equidistant from the wires, is given by

$$B = \frac{2\mu_0 id}{\pi(4R^2 + d^2)}.$$

(b) In what direction does **B** point?

11. Figure 34-25 shows a hollow cylindrical conductor of radii a and b which carries a current i uniformly spread over its cross section. (a) Show that the magnetic field B for points inside the body of the conductor (that is, a < r < b) is given by

$$B = \frac{\mu_0 i}{2\pi(b^2 - a^2)}\frac{r^2 - a^2}{r}.$$

Check this formula for the limiting case of a = 0. (b) Make a rough plot of the general behavior of B(r) from r = 0 to r → ∞.

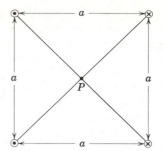

figure 34-22
Problem 8

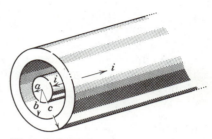

figure 34-23
Problem 9

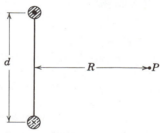

figure 34-24
Problem 10

figure 34-25
Problem 11

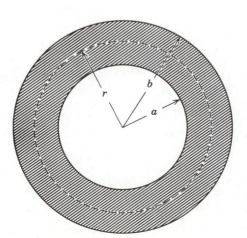

12. Two long straight wires a distance d (10 cm) apart each carry a current i (100 A). Figure 34-26 shows a cross section, with the wires running perpendicular to the page and point P lying on the perpendicular bisector of d. Find the magnitude and direction of the magnetic field at P when the current in the left-hand wire is out of the page and the current in the right-hand wire is (a) in the same direction and (b) in the opposite direction.

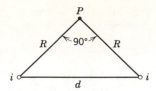

figure 34-26
Problem 12

13. A conductor consists of an infinite number of adjacent wires, each infinitely long and carrying a current i. Show that the lines of **B** will be as represented in Fig. 34-27 and that B for all points in front of the infinite current sheet will be given by

$$B = \tfrac{1}{2}\mu_0 ni,$$

where n is the number of conductors per unit length. Derive both by direct application of Ampère's law and by considering the problem as a limiting case of Example 2.

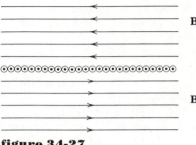

figure 34-27
Problem 13

14. A long straight conductor has a circular cross section of radius R and carries a current i. Inside the conductor there is a cylindrical hole of radius a whose axis is parallel to the conductor axis at a distance b from it. Use superposition ideas and obtain an expression for the magnetic field **B** inside the hole.

SECTION 34-3

15. A long wire carrying a current of 100 A is placed in a uniform external magnetic field of 50 gauss. The wire is at right angles to this external field. Locate the points at which the *resultant* magnetic field is zero.
Answer: $B = 0$ along a line parallel to the wire and 4.0 mm from it.

SECTION 34-4

16. In Problem 8 what is the force per unit length (N/m) acting on the lower left wire, in magnitude and direction?

17. Figure 34-28 shows a long wire carrying a current of 30 A. The rectangular loop carries a current of 20 A. Calculate the resultant force acting on the loop. Assume that $a = 1.0$ cm, $b = 8.0$ cm, and $l = 30$ cm.
Answer: 3.2×10^{-3} N, toward the wire.

18. Suppose, in Fig. 34-22, that the currents are all in the same direction. What is the force per unit length (N/m, magnitude and direction) on any one wire? In the analogous case of parallel motion of charged particles in a plasma, this is known as the pinch effect.

19. Two long parallel wires of negligible radius are a distance d apart. Let there be a current i in each wire (a) in the same direction and (b) in opposite directions. Letting r be the perpendicular distance from the center of one wire, find the magnitude B of the magnetic field in the region between the wires at points in the plane of the two wires.

Answer: (a) $\dfrac{\mu_0 i}{2\pi}\left[\dfrac{d-2r}{r(d-r)}\right]$. (b) $\dfrac{\mu_0 i}{2\pi}\left[\dfrac{d}{r(d-r)}\right]$.

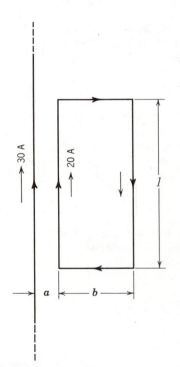

figure 34-28
Problem 17

SECTION 34-5

20. A 200-turn solenoid having a length of 25 cm and a diameter of 10 cm carries a current of 0.30 A. (a) What is the magnitude of the magnetic field **B** near the center of the solenoid? (b) What is the magnetic flux through an annular ring having an inside diameter of 2.0 cm and an outside diameter of 8.0 cm if the plane of the ring is perpendicular to the axis of the solenoid?

21. Express (a) the magnetic field **B** and (b) the magnetic flux Φ_B in terms of mass, length, time, and charge (M, L, T, Q).
Answer: (a) M/QT. (b) ML^2/QT.

22. A toroid having a 5.0 cm $\times$ 5.0 cm cross section and an inside radius of 15 cm has 500 turns of wire and carries a current of 0.80 A. (a) What is the magnitude of the magnetic field **B** in the center of the toroid (that is, at a radius of 17.5 cm)? (b) What is the magnetic flux through the cross section?

23. A long straight wire of radius a carries a constant current i (a) Consider a concentric hypothetical circle of radius $2a$, its plane being at right angles to the wire. What magnetic flux Φ_B passes through this circle? (b) If we were to double the current i, what do you imagine would happen to this flux?
Answer: (a) Zero. (b) Still zero; no lines of **B** pierce the circle in either case.

24. Derive the solenoid equation (Eq. 34-7) starting from the expression for the field on the axis of a circular loop (Example 8). (Hint: Subdivide the solenoid into a series of current loops of infinitesimal thickness and integrate.)

25. A long copper wire carries a current of 10 A. Calculate the magnetic flux per meter of wire through a plane surface S inside the wire, as in Fig. 34-29.
Answer: 1.0×10^{-6} Wb/m.

26. Two long, parallel #10 copper wires (diameter = 0.10 in.) carry currents of 10 A in opposite directions. (a) If their centers are 2.0 cm apart, calculate the flux per meter of wire length that exists in the space between the axes of the wires. (b) What fraction of the flux in (a) lies inside the wires? (c) Repeat the calculation of (a) for parallel currents.

SECTION 34-6

27. A long "hairpin" is formed by bending a piece of wire as shown in Fig. 34-19. If a 10-A current is set up, what are the direction and magnitude of **B** at point a? At point b? Take $R = 0.50$ cm.
Answer: (a) 1.0×10^{-3} T, out of figure. (b) 8.0×10^{-4} T, out of figure.

28. Use the Biot-Savart law to calculate the magnetic field **B** at C, the common center of the semicircular arcs AD and HJ, of radii R_2 and R_1 respectively, forming part of the circuit $ADJHA$ carrying current i, as shown in Fig. 34-30.

29. Consider the circuit of Fig. 34-31. The curved segments are part of circles of radii a and b. The straight segments are along the radii. Find the magnetic field **B** at P, assuming a current i in the circuit.
Answer: $B = \dfrac{\mu_0 i \theta}{4\pi} \left(\dfrac{1}{b} - \dfrac{1}{a} \right)$, out of page.

30. The wire shown in Fig. 34-32 carries a current i. What is the magnetic field **B** at the center C of the semicircle arising from (a) each straight segment of length l, (b) the semicircular segment of radius R, and (c) the entire wire?

31. A straight conductor is split into identical semicircular turns as shown in Fig. 34-33. What is the magnetic field at the center C of the circular loop so formed?
Answer: Zero.

32. A circular loop of radius 10 cm carries a current of 15 A. At its center is placed a second loop of radius 1.0 cm, having 50 turns and a current of 1.0 A. (a) What magnetic field **B** does the large loop set up at its center? (b) What torque acts on the small loop? Assume that the planes of the two loops are at right angles and that **B** provided by the large loop is essentially uniform throughout the volume occupied by the small loop.

33. You are given a closed circuit with radii a and b, as shown in Fig. 34-34, carrying a current i. (a) What are the magnitude and direction of **B** at point P? (b) Find the dipole moment of the circuit.
Answer: (a) $\dfrac{\mu_0 i}{4} \left(\dfrac{1}{a} + \dfrac{1}{b} \right)$, into the page. (b) $\dfrac{i\pi}{2} (a^2 + b^2)$, into the page.

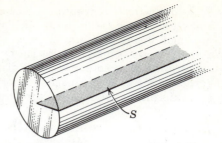

figure 34-29
Problem 25

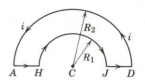

figure 34-30
Problem 28

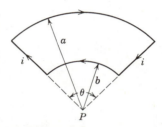

figure 34-31
Problem 29

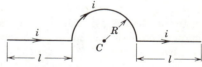

figure 34-32
Problem 30

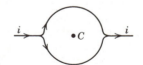

figure 34-33
Problem 31

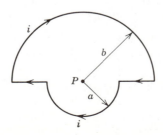

figure 34-34
Problem 33

34. (a) A long wire is bent into the shape shown in Fig. 34-35, without cross-contact at P. The radius of the circular section is R. Determine the magnitude and direction of **B** at the center C of the circular portion when the current i flows as indicated. (b) The circular part of the wire is rotated without distortion about its (dashed) diameter perpendicular to the straight portion of the wire. The magnetic moment associated with the circular loop is now in the direction of the current in the straight part of the wire. Determine **B** at C in this case.

35. *Helmholtz coils.* Two 300-turn coils are arranged a distance apart equal to their radius, as in Fig. 34-36. For $R = 5.0$ cm and $i = 50$ A, plot B as a function of distance x along the common axis over the range $x = -5$ cm to $x = +5$ cm, taking $x = 0$ at point P. (Such coils provide an especially uniform field of B near point P.)

36. In Problem 35 let the separation of the coils be a variable z. Show that if $z = R$, then not only the first derivative (dB/dx) but also the second (d^2B/dx^2) of B is zero at point P. This accounts for the uniformity of B near point P for this particular coil separation.

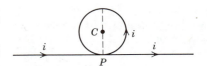

figure 34-35
Problem 34

37. A plastic disk of radius R has a charge q uniformly distributed over its surface. If the disk is rotated at an angular frequency ω about its axis, show that (a) the magnetic field at the center of the disk is

$$B = \frac{\mu_0 \omega q}{2\pi R}$$

and (b) the magnetic dipole moment of the disk is

$$\mu = \frac{\omega q R^2}{4}.$$

(Hint: The rotating disk is equivalent to an array of current loops; see Example 8.)

38. A messy loop of limp wire is placed on a table and anchored at points a and b as shown in Fig. 34-37. If a current i is now passed through the wire, will it try to form a circular loop or will it try to bunch up further?

39. A straight wire segment of length l carries a current i. (a) Show that the magnetic field **B** associated with this segment, at a distance R from the segment along a perpendicular bisector (see Fig. 34-38), is given in magnitude by

$$B = \frac{\mu_0 i}{2\pi R} \frac{l}{(l^2 + 4R^2)^{1/2}}.$$

(b) Does this expression reduce to an expected result as $l \to \infty$?
Answer: (b) Yes.

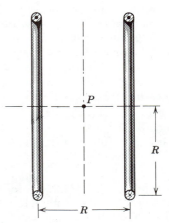

figure 34-36
Problems 35, 36

40. A square loop of wire of edge a carries a current i. Show that the value of B at the center is given by

$$B = \frac{2\sqrt{2}\ \mu_0 i}{\pi a}.$$

41. A square loop of wire of edge a carries a current i. (a) Show that B for a point on the axis of the loop and a distance x from its center is given by

$$B = \frac{4\mu_0 i a^2}{\pi (4x^2 + a^2)(4x^2 + 2a^2)^{1/2}}.$$

(b) Does this reduce to the result of Problem 40 for $x = 0$? (c) Does the square loop behave like a dipole for points such that $x \gg a$? If so, what is its dipole moment?   *Answer:* (b) Yes. (c) Yes; $\mu = ia^2$.

42. (a) Show that B at the center of a rectangle of length l and width d, carrying a current i, is given by

$$B = \frac{2\mu_0 i}{\pi} \frac{(l^2 + d^2)^{1/2}}{ld}.$$

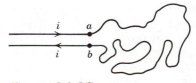

figure 34-37
Problem 38

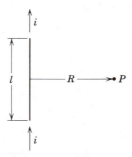

figure 34-38
Problem 39.

(b) What does B reduce to for $l \gg d$? Is this a result that you expect?

43. (a) A wire in the form of a regular polygon of n sides is just enclosed by a circle of radius a. If the current in this wire is i, show that the magnetic field **B** at the center of the circle is given in magnitude by

$$B = \frac{\mu_0 ni}{2\pi a} \tan (\pi/n).$$

(b) Show that as $n \to \infty$ this result approaches that of a circular loop.

44. Calculate **B** approximately at point P in Fig. 34-39. Assume that $i = 10$ A and $a = 8.0$ cm.

45. You are given a length l of wire in which a current i may be established. The wire may be formed into a circle or a square. Which yields the larger value for B at the central point? See Problem 40.   *Answer:* The square.

46. (a) A current i flows in a straight wire of length L in the direction shown in Fig. 34-40a. Starting from the Biot-Savart law, determine the resulting magnetic field ($\mathbf{B}_P$, $\mathbf{B}_Q$, $\mathbf{B}_R$, $\mathbf{B}_S$, respectively-direction and magnitude in each case) at each of the four points P, Q, R, S (all coplanar with the wire). (b) Using the results of part (a), compute the magnetic field **B** (magnitude and direction) resulting at the point T from the current flowing as indicated in the six-sided rectilinear closed loop shown in Fig. 34-40b. (Everything drawn is meant to lie in the same plane and all angles are 90°.)

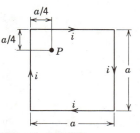

figure 34-39
Problem 44.

figure 34-40
Problem 46.

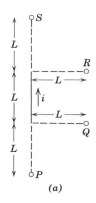

(a)

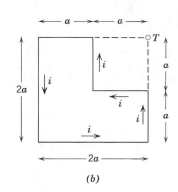

(b)

35
faraday's law of induction

For some physical laws it is hard to find experiments that lead in a direct and convincing way to the formulation of the law. Gauss's law, for example, emerged only slowly as the common factor with whose aid all electrostatic experiments could be interpreted and correlated. In Chapter 28 we found it best to state Gauss's law first and then to show that the underlying experiments were consistent with it.

Faraday's law of induction, which is one of the basic equations of electromagnetism (see Table 40-2), is different in that there are a number of simple experiments from which the law can be—and was—deduced directly. Such experiments were carried out by Michael Faraday in England in 1831 and by Joseph Henry in the United States at about the same time.

Faraday and Henry had several parallels in their lives. Both were apprentices at an early age. Faraday, at age 14, was apprenticed to a London bookbinder. He wrote, "There were plenty of books there and I read them." Henry, at age 13, was apprenticed to a watchmaker in Albany, New York.

In later years Faraday was appointed director of the Royal Institution in London, whose founding was due in large part to the American, Benjamin Thompson (Count Rumford). Henry, on the other hand, became secretary of the Smithsonian Institution in Washington, D.C., which was founded by an endowment from an Englishman, James Smithson.

Their greatest scientific overlap was that each discovered the law of electromagnetic induction, with which this chapter deals, independently and at about the same time. Even though Faraday published his results first, which gives him priority of discovery, the SI unit of inductance (see Chapter 36) is called the *henry* (abbr. H). On the other hand the SI unit of capacitance is, as we have seen, called the *farad* (abbr. F). In Section 38-1, in which we discuss oscillations in capacitative-inductive circuits, we shall see how appropriate it is to link the names of these two talented contemporaries in a single context.

Figure 35-1 shows the terminals of a coil connected to a galvanometer. Normally we would not expect this instrument to deflect because there seems to be no electromotive force in this circuit; but if we push a bar magnet toward the coil, with its north pole facing the coil, a remarkable thing happens. *While the magnet is moving,* the galvanometer deflects, showing that a current has been set up in the coil. If we hold the magnet stationary with respect to the coil, the galvanometer does not deflect. If we move the magnet *away* from the coil, the galvanometer again deflects, but in the opposite direction, which means that the current in the coil is in the opposite direction. If we use the south pole end of a magnet instead of the north pole end, the experiment works as described but the deflections are reversed. Further experimentation shows that *what matters is the relative motion of the magnet and the coil.* It makes no difference whether we move the magnet toward the coil or the coil toward the magnet.

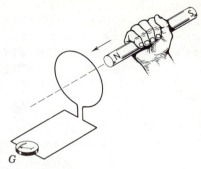

figure 35-1
Galvanometer G deflects while the magnet is moving with respect to the coil. Only their relative motion counts.

The current that appears in this experiment is called an *induced current* and is said to be set up by an *induced electromotive force.* Note that there are no batteries anywhere in the circuit. Faraday was able to deduce from experiments like this the law that gives their magnitude and direction. Such emfs are very important in practice. The chances are good that the lights in the room in which you are reading this book are operated from an induced emf produced in a commercial electric generator.

In another experiment the apparatus of Fig. 35-2 is used. The coils are placed close together but at rest with respect to each other. When we close the switch S, thus setting up a steady current in the right-hand coil, the galvanometer deflects momentarily; when we open the switch, thus interrupting this current, the galvanometer again deflects momentarily, but in the opposite direction. No gross objects are moving in this experiment. In Faraday's words:

When the contact was made, there was a sudden and very slight effect at the galvanometer, and there was also a similar slight effect when the contact with the battery was broken. But whilst the voltaic current was continuing to pass through the one helix, no galvanometrical appearances nor any effect like induction upon the other helix could be perceived, although the active power of the battery was proved to be very great. . . .

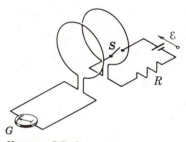

figure 35-2
Galvanometer G deflects momentarily when switch S is closed or opened. No motion is involved.

Experiment shows that there will be an induced emf in the left coil of Fig. 35-2 whenever the current in the right coil is *changing.* It is the *rate at which the current is changing* and *not the size of the current* that is significant.

Faraday had the insight to perceive that the change in the flux Φ_B for the left coil in the preceding experiments is the important common factor. This flux may be set up by a bar magnet or a current loop. Faraday's law of induction says that the induced emf ε in a circuit is equal (except for a minus sign) to the rate at which the flux through the circuit is changing. If the rate of change of flux is in webers/second, the emf ε will be in volts. In equation form

35-2
FARADAY'S LAW OF INDUCTION

$$\varepsilon = -\frac{d\Phi_B}{dt}. \qquad (35\text{-}1)$$

This is *Faraday's law of induction.* The minus sign is an indication of the *direction* of the induced emf, a matter we discuss in Section 35-3.*

If we apply Eq. 35-1 to a coil of N turns, an emf appears in every turn and these emfs are to be added. If the coil is so tightly wound that each turn can be said to occupy the same region of space, the flux through each turn will then be the same. The flux through each turn is also the same for (ideal) toroids and solenoids (see Section 34-5). The induced emf in all such devices is given by

$$\varepsilon = -N\frac{d\Phi_B}{dt} = -\frac{d(N\Phi_B)}{dt},$$ (35-2)

where $N\Phi_B$ measures the so-called *flux linkages* in the device.

Figures 35-1 and 35-2 suggest that there are at least two ways in which we can make the flux through a circuit change and thus induce an emf in that circuit. Speaking loosely, the coil that is connected to the galvanometer cannot tell in which of these experiments it is participating; it is aware only that the flux passing through its cross-sectional area is changing. The flux through a circuit can also be changed by changing its shape, that is, by squeezing or stretching it.

EXAMPLE 1

A long solenoid has 200 turns/cm and carries a current of 1.5 A; its diameter is 3.0 cm. At its center we place a 100-turn, close-packed coil of diameter 2.0 cm. This coil is arranged so that **B** at the center of the solenoid is parallel to its axis. The current in the solenoid is reduced to zero and then raised to 1.5 A in the other direction at a steady rate over a period of 0.050 s. What induced emf appears in the coil while the current is being changed?

The field B at the center of the solenoid is given by Eq. 34-7, or

$$B = \mu_0 ni = (4\pi \times 10^{-7}\,\text{T·m/A})(200 \times 10^2\,\text{turns/m})(1.5\,\text{A})$$
$$= 3.8 \times 10^{-2}\,\text{T}.$$

The area of the coil (not of the solenoid) is $3.1 \times 10^{-4}\,\text{m}^2$. The initial flux Φ_B through each turn of the coil is given by

$$\Phi_B = BA = (3.8 \times 10^{-2}\,\text{T})(3.1 \times 10^{-4}\,\text{m}^2) = 1.2 \times 10^{-5}\,\text{Wb}.$$

The flux goes from an initial value of 1.2×10^{-5} Wb to a final value of -1.2×10^{-5} Wb. The *change* in flux $\Delta\Phi_B$ for each turn of the coil during the 0.050-s period is thus twice the initial value. The induced emf is given by

$$\varepsilon = -\frac{N\Delta\Phi_B}{\Delta t} = -\frac{(100)(2 \times 1.2 \times 10^{-5}\,\text{Wb})}{0.050\,\text{s}} = -4.8 \times 10^{-2}\,\text{V} = -48\,\text{mV}.$$

The minus sign deals with the *direction* of the emf, as we explain below.

35-3
LENZ'S LAW

So far we have not specified the directions of the induced emfs. Although we can find these directions from a formal analysis of Faraday's law, we prefer to find them from the conservation-of-energy principle which, in this context, takes the form of Lenz's law, deduced by Heinrich Friedrich Lenz (1804–1865) in 1834: *The induced current will appear in such a direction that it opposes the change that produced it.* The minus sign in Faraday's law suggests this opposition. In mechanics

* Faraday, being untrained in mathematics, did not express his law of induction in equation form. In fact, in his three-volume "Experimental Researches in Electricity," a landmark work in the development of physics and chemistry, not a single equation appears!

the energy principle often allows us to draw conclusions about mechanical systems without analyzing them in detail. We use the same approach here.

Lenz's law refers to induced *currents*, which means that it applies only to closed conducting circuits. If the circuit is open, we can usually think in terms of what would happen if it *were* closed and in this way find the direction of the induced emf.

Consider the first of Faraday's experiments described in Section 35-1. Figure 35-3 shows the north pole of a magnet and a cross section of a nearby conducting loop. As we push the magnet toward the loop (or the loop toward the magnet) an induced current is set up in the loop. What is its direction?

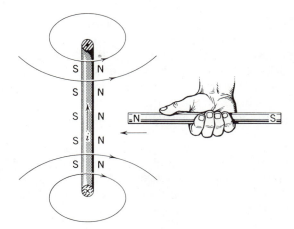

figure 35-3
If we move the magnet toward the loop, the induced current points as shown, setting up a magnetic field that opposes the motion of the magnet.

A current loop sets up a magnetic field at distant points like that of a magnetic dipole, one face of the loop being a north pole, the opposite face being a south pole. The north pole, as for bar magnets, is that face *from* which the lines of **B** emerge. If, as Lenz's law predicts, the loop in Fig. 35-3 is to oppose the motion of the magnet toward it, the face of the loop toward the magnet must become a north pole. The two north poles —one of the current loop and one of the magnet—will repel each other. The right-hand rule shows that for the magnetic field set up by the loop to emerge from the right face of the loop, the induced current must be as shown. The current will be counterclockwise as we sight along the magnet toward the loop.

When we push the magnet toward the loop (or the loop toward the magnet), an induced current appears. In terms of Lenz's law this pushing is the "change" that produces the induced current, and, according to this law, the induced current will oppose the "push." If we pull the magnet away from the coil, the induced current will oppose the "pull" by creating a *south* pole on the right-hand face of the loop of Fig. 35-3. To make the right-hand face a south pole, the current must be opposite to that shown in Fig. 35-3. Whether we pull or push the magnet, its motion will always be automatically opposed.

The agent that causes the magnet to move, either toward the coil or away from it, will always experience a resisting force and will thus be required to do work. From the conservation-of-energy principle this work done on the system must be exactly equal to the thermal energy produced in the coil, since these are the only two energy transfers that occur in the system. If we move the magnet more rapidly, we will have to do work at a faster rate and the rate of production of thermal energy will increase correspondingly. If we cut the loop and then perform the experiment, there will be no induced current, no thermal energy, no force on the magnet, and no work required to move it. There will still be an emf in the loop, but, like a battery connected to an open circuit, it will not set up a current.

If the current in Fig. 35-3 were in the *opposite* direction to that shown, the face of the loop toward the magnet would be a south pole, which would *pull* the bar magnet toward the loop. We would only need to push the magnet slightly to start the process and then the action would be self-perpetuating. The magnet would accelerate toward the loop, increasing its kinetic energy all the time. At the same time thermal energy would appear in the loop at a rate that would increase with time. This would indeed be a something-for-nothing situation! Needless to say, it does not occur.

Let us apply Lenz's law to Fig. 35-3 in a different way. Figure 35-4 shows the lines of **B** for the bar magnet.* On this point of view the "change" is the increase in Φ_B through the loop caused by bringing the magnet nearer. The induced current opposes this change by setting up a field that tends to oppose the increase in flux caused by the moving magnet. Thus the field due to the induced current must point from left to right through the plane of the coil, in agreement with our earlier conclusion.

It is not significant here that the induced field opposes the *field* of the magnet but rather that it opposes the *change*, which in this case is the *increase* in Φ_B through the loop. If we withdraw the magnet, we reduce Φ_B through the loop. The induced field will now oppose this decrease in Φ_B (that is, the change) by *reenforcing* the magnet field. In each case the induced field opposes the change that gives rise to it.

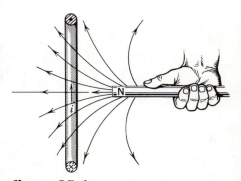

figure 35-4
If we move the magnet toward the loop, we increase Φ_B through the loop.

figure 35-5
A rectangular loop is pulled out of a magnetic field with velocity **v**.

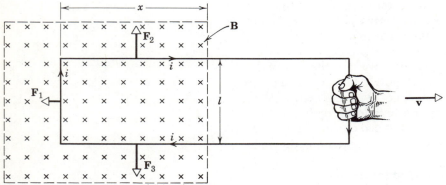

The example of Fig. 35-4, although easy to understand qualitatively, does not lend itself to quantitative calculations. Consider then Fig. 35-5, which shows a rectangular loop of wire of width *l*, one end of which is in a uniform field **B** pointing at right angles to the plane of the loop. This field **B** may be produced in the gap of a large electromagnet like that of Fig. 33-2. The dashed lines show the assumed limits of the magnetic field. The experiment consists in pulling the loop to the right at a constant speed *v*.

Note that the situation described by Fig. 35-5 does not differ in any

35-4
INDUCTION—A QUANTITATIVE STUDY

* There are two fields of **B** in this problem—one connected with the current loop and one with the bar magnet. You must always be certain which one is meant.

essential particular from that of Fig. 35-4. In each case a conducting loop and a magnet are in relative motion; in each case the flux of the field of the magnet through the loop is being caused to change with time.

The flux Φ_B enclosed by the loop in Fig. 35-5 is

$$\Phi_B = Blx,$$

where lx is the area of that part of the loop in which B is not zero. We find the emf ε, from Faraday's law, or

$$\varepsilon = -\frac{d\Phi_B}{dt} = -\frac{d}{dt}(Blx) = -Bl\frac{dx}{dt} = Blv, \qquad (35\text{-}3)$$

where we have set $-dx/dt$ equal to the speed v at which the loop is pulled out of the magnetic field. Note that the only dimension of the loop that enters into Eq. 35-3 is the length l of the left end conductor. As we shall see later, the induced emf in Fig. 35-5 may be regarded as localized here. An induced emf such as this, produced by pulling a conductor through a magnetic field (or conversely), is sometimes called a *motional emf*.

The emf Blv sets up a current in the loop given by

$$i = \frac{\varepsilon}{R} = \frac{Blv}{R}, \qquad (35\text{-}4)$$

where R is the loop resistance. From Lenz's law, this current (and thus ε) must be clockwise in Fig. 35-5; it opposes the "change" (the decrease in Φ_B) by setting up a field that is parallel to the external field within the loop.

The current in the loop will cause forces $\mathbf{F}_1$, $\mathbf{F}_2$, and $\mathbf{F}_3$ to act on the three conductors, in accord with Eq. 33-6a, or

$$\mathbf{F} = i\mathbf{l} \times \mathbf{B}. \qquad (35\text{-}5)$$

Because $\mathbf{F}_2$ and $\mathbf{F}_3$ are equal and opposite, they cancel each other; $\mathbf{F}_1$, which is the force that opposes our effort to move the loop, is given in magnitude from Eqs. 35-4 and 35-5 as

$$F_1 = ilB \sin 90° = \frac{B^2l^2v}{R}.$$

The agent that pulls the loop must do work at the steady rate of

$$P = F_1 v = \frac{B^2l^2v^2}{R}. \qquad (35\text{-}6)$$

From the principle of the conservation of energy, thermal energy must appear in the resistor at this same rate. We introduced the conservation-of-energy principle into our derivation when we wrote down the expression for the current (Eq. 35-4); recall that the relation $i = \varepsilon/R$ for single-loop circuits is a direct consequence of this principle. Thus we should be able to write down the expression for the rate of thermal energy production in the loop with the expectation that we will obtain a result identical with Eq. 35-6. Recalling Eq. 35-4, we put

$$P_J = i^2R = \left(\frac{Blv}{R}\right)^2 R = \frac{B^2l^2v^2}{R},$$

which is indeed the expected result. This example provides a quantitative illustration of the conversion of mechanical energy (the work done by an external agent) into electrical energy (associated with the induced emf) and finally into thermal energy.

Figure 35-6 shows a side view of the coil in the field. In Fig. 35-6a the coil is stationary; in Fig. 35-6b we are moving it to the right; in Fig. 35-6c we are moving it to the left. The lines of **B** in these figures represent the *resultant field* produced by the vector addition of the field **B₀** due to the magnet and the field **Bᵢ** due to the induced current, if any, in the coil. These lines suggest convincingly that the agent moving the coil always experiences an opposing force.

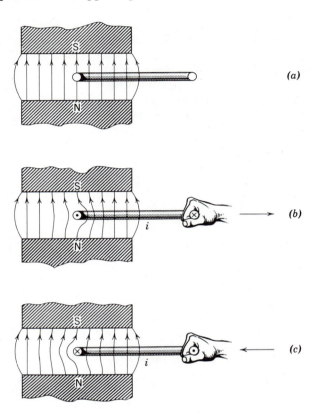

(a)

(b)

(c)

figure 35-6
Side view of a rectangular loop in a magnetic field showing the loop (a) at rest, (b) being pulled out, and (c) being pushed in. The configuration of the lines suggests an opposing force in both (b) and (c).

Figure 35-7 shows a rectangular loop of resistance R, width l, and length a being pulled at constant speed v through a region of thickness d in which a uniform magnetic field **B** is set up by a magnet.

(a) Plot the flux Φ_B through the loop as a function of the coil position x. Assume that $l = 4$ cm, $a = 10$ cm, $d = 15$ cm, $R = 16$ Ω, $B = 2.0$ T, and $v = 1.0$ m/s.

The flux Φ_B is zero when the loop is not in the field; it is Bla when the loop is entirely in the field; it is Blx when the loop is entering the field and $Bl[a - (x - d)]$ when the loop is leaving the field. These conclusions, which you should verify, are shown graphically in Fig. 35-8a.

(b) Plot the induced emf ε.

The induced emf ε is given by $\varepsilon = -d\Phi_B/dt$, which we can write as

$$\varepsilon = -\frac{d\Phi_B}{dt} = -\frac{d\Phi_B}{dx}\frac{dx}{dt} = -\frac{d\Phi_B}{dx}v,$$

where $d\Phi_B/dx$ is the slope of the curve of Fig. 35-8a. $\varepsilon(x)$ is plotted in Fig. 35-8b. Lenz's law, from the same type of reasoning as that used for Fig. 35-5, shows that when the coil is entering the field, the emf ε acts counterclockwise as seen from above. Note that there is no emf when the coil is entirely in the magnetic field because the flux Φ_B through the coil is not changing with time, as Fig. 35-8a shows.

(c) Plot the rate P of thermal energy production in the loop.

EXAMPLE 2

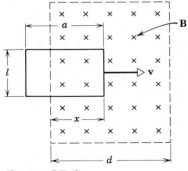

figure 35-7
Example 2. A rectangular loop is caused to move with a velocity **v** through a magnetic field. The position of the loop is measured by x, the distance between the effective left edge of the field **B** and the right end of the loop.

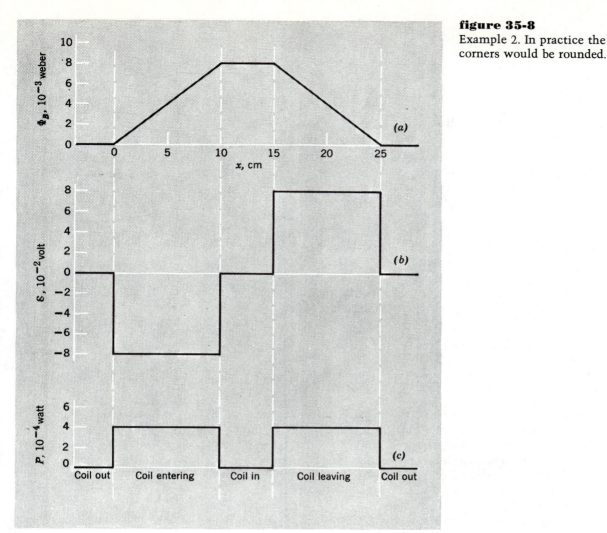

figure 35-8
Example 2. In practice the sharp
corners would be rounded.

This is given by $P = \varepsilon^2/R$. It may be calculated by squaring the ordinate of the curve of Fig. 35-8b and dividing by R. The result is plotted in Fig. 35-8c.

If the fringing of the magnetic field, which cannot be avoided in practice (see Problem 34-2), is taken into account, the sharp bends and corners in Fig. 35-8 will be replaced by smooth curves. What changes would occur in the curves of Fig. 35-8 if the coil were open circuited?

EXAMPLE 3

A copper rod of length L rotates at angular frequency ω in a uniform magnetic field **B** as shown in Fig. 35-9. Find the emf ε developed between the two ends of the rod. (We might measure this emf if we place a conducting rail along the dashed circle in the figure and connect a voltmeter between the rail and point O.

If a wire of length dl is moved at velocity **v** at right angles to a field **B**, a motional emf $d\varepsilon$ will be developed (see Eq. 35-3) given by

$$d\varepsilon = Bv \, dl.$$

The rod of Fig. 35-9 may be divided into elements of length dl, the linear speed v of each element being ωl. Each element is perpendicular to **B** and is also moving in a direction at right angles to **B** so that, since the $d\varepsilon$'s are "in series",

$$\varepsilon = \int d\varepsilon = \int_0^L Bv \, dl = \int_0^L B(\omega l) \, dl = \tfrac{1}{2} B\omega L^2.$$

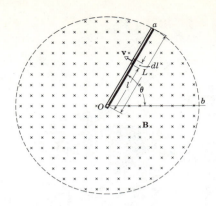

figure 35-9
Example 3

For a second approach, consider that at any instant the flux enclosed by the sector aOb in Fig. 35-9 is given by

$$\Phi_B = BA = B(\tfrac{1}{2}L^2\theta),$$

where $\tfrac{1}{2}L^2\theta$ can be shown to be the area of the sector. Differentiating gives

$$\frac{d\Phi_B}{dt} = \tfrac{1}{2}BL^2\frac{d\theta}{dt} = \tfrac{1}{2}B\omega L^2.$$

From Faraday's law, this is precisely the magnitude of $\mathcal{E}$, and agrees with the result just derived.

35-5
TIME-VARYING MAGNETIC FIELDS

So far we have considered emfs induced by the relative motion of magnets and coils. In this section we assume that there is no physical motion of gross objects but that the magnetic field may vary with time. If we place a conducting loop in such a time-varying field, the flux through the loop will change and an induced emf will appear in the loop. This emf will set the charge carriers in motion, that is, it will induce a current.

From a microscopic point of view we can say, equally well, that the changing flux of **B** sets up an induced electric field **E** at various points around the loop. These induced electric fields are just as real as electric fields set up by static charges and will exert a force **F** on a test charge q_0 given by $\mathbf{F} = q_0\mathbf{E}$. Thus we can restate Faraday's law of induction in a loose but informative way as: *A changing magnetic field produces an electric field.*

To fix these ideas, consider Fig. 35-10, which shows a uniform magnetic field **B** at right angles to the plane of the page. We assume that **B** is increasing in magnitude at the same constant rate dB/dt at every point. This could be done by causing the current in the windings of the electromagnet that establishes the field to increase with time in the proper way.

The circle of arbitrary radius r shown in Fig. 35-10 encloses, at any instant, a flux Φ_B. Because this flux is changing with time, an induced emf given by $\mathcal{E} = -d\Phi_B/dt$ will appear around the loop. The electric fields **E** induced at various points of the loop must, from symmetry, be tangent to the loop. Thus the electric lines of force that are set up by the changing magnetic field are in this case concentric circles.

If we consider a test charge q_0 moving around the circle of Fig. 35-10, the work W done on it per revolution is, in terms of the definition of an emf, simply $\mathcal{E}q_0$. From another point of view it is $(q_0E)(2\pi r)$, where q_0E is the force that acts on the charge and $2\pi r$ is the distance over

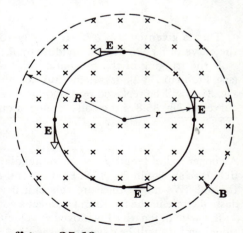

figure 35-10
The induced electric fields at four points produced by an increasing magnetic field. (The magnetic field cannot end abruptly at radius R but must approach zero gradually. This "fringing" does not change any of the arguments of this section.)

which the force acts. Setting the two expressions for W equal and canceling q_0 yields

$$\varepsilon = E2\pi r. \qquad (35\text{-}7)$$

In a more general case than that of Fig. 35-10 we must write

$$\varepsilon = \oint \mathbf{E} \cdot d\mathbf{l}. \qquad (35\text{-}8)$$

If this integral is evaluated for the conditions of Fig. 35-10, we obtain Eq. 35-7 at once. If Eq. 35-8 is combined with Eq. 35-1 ($\varepsilon = -d\Phi_B/dt$), we can write Faraday's law of induction as

$$\oint \mathbf{E} \cdot d\mathbf{l} = -\frac{d\Phi_B}{dt}, \qquad (35\text{-}9)$$

which is the form in which this law is expressed in Table 40-2.

EXAMPLE 4

779

TIME-VARYING MAGNETIC FIELDS   SEC. 35-5

Let B in Fig. 35-10 be increasing at the rate dB/dt. Let R be the radius of the cylindrical region in which the magnetic field is assumed to exist. What is the magnitude of the electric field $\mathbf{E}$ at any radius r? Assume that $dB/dt = 0.10$ T/s and $R = 10$ cm.

(a) For $r < R$, the flux Φ_B through the loop is

$$\Phi_B = B(\pi r^2).$$

Substituting into Faraday's law (Eq. 35-9),

$$\oint \mathbf{E} \cdot d\mathbf{l} = -\frac{d\Phi_B}{dt}$$

yields

$$(E)(2\pi r) = -\frac{d\Phi_B}{dt} = -(\pi r^2)\frac{dB}{dt}.$$

Solving for E yields

$$E = -\tfrac{1}{2}r\frac{dB}{dt}.$$

The minus sign is retained to suggest that the induced electric field $\mathbf{E}$ acts to *oppose* the change of the magnetic field. Note that $E(r)$ depends on dB/dt and not on B. Substituting numerical values, assuming $r = 5.0$ cm, yields, for the magnitude of E,

$$E = \tfrac{1}{2}r\frac{dB}{dt} = (\tfrac{1}{2})(5 \times 10^{-2}\text{ m})(0.10\text{ T/s}) = 2.5 \times 10^{-3}\text{ V/m}.$$

(b) For $r > R$ the flux through the loop is

$$\Phi_B = \int \mathbf{B} \cdot d\mathbf{S} = B(\pi R^2).$$

This equation is true because $\mathbf{B} \cdot d\mathbf{S}$ is zero for those points of the loop that lie outside the effective boundary of the magnetic field.

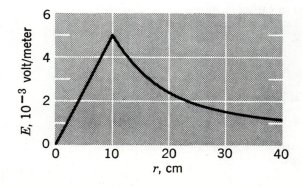

figure 35-11
Example 4. If the fringing of the field in Fig. 35-10 were to be taken into account, the result would be a rounding of the sharp cusp at $r = R$ (= 10 cm).

From Faraday's law (Eq. 35-9),

$$(E)(2\pi r) = -\frac{d\Phi_B}{dt} = -(\pi R^2)\frac{dB}{dt}.$$

Solving for E yields

$$E = -\frac{1}{2}\frac{R^2}{r}\frac{dB}{dt}.$$

These two expressions for $E(r)$ yield the same result, as they must, for $r = R$. Figure 35-11 is a plot of the magnitude of $\mathbf{E}(r)$ for the numerical values given.

In applying Lenz's law to Fig. 35-10, imagine that a circular conducting loop is placed concentrically in the field. Since Φ_B through this loop is increasing, the induced current in the loop will tend to oppose this "change" by setting up a magnetic field of its own that points up within the loop. Thus the induced current i must be counterclockwise, which means that the lines of the induced electric field $\mathbf{E}$, which is responsible for the current, must also be counterclockwise. If the magnetic field in Fig. 35-10 were *decreasing* with time, the induced current and the lines of force of the induced electric field $\mathbf{E}$ would be clockwise, again opposing the *change* in Φ_B.

Figure 35-12 shows four of many possible loops to which Faraday's law may be applied. For loops 1 and 2, the induced emf ε is the same because these loops lie entirely within the changing magnetic field and, having the same area, thus have the same value of $d\Phi_B/dt$. Note that even though the emf ε $(= \oint \mathbf{E}\cdot d\mathbf{l})$ is the same for these two loops, the distribution of electric fields $\mathbf{E}$ around the perimeter of each loop, as indicated by the electric lines of force, is different. For loop 3 the emf is less because Φ_B and $d\Phi_B/dt$ for this loop are less, and for loop 4 the induced emf is zero.

The induced electric fields that are set up by the induction process are not associated with charges but with a changing magnetic flux. Although both kinds of electric fields exert forces on charges, there is a difference between them. The simplest manifestation of this difference is that lines of $\mathbf{E}$ associated with a changing magnetic flux can form closed loops (see Fig. 35-12); lines of $\mathbf{E}$ associated with charges cannot form closed loops but can always be drawn to start on a positive charge and end on a negative charge.

Equation 29-5, which defined the potential difference between two points a and b, is

$$V_b - V_a = \frac{W_{ab}}{q_0} = -\int_a^b \mathbf{E}\cdot d\mathbf{l}.$$

We have insisted that if potential is to have any useful meaning this integral (and W_{ab}) must have the same value for every path connecting a and b. This proved to be true for every case examined in earlier chapters.

An interesting special case comes up if a and b are the same point. The path connecting them is now a closed loop; V_a must be identical with V_b and this equation reduces to

$$\oint \mathbf{E}\cdot d\mathbf{l} = 0. \tag{35-10}$$

However, when changing magnetic flux is present, $\oint \mathbf{E}\cdot d\mathbf{l}$ is precisely *not* zero but is, according to Faraday's law (see Eq. 35-9), $-d\Phi_B/dt$. Electric fields associated with stationary charges are *conservative*, but those associated with changing magnetic fields are *nonconservative*; see Section 8-2. Electric potential, which can be defined only for a conservative force, *has no meaning for electric fields produced by induction.*

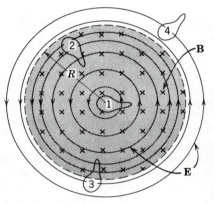

figure 35-12
Showing the circular lines of $\mathbf{E}$ from an increasing magnetic field. The four loops are imaginary paths around which an emf can be calculated.

The betatron is a device used to accelerate electrons to high speeds by allowing them to be acted upon by induced electric fields that are set up by a changing magnetic flux. It provides an excellent illustration of the "reality" of such induced fields and it is in this context that we discuss it. The energetic electrons can be used for fundamental research in physics or to produce penetrating X-rays which are useful in cancer therapy and in industry.

Figure 35-13 shows a 100-MeV betatron that was constructed at the General Electric Company. At this energy the electron speed is $0.999986c$, where c is the speed of light, so that relativistic mechanics must certainly be used in the analysis of its operation. Figure 35-14 shows a vertical cross section through the central part of the betatron to which the man in Fig. 35-13 is pointing.

The magnetic field in the betatron has several functions: (a) it guides the electrons in a circular path; (b) the changing magnetic field generates an electric field (Section 35-5) that accelerates the electrons in this path; (c) it keeps the radius of the orbit in which the electrons are moving essentially constant; (d) it introduces the electrons into the or-

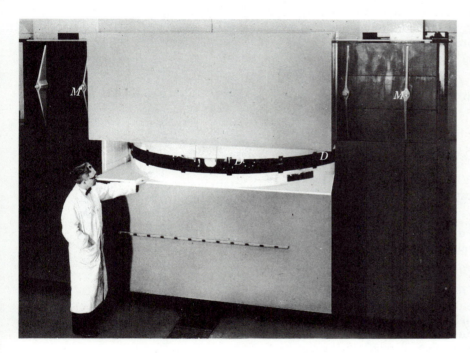

figure 35-13
A 100-MeV betatron. M shows the magnet, C the magnetizing coils, and D the region in which the "doughnut" is located. (Courtesy General Electric Company.)

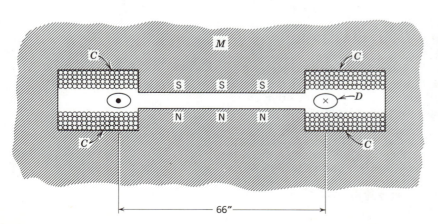

figure 35-14
Cross section of a betatron, showing magnet M, coils C, and "doughnut" D. Electrons emerge from the page at the left and enter it at the right. See Problem 33.

bit initially and removes them from the orbit after they have reached full energy; and finally (e) it provides a restoring force that resists any tendency for the electrons to leave their orbit, either vertically or radially. It is remarkable that it is possible to do all these things by proper shaping and control of the magnetic field.

The object marked D in Fig. 35-14 is an evacuated glass "doughnut" inside which the electrons travel. Their orbit is a circle at right angles to the plane of the figure. The electrons emerge from the plane at the left (·) and enter it at the right (×). In the General Electric machine the radius of the electron path is 33 in. The coils C and the 130-ton steel magnet shown in Fig. 35-13 provide the magnetic flux that passes through the plane of this orbit.

The current in coils C is made to alter periodically, 60 times/second to produce a changing flux through the orbit, shown in Fig. 35-15. Here Φ_B is taken as positive when **B** is pointing up, as in Fig. 35-14. If the electrons are to circulate in the direction shown, they must do so during the positive half-cycle, marked ac in Fig. 35-15. You should verify this (see Section 33-6). The electrons are accelerated by electric fields set up by the changing flux. The direction of these induced fields depends on the sign of $d\Phi_B/dt$ and must be chosen to accelerate, and not to decelerate, the electrons. Thus only half the positive half-cycle in Fig. 35-15 can be used for acceleration; it will prove to be ab.

The average value of $d\Phi_B/dt$ during the quarter-cycle ab is the slope of the dashed line, or

$$\frac{d\Phi_B}{dt} = \frac{1.8 \text{ Wb}}{4.2 \times 10^{-3} \text{ s}} = 430 \text{ V}.$$

From Faraday's law (Eq. 35-1), this is also the emf in volts. The electron will thus increase its energy by 430 eV every time it makes a trip around the orbit in the changing flux. If the electron gains only 430 eV of energy per revolution, it must make about 230,000 rev to gain its full 100 MeV. For an orbit radius of 33 in., this corresponds to a length of path of some 750 miles.

The betatron provides a good example of the fact that electric potential has no meaning for electric fields produced by induction. If a potential exists, it must be true that, as Eq. 35-10 shows, $\oint \mathbf{E} \cdot d\mathbf{l} = 0$ for any closed path. In the betatron, however, this integral, evaluated around the orbit, is precisely *not* zero but is, in our example, 430 volts. It must not be thought, of course, that the betatron violates the conservation-of-energy principle. The gain in kinetic energy of the circulating electron (430 eV/rev) must be supplied by an identifiable energy source. It comes, in fact, from the generator that energizes the magnet coils, thus providing the changing magnetic field. The energy is transmitted to the electron through the intermediary of this changing field.

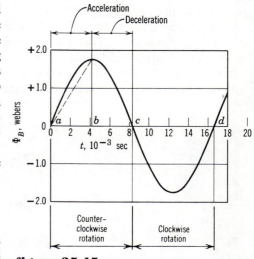

figure 35-15
The flux through the orbit of a betatron, during one cycle. Rotation of the electrons in the desired direction (counterclockwise as viewed from above in Fig. 35-14) is possible only during half-cycle ac.

EXAMPLE 5

In the betatron of Fig. 35-13, which is the "accelerating" quarter-cycle?

Let us assume that it is ab in Fig. 35-15, during which Φ_B through the orbit is *increasing*. If a conducting loop were placed to coincide with the orbit, an induced current would appear in the loop to oppose the tendency of Φ_B to increase. This means that a magnetic field would be set up that would oppose the field of the large magnet. Thus **E** would point outward at the right side of "doughnut" D in Fig. 35-14 and inward on the left side. The force $(-e\mathbf{E})$ acting on the electron is in the opposite direction to **E** because of the negative charge of the electron. Thus the tangential force acting on the electron is in the same direction as that at which it circulates in its orbit; this means that the speed of the electron will increase, as desired. You should go through this same analysis carefully, assuming (incorrectly, as it will turn out) that the accelerating half-cycle is *bc* in Fig. 35-15 rather than *ab*.

Faraday's law, in the form $\varepsilon = -d\Phi_B/dt$, gives correctly the induced emf no matter whether the change in Φ_B is produced by moving a coil, moving a magnet, changing the strength of a magnetic field, changing the shape of a conducting loop, or in other ways. However, observers who are in relative motion with respect to each other, even though they would all agree on the numerical value of the emf, would give different microscopic descriptions of the induction process. In electromagnetic systems, as well as in mechanical systems, it is important that the state of motion of the observer with respect to his environment be made perfectly clear.

Figure 35-16 shows a closed loop which is caused to move at velocity **v** with respect to a magnet that provides a uniform field **B** in the region shown. We consider first an observer, identified as *S,* who is *at rest with respect to the magnet* used to establish the field **B;** see Fig. 35-16a. The induced emf in this case is called a *motional emf* because the conducting loop is moving with respect to this observer.

Consider a positive charge carrier at the center of the left end of the conducting loop. To observer *S,* this charge, constrained to move to the right along with the loop, is a charge q moving with a velocity **v** in the magnetic field **B** and as such it experiences a sideways magnetic deflecting force given by Eq. 33-2 $(\mathbf{F} = q\mathbf{v} \times \mathbf{B})$. This force causes the carriers to move upward along the conductor, so that they acquire a drift velocity $\mathbf{v}_d$ also, as shown in Fig. 35-16a.

The resultant equilibrium speed of the carrier is now **V,** which we find by adding **v** and $\mathbf{v}_d$ vectorially in Fig. 35-16a. The magnetic deflecting force $\mathbf{F}_m$ is, as always, at right angles to the resultant velocity **V** of the carrier and is given by

$$\mathbf{F}_m = q\mathbf{V} \times \mathbf{B}. \qquad (35\text{-}13)$$

35-7
INDUCTION AND RELATIVE MOTION

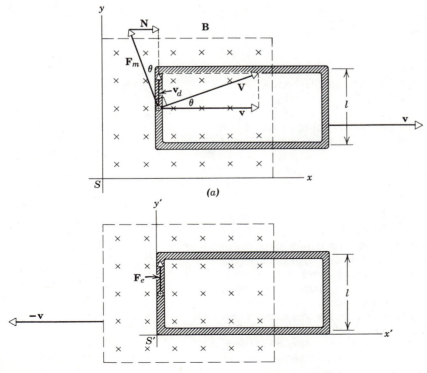

(a)

(b)

figure 35-16
A closed conducting loop is in relative motion with respect to a magnet. (*a*) An observer *S,* fixed with respect to the magnet that produces the field **B,** sees the loop moving to the right. He also sees (see text) a *magnetic* force (Eq. 33-2) equal to $F_m \cos \theta$ acting upward on the positive charge carriers. (*b*) An observer *S',* fixed with respect to the loop, sees the magnet moving toward the left. He also sees an *electric* force (see text) acting upward on the positive charge carriers. In both figures there is, of course, an average internal collision force (not shown; see Sections 31-4 and 31-5) that keeps the charge carriers from accelerating.

Now $\mathbf{F}_m$ acting alone would tend to push the carriers through the left wall of the conductor. Because this does not happen the conductor wall must exert a normal force $\mathbf{N}$ on the carriers (see Fig. 35-16a) of magnitude such that $\mathbf{v}_d$ lies parallel to the axis of the wire; in other words, $\mathbf{N}$ exactly cancels the horizontal component of $\mathbf{F}_m$, leaving only the component $F_m \cos \theta$ that lies along the direction of the conductor. This component of force on the carrier is also canceled out, this time by the average force $\overline{\mathbf{F}}_i$ associated with the internal collisions that the carrier experiences as it drifts with (constant) speed v_d through the wire.

The kinetic energy of the charge carrier as it drifts through the wire remains constant. This is consistent with the fact that the resultant force acting on the charge carrier $(= \mathbf{F}_m + \overline{\mathbf{F}}_i + \mathbf{N})$ is zero. The work done by $\mathbf{F}_m$ is zero because magnetic forces, acting at right angles to the velocity of a moving charge, can do no work on that charge. Thus the (negative) work done on the carrier by the average internal collision force $\mathbf{F}_i$ must be exactly canceled by the (positive) work done on the carrier by the force $\mathbf{N}$. In the last analysis $\mathbf{N}$ is exerted by the agent who pulls the loop through the magnetic field, and the mechanical energy expended by this agent appears as thermal energy in the loop, as we have seen in Section 35-4.

Let us then calculate the work dW done on the carrier in time dt by the force $\mathbf{N}$; it is

$$dW = N(v \, dt) \qquad (35\text{-}14)$$

in which $v \, dt$ is the distance that the loop (and the carrier) have moved to the right in Fig. 35-16a in time dt. We can write for N (see Eq. 35-13 and Fig. 35-16a).

$$N = F_m \sin \theta = (qVB)(v_d/V) = qBv_d. \qquad (35\text{-}15)$$

Substituting Eq. 35-15 into Eq. 35-14 yields

$$\begin{aligned} dW &= (qBv_d)(v \, dt) \\ &= (qBv)(v_d \, dt) = qBv \, dl \end{aligned} \qquad (35\text{-}16)$$

in which $dl \,(= v_d dt)$ is the distance the carrier drifts along the conductor in time dt.

The work done on the carrier as it makes a complete circuit of the loop is found by integrating Eq. 35-16 around the loop and is

$$W = \oint dW = qBvl. \qquad (35\text{-}17)$$

This follows because work contributions for the top and the bottom of the loops are opposite in sign and cancel and no work is done in those portions of the loop that lie outside the magnetic field.

An agent that does work on charge carriers, thus establishing a current in a closed conducting loop, can be viewed as an emf. We can write, making use of Eq. 35-17,

$$\varepsilon = \frac{W}{q} = \frac{qBvl}{q} = Blv, \qquad (35\text{-}18)$$

which is, of course, the same result that we derived from Faraday's law of induction; see Eq. 35-3. Thus a motional emf is intimately connected with the sideways deflection of a charged particle moving through a magnetic field.

We now consider how the situation of Fig. 35-16 would appear to an observer S' who is *at rest with respect to the loop.* To this observer, the magnet is moving to the left in Fig. 35-16b with velocity $-\mathbf{v}$, and the charge q is at rest as far as its left-to-right motion is concerned. However S', like S, observes that the charge drifts clockwise around the loop and he measures the same emf ε that S measures. S' accounts for this, at the microscopic level, by postulating that an electric field $\mathbf{E}$ is induced in the loop by the action of the moving magnet. This induced field $\mathbf{E}$, which has the same origin as the induced fields that we discussed in Section 35-5, exerts a force on the charge carrier given by $q\mathbf{E}$.

The induced field $\mathbf{E}$, which exists in the end of the loop only, is associated with an emf ε and generates a current in the closed loop. Note that, in any closed loop in which there is a current, an internal electric field must exist at every point at which charges are moving. These electric fields, however, are set up by the emf, as in the case of a closed loop connected to a battery, and are *not* induced by the motion of the magnet. It is only this induced field **E** that we associate with the emf, through the relation (Eq. 35-8)

$$\varepsilon = \oint \mathbf{E} \cdot d\mathbf{l}$$

which reduced to

$$\varepsilon = El \tag{35-19}$$

in this case. This is so because no *induced* electric field appears in the upper and lower bars (because of the nature of their motions), and none appears in the part of the loop outside the magnetic field.

The emfs given by Eqs. 35-18 and 35-19 must be identical because the relative motion of the loop and the magnet is identical in the two cases shown in Fig. 35-16. Equating these relations yields

$$El = Blv,$$

or

$$E = vB. \tag{35-20a}$$

In Fig. 35-16*b* the vector **E** points upward along the axis of the left end of the conducting loop because this is the direction in which positive charges are observed to drift. The directions of **v** and **B** are clearly shown in this figure. We see, then, that Eq. 35-20*a* is consistent with the more general vector relation

$$\mathbf{E} = \mathbf{v} \times \mathbf{B}. \tag{35-20b}$$

We have not proved Eq. 35-20*b* except for the special case of Fig. 35-16; nevertheless it proves to be true in general, that is, no matter what the angle between **v** and **B**.

We interpret Eq. 35-20*b* in the following way: Observer *S* fixed with respect to the magnet is aware only of a magnetic field. The force to him arises from the motion of the charges through **B**. Observer *S'* fixed on the charge carrier is aware of an electric field **E** also and attributes the force on the charge (at rest with respect to him initially) to the electric field. *S* says the force is of purely magnetic origin and *S'* says the force is of purely electric origin. From the point of view of *S*, the induced emf is given by $\oint (\mathbf{v} \times \mathbf{B}) \cdot d\mathbf{l}$. From the point of view of *S'*, the same induced emf is given by $\oint \mathbf{E} \cdot d\mathbf{l}$, where **E** is the (induced) electric vector that he observes at points along the circuit.

For a third observer *S''* who judges that both the magnet and the loop are moving, the force tending to move charges around the loop is neither purely electric nor purely magnetic, but a bit of each. In summary, in the equation

$$\mathbf{F}/q = \mathbf{E} + \mathbf{v} \times \mathbf{B},$$

different observers form different assessments of **E, B,** and **v** but, when these are combined, all observers form the same assessment of $\mathbf{F}/q$ and all obtain the same value for the induced emf in the loop (this depends only on the relative motion). That is, the total force is the same for all observers, but each observer forms a different estimate of the separate electric and magnetic forces contributing to the same total force.

The essential point is that what seems like a magnetic field to one observer may seem like a mixture of an electric field and a magnetic field to a second observer in a different inertial reference frame. Both observers would agree, however, on the gross measurable result, in the case of Fig. 35-16, the current in the loop. We are forced to conclude that magnetic and electric fields are *not* independent of each other and have no separate unique existence; they depend on the inertial frame.

Einstein started to think about relative motion at age 16 and published his famous paper on the theory of special relativity in 1905, when he was a 26-year-old patent examiner in the Swiss Patent Office in Berne. He was *not* led to this theory by early considerations of the nature of space and time but precisely by the problems raised in this section.

This is not only made clear by the title of his paper (translated from the German), "On the Electromagnetic Forces Acting on Moving Bodies," but is strongly reenforced by the opening lines of his paper, which are:

It is known that Maxwell's electrodynamics—as usually understood at the present time—when applied to moving bodies, leads to asymmetries which do not appear to be inherent in the phenomena. Take, for example, the reciprocal electrodynamic action of a magnet and a conductor. The observable phenomenon here depends only on the relative motion of the conductor and the magnet, whereas the customary view draws a sharp distinction between the two cases in which either the one or the other of these bodies is in motion. For if the magnet is in motion and the conductor at rest, there arises in the neighbourhood of the magnet an electric field with a certain definite energy, producing a current at the places where parts of the conductor are situated. But if the magnet is stationary and the conductor in motion, no electric field arises in the neighbourhood of the magnet. In the conductor, however, we find an electromotive force, to which in itself there is no corresponding energy, but which gives rise— assuming equality of relative motion in the two cases discussed—to electric currents of the same path and intensity as those produced by the electric forces in the former case.

If you want to pursue this matter further, please read Supplementary Topic V carefully and then refer to *Introduction to Special Relativity* (Chapter IV) by Robert Resnick, John Wiley & Sons (1968).

EXAMPLE 6

In Fig. 35-16 assume that $B = 2.0$ T, $l = 10$ cm, and $v = 1.0$ m/s. Calculate (a) the induced electric field observed by S' and (b) the emf induced in the loop.

(a) The electric field, which is apparent only to observer S', is associated with the moving magnet and is given in magnitude (see Eq. 35-20a) by

$$E = vB$$
$$= (1.0 \text{ m/s})(2.0 \text{ T})$$
$$= 2.0 \text{ V/m}.$$

(b) Observer S would calculate the induced (motional) emf from

$$\varepsilon = Blv$$
$$= (2.0 \text{ T})(1.0 \times 10^{-1} \text{m})(1.0 \text{ m/s})$$
$$= 0.20 \text{ V}.$$

Observer S' would not regard the emf as motional and would use the relationship

$$\varepsilon = El$$
$$= (2.0 \text{ V/m})(1.0 \times 10^{-1} \text{ m})$$
$$= 0.20 \text{ V}.$$

As must be the case, both observers agree as to the numerical value of the emf.

---

questions

1. In Figs. 35-1, 35-2, and 35-3, etc. we show, for simplicity, coils of only one turn. Explain the advantage of increasing the number of turns.

2. Are "induced" emfs and currents different in any way from emfs and currents provided by a battery connected to a conducting loop? Discuss.

3. Although we discussed these matters in earlier chapters, can you now explain more clearly in your own words the difference between a magnetic field **B** and the flux of a magnetic field Φ_B? Are they vectors or scalars? In what units may each be expressed? How are these units related? Are either or both (or neither) properties of a given point in space? Discuss fully.

4. A magnet is dropped from the ceiling along the axis of a copper loop lying flat on the floor. If the falling magnet is photographed with a time sequence

camera, what differences, if any, will be noted if (*a*) the loop is at room temperature and (*b*) the loop is packed in dry ice?

5. Two conducting loops face each other a distance *d* apart (Fig. 35-17). An observer sights along their common axis from left to right. If a clockwise current *i* is suddenly established in the larger loop, (*a*) what is the direction of the induced current in the smaller loop? (*b*) What is the direction of the force (if any) that acts on the smaller loop?

6. What is the direction of the induced emf in coil *Y* of Fig. 35-18 when (*a*) coil *Y* is moved toward coil *X* and (*b*) the current in coil *X* is decreased, without any change in the relative positions of the coils?

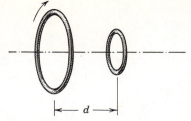

figure 35-17
Question 5

figure 35-18
Question 6

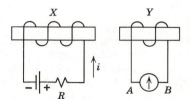

7. The north pole of a magnet is moved away from a metallic ring, as in Fig. 35-19. In the part of the ring farthest from the reader, which way does the current point?

8. *Eddy currents.* A sheet of copper is placed in a magnetic field as shown in Fig. 35-20. If we attempt to pull it out of the field or push it further in, an automatic resisting force appears. Explain its origin. (Hint: Currents, called eddy currents, are induced in the sheet in such a way as to oppose the motion.)

9. A current-carrying solenoid is moved toward a conducting loop as in Fig. 35-21. What is the direction of circulation of current in the loop as we sight toward it as shown?

10. If the resistance *R* in the left-hand circuit of Fig. 35-22 is increased, what is the direction of the induced current in the right-hand circuit?

figure 35-19
Question 7

figure 35-20
Question 8

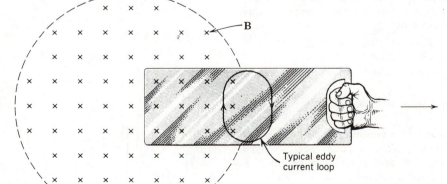

Typical eddy current loop

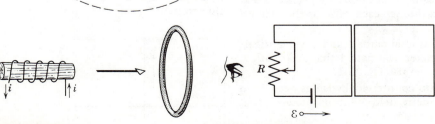

figure 35-21
Question 9

figure 35-22
Question 10

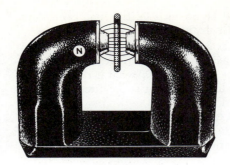

figure 35-23
Question 11

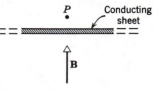

figure 35-24
Question 12

11. A loop, shown in Fig. 35-23, is removed from the magnet by pulling it vertically upward. (a) What is the direction of the induced current? (b) Is a force required to remove the loop? (c) Does the total amount of thermal energy produced in removing the loop depend on the time taken to remove it?

12. *Electromagnetic shielding.* Consider a conducting sheet lying in a plane perpendicular to a magnetic field **B**, as shown in Fig. 35-24. (a) If **B** suddenly changes, the full *change* in **B** is not immediately detected in region P. Explain. (b) If the resistivity of the sheet is zero, the *change* is not ever detected at P. Explain. (c) If **B** changes periodically at high frequency and the conductor is made of a material of low resistivity, the region near P is almost completely shielded from the *changes* in flux. Explain. (d) Is such a conductor useful as a shield from *static* magnetic fields? Explain.

13. *Magnetic damping.* A strip of copper is mounted as a pendulum about O in Fig. 35-25. It is free to swing through a magnetic field normal to the page. If the strip has slots cut in it as shown, it can swing freely through the field. If a strip without slots is substituted, the vibratory motion is strongly damped. Explain. (Hint: Use Lenz's law; consider the paths that the charge carriers in the strip must follow if they are to oppose the motion.)

14. What is the direction, if any, of the conventional current through resistor R in Fig. 35-26 (a) immediately after switch S is closed, (b) some time after switch S was closed, and (c) immediately after switch S is opened. (d) When switch S is held closed, which end of the long coil acts as a north pole? (e) How do the free charges in the coil containing R know about the flux within the long coil? What really gets them moving?

15. In Fig. 35-27 the movable wire is moved to the right, causing an induced current as shown. What is the direction of **B** in region A?

16. Account qualitatively for the configurations of the lines of **B** in Figs. 35-6 b and c.

17. How would Fig. 35-8 be changed if we took into account the necessary fringing of the magnetic field **B** in Fig. 35-7?

18. (a) In Fig. 35-10, need the circle of radius r be a conducting loop for **E** and ε to be present? (b) If the circle of radius r were not concentric (moved slightly to the left, say), would ε change? Would the configuration of **E** around the circle change? (c) For a concentric circle of radius r, with r > R, does an emf exist? Do electric fields exist?

19. A copper ring and a wooden ring of the same dimensions are placed so that there is the same changing magnetic flux through each. How do the induced electric fields in each ring compare?

20. In Fig. 35-12 how can the induced emfs around paths 1 and 2 be identical? The induced electric fields are much weaker near path 1 than near path 2, as the spacing of the lines of force shows. See also Fig. 35-11.

21. In a certain betatron the electrons rotate counterclockwise as seen from above. In what direction must the magnetic field point and how must it change with time while the electron is being accelerated?

22. Why can a betatron be used for acceleration only during one-quarter of a cycle?

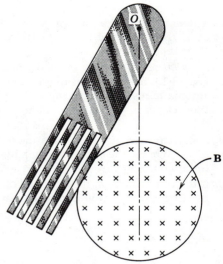

figure 35-25
Question 13

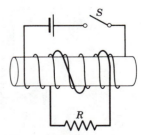

figure 35-26
Question 14

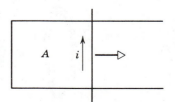

figure 35-27
Question 15

23. To make the electrons in a betatron orbit spiral outward, would it be necessary to increase or to decrease the central flux? Assume that **B** at the orbit remains essentially unchanged.

24. A cyclotron (see Section 33-7) is a so-called *resonance device*. Does a betatron depend on resonance? Discuss.

25. In Fig. 35-16a we can see that a force ($F_m \cos \theta$) acts on the charge carriers in the left branch of the loop. However, if there is to be a continuous current in the loop, and there is, a force of some sort must act on charge carriers in the other three branches of the loop to maintain the same drift speed v_d in these branches. What is its source? (Hint: Consider that the left branch of the loop was the only conducting element, the other three being nonconducting. Would not positive charge pile up at the top of the left half and negative charge at the bottom?)

26. Show that one volt = one weber/second.

problems

SECTION 35-2

1. A uniform magnetic field **B** is normal to the plane of a circular ring 10 cm in diameter made of #10 copper wire (diameter = 0.10 in.). At what rate must B change with time if an induced current of 10 A is to appear in the ring? *Answer:* 1.3 T/s.

2. You are given 50 cm of #18 copper wire (diameter = 0.040 in.). It is formed into a circular loop and placed at right angles to a uniform magnetic field that is increasing with time at the constant rate of 0.010 T/s. At what rate (in watts) is thermal energy generated in the loop?

3. A hundred turns of insulated copper wire are wrapped around an iron cylinder of cross-sectional area 0.001 m² and are connected to a resistor. The total resistance in the circuit is 10 Ω. If the longitudinal magnetic field in the iron changes from 1.0 T in one direction to 1.0 T in the opposite direction, how much charge flows through the circuit?   *Answer:* 2.0×10^{-2} C.

4. The current in the solenoid of Example 1 changes, not as in that Example, but according to $i = 3.0t + 1.0t^2$, where i is in amperes and t in seconds. (a) Plot quantitatively the induced emf in the coil from $t = 0$ to $t = 4.0$ s. (b) The resistance of the coil is 0.15 Ω. What is the instantaneous current in the coil at $t = 2.0$ s?

5. A small loop of area A is inside of, and has its axis in the same direction as, a long solenoid of n turns per unit length and current i. If $i = i_0 \sin \omega t$, find the emf ε in the loop.   *Answer:* $-\mu_0 n A i_0 \omega \cos \omega t$.

6. *Alternating current generator.* A rectangular loop of N turns and of length a and width b is rotated at a frequency ν in a uniform magnetic field **B**, as in Fig. 35-28. (a) Show that an induced emf given by

$$\varepsilon = 2\pi\nu NbaB \sin 2\pi\nu t = \varepsilon_0 \sin 2\pi\nu t$$

appears in the loop. This is the principle of the commercial alternating-current generator. (b) Design a loop that will produce an emf with $\varepsilon_0 = 150$ V when rotated at 60 rev/s in a magnetic field of 0.50 T.

7. In Fig. 35-29 a closed single turn with a copper coil resistance of 5.0 Ω is placed *outside* a solenoid like that of Example 1. If the current in the solenoid is changed as in that example, (a) what current appears in the loop while the solenoid current is being changed? (b) How do the free charges

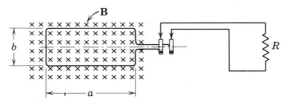

figure 35-28

Problem 6

in the loop "get the message" from the solenoid that they should start moving (to establish a current)? After all, the magnetic flux is entirely confined to the interior of the solenoid.
Answer: (a) 2.1×10^{-4} A.

8. Figure 35-30 shows a copper rod moving on conducting rails with velocity **v** parallel to a long straight wire carrying a current i. Calculate the induced emf ε in the rod, assuming that $v = 5.0$ m/s, $i = 100$ A, $a = 1.0$ cm, and $b = 20$ cm.

9. A circular loop of wire 10 cm in diameter is placed with its normal making an angle of 30° with the direction of a uniform 0.50-T magnetic field. The loop is "wobbled" so that its normal rotates about the field direction at the constant rate of 100 rev/min; the angle between the normal and the field direction (= 30°) remains unchanged during this process. What emf appears in the loop?   *Answer:* Zero.

10. A stiff wire bent into a semicircle of radius R is rotated with a frequency ν in a uniform magnetic field **B,** as shown in Fig. 35-31. What are the amplitude and frequency of the induced emf and of the induced current when the internal resistance of the meter M is R_M and the remainder of the circuit has negligible resistance?

11. Figure 35-32 shows a copper rod moving with velocity **v** parallel to a long straight wire carrying a current i. Calculate the induced emf in the rod, assuming that $v = 5.0$ m/s, $i = 100$ A, $a = 1.0$ cm, and $b = 20$ cm.
Answer: 3.0×10^{-4} V.

12. A uniform magnetic field **B** is changing in magnitude at a constant rate dB/dt. You are given a mass m of copper which is to be drawn into a wire of radius r and formed into a circular loop of radius R. Show that the induced current in the loop does not depend on the size of the wire or of the loop and, assuming **B** perpendicular to the loop, is given by

$$i = \frac{m}{4\pi\rho\delta} \frac{dB}{dt},$$

where ρ is the resistivity and δ the density of copper.

13. A circular loop of radius r (10 cm) is placed in a uniform magnetic field **B** (0.80 T) normal to the plane of the loop. The radius of the loop begins shrinking at a constant rate dr/dt (80 cm/s). (a) What is the emf ε induced in the loop? (b) At what constant rate would the area have to shrink to induce this same emf?   *Answer:* (a) 0.40 V. (b) 0.50 m²/s.

14. Prove that the electric field **E** in a charged parallel-plate capacitor cannot drop abruptly to zero as one moves at right angles to it, as suggested by the arrow in Fig. 35-33 (see point a). In actual capacitors fringing of the lines of force always occurs, which means that **E** approaches zero in a continuous and gradual way; see Problem 34-2. (Hint: Apply Faraday's law to the rectangular path shown by the dashed lines.)

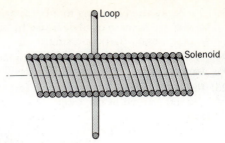

figure 35-29
Problem 7

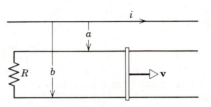

figure 35-30
Problem 8

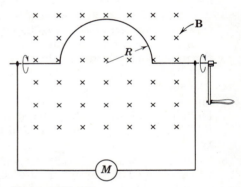

figure 35-31
Problem 10

figure 35-32
Problem 11

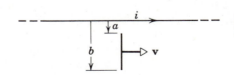

figure 35-33
Problem 14

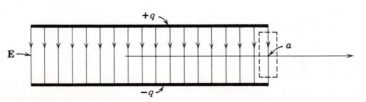

15. An electromagnetic "eddy current" brake consists of a disk of conductivity σ and thickness t rotating about an axis through its center with a magnetic field **B** applied perpendicular to the plane of the disk over a small area a^2 (see Fig. 35-34). If the area a^2 is at a distance r from the axis, find an approximate expression for the torque tending to slow down the disk at the instant its angular velocity equals ω. *Answer:* $\tau = B^2 a^2 r^2 \, \omega \, \sigma \, t$.

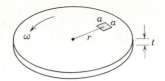

figure 35-34
Problem 15

SECTION 35-3

16. A small bar magnet is pulled rapidly through a conducting loop, along its axis. Sketch qualitatively (*a*) the induced current and (*b*) the rate of thermal energy production as a function of the position of the center of the magnet. Assume that the north pole of the magnet enters the loop first and that the magnet moves at constant speed. Plot the induced current as positive if it is clockwise as viewed along the path of the magnet.

17. A metal wire of mass m slides without friction on two rails spaced a distance d apart, as in Fig. 35-35. The track lies in a vertical uniform magnetic field **B**. (*a*) A *constant current* i flows from generator G along one rail, across the wire, and back down the óther rail. Find the velocity (speed and direction) of the wire as a function of time, assuming it to be at rest at $t = 0$. (*b*) The generator is replaced by a battery with *constant emf* ε. The velocity of the wire now approaches a constant final value. What is this terminal speed? (*c*) What is the current in part (*b*) when the terminal speed has been reached? *Answer:* (*a*) $Bidt/m$, away from G. (*b*) ε/Bd. (*c*) Zero.

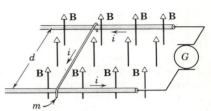

figure 35-35
Problem 17

18. In Fig. 35-36 the magnetic flux through the loop perpendicular to the plane of the coil and directed into the paper is varying according to the relation

$$\Phi_B = 6t^2 + 7t + 1,$$

where Φ_B is in milliwebers (1 milliweber $= 10^{-3}$ weber) and t is in seconds. (*a*) What is the magnitude of the emf induced in the loop when $t = 2.0$s? (*b*) What is the direction of the current through R?

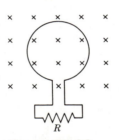

figure 35-36
Problem 18

SECTION 35-4

19. In the arrangement of Example 3 put $B = 1.2$ T and $L = 5.0$ cm. If $\varepsilon = 1.0$ V, what acceleration will a point at the end of the rotating rod experience? *Answer:* 2300 "g".

20. In Fig. 35-37 a conducting rod AB makes contact with the metal rails AD and BC which are 50 cm apart in a uniform magnetic field of 1.0 T perpendicular to the plane of the paper as shown. The total resistance of the circuit $ABCD$ is 0.4 Ω (assumed constant). (*a*) What is the magnitude and direction of the emf induced in the rod when it is moved to the left with a velocity of 8.0 m/s? (*b*) What force is required to keep the rod in motion? (*c*) Compare the rate at which mechanical work is done by the force **F** with the rate of development of thermal energy in the circuit.

21. In Fig. 35-38, $l = 2.0$ m and $v = 50$ cm/s. **B** is the earth's magnetic field, directed perpendicularly out of the page and having a magnitude 6.0×10^{-5} T at that place. The resistance of the circuit $ADCB$, assumed constant (explain how this may be achieved approximately), is $R = 1.2 \times 10^{-5}\,\Omega$. (*a*) What is the emf induced in the circuit? (*b*) What is the electric field in the wire AB? (*c*) What force does each electron in the wire experience due to the motion of the wire in the magnetic field? (*d*) What is the magnitude and direction of the current in the wire? (*e*) What force must an external agency exert in order to keep the wire moving with this constant velocity? (*f*) Compute the rate at which the external agency is doing work. (*g*) Compute the rate at which electrical energy is being converted into thermal energy. *Answer:* (*a*) 6.0×10^{-5} V. (*b*) 3.0×10^{-5} V/m. (*c*) 4.8×10^{-24} N. (*d*) 5.0 A. (*e*) 6.0×10^{-4} N. (*f*) 3.0×10^{-4} W. (*g*) 3.0×10^{-4} W.

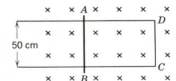

figure 35-37
Problem 20

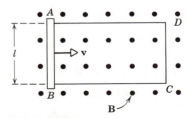

figure 35-38
Problem 21

22. A square wire of length l, mass m, and resistance R slides without friction down parallel conducting rails of negligible resistance, as in Fig. 35-39. The rails are connected to each other at the bottom by a resistanceless rail parallel to the wire, so that the wire and rails form a closed rectangular conduct-

ing loop. The plane of the rails makes an angle θ with the horizontal, and a uniform vertical magnetic field **B** exists throughout the region. (a) Show that the wire acquires a steady-state velocity of magnitude

$$v = \frac{mgR \sin \theta}{B^2 l^2 \cos^2 \theta}.$$

(b) Prove that this result is consistent with the conservation-of-energy principle. (c) What change, if any, would there be if **B** were directed down instead of up?

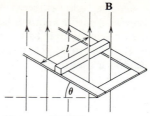

figure 35-39
Problem 22

SECTION 35-5

23. A long solenoid of radius r (2.5 cm) and turns per unit length n (100/cm) carries an initial current of i_0 (1.0 A). A single loop of wire of diameter D (10 cm) is placed around the solenoid, the axes coinciding. The current in the solenoid is reduced uniformly to i (0.50 A) over a period T (0.010 s). While the current is changing what is the induced emf ε in the surrounding loop? *Answer:* 1.2×10^{-3} V.

24. A circular coil of radius r (10 cm) is made of wire of resistance R (10 Ω). Perpendicular to the plane of the coil is a uniform magnetic field **B**. (a) At what constant rate must B increase so that there is a steady current i (0.010 A) in the circuit? (b) What power is dissipated in the resistor?

25. (a) For the arrangement of Fig. 34-28, what would be the current induced around the rectangular loop if the current in the wire decreased uniformly from 30 A to zero in 1.0 s? Assume no initial current in the loop and a resistance for the loop of 0.020 Ω. Take $a = 1.0$ cm, $b = 8.0$ cm, and $l = 30$ cm. (b) How much energy would be transferred to the loop in the 1.0-s interval? *Answer:* (a) 1.9×10^{-4} A. (b) 7.2×10^{-10} J.

26. The perpendicular field B through a one-turn circular loop of wire of negligible resistance changes with time as shown in Fig. 35-40. The loop is of radius r (10 cm) and is connected to a resistor R (10 Ω). (a) Plot the emf appearing across the resistor. (b) Plot the current i through the resistor R. (c) Plot the rate of thermal energy production in the resistor.

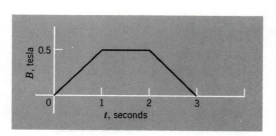

figure 35-40
Problem 26

27. A loop of wire of area A is connected to a resistance R. The loop is exposed to a time-varying **B** field (see Fig. 35-41). (a) Derive an expression for the net charge transferred through the resistor between $t = t_1$ and $t = t_2$. Show that your answer is proportional to the difference $\Phi_B(t_2) - \Phi_B(t_1)$, and is otherwise independent of the manner in which **B** is changing. (b) Suppose the change in flux, $\Phi_B(t_2) - \Phi_B(t_1)$ is zero. Does it then follow that no thermal energy production occurred during this time interval?

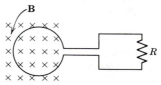

figure 35-41
Problem 27

Answer: (a) $Q = -\frac{1}{R} [\Phi_B(t_2) - \Phi_B(t_1)]$. (b) No.

28. Figure 35-42 shows two loops of wire having the same axis. The smaller loop is above the larger one, by a distance x which is large compared to the radius R of the larger loop. Hence, with current i flowing as indicated in the

larger loop, the consequent magnetic field is nearly constant throughout the plane area πr^2 bounded by the smaller loop. Suppose now that x is not constant but is changing at the constant rate $dx/dt = v$ (x increasing). (a) Determine the magnetic flux across the area bounded by the smaller loop as a function of x. (b) Compute the emf generated in the smaller loop at the instant when $x = NR$. (c) Determine the direction of the induced current flowing in the smaller loop if $v > 0$.

29. A wire is bent into three circular segments of radius r (10 cm) as shown in Fig. 35-43. Each segment is a quadrant of a circle, ab lying in the x-y plane, bc lying in the y-z plane, and ca lying in the z-x plane. (a) If a spatially uniform magnetic field $\mathbf{B}$ points in the x-direction, what is the magnitude of the emf ε developed in the wire when $\mathbf{B}$ increases at the rate of 3.0×10^{-3} T/s? (b) What is the direction of the current in the segment bc?
    Answer: (a) 2.4×10^{-5} V. (b) From c to b.

30. Figure 35-44 shows a uniform magnetic field $\mathbf{B}$ confined to a cylindrical volume of radius R. $\mathbf{B}$ is decreasing in magnitude at a constant rate of 0.010 T/s. What is the instantaneous acceleration (direction and magnitude) experienced by an electron placed at a, at b, and at c? Assume $r = 5.0$ cm. (The necessary fringing of the field beyond R will not change your answer as long as there is axial symmetry about a perpendicular axis through b.)

31. A closed loop of wire consists of a pair of equal semicircles, radius 3.70 cm, lying in mutually perpendicular planes. The loop was formed by folding a circular loop along a diameter until the two halves became perpendicular. A uniform magnetic field $\mathbf{B}$ of magnitude 0.076 T is directed perpendicular to the fold diameter and makes equal angles (45°) with the planes of the semicircles as shown in Fig. 35-45a. (a) The magnetic field is reduced at a uniform rate to zero during a time interval 4.50×10^{-3} s. Determine the magnitude of the induced emf and the sense of the induced current in the loop during this interval. (b) How would the answers change if $\mathbf{B}$ is directed as shown in Fig. 35-45b, perpendicular to the direction first given for it but still perpendicular to the "fold-diameter?"
    Answer: (a) 51×10^{-3} V. (b) $\varepsilon = 0$.

32. A uniform magnetic field $\mathbf{B}$ fills a cylindrical volume of radius R. A metal rod of length l is placed as shown in Fig. 35-46. If B is changing at the rate dB/dt, show that the emf that is produced by the changing magnetic field and that acts between the ends of the rod is given by

$$\varepsilon = \frac{dB}{dt}\frac{l}{2}\sqrt{R^2 - \left(\frac{l}{2}\right)^2}.$$

SECTION 35-6

33. Some measurements of the maximum magnetic field as a function of radius for the betatron described on page 781 are as follows:

r, cm	B, tesla	r, cm	B, tesla
0	0.400	81.2	0.409
10.2	0.950	83.7	0.400
68.2	0.950	88.9	0.381
73.2	0.528	91.4	0.372
75.2	0.451	93.5	0.360
77.3	0.428	95.5	0.340

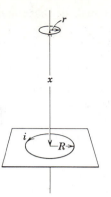

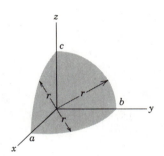

figure 35-42
Problem 28

figure 35-43
Problem 29

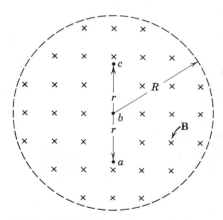

figure 35-44
Problem 30

figure 35-45
Problem 31

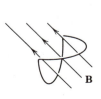

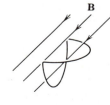

(a)   (b)

Show by graphical analysis that the relation $\overline{B} = 2B_R$ is satisfied at the orbit radius, $R = 84$ cm. (Hint: Note that $\overline{B} = \dfrac{1}{\pi R^2} \displaystyle\int_0^R B(r)(2\pi r)\, dr$ and evaluate the integral graphically.)

SECTION 35-7

34. (a) Estimate θ in Fig. 35-16a. Recall (see Section 31-1) that $v_d = 4 \times 10^{-2}$ cm/s in a typical case. Assume $v = 10$ cm/s. (b) It is clear that θ will be small. However, must we have $\theta \neq 0$ for the arguments presented in connection with this figure to be valid?

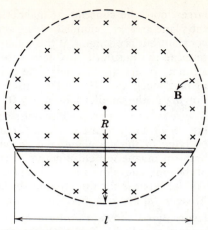

figure 35-46
Problem 32

36
inductance

If two coils are near each other, a current i in one coil will set up a flux Φ_B through the second coil. If this flux is changed by changing the current, an induced emf will appear in the second coil according to Faraday's law; see Fig. 35-2. However, two coils are not needed to show an inductive effect. An induced emf appears in a coil if the current *in that same coil* is changed. This is called *self-induction* and the electromotive force produced is called a *self-induced emf.* It obeys Faraday's law of induction just as other induced emfs do.

Consider first a "close-packed" coil, a toroid, or the central section of a long solenoid. In all three cases the flux Φ_B set up in each turn by a current i is essentially the same for every turn. Faraday's law for such coils (Eq. 35-2)

$$\varepsilon = -\frac{d(N\Phi_B)}{dt} \tag{36-1}$$

shows that the number of *flux linkages* $N\Phi_B$ (N being the number of turns) is the important characteristic quantity for induction. For a given coil, provided no magnetic materials such as iron are nearby, this quantity is proportional to the current i, or

$$N\Phi_B = Li, \tag{36-2}$$

in which L, the proportionality constant, is called the *inductance* of the device.

From Faraday's law (see Eq. 36-1) we can write the induced emf as

$$\varepsilon = -\frac{d(N\Phi_B)}{dt} = -L\frac{di}{dt}. \tag{36-3a}$$

Written in the form

$$L = -\frac{\mathcal{E}}{di/dt},\qquad(36\text{-}3b)$$

this relation may be taken as the defining equation for inductance for coils of all shapes and sizes, whether or not they are close-packed and whether or not iron or other magnetic material is nearby. It is analogous to the defining relation for capacitance, namely,

$$C = \frac{q}{V}.$$

If no iron or similar materials are nearby, L depends only on the geometry of the device. In an *inductor* (symbol ) the presence of a *magnetic field* is the significant feature, corresponding to the presence of an *electric* field in a *capacitor*.

The SI unit of inductance, from Eq. 36-3b, is the volt·second/ampere. A special name, the *henry* (abbr. H), has been given to this combination of units, or

$$1 \text{ henry} = 1 \text{ volt}\cdot\text{second/ampere}.$$

As we saw in Section 35-1, the SI unit of inductance is named after Joseph Henry (1797–1878), an American physicist and a contemporary of Faraday.

We can find the direction of a self-induced emf from Lenz's law; see Section 35-3. Suppose that a steady current i, produced by a battery, exists in a coil. Let us suddenly *decrease* the (battery) emf in the circuit. The current i will start to *decrease* at once. This decrease in current, in the language of Lenz's law, is the "change" which the self-induction must oppose. To oppose the falling current, the induced emf must point in the *same* direction as the current, as shown in Fig. 36-1a.

However, if we suddenly *increase* the (battery) emf, the current i will start to *increase* at once. This increase in current is now the "change" which self-induction must oppose. To oppose the rising current the induced emf must point in the *opposite* direction to the current, as Fig. 36-1b shows. In each case the self-induced emf acts to oppose the *change* in the current. The minus sign in Eq. 36-3 shows that $\mathcal{E}$ and di/dt are opposite in sign, since L is always a positive quantity.

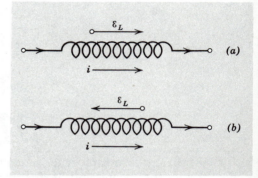

figure 36-1
In (*a*) the current i is made to *decrease* and in (*b*) it is made to *increase*. The self-induced emf $\mathcal{E}_L$ opposes the *change* in each case.

36-2
CALCULATION OF INDUCTANCE

We were able to make a direct calculation of capacitance in terms of geometrical factors for a few special cases, such as the parallel-plate capacitor. In the same way, we can calculate the self-inductance L for a few special cases.

For a close-packed coil with no iron nearby, we have, from Eq. 36-2,

$$L = \frac{N\Phi_B}{i}.\qquad(36\text{-}4)$$

Let us apply this equation to calculate L for a section of length l near the center of a long solenoid. The number of flux linkages in the length l of the solenoid is

$$N\Phi_B = (nl)(BA),$$

where n is the number of turns per unit length, B is the magnetic field inside the solenoid, and A is the cross-sectional area. From Eq. 34-7, B is given by

$$B = \mu_0 ni.$$

Combining these equations gives

$$N\Phi_B = \mu_0 n^2 liA.$$

Finally, the inductance, from Eq. 36-4, is

$$L = \frac{N\Phi_B}{i} = \mu_0 n^2 lA. \tag{36-5}$$

The inductance of a length l of a solenoid is proportional to its volume (lA) and to the square of the number of turns per unit length. Note that it depends on geometrical factors only. The proportionality to n^2 is expected. If we double the number of turns per unit length, not only is the *total* number of turns N doubled but the flux *through each turn* Φ_B is also doubled, an over-all factor of four for the flux linkages $N\Phi_B$, hence also a factor of four for the inductance (Eq. 36-4).

Derive an expression for the inductance of a toroid of rectangular cross section as shown in Fig. 36-2. Evaluate for $N = 10^3$, $a = 5.0$ cm, $b = 10$ cm, and $h = 1.0$ cm.

EXAMPLE 1

The lines of **B** for the toroid are concentric circles. Applying Ampère's law,

$$\oint \mathbf{B} \cdot d\mathbf{l} = \mu_0 i,$$

to a circular path of radius r yields

$$(B)(2\pi r) = \mu_0 i_0 N,$$

where N is the number of turns and i_0 is the current in the toroid windings; recall that i in Ampère's law is the *total* current that passes through the path of integration. Solving for B yields

$$B = \frac{\mu_0 i_0 N}{2\pi r}.$$

The flux Φ_B for the cross section of the toroid is

$$\Phi_B = \int \mathbf{B} \cdot d\mathbf{S} = \int_a^b (B)(h \, dr) = \int_a^b \frac{\mu_0 i_0 N}{2\pi r} h \, dr$$

$$= \frac{\mu_0 i_0 Nh}{2\pi} \int_a^b \frac{dr}{r} = \frac{\mu_0 i_0 Nh}{2\pi} \ln \frac{b}{a},$$

where $h \, dr$ is the area of the elementary strip shown (dashed lines) in the figure.

The inductance follows from Eq. 36-4, or

$$L = \frac{N\Phi_B}{i_0} = \frac{\mu_0 N^2 h}{2\pi} \ln \frac{b}{a}.$$

Substituting numerical values yields

$$L = \frac{(4\pi \times 10^{-7} \text{ Wb/A} \cdot \text{m})(10^3)^2(1.0 \times 10^{-2} \text{ m})}{2\pi} \ln \frac{10 \times 10^{-2} \text{ m}}{5 \times 10^{-2} \text{ m}}$$

$$= 1.4 \times 10^{-3} \text{ Wb/A} = 1.4 \text{ mH}.$$

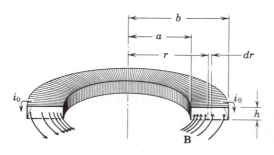

figure 36-2
Example 1. A cross section of a toroid, showing the current in the windings and the magnetic field.

In Section 32-8 we saw that if we suddenly introduce an emf ε, perhaps by using a battery, into a single loop circuit containing a resistor R and a capacitor C, the charge does not build up immediately to its final equilibrium value ($= C\varepsilon$) but approaches it in an exponential fashion described by Eq. 32-15, or

$$q = C\varepsilon\,(1 - e^{-t/\tau_C}). \tag{36-6}$$

The delay in the rise of the charge is described by the *capacitive time constant* τ_C, defined from

$$\tau_C = RC. \tag{36-7}$$

If in this same circuit the battery emf ε is suddenly removed, the charge does not immediately fall to zero but approaches zero in an exponential fashion, described by Eq. 32-18b, or

$$q = C\varepsilon\,e^{-t/\tau_C}. \tag{36-8}$$

The same time constant τ_C describes the fall of the charge as well as its rise.

An analogous delay in the rise (or fall) of the current occurs if we suddenly introduce an emf ε into (or remove it from) a single-loop circuit containing a resistor R and an inductor L. When the switch S in Fig. 36-3 is closed on a, for example, the current in the resistor starts to rise. If the inductor were not present, the current would rise rapidly to a steady value ε/R. Because of the inductor, however, a self-induced emf ε_L appears in the circuit and, from Lenz's law, this emf opposes the *rise* of current, which means that it opposes the battery emf ε in polarity. Thus the resistor responds to the difference between two emfs, a constant one ε due to the battery and a variable one ε_L ($= -L\,di/dt$) due to self-induction. As long as this second emf is present, the current in the resistor will be less than ε/R.

As time goes on, the rate at which the current increases becomes less rapid and the self-induced emf ε_L, which is proportional to di/dt, becomes smaller. Thus a time delay is introduced, and the current in the circuit approaches the value ε/R asymptotically.

When the switch S in Fig. 36-3 is thrown to a, the circuit reduces to that of Fig. 36-4. Let us apply the loop theorem, starting at x in this figure and going clockwise around the loop. For the direction of current shown, x will be higher in potential than y, which means that we encounter a drop in potential of $-iR$ as we traverse the resistor. Point y is higher in potential than point z because, for an increasing current, the induced emf will oppose the *rise* of the current by pointing as shown. Thus as we traverse the inductor from y to z we encounter a drop in potential of $-L(di/dt)$. We encounter a rise in potential of $+\varepsilon$ in traversing the battery from z to x. The loop theorem thus gives

$$-iR - L\frac{di}{dt} + \varepsilon = 0$$

or

$$L\frac{di}{dt} + iR = \varepsilon. \tag{36-9}$$

Equation 36-9 is a *differential equation* involving the variable i and its first derivative di/dt. We seek the function $i(t)$ such that when it and its first derivative are substituted in Eq. 36-9 the equation is satisfied.

Although there are formal rules for solving various classes of differential equations (and Eq. 36-9 can, in fact, be easily solved by direct

36-3
AN LR CIRCUIT

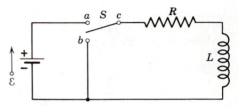

figure 36-3
An *LR* circuit.

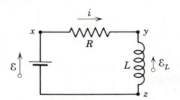

figure 36-4
The circuit of Fig. 36-3 just after switch S is closed on a.

integration, after rearrangement) we often find it simpler to guess at the solution, guided by physical reasoning and by previous experience. We can test any proposed solution by substituting it in the differential equation and seeing whether this equation reduces to an identity.

The solution to Eq. 36-9 is, we assert,

$$i = \frac{\varepsilon}{R}(1 - e^{-Rt/L}). \tag{36-10}$$

To test this solution by substitution, we find the derivative di/dt, which is

$$\frac{di}{dt} = \frac{\varepsilon}{L}e^{-Rt/L}. \tag{36-11}$$

Substituting i and di/dt into Eq. 36-9 leads to an identity, as you can easily verify. Thus Eq. 36-10 is a solution of Eq. 36-9. Figure 36-5 shows how the potential difference V_R across the resistor $(= iR;$ see Eq. 36-10) and V_L across the inductor $(\stackrel{.}{=} L\,di/dt;$ see Eq. 36-11) vary with time for particular values of ε, L, and R. Compare this figure carefully with the corresponding figure for an RC circuit (Fig. 32-11).

We can rewrite Eq. 36-10 as

$$i = \frac{\varepsilon}{R}(1 - e^{-t/\tau_L}), \tag{36-12}$$

in which τ_L, the *inductive time constant*, is given by

$$\tau_L = L/R. \tag{36-13}$$

Note the correspondence between Eqs. 36-6 and 36-12.

To show that the quantity τ_L $(= L/R)$ has the dimensions of time, we put

$$\frac{1\ \text{henry}}{\text{ohm}} = \frac{1\ \text{henry}}{\text{ohm}}\left(\frac{1\ \text{volt}\cdot\text{second}}{1\ \text{henry}\cdot\text{ampere}}\right)\left(\frac{1\ \text{ohm}\cdot\text{ampere}}{1\ \text{volt}}\right) = 1\ \text{second}.$$

The first quantity in parentheses is a conversion factor based on the defining equation for inductance $[L = -\varepsilon/(di/dt);$ see Eq. 36-3b]. The second conversion factor is based on the relation $V = iR$.

The physical significance of the time constant follows from Eq. 36-12. If we put $t = \tau_L = L/R$ in this equation, it reduces to

$$i = \frac{\varepsilon}{R}(1 - e^{-1}) = (1 - 0.37)\frac{\varepsilon}{R} = 0.63\,\frac{\varepsilon}{R}.$$

Thus the time constant τ_L is that time at which the current in the circuit will reach a value within $1/e$ (about 37%) of its final equilibrium value (see Fig. 36-5).

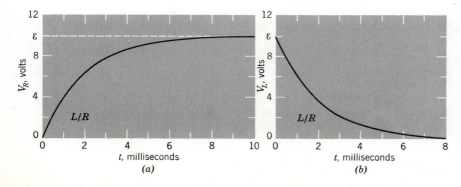

figure 36-5
If in Fig. 36-3 we assume that $R = 2000\ \Omega$, $L = 4$ H, and $\varepsilon = 10$ V, then (a) shows the variation of V_R with t during the current buildup after switch S is closed on a, and (b) the variation of V_L with t. The time constant is $L/R = 2.0 \times 10^{-3}$ s. Compare this figure carefully with Fig. 32-11, the corresponding figure for an RC circuit.

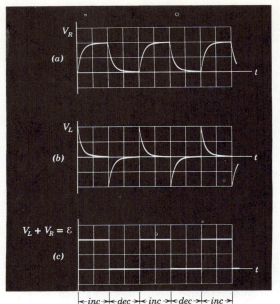

figure 36-6
Oscilloscope photograph showing the variation with time of (a) the potential drop V_R across the resistor, (b) the potential drop V_L across the inductor, and (c) the applied emf ε. During the intervals marked *inc* the current is increasing; during those marked *dec* it is decreasing. Compare with Fig. 32-13. (Courtesy E. K. Hege, Rensselaer Polytechnic Institute.)

If the switch S in Fig. 36-3, having been left in position a long enough for the equilibrium current ε/R to be established, is thrown to $b*$, the effect is to remove the battery from the circuit. The differential equation that governs the subsequent decay of the current in the circuit can be found by putting $\varepsilon = 0$ in Eq. 36-9, or

$$L \frac{di}{dt} + iR = 0. \tag{36-14}$$

You can show by the test of substitution that the solution of this differential equation is

$$i = \frac{\varepsilon}{R} e^{-t/\tau_L}. \tag{36-15}$$

Just as for the RC circuit, the behavior of the circuit of Fig. 36-3 can be investigated experimentally, using a cathode-ray oscilloscope. If switch S in this figure is thrown periodically between a and b, the applied (battery) emf alternates between the values ε and zero. If the terminals of an oscilloscope are connected across b and c in Fig. 36-3, the oscilloscope will display the waveform of this applied emf on its screen, as in Fig. 36-6c.

If the terminals of the oscilloscope are connected across the resistor, the waveform displayed (Fig. 36-6a) will be that of the current in the circuit, since the potential drop across R, which determines the oscilloscope deflection, is given by $V_R = iR$. During the intervals marked *inc* in Fig. 36-6, the current is increasing and the waveform (see Eq. 36-12) is given by

$$V_R \; (= iR) = \varepsilon(1 - e^{-t/\tau_L}).$$

During the intervals marked *dec*, the current is decreasing and V_R (see Eq. 36-15) is given by

$$V_R \; (= iR) = \varepsilon e^{-t/\tau_L}.$$

Note that both the growth and the decay of the current are delayed.

If the oscilloscope terminals are connected across the inductor, the screen will show a plot of the potential difference across it as a function of time (Fig. 36-6b). While the current is increasing, the equation of the trace (see Eq. 36-11) should be

---

* The connection to b must be made before the connection to a is broken. A switch which does this is called a "make before break" switch.

$$V_L \left(= L \frac{di}{dt}\right) = \varepsilon e^{-t/\tau_L}.$$

When the current is decreasing, V_L is given in terms of the time derivative of Eq. 36-15 and is

$$V_L \left(= L \frac{di}{dt}\right) = -\varepsilon e^{-t/\tau_L}.$$

Note that V_L is opposite in sign when the current is increasing (di/dt positive) and when it is decreasing (di/dt negative), as is true also for the induced emf $\varepsilon_L[=-L(di/dt)=-V_L]$.

Examination of Fig. 36-6 shows that at any instant the sum of curves a and b always yields curve c. This is an expected consequence of the loop theorem, as Eq. 36-9 shows.

EXAMPLE 2

A solenoid has an inductance of 50 H and a resistance of 30 Ω. If it is connected to a 100-V battery, how long will it take for the current to reach one-half its final equilibrium value?

The equilibrium value of the current is reached as $t \to \infty$; from Eq. 36-12 it is ε/R. If the current has half this value at a particular time t_0, this equation becomes

$$\frac{1}{2}\frac{\varepsilon}{R} = \frac{\varepsilon}{R}(1 - e^{-t_0/\tau_L}).$$

Solving for t_0 yields

$$t_0 = \tau_L \ln 2 = 0.69 \frac{L}{R}.$$

Using the values given, this reduces to

$$t_0 = 0.69\tau_L = 0.69 \left(\frac{50 \text{ H}}{30 \text{ }\Omega}\right) = 1.2 \text{ s}.$$

36-4
ENERGY AND THE MAGNETIC FIELD

When we lift a stone we do work, which we can get back again by lowering the stone. It is convenient to think of the work done to lift the stone being stored temporarily in the gravitational field between the earth and the lifted stone, and being withdrawn from this field when we lower the stone.

When we pull two unlike charges apart we like to say that the work we do is stored in the electric field between the charges. We can get it back from the field by letting the charges move closer together again.

In the same way we can store energy in a magnetic field. For example, two long, rigid, parallel wires carrying current in the same direction attract each other and we must do work to pull them apart. We can get this stored energy back at any time by letting the wires move back to their original positions.

To derive a quantitative expression for the storage of energy in the magnetic field, consider Fig. 36-4, which shows a source of emf ε connected to a resistor R and an inductor L.

$$\varepsilon = iR + L \frac{di}{dt}, \qquad (36\text{-}9)$$

is the differential equation that describes the growth of current in this circuit. We stress that this equation follows immediately from the loop theorem and that the loop theorem in turn is an expression of the principle of conservation of energy for single-loop circuits. If we multiply each side of Eq. 36-9 by i, we obtain

$$\varepsilon i = i^2 R + Li \frac{di}{dt}, \qquad (36\text{-}16)$$

which has the following physical interpretation in terms of work and energy:

1. If a charge dq passes through the seat of emf ε in Fig. 36-4 in time dt, the seat does work on it in amount $\varepsilon\, dq$. The *rate* of doing work is $(\varepsilon\, dq)/dt$, or εi. Thus the left term in Eq. 36-16 is the *rate at which the seat of emf delivers energy to the circuit.*

2. The second term in Eq. 36-16 is the *rate at which energy appears as thermal (Joule) energy in the resistor.*

3. Energy that does not appear as thermal (Joule) energy must, by our hypothesis, be stored in the magnetic field. Since Eq. 36-16 represents a statement of the conservation of energy for LR circuits, the last term must represent the *rate dU_B/dt at which energy is stored in the magnetic field,* or

$$\frac{dU_B}{dt} = Li \frac{di}{dt}. \qquad (36\text{-}17)$$

We can write this as $\qquad dU_B = Li\, di.$
Integrating yields

$$U_B = \int_0^{U_B} dU_B = \int_0^i Li\, di = \tfrac{1}{2} Li^2, \qquad (36\text{-}18)$$

which represents the total stored magnetic energy in an inductance L carrying a current i.

We can compare this relation with the expression for the energy associated with a capacitor C carrying a charge q, namely,

$$U_E = \frac{1}{2}\frac{q^2}{C}.$$

Here the energy is stored in an electric field. In each case the expression for the stored energy was derived by setting it equal to the work that must be done to set up the field.

EXAMPLE 3

A coil has an inductance of 5.0 H and a resistance of 20 Ω. If a 100-V emf is applied, what energy is stored in the magnetic field after the current has built up to its maximum value ε/R?

The maximum current is given by

$$i = \frac{\varepsilon}{R} = \frac{100\ \text{V}}{20\ \Omega} = 5.0\ \text{A}.$$

The stored energy is given by Eq. 36-18:

$$U_B = \tfrac{1}{2} Li^2 = \tfrac{1}{2}(5.0\ \text{H})(5.0\ \text{A})^2 = 63\ \text{J}.$$

Note that the time constant for this coil $(= L/R)$ is 0.25 s. After how many time constants will *half* of this equilibrium energy be stored in the field?

EXAMPLE 4

A 3.0 H inductor is placed in series with a 10-Ω resistor, an emf of 3.0 V being suddenly applied to the combination. At 0.30 s (which is one inductive time constant) after the contact is made, (a) what is the rate at which energy is being delivered by the battery?

The current is given by Eq. 36-12, or

$$i = \frac{\varepsilon}{R}(1 - e^{-t/\tau_L}),$$

which at $t = 0.30$ s $(= \tau_L)$ has the value

$$i = \left(\frac{3.0 \text{ V}}{10 \text{ }\Omega}\right)(1 - e^{-1}) = 0.189 \text{ A}.$$

The rate P_ε at which energy is delivered by the battery is

$$P_\varepsilon = \varepsilon i$$
$$= (3.0 \text{ V})(0.189 \text{ A})$$
$$= 0.567 \text{ W}.$$

(b) At what rate does energy appear as thermal (Joule) energy in the resistor? This is given by

$$P_J = i^2 R$$
$$= (0.189 \text{ A})^2 (10 \text{ }\Omega)$$
$$= 0.357 \text{ W}.$$

(c) At what rate P_B is energy being stored in the magnetic field?
This is given by the last term in Eq. 36-16, which requires that we know di/dt. Differentiating Eq. 36-12 yields

$$\frac{di}{dt} = \left(\frac{\varepsilon}{R}\right)\left(\frac{R}{L}\right) e^{-t/\tau_L}$$
$$= \frac{\varepsilon}{L} e^{-t/\tau_L}.$$

At $t = \tau_L$ we have

$$\frac{di}{dt} = \left(\frac{3.0 \text{ V}}{3.0 \text{ H}}\right) e^{-1} = 0.37 \text{ A/s}.$$

From Eq. 36-17, the desired rate is

$$P_B = \frac{dU_B}{dt} = Li \frac{di}{dt}$$
$$= (3.0 \text{ H})(0.189 \text{ A})(0.37 \text{ A/s})$$
$$= 0.210 \text{ W}.$$

Note that as required by the principle of conservation of energy (see Eq. 36-16)

$$P_\varepsilon = P_J + P_B,$$

or

$$0.567 \text{ W} = 0.357 \text{ W} + 0.210 \text{ W}$$
$$= 0.567 \text{ W}.$$

36-5
ENERGY DENSITY AND THE MAGNETIC FIELD

We now derive an expression for the *density* of energy u_B in a magnetic field. Consider a length l near the center of a very long solenoid; Al is the volume associated with this length. The stored energy must lie entirely within this volume because the magnetic field outside such a solenoid is essentially zero. Moreover, the stored energy must be uniformly distributed throughout the volume of the solenoid because the magnetic field is uniform everywhere inside.

Thus, we can write

$$u_B = \frac{U_B}{Al}$$

or, since

$$U_B = \tfrac{1}{2}Li^2,$$

we have

$$u_B = \frac{\tfrac{1}{2}Li^2}{Al}.$$

To express this in terms of the magnetic field, we can substitute for L in this equation, using the relation $L = \mu_0 n^2 lA$ (Eq. 36-5). Also we can solve Eq. 34-7 ($B = \mu_0 in$) for i and substitute in this equation. Doing so yields finally

$$u_B = \frac{1}{2}\frac{B^2}{\mu_0}. \qquad (36\text{-}19)$$

This equation gives the energy density stored at any point (in a vacuum or in a nonmagnetic substance) where the magnetic field is **B**. The equation is true for all magnetic field configurations, even though we derived it by considering a special case, the solenoid. Equation 36-19 is to be compared with Eq. 30-9,

$$u_E = \tfrac{1}{2}\epsilon_0 E^2, \qquad (36\text{-}20)$$

which gives the energy density (in a vacuum) at any point in an electric field. Note that both u_B and u_E are proportional to the square of the appropriate field quantity, B or E.

The solenoid plays a role with relationship to magnetic fields similar to the role the parallel-plate capacitor plays with respect to electric fields. In each case we have a simple device that can be used for setting up a uniform field throughout a well-defined region of space and for deducing, in a simple way, some properties of these fields.

EXAMPLE 5

A long *coaxial cable* (Fig. 36-7) consists of two concentric cylinders with radii a and b. Its central conductor carries a steady current i, the outer conductor providing the return path. (a) Calculate the energy stored in the magnetic field for a length l of such a cable.

In the space between the two conductors Ampère's law,

$$\oint \mathbf{B} \cdot d\mathbf{l} = \mu_0 i,$$

leads to

$$(B)(2\pi r) = \mu_0 i$$

or

$$B = \frac{\mu_0 i}{2\pi r}.$$

Ampère's law shows further that the magnetic field is zero for points outside the outer conductor (why?). Magnetic fields exist *inside* each of the conductors; although we can find their values from Ampère's law, we choose to ignore them, on the assumption that the cable dimensions are chosen so that most of the stored magnetic energy is in the space between the conductors.

The energy density for points between the conductors, from Eq. 36-19, is

$$u_B = \frac{1}{2\mu_0}B^2 = \frac{1}{2\mu_0}\left(\frac{\mu_0 i}{2\pi r}\right)^2 = \frac{\mu_0 i^2}{8\pi^2 r^2}.$$

Consider a volume element dV consisting of a cylindrical shell whose radii are r and $r + dr$ and whose length is l. The energy dU contained in it is

$$dU = u_B dV = \frac{\mu_0 i^2}{8\pi^2 r^2}(2\pi rl)(dr) = \frac{\mu_0 i^2 l}{4\pi}\frac{dr}{r}.$$

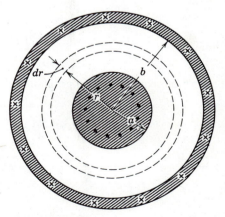

figure 36-7
Example 5. Cross section of a coaxial cable, showing steady equal but opposite currents in the central and outer conductors.

The total stored magnetic energy is found by integration, or

$$U = \int dU = \frac{\mu_0 i^2 l}{4\pi} \int_a^b \frac{dr}{r} = \frac{\mu_0 i^2 l}{4\pi} \ln \frac{b}{a},$$

which is the desired expression.

(b) What is the inductance of a length l of coaxial cable?

We can find the inductance L from Eq. 36-18 ($U = \frac{1}{2} L i^2$), which leads to

$$L = \frac{2U}{i^2} = \frac{\mu_0 l}{2\pi} \ln \frac{b}{a}.$$

You should also derive this expression directly from the definition of inductance, using the procedures of Example 1.

Compare the energy required to set up, in a cube 10 cm on edge (a) a uniform electric field of 1.0×10^5 V/m and (b) a uniform magnetic field of 1.0 T ($= 10^4$ gauss). Both these fields would be judged reasonably large but they are readily available in the laboratory.

EXAMPLE 6

(a) In the electric case we have, where V_0 is the volume of the cube,

$$U_E = u_E V_0 = \frac{1}{2} \epsilon_0 E^2 V_0$$

$$= (0.5)(8.9 \times 10^{-12} \text{ C}^2/\text{N}\cdot\text{m}^2)(10^5 \text{ V/m})^2(0.1 \text{ m})^3$$

$$= 4.5 \times 10^{-5} \text{ J}.$$

(b) In the magnetic case, from Eq. 36-19, we have

$$U_B = u_B V_0 = \frac{B^2}{2\mu_0} V_o = \frac{(1.0 \text{ T})^2(0.1 \text{ m})^3}{(2)(4\pi \times 10^{-7} \text{ T}\cdot\text{m/A})}$$

$$= 400 \text{ J}.$$

In terms of fields normally available in the laboratory, much larger amounts of energy can be stored in a magnetic field than in an electric one, the ratio being about 10^7 in this example. Conversely, much more energy is required to set up a magnetic field of reasonable laboratory magnitude than is required to set up an electric field of similarly reasonable magnitude.

In Section 35-2 we saw that if two coils are close together, as in Fig. 35-2, a steady current i in one coil will set up a magnetic flux Φ linking the other coil. If we change i with time, an emf ε given by Faraday's law (Eq. 35-2) appears in the second coil; we called this process *induction*. We could better have called it *mutual induction*, to suggest the mutual interaction of the two coils and to distinguish it from *self-induction*, in which only one coil is involved as described in the preceding sections of this chapter.

Let us look a little more quantitatively at mutual induction. Figure 36-8 shows two circular close-packed coils near each other and sharing a common axis. There is a current i_1 in coil 1, set up by an external circuit not shown. This current produces a magnetic field suggested by the lines of $\mathbf{B}_1$ in the figure. Coil 2 stands alone as a closed circuit, with no external connections to a battery; a magnetic flux Φ_{21} passes through (links) it.

We define the *mutual inductance* M_{21} of coil 2 with respect to coil 1 as

$$M_{21} = \frac{N_2 \Phi_{21}}{i_1}. \tag{36-21a}$$

36-6
MUTUAL INDUCTANCE

Compare this with Eq. 36-4 ($L = N\Phi/i$), the definition of (self-) inductance. We can recast Eq. 36-21a as

$$M_{21} i_1 = N_2 \Phi_{21}. \tag{36-21b}$$

If, by external means, we cause i_1 to vary with time, we have

$$M_{21} \frac{di_1}{dt} = N_2 \frac{d\Phi_{21}}{dt}. \tag{36-22}$$

The right side of this equation, from Faraday's law (Eq. 35-2), is, apart from a change in sign, just the emf ε_2 appearing in coil 2 due to the changing current in coil 1, or

$$\varepsilon_2 = -M_{21} \frac{di_1}{dt} \tag{36-23a}$$

which you should compare with Eq. 36-3a ($\varepsilon = -L\, di/dt$) for self-inductance.

Let us now interchange the roles of coils 1 and 2 in Fig. 36-8. That is, we set up a current i_2 in coil 2, by external circuitry not shown, and this produces a magnetic flux Φ_{12} that links coil 1 (from which the external circuitry has now been removed). If we change i_2 with time, we have, by the same argument given above,

$$\varepsilon_1 = -M_{12} \frac{di_2}{dt}; \tag{32-23b}$$

compare Eq. 36-23a.

Thus we see that the emf in *either* coil is proportional to the rate of change of current in the *other* coil. The proportionality constants M_{21} and M_{12} seem to be different, but we assert, without proof, that they are in fact the same so that no subscripts are needed. This conclusion is in no way obvious. Thus we have

$$M_{21} = M_{12} = M \tag{32-24}$$

and we can rewrite Eqs. 32-23 as

$$\varepsilon_2 = -M\, di_1/dt \quad \text{and} \quad \varepsilon_1 = -M\, di_2/dt. \tag{32-25}$$

The induction is indeed *mutual*. The SI unit for M (compare Eqs. 36-3 and 36-23) is the henry.

The calculation of M, like that of L, depends on the geometry of the system. The simplest case is that in which *all* of the flux from one coil links the other coil. Example 7 shows such a situation.

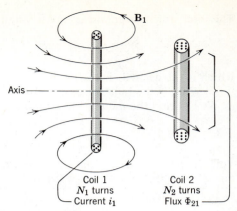

figure 36-8
The arrangement used to define the mutual inductance M_{21} of coil 2 with respect to coil 1.

In the toroid of Example 1 (see Fig. 36-2) let us relabel the turns in the winding shown (coil 1) as N_1 and the current in this winding as i_1. Now let us superimpose a second winding (coil 2) over the first, having N_2 turns. The coils are electrically insulated from each other. In terms of the geometrical factors of Example 1, what is the mutual inductance M for the two windings?

From Eq. 36-21a we have

$$M = \frac{N_2 \Phi_{21}}{i_1}$$

in which Φ_{21} is here identical with Φ, the total common flux in coils 1 and 2 due to i_1. From Example 1 we see that, with the noted changes in notation,

$$\Phi = \frac{\mu_0 i_1 N_1 h}{2\pi} \ln \frac{b}{a},$$

EXAMPLE 7

so that

$$M = \frac{\mu_0 N_1 N_2 h}{2\pi} \ln \frac{b}{a}.$$

Note that if $N_1 = N_2$, the inductances L_1 and L_2 of the two windings are virtually the same ($= L$). A comparison with Example 1 shows that, in this case, $M = L$; see Problem 36 for a more general result.

questions

1. Under what conditions can we write Eq. 36-1 [$\varepsilon = d(N\Phi_B/dt)$] as $\varepsilon = N(d\Phi_B/dt)$? Can you think of a physical situation in which changes in N alone with time would produce an induced emf?

2. In a case of mutual induction, as in Fig. 35-2, is self-induction also present? Discuss.

3. Is the inductance per unit length for a solenoid near its center (a) the same as, (b) less than, or (c) greater than the inductance per unit length near its ends?

4. Two solenoids, A and B, have the same diameter and length and contain only one layer of copper windings, with adjacent turns touching, insulation thickness being negligible. Solenoid A contains many turns of fine wire and solenoid B contains fewer turns of heavier wire. (a) Which solenoid has the larger inductance? (b) Which solenoid has the larger inductive time constant?

5. If the flux passing through each turn of a coil is the same, the inductance of the coil may be computed from $L = N\Phi_B/i$ (Eq. 36-4). How might one compute L for a coil for which this assumption is not valid?

6. Show that the dimensions of the two expressions for L, $N\Phi_B/i$ (Eq. 36-4) and $\varepsilon/(di/dt)$ (Eq. 36-3b), are the same.

7. You are given a length l of copper wire. How would you arrange it to obtain the maximum self-inductance?

8. You want to wind a coil so that it has resistance R but essentially no inductance. How would you do it?

9. Does the time required for the current in a particular LR circuit to build up to any given fraction of its equilibrium value depend on the value of the applied constant emf?

10. A steady current is set up in a coil with a very large inductive time constant. When the current is interrupted with a switch, a heavy arc tends to appear at the switch blades. Explain. (Note: Interrupting currents in highly inductive circuits can be dangerous.)

11. In an LR circuit like that of Fig. 36-4, can the self-induced emf ever be larger than the battery emf?

12. In an LR circuit like that of Fig. 36-4, is the current in the resistance *always* the same as the current in the inductance?

13. In the circuit of Fig. 36-3 the self-induced emf is a maximum at the instant the switch is closed on a. How can this be since there is no current in the inductance at this instant?

14. The switch in Fig. 36-3, having been closed on a for a "long" time, is thrown to b. What happens to the energy initially stored in the inductor?

15. A coil has a (measured) inductance L and a (measured) resistance R. Is its inductive time constant given by Eq. 36-13? Bear in mind that we derived that equation (see Fig. 36-3) for a situation in which the inductive and resistive elements were physically separated. Discuss.

16. In Section 36-3 we show that Eq. 36-10 is a solution of Eq. 36-9. Can you be sure that it is the *only* solution?

17. If a current in a source of emf is in the direction of the emf, the energy of the source decreases; if a current is in a direction opposite to the emf (as in

charging a battery), the energy of the source increases. Do these statements apply to the inductor in Fig. 36-1a and 36-1b?

18. Can you make an argument based on the manipulation of bar magnets to suggest that energy may be stored in a magnetic field?

19. Does Eq. 36-18 $(U = \frac{1}{2}Li^2)$ shed any light on the fact that (see Eq. 36-5) the inductance of a length l of a long solenoid is proportional to its volume?

20. Draw all the formal analogies that you can between a parallel-plate capacitor (for electric fields) and a long solenoid (for magnetic fields).

21. In a toroid is the energy density larger near the inner radius or near the outer radius?

22. Two coils are connected in series. Does their equivalent inductance depend on their geometrical relationship to each other?

23. You are given two similar flat circular coils of N turns each. For what geometry will their mutual inductance M be the greatest? For what geometry will it be the least? Assume that the coils are close together.

24. You are given two coils geometrically close together. Need they be electrically connected for them to exhibit mutual inductance? If they *are* electrically connected, can they still exhibit mutual inductance?

25. A flat circular coil is placed completely outside a long solenoid, near its center, the axes of the coil and the solenoid being parallel. Is there a mutual induction effect? Suppose the coil surrounds the solenoid? In each case justify your answer.

26. A circular coil of N turns surrounds a long solenoid. Is the mutual inductance greater when the coil is near the center of the solenoid or when it is near its end? Explain.

problems

SECTION 36-1

1. The inductance of a close-packed coil of 400 turns is 8.0 mH. What is the magnetic flux through the coil when the current is 5.0×10^{-3} A?
Answer: 1.0×10^{-7} Wb.

2. Each item (a) coulomb·ohm·meter/weber, (b) volt·second, (c) coulomb·ampere/farad, (d) kilogram·volt·meter²/(henry·ampere)², (e) (henry/farad)$^{1/2}$ is equal to one of the items in the following list: meter, second, kilogram, dimensionless number, newton, joule, volt, ohm, watt, coulomb, ampere, weber, henry, farad. Give the equalities.

3. A 10-H inductor carries a steady current of 2.0 A. How can a 100-V self-induced emf be made to appear in the inductor?
Answer: Let the current change at 10 A/s.

SECTION 36-2

4. A solenoid is wound with a single layer of #10 copper wire (diameter, 0.10 in.). It is 4.0 cm in diameter and 2.0 m long. What is the inductance per unit length for the solenoid near its center? Assume that adjacent wires touch and that insulation thickness is negligible.

5. A long thin solenoid can be bent into a ring to form a toroid. Show that if the solenoid is long and thin enough, the equation for the inductance of a toroid (see Example 1) reduces to that for a solenoid of comparable length (Eq. 36-5).

6. *Inductors in Series* Two inductors L_1 and L_2 are connected in series and are separated by a large distance. (a) Show that the equivalent inductance L is $L_1 + L_2$. (b) Why must their separation be large?

7. Show that the self-inductance for a length l of a long wire associated with the flux *inside* the wire only is $\mu_0 l/8\pi$. Assume a uniform distribution of the current over the cross section of the wire.

8. *Inductors in Parallel* Two inductors L_1 and L_2 are connected in parallel and separated by a large distance. (*a*) Show that the equivalent inductance L is given by

$$\frac{1}{L} = \frac{1}{L_1} + \frac{1}{L_2}.$$

(*b*) Why must their separation be large for this relationship to hold?

9. Two long parallel wires whose centers are a distance d apart carry equal currents in opposite directions. Show that, neglecting the flux within the wires themselves, the inductance of a length l of such a pair of wires is given by

$$L = \frac{\mu_0 l}{\pi} \ln \frac{d-a}{a},$$

where a is the wire radius. See Example 4, Chapter 34.

10. Calculate the self-inductance of two concentric hollow cylinders of radii a and b, and of length $l \gg a,b$. At one end the cylinders are connected by a flat conducting plate so that the current travels down in the inner cylinder and back in the outer. See Example 5 for hints.

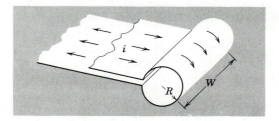

figure 36-9
Problem 11

11. A very wide copper strip of width W is bent into a piece of slender tubing of radius R with two plane extensions, as shown in Fig. 36-9. A current i flows through the strip, distributed uniformly over its width. In this way a "one-turn solenoid" has been formed. (*a*) Find the magnitude of the magnetic field **B** in the tubular part (far away from the edges). (Hint: Assume that the field outside this one-turn solenoid is negligibly small.) (*b*) Find the inductance of this one-turn solenoid, neglecting the two plane extensions. *Answer:* (*a*) $\mu_0 i/W$. (*b*) $\pi \mu_0 R^2/W$.

SECTION 36-3

12. The current in an LR circuit builds up to one-third of its steady-state value in 5.0 s. What is the inductive time constant?

13. How many "time constants" must we wait for the current in an LR circuit to build up to within 0.10 percent of its equilibrium value? *Answer:* 6.9.

14. A 50-V potential difference is suddenly applied to a coil with $L = 50$ mH and $R = 180\ \Omega$. At what rate is the current increasing after 0.001 s?

15. A wooden toroidal core with a square cross section has an inner radius of 10 cm and an outer radius of 12 cm. It is wound with one layer of #18 wire (diameter, 0.040 in.; "resistance", 160 ft/Ω). What are (*a*) the inductance and (*b*) the inductive time constant? Ignore the thickness of the insulation. *Answer:* (*a*) 2.8×10^{-4} H. (*b*) 2.7×10^{-4} s.

16. How long would it take for the potential difference across the resistor in an LR circuit ($L = 1.0$ H, $R = 1.0\ \Omega$) to drop to 10% of its initial value?

17. A solenoid having an inductance of 6.0×10^{-6} H is connected in series to a 1.0×10^{3}-Ω resistor. (a) If a 10-V battery is switched across the pair, how long will it take for the current through the resistor to reach 80% of its final value? (b) What is the current through the resistor after one time constant? *Answer:* (a) 9.7×10^{-9} s. (b) 6.3×10^{-3} A.

18. The current in an LR circuit drops from 1.0 A at $t = 0$ to 0.010 A one second later. If L is 10 H, find the resistance R in the circuit.

19. In the circuit shown in Fig. 36-10, $\varepsilon = 10$ V, $R_1 = 5.0 \, \Omega$, $R_2 = 10 \, \Omega$, and $L = 5.0$ H. For the two separate conditions (I) switch S just closed and (II) switch S closed for a very long time, calculate (a) the current i_1 through R_1, (b) the current i_2 through R_2, (c) the current i through the switch, (d) the potential difference across R_2, (e) the potential difference across L, and (f) di_2/dt.
Answer: I. (a) 2.0 A. (b) Zero. (c) 2.0 A. (d) Zero. (e) 10 V. (f) 2.0 A/s II. (a) 2.0 A. (b) 1.0 A. (c) 3.0 A. (d) 10 V. (e) Zero. (f) Zero.

20. In Fig. 36-11, $\varepsilon = 100$ V, $R_1 = 10 \, \Omega$, $R_2 = 20 \, \Omega$, $R_3 = 30 \, \Omega$ and $L = 2.0$ H. Find the values of i_1 and i_2 (a) immediately after switch S is closed; (b) a long time later; (c) immediately after switch S is opened again; (d) a long time later.

21. Show that the inductive time constant τ_L can also be defined as the time that would be required for the current in an LR circuit to reach its equilibrium value *if it continued to increase at its initial rate.*

22. The switch S in Fig. 36-3 is thrown from b to a. After one inductive time constant show that (a) the total energy transformed to thermal energy in the resistor is $0.168\varepsilon^2\tau_L/R$ and that (b) the energy stored in the magnetic field is $0.200\varepsilon^2\tau_L/R$. (c) Show that the equilibrium energy stored in the magnetic field is $0.500\varepsilon^2\tau_L/R$.

23. A coil with an inductance of 2.0 H and a resistance of 10 Ω is suddenly connected to a resistanceless battery with $\varepsilon = 100$ V. (a) What is the equilibrium current? (b) How much energy is stored in the magnetic field when this current exists in the coil? *Answer:* (a) 10 A (b) 100 J.

24. A coil with an inductance of 2.0 H and a resistance of 10 Ω is suddenly connected to a resistanceless battery with $\varepsilon = 100$ V. At 0.10 s after the connection is made, what are the rates at which (a) energy is being stored in the magnetic field, (b) thermal energy is appearing, and (c) energy is being delivered by the battery?

25. A given coil is connected in series with a 10,000-Ω resistor. When a 50-V battery is applied to the two, the current reaches a value of 2.0 mA after 5.0 ms. (a) Find the inductance of the coil: (b) What is the energy stored in the inductance at this same moment? *Answer:* (a) 98 H. (b) 2.0×10^{-4} J.

26. A long wire carries a current i uniformly distributed over a cross section of the wire. Show that the magnetic energy per unit length stored *within* the wire equals $\mu_0 i^2/16\pi$. Note that it does not depend on the wire diameter.

27. The coaxial cable of Example 5 has $a = 1.0$ mm, $b = 4.0$ mm, and $c = 5.0$ mm (c is the radius of the outer surface of the outer conductor). It carries a current of 10 A in the inner conductor and an equal but oppositely directed return current in the outer conductor. Calculate and compare the stored magnetic energy per meter of cable length (a) within the central conductor, (b) in the space between the conductors, and (c) within the outer conductor. *Answer:* (a) 2.5×10^{-6} J/m. (b) 14×10^{-6} J/m. (c) 0.80×10^{-6} J/m.

28. Prove that when switch S in Fig. 36-3 is thrown from a to b all the energy stored in the inductor appears as thermal energy in the resistor.

SECTION 36-5

29. What is the energy density in the magnetic field near the center of the solenoid of Problem 23, Chapter 35? *Answer:* 63 J/m^3.

30. A circular loop of wire 5.0 cm in radius carries a current of 100 A. What is the energy density at the center of the loop?

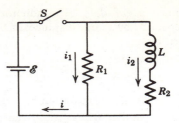

figure 36-10
Problem 19

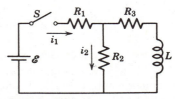

figure 36-11
Problem 20

31. A length of #10 copper wire carries a current of 10 A. Calculate (a) the magnetic energy density and (b) the electric energy density at the surface of the wire. The wire diameter is 0.10 in. and its resistance per unit length is 1.0 ohm/1000 ft.
Answer: (a) 0.99 J/m³. (b) 4.8×10^{-15} J/m³.

32. (a) What is the magnetic energy density of the earth's magnetic field of 5.0×10^{-5} T? (b) Assuming this to be relatively constant over distances small compared with the earth's radius and neglecting the variations near the magnetic poles, how much energy would be stored in a shell between the the earth's surface and 16 km above the surface?

33. (a) Find an expression for the energy density as a function of the radius for the toroid of Example 1. (b) Integrating the energy density over the volume of the toroid, find the total energy stored in the field of the toroid; assume $i = 0.50$ A. (c) Using Eq. 36-18 evaluate the energy stored in the toroid directly from the inductance and compare with (b).

Answer: (a) $\dfrac{\mu_0 i^2 N^2}{8\pi^2 r^2}$. (b) 1.8×10^{-4} J. (c) 1.8×10^{-4} J.

34. What must be the magnitude of a uniform electric field if it is to have the same energy density as that possessed by a 0.50-tesla magnetic field?

35. What is the magnetic energy density at the center of a circulating electron in the hydrogen atom (see Example 9, Chapter 34)?
Answer: 5.7×10^7 J/m³.

SECTION 36-6
36. In Example 7 show that, if $N_1 \neq N_2$, the mutual inductance is given by
$$M = \sqrt{L_1 L_2}.$$

Do you suppose that this relation holds even if the situation is not like that of Example 7, that is, if it is *not* true that *all* the flux from one coil links *all* the turns of the other coil?

37. Two short cylindrical coils are connected in series; the coils are reasonably close together and share the same axis. (a) Show that the effective inductance of the combination is
$$L = L_1 + L_2 \pm 2M.$$

(b) What is the significance of the ± sign? Does it have anything to do with the relative sense (clockwise or counterclockwise) in which the coils are wound?

37

magnetic
properties
of matter

In electricity the *isolated charge q* is the simplest structure that can exist. If two such charges of opposite sign are placed near each other, they form an *electric dipole,* characterized by an electric dipole moment **p.** In magnetism isolated magnetic poles (usually called *magnetic monopoles*) which would correspond to isolated electric charges, apparently do not exist. The simplest magnetic structure is the *magnetic dipole,* characterized by a magnetic dipole moment **μ.** Table 34-1 summarizes some characteristics of electric and magnetic dipoles.

A current loop, a bar magnet, and a solenoid of finite length are examples of magnetic dipoles. We can identify their north poles (from which the lines of **B** *emerge*) by suspending them as compass needles and observing which end points north. We can find their magnetic dipole moments by placing the dipole in an external magnetic field **B,** measuring the torque τ that acts on it, and computing μ from Eq. 33-11, or

$$\tau = \mu \times \mathbf{B}. \qquad (37\text{-}1)$$

Alternatively, we can measure **B** due to the dipole at a point along its axis a (large) distance r from its center and compute μ from the expression in Table 34-1, or

$$B = \frac{\mu_0}{2\pi} \frac{\mu}{r^3}. \qquad (37\text{-}2)$$

Figure 37-1, which shows iron filings sprinkled on a sheet of paper under which there is a bar magnet, suggests that this dipole might be viewed as two "poles" separated by a distance d. However, all attempts

figure 37-1
A bar magnet is a magnetic dipole. The iron filings suggest the pattern of lines of force in Fig. 37-4a. See Fig. 27-6b for the electrostatic analog. (Courtesy Physical Science Study Committee.)

to isolate these poles fail. If we break the magnet, as in Fig. 37-2, the fragments prove to be dipoles and not isolated poles. Where one north pole and one south pole existed there are now three of each. If we break up a magnet into the electrons, protons, and neutrons that make up its atoms, we will find that even these elementary particles are magnetic dipoles. Figure 37-3 contrasts the electric and the magnetic characters of the free electron.

All electrons have a characteristic *"spin" angular momentum* about a certain axis, which has the value of

$$L_s = 0.5272943 \times 10^{-34} \text{ joule} \cdot \text{second}.$$

This is suggested by the vector $\mathbf{L}_s$ in Fig. 37-3b. Such a spinning charge can be viewed classically as being made up of infinitesimal current

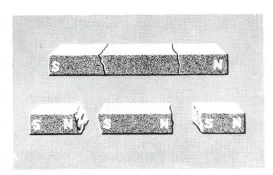

figure 37-2
If we break a bar magnet, each fragment becomes a dipole. There will always be equal numbers of north and south poles, associated in pairs.

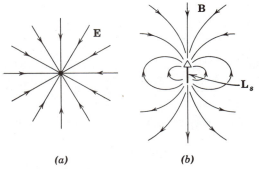

(a)   *(b)*

figure 37-3
(a) The lines of **E** and (b) the lines of **B** for an electron. The magnetic dipole moment of the electron, $\boldsymbol{\mu}_l$, is directed opposite to the spin angular momentum vector, $\mathbf{L}_s$.

loops. Each such loop is a tiny magnetic dipole, its moment being given by (see Eq. 33-10)

$$\mu = NiA, \qquad (37\text{-}3)$$

where i is the equivalent current in each infinitesimal loop and A is the loop area. The number of turns, N, is unity for each loop. The magnetic dipole moment of the spinning charge can be found by integrating over the moments of the infinitesimal current loops that make it up; see Problem 8.

Although this model of the spinning electron is too mechanistic and is not in accord with modern quantum physics, it remains true that the magnetic dipole moments of elementary particles are closely connected with their intrinsic angular momenta, or spins. Those particles and nuclei whose spin angular momentum is zero (the α-particle, the pion, the O^{16} nucleus, etc.) have no magnetic dipole moment. We must distinguish the "intrinsic" or "spin" magnetic moment of the electron from any additional magnetic moment it may have because of its *orbital* motion in an atom; see Example 2.

EXAMPLE 1

Devise a method for measuring μ for a bar magnet.

(a) Place the magnet in a uniform external magnetic field **B**, with **μ** making an angle θ with **B**. The magnitude of the torque acting on the magnet (see Eq. 37-1) is given by

$$\tau = \mu B \sin \theta.$$

Clearly we can find μ if we measure τ, B, and θ.

(b) A second technique is to suspend the magnet from its center of mass and to allow it to oscillate about its stable equilibrium position in the external field **B**. For *small* oscillations, $\sin \theta$ can be replaced by θ and the equation just given becomes

$$\tau = -(\mu B)\theta = -\kappa\theta,$$

where κ is a constant. The minus sign has been inserted to show that τ is a *restoring torque*. Since τ is proportional to θ, the condition for simple angular harmonic motion is met. The frequency ν is given by the reciprocal of Eq. 15-24, or

$$\nu = \frac{1}{2\pi}\sqrt{\frac{\kappa}{I}} = \frac{1}{2\pi}\sqrt{\frac{\mu B}{I}},$$

in which I is the rotational inertia. With this equation μ can be found from the measured quantities ν, B, and I.

EXAMPLE 2

An electron in an atom circulating in an assumed circular orbit of radius r behaves like a tiny current loop and has an *orbital magnetic dipole moment* usually represented by μ_l.*,† Derive a connection between μ_l and the *orbital angular momentum* L_l.

Newton's second law $(F = ma)$ yields, if we substitute Coulomb's law for F,

$$\frac{1}{4\pi\epsilon_0}\frac{e^2}{r^2} = ma = \frac{mv^2}{r}$$

or

$$v = \sqrt{\frac{e^2}{4\pi\epsilon_0 mr}} . \qquad (37\text{-}4)$$

* This must not be confused with the magnetic dipole moment μ_s of the electron spin, which is also present.
† Although this model is too mechanistic and not in the spirit of modern quantum physics, it is nevertheless instructive to examine it.

The angular velocity ω is given by

$$\omega = \frac{v}{r} = \sqrt{\frac{e^2}{4\pi\epsilon_0 mr^3}}.$$

The current for the orbit is the rate at which charge passes any given point, or

$$i = ev = e\left(\frac{\omega}{2\pi}\right) = \sqrt{\frac{e^4}{16\pi^3\epsilon_0 mr^3}}.$$

The orbital dipole moment μ_l is given from Eq. 37-3 if we put $N = 1$ and $A = \pi r^2$, or

$$\mu_l = NiA = (1)\sqrt{\frac{e^4}{16\pi^3\epsilon_0 mr^3}}\,(\pi r^2) = \frac{e^2}{4}\sqrt{\frac{r}{\pi\epsilon_0 m}}. \qquad (37\text{-}5)$$

The orbital angular momentum L_l is

$$L_l = (mv)r.$$

Combining with Eq. 37-4 leads to

$$L_l = \sqrt{\frac{e^2 mr}{4\pi\epsilon_0}}.$$

Finally, eliminating r between this equation and Eq. 37-5 yields

$$\mu_l = L_l\left(\frac{e}{2m}\right),$$

which shows that the orbital magnetic moment of an electron is proportional to its orbital angular momentum. Convince yourself that the vectors $\boldsymbol{\mu}_l$ and $\mathbf{L}_l$ point in opposite directions.

For $r = 5.3 \times 10^{-11}$ m, which corresponds to hydrogen in its normal state, we have, from Eq. 37-5,

$$\mu_l = \frac{e^2}{4}\sqrt{\frac{r}{\pi\epsilon_0 m}}$$

$$= \frac{(1.6 \times 10^{-19}\text{ C})^2}{4}\sqrt{\frac{5.3 \times 10^{-11}\text{ m}}{(\pi)(8.9 \times 10^{-12}\text{ C}^2/\text{N}\cdot\text{m}^2)(9.1 \times 10^{-31}\text{ kg})}}$$

$$= 9.2 \times 10^{-24}\text{ A}\cdot\text{m}^2.$$

It is clear in Fig. 37-4 that, for the bar magnet and the solenoid, lines of $\mathbf{B}$ emerge from the top end and enter the bottom end. These are localized regions called the "north pole" (top end) and the "south pole" (bottom end). There are no sharply defined points so defined, as there are for the electrostatic dipole of Fig. 37-4c. Note that the lines of $\mathbf{B}$ and $\mathbf{E}$ in Fig. 37-4 are much alike (dipole field) at distances from the device that are much larger than its dimensions but that the behavior is much different for points very close to the device, including internal points.

The symmetry of nature has always been a guiding principle for physicists. For example, the existence of the (negative) electron suggested the existence of a positive electron, or positron, which was eventually discovered. In the same way the existence of a (positive) proton suggested that there might also be a negative proton and a large accelerator was built at the University of California (Berkeley) primarily to search for this particle; it was found. The discoveries of the positron and of the negative proton each were associated with Nobel prizes; see Appendix K.

With this motivation it is not surprising that physicists have long sought experimental evidence for the existence of magnetic monopoles.

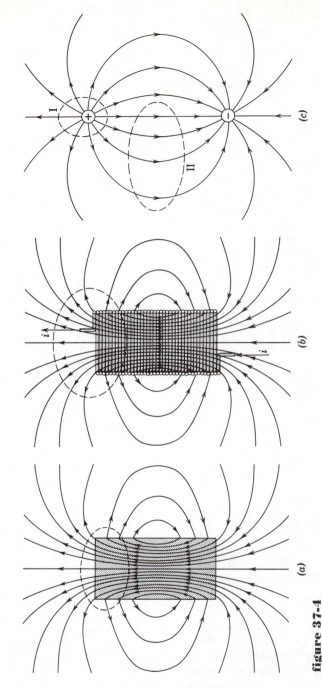

(a)   *(b)*   *(c)*

figure 37-4
Lines of **B** (*a*) for a bar magnet and (*b*) for a short solenoid. In each case the top of the figure is a north pole, the bottom being a south pole. (*c*) Lines of **E** for an electric dipole. At large enough distances all three fields vary like those for a dipole. The four dashed curves are intersections with the plane of the figure of closed Gaussian surfaces. Note that Φ_B equals zero for (*a*) or (*b*). Φ_E equals zero for surfaces like II in (*c*), which do not contain any charge, but Φ_E is not zero for surfaces like I. See Section 28-3.

Their absence, as we shall see in more detail in Chapter 40, represents a serious lack of symmetry between electricity and magnetism. Actually, magnetic monopoles were predicted to exist, on the basis of theory, by P. A. M. Dirac in 1931 and physicists have been searching for them constantly ever since.*

Gauss's law for magnetism, which is one of the basic equations of electromagnetism (see Table 40-2), is a formal way of stating a conclusion that seems to be forced on us by the facts of magnetism, namely, that *isolated magnetic poles do not seem to exist.* This equation asserts that the flux Φ_B through any *closed* Guassian surface must be zero, or

37-2
GAUSS'S LAW FOR MAGNETISM

$$\Phi_B = \oint \mathbf{B} \cdot d\mathbf{S} = 0, \qquad (37\text{-}6)$$

where the integral is to be taken over the entire closed surface. We contrast this with Gauss's law for electricity, which is

$$\epsilon_0 \oint \mathbf{E} \cdot d\mathbf{S} = q. \qquad (37\text{-}7)$$

The fact that a zero appears at the right of Eq. 37-6, but not at the right of Eq. 37-7, means that in magnetism there seems to be no counterpart to the free charge q in electricity.

Figure 37-4a shows a Gaussian surface that encloses one end of a bar magnet. Note that the lines of **B** enter the surface inside the magnet and leave it outside the magnet. There is thus an inward (or negative) flux inside the magnet and an outward (or positive) flux outside it. The total flux for the whole surface is zero.

Figure 37-4b shows a similar surface for a solenoid of finite length which, like a bar magnet, is also a magnetic dipole. Here, too, Φ_B equals zero. Figures 37-4a and b show clearly that there are no "sources" of **B**; that is, there are no points from which lines of **B** emanate. Also, there are no "sinks" of **B**; that is, there are no points toward which **B** converges. In other words, *there are no free magnetic poles.*

Figure 37-4c shows a Gaussian surface (I) surrounding the positive end of an *electric* dipole. Here there *is* a net flux of the lines of **E**. There is a "source" of **E**; it is the charge q. If q is negative, we have a "sink" of **E** because the lines of **E** end on negative charges. For surfaces like surface II in Fig. 37-4c for which the charge inside is zero, the flux of **E** over the surface is also zero.

The magnet with which we are most familiar is that on which we live, the earth. The supposition that the earth is a large magnet, with magnetic poles and a magnetic equator, was first made by Sir William Gilbert (1544–1603), physician to Queen Elizabeth I. Gilbert made a small spherical *terrella* ("little earth") out of a naturally occurring magnetic loadstone (literally "leading stone" or compass) and traced its lines of magnetism. In those days of navigation and exploration there was a natural interest in the compass and in the earth's magnetism.

37-3
THE MAGNETISM OF THE EARTH

* See "Quest for the Magnetic Monopole" by Richard A. Carrigan, Jr., *The Physics Teacher,* October 1975, for more information.

Figure 37-5a is an idealized sketch of the lines of **B** associated with the earth's magnetic field, both at and above its surface. To a first approximation we can represent this field by imagining a strong bar magnet located at the center of the earth, as in Fig. 37-5b. Note that the earth's magnetic axis and its rotational axis RR do not coincide, being separated by about 15°.

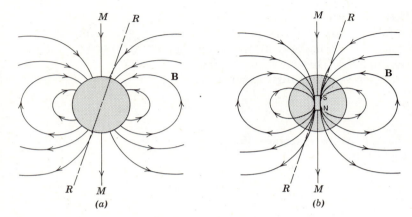

(a)   (b)

figure 37-5
(a) An idealized representation of the lines of **B** associated with the earth's magnetic field. MM is the earth's magnetic axis and RR is its rotational axis. (b) We can approximate the earth's external magnetic field by imagining that a srong bar magnet is located at the center of the earth.

The magnetic pole in the northern hemisphere is in arctic Canada. Note that it is a *south* magnetic pole because lines of **B** *converge* toward it, as in Fig. 37-4. There is a *north* magnetic pole in the southern hemisphere, in Antarctica; lines of **B** *emerge* from it.

There is, of course, no bar magnet buried at the center of the earth. The earth's magnetism must be related to the facts that the central core of the earth, whose radius is 55% of the earth's radius, is (a) liquid, (b) highly conducting, and (c) partakes of the earth's rotation. A dynamo effect, involving circulating currents in the earth's core whose mechanism is not yet fully understood, is almost certainly operating.*

Several other planets in our solar system, Mercury and Jupiter among them, also have magnetic fields. So do the sun and many other stars. There is also a magnetic field associated with our galaxy, that is, with the family of stars whose plane of symmetry is defined by the Milky Way. The galactic magnetic field is relatively weak (about 2 pT on the average) but its effects can be important because it extends over such large distances.

Two simple tools for exploring the earth's magnetic field are the compass needle and the dip needle, the latter being a gravitationally balanced magnetized needle whose axis of rotation is horizontal rather than vertical. In Tucson, Arizona, for example, the north pole of a compass needle (as of 1964) pointed about 13° east of geographic north. Such *declinations* must be known and taken into account when using the compass for navigation or direction finding. The horizontal component of the earth's magnetic field B_h, to which the compass needle responds, is, at Tucson, 26 μT (=0.26 gauss).

Now let us orient the (horizontal) rotational axis of a dip needle at Tucson until it is at right angles to the horizontal component of **B.** Can you convince yourself that the needle will now point in the direction of **B?** It does, and we find that, at this site and at this date, the north end of the needle points downward toward the earth, the needle making an angle φ_i of about 59° (the *inclination*) with a horizontal plane containing

* See "The Earth's Magnetism" by S. K. Runcorn, *Scientific American*, September 1955.

Color Plate 1

At the right we see an aurora viewed from Sky Lab 3, from somewhere over the Indian Ocean. The bright band extending to the left is the normal "airglow", a phenomenon not visible from earth to the naked eye. It represents the emission of radiation from atoms and molecules in the upper atmosphere that absorb energy from sunlight during the daytime. The full moon illuminates a largely clouded earth. A piece of Sky Lab barely shows at the upper left. (This photograph was taken by astronaut Owen K. Garriott by a hand-held camera in Sky Lab. We are indebted to him and to the National Aeronautics and Space Administration for permission to publish the photograph.)

its rotational axis. This shows that, as we expect from Fig. 37-5*a*, lines of **B** are *entering* the earth's surface at this point.

The earth's magnetic field is neither as regular nor as static as the idealized field in Fig. 37-5 suggests. Also, there are observable phenomena, going far beyond deflections of the compass needle, that would not occur if the earth had no magnetic field. Consider the following:

1. *Local variations.* The earth's magnetic field has important local variations, caused by differences in the magnetic properties of the rocks that make up the earth's crust and by the presence of concentrated magnetic ore bodies.

2. *Changes with time.* The mean magnetic declination and inclination vary measurably from year to year at any given location. For example, between the years 1600 and 1800 the measured magnetic declination at London varied in a continuous way from 11° east to 24° west. The north magnetic pole (as of 1948) was measured to be moving north west at about 8 km/y.

This wandering of the earth's magnetic axis, and thus the variation with time of local declinations and inclinations, has resulted in a new archeological specialty, *archeomagnetism*, by means of which dates may be assigned to ancient kilns, ovens, and hearths. The principle is based on the fact that most clays, from which such structures were made, contain small amounts of magnetic materials, whose orientation is frozen into position by heating during normal use. By comparing the present direction of the earth's magnetic field with that of the "frozen in" direction of the magnetization, an approximate archeological date may be established.

On a longer (geological) time scale there is evidence that the earth's magnetic axis has actually completely reversed in direction as many as nine times during the last 4×10^6 years.* The evidence rests on measurements of the (weak) magnetism frozen into rocks of known geological age at the time of their formation.

3. *Interactions with the solar wind.* The sun emits a steady stream of ionized hydrogen atoms (protons) and electrons, that sweeps through the solar system at supersonic speeds. This ever present "solar wind" interacts strongly in several ways with the earth's magnetic field.† (*a*) Occasional sharp increases in the intensity of the solar wind produce terrestrial *magnetic storms*, which seriously interfere with long-distance radio communication. (*b*) The protons and electrons of the solar wind, acted upon by forces given by Eq. 33-2 ($\mathbf{F} = q_0 \mathbf{v} \times \mathbf{B}$), spiral along the lines of **B**, moving back and forth between the magnetic north and south polar regions. These "trapped" electrons and protons form the so called *Van Allen radiation belts*. They were discovered by James A. Van Allen, of the State University of Iowa, in early satellite experiments.‡ (*c*) The trapped solar wind particles, as they interact with the earth's atmosphere, produce the dazzling spectacle of the *aurora*, which is most prominent at about ±75° geomagnetic latitude; see Plate I, opposite.**

---

* "Reversals of the Earth's Magnetic Field" by Allan Cox, G. Brent Dalrymple, and Richard R. Doell, *Scientific American*, February 1967.
† "The Solar Wind" by E. N. Parker, *Scientific American*, April 1964.
‡ See "Radiation Belts" by Brian J. O'Brien, *Scientific American*, May 1963. See also "The Magnetosphere" by J. A. Ratcliffe, *Contemporary Physics*, vol. 18, 1977.
** "Aurora and Airglow" by C. T. Elvey and Franklin E. Roach, *Scientific American*, September 1955.

From data given earlier in this section find (a) the vertical component $\mathbf{B}_v$ of the earth's magnetic field and (b) the resultant magnetic field $\mathbf{B}$ at Tucson.

Figure 37-6 shows the situation. We have

(a)
$$B_v = B_h \tan \varphi_i$$
$$= (26 \ \mu\text{T})(\tan 59°)$$
$$= 44 \ \mu\text{T} \ (= 0.44 \ \text{gauss}),$$

and

(b)
$$B = \sqrt{B_h{}^2 + B_v{}^2}$$
$$= \sqrt{(26 \ \mu\text{T})^2 + (44 \ \mu\text{T})^2}$$
$$= 51 \ \mu\text{T} \ (= 0.51 \ \text{gauss}).$$

Note that the magnetic declination at Tucson plays no role in this problem.

EXAMPLE 3

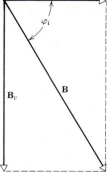

figure 37-6
Example 3.

37-4
PARAMAGNETISM

Magnetism as we know it in our daily experience is an important but special branch of the subject called *ferromagnetism*; we discuss this in Section 37-6. Here we discuss a weaker form of magnetism called *paramagnetism*.

For most atoms and ions, the magnetic effects of the electrons, including both their spins and orbital motions, exactly cancel so that the atom or ion is not magnetic. This is true for the rare gases such as neon and for ions such as Cu^+, which makes up ordinary copper.* For other atoms or ions the magnetic effects of the electrons do *not* cancel, so that the atom as a whole has a magnetic dipole moment $\boldsymbol{\mu}$. Examples are found among the transition elements, such as Mn^{++}; the rare earths, such as Gd^{+++}; and the actinide elements, such as U^{++++}.

If we place a sample of N atoms, each of which has a magnetic dipole moment $\boldsymbol{\mu}$, in a magnetic field, the elementary atomic dipoles tend to line up with the field. This tendency to align is called *paramagnetism*. For perfect alignment, the sample as a whole would have a magnetic dipole moment of $N\boldsymbol{\mu}$. However, the aligning process is seriously interfered with by thermal agitation effects. The importance of thermal agitation may be measured by comparing two energies: one $(= \frac{3}{2}kT)$ is the mean kinetic energy of translation of a gas atom at temperature T; the other $(= 2\mu B)$ is the difference in energy between an atom lined up with the magnetic field and one pointing in the opposite direction. As Example 4 shows, the effect of the collisions at ordinary temperatures and fields is very great. The sample acquires a magnetic moment when placed in an external magnetic field, but this moment is usually very much smaller than the maximum possible moment $N\mu$.

EXAMPLE 4

A paramagnetic gas, whose atoms (see Example 2) have a magnetic dipole moment of about 10^{-23} A·m², is placed in an external magnetic field of magnitude 1.0 T $(= 10^4$ gauss). At room temperature $(T = 300$ K) calculate and compare U_T, the mean kinetic energy of translation $(= \frac{3}{2}kT)$, and U_B, the magnetic energy $(= 2\mu B)$:

* Cu^+ indicates a copper atom from which one electron has been removed; Al^{+++} indicates an aluminum atom from which three electrons have been removed, etc.

$$U_T = \tfrac{3}{2}kT = (\tfrac{3}{2})(1.38 \times 10^{-23} \text{ J/K})(300 \text{ K}) = 6 \times 10^{-21} \text{ J},$$

$$U_B = 2\mu B = (2)(10^{-23} \text{ A} \cdot \text{m}^2)(1.0 \text{ T}) = 2 \times 10^{-23} \text{ J}.$$

Because U_T equals $300\, U_B$, we see that energy exchanges in collisions can interfere seriously with the alignment of the dipoles with the external field.

If we place a specimen of a paramagnetic substance in a nonuniform magnetic field, such as that near the pole of a strong magnet, it will be attracted toward the region of higher field, that is, toward the pole. We can understand this by drawing an analogy with the corresponding electric case of Fig. 37-7, which shows a dielectric specimen (a sphere) in a nonuniform electric field. The net electric force points to the right in the figure and is

$$F_e = q(E_0 + \Delta E) - q(E_0 - \Delta E) = q(2\Delta E),$$

which we can write as

$$F_e = \frac{(q\,\Delta x)}{\Delta x}2\Delta E = p\left(\frac{2\Delta E}{\Delta x}\right) \cong p\left(\frac{dE}{dx}\right)_{\max}$$

Here $p\ (= q\,\Delta x)$ is the induced electric dipole moment of the sphere. In the differential limit of a very small sphere $(2\Delta E/\Delta x)$ approaches $(dE/dx)_{\max}$, the gradient of the electric field at the center of the sphere.

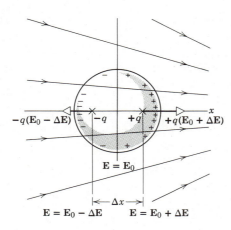

$$-q(\mathbf{E_0} - \Delta \mathbf{E}) \qquad -q \qquad +q \qquad +q(\mathbf{E_0} + \Delta \mathbf{E})$$

$$\mathbf{E} = \mathbf{E_0}$$

$$\mathbf{E} = \mathbf{E_0} - \Delta \mathbf{E} \qquad \mathbf{E} = \mathbf{E_0} + \Delta \mathbf{E}$$

figure 37-7
A dielectric sphere in a nonuniform electric field. The effective induced charges are represented by the point charges $+q$ and $-q$

In the corresponding magnetic case we have, by analogy,

$$F_m = \mu\left(\frac{dB}{dx}\right)_{\max} \qquad (37\text{-}8)$$

Thus, by measuring the magnetic force F_m that acts on a small paramagnetic specimen when we place it in a nonuniform magnetic field whose field gradient $(dB/dx)_{\max}$ is known, we can learn its magnetic dipole moment μ. The *magnetization* $\mathbf{M}$ of the specimen is defined as the magnetic moment per unit volume, or

$$\mathbf{M} = \frac{\mu}{V},$$

where V is the volume of the specimen. It is a vector because $\boldsymbol{\mu}$, the dipole moment of the specimen, is a vector.

In 1895 Pierre Curie (1859–1906) discovered experimentally that the magnetization M of a paramagnetic specimen is directly proportional to

B, the effective value of magnetic field in which the specimen is placed, and inversely proportional to the temperature, or

$$M = C \frac{B}{T}, \qquad (37\text{-}9)$$

in which C is a constant. This equation is known as *Curie's law*. The law is physically reasonable in that increasing B tends to align the elementary dipoles in the specimen, that is, to increase M, whereas increasing T tends to interfere with this alignment, that is, to decrease M. Curie's law is well verified experimentally, provided that the ratio B/T does not become too large.

M cannot increase without limit, as Curie's law implies, but must approach a value M_{max} ($= \mu N/V$) corresponding to the complete alignment of the N dipoles contained in the volume V of the specimen. Figure 37-8 shows this saturation effect for a sample of $CrK(SO_4)_2 \cdot 12H_2O$. The chromium ions are responsible for all the paramagnetism of this salt, all the other elements being paramagnetically inert. To achieve 99.5% saturation, it is necessary to use applied magnetic fields as high as 50,000 gauss ($= 5.0$ T) and temperatures as low as 1.3 K. Note that for more readily achievable conditions, such as $B = 10,000$ gauss ($= 1.0$ T) and $T = 10$ K, the abscissa in Fig. 37-8 is only 1.0 so that Curie's law would appear to be well obeyed for this and for all lower values of B/T.

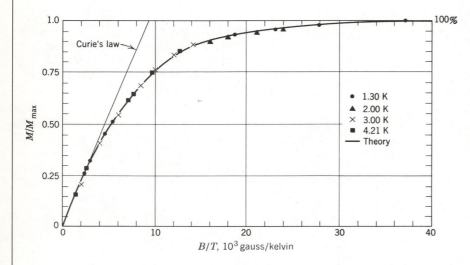

figure 37-8
The ratio M/M_{max} for a paramagnetic salt (chromium potassium alum) in various magnetic fields and at various temperatures. The curve through the experimental points is a theoretical curve calculated from modern quantum physics. (From measurements by W. E. Henry.)

The curve that passes through the experimental points in this figure is calculated from a theory based on modern quantum physics; it is in excellent agreement with experiment.

37-5
DIAMAGNETISM

In 1846 Michael Faraday discovered that a specimen of bismuth brought near to the pole of a strong magnet is *repelled*. He called such substances *diamagnetic* (in contrast with paramagnetic specimens, which are *attracted*). Diamagnetism, present in all substances, is such a feeble effect that its presence is masked in substances made of atoms that have a net magnetic dipole moment, that is, in paramagnetic or ferromagnetic substances.

Figures 37-9a and b show an electron circulating in a diamagnetic atom at angular frequency ω_0 in an assumed circular orbit of radius r.

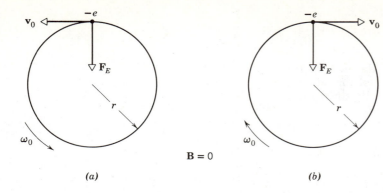

(a)   *(b)*

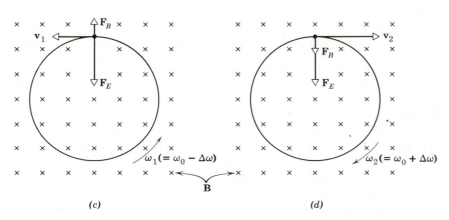

(c)   *(d)*

figure 37-9
figure 37-9
(*a*) An electron circulating in an atom. (*b*) An electron circulating in the opposite direction. (*c*) A magnetic field is introduced, *decreasing* the linear speed of the electron in (*a*), that is, $v_1 < v_0$. (*d*) The magnetic field *increases* the linear speed of the electron in (*b*), that is, $v_2 > v_0$. See problem 15 for a different way of looking at diamagnetism. Our treatment of diamagnetism, though mechanistic and classical, nevertheless yields results in reasonable agreement with experiment.

Each electron is moving under the action of a centripetal force $\mathbf{F}_E$ of electrostatic origin where, from Newton's second law,

$$F_E = ma = m\omega_0^2 r. \qquad (37\text{-}10)$$

Each revolving electron has an orbital magnetic moment, but for the atom as a whole the orbits are randomly oriented so that there is no *net* magnetic effect. In Fig. 37-9*a*, for example, the magnetic dipole moment $\boldsymbol{\mu}_l$ points into the page; in Fig. 37-9*b* it points out and the suggested net effect for the two orbits shown is cancellation. This cancellation is also suggested at the left in Fig. 37-10.

If we apply an external field **B** as in Figs. 37-9*c* and *d*, an *additional* force, given by $-e(\mathbf{v} \times \mathbf{B})$, acts on the electron. This magnetic force acts always at right angles to the direction of motion; its magnitude is

$$F_B = evB = e(\omega r)B. \qquad (37\text{-}11)$$

Show that in Fig. 37-9*c* $\mathbf{F}_B$ and $\mathbf{F}_E$ point in opposite directions and that in Fig. 37-9*d* they point in the same direction. Note that since the centripetal force changes when we turn on the magnetic field (the radius can be shown to remain constant), the angular velocity must also change; thus ω in Eq. 37-11 differs from ω_0 in Eq. 37-10.

Applying Newton's second law to Figs. 37-9*c* and *d*, and allowing for both directions of circulation, yields for the *resultant* forces on the electrons

$$F_E \pm F_B = ma = m\omega^2 r.$$

Substituting Eqs. 37-10 and 37-11 into this equation yields

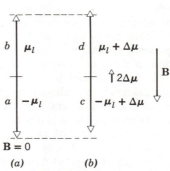

(a)   *(b)*

figure 37-10
The magnetic moments of the two oppositely circulating electrons in an atom cancel when there is no external magnetic field, as in (*a*), but do *not* cancel when a field is applied, as in (*b*). Note that the resultant moment in (*b*) points in the *opposite* direction to **B**. Compare carefully with Fig. 37-9.

$$m\omega_0^2 r \pm e\omega r B = m\omega^2 r$$

or
$$\omega^2 \mp \left(\frac{eB}{m}\right)\omega - \omega_0^2 = 0. \qquad (37\text{-}12)$$

We could solve this quadratic equation for ω, the new angular velocity. Rather than doing this, we take advantage of the fact (presented without proof; see Problem 14) that ω differs only slightly from ω_0, even in the strongest external magnetic fields. Thus

$$\omega = \omega_0 + \Delta\omega, \qquad (37\text{-}13)$$

where $\Delta\omega \ll \omega_0$. Substituting this equation into Eq. 37-12 yields

$$[\omega_0^2 + 2\omega_0\,\Delta\omega + (\Delta\omega)^2] \mp [\beta\omega_0 + \beta\Delta\omega] - \omega_0^2 = 0,$$

where β is a convenient abbreviation for eB/m. The two terms ω_0^2 cancel each other; the terms $(\Delta\omega)^2$ and $\beta\Delta\omega$ are small compared to the remaining terms and may be set equal to zero with only small error. This leads, as an excellent approximation, to

$$\Delta\omega \cong \mp \tfrac{1}{2}\beta = \pm \frac{eB}{2m} \qquad (37\text{-}14)$$

or, from Eq. 37-13,
$$\omega = \omega_0 \pm \frac{eB}{2m}.$$

Thus the effect of applying a magnetic field is to increase or decrease (depending on the direction of circulation) the angular velocity. This, in turn, increases or decreases the orbital magnetic moment of the circulating electron (see Example 2).

In Fig. 37-9c the angular velocity is reduced (because the centripetal force is reduced) so that the magnitude of the magnetic moment is reduced. In Fig. 37-9d, however, the angular velocity is increased so that the magnitude of $\boldsymbol{\mu}_l$ is increased. These effects are shown on the right in Fig. 37-10, where it will be noted that the two magnetic moments *no longer cancel*.

We see that if we apply a magnetic field **B** to a diamagnetic substance, a magnetic moment will be *induced* whose direction (out of the plane of Fig. 37-9) is *opposite* to **B**; see also Fig. 37-10. This is precisely the reverse of paramagnetism, in which the (*permanent*) magnetic dipoles tend to point in the *same* direction as the applied field.

We can now understand why a diamagnetic specimen is repelled when brought near to the pole of a strong magnet. If the pole is a north pole, there exists a nonuniform magnetic field **B** pointing away from the pole. If a sphere made of a diamagnetic material (bismuth, say) is brought near to this pole, the magnetization **M** that is induced in it points toward the pole, that is, *opposite to* **B.** Thus the side of the sphere closest to the magnet behaves like a north pole and is *repelled* by the nearby north pole of the magnet. For a paramagnetic sphere, the vector **M** points *along the direction of* **B** and the side of the sphere nearest to the magnet is a south pole, which is *attracted* to the north pole of the magnet.

EXAMPLE 5

Calculate the *change* in magnetic moment for a circulating electron, as described in Example 2, if a magnetic field **B** of 2.0 T (= 20,000 gauss) acts at right angles to the plane of the orbit.

We obtain μ from Eq. 37-3, or

$$\mu = NiA = (1)(ev)(\pi r^2) = (1)\left(\frac{e\omega}{2\pi}\right)(\pi r^2) = \tfrac{1}{2}er^2\omega.$$

The *change* in μ is

$$\Delta\mu = \tfrac{1}{2}er^2\,\Delta\omega$$

or, from Eq. 37-14,

$$\Delta\mu = \pm\tfrac{1}{2}er^2\left(\frac{eB}{2m}\right) = \pm\frac{e^2Br^2}{4m}.$$

Substituting numbers yields

$$\Delta\mu = \pm\frac{(1.6\times10^{-19}\ \text{C})^2(2.0\ \text{T})(5.3\times10^{-11}\ \text{m})^2}{(4)(9.1\times10^{-31}\ \text{kg})}$$

$$= \pm4.0\times10^{-29}\ \text{A}\cdot\text{m}^2.$$

In Example 2 the moment μ_l was $9.2\times10^{-24}\ \text{A}\cdot\text{m}^2$, so that the change induced by even a strong external magnetic field is rather small, the ratio $\Delta\mu/\mu_l$ being about 4×10^{-6}.

37-6
FERROMAGNETISM

For three elements (Fe, Co, and Ni) and for a variety of alloys of these and other elements, a special effect occurs which permits a specimen to achieve a high degree of magnetic alignment in spite of the randomizing tendency of the thermal motions of the atoms. In such materials, described as *ferromagnetic*, a special form of interaction called *exchange coupling* occurs between adjacent atoms, coupling their magnetic moments together in rigid parallelism.* If the temperature is raised above a certain critical value, called the *Curie temperature*, the exchange coupling suddenly disappears and the materials become simply paramagnetic. For iron the Curie temperature is 1043 K. Ferromagnetism is evidently a property not only of the individual atom or ion but also of the interaction of each atom or ion with its neighbors in the crystal lattice (see Fig. 21-5) of the solid.

Figure 37-11 shows a *magnetization curve* for a specimen of iron. To obtain such a curve, we form the specimen, assumed initially unmagnetized, into a ring and wind a toroidal coil around it, as in Fig. 37-12, to form a so-called *Rowland ring.*† When a current i is set up in the coil, *if the iron core is not present*, a magnetic field is set up within the toroid given by (see Eq. 34-7)

$$B_0 = \mu_0 ni, \qquad (37\text{-}15)$$

where n is the number of turns per unit length for the toroid. Although this formula was derived for a long solenoid, we can apply it to a toroid

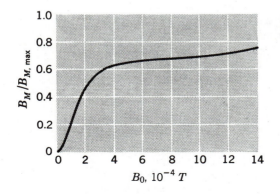

figure 37-11
A magnetization curve for iron.

* Exchange coupling, a purely quantum effect, cannot be "explained" in terms of classical physics.
† See Section 33-1 for further information about H. A. Rowland.

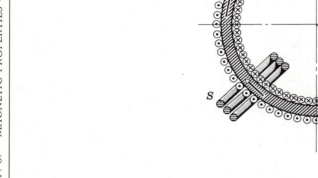

figure 37-12
A Rowland ring, showing a
secondary coil S.

if $d \ll r$ in Fig. 37-12. Because of the iron core, the actual value of **B** in the toroidal space will exceed **B**$_0$, by a large factor in many cases, since the elementary atomic dipoles in the core line up with the applied field **B**$_0$, thereby setting up their own magnetic field. Thus we can write

$$B = B_0 + B_M \qquad (37\text{-}16)$$

where B_M is the magnetic field contribution due to the core; it is proportional to the magnetization M of the core. Often $B_M \gg B_0$.

The field B_0 is proportional to the current in the toroid and we can calculate it readily, using Eq. 37-15; B can be measured in a way that we describe below. An experimental value for B_M can be derived from Eq. 37-16. It has a maximum value $B_{M,\max}$ corresponding to complete alignment of the atomic dipoles in the iron. Thus we can plot, as in Fig. 37-11, the fractional degree of alignment $(= B_M/B_{M,\max})$ as a function of B_0. For this specimen a value of 96.5% saturation is achieved at $B_0 = 0.13$ T $(= 1300$ gauss; this point is about 16 ft to the right of the origin in the figure); increasing B_0 to 1.0 T $(= 10,000$ gauss; about 120 ft to the right in Fig. 37-11) increases the fractional saturation only to 97.7%.

The use of iron in transformers, electromagnets, etc., greatly increases the strength of the magnetic field that can be generated by a given current in a given set of windings. That is, very often, $B_M \gg B_0$ in Eq. 37-16. However, the presence of iron also sets a limit to the maximum magnetic field that can be produced because of the saturation effect suggested in Fig. 37-11. To generate magnetic fields greater than this saturation limit, it is necessary to abandon the use of iron and to rely on the "brute force" application of very large (and often transient) currents.*

To measure B in the system of Fig. 37-12, let the current in the toroid windings be increased from zero to i. The flux through the secondary coil S will change by BA, where A is the area of the toroid. While the flux is changing, an induced emf will appear in coil S, according to Faraday's law. For simplicity, we assume that the current in the toroid is so adjusted that B increases linearly with time for an interval Δt, as shown in Fig. 37-13a. The emf in coil S during this interval, from Faraday's law,† will then be

---

* See "Megagauss Physics" by C. M. Fowler, *Science*, April 1973.
† We ignore the minus sign because we are concerned only with the magnitude of ε.

$$\varepsilon = N \frac{\Delta \Phi_B}{\Delta t} = \frac{NBA}{\Delta t},$$

where N is the number of turns in coil S. This emf will set up a current i_s in coil S given by

$$i_s = \frac{\varepsilon}{R} = \frac{NBA}{R \, \Delta t}$$

or

$$B = \frac{(i_s \, \Delta t)R}{NA} = \frac{qR}{NA},$$

in which R is the resistance of coil S and $i_s \, \Delta t$ is the charge q that passes through this coil during time Δt. If a so-called *ballistic galvanometer* is connected to S, its deflection will be a measure of the charge q. Thus it is possible to find B for any value of the current i in the toroid windings. A more detailed analysis shows that it is not necessary that the curve $B(t)$ in Fig. 37-13a be linear during the interval Δt.

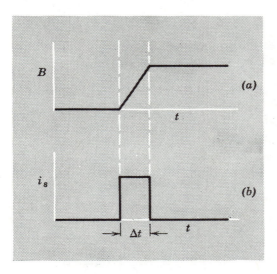

figure 37-13
(a) B for a Rowland ring as the current in the windings is increased from zero during an interval Δt. (b) The corresponding induced current in the secondary coil. Both curves are idealized; in practice, the sharp corners would be rounded off.

Magnetization curves for ferromagnetic materials *do not retrace themselves* as we increase and then decrease the toroid current. Figure 37-14 shows the following operations with a Rowland ring: (1) starting with the iron unmagnetized (point a), increase the toroid current until B_0 (= $\mu_0 n i$) has the value corresponding to point b; (2) reduce the current in the toroid winding back to zero (point c); (3) reverse the toroid current and increase it in magnitude until point d is reached; (4) reduce the current to zero again (point e); (5) reverse the current once more until point b is reached again. The lack of retraceability shown in Fig. 37-14 is called *hysteresis*. Note that at points c and e the iron core is magnetized, even though there is no current in the toroid windings; this is the familiar phenomenon of *permanent magnetism*.

We explained the magnetization curve for paramagnetism (see Fig. 37-8) in terms of the mutually opposing tendencies of alignment with the external field and of randomization because of thermal agitation. In ferromagnetism, however, we have assumed that adjacent atomic dipoles are locked in rigid parallelism. Why, then, does the magnetic moment of the specimen not reach its saturation value for very low—even zero—values of B_0? The current interpretation is to assume the existence within the specimen of *domains*, that is, of local regions

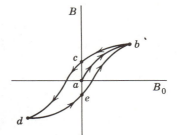

figure 37-14
A magnetization curve (*ab*) for a specimen of iron and an associated hysteresis loop (*ebcde*).

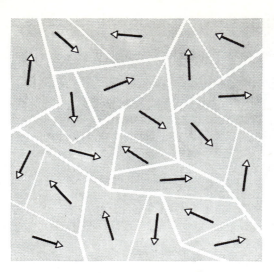

figure 37-15
The separate magnetic domains in
an unmagnetized polycrystalline
ferromagnetic sample are oriented
to produce little external effect.
Each domain, however, is made up
of completely aligned atomic
dipoles, as suggested by the arrows.
The white boundaries define the
crystals that make up the solid; the
light boundaries define the domains
within the crystals.

within which there is essentially perfect alignment. The domains themselves, however, as Fig. 37-15 suggests, are not parallel at moderately low values of B_0.

Figure 37-16 shows some domain photographs, taken by sprinkling a colloidal suspension of finely powdered iron oxide on a properly etched single crystal of iron. The domain boundaries, which are thin regions in which the alignment of the elementary dipoles changes from a certain direction in one domain to an entirely different direction in the other, are the sites of intense but highly localized and nonuniform magnetic fields. The suspended colloidal particles are attracted to these regions. Although the atomic dipoles in the individual domains are completely aligned, the specimen as a whole may have a very small resultant magnetic moment. This is the state of affairs in an unmagnetized iron nail.

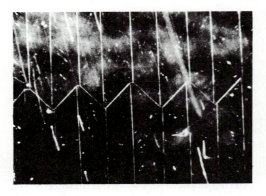

figure 37-16
Domain patterns for a single
crystal of iron containing 3.8%
silicon. The white lines show the
boundaries between the domains.
These boundaries are regular rather
than irregular, as in Fig. 37-15,
because the specimen is a single
crystal. In Fig. 37-15, the specimen
is not a single crystal but is
made up of many crystallites or
grains. (Courtesy H. J. Williams,
Bell Telephone Laboratories.)

As we magnetize a piece of iron by placing it in an external magnetic field, two effects take place. One is a growth in size of the domains that are favorably oriented at the expense of those that are not, as in Fig. 37-17. Second, the direction of orientation of the dipoles within a domain may swing around as a unit, becoming closer to the field direction. Hysteresis comes about because the domain boundaries do not move completely back to their original positions when the external field B_0 is removed.

Two other types of magnetism, closely related to ferromagnetism, are *antiferromagnetism* and *ferrimagnetism* (note spelling). In antiferro-

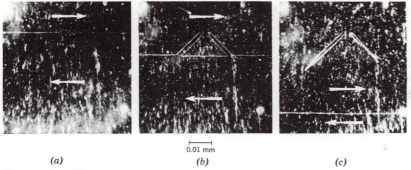

(a)   (b)   (c)

0.01 mm

figure 37-17
(a) A boundary between two domains, with the magnetization in each domain shown by the white arrows. (b) If an external magnetic field pointing from left to right is imposed on the specimen, the upper domain will grow at the expense of the lower. The domain boundary will move down as the elementary dipoles reverse themselves. (c) The process continues. The boundary has moved across a region in which there is a crystal imperfection. (Courtesy H. J. Williams, Bell Telephone Laboratories.)

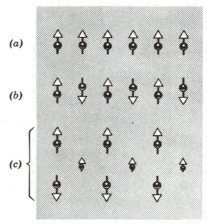

figure 37-18
Showing how elementary magnetic dipoles are oriented by the interatomic exchange coupling in (a) ferromagnetism, (b) antiferromagnetism, and (c) ferrimagnetism. The actual arrangements are, of course, three-dimensional.

magnetic substances, of which MnO_2 is an example, the exchange coupling to which we referred on page 825 serves to lock adjacent ions into rigid *antiparallelism* (see Fig. 37-18). Such materials exhibit very little gross external magnetism. However, if they are heated above a certain temperature, called the *Néel temperature,* the exchange coupling ceases to act and the material becomes paramagnetic. In ferrimagnetic substances, of which iron ferrite is an example, two different kinds of magnetic ions are present. In iron ferrite the two ions are Fe^{++} and Fe^{+++}. The exchange coupling locks the ions into a pattern like that of Fig. 37-18c, in which the external effects are intermediate between ferromagnetism and antiferromagnetism. Here, too, the exchange coupling disappears if the material is heated above a certain characteristic temperature.

37-7
NUCLEAR MAGNETISM

Many nuclei have magnetic dipoles, and the possibility arises that a specimen of matter may exhibit gross external magnetic effects associated with its nuclei. However, nuclear magnetic moments are several orders of magnitude smaller than those associated with the electronic motions in an atom or ion. The magnetic moment of an electron associated with its spin, for example, exceeds that of the proton (the nucleus of hydrogen) by a factor of 660.

Gross external effects for nuclear magnetism are smaller than the corresponding (ionic) paramagnetic effects by the *square* of ratios of this order of magnitude, because (a), *all else being equal,* the external magnetism is reduced by such a ratio, but (b) the very fact that the magnetic dipole moment of the nucleus is smaller means that (see Example 4) the thermal vibrations are proportionally (to a good approximation) more effective in reducing the degree of alignment of the elementary dipoles in an external magnetic field; thus all else is *not* equal and the ratio enters twice.

Techniques such as the Rowland ring (see Fig. 37-12) are far too insensitive to detect nuclear magnetism. We describe here a *nuclear resonance technique* by means of which nuclear magnetism can readily reveal itself. This method is also vastly useful for studying paramagnetism, ferromagnetism, antiferromagnetism, and ferrimagnetism, in all of which cases the magnetic effects are associated not with the nuclei but with the atomic electrons. The nuclear-resonance technique was developed in 1946 by E. M. Purcell and his co-workers at Harvard. Simultaneously and independently, F. Bloch and his co-workers at Stanford discovered a very similar method. For these achievements the two physicists shared a Nobel prize.

We focus our attention on the problem of measuring the magnitude μ of the magnetic moment of the proton. In principle, this can be done by placing a specimen containing protons in an external magnetic field **B** and by measuring the energy $(= 2\mu B)$ required to turn the protons end for end. A rigorously correct description of the procedures cannot be given without using quantum physics. The description given, although based entirely on classical physics, nevertheless leads to the correct conclusions.

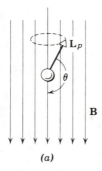
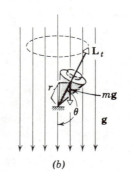

(a)   (b)

figure 37-19

(a) A spinning proton precessing in an external magnetic field and (b) a spinning top precessing in an external gravitational field. L_p and L_t are the two angular momentum vectors.

Figure 37-19a shows a spinning proton with its axis making an angle θ with a uniform external magnetic field **B.** Figure 37-19b shows a spinning top with its axis making an angle θ with a uniform external gravitational field **g.** In each case there is a torque that tends to align the axis of the spinning object with the field. For the proton (see Eq. 33-11) it is given by

$$\tau_p = \mu B \sin\theta. \qquad (37\text{-}17a)$$

For the top it is given by $\tau_t = mgr \sin\theta,$ (37-17b)

where r locates the center of mass of the top and m is its mass.

In Section 13-2, we saw that the spinning top precesses about a vertical axis with an angular frequency given by

$$\omega_t = \frac{mgr}{L_t}, \qquad (37\text{-}18a)$$

in which L_t is the spin angular momentum of the top.

The proton, which has a quantized spin angular momentum L_p, will also precess about the direction of the (magnetic) field because of the action of the (magnetic) torque. You should derive the expression for the frequency of precession, being guided by the derivation of Section 13-2, but using the magnetic torque (see Eq. 37-17a) instead of the gravitational torque (see Eq. 37-17b). The relation is

$$\omega_p = \frac{\mu B}{L_p}. \qquad (37\text{-}18b)$$

EXAMPLE 6

831 NUCLEAR MAGNETISM SEC. 37-7

What is the precession frequency of a proton in a magnetic field of 0.50 T (= 5000 gauss)?

The quantities μ and L_p in Eq. 37-18b are 1.4×10^{-26} A·m² and 0.53×10^{-34} J·s. This equation then yields

$$\nu_p = \frac{\omega_p}{2\pi} = \frac{\mu B}{2\pi L_p} = \frac{(1.4 \times 10^{-26} \text{ A·m}^2)(0.50 \text{ T})}{(2\pi)(0.53 \times 10^{-34} \text{ J·s})} = 2.1 \times 10^7 \text{ Hz.}$$

This frequency (=21 MHz) is in the radio-frequency range.

It is possible to change the energy of any system in periodic motion if we allow an external influence to act on it at the same frequency as that of its motion. This is the familiar *resonance* condition. As an "external influence" for the precessing proton, we use a small alternating magnetic field $\mathbf{B}_{osc}$ arranged to be at right angles to the steady field $\mathbf{B}$. This oscillating field combines vectorially with the steady field so that the *resultant* field rocks back and forth between the limits shown by the dashed lines in Fig. 37-20. Typical values for B and for the amplitude of $\mathbf{B}_{osc}$ are 5000 gauss and one gauss, respectively, so that the rocking angle α in the figure is quite small. If the angular frequency ω_0 of the oscillating field is chosen equal to the angular precession frequency ω_p of the proton, it turns out that the precessing proton can absorb energy. An increase in energy means an increase in θ in Fig. 37-19a.

The resonance condition

$$\omega_0 = \frac{\mu B}{L_p} \qquad (37\text{-}19)$$

can be used to measure μ. We place the spinning proton in a known field $\mathbf{B}$, apply a "perturbing field" at right angles to it, and vary the angular frequency ω_0 of this perturbing field until resonance occurs. It is possible to tell when Eq. 37-19 is satisfied because, at resonance, many spinning protons will tend to turn end for end in the field, absorbing energy which can be detected by appropriate electronic techniques.

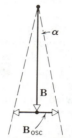

figure 37-20
In the nuclear magnetic resonance method a small oscillating magnetic field $\mathbf{B}_{osc}$ is placed at right angles to a steady field $\mathbf{B}$.

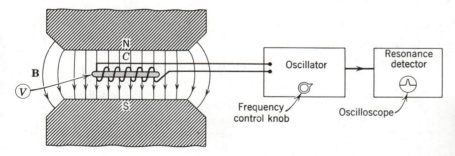

figure 37-21
An arrangement to observe nuclear resonance. The oscillating field is horizontal within the coil.

Figure 37-21 is a schematic diagram of an experimental arrangement. The protons, present as hydrogen nuclei in a small vial V of water, are immersed in a strong steady magnetic field caused by the electromagnet whose pole faces N and S are shown. A rapidly alternating current in the small coil C provides the (horizontal) weak, perturbing magnetic field $\mathbf{B}_{osc}$. This current is provided by a radio-frequency oscillator whose angular frequency ω_0 can be varied; an electronic "resonance detector," also connected to the oscillator, serves to indicate when energy is being drained from the oscillator and used to "flip the protons." In principle, the oscillator angular frequency ω_0 is varied until the resonance detector shows that Eq. 37-19 is satisfied (see Fig. 37-22). The magnetic moment μ can then be determined by measuring B and ω_0. Surprisingly enough, magnetic moments can be measured in this and similar ways to a much greater accuracy than we can measure μ for a bar magnet. For the proton we have

$$\mu_p = 1.410617 \times 10^{-26} \text{ A·m}^2.$$

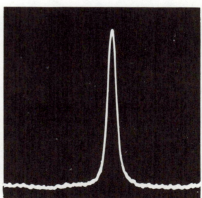

figure 37-22
An oscilloscope photograph of a proton resonance peak showing energy absorbed from the oscillator (vertical axis) versus oscillator frequency (horizontal axis). (From Bloembergen et al.

In Chapter 30 we saw that if a dielectric is placed in an electric field, polarization charges will appear on its surface. These surface charges, which find their origin in the elementary electric dipoles (permanent or induced) that make up the dielectric, set up a field of their own that modifies the original field. For the simple case discussed in Chapter 30 – a dielectric slab in a parallel-plate capacitor – this complication can readily be handled in terms of the electric field vector **E** and some knowledge of the electric properties of the slab material, such as the dielectric constant. For more complex problems we asserted that it was useful to introduce two other (subsidiary) electric vectors, the *electric polarization* **P** and the *electric displacement* **D.** Table 30-2 shows some of the characteristics of these three vectors.

In magnetism we find a similar situation. If magnetic materials are placed in an external magnetic field, the elementary magnetic dipoles (permanent or induced) will act to set up a field of their own that will modify the original field. For the simple case discussed in this chapter – a Rowland ring with a ferromagnetic core – this complication can readily be handled in terms of the magnetic field vector **B** and some knowledge of the magnetic properties of the ring material, such as that provided by Fig. 37-11. For more complex problems we find it useful to introduce two other (subsidiary) magnetic vectors, the *magnetization* **M** and the *magnetic field strength* **H.** We do so largely so that the student who takes a second course in electromagnetism will have some familiarity with them.

Consider a Rowland ring carrying a current i_0 in its windings and designed so that its core, assumed to be iron, can be removed. The magnetic field **B,** measured by the methods of Section 37-6, will be much greater when the core is in place than when it is not, assuming that the current in the windings remains unchanged.

Physically we can understand the large value of B in the iron core in terms of the alignment of the elementary dipoles in the iron. A hypothetical slice out of the iron core, as in Fig. 37-23b, has a magnetic moment $d\mu$ equal to the vector sum of all of the elementary dipoles contained in it. We define our first subsidiary vector, the *magnetization* **M,** as the magnetic moment per unit volume of the core material. For the slice of Fig. 37-23b we have

$$d\mu = \mathbf{M}(A\ dl),$$

where $(A\ dl)$ is the volume of the slice, A being the cross section of the core.

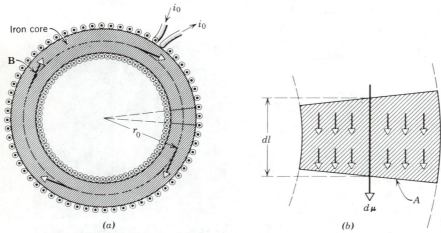

figure 37-23
(*a*) A Rowland ring with an iron core. (*b*) A slice of the core, showing its magnetic moment $d\mu$ caused by the alignment of the elementary magnetic dipoles in the iron.

When we discussed Ampère's law in Chapter 34, we assumed that no magnetic materials were present. If we apply this law, namely,

$$\oint \mathbf{B} \cdot d\mathbf{l} = \mu_0 i, \qquad (37\text{-}20)$$

to the circular path of integration shown in Fig. 37-23a, we obtain

$$(B)(2\pi r_0) = \mu_0(N_0 i_0), \qquad (37\text{-}21)$$

has when the core is in place if we increase the current in the windings by an amount $i_{M,0}$. The magnetization of the iron core is thus *equivalent in its effect on* **B** to such a hypothetical current increase. We choose to modify Ampère's law by arbitrarily inserting a *magnetizing current* term i_M on the right, obtaining

$$\oint \mathbf{B} \cdot d\mathbf{l} = \mu_0(i + i_M). \tag{37-22}$$

If we give i_M a suitable value when the iron core is in place, it is clear that Ampère's law, in this new form, can remain valid. It remains to relate this (largely hypothetical) magnetizing current to something more physical, the magnetization **M**.*

Applying Eq. 37-22 to the iron ring of Fig. 37-23a yields

$$(B)(2\,\pi r_0) = \mu_0(N_0 i_0 + N_0 i_{0M,0}). \tag{37-23}$$

We can relate $i_{M,0}$ to the magnetization **M** if we recall (Eq. 33-10) that the magnetic moment of a magnetic dipole in the form of a current loop is given by

$$\mu = NiA,$$

where N is the number of turns in the loop, i is the loop current, and A is the loop area. Let us use this equation to find what increase $i_{M,0}$ in current in the windings around the slice of Fig. 37-23b would produce a magnetic moment equivalent to that actually produced by the alignment of elementary dipoles in the slice. We have

$$M(A\ dl) = \left(N_0 \frac{dl}{2\pi r_0}\right)(i_{M,0})(A),$$

the quantity in the first parentheses on the right being the number of turns associated with the slice of thickness dl. This reduces to

$$N_0 i_{M,0} = M(2\pi r_0). \tag{37-24}$$

Substituting this into Eq. 37-23 yields

$$(B)(2\pi r_0) = \mu_0(N_0 i_0) + \mu_0(M)(2\pi r_0). \tag{37-25}$$

We now choose to generalize from the special case of the Rowland ring by writing Eq. 37-25 as

$$\oint \mathbf{B} \cdot d\mathbf{l} = \mu_0 i + \mu_0 \oint \mathbf{M} \cdot d\mathbf{l}$$

or

$$\oint \left(\frac{\mathbf{B} - \mu_0 \mathbf{M}}{\mu_0}\right) \cdot d\mathbf{l} = i.$$

The quantity $(\mathbf{B} - \mu_0 \mathbf{M})/\mu_0$ occurs so often in magnetic situations that we give it a special name, the *magnetic field strength* **H**, or

$$\mathbf{H} = \frac{\mathbf{B} - \mu_0 \mathbf{M}}{\mu_0}$$

which we write as $\mathbf{B} = \mu_0 \mathbf{H} + \mu_0 \mathbf{M}.$ $\qquad$ (37-26)

---

* It is possible to give reality to the magnetizing current by viewing it as a real current that flows around the magnet at its surface, being the resultant macroscopic effect of all the microscopic current loops that constitute the atomic electron orbits. This *Amperian current* viewpoint, however, does not take the magnetization due to electron spin readily into account. Since we do not attempt to measure magnetizing currents experimentally, other than through their (postulated) magnetic effects, we prefer to view the magnetizing current as a convenient formalism.

Ampère's law can now be written in the simple form

$$\oint \mathbf{H} \cdot d\mathbf{l} = i, \tag{37-27}$$

which holds in the presence of magnetic materials and in which i is the *true current only*, that is, it does not include the magnetizing current. This reminds us that the electric displacement vector $\mathbf{D}$ permitted us to write Gauss's law for the case in which dielectric materials are present, in a form involving *free charges only*, that is, not polarization charges; see Table 30-2.

We state without proof (see Problems 23 and 24) that at a boundary between two media (1) the component of $\mathbf{H}$ tangential to the surface has the same value on each side of the surface* and (2) the component of $\mathbf{B}$ perpendicular to the surface has the same value on each side of the surface. These *boundary conditions* are of great value in solving complex problems.

To find H in our Rowland ring, let us apply Ampère's law in the generalized form of Eq. 37-27. We have

$$(H)(2\pi r_0) = N_0 i_0,$$

where i_0 is the (true) current in the windings. This gives

$$H = \left(\frac{N_0}{2\pi r_0}\right) i_0 = n i_0, \tag{37-28}$$

in which n is the number of turns per unit length. Since we have not introduced any information describing the core into Eq. 37-27, the value of H computed from Eq. 37-28 is independent of the core material.

B can be measured experimentally by the method of Section 37-6 and M can then be calculated from Eq. 37-26. Note in passing (see Eq. 37-15) that the abscissa B_0 in Fig. 37-11 is proportional to $H(= \mu_0 H)$, the ordinate being proportional to B. Curves such as this and that of Fig. 37-14 are called *B-H* curves.

Let us assume that we have made measurements of $\mathbf{H}$, $\mathbf{B}$, and $\mathbf{M}$ for a wide variety of magnetic materials, using either the technique described or an equivalent one. For *paramagnetic* and *diamagnetic* materials we would find, as an experimental result, that $\mathbf{B}$ is directly proportional to $\mathbf{H}$, or

$$\mathbf{B} = \kappa_m \mu_0 \mathbf{H}, \tag{37-29}$$

in which κ_m, the *permeability* of the magnetic medium, is a constant for a given temperature and density of the material. Eliminating $\mathbf{B}$ between Eqs. 37-29 and 37-26 allows us to write

$$\mathbf{M} = (\kappa_m - 1)\mathbf{H}, \tag{37-30}$$

which is another expression of the linear or proportional character of paramagnetic and diamagnetic materials.

For a vacuum, in which there are no magnetic dipoles present to be aligned, the magnetization $\mathbf{M}$ must be zero. Putting $\mathbf{M} = 0$ in Eq. 37-26 leads to

$$\mathbf{B} = \mu_0 \mathbf{H} \quad \text{(in vacuum)}. \tag{37-31}$$

Comparison with Eq. 37-29 shows that a vacuum must be described by $\kappa_m = 1$. Equation 37-30 verifies that the magnetization vanishes if we

---

* Assuming that there are no true currents at the surface, as there are in the Rowland ring in Fig. 37-23a, for example.

Table 37-1
Three magnetic vectors

Name	Symbol	Associated with	Boundary Condition
Magnetic induction*	**B**	All currents	Normal component continuous
Magnetic field strength	**H**	True currents only	Tangential component continuous†
Magnetization (magnetic dipole moment per unit volume)	**M**	Magnetization currents only	Vanishes in a vacuum

Defining equations for **B**	$\mathbf{F} = q\mathbf{v} \times \mathbf{B}$	Eq. 33-2
	or $= i\mathbf{l} \times \mathbf{B}$	Eq. 33-6a
General relation among the three vectors	$\mathbf{B} = \mu_0\mathbf{H} + \mu_0\mathbf{M}$	Eq. 37-26
Ampère's law when magnetic materials are present	$\oint \mathbf{H} \cdot d\mathbf{l} = i$ (i = true current only)	Eq. 37-27
Empirical relations for certain magnetic materials**	$\mathbf{B} = \kappa_m\mu_0\mathbf{H}$	Eq. 37-29
	$\mathbf{M} = (\kappa_m - 1)\mathbf{H}$	Eq. 37-30

* We have usually called **B** simply "the magnetic field." Here we call it by its alternative name, "magnetic induction," to avoid confusion with "magnetic field strength," the name for **H**.
** For paramagnetic and diamagnetic materials only, if κ_m is to be independent of **H**.
† Assuming no true currents exist at the boundary.

put κ_m equal to unity. For paramagnetic materials κ_m is slightly greater than unity. For diamagnetic materials it is slightly less than unity; Eq. 37-30 shows that this requires **M** and **H** to be oppositely directed, a fact implicit in Section 37-5.

In ferromagnetic materials the relationship between **B** and **H** is far from linear, as Figs. 37-11 and 37-14 show. Experimentally, κ_m proves to be a function not only of the value of H but also, because of hysteresis, of the magnetic and thermal history of the specimen.*

An interesting special case of ferromagnetism is the permanent magnet, for which **H**, **M**, and **B** all have nonvanishing values inside the magnet even though there is no true current. Figure 37-24 shows typical lines of **B** and **H** associated with such a magnet. The lines of **B** may be drawn as continuous loops, the boundary condition (2), mentioned above, being satisfied where the lines enter and leave the magnet.

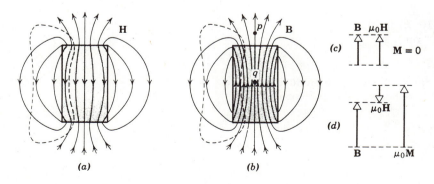

(a)   (b)   (c)   (d)

figure 37-24
(a) The lines of **H** and (b) the lines of **B** for a permanent magnet. Note that the lines of **H** change direction at the boundary. The closed dashed curves are paths of integration around which Ampère's law may be applied. The relation $\mathbf{B} = \mu_0\mathbf{H} + \mu_0\mathbf{M}$ is shown to be satisfied for (c) a particular outside point p and (d) a particular inside point q.

* In the dielectric case there are waxy materials, called *ferroelectrics*, for which the relationship between **D** and **E** is nonlinear, which exhibit hysteresis, and from which quasi-permanent electric dipoles (*electrets*) can be constructed. However, most commonly useful dielectric materials are linear, whereas most commonly useful magnetic materials are nonlinear.

Equation 37-22 shows that the vector **B** is associated with the *total* current, both true and magnetizing. In Fig. 37-24b $\oint \mathbf{B} \cdot d\mathbf{l}$ around any loop such as that shown by the dashed curve is not zero and must be associated with a hypothetical magnetizing current i_M, imagined to circulate around the magnet at its surface; actual or true currents (i) do not exist in this problem. Figure 37-24a shows that **H** reverses direction at the boundary. Since **H** (see Eq. 37-27) is associated with true currents only, we must have $\oint \mathbf{H} \cdot d\mathbf{l} = 0$ around any loop such as that shown by the dashed lines. The reversal of **H** at the boundary makes this possible. Note that **M** and **H** point in opposite directions within the magnet. Table 37-1 summarizes the properties of the three vectors **B**, **H**, and **M**.

EXAMPLE 7

In the Rowland ring the (true) current i_0 in the windings is 2.0 A and the number of turns per unit length (n) in the toroid is 10 turns/cm. B, measured by the technique of Section 37-6, is 1.0 T. Calculate (a) H, (b) M, and (c) the magnetizing current $i_{M,0}$ both when the core is in place and when it is removed. (d) For these particular operating conditions, what is κ_m?

(a) H is independent of the core material and may be found from Eq. 37-28:

$$H = ni$$

$$= (10 \text{ turns/cm})(2.0 \text{ A})$$

$$= 2.0 \times 10^3 \text{ A/m}.$$

(b) M is zero when the core is removed. With the core in place, we may solve Eq. 37-26 for M, obtaining for the magnitude of **M**,

$$M = \frac{B - \mu_0 H}{\mu_0}$$

$$= \frac{(1.0 \text{ T}) - (4\pi \times 10^{-7} \text{ T} \cdot \text{m/A})(2.0 \times 10^3 \text{ A/m})}{(4\pi \times 10^{-7} \text{ T} \cdot \text{m/A})}$$

$$= 7.9 \times 10^5 \text{ A/m}.$$

(c) The effective magnetizing current follows from Eq. 37-24:

$$i_{M,0} = M \left(\frac{2\pi r_0}{N_0}\right) = \frac{M}{n}$$

$$= \frac{7.9 \times 10^5 \text{ A/m}}{2.0 \times 10^3 \text{ turns/m}}$$

$$= 390 \text{ A}.$$

An *additional* current of this amount in the windings would produce the same value of B in the absence of a core as that obtained by alignment of the elementary dipoles, with the core in place.

(d) The permeability can be found from Eq. 37-29, or

$$\kappa_m = \frac{B}{\mu_0 H}$$

$$= \frac{1.0 \text{ T}}{(4\pi \times 10^{-7} \text{ T} \cdot \text{m/A})(2.0 \times 10^3 \text{ A/m})}$$

$$= 397.$$

We emphasize that this value of κ_m holds only for the special conditions of this experiment.

questions

1. Two iron bars are identical in appearance. One is a magnet and one is not. How can you tell them apart? You are not permitted to suspend either bar as a compass needle or to use any other apparatus.

2. Two iron bars always attract, no matter the combination in which their ends are brought near each other. Can you conclude that one of the bars must be unmagnetized?

3. The neutron, which has no charge, has a magnetic dipole moment. Is this possible on the basis of classical electromagnetism, or does this evidence alone indicate that classical electromagnetism has broken down?

4. Do all permanent magnets have to have identifiable north and south poles? Consider geometries other than the straight bar magnet.

5. If magnetic monopoles are shown to exist, is it conceivable that there might be a whole family of them, with different masses, magnetic pole strengths, intrinsic angular momenta, etc.? Compare Appendix F.

6. A certain short iron rod is found, by test, to have a north pole at each end. You sprinkle iron filings over the rod. Where (in the simplest case) will they cling? Make a rough sketch of what the lines of B must look like, both inside and outside the rod. See "A Three-Pole Bar Magnet?" by Jerry D. Wilson, *The Physics Teacher,* September 1976.

7. Consider these two situations. (*a*) A (hypothetical) magnetic monopole is pulled through a single-turn conducting loop along its axis, at a constant speed. (*b*) A short bar magnet (a magnetic dipole) is similarly pulled. Compare qualitatively the net amounts of charge transferred through any cross section of the loop during these two processes.

8. Cosmic rays are charged particles that strike our atmosphere from some external source. We find that more low-energy cosmic rays reach the earth near the north and south magnetic poles than at the (magnetic) equator. Why is this so?

9. How might the magnetic dipole moment of the earth be measured?

10. Give three reasons for believing that the flux Φ_B of the earth's magnetic field is greater through the boundaries of Alaska than through those of Texas.

11. You are a manufacturer of compasses. (*a*) Describe ways in which you might magnetize the needle. (*b*) The end of the needle that points north is usually painted a characteristic color. Without suspending the needle in the earth's field, how might you find out which end of the needle to paint? (*c*) Is the painted end a north or a south magnetic pole?

12. Can you think of a mechanism by which a magnetic storm, that is, a violent perturbation of the earth's magnetic field, can interfere with radio communication?

13. Convince yourself, in terms of the relation $\mathbf{F} = q_0\mathbf{v} \times \mathbf{B}$ (Eq. 33-2) that electrons and protons of the solar wind that are trapped in the earth's Van Allen radiation belts will indeed spiral around the lines of **B** and be reflected backward near the earth's north and south magnetic poles (the magnetic mirror effect). In what way do the motions of the trapped electrons and protons differ?

14. Aurorae are most frequently observed, not at the north and south magnetic poles, but at magnetic latitudes about 23° away from these poles (passing through Hudson Bay, for example, in the northern geomagnetic hemisphere). Can you think of any reason, however qualitative, why the auroral activity should not be strongest at the poles themselves?

15. Is the magnetization at saturation for a paramagnetic substance very much different from that for a saturated ferromagnetic substance of about the same size?

16. The magnetization induced in a given diamagnetic sphere by a given external magnetic field does not vary with temperature, in sharp contrast to the situation in paramagnetism. Is this understandable in terms of the description that we have given of the origin of diamagnetism?

17. Explain why a magnet attracts an unmagnetized iron object such as a nail.

18. Does any net force or torque act on (a) an unmagnetized iron bar or (b) a permanent bar magnet when placed in a uniform magnetic field?

19. A nail is placed at rest on a smooth table top near a strong magnet. It is released and attracted to the magnet. What is the source of the kinetic energy it has just before it strikes the magnet?

20. Compare the magnetization curves for a paramagnetic substance (see Fig. 37-8) and for a ferromagnetic substance (see Fig. 37-11). What would a similar curve for a diamagnetic substance look like? Do you think that it would show saturation effects in strong applied fields (say 10 T)?

21. Why do iron filings line up with a magnetic field, as in Fig. 37-1? After all, they are not intrinsically magnetized.

22. Distinguish between the precession frequency and the cyclotron frequency of a proton in a magnetic field.

23. In our discussion of nuclear magnetism we said that energy absorption occurs because the dipoles are turned end for end. However, a given dipole might initially be lined up either with the field or against it. In the first case there would be an *absorption* of energy, but in the second case there would be a *release* of energy, each amount being $2\mu B$. Why do we observe a *net* absorption? These two events would seem to cancel.

24. Discuss similarities and differences in Tables 30-2 and 37-1.

25. In what sense do a parallel plate capacitor filled with a dielectric and a Rowland ring (see Fig. 37-12) with an iron core show formal similarities as far as E and B (and their related vectors) are concerned? Discuss in terms of Tables 30-2 and 37-1.

26. A Rowland ring (see Fig. 37-12) carries a constant current. If we cut a small slot in the iron core, leaving an air gap, what changes occur in **B**, **H**, and **M** within the core?

27. You are given a bar magnet, with its north pole pointing up and its south pole pointing down. What are the directions of **B**, **H**, and **M** for points (a) inside the magnet near its center, (b) outside and just above the magnet, and (c) outside and just below the magnet?

problems

SECTION 37-1

1. The dipole moment of a small, single-turn, current loop is 2.0×10^{-4} A·m². What is the magnetic field (due to the dipole) on the axis of the dipole 8.0 cm away from the loop?   *Answer:* 7.8×10^{-8} T, pointing along the axis.

2. A simple bar magnet is suspended by a string, as shown in Fig. 37-25. If a uniform magnetic field **B** directed parallel to the ceiling is then established, sketch the resulting orientation of string and magnet.

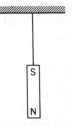

figure 37-25
Problem 2

3. Calculate (a) the electric field **E** and (b) the magnetic field **B** at a point 0.10 nm away from a proton, measured along its axis of spin. The magnetic moment of the proton is 1.4×10^{-26} A·m².
Answer: (a) 1.4×10^{11} V/m. (b) 2.8×10^{-3} T.

4. Show that, classically, a spinning positive charge will have a spin magnetic moment that points in the same direction as its spin angular momentum.

5. An electron has a spin angular momentum L of 0.53×10^{-34} J·s and a magnetic moment μ of 9.3×10^{-24} A·m². Compare μ/L and e/m for the electron. *Answer:* They are equal.

6. A total charge q is distributed uniformly on a dielectric ring of radius r. If the ring is rotated about an axis perpendicular to its plane and through its center at an angular speed ω, find the magnitude and direction of its resulting magnetic moment.

7. Show, by sketching the magnetic field of a magnetic dipole, that (a) if the dipole moments of two nearby dipoles are parallel, they will not tend to remain that way and (b) if they are anti-parallel, they *will* tend to remain that way.

In each case consider the torques on the second dipole in the field of the first.

8. Assume that the electron is a small sphere of radius R, its charge and mass being spread uniformly throughout its volume. Such an electron has a "spin" angular momentum L of 0.53×10^{-34} J·s and a magnetic moment μ of 9.3×10^{-24} A·m². Show that $e/m = 2\mu/L$. Is this prediction in agreement with experiment? (Hint: The spherical electron must be divided into infinitesimal current loops and an expression for the magnetic moment found by integration. This model of the electron is too mechanistic to be in the spirit of quantum physics.)

SECTION 37-3

9. From the data given in the text find (a) the average vertical component of the earth's magnetic field at Tucson in 1964 and (b) the average magnitude of **B**.
   Answer: (a) 44μT $(=0.44$ gauss). (b) 51μT $(=0.51$ gauss).

10. The earth has a magnetic dipole moment of 8.0×10^{22} A·m². (a) What current would have to be set up in a single turn of wire going around the earth at its magnetic equator if we wished to set up such a dipole? (b) Could such an arrangement be used to cancel out the earth's magnetism at points in space well above the earth's surface? (c) On the earth's surface?

SECTION 37-4

11. (a) What is the magnetic moment due to the orbital motion of an electron in an atom when the orbital angular momentum is one quantum unit $(= h/2\pi = 1.05 \times 10^{-34}$ J · s). (b) The intrinsic spin magnetic moment of an electron is 0.928×10^{-23} A · m². What is the difference in the magnetic potential energy U between the states in which the magnetic moment is aligned with and aligned in the opposite direction to an external magnetic field of 1.2 T? (c) What absolute temperature would be required so that the energy difference in (b) would equal the mean thermal energy $kT/2$?
   Answer: (a) 9.2×10^{-24} A · m². (b) 2.2×10^{-23} J. (c) 3.2 K.

12. At what temperature will the average thermal energy of a paramagnetic gas be equal to the magnetic energy in a field of 0.50 T $(=5000$ gauss) if the dipole moments of the atoms are about 10^{-23} A·m²?

SECTION 37-5

13. An electron travels in a circular orbit about a fixed positive point charge in the presence of a uniform magnetic field **B** directed normal to the plane of its motion. The electric force has precisely N times the magnitude of the magnetic force on the electron. (a) Determine the two possible angular speeds of the electron's motion. (b) Evaluate these speeds numerically if $B = 0.427$ T $(=4.27 \times 10^3$ gauss) and $N = 100$.
   Answer: (a) $(N \pm 1) \dfrac{eB}{m}$. (b) 7.43×10^{12} rad/s; 7.57×10^{12} rad/s.

14. Prove that $\Delta\omega \ll \omega_0$ in Eq. 37-13.

15. Can you give an explanation of diamagnetism based on Faraday's law of induction? In Figs. 37-9a and b, for example, what inductive effects can be expected as the magnetic field is built up from zero to the final value **B**?

SECTION 37-6

16. The dipole moment associated with an atom of iron in an iron bar is 1.8×10^{-23} A·m². Assume that all the atoms in the bar, which is 5.0 cm long and has a cross-sectional area of 1.0 cm², have their dipole moments aligned. (a) What is the dipole moment of the bar? (b) What torque must be exerted to hold this magnet at right angles to an external field of 1.5 T $(=15,000$ gauss)? The density of iron is 7.9 g/cm³.

17. A Rowland ring is formed of ferromagnetic material. It is circular in cross section, with an inner radius of 5.0 cm and an outer radius of 6.0 cm and is wound with 400 turns of wire. (a) What current must be set up in the windings to attain $B_0 = 2.0 \times 10^{-4}$ T in Fig. 37-11? (b) A secondary coil wound around the toroid has 50 turns and has a resistance of 8.0 Ω. If, for this value

of B_0, we have $B_M = 800B_0$, how much charge moves through the secondary coil when the current in the toroid windings is turned on?
Answer: (a) 0.14 A. (b) 7.9×10^{-5} C.

18. *Dipole-dipole interaction.* The exchange coupling mentioned in Section 37-6 as being responsible for ferromagnetism is *not* the mutual magnetic interaction energy between two elementary magnetic dipoles. To show this (a) compute B a distance a (=10 nm) away from a dipole of moment μ (= 1.8×10^{-23} A·m²); (b) compute the energy (= $2\mu B$) required to turn a second similar dipole end for end in this field. What do you conclude about the strength of this dipole-dipole interaction? Compare with the results of Example 4. (Note: For the same distance, the field in the median plane of a dipole is only half as large as on the axis; see Eq. 37-2.)

SECTION 37-7

19. Assume that the hydrogen nuclei (protons) in 1.0 g of water could all be aligned. What magnetic field B would be produced 5.0 cm from the sample, along its alignment axis?   *Answer:* 7.5×10^{-6} T.

20. It is possible to measure e/m for the electron by measuring (a) the cyclotron frequency ν_c of electrons in a given magnetic field and (b) the precession frequency ν_p of protons in the same field. Show that the relation is

$$\frac{e}{m} = \frac{\nu_c}{\nu_p}\frac{\mu_s}{L_s}.$$

SECTION 37-8

21. The magnetic energy density can be shown to be given in its most general form as

$$\mu_B = \tfrac{1}{2}\mathbf{B}\cdot\mathbf{H}.$$

Does this reduce to a familiar result for a vacuum?   *Answer:* Yes.

22. An iron magnet containing iron of relative permeability 5000 has a flux path 1.0 m long in the iron and an air gap 0.01 m long each with cross-sectional areas of 0.02 m². What current is necessary in a 500 turn coil wrapped around the iron to give a flux density in the air gap of 1.8 T?

23. *Boundary condition for* **H.** Prove that at the boundary between two media the tangential component of **H** has the same value on each side of the surface, assuming that there is no current at the surface. (Hint: Construct a closed rectangular loop, the two opposite longer sides being parallel to the surface, with one side in each medium. Use Ampère's law in the form that applies when magnetic materials are present.)

24. *Boundary condition for* **B.** Prove that at the boundary between two media the normal component of **B** has the same value on each side of the surface. (Hint: Construct a closed Gaussian surface shaped like a flat pillbox with one face in each medium and apply Gauss's law for magnetism.)

38
electromagnetic
oscillations

The LC system in Fig. 38-1 (assumed resistanceless*) resembles the mass-spring system of Fig. 8-4 (assumed frictionless*) in that, among other things, each system has a characteristic frequency of oscillation. The analogy actually goes far beyond this, as we will see quantitatively in Section 38-3. Let us first, however, treat LC oscillations from a physical but semiquantitative point of view.

Assume that initially the capacitor C in Fig. 38-1a carries a charge q_m and the current i in the inductor is zero. At this instant the energy stored in the capacitor is given by Eq. 30-7, or

$$U_E = \frac{1}{2}\frac{q_m^2}{C}. \qquad (38\text{-}1)$$

The energy stored in the inductor, given by

$$U_B = \tfrac{1}{2}Li^2, \qquad (38\text{-}2)$$

is zero because the current is zero. The capacitor now starts to discharge through the inductor, positive charge carriers moving counterclockwise, as shown in Fig. 38-1b. This means that a current i, given by dq/dt and pointing down in the inductor, is established.

As q decreases, the energy stored in the electric field in the capacitor also decreases. This energy is transferred to the magnetic field that appears around the inductor because of the current i that is building up

38-1
LC OSCILLATIONS

---

* A common term to describe the closely related situations of Figs. 8-4 and 38-1 is "undamped." If friction is present (as in Fig. 15-19) or resistance (as in Fig. 38-3), we describe the situation as "damped."

there. Thus the electric field decreases, the magnetic field builds up, and energy is transferred from the former to the latter.

At a time corresponding to Fig. 38-1c, all the charge on the capacitor will have disappeared. The electric field in the capacitor will be zero, the energy stored there having been transferred entirely to the magnetic field of the inductor. According to Eq. 38-2, there must then be a current—and indeed one of maximum value—in the inductor. Note that even though q equals zero, the current (which is dq/dt) is *not* zero at this time.

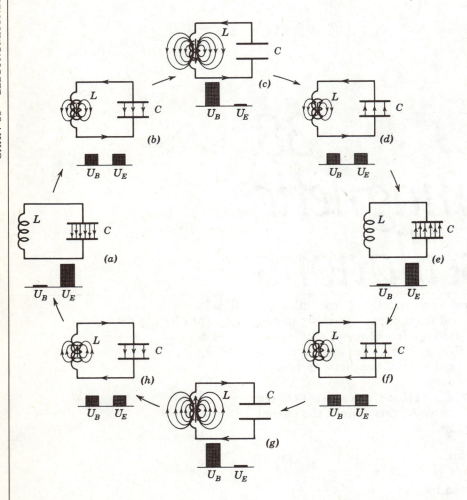

figure 38-1
Showing eight stages in a cycle of oscillation of a resistanceless *LC* circuit. The bar graphs below each figure show the stored magnetic and electric energy. The vertical arrows on the inductor axis show the current. Compare this figure in detail with Fig. 8-4, to which it exactly corresponds.

The large current in the inductor in Fig. 38-1c continues to transport positive charge from the top plate of the capacitor to the bottom plate, as shown in Fig. 38-1d; energy now flows from the inductor back to the capacitor as the electric field builds up again. Eventually, the energy will have been transferred completely back to the capacitor, as in Fig. 38-1e. The situation of Fig. 38-1e is like the initial situation, except that the capacitor is charged oppositely.

The capacitor will start to discharge again, the current now being clockwise, as in Fig. 38-1f. Reasoning as before, we see that the circuit eventually returns to its initial situation and that the process continues at a definite frequency ν (measured, say, in hertz, or cycles/second) to which corresponds a definite *angular* frequency ω ($= 2\pi\nu$ and measured, say, in radians/second). Once started, such LC oscillations (in the ideal case described, in which the circuit contains no resistance) continue indefinitely, energy being shuttled back and forth between the electric field in the capacitor and the magnetic field in the inductor. Any configuration in Fig. 38-1 can be set up as an initial condition. The oscillations will then continue from that point, proceeding clockwise around the figure. Compare these oscillations carefully with those of the mass-spring system described in Fig. 8-4.

To measure the charge q as a function of time, we can measure the variable potential difference $V_C(t)$ that exists across capacitor C. The relation

$$V_C = \left(\frac{1}{C}\right) q$$

shows that V_C is proportional to q. To measure the current we can insert a small resistance R in the circuit and measure the potential difference across it. This is proportional to i through the relation

$$V_R = (R)i.$$

We assume here that R is so small that its effect on the behavior of the circuit is negligible. Both q and i, or more correctly V_C and V_R, which are proportional to them, can be displayed on a cathode-ray oscilloscope. This instrument can plot automatically on its screen graphs proportional to $q(t)$ and $i(t)$, as in Fig. 38-2.

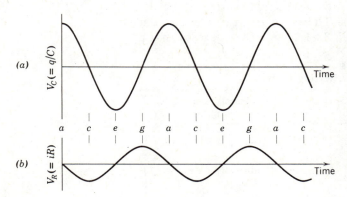

figure 38-2
A drawing of an oscilloscope screen showing potential differences proportional to (a) the charge and (b) the current, in the circuit of Fig. 38-1, as a function of time. The letters indicate corresponding phases of oscillation in that figure. Note that because $i = dq/dt$ the lower curve is proportional to the derivative of the upper. Can you verify this?

EXAMPLE 1

A 1.0-μF capacitor is charged to 50 V. The charging battery is then disconnected and a 10-mH coil is connected across the capacitor, so that LC oscillations occur. What is the maximum current in the coil? Assume that the circuit contains no resistance.

The maximum stored energy in the capacitor must equal the maximum stored energy in the inductor, from the conservation-of-energy principle. This leads, from Eqs. 38-1 and 38-2, to

$$\frac{1}{2}\frac{q_m^2}{C} = \tfrac{1}{2}Li_m^2,$$

where i_m is the *maximum* current and q_m is the *maximum* charge. Note that the maximum current and the maximum charge do not occur at the same time but one-fourth of a cycle apart; see Figs. 38-1 and 38-2. Solving for i_m and substituting CV_0 for q_m gives

$$i_m = V_0 \sqrt{\frac{C}{L}} = (50 \text{ V}) \sqrt{\frac{1.0 \times 10^{-6} \text{ F}}{10 \times 10^{-3} \text{ H}}} = 0.50 \text{ A}.$$

In an actual *LC* circuit the oscillations will not continue indefinitely because there is always some resistance present that will drain energy from the electric and magnetic fields and dissipate it as thermal (Joule) energy; the circuit will warm up. The oscillations, once started, will die away, as in Fig. 38-3. Compare this figure with Fig. 15-19, which shows the decay of the mechanical oscillations of a mass-spring system caused by frictional damping.

It is possible to have sustained electromagnetic oscillations if arrangements are made to supply, automatically and periodically (once a cycle, say), enough energy from an outside source to compensate for that dissipated as thermal energy.* We are reminded of a clock escapement, which is a device for feeding energy from a spring or a falling weight into an oscillating pendulum, thus compensating for frictional losses that would otherwise cause the oscillations to die away. *LC* oscillators, whose frequency may be varied between certain limits, are commercially available as packaged units over a wide range of frequencies, extending from low audio-frequencies (lower than 10 Hz) to microwave frequencies (higher than 10 GHz).

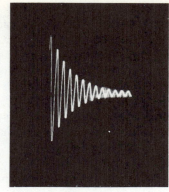

figure 38-3
A photograph of an oscilloscope trace showing how the oscillations in an *RCL* circuit die away because energy is dissipated as thermal energy in the resistor. The figure is a plot of the potential difference across the resistor as a function of time.

38-2
ANALOGY TO SIMPLE HARMONIC MOTION

Figure 8-4 shows that in a mass-spring system performing simple harmonic motion, as in an oscillating *LC* circuit, two kinds of energy occur. One is potential energy of the compressed or extended spring; the other is kinetic energy of the moving mass. These are given by the familiar formulas in the first column of Table 38-1. The table suggests that a

Table 38-1
Energy in oscillating systems

Mechanical (Fig. 8-4)		Electromagnetic (Fig. 38-1)	
spring	$U_P = \frac{1}{2}kx^2$	capacitor	$U_E = \frac{1}{2}\frac{q^2}{C}$
mass	$U_K = \frac{1}{2}mv^2$	inductor	$U_B = \frac{1}{2}Li^2$
	$v = dx/dt$		$i = dq/dt$

capacitor is in some formal way like a spring and an inductor is like a mass and that certain electromagnetic quantities "correspond" to certain mechanical ones, namely,

q corresponds to x,
i corresponds to v,
C corresponds to $1/k$,
L corresponds to m.

Comparison of Fig. 38-1, which shows the oscillations of a resistanceless *LC* circuit, with Fig. 8-4, which shows the oscillations in a frictionless mass-spring system, indicates how close the correspondence is. Note how v and i correspond in the two figures; also x and q. Note, too, how in each case the energy alternates between two forms, magnetic

---

* From another point of view it is possible to supply an electronically generated "negative resistance" to the circuit, large enough in magnitude to just cancel the actual resistance. See "Negative Resistor to Provide Self-Oscillation in RLC Circuits" by Edwin A. S. Lewis, *American Journal of Physics,* December 1976.

and electric for the LC system, and kinetic and potential for the mass-spring system.

In Section 15-3 we saw that the natural angular frequency of oscillation of a (frictionless) mass-spring system is

$$\omega = 2\pi\nu = \sqrt{\frac{k}{m}}.$$

The method of correspondences suggests that to find the natural frequency for a (resistanceless) LC circuit, k should be replaced by $1/C$ and m by L, obtaining

$$\omega = 2\pi\nu = \sqrt{\frac{1}{LC}}. \tag{38-3}$$

This formula is indeed correct, as we show in the next section.

We now derive an expression for the frequency of oscillation of a (resistanceless) LC circuit, our derivation being based on the conservation-of-energy principle. The total energy U present at any instant in an oscillating LC circuit is given by

$$U = U_B + U_E = \tfrac{1}{2}Li^2 + \frac{1}{2}\frac{q^2}{C},$$

which expresses the fact that at any arbitrary time the energy is stored partly in the magnetic field in the inductor and partly in the electric field in the capacitor. If we assume the circuit resistance to be zero, there is no energy transfer to thermal energy and U remains constant with time, even though i and q vary. In more formal language, dU/dt must be zero. This leads to

$$\frac{dU}{dt} = \frac{d}{dt}\left(\tfrac{1}{2}Li^2 + \frac{1}{2}\frac{q^2}{C}\right) = Li\frac{di}{dt} + \frac{q}{C}\frac{dq}{dt} = 0. \tag{38-4}$$

Now, q and i are not independent variables, being related by

$$i = \frac{dq}{dt}.$$

Differentiating yields
$$\frac{di}{dt} = \frac{d^2q}{dt^2}.$$

Substituting these two expressions into Eq. 38-4 leads to

$$L\frac{d^2q}{dt^2} + \frac{1}{C}q = 0. \tag{38-5}$$

This is the differential equation that describes the oscillations of a (resistanceless) LC circuit. To solve it, note that Eq. 38-5 is mathematically of *exactly* the same form as Eq. 15-6,

$$m\frac{d^2x}{dt^2} + kx = 0, \tag{15-6}$$

which is the differential equation for the mass-spring oscillations. Fundamentally, it is by comparing these two equations that the correspondences on p. 844 arise.

The solution of Eq. 15-6 proved to be

$$x = A\cos(\omega t + \phi), \tag{15-8}$$

where A $(= x_m)$ is the amplitude of the motion and ϕ is an arbitrary

38-3
ELECTROMAGNETIC OSCILLATIONS– QUANTITATIVE

phase angle. Since q corresponds to x, we can write the solution of Eq. 38-5 as

$$q = q_m \cos{(\omega t + \phi)}, \qquad (38\text{-}6)$$

where ω is the still unknown angular frequency of the electromagnetic oscillations.

We can test whether Eq. 38-6 is indeed a solution of Eq. 38-5 by substituting it and its second derivative in that equation. To find the second derivative, we write

$$\frac{dq}{dt} = i = -\omega q_m \sin{(\omega t + \phi)} \qquad (38\text{-}7a)$$

and

$$\frac{d^2q}{dt^2} = -\omega^2 q_m \cos{(\omega t + \phi)}. \qquad (38\text{-}7b)$$

Substituting q and d^2q/dt^2 into Eq. 38-5 yields

$$-L\omega^2 q_m \cos{(\omega t + \phi)} + \frac{1}{C} q_m \cos{(\omega t + \phi)} = 0.$$

Canceling $q_m \cos{(\omega t + \phi)}$ and rearranging leads to

$$\omega = \sqrt{\frac{1}{LC}}.$$

Thus, if ω is given the constant value $1/\sqrt{LC}$, Eq. 38-6 is indeed a solution of Eq. 38-5. This expression for ω agrees with Eq. 38-3, which we arrived at by the method of correspondences.

The phase angle ϕ in Eq. 38-6 is determined by the conditions that prevail at $t = 0$. If the initial condition is as represented by Fig. 38-1a, then we put $\phi = 0$ in order that Eq. 38-6 may predict $q = q_m$ at $t = 0$. What initial condition in Fig. 38-1 is implied if we select $\phi = 90°$?

EXAMPLE 2

(*a*) In an oscillating LC circuit, what value of charge, expressed in terms of the maximum charge, is present on the capacitor when the energy is shared equally between the electric and the magnetic field? (*b*) How much time is required for this condition to arise, assuming the capacitor to be fully charged initially? Assume that $L = 10$ mH and $C = 1.0$ μF.

(*a*) The stored energy and the *maximum* stored energy in the capacitor are, respectively,

$$U_E = \frac{q^2}{2C} \qquad \text{and} \qquad U_{E,m} = \frac{q_m^2}{2C}.$$

Substituting $U_E = \frac{1}{2}U_{E,m}$ yields

$$\frac{q^2}{2C} = \frac{1}{2}\frac{q_m^2}{2C} \qquad \text{or} \qquad q = \frac{1}{\sqrt{2}}q_m.$$

(*b*) To find the time, we write, assuming $\phi = 0$ in Eq. 38-6,

$$q = q_m \cos{\omega t} = \frac{1}{\sqrt{2}}q_m,$$

which leads to $\qquad \omega t = \cos^{-1}\frac{1}{\sqrt{2}} = \frac{\pi}{4} \qquad \text{or} \qquad t = \frac{\pi}{4\omega}.$

The angular frequency ω is found from Eq. 38-3, or

$$\omega = \sqrt{\frac{1}{LC}} = \sqrt{\frac{1}{(10 \times 10^{-3} \text{ H})(1.0 \times 10^{-6} \text{ F})}} = 1.0 \times 10^4 \text{ rad/s}.$$

The time t is then

$$t = \frac{\pi}{4\omega} = \frac{\pi}{(4)(1.0 \times 10^4 \text{ rad/s})} = 7.9 \times 10^{-5} \text{ s } (=79 \ \mu\text{s}).$$

What is the frequency ν in Hz?

The stored electric energy in the LC circuit, using Eq. 38-6, is

$$U_E = \frac{1}{2}\frac{q^2}{C} = \frac{q_m^2}{2C}\cos^2(\omega t + \phi), \qquad (38\text{-}8)$$

and the magnetic energy, using Eq. 38-7a, is

$$U_B = \tfrac{1}{2}Li^2 = \tfrac{1}{2}L\omega^2 q_m^2 \sin^2(\omega t + \phi).$$

Substituting the expression for ω (see Eq. 38-3) into this last equation yields

$$U_B = \frac{q_m^2}{2C}\sin^2(\omega t + \phi). \qquad (38\text{-}9)$$

Figure 38-4 shows plots of $U_E(t)$ and $U_B(t)$ for the case of $\phi = 0$. Note that (a) the maximum values of U_E and U_B are the same $(= q_m^2/2C)$; (b) at any instant the sum of U_E and U_B is a constant $(= q_m^2/2C)$; and (c) when U_E has its maximum value, U_B is zero and conversely. This analysis supports the qualitative analysis of Section 38-1. Compare this discussion with that given in Section 15-4 for the energy transfers in a mass-spring system.

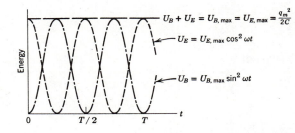

$U_B + U_E = U_{B,\max} = U_{E,\max} = \frac{q_m^2}{2C}$

$U_E = U_{E,\max}\cos^2 \omega t$

$U_B = U_{B,\max}\sin^2 \omega t$

figure 38-4
The stored magnetic (—·—·—) and electric (————) energies in the circuit of Fig. 38-1. Note that their sum is a constant. $T \ (=1/\nu)$ is the period of the oscillation.

EXAMPLE 3

The RCL circuit. (a) Derive an expression for the quantity $q(t)$ for a single-loop circuit containing a resistance R as well as an inductance L and a capacitance C. (b) After how long a time will the charge oscillations decay to half-amplitude if $L = 10$ mH, $C = 1.0 \ \mu$F, and $R = 0.10 \ \Omega$?

(a) If U is the total stored field energy, we have, as before,

$$U = \tfrac{1}{2}Li^2 + \frac{1}{2}\frac{q^2}{C}.$$

U is no longer constant but rather

$$\frac{dU}{dt} = -i^2 R,$$

the minus sign signifying that the stored energy U *decreases* with time, being converted to thermal (Joule) energy at the rate i^2R. Combining these two equations leads to

$$Li\frac{di}{dt} + \frac{q}{C}\frac{dq}{dt} = -i^2R.$$

Substituting dq/dt for i and d^2q/dt^2 for di/dt leads finally to

$$L \frac{d^2q}{dt^2} + R \frac{dq}{dt} + \frac{1}{C} q = 0,$$

which is the differential equation that describes the damped LC oscillations. If we put $R = 0$, this equation reduces, as expected, to Eq. 38-5.

Compare this differential equation for damped LC oscillations with Eq. 15-37, or

$$m \frac{d^2x}{dt^2} + b \frac{dx}{dt} + kx = 0, \tag{15-37}$$

which describes damped mass-spring oscillations. Once again the equations are mathematically identical, the resistance R corresponding to the mechanical damping constant b and q corresponding to x.

The solution of the RCL circuit follows at once, by correspondence, from the solution of Eq. 15-37. It is (see Eqs. 15-38 and 15-39) for R *reasonably small*, and for an initial condition in which the capacitor has a maximum charge (that is, $\phi = 0$),

$$q = q_m e^{-Rt/2L} \cos \omega't, \tag{38-10}$$

where

$$\omega' = \sqrt{\frac{1}{LC} - \left(\frac{R}{2L}\right)^2}. \tag{38-11}$$

Note that Eq. 38-10, which can be described as a cosine function with an exponentially decreasing amplitude, is the equation of the decay curve of Fig. 38-3. Note, too (see Eq. 38-11), that the presence of resistance reduces the oscillation frequency. These two equations reduce to familiar results as $R \rightarrow 0$.

(b) The oscillation amplitude will have decreased to half the initial amplitude when the amplitude factor $e^{-Rt/2L}$ in Eq. 38-10 has the value one-half, or

$$\tfrac{1}{2} = e^{-Rt/2L}.$$

At what time t will this occur? What is ω'?

We have

$$t = \frac{2L}{R} \ln 2 = \frac{(2)(10 \times 10^{-3} \text{ H})(0.69)}{0.10 \ \Omega} = 0.14 \text{ s}.$$

The angular frequency, from Eq. 38-11, is

$$\omega' = \sqrt{\frac{1}{(10 \times 10^{-3} \text{ H})(1.0 \times 10^{-6} \text{ F})} - \left(\frac{0.10 \ \Omega}{2 \times 10 \times 10^{-3} \text{ H}}\right)^2}$$

$$= \sqrt{10^8 \ (\text{rad/s})^2 - 25 \ (\text{rad/s})^2} = 1.0 \times 10^4 \text{ rad/s}.$$

Note that the second term is quite small, so that in this case, as in many practical cases, the resistance has a negligible effect on the frequency. Can you show that 0.14 s, the time at which the oscillations decrease to half-amplitude, corresponds to about 220 cycles of oscillation? The damping is much less severe than that shown in Fig. 38-3.

In this section we put electromagnetic oscillations aside for the moment and return to mechanical oscillating systems, so that we can develop the concepts of *lumped* and *distributed* elements. In Section 38-5 we will return to electromagnetic systems and make a similar comparison.

In the oscillating mass-spring system in Fig. 8-4 the two kinds of energy involved appear in quite separate parts of the system, the potential energy being associated with the spring and the kinetic energy with the moving mass. A closed organ pipe (see Fig. 20-6b) is a mechanical oscillating system in which the two kinds of energy are *not* separated in

38-4
LUMPED AND DISTRIBUTED ELEMENTS

space. Kinetic energy, associated with moving elements of air, and potential energy, associated with compressions and rarifications of elements of air, are both present throughout the volume of the pipe. We say that such an oscillating system has *distributed* rather than *lumped* (as in the mass-spring system) elements.

We see at once one difference between lumped and distributed mechanical oscillating systems. A lumped system has a single frequency of oscillation, given for the mass-spring system by Eq. 15-11; see Section 15-5 for other examples of lumped oscillating mechanical systems. On the other hand, distributed systems, such as the organ pipes in Fig. 20-6 and the vibrating string of Fig. 20-4, have a number of discrete oscillation frequencies (harmonics), the values for these two cases being given in Section 20-5.

Now let us examine in more detail, from the point of view of distributed elements, the behavior of a closed organ pipe, which we will idealize as an *acoustic resonant cavity*.

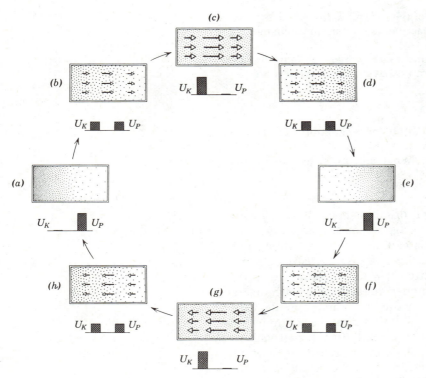

figure 38-5
Showing eight stages in a cycle of oscillation of a cylindrical acoustic resonant cavity, suggestive of a closed organ pipe. The bar graphs below each figure show the kinetic and potential energy. The arrows represent the directed velocities of small volume elements of the gas.

Figure 38-5, a series of "snapshots" taken one-eighth of a cycle apart, shows the pressure and velocity variations in the fundamental mode of a particular acoustic resonant cavity. There is a pressure node at the center and a pressure antinode at each end. There is a velocity* node at each end and a velocity antinode at the center. When the pressure variation is the greatest, the velocity is zero (Figs. 38-5a and e). When the pressure is uniform, the velocities have their maximum values (Figs. 38-5c, and g).

The energy in the acoustic resonator alternates between kinetic

---

* The velocity of interest here is the directed velocity $\mathbf{v}_{gas}$ of small volume elements of the gas which are, however, large enough to contain a great number of molecules. The thermal velocities of the molecules have no directional preference and are ignored.

energy associated with the moving gas and potential energy associated with the compression and rarefaction of the gas. In Figs. 38-5c and g the energy is all kinetic and in Figs. 38-5a and e it is all potential. In intermediate phases it is part kinetic and part potential.

The kinetic energy of a small mass Δm of the gas, which is moving parallel to the cylinder axis with a speed v_{gas}, is $\frac{1}{2}\Delta m v_{gas}^2$. The *kinetic energy density*, that is, the kinetic energy per unit volume, is

$$u_K = \frac{1}{2}\frac{\Delta m}{\Delta V}v_{gas}^2 = \frac{1}{2}\rho_0 v_{gas}^2,$$

in which ΔV is the volume of the gas element and ρ_0 is the mean density of the gas.

The potential energy per unit volume in the gas, that is, the *potential energy density*, associated with the compressions and rarefactions of the gas is given (we state) by

$$u_P = \frac{1}{2}B\left(\frac{\Delta\rho}{\rho_0}\right)^2.$$

Here B is the bulk modulus* of elasticity of the gas and $\Delta\rho/\rho_0$, which is positive for a compression and negative for a rarefaction, is the fractional change in gas density at any point.

The angular frequency of oscillation for the cavity of Fig. 38-5, in the fundamental (or lowest frequency) mode, shown in that figure, is found from

$$\omega_1 = 2\pi\nu = \frac{2\pi v}{\lambda} = \frac{\pi v}{l'}$$

where v is the speed of sound in the gas and l is the length of the cavity. From Eq. 20-1 we may write v as $\sqrt{B/\rho_0}$. Note that in the above we have put $\lambda = 2l$, corresponding to the fundamental mode. What are the angular frequencies ω_2, ω_3, etc., of the higher frequency modes?

38-5
ELECTROMAGNETIC RESONANT CAVITY

Consider now a second closed cylinder, of radius a and length l, whose walls are made of copper or some other good conductor. A system of oscillating electric and magnetic fields can be set up in such a cavity. Such an *electromagnetic resonant cavity* is a distributed electromagnetic oscillator, in contrast to an LC circuit, which is a lumped system. As for the acoustic resonator, many modes of oscillation with discrete frequencies are possible; we describe only the fundamental mode, which has the lowest frequency. The cavity oscillations can be set up by suitably connecting the cavity, through a small hole in its side wall, to a source of electromagnetic radiation such as a magnetron. If the cavity dimensions are of the order of a few centimeters, the resonant frequencies will be of the order of 10^{10} Hz. This is far higher than the acoustic frequencies in cavities of the same size, reflecting the fact that the speed of electromagnetic waves in free space ($= 3 \times 10^8$ m/s) is much greater than the speed of sound in air (about 350 m/s).

Figure 38-6, which is a series of "snapshots" taken one-eighth of a cycle apart, shows, by the horizontal lines, how the electric field **E** varies with time in the cavity. The electric lines of force originate on positive charges at one end of the cylinder and terminate on negative charges at the other end. As **E** changes with time, eventually reversing itself, currents flow from one end of the cylinder to the other on the inner cylinder wall. At any point in the cavity, energy is stored in the electric field in an amount per unit volume given by Eq. 30-9, or

$$u_E = \frac{1}{2}\epsilon_0 E^2. \qquad (38\text{-}12)$$

* See p. 435.

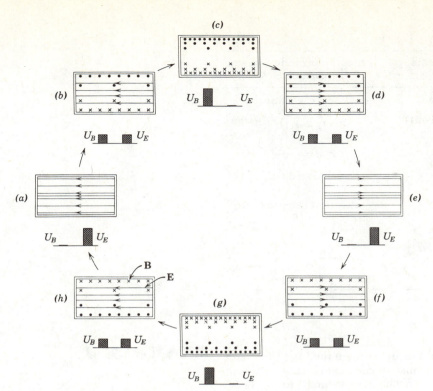

figure 38-6
Showing eight stages in a cycle of oscillation of a cylindrical electromagnetic resonant cavity. The bar graphs below each figure show the stored electric and magnetic energy. The dots and crosses represent circular lines of **B**; the horizontal lines represent **E**. Compare with Fig. 38-5; both figures are examples of distributed elements.

Figure 38-6 also shows, by the dots and crosses, how the magnetic field **B** varies with time. The magnetic lines form circles about the cylinder axis. Note that the magnetic field has a maximum value when the electric field is zero, and conversely. At any point in the cavity energy is stored in the magnetic field in an amount per unit volume given by Eq. 36-19, or

$$u_B = \frac{1}{2\mu_0} B^2. \qquad (38\text{-}13)$$

Thus, as in the *LC* circuit, energy is shuttled back and forth between the electric and the magnetic fields. However, these fields no longer occupy completely separate regions of space. In Section 40-5 we clarify just *why* the electric and magnetic fields in Fig. 38-6 interact in the way that we have described.

We state without proof that the angular frequency of oscillation for the electromagnetic cavity in Fig. 38-6 is, in the fundamental mode shown in that figure,

$$\omega_1 = \frac{2.41c}{a},$$

in which a is the cavity radius and c is the speed of electromagnetic radiation in free space. We will see in Section 41-8 that c may be written in terms of electromagnetic quantities as $1/\sqrt{\mu_0\epsilon_0}$. As the field patterns in Fig. 38-6 suggest, the resonant frequency of oscillation of the cavity, for the mode of oscillation shown, depends only on the radius of the cavity and not on its length.

Table 38-2 summarizes some characteristics of the four oscillating systems that we have discussed so far. For lumped systems it gives expressions for the two kinds of energy involved and for the (single) oscillation frequency. For the distributed systems it gives expressions for the two kinds of energy density involved and for the oscillation frequency in the fundamental mode. You should study carefully all the correspondences, similarities, and differences that occur in the table.

Table 38-2
Four oscillating systems

	Mechanical Systems (frictionless)	Electromagnetic Systems (resistanceless)
Lumped systems	Mass + spring $U_K = \frac{1}{2}mv^2$ $U_P = \frac{1}{2}kx^2$ $\omega = \sqrt{\dfrac{k}{m}}$ (Fig. 8-4)	LC circuit $U_B = \frac{1}{2}Li^2$ $U_E = \frac{1}{2}(1/C)q^2$ $\omega = \sqrt{\dfrac{1}{LC}}$ (Fig. 38-1)
Distributed systems	Acoustic cavity $u_K = \frac{1}{2}\rho_0 v_{gas}^2$ $u_P = \frac{1}{2}B(\Delta\rho/\rho_0)^2$ $\omega_1 = \dfrac{\pi v}{l};\quad v = \sqrt{\dfrac{B}{\rho_0}}$ (Fig. 38-5)	Electromagnetic cavity $u_B = \frac{1}{2}(1/\mu_0)B^2$ $u_E = \frac{1}{2}\epsilon_0 E^2$ $\omega_1 = \dfrac{2.41c}{a};\quad c = \sqrt{\dfrac{1}{\epsilon_0\mu_0}}$ (Fig. 38-6)

EXAMPLE 4

In the cavity of Fig. 38-6, what is the relationship between the "average" value of E throughout the cavity, measured at the instant corresponding to Fig. 38-6a, to the "average" value of B, measured at the instant corresponding to Fig. 38-6c?

At the first instant the energy is all electric and at the second it is all magnetic. The total energy U, found by integrating the energy density over the volume of the cavity, must be the same at these two instants, or

$$U = \int u_{E,m}\, dV = \int u_{B,m}\, dV,$$

where dV is a volume element in the cavity and $u_{E,m}$ and $u_{B,m}$ are the *maximum* values of u_E and of u_B at the site of this volume element; these maximum values occur one-fourth of a cycle apart, as Fig. 38-6 shows. Substituting Eqs. 38-12 and 38-13 leads to

$$\int \frac{\epsilon_0 E_m^2}{2}\, dV = \int \frac{B_m^2}{2\mu_0}\, dV$$

or

$$\mu_0\epsilon_0 \int E_m^2\, dV = \int B_m^2\, dV.$$

The quantity $\int E_m^2\, dV$ can be written as $\overline{E_m^2}V$, where V is the cavity volume and $\overline{E_m^2}$ is the average value of E_m^2 throughout the cavity. Treating B_m in the same way leads to

$$\mu_0\epsilon_0 \overline{E_m^2} = \overline{B_m^2}$$

or, taking square roots,

$$\sqrt{\overline{B_m^2}} = \sqrt{\mu_0\epsilon_0}\,\sqrt{\overline{E_m^2}}.$$

We can represent $\sqrt{\overline{B_m^2}}$ by B_{rms}, a so-called "root-mean-square" value of B_m. In computing B_{rms}, note that the averaging is done throughout the volume of the cavity, at the instant corresponding to Fig. 38-6c. It is not a time average for a particular point in the cavity. Doing the same for E yields

$$E_{rms}/B_{rms} = \frac{1}{\sqrt{\mu_0\epsilon_0}}$$

$$= [(4\pi \times 10^{-7}\ \text{T·m/A})(8.9 \times 10^{-12}\ \text{C}^2/\text{N·m}^2)]^{-1/2}$$

$$= 3.00 \times 10^8\ \text{m/s}.$$

If this reminds you of the speed of light, it is no coincidence because that is exactly what it is; we will explore this in detail in Section 41-8.

If E_{rms} above equals 1.0×10^4 V/m, a reasonable value, then

$$B_{rms} = (1.0 \times 10^4 \text{ V/m})/(3.00 \times 10^8 \text{ m/s})$$

$$= 3.3 \times 10^{-5} \text{ T} = 0.33 \text{ gauss.}$$

What is the total stored energy in the cavity under these conditions, assuming the cavity to be 10 cm long and 3.0 cm in diameter?

Recall that in Example 6, Chapter 36, we showed the energy density for a magnetic field of "ordinary" laboratory magnitude (say, 1 T) to be enormously greater than that for an electric field of "ordinary" magnitude (say, 10^5 V/m). This fact is consistent with the present example.

questions

1. Why doesn't the LC circuit of Fig. 38-1 simply stop oscillating when the capacitor has been completely discharged?

2. How might you start an LC circuit into oscillation with its initial condition being represented by Fig. 38-1c? Devise a switching scheme to bring this about.

3. In an oscillating LC circuit, assumed resistanceless, what determines (a) the frequency and (b) the amplitude of the oscillations?

4. In connection with Figs. 38-1c and g, explain how there can be a current in the inductor even though there is no charge on the capacitor.

5. In Fig. 38-1 is it possible to have (a) an LC circuit without resistance, (b) an inductor without inherent capacitance, or (c) a capacitor without inherent inductance? Discuss the practical validity of the LC circuit of Fig. 38-1, in which each of the above possibilities is ignored. See "Self-resonant Effects in Coils and Capacitors: an Experiment" by Samuel Derman, *The Physics Teacher*, September 1976.

6. In Fig. 38-1, what changes are called for if the oscillations are to proceed *counterclockwise* around the figure?

7. In Fig. 38-1, what phase angles ϕ in Eq. 38-6 correspond to the eight circuit situations shown?

8. All practical LC circuits have to contain *some* resistance and thus are RCL circuits. However, one can buy a packaged audio oscillator in which the output maintains a constant amplitude indefinitely and does not decay, as in Fig. 38-3. How can this happen? (*Hint:* Consider the analogy to the pendulum clock, in which falling weights are involved.)

9. Can you see any physical reason for assuming that R is "small" in Eqs. 38-10 and 38-11? (*Hint:* Consider what might happen if the damping R were so large that Eq. 38-10 could not even go through one cycle of oscillation before q was reduced to essentially zero. Could this happen? If so, what do you imagine Fig. 38-3 would look like?)

10. Tabulate as many mechanical or electric systems as you can think of that possess a natural frequency, along with the formula for that frequency if given in the text.

11. Discuss the periodic flow of energy, if any, from point to point in an acoustic resonant cavity.

12. Can a given circuit element (a capacitor, say) behave like a "lumped" element under some circumstances and like a "distributed" element under others?

13. List as many (a) lumped and (b) distributed mechanical oscillating systems as you can.

14. Are oscillating systems (mechanical, say) *either* lumped *or* distributed? That is, is there no middle ground? (a) Consider a lumped system such as an

idealized mass-spring arrangement. How might you change it physically to make it more distributed? (b) Consider a distributed system such as a vibrating string. How might you change it physically to make it more lumped?

15. A coil has a measured inductance L. In a practical case it also has a capacitance C, adjacent windings behaving as "plates." It can be made to oscillate at a certain frequency without attaching it to an external capacitance. Is this a case of distributed elements? Do you suppose that it can oscillate at more than one frequency? Discuss.

16. A violin string is an oscillating mechanical system with distributed elements. Give some qualitative details as to why this is so. For example, where are the kinetic and potential energies to be found? (Compare Figs. 19-16 and 38-5.)

17. An air-filled acoustic resonant cavity and an electromagnetic resonant cavity of the same size have resonant frequencies that are in the ratio of 10^6 or so. Which has the higher frequency and why?

18. What constructional difficulties would you encounter if you tried to build an LC circuit of the type shown in Fig. 38-1 to oscillate (a) at 0.01 Hz, or (b) at 10^{10} Hz?

19. Electromagnetic cavities are often silver-plated on the inside. Why?

problems

SECTION 38-1

1. Find the capacitance of an LC circuit if the maximum charge on the capacitor is 1.0 μC and the total energy is 1.4×10^{-4} J.   *Answer:* 3.6×10^{-9} F.

2. A 1.5-mH inductor in an LC circuit stores a maximum energy of 1.0×10^{-5} J. What is the peak current?

3. In an oscillating LC circuit $L = 1.0$ mH, $C = 4.0$ μF and the maximum charge on C is 3.0 μC. Find the maximum current.   *Answer:* 47 mA.

4. An oscillating LC circuit consisting of a 1.0 nF (= 1.0 nanofarad = 1.0×10^{-9} F) capacitor and a 3.0-mH coil carries a peak voltage of 3.0 V. (a) What is the maximum charge on the capacitor? (b) What is the peak current through the circuit? (c) What is the maximum energy stored in the magnetic field of the coil?

5. In an oscillating LC circuit, (a) in terms of the maximum charge on the capacitor, what value of charge is present when the energy is shared equally between the electric and the magnetic fields? (b) What fraction of a period must elapse following the time the capacitor is fully charged for this condition to arise?   *Answer:* (a) $q = q_m/\sqrt{2}$. (b) $t = T/8$.

6. At some instant in an oscillating LC circuit, three-fourths of the total energy is stored in the magnetic field of the inductor. (a) In terms of the maximum charge on the capacitor, what is the charge on the capacitor at this instant? (b) In terms of the maximum current in the inductor, what is the current in the inductor at this instant?

SECTION 38-2

7. Given a 1.0-mH inductor, how would you make it oscillate at 1.0 MHz (= 1.0×10^6 Hz)?
   Answer: Connect a 25-pF capacitor across it and use it as the resonant element in an oscillator.

8. You are given a 10-mH inductor and two capacitors, of 5.0- and 2.0-μF capacitance. (a) Can you find four resonant frequencies that can be obtained by connecting these elements in various ways? (b) Are there more than four such frequencies?

9. An LC circuit has an inductance $L = 3.0$ mH and a capacitance $C = 10$ μF. (a) Calculate the angular frequency ω of oscillation. (b) Find the period T of the oscillation. (c) At time $t = 0$ the capacitor is charged to 200 μC, and the current is zero. Sketch roughly the charge on the capacitor as a function of time.   *Answer:* (a) 5.8×10^3 rad/s. (b) 1.1×10^{-3} s.

10. How long will it take for an uncharged 4.0-pF capacitor in an LC circuit to charge if its final voltage is 1.0 mV and the maximum current is 50 mA?

11. An inductor is connected across a capacitor whose capacitance can be varied by turning a knob. We wish to make the frequency of the LC oscillations vary linearly with the angle of rotation of the knob, going from 2×10^5 Hz to 4×10^5 Hz as the knob turns through 180°. If $L = 1.0$ mH, plot C as a function of angle for the 180° rotation.
   Answer: 0°, 45°, 90°, 135°, and 180° correspond respectively to 6.4, 4.1, 2.8, 2.1, and 1.6×10^{-10} F.

12. A variable capacitor with a range from 10 to 365 pF is used with a coil to tune the input to a radio. (a) What ratio of maximum to minimum frequencies may be tuned with such a capacitor? (b) If this capacitor is to tune from 0.54 to 1.60 MHz, the ratio computed in (a) is too large. By adding a capacitor in parallel to the variable capacitor this range may be adjusted. How large should this capacitor be and what inductance should be chosen in order to tune the desired range of frequencies?

13. An oscillating LC circuit is designed to operate at a peak current i (30 mA). The inductance L (0.042 H) is fixed and the frequency is varied by changing C. (a) If the capacitor has a maximum peak voltage V_m (50 V), can the circuit safely operate at a frequency ν of 1.0 MHz? (b) What is the maximum safe operating frequency? (c) What is the minimum capacitance?
   Answer: (a) 7 900 V; no. (b) 6300 Hz. (c) 1.5×10^{-8} F.

14. A 10-kg mass oscillates on a spring that, when extended 2.0 cm from equilibrium, has a restoring force of 5.0 N. (a) Find the capacitance of the analogous LC system with $L = 1.0 \times 10^{-3}$ H. (b) Would it be a simple matter to construct the analogous circuit?

15. Initially the 900-μF capacitor is charged to 100 V, and the 100-μF capacitor is uncharged in Fig. 38-7. (a) Describe in detail how one may charge the 100-μF capacitor to 300 V using S_1 and S_2 appropriately. (b) Describe in detail the mass + spring mechanical analogy of this problem.
   Answer: Let T_2 be the period of the inductor and 900 μF capacitor and T_1 the period of inductor and 100 μF capacitor. Then (a) close S_2, wait $T_2/4$; quickly close S_1, then open S_2; wait $T_1/4$ and then open S_1.

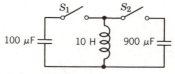

figure 38-7
Problem 15

SECTION 38-3

16. A circuit has $L = 10$ mH and $C = 1.0$ μF. How much resistance must be inserted in the circuit to reduce the (undamped) resonant frequency by 0.01%?

17. In an oscillating LC circuit $L = 3.0$ mH and $C = 2.7$ μF. At $t = 0$ the charge $q = 0$ and the current $i = 2.0$ A. (a) What is the maximum charge that will appear on the capacitor? (b) In terms of the period T of oscillation, how much time after $t = 0$ will elapse until the energy stored in the capacitor will be increasing at its greatest rate? (c) What is this greatest rate at which the energy stored in the capacitor increases?
   Answer: (a) 1.8×10^{-4} C. (b) $T/8$. (c) 67 W.

18. In a damped LC circuit, find the time required for the maximum energy present in the capacitor during one oscillation to fall to half the maximum energy present during the first oscillation. Assume $q = q_m$ at $t = 0$ (i.e., use Eq. 38-10).

19. Derive the differential equation for an LC circuit (Eq. 38-5) using the loop theorem.

20. Show that, for low damping, the current in a damped LC circuit is given approximately by

$$i = -q_m\omega' e^{-Rt/2L} \sin(\omega' t + \phi),$$

in which

$$\phi = \tan^{-1} \frac{R}{2L\omega'}.$$

Start from Eq. 38-10.

21. Suppose that in an oscillating *RCL* circuit the amplitude of the charge oscillations drops to one-half its initial value after n cycles. Show that the fractional reduction in the frequency of resonance, caused by the presence of the resistor, is given to a close approximation by

$$\frac{\omega - \omega'}{\omega} = \frac{0.0061}{n^2},$$

which is independent of L, C, or R. Apply to the decay curve of Fig. 38-3.

22. *"Q" for a circuit.* In the damped *LC* circuit of Example 3 show that the fraction of the energy lost per cycle of oscillation, $\Delta U/U$, is given to a close approximation by $2\pi R/\omega L$. The quantity $\omega L/R$ is often called the "Q" of the circuit (for "quality"). A "high-Q" circuit has low resistance and a low fractional energy loss per cycle $(= 2\pi/Q)$.

SECTION 38-5

23. What would be the dimensions of a cylindrical electromagnetic resonant cavity (like that described in the text) operating, in the fundamental mode, at 60 Hz, the frequency of household alternating current?
Answer: Radius $= 1.9 \times 10^3$ km, independent of its length.

24. Sketch diagrams like those shown in Figure 38-6 showing a cycle of oscillation of a cylindrical electromagnetic resonant cavity operating, not in the fundamental mode, but in the first overtone.

39
alternating currents

Thus far we have discussed the response, that is, the generated currents, when emfs that vary with time in different ways are applied to circuits containing elements of resistance R, capacitance C, and inductance L, in various combinations.

In Chapter 32 we discussed the (steady) currents generated when we apply (steady) emfs to purely resistive networks. In Section 32-8 we discussed the response of a single-loop RC circuit when a "square wave" emf is applied, as in Fig. 32-13. In Section 36-3 we described the behavior of a single-loop LR network when a similar square wave is applied, as in Fig. 36-6. Finally, in Chapter 38, we analyzed the responses of single-loop LC and RCL circuits in which *no* emf is applied other than the transient emf needed to start the "free oscillations"; see Figs. 38-1, 38-2*b*, and 38-3.

This chapter deals with *alternating currents* set up in single-loop RCL circuits, such as that of Fig. 39-1, by an emf that varies with time as

$$\varepsilon = \varepsilon_m \sin \omega t, \qquad (39\text{-}1)$$

$\omega (= 2\pi\nu$, where ν is measured in hertz) being a fixed angular frequency. An emf of this type might be established, for example, by an alternating current generator in a commercial power plant, where usually $\nu = 60$ Hz. (See Problem 6, Chapter 35.) The symbol for a source of alternating emf such as that described by Eq. 39-1 is ———◯⁓◯——— Such a device is called an alternating current generator or an *ac generator*.*

Alternating currents are important for two reasons: (1) On the practical side, modern technology and indeed the life-style in technologically

---

* The term 'alternating current' is used loosely here: strictly this is an alternating voltage.

857

advanced nations, would be very different if emfs such as those given by Eq. 39-1, and the alternating currents to which they correspond, were not available. Power distribution systems, radio, television, satellite communication systems, computor systems, and so on, would be either much less effective or impossible without alternating emfs and the alternating currents generated by them. (2) On the theoretical side, if we know the response of *any RCL* circuit (no matter how many elements or loops are involved) to the emf of Eq. 39-1, we can find the response, that is, the currents generated, to *any* arbitrary emf, no matter how complicated its waveform. We rely here on the facts that we can write any complex waveform as the sum of separate sine (and cosine) terms in a Fourier series and that we can apply the principle of superposition. See Section 19-4 and Fig. 19-5 for an analogy.

A central aim of this chapter is to find the alternating current i in the circuit of Fig. 39-1 in terms of ε_m, ω, R, C, and L. Note that, for the conditions assumed in Fig. 39-1,* the current in all parts of the loop is the same (as for single-loop direct current circuits in Chapter 32) and we may safely assume that i is given by

$$i = i_m \sin(\omega t - \phi) \tag{39-2}$$

in which ω is the angular frequency of the applied alternating emf of Eq. 39-1.

In Eq. 39-2 i_m is the current amplitude and ϕ is the phase angle between the alternating current of Eq. 39-2 and the alternating emf of Eq. 39-1. It is our job to express i_m and ϕ in Eq. 39-2 in terms of ε_m, ω, R, C, and L.†

To do so, let us break down the problem suggested by Fig. 39-1 into three separate problems, in which we consider R, C, and L separately and in turn.‡ We start with R.

1. A resistive circuit. Figure 39-2a shows a circuit containing a resistive element only, acted on by the alternating emf of Eq. 39-1. From the loop theorem and from the definition of resistance we can write

$$V_R = \varepsilon_m \sin \omega t \text{ (loop theorem)} \tag{39-3a}$$
$$V_R = i_R R \text{ (definition of } R) \tag{39-3b}$$

or

$$i_R = \left(\frac{\varepsilon_m}{R}\right) \sin \omega t. \tag{39-3c}$$

Comparison of Eqs. 39-3a and 39-3c shows that the time-varying quantities V_R and i_R are *in phase*, that is, they reach their maximum values at the same time. As expected from Eq. 39-2 they also have the same angular frequency ω. We show both of these things in Fig. 39-2b, a plot of Eqs. 39-3a and 39-3c.

39-2
RCL ELEMENTS, CONSIDERED SEPARATELY

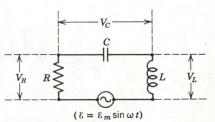

$(\varepsilon = \varepsilon_m \sin \omega t)$

figure 39-1
A single-loop *RCL* circuit contains an ac generator. V_R, V_C, and V_L are the time-varying potential differences across the resistor, the capacitor, and the inductor, respectively.

---

* The condition is, as Fig. 39-1 implies, that we can localize R, C, and L in separate, physically identifiable parts of the circuit. We discuss the basis of this assumption (that of *lumped impedances*) fully in Chapter 40. The lower the frequency the more justifiable this assumption becomes.

† As we shall see (Eq. 39-14), ϕ actually does not depend on ε_m.

‡ We could apply the loop theorem (see Section 32-2) to the circuit of Fig. 39-1 and solve the differential equation that results but this would involve us in mathematical difficulties beyond the scope of this book. We adopt an indirect but equally valid and more physically informative approach.

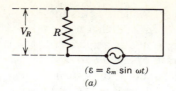

(a)

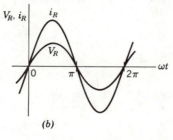

(b)

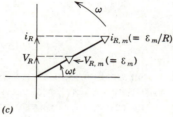

(c)

figure 39-2
(a) A single-loop resistive circuit containing an ac generator. (b) The current and the potential difference across the resistor are in phase. In Eq. 39-2, $\phi = 0$. (c) A phasor diagram shows the same thing. The arrows on the vertical axis are instantaneous values.

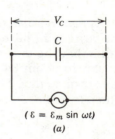

(a)

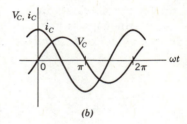

(b)

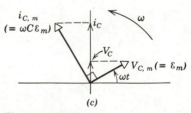

(c)

figure 39-3
(a) A single-loop capacitive circuit containing an ac generator. (b) The potential difference across the capacitor *lags* the current by a quarter-cycle. In Eq. 39-2, $\phi = -90°$. (c) A phasor diagram shows the same thing. The arrows on the vertical axis are instantaneous values.

Figure 39-2c shows another way of looking at the situation. It is sometimes called a *phasor diagram*, in which the phasors, represented by the open arrows, rotate counterclockwise with an angular frequency ω about the origin. The phasors have these properties: (a) The length of the phasor is proportional to the *maximum* value of the alternating quantity involved, thus ε_m for V_R (Eq. 39-3a) and (ε_m/R) for i_R (Eq. 39-3c), and (b) the projection of the phasors on the vertical axis gives the *instantaneous* values of the alternating currents involved. Thus the arrows on the vertical axis represent the time-varying quantities V_R and i_R, as in Eqs. 39-3a and 3c, respectively. That V_R and i_R are in phase follows from the fact that their phasors lie along the same line in Fig. 39-2c.

Follow the rotation of the phasors in Fig. 39-2c and convince yourself that this phasor diagram completely and correctly describes Eqs. 39-3a and 3c.

2. *A capacitive circuit.* Figure 39-3a shows a circuit containing a capacitive element only, acted on by the alternating emf of Eq. 39-1. From the loop theorem and from the definition of capacitance we can write

$$V_C = \varepsilon_m \sin \omega t \quad \text{(loop theorem)} \qquad (39\text{-}4a)$$

and

$$V_C = q/C \quad \text{(definition of } C\text{)}. \qquad (39\text{-}4b)$$

From these relations we have

$$q = \varepsilon_m C \sin \omega t$$

or

$$i_C \left(= \frac{dq}{dt}\right) = \omega C \varepsilon_m \cos \omega t. \qquad (39\text{-}4c)$$

Comparison of Eqs. 39-4a and 39-4c shows that the time-varying quantities V_C and i_C are one-quarter cycle out of phase. We show this in Fig. 39-3b, a plot of Eqs. 39-4a and 39-4c. We see that V_C lags i_C, that is, as time goes on, V_C reaches its maximum *after* i_C does, by one-quarter cycle.

We show this with equal clarity in the phasor diagram of Fig. 39-3c. As the phasors rotate in the assumed counterclockwise direction it is clear that the $V_{C,m}$ phasor *lags* behind the $i_{C,m}$ phasor by one-quarter cycle.

The phase angle ϕ between V_C and i_C in Fig. 39-3 is $-90°$. To show this put $\phi = -90°$ in Eq. 39-2 and expand. We obtain

$$i = i_m \cos \omega t,$$

in agreement with Eq. 39-4c.

For reasons of symmetry of notation we choose to rewrite Eq. 39-4c as

$$i_C = \left(\frac{\varepsilon_m}{X_C}\right) \cos \omega t \qquad (39\text{-}5a)$$

in which we must have

$$X_C = \frac{1}{\omega C}. \qquad (39\text{-}5b)$$

We call X_C the *capacitive reactance*. Comparing Eqs. 39-3c and 39-5a we see that the current amplitudes are (ε_m/R) and (ε_m/X_C), respectively. We see from this that the unit of X_C must be the ohm. Can you prove this explicitly, starting from Eq. 39-5b?

Also, let us recognize that ε_m in Eq. 39-4a is the maximum value of V_C $(= V_{C,m})$. Thus, from Eq. 39-5a we can write

$$V_{C,m} = i_{C,m} X_C. \qquad (39\text{-}6a)$$

This suggests that when *any* alternating current, of amplitude i_m and angular frequency ω, exists in a capacitor the maximum potential difference across the capacitor (*regardless of the complexity of the circuit in which the capacitor is involved*) is given by

$$V_{C,m} = i_m X_C. \qquad (39\text{-}6b)$$

EXAMPLE 1

In Fig. 39-3a, let $C = 150 \ \mu F$, $\nu = 60$ Hz, and $\varepsilon_m = 300$ V. Find (a) $V_{C,m}$, (b) X_C, and (c) $i_{C,m}$.

(a) From Eq. 39-4a,

$$V_{C,m} = 300 \text{ V}.$$

(b) From Eq. 39-5b,.

$$X_C = \frac{1}{\omega C} = \frac{1}{2\pi\nu C}$$
$$= \frac{1}{(2\pi)(60 \text{ Hz})(150 \times 10^{-6} \text{ F})}$$
$$= 18 \ \Omega.$$

(c) From Eq. 39-6a,

$$i_{C,m} = \frac{V_{C,m}}{X_C}$$
$$= \frac{300 \text{ V}}{18 \ \Omega} = 17 \text{ A}.$$

How would $i_{C,m}$ change if you doubled the frequency? Does this seem intuitively reasonable?

3. *An inductive circuit.* Figure 39-4*a* shows a circuit containing an inductive element only, acted on by the alternating emf of Eq. 39-1. From the loop theorem and the definition of inductance we can write

$$V_L = \mathcal{E}_m \sin \omega t \text{ (loop theorem)} \qquad (39\text{-}7a)$$

and

$$V_L = L(di/dt) \text{ (definition of } L). \qquad (39\text{-}7b)$$

From these relations we see that

$$di = (\mathcal{E}_m/L) \sin \omega t \, dt$$

or

$$i_L = \int di = -(\mathcal{E}_m/\omega L) \cos \omega t. \qquad (39\text{-}7c)$$

Comparison of Eqs. 39-7*a* and 7*c* shows that the time-varying quantities V_L and i_L are a quarter-cycle out of phase. We show this in Fig. 39-4*b*, a plot of Eqs. 39-7*a* and 39-7*c*. We see that V_L leads i_L, that is, as time goes on V_L reaches its maximum *before* i_L does, by a quarter-cycle.

We show this with equal clarity in the phasor diagram of Fig. 39-4*c*. As the phasors rotate in the counterclockwise direction it is clear that the $V_{L,m}$ phasor precedes (that is, *leads*) the $i_{L,m}$ phasor by a quarter-cycle.

The phase angle ϕ between V_L and i_L in Fig. 39-4 is $+90°$. To show this put $\phi = +90°$ in Eq. 39-2 and expand. We obtain

$$i = -i_m \cos \omega t,$$

in agreement with Eq. 39-7*c*.

Again, for reasons of compactness of notation we choose to rewrite Eq. 39-7*c* as

$$i_L = -(\mathcal{E}_m/X_L) \cos \omega t \qquad (39\text{-}8a)$$

in which we must have

$$X_L = \omega L. \qquad (39\text{-}8b)$$

We call X_L the *inductive reactance.* As for the capacitive reactance (Eq. 39-5) we see that the SI unit for X_L must also be the ohm. Can you prove this explicitly, starting from Eq. 39-8*b*?

Also, let us recognize that $\mathcal{E}_m$ in Eq. 39-7*a* is the maximum value of $V_L \ (= V_{L,m})$. Thus from Eq. 39-7*a* we can write

$$V_{L,m} = i_{L,m} X_L. \qquad (39\text{-}9a)$$

This suggests that when *any* alternating current of amplitude i_m and angular frequency ω exists in an inductor, the maximum potential difference across the inductor (*regardless of the complexity of the circuit in which the inductor is involved*) is given by

$$V_{L,m} = i_m X_L. \qquad (39\text{-}9b)$$

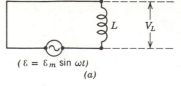

($\mathcal{E} = \mathcal{E}_m \sin \omega t$)
(a)

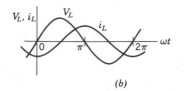

(b)

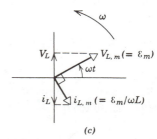

(c)

figure 39-4
(*a*) A single-loop inductive circuit containing an ac generator. (*b*) The potential difference across the inductor *leads* the current by a quarter-cycle. In Eq. 39-2, $\phi = +90°$. (*c*) A phasor diagram shows the same thing. The arrows on the vertical axis are instantaneous values.

In Fig. 39-4 let $L = 60$ mH, $\nu = 60$ Hz, and $\mathcal{E}_m = 300$ V. Find (*a*) $V_{L,m}$, (*b*) X_L, and (*c*) $i_{L,m}$.

EXAMPLE 2

(*a*) From Eq. 39-7*a*,

$$V_{L,m} = \mathcal{E}_m = 300 \text{ V}.$$

(*b*) From Eq. 39-8*b*,

$$\begin{aligned} X_L &= \omega L = 2\pi\nu L \\ &= (2\pi)(60 \text{ Hz})(60 \times 10^{-3} \text{ H}) \\ &= 23 \ \Omega. \end{aligned}$$

(c) From Eq. 39-9a,

$$i_{L,m} = \frac{V_{L,m}}{X_L}$$

$$= \frac{300 \text{ V}}{23 \ \Omega} = 13 \text{ A}.$$

How would $i_{L,m}$ change if you doubled the frequency? Does your answer seem intuitively reasonable?

Having finished our analysis of separate single-loop R, C, and L circuits, we now return to Fig. 39-1, in which all three elements are present. The emf is given by Eq. 39-1

$$\varepsilon = \varepsilon_m \sin \omega t \qquad (39\text{-}1)$$

and the (single) current in the circuit has the form shown in Eq. 39-2

$$i = i_m \sin (\omega t - \phi) \qquad (39\text{-}2)$$

in which we have yet to determine i_m and ϕ.

We start by applying the loop theorem to the circuit of Fig. 39-1, obtaining

$$\varepsilon = V_R + V_C + V_L \qquad (39\text{-}10)$$

These are all sinusoidally time-varying quantities, their maximum values being, in order, ε_m (see Eq. 39-1), $V_{R,m}$ $(= i_m R)$, $V_{C,m}$ $(= i_m X_C$; see Eq. 39-6b), and $V_{L,m}$ $(= i_m X_L$; see Eq. 39-9b).

Although Eq. 39-10 is true at any instant of time we cannot easily use it to find i_m and ϕ in Eq. 39-2 because of the phase differences that exist among the separate terms. We therefore turn to the phasor diagram of Fig. 39-5a, which shows the maximum values of i, V_R, V_C, and V_L. Check Section 39-2 to see that the phase differences are correct, that is, V_R is *in phase* with the current (Fig. 39-2), V_C *lags* the current by a quarter-cycle (Fig. 39-3), and V_L *leads* the current by a quarter-cycle (Fig. 39-4).

As in Section 39-2, the projections of the phasors on the vertical axis give the *instantaneous* values of the quantities involved. Thus the *algebraic* sum of V_R, V_C, and V_L on the vertical axis equals ε, as Eq. 39-10 requires.

On the other hand, we assert that the *vector* sum of the phasor amplitudes $V_{R,m}$, $V_{C,m}$, and $V_{L,m}$ yields a phasor whose amplitude is the ε_m of Eq. 39-1. The projection of ε_m on the vertical axis would, of course, be the time-varying ε of Eq. 39-1, that is, it would be $V_R + V_C + V_L$ as Eq. 39-10 asserts. Note that, in vector operations, the (algebraic) sum of the projections of any number of vectors on a given straight line is equal to the projection on that line of the (vector) sum of those vectors.

We can find ε_m from Fig. 39-5b, in which we have formed the phasor $V_{L,m} - V_{C,m}$. This is at right angles to $V_{R,m}$ and we have

$$\varepsilon_m = \sqrt{V_{R,m}^2 + (V_{L,m} - V_{C,m})^2}$$

$$= \sqrt{(i_m R)^2 + (i_m X_L - i_m X_C)^2}$$

$$= i_m \sqrt{R^2 + (X_L - X_C)^2} \qquad (39\text{-}11)$$

We call the quantity multiplying i_m the *impedance* Z of the circuit of Fig. 39-1. Thus we can write

39-3
THE SINGLE-LOOP RCL CIRCUIT

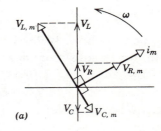

(a)

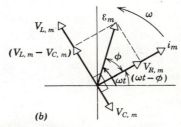

(b)

figure 39-5

(a) A phasor diagram corresponding to the circuit of Fig. 39-1. Note the phasor amplitudes and the instantaneous values shown on the vertical axis. (b) The same diagram, showing the relationship between the emf ξ and the current i of Eqs. 39-1 and 39-2, respectively. The phase angle ϕ of Eq. 39-2 is clearly shown.

$$i_m = \frac{\varepsilon_m}{Z},$$

(39-12)

which reminds us of the relation $i = \varepsilon/R$ (Eq. 32-2) for single-loop resistive networks with steady emfs. The SI unit of impedance is evidently the ohm.

We can write out Eq. 39-12 in full detail (see Eqs. 39-11, 39-5b, and 39-8b) as

$$i_m = \frac{\varepsilon_m}{\sqrt{R^2 + (\omega L - 1/\omega C)^2}}.$$

(39-13)

Thus we have solved the first problem posed in the second paragraph of Section 39-2; we have expressed i_m in terms of ε_m, ω, R, C, and L. It remains to find the phase angle ϕ between i and ε; compare Eqs. 39-1 and 39-2.

We show the angle ϕ in Fig. 39-5b and we can write

$$\tan \phi = \frac{V_{L,m} - V_{C,m}}{V_{R,m}}$$

$$= \frac{i_m(X_L - X_C)}{i_m R}$$

$$= \frac{X_L - X_C}{R}.$$

(39-14)

Thus we have solved the second problem posed in the second paragraph of Section 39-2; we have expressed ϕ in terms of ω, R, C, and L. As we said earlier, ϕ does not depend on ε_m. Increasing ε_m increases i_m (see Eq. 39-13) but it does not change ϕ; the *scale* of the operation is changed but not its *nature*.

We drew Fig. 39-5b arbitrarily with $X_L > X_C$; that is, we assumed the circuit of Fig. 39-1 to be more inductive than capacitive. For this assumption ε_m *leads* i_m (although not by so much as a quarter-cycle but less than $+90°$, as it did in the purely inductive circuit of Fig. 39-4). The phase angle ϕ in Eq. 39-14 (and thus in Eq. 39-2) is positive.

If, on the other hand, we had $X_C > X_L$, the circuit would be more capacitive than inductive and ε_m would *lag* i_m (although not by as much as a quarter-cycle, as it did in the purely capacitive circuit of Fig. 39-3). Consistent with this change from leading to lagging, the angle ϕ in Eq. 39-14 (and thus in Eq. 39-2) would automatically become negative.

EXAMPLE 3

In Fig. 39-1 let $R = 4.0\ \Omega$, $C = 150\ \mu\text{F}$, $L = 60\ \text{mH}$, $\nu = 60\ \text{Hz}$, and $\varepsilon_m = 300\ \text{V}$. Find (a) X_C, (b) X_L, (c) Z, (d) i_m, and (e) ϕ.

(a) $X_C = 18\ \Omega$ as in Example 1. Note that X_C depends only on C and ω (Eq. 39-5b) and *not* on the nature of the circuit in which C is an element.

(b) $X_L = 23\ \Omega$, as in Example 2. Note that X_L depends only on L and ω (Eq. 39-8b) and *not* on the nature of the circuit in which L is an element.

(c) From Eq. 39-11 we have

$$Z = \sqrt{R^2 + (X_L - X_C)^2}$$
$$= \sqrt{(4.0 \ \Omega)^2 + (23 \ \Omega - 18 \ \Omega)^2}$$
$$= 6.4 \ \Omega.$$

Note that this circuit is more inductive than capacitive, that is, $X_L > X_C$, as in Fig. 39-5.

(d) From Eq. 39-12 we have

$$i_m = \frac{\mathcal{E}_m}{Z} = \frac{300 \ \text{V}}{6.4 \ \Omega} = 47 \ \text{A}.$$

(e) From Eq. 39-14 we have

$$\tan \phi = \frac{X_L - X_C}{R}$$
$$= \frac{23 \ \Omega - 18 \ \Omega}{4.0 \ \Omega} = 1.25$$

or $\phi = +51°$. Because $X_L > X_C$, ϕ is positive and $\mathcal{E}_m$ *leads* i_m, as Fig. 39-5b suggests, but, as we expect, by less than $+90°$.

Power is an important practical matter. If you turn on the lights in a room, you are dealing directly with the subject matter of this section. We start by realizing that power dissipation in *RCL* circuits (Fig. 39-1 is an example on which we will focus), occurs *only* in the resistive element *R*; there is no mechanism for dissipating power in purely capacitive or purely inductive elements.

Let us therefore consider the purely resistive single-loop circuit of Fig. 39-2a, from two points of view. First let us replace the alternating emf by a *steady* emf of magnitude $\mathcal{E}_m$. The (steady) power dissipated in *R* (see Eq. 31-17) is given by

$$P = \frac{\mathcal{E}_m^2}{R} \text{ (steady emf).} \tag{39-15}$$

If the alternating emf of Eq. 39-1 applies, we have, however

$$P(t) = \frac{(\mathcal{E}_m \sin \omega t)^2}{R} \text{ (alternating emf).} \tag{39-16}$$

Our real concern is not so much for $P(t)$, which is sometimes zero, but for the average power $P(t)$ ($= P_{\text{av}}$), or

$$P_{\text{av}} = \frac{\mathcal{E}_m^2}{R} \overline{(\sin \omega t)^2}. \tag{39-17}$$

Figure 39-6 shows (a) a plot of $\sin \omega t$ and (b) a plot of $(\sin \omega t)^2$. Note that $\overline{\sin \omega t} = 0$ (the positive parts just cancel the negative parts). However, $\overline{(\sin \omega t)^2} = \frac{1}{2}$ (the parts above the line "$\frac{1}{2}$" just cancel those below that line; there are no negative values). Thus,

$$P_{\text{av}} = \frac{1}{2} \frac{\mathcal{E}_m^2}{R} = \left(\frac{\mathcal{E}_m}{\sqrt{2}} \right)^2 \frac{1}{R}. \tag{39-18}$$

We choose to call $\mathcal{E}_m/\sqrt{2}$ the *root-mean-square* (abbr. rms) value of $\mathcal{E}$ ($= \mathcal{E}_{\text{rms}}$). The notation is justified. First we *squared* $\mathcal{E}_m$, then we averaged it (or took the *mean* of it) over an integral number of cycles (the averaging factor, in this case of sinusoidal functions, being $\frac{1}{2}$), and

39-4
POWER IN ALTERNATING CURRENT CIRCUITS

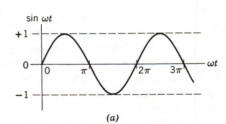

(a)

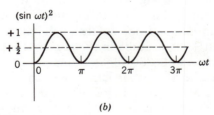

(b)

figure 39-6

(a) A plot of $\sin \omega t$, showing that its average value over an integral number of cycles is zero. The parts of the curve above the horizontal axis just cancel those below. (b) A plot of $(\sin \omega t)^2$, showing that its average value over an integral number of cycles is $\frac{1}{2}$. The parts of the curve above the horizontal line "$\frac{1}{2}$" just cancel those below it; the curve has no negative values.

finally we took the *square root*.* We can thus write Eq. 39-18 as

$$P_{av} = \frac{\mathcal{E}_{rms}^2}{R}. \qquad (39\text{-}19)$$

Equation 39-19 looks much like Eq. 31-17 and the message is that if we use rms quantities for $\mathcal{E}$, for V, and for i, the *average* power dissipation, which is all that usually matters, will be the same for alternating current circuits as for direct current circuits with a constant emf. Alternating current instruments, such as ammeters and voltmeters, are almost always too sluggish mechanically to follow the actual cycles of oscillation. They are deliberately calibrated to read $\mathcal{E}_{rms}$, V_{rms}, and i_{rms}. Thus, if you plug an alternating current voltmeter into a household electric outlet and it reads "120 V," the maximum value of the potential difference at the outlet is $\sqrt{2}$ (120 V) or 170 V. The sole reason for using rms values in alternating current circuits is to let us use the familiar direct current power relationships of Section 31-5. Thus, we summarize:

$$\mathcal{E}_{rms} = \mathcal{E}_m/\sqrt{2}, \quad V_{rms} = V_m/\sqrt{2}, \quad \text{and} \quad i_{rms} = i_m/\sqrt{2}. \qquad (39\text{-}20)$$

Because the proportionality factor $(= 1/\sqrt{2})$ is the same in each case, we can write the important Eq. 39-13 as

$$i_{rms} = \frac{\mathcal{E}_{rms}}{\sqrt{R^2 + (\omega L - 1/\omega C)^2}} \qquad (39\text{-}21)$$

and, indeed, this is the form that we almost always use.

EXAMPLE 4

(a) A steady emf ($\mathcal{E}_0 = 120$ V) is applied to a single-loop resistive circuit with $R = 150\ \Omega$. What is the power dissipation? (b) The steady emf $\mathcal{E}_0$ is replaced by an alternating emf ($\mathcal{E} = \mathcal{E}_m \sin \omega t$). If the average power is to remain unchanged, what must $\mathcal{E}_m$ be?

(a)
$$P = \frac{\mathcal{E}_0^2}{R} = \frac{(120\ \text{V})^2}{150\ \Omega} = 96\ \text{W}.$$

(b) From Eq. 39-18 and the conditions of the problem,

$$P_{av}\ (= 96\ \text{W}) = \left(\frac{\mathcal{E}_m}{\sqrt{2}}\right)^2 \frac{1}{R}$$

or

$$\mathcal{E}_m = \sqrt{2 P_{av} R}$$
$$= \sqrt{(2)(96\ \text{W})(150\ \Omega)}$$
$$= 170\ \text{V}.$$

From Eq. 39-20 we have

$$\mathcal{E}_{rms} = \mathcal{E}_m/\sqrt{2} = (170\ \text{V})/\sqrt{2}$$
$$= 120\ \text{V},$$

just as we expect.

Now we turn from power considerations for the resistive circuit of Fig. 39-2 to those for the more general circuit of Fig. 39-1, in which all three elements, R, C, and L, appear. For the instantaneous power we can write (compare Eq. 31-15)

*We have encountered rms quantities before, namely, in the rms speeds of molecules (see Section 23-4; also, see p. 852).

$$P(t) = \varepsilon(t)\, i\,(t)$$

$$= [\varepsilon_m \sin \omega t]\,[i_m \sin (\omega t - \phi)]. \qquad (39\text{-}22)$$

If we expand the factor $\sin (\omega t - \phi)$ by a trignometric identity that is evident by comparison with Eq. 39-23, we obtain

$$P(t) = (\varepsilon_m\, i_m)(\sin\, \omega t)(\sin \omega t \cos \phi - \cos \omega t \sin \phi)$$

$$= \varepsilon_m\, i_m \sin^2 \omega t \cos \phi - \varepsilon_m\, i_m \sin \omega t \cos \omega t \sin \phi. \qquad (39\text{-}23)$$

If we now find $\overline{P(t)}$ ($= P_{av}$), which is our principle practical concern, we have

$$P_{av} = \tfrac{1}{2}\varepsilon_m\, i_m \cos \phi + 0. \qquad (39\text{-}24)$$

This comes about because, as we have seen, $\overline{\sin^2 \omega t} = \tfrac{1}{2}$ and we assert (see Problem 9) that $\overline{\sin \omega t \cos \omega t} = 0$. Because $\varepsilon_{rms} = \varepsilon_m/\sqrt{2}$ and $i_{rms} = i_m/\sqrt{2}$ (Eq. 39-20) we can write Eq. 39-24 as

$$P_{av} = \varepsilon_{rms}\, i_{rms} \cos \phi \qquad (39\text{-}25)$$

in which we call $\cos \phi$ the *power factor*. Figure 39-2, which shows a purely resistive load, is characterized by $\phi = 0$ (and thus $\cos \phi = 1$)so that Eq. 39-24 becomes

$$P_{av} = \varepsilon_{rms}\, i_{rms} \quad \text{(resistive load)}. \qquad (39\text{-}26)$$

Subscripts aside, this is just the relation that we would write for direct current circuits (compare Eq. 31-15).

EXAMPLE 5

Let us use the same parameters for Fig. 39-1 that we did in Example 3, namely, $R = 4.0\ \Omega$, $C = 150\mu F$, $L = 60$ mH, $\nu = 60$ Hz, and $\varepsilon_m = 300$ V. Find (a) ε_{rms}, (b) i_{rms}, (c) ϕ, (d) $\cos\phi$, and (e) P_{av}.

(a) $\varepsilon_{rms} = \varepsilon_m/\sqrt{2} = (300\ \text{V})/\sqrt{2} = 210$ V.

(b) $i_{rms} = i_m/\sqrt{2} = (47\ \text{A})/\sqrt{2} = 33$ A; see Example 3, part (d).

(c) $\phi = 51°$, as in Example 3, part (e).

(d) The power factor, $\cos \phi$, is $\cos 51°$ or 0.63.

(e) $P_{av} = \varepsilon_{rms}\, i_{rms} \cos \phi$

$$= (210\ \text{V})(33\ \text{A})(0.63)$$

$$= 4.4\ \text{kW}.$$

39-5
RESONANCE IN ALTERNATING CURRENT CIRCUITS

In this section we return to the circuit of Fig. 39-1 and consider the effect, as far as the current i_{rms} is concerned, of varying the angular frequency ω of the applied emf of Eq. 39-1, assuming that ε_m, R, C, and L remain fixed.

We have stressed the analogy between mechanical systems, such as mass-spring arrangements (Fig. 8-4) and electromagnetic systems (Fig. 38-1). With this in mind we are entitled to look upon the *RCL* circuit of Fig. 39-1 as possessing a "natural" frequency of oscillation, and to view the circuit as acted upon by an external influence, in this case the applied alternating emf, given by $\varepsilon = \varepsilon_m \sin \omega t$ (Eq. 39-1), in which ω is the angular frequency of the "driving force." We expect a maximum "response," defined here by the i_{rms} of Eq. 39-21, when the angular frequency ω of the driving force just equals the natural frequency of oscillation ω_0 for the free oscillations of the circuit. Let us see whether these analogies and expectations turn out to be true.

From Eq. 39-21 we see that the maximum value of i_{rms} occurs when $X_L = X_C$ and has the value

$$i_{rms,max} = \frac{\mathcal{E}_{rms}}{R}, \qquad (39\text{-}27)$$

that is, i_{rms} is limited only by the circuit resistance. If $R \rightarrow 0$, $i_{rms,max} \rightarrow \infty$. Putting $X_L = X_C$ yields

$$\omega L = \frac{1}{\omega C}$$

or

$$\omega = \frac{1}{\sqrt{LC}}. \qquad (39\text{-}28)$$

The quantity on the right (see Eq. 38-3) is just the natural angular frequency ω_0 of the circuit of Fig. 39-1, with $\mathcal{E}$ and R removed. That is:

the maximum value of i_{rms} occurs when the frequency ω of the driving force [the $\mathcal{E}(t)$ of Eq. 39-1], is exactly equal to the natural frequency ω_0 of the *undamped* ($R = 0$) circuit of Fig. 39-1.

This condition

$$\omega = \omega_0 \qquad (39\text{-}29)$$

is called *resonance.* We met it earlier in connection with the cyclotron (see Section 33-7) and in many equivalent mechanical situations.

Figure 39-7 is a plot of i_{rms} versus ω (see Eq. 39-21), for fixed values of $\mathcal{E}_m$, C, and L, but for three different values of R. Note how rapidly the sharpness of the resonance peak broadens as R increases.

Figure 39-7 suggests to us the common experience of tuning a radio set. What we do here when we turn the tuning knob is to adjust the natural frequency ω_0 of an internal circuit to the frequency ω of the signal transmitted by the station antenna, until Eq. 39-29 is satisfied. In a metropolitan area, where there are many incoming signals whose fre-

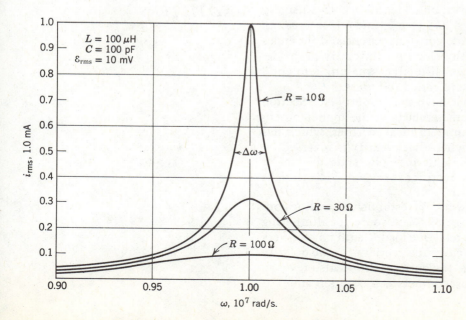

figure 39-7
Showing resonance in the *RCL* circuit of Fig. 39-1 for three different values of R. Along the horizontal axis we plot the (variable) angular frequency ω of the impressed emf. The arrow marked "$\Delta\omega$" on the curve for $R = 10\ \Omega$ is called the "full width of the curve at half maximum" or, more concisely, the *half width.*

quencies are not too far apart, the sharpness (or "quality"; see Problem 17) of tuning becomes very important.

Now let us consider the phenomenon of resonance from the point of view of the phasor diagram of Fig. 39-5b. If $X_L = X_C$, as resonance requires, the phasors $V_{L,m}$ ($=i_m X_L$) and $V_{C,m}$ ($=i_m X_C$) just cancel each other and the phase angle $\phi = 0$. The fact that $\phi = 0$ if $X_L = X_C$ is explicitly shown to be true by Eq. 39-14. If $\phi > 0$, as in Fig. 39-5b, we have $X_L > X_C$ and the circuit is predominantly inductive. On the other hand, if $\phi < 0$, we have $X_C > X_L$ and the circuit is predominantly capacitive.

EXAMPLE 6

In Fig. 39-7 (a) show that the angular frequency of the impressed emf at which resonance occurs is indeed 1.0×10^7 rad/s; (b) show also that the maximum value of i_{rms} at resonance, for $R = 10 \ \Omega$, is 1.0 mA.

(a) At resonance $X_L = X_C$ or

$$\omega L = \frac{1}{\omega C}$$

which leads to

$$\omega = \frac{1}{\sqrt{LC}} = \frac{1}{\sqrt{(100 \ \mu\text{H})(100 \ \text{pF})}}$$

$$= 1.0 \times 10^7 \ \text{rad/s}.$$

(b) At resonance Eq. 39-27 applies, or

$$i_{\text{rms,max}} = \frac{\mathcal{E}_{\text{rms}}}{R}$$

$$= \frac{10 \ \text{mV}}{10 \ \Omega} = 1.0 \ \text{A}.$$

39-6
ALTERNATING CURRENT RECTIFIERS AND FILTERS

We often have available a source of alternating emf (given, say, by Eq. 39-1, $\mathcal{E} = \mathcal{E}_m \sin \omega t$) and we want to derive from it, by electronic means, a constant potential difference. For example, in television sets, sound reproducing systems, etc., the available electric input is usually an alternating emf, often described by 120 V ($= \mathcal{E}_{\text{rms}}$) and 60 Hz ($= \omega/2\pi$). From this we need to derive one or more constant potential differences (50 V, 300 V, 1500 V, etc.) to operate the electronic circuitry of the device. This process is called *rectification* (literally, "making straight") and the devices that make it possible are called *rectifiers*.

Rectifiers are nonohmic devices that have the property (see Section 31-3) that their resistance depends on the polarity of the applied potential difference. In this section we assume an ideal rectifier, which has $R = 0$ for a given polarity and $R \to \infty$ when that polarity is reversed.

Physically, rectifiers may be semiconducting solid-state devices or vacuum tube diodes. The symbol for a rectifier is ——▷⊢——, left to right being the direction of "easy conduction."

Figure 39-8 suggests how rectifiers work. To establish a baseline consider Fig. 39-8a, which is identical to Fig. 39-2a. A source of alternating emf $\mathcal{E}$ ($= \mathcal{E}_m \sin \omega t$) is connected with a resistive load R with no rectifiers in the circuit. The arrows indicate the current directions for the polarity shown. When the polarity of $\mathcal{E}$ reverses, the current arrows also reverse.

If we connect a cathode ray oscilloscope between points b and c, it

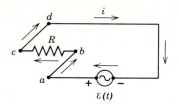

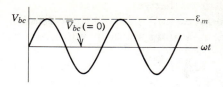

(a)

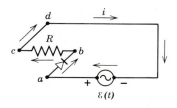

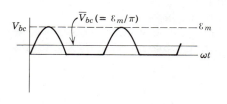

(b)

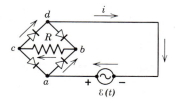

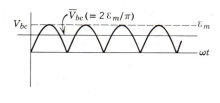

(c)

figure 39-8
An ac generator $\varepsilon \ (= \varepsilon_m \sin \omega t)$ acts on the three circuits shown. The potential difference V_{bc} is displayed on the right. Note that $\overline{V}_{bc}$ increases from zero to ε_m/π to $2\varepsilon_m/\pi$.

displays the waveform shown on the right. Note that $\overline{V}_{bc} = 0$ in this case, the positive halves of the sine wave just cancelling the negative halves. No rectification occurs, which is not surprising because there are no rectifiers in the circuit.

If we connect a dc voltmeter* between points b and c in Fig. 39-8*a*, it would read zero because the coil assembly is too sluggish to follow the waveform shown in Fig. 39-8*a*. For a typical dc voltmeter, at 60 Hz, say, the pointer will oscillate rapidly with small amplitude about its zero position. Why?

In Fig. 39-8*b* we insert a rectifier in the path ab, the direction of "easy conduction" being from a to b. When the polarity reverses, the circuit is virtually open because the (ideal) rectifier has infinite resistance.

If we connect a cathode ray oscilloscope between points b and c, it displays the waveform shown on the right. It is clear that $\overline{V}_{bc} > 0$. In fact,

$$\overline{V}_{bc} = \frac{\varepsilon_m}{2\pi} \left\{ \int_0^{\pi} \sin \omega t \ d(\omega t) + \int_{\pi}^{2\pi} 0 \ d(\omega t) \right\} = \frac{\varepsilon_m}{\pi}. \qquad (39\text{-}30)$$

The arrangement of Fig. 39-8*b* is called a *half-wave rectifier*. It does not yet produce a truly constant potential difference but it is a step in

---

* We can construct such a voltmeter by putting a large resistor R_0 in series with the coil of the sensitive galvanometer shown in Fig. 33-9. The large series resistor R_0 ($\gg R$ in Fig. 39-8*a*) is needed to insure that touching the voltmeter terminals to points b and c in Fig. 39-8*a* will not appreciably change the current distribution in that circuit.

that direction. If we connect a dc voltmeter between points b and c in Fig. 39-8b, it will give a definite reading. As for Fig. 39-8a there will be rapid small-amplitude oscillations of the pointer about its equilibrium position.

In Fig. 39-8c we introduce four rectifiers. For the polarity shown the current follows the path indicated by the arrows. When the polarity of the emf reverses, the current becomes counterclockwise and its path through the rectifier network is $dbca$. Note that the direction of the current through R remains unchanged; it is always from b to c, regardless of the polarity of the source of emf.

If we connect a cathode ray oscilloscope between points b and c, it displays the waveform shown on the right. Convince yourself that $\overline{V_{bc}}$ is just twice the value given for the half-wave rectifier of Fig. 39-8b, that is,

$$\overline{V_{bc}} = 2 \ \varepsilon_m/\pi. \qquad (39\text{-}31)$$

The arrangement of Fig. 39-8c is called a *full-wave rectifier*. If we connect a dc voltmeter between points b and c, it will deflect twice as far as it did in connection with Fig. 39-8b. Oscillations of the pointer will still occur but they will have smaller amplitude. Why?

In Fig. 39-8 the emf is $\varepsilon = \varepsilon_m \sin \omega t$, in which $\varepsilon_m = 300$ V and $R = 15 \ \Omega$. What average power P_{av} is dissipated in the resistor for the three cases shown?

EXAMPLE 7

(*a*) For Fig. 39-8a we have, from Eqs. 39-18 and 39-19,

$$P_{av} = \frac{\varepsilon_m{}^2}{2R} \left(= \frac{\varepsilon_{rms}{}^2}{R} \right)$$

$$= \frac{(300 \text{ V})^2}{(2)(15 \ \Omega)} = 3.0 \text{ kW}.$$

(*b*) Because the circuit of Fig. 39-8b is nonconducting (open circuit) for half the time, P_{av} is just half the value shown in (a) above, or 1.5 kW.

(*c*) In Fig. 39-8c P_{av} has the same value ($= 3.0$ kW) as in (a) above. Note that P_{av} does not depend on the *direction* of i or V; both quantities occur as squares.

The waveform of Fig. 39-8c is *still* not a nontime-varying potential difference. We can, however, resolve it into a steady component and a series of sinusoidally alternating potential differences with various angular frequencies ω, amplitudes V, and relative phases ϕ. We often call these ac components *the ripple*. Qualitatively we can see that this division of the waveform of Fig. 39-8c is reasonable, just by inspection. Quantitatively we could (but won't) derive it by Fourier analysis of the waveform.

What we need to do now is to connect the potential difference between points b and c of the circuit of Fig. 39-8 c, that is, the waveform of that figure, through a *filter circuit*, which has these properties: (1) it passes the dc component from input to output with negligible reduction in value, and (2) it greatly decreases the amplitude of the time-varying components, that is, of the ripple.

Figure 39-9 shows a simple filter circuit. It contains an ideal inductor L (that is, it has no resistive or capacitive properties) and an ideal capacitor C (that is, it has no resistive or inductive properties). The input V_{in} to the filter may be steady or sinusoidally oscillating. To explore the behavior of the filter we will look at these two cases separately.

For V_{in} = a constant we see that,

$$V_{out} = V_{in} = \text{the same constant.}$$

Neither L nor C has any effect. In fact, L might be replaced by a straight wire (shorted out) and C removed from (clipped out of) the circuit, with no noticeable effect on V_{out}.

For an ac input, however, the situation is quite different. We assume from the beginning that both L and C are "large" so that $X_L (=\omega L) \gg X_C (=1/\omega C)$. If ω and C are large enough, $X_C \to 0$ and the capacitor acts as a virtual short circuit for ac components, even though it has no effect on the dc component.

Assume

$$V_{in} = V_{in,m} \sin \omega t. \tag{39-32}$$

For the current we can put

$$i = i_m \sin (\omega t + \phi). \tag{39-2}$$

From Eq. 39-13, with $R = 0$ and ε_m replaced $V_{in.m}$, we have

$$i = \frac{V_{in,m}}{X_L - X_C} \sin (\omega t + \phi). \tag{39-33}$$

Because we have already assumed that $X_L \gg X_C$ we can write this as

$$i \cong \frac{V_{in,m}}{X_L} \sin (\omega t + \phi). \tag{39-34}$$

To find the phase angle ϕ we turn to Eq. 39-14, or

$$\tan \phi = \frac{X_L - X_C}{R}. \tag{39-14}$$

With $X_L \gg X_C$ and $R \to 0$ we have $\tan \phi \to +\infty$ or $\phi = +90°$. Thus Eq. 39-33 becomes (see Eq. 39-14)

$$i = \frac{V_{in,m}}{X_L} \cos \omega t. \tag{39-35}$$

Note that a cosine function is just 90° out of phase with a sine function.

The output V_{out} is given by q/C so that we must find q for the capacitor. From Eq. 39-35 we have

$$q = \int_0^t i \,(dt) = \frac{1}{\omega} \int_0^{\omega t} i \, d(\omega t)$$

$$= \left(\frac{1}{\omega}\right)\left(\frac{V_{in,m}}{X_L}\right) \int_0^{\omega t} \cos \omega t \, d(\omega t)$$

$$= \left(\frac{1}{\omega}\right)\left(\frac{V_{in,m}}{X_L}\right) \Big| \sin \omega t \Big|_0^{\omega t}$$

$$= \left(\frac{1}{\omega}\right)\left(\frac{V_{in,m}}{X_L}\right) \sin \omega t. \tag{39-36}$$

Because $X_C = 1/\omega C$ we have

$$V_{out} = \frac{q}{C} = (X_C)\left(\frac{V_{in,m}}{X_L}\right) \sin \omega t. \tag{39-37}$$

Comparison with Eq. 39-32 gives for the ac attenuation

$$V_{out,m}/V_{in,m} = X_C/X_L. \tag{39-38}$$

Because we have chosen $X_L \gg X_C$ we see that the attenuation of the "ripple components" is large. Much more effective filters than those of Fig. 39-9 are possible, as amateur builders of sound reproducing equipment are well aware (see Problem 23).

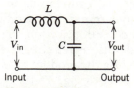

figure 39-9

A simple filter circuit designed to pass dc potential differences from input to output without appreciable attenuation but to attenuate time-varying components greatly. Convince yourself that if L, C, and ω all have "large" values, the filter element will be more effective in reducing the amplitudes of ac input components.

In Fig. 39-9 let $V_{in} = V_{in,m} \sin \omega t$, in which $V_{in,m} = 300$ V, and $\omega/2\pi = 60$ Hz. Let $L = 10$ H and $C = 300$ μF. Find $V_{out,m}$. **EXAMPLE 8**

From Eq. 39-37, we have

$$V_{out,m} = (V_{in,m})(X_C)\left(\frac{1}{X_L}\right)$$

$$= (V_{in,m})\left(\frac{1}{\omega C}\right)\left(\frac{1}{\omega L}\right)$$

$$= \frac{V_{in,m}}{\omega^2 LC}$$

$$= \frac{(300 \text{ V})}{(2\pi \times 60 \text{ Hz})^2(10 \text{ H})(300 \times 10^{-6} \text{ F})}$$

$$= 0.70 \text{ V}.$$

In this case the attenuation factor is

$$(V_{out,m}/V_{in,m}) = (0.70 \text{ V})/(300 \text{ V})$$

or 2.3×10^{-3}. Convince yourself that this is the ratio X_C/X_L, in which we have assumed $X_L \gg X_C$.

In dc circuits the power dissipation in a resistive load is given by Eq. 31-15 $(P = iV)$. This means that, for a given power requirement, we have our choice of a relatively large current i and a relatively small potential difference V or just the reverse, provided that their product remains constant. In the same way, for ac circuits the average power dissipation is given by Eq. 39-25 $(P_{av} = i_{rms} V_{rms})$ and we have the same choice as to the relative values of i_{rms} and V_{rms}.*

39-7
THE TRANSFORMER

In electric power distribution systems it is clear that at both the generating end (the electric power plant) and the receiving end (the home or factory) it is desirable, both for reasons of safety and the efficient design of equipment, to deal with relatively low voltages. For example, no one wants an electric toaster or a child's electric train to operate at, say, 10 kV.

On the other hand, in the transmission of electric energy from the generating plant to the consumer, we want the *lowest* possible current (and thus the *largest* possible potential difference) so as to minimize the $i^2 R$ ohmic losses in the transmission line. $V_{rms} = 350$ kV is not uncommon. Thus there is a fundamental "mismatch" between the requirements for efficient transmission on the one hand and efficient and safe generation and consumption on the other. We need a device that can, as design considerations require, raise (or lower) the potential difference in a circuit, keeping the product iV essentially constant. The alternating current *transformer* of Fig. 39-10 is such a device. It has no direct current counterpart of equivalent simplicity, which is why dc distribution systems, strongly advocated by Edison, have now been essentially totally replaced by ac systems, strongly advocated by Tesla and others.

Within the last few decades the use of dc transmission over very long distances or for special situations such as underwater or underground transmission cables has undergone a revival. In the Soviet Union, for example, there is a 300-mile line from Volgograd to the Donbass operating at a dc potential difference of 800 kV. The power is generated as ac, stepped up to the transmission voltage, converted to dc, reconverted

---

* We assume a resistive load so that $\cos \phi = 1$ in Eq. 39-25.

to ac at the receiving point, and stepped down to lower voltages for transmission. The advantage of high-voltage dc transmission under these circumstances is that, for dc, the capacitive and inductive properties of the transmission line can be ignored. Note that transformers, the subject of this section, are still very much required.*

In Fig. 39-10 two coils are shown wound around a soft iron core. The *primary* winding, of N_1 turns, is connected with an alternating current generator whose emf $\varepsilon_1(t)$ is given by $\varepsilon_1 = \varepsilon_m \sin \omega t$. The *secondary* winding, of N_2 turns, is on open circuit as long as switch S is open, which we assume for the present. Thus there is no secondary current. We assume further that the resistances of the primary and secondary windings and also the magnetic "losses" in the iron core are negligible. Actually, well-designed, high-capacity transformers can have energy losses as low as one percent so that our assumption of an ideal transformer is not unreasonable.

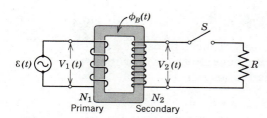

figure 39-10
An ideal transformer, showing two coils wound on the same soft iron core. An ac generator is connected with the primary winding. The secondary winding may be on open circuit (if switch S is open) or connected with a resistive load R (if switch S is closed).

For the above conditions the primary winding is a pure inductance; compare Fig. 39-4. Thus the (very small) primary current, called the *magnetizing current* $i_{mag}(t)$, lags the primary potential difference $V_1(t)$ by 90°; the power factor ($=\cos\phi$ in Eq. 39-25) is zero, and thus no power is delivered from the generator to the transformer.

However, the small alternating primary current $i_{mag}(t)$ induces an alternating magnetic flux $\Phi_B(t)$ in the iron core and we assume that all this flux links the turns of the secondary windings. From Faraday's law of induction (see Eq. 35-2) the emf *per turn* ε_T is the same for both the primary and secondary windings. Thus, on an rms basis, we can write

$$(\varepsilon_T)_{rms} = \left(-\frac{d\Phi_B}{dt}\right)_{rms} = \frac{V_{1,rms}}{N_1} = \frac{V_{2,rms}}{N_2}$$

or

$$V_{2,rms} = V_{1,rms}(N_2/N_1). \qquad (39-39)$$

If $N_2 > N_1$, we speak of a *step-up transformer*; if $N_2 < N_1$, we speak of a *step-down transformer*.

In all of the above we have assumed an open circuit secondary so that no power is transmitted through the transformer. If we now close switch S in Fig. 39-10, however, we have a more practical situation in which the secondary winding is connected with a resistive load R. In the general case the load would also contain inductive and capacitive elements but we confine ourselves to this special case.

Several things happen when we close switch S. (1) A current $i_2(t)$ appears in the secondary circuit, with a corresponding power dissipation $i_{2,rms}^2 R$ ($=V_{2,rms}^2/R$) in the resistive load. (2) This current induces its

*See L. O. Barthold and H. G. Pfeiffer, "High-Voltage Power Transmission," *Scientific American*, May 1964.

own alternating magnetic flux in the iron core and this flux induces (from Faraday's law and Lenz's law) an *opposing* emf in the primary windings. The two windings appear now as a fully coupled mutual inductance; see Section 36-6. (3) $V_1(t)$, however, cannot change in response to this opposing emf because it must *always* equal $\varepsilon(t)$ as provided by the generator; closing switch S cannot change this fact. (4) For this reason a *new* resultant current $i_1(t)$ must appear in the primary windings, its magnitude and phase angle being just that needed to cancel the opposing emf generated in the primary windings by $i_2(t)$. In particular the phase angle ϕ between $i_1(t)$ and $\varepsilon(t)$ for an ideal transformer must approach $0°$, so that the power factor in Eq. 39-25, $\cos \phi$, must approach unity.

All of the above is consistent with energy conservation. When we close switch S, power is dissipated in the resistive load. This requires that the generator provide an equal power to the (ideal) transformer, or, assuming $\phi = 0°$ ($\cos \phi = 1$),

$$\varepsilon_{rms} i_{1,rms} = V^2_{2,rms}/R. \qquad (39\text{-}40)$$

This relation expresses the fact that, for an ideal transformer with a resistive load, the power provided by the generator on the primary side just equals that dissipated in the resistive load on the secondary side.

EXAMPLE 9

A transformer on a utility pole operates at $V_{1,rms} = 8.0$ kV (see Fig. 39-10) on the primary side and supplies electric energy to a number of nearby houses at $V_{2,rms} = 120$ V.

(*a*) What is the turns ratio N_1/N_2? (*b*) If the average power consumption in the houses for a given time interval is 70 kW, what are the rms currents in the primary and secondary windings of the transformer? Assume an ideal transformer, a resistive load, and a power factor of unity. (*c*) What is the equivalent resistive load R in the secondary circuit?

(*a*) From Eq. 39-39,

$$N_1/N_2 = V_{1,rms}/V_{2,rms}$$

$$= 8.0 \text{ kV}/120 \text{ V}$$

$$= 67.$$

(*b*) From Eq. 39-25 we have, with $\cos \phi = 1$

$$i_{2,rms} = P_{av}/(V_{2,rms})(\cos \phi)$$

$$= (70 \text{ kW})/(120 \text{ V})(1)$$

$$= 580 \text{ A},$$

and also

$$i_{1,rms} = P_{av}/(V_{1,rms})(\cos \phi)$$

$$= (70 \text{ kW})/(8.0 \text{ kV})(1)$$

$$= 8.8 \text{ A}.$$

Note that, as required for an ideal transformer,

$$i_{1,rms} V_{1,rms} = i_{2,rms} V_{2,rms} = 70 \text{ kW}.$$

(*c*) Here we have

$$R = (V_{2,rms})^2/P_{av}$$

$$= (120 \text{ V})^2/70 \text{ kW}$$

$$= 0.21 \ \Omega.$$

Show that the same result follows by solving Eq. 39-40 for R.

1. In the relation $\omega = 2\pi\nu$ we measure ω in radians/second and ν in hertz or cycles/seconds. The radian is a measure of angle. What connection do angles have with alternating currents?

2. Problem 35-6 suggests how an alternating emf such as that described by Eq. 39-1 can be generated. If the output of such an ac generator is connected to an *RCL* circuit such as that of Fig. 39-1, what is the ultimate source of the power dissipated in the circuit? In what part of the circuit does the power dissipation occur?

3. In the circuit of Fig. 39-1, why is it safe to assume that (a) the alternating current of Eq.39-2 has the same angular frequency $\mathcal{E}_m$ as the alternating emf of Eq. 39-1, and (b) that the phase angle ϕ in Eq. 39-2 does not vary with time? What would happen if either of these (true) statements were false?

4. Is there an analogy, however loose, between the facts that (a) the phase angle ϕ in Eq. 39-2 does not depend on the value of ω in Eq. 39-1, and (b) the oscillation frequency of a mass-spring system does not depend on the amplitude of the oscillation? Distinguish between the *nature* of a physical event and its *scale*.

5. How does a phasor differ from a vector? We know, for example, that emfs, potential differences, and currents are *not* vectors. How then can we justify constructions such as Fig. 39-5?

6. In the purely resistive circuit of Fig. 39-2 how does the maximum value of the alternating current i_m vary with the angular frequency of the applied emf?

7. Would any of the discussion of Section 39-2 be invalid if the phasor diagrams were to rotate in the clockwise direction, rather than the counterclockwise direction which we assumed?

8. Does it seem intuitively reasonable that the capacitive reactance $(= 1/\omega C)$ should vary inversely with the angular frequency, whereas the inductive reactance $(=\omega L)$ varies directly with this quantity?

9. During World War II, at a large research laboratory in this country, an alternating current generator was located a mile or so from the laboratory building which it served. A technician increased the speed of the generator to compensate for what he called "the loss of frequency along the transmission line" connecting the generator with the laboratory building. Comment.

10. Discuss in your own words what it means to say that a potential difference "leads" or "lags" an alternating current.

11. If, as we stated in Section 39-3, a given circuit is "more inductive than capacitive," that is, that $X_L > X_C$, does this mean, for a fixed angular frequency, that (a) L is relatively "large" and C is relatively "small," or (b) L and C are both relatively "large"? (c) For fixed values of L and C does this mean that ω is relatively "large" or relatively "small"?

12. Assume that in Fig. 39-1 we let $\omega \rightarrow 0$. Does Eq. 39-13 approach an expected value? Discuss.

13. How is it that in Fig. 39-2 the current phasor points in the direction of the voltage phasor but in Figs. 39-3, 39-4, and 39-5 the current phasors point in three different directions? The applied emf (Eq. 39-1) is the same in all four cases.

14. Consider this statement: "If $X_L > X_C$, then, regardless of the frequency, we must have $L > 1/C$." What is wrong with this statement?

15. Do Kirchoff's loop theorem (see Section 32-2) and Kirchoff's junction theorem (see Section 32-5) apply to multiloop ac circuits as well as to multiloop dc circuits?

16. In Example 5 what would be the effect on P_{av} if you increased (a) R, (b) C, and (c) L? How would ϕ in Eq. 39-25 change in these three cases?

17. Do commercial power station engineers like to have a low power factor (see Eq. 39-25) or a high one, or does it make any difference to them? Between what values can the power factor fluctuate? What determines the power factor; is it characteristic of the generator, of the transmission line,

of the circuit to which the transmission line is connected, or of some combination of these?

18. If you know the power factor $(=\cos\phi$ in Eq. 39-25) for a given RCL circuit, can you tell whether or not the applied alternating emf is leading or lagging the current? If so, how? If not, why not?

19. If $R = 0$ in the circuit of Fig. 39-1, there can be no Joule internal energy dissipation in the circuit. However, an alternating emf and an alternating current are still present. Discuss the energy flow in the circuit under these conditions.

20. If you want to reduce your electric bill, do you hope for a small or a large power factor $(=\cos\phi$ in Eq. 39-25) or does it make any difference? If it does, is there anything that you can do about it? Discuss.

21. If $R = 0$ in Fig. 39-1, how would the instantaneous current vary with time? What value would the phase angle ϕ have? What would be the average power dissipation?

22. In Eq. 39-25 is ϕ the phase angle between $\varepsilon(t)$ and $i(t)$ or between ε_{rms} and i_{rms}? Discuss.

23. Resonance in RCL circuits, as judged by Eq. 39-29 and Fig. 39-7, occurs when the angular frequency of the alternating emf (the driving force) is exactly equal to the natural angular frequency of the (undamped) LC circuit. In Section 15-10, however, we saw that resonance for damped mass-spring systems, as judged by Eq. 15-41 and Fig. 15-20, occurs when the angular frequency of the driving force is close to, *but not exactly equal to,* the natural angular frequency of the (undamped) mass-spring system. Is there a failure of the principle of correspondence here? (*Hint:* Do the quantities displayed on the vertical axes of Figs. 15-20 and 39-7 truly "correspond"? See Problem 15.)

24. Convince yourself that filter circuits, such as those shown in Fig. 39-9 become, for fixed values of L and C, more effective as the angular frequency ω of V_{in} increases. What does "more effective" mean in this connection?

25. In Fig. 39-8c why does the current approaching junction c not divide equally between paths cd and ca? Both are potentially conducting paths.

26. Sketch roughly the waveforms of Fig. 39-8b,c if the rectifiers are not "ideal," that is, if the resistance in the forward direction, though small, is not actually zero, and if the resistance in the backward direction, though large, is not infinitely large.

27. Sketch an $i - V$ curve (see Figs. 31-4, 31-5, and 31-6) for the idealized rectifier assumed in Section 39-6. Include both polarities of the applied potential difference V (and of the corresponding current i) in your plot.

28. A doorbell transformer is designed for a primary rms input of 120 V and a secondary rms output of 6.0 V. What would happen if the primary and secondary connections were accidentally interchanged during installation? Would you have to wait for someone to push the doorbell to find out? Discuss.

29. You are given a transformer enclosed in a wooden box, its primary and secondary terminals being available at two opposite faces of the box. How could you find its turns ratio without opening the box?

problems

SECTION 39-1

1. A commercial alternating current generator is characterized by $\nu = 60$ Hz. What is the angular frequency ω and in what units is it expressed?
   Answer: $\omega = 2\pi\nu = 380$ rad/s.

SECTION 39-2

2. An 0.50-μF capacitor is connected, as shown in Fig. 39-3a, to an ac generator with $\varepsilon_m = 300$ V. What is the amplitude i_m of the resulting alternating current if the angular frequency ω is (a) 100 rad/s, and (b) 1000 rad/s?

3. A 45-mH inductor has a reactance X_L of 1300Ω. What must be (a) the applied angular frequency ω and (b) the applied frequency ν for this to be true? (c) If, as in Fig. 39-4a, an alternating emf with $\mathcal{E}_m = 300$ V is applied, what is the amplitude i_m of the alternating current that results? *Answer:* (a) 2.9×10^{-4} rad/s. (b) 4.6 kHz. (c) 0.23A.

4. A 1.5-μF capacitor has a reactance X_C of 12 Ω. What must be (a) the applied frequency ν and (b) the applied angular frequency ω for this to be true? (c) If, as in Fig. 39-3a, an alternating emf with $\mathcal{E}_m = 300$ V is applied, what is the amplitude i_m of the alternating current that results?

5. (a) At what frequency ν would a 6.0-mH inductor and a 10-μF capacitor have the same reactance? (b) What would this reactance be? (c) How would this frequency compare with the natural resonant frequency of free oscillations if the components were connected as a (resistanceless) LC oscillator, as in Fig. 38-1? *Answer:* (a)650 Hz. (b) 24 Ω. (c) They are equal.

SECTION 39-3

6. Recalculate Example 3 for $C = 20$ μF, all other given quantities remaining the same. Will $\mathcal{E}_m$ lead or lag i_m in this new situation? Draw a rough diagram corresponding to Fig. 39-5.

7. Redraw (roughly) Figs. 39-5a and 5b for the cases of $X_C > X_L$ and $X_C = X_L$.

SECTION 39-4

8. We have seen that $\overline{\sin^2 \omega t} = \frac{1}{2}$ (average value). Find the average value of $\sin^2 (\omega t + \phi)$ where ϕ is a (constant) phase angle?

9. Show that (see Eq. 39-23) $\overline{\sin \omega t \cos \omega t} = 0$ by using the trignometric identity $2 \sin \omega t \cos \omega t = \sin 2\omega t$. Also, plot $\sin \omega t \cos \omega t$ roughly and show graphically that its average value is zero.

10. The average value of $\mathcal{E}_m \sin \omega t$ over many cycles (see Fig. 39-6a) is zero. We have also seen that $\mathcal{E}_{\text{rms}} = \mathcal{E}_m / \sqrt{2}$. How does the average value of $\mathcal{E}_m \sin \omega t$, averaged over one-half of a cycle only, compare with $\mathcal{E}_m$ and with $\mathcal{E}_{\text{rms}}$?

11. In an RCL circuit such as that of Fig. 39-1, assume $R = 5.0$ Ω, $L = 60$ mH, $\nu = 60$ Hz, and $\mathcal{E}_m = 300$ V. For what values of C would P_{av} for the circuit be (a) a maximum and (b) a minimum? (c) What are these maximum and minimum values and what are the corresponding phase angles and power factors? (d) If $\cos \phi$ in Eq. 39-25 is negative, it would seem that P_{av} in Eq. 39-25 would be negative. What can this mean? Can you find a value of C above that will give a negative value for $\cos \phi$? Discuss. *Answer:* (a) 120 μF. (b) Zero. (c) 9000W, 420W; 0°, 78°; 1.0, 0.22.

SECTION 39-5

12. An RCL circuit, such as that of Fig. 39-1, has $R = 20$ Ω, $C = 20$ μF, and $L = 1.0$ H. (a) At what angular frequency ω of the ac generator will the circuit resonate with maximum response? (b) At what angular frequencies will the response be one-half the maximum value? We define "response" to be measured by the rms current in the circuit, as in Fig. 39-7.

13. A resistor-capacitor-inductor circuit R_1, C_1, L_1, connected as in Fig. 39-1, exhibits resonance at the same frequency as a second, separate, combination R_2, C_2, L_2. If the two combinations are connected in series in a single circuit, at what frequency would the combined circuit resonate?

Answer: $\omega = \sqrt{\dfrac{1}{L_1 C_1}} = \sqrt{\dfrac{1}{L_2 C_2}}$, independent of R_1 and R_2.

14. In Fig. 39-7 show that for frequencies higher than the resonant frequency the circuit is predominantly inductive and for frequencies lower than the resonant frequency it is predominantly capacitive. What does this statement mean? How do you interpret it in terms of Fig. 39-5b?

15. Show that the amplitude of the *charge* (not current) oscillations in an RCL circuit such as that of Fig. 39-1 is

$$q_m = \frac{\mathcal{E}_m}{\sqrt{(\omega^2 L - 1/C)^2 + (\omega R)^2}}.$$

(a) For what value of ω will q_m be a maximum? (b) Does this result shed any light on the comparison of Figs. 15-20 and 39-7, suggested in Question 23?

Answer: (a) $\omega = \sqrt{\dfrac{1}{LC} - \dfrac{R^2}{2L^2}}$.

16. Figure 39-11 shows an ac generator connected, through terminals a and b, to a "black box" containing an RCL circuit whose elements and arrangements we do not know. An alternating current, given by $i = i_m \sin(\omega t + \phi)$ appears in the lead-in wires. (a) What is the power factor? (b) Is the circuit in the box capacitive or inductive in nature? (c) Does the emf lag the current or lead it? (d) What average power P_{av} is delivered to the box by the source of emf if $\varepsilon_m = 750$ V and $i_m = 12$ A? (e) Why don't you need to know the angular frequency ω to answer the preceding question? (f) If you wanted the circuit to resonate, in the sense of Fig. 39-7, what is the nature of the circuit element that you would connect between terminals a and b? (g) At resonance, what values would ϕ and P_{av} have?

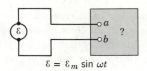

$\varepsilon = \varepsilon_m \sin \omega t$

figure 39-11
Problem 16

17. Show that the fractional half-width of the resonance curves of Fig. 39-7 is given, to a close approximation, by

$$\frac{\Delta \omega}{\omega} = \frac{\sqrt{3}R}{\omega L},$$

in which ω is the resonant frequency and $\Delta \omega$ is the width of the resonance peak at $i = \frac{1}{2}i_m$. Note (see Problem 38-22) that this expression may be written as $\sqrt{3}/Q$ which shows clearly that a "high-Q" circuit has a sharp resonance peak, that is, a small $\Delta \omega/\omega$.

SECTION 39-6

18. In Fig. 39-8 what are (a) $V_{bc,rms}$ and (b) P_{av} for the resistor R in each of the cases shown?

19. In Fig. 39-8, shade the areas, in all three cases, that cancel to form $\overline{V_{bc}}$.

20. For the full-wave rectified waveform of Fig. 39-8c, what is the fundamental (lowest) frequency ν_0 of the "ac ripple"? Assume that, for the ac generator, $\nu = \omega/2\pi = 60$ Hz.

21. In Fig. 39-12 in which $R_1 \gg R_2$ and both V_{in} and V_{out} are constant potential differences, show that the attenuation factor $V_{out}/V_{in} \cong R_2/R_1$. Compare with Fig. 39-9 and Eq. 39-38, in which the (ac) attenuation factor is (approximately) X_C/X_L. Discuss analogies and differences.

22. (a) Show that the attenuation factor $V_{out,m}/V_{in,m}$ (see Eq. 39-38 and Fig. 39-9) can be written, for $\omega \gg \omega_0$, as $(\omega_0/\omega)^2$. Here ω is the input angular frequency and ω_0 is the resonant angular frequency $(= \sqrt{1/LC})$ of the LC filter combination. (b) Show that $\omega \gg \omega_0$ corresponds to $X_L \gg X_C$. Is this reasonable?

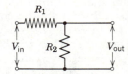

figure 39-12
Problem 21

23. Show by qualitative argument that the three-stage filter of Fig. 39-13 is more effective than the one-stage filter of Fig. 39-9. Sketch its dc equivalent, that is, replace L by R_1 and C by R_2, with $R_1 \gg R_2$. Derive the dc attenuation factor V_{out}/V_{in}, both of these quantities being dc potentials. Answer: $(R_2/R_1)^3$.

24. Show that, in low current, high-voltage applications (such as power supplies for television picture tubes) we can replace the inductor L of Fig. 39-9 by a "large" resistor R and still achieve substantial reduction of the ac component of V_{in} without too much reduction of the dc component.

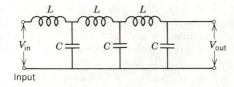

Input

figure 39-13
Problem 23

SECTION 39-7

25. An ac transmission line transfers energy at the rate $P_{av} = 5.0$ MW from a generating plant to a factory. (a) What current i_{rms} is present in the line if the transmission voltage V_{rms} is 120 V? (b) If $V_{rms} = 80$ kV? (c) What is the ratio of the thermal (Joule) energy losses in the line for these two cases? Assume that the power factor $\cos \phi = 1$.
Answer: (a) 42 kA, (b) 63 A, (c) 4.5×10^5.

26. A transformer has 500 primary turns and 10 secondary turns. (a) If $V_{1,rms}$ for the primary is 120 V, what is $V_{2,rms}$ for the secondary, assumed on open

circuit? (b) If the secondary now has a resistive load R of 15Ω, what are $i_{1,\text{rms}}$ and $i_{2,\text{rms}}$? Assume an ideal transformer with $\phi = 0$.

27. The output of a full-wave rectifier (see Fig. 39-8c) is fed into an (ideal) transformer with a step-up ratio of 2:1. Sketch roughly the waveform appearing at the secondary, assumed to be on open circuit.

28. In Fig. 39-10 compare these quantities for (a) switch S open, and (b) switch S closed: $\Phi_B(t)$, $\mathcal{E}(t)$, $V_1(t)$, $V_2(t)$, $V_{1,\text{rms}}$, $V_{2,\text{rms}}$, $i_1(t)$, $i_2(t)$, $i_{1,\text{rms}}$, and $i_{2,\text{rms}}$. Assume an ideal transformer, with $\phi = 0$.

29. In Fig. 39-10 show that $i_1(t)$ in the primary circuit remains unchanged if a resistance $R'[=R(N_1/N_2)^2]$ is connected directly across the generator, the transformer and the secondary circuit being removed. That is,

$$i_1(t) = \frac{\mathcal{E}(t)}{R'}.$$

In this sense we see that a transformer not only "transforms" potential differences and currents but also resistances. In the more general case, in which the secondary load in Fig. 39-10 contains capacitive and inductive elements as well as resistive, we say that a transformer transforms *impedances*. See Problem 31 for an example.

30. In Problem 7 of Chapter 32 we asserted (see Fig. 32-3a) that, for dc circuits, the power dissipated in an external resistance R is a maximum when $R = r$, where r is the internal resistance of the (dc) source of emf. In Fig. 39-14 show that, in the same way, the average power dissipation P_{av} in R is a maximum when $R = r$, in which r is the internal resistance of the ac generator. In the text we have tacitly assumed that $r = 0$; here we assume that $r \neq 0$.

31. *Impedance Matching.* We have seen in Problem 29 that a transformer can serve as a resistance (generally, impedance) transforming device. We saw also, in Problem 30 that (see Fig. 39-14) the transfer of power from an ac generator (internal resistance r) to a resistive load R is a maximum when $R = r$. Suppose that, in Fig. 39-14, $r = 1.0$ kΩ, $R = 10$ Ω, $\omega/2\pi = 60$ Hz, and $\mathcal{E}_{\text{rms}} = 120$ V. Design a transformer, to be interposed between the ac generator and the load, that will insure maximum power transfer to R. Assume an ideal transformer with $\phi = 0$. Such a technique is used, for example, when it is necessary to transfer power efficiently from a (high-impedance) audio amplifier to a (low-impedance) loudspeaker.
Answer: $N_1/N_2 = 10$.

figure 39-14
Problem 30

40
maxwell's equations

In classical mechanics and thermodynamics we sought to identify the smallest, most compact set of equations or laws that would define the subject as completely as possible. In mechanics we found this in Newton's three laws of motion (Sections 5-2, 5-4, and 5-5) and in the associated force laws, such as Newton's law of gravitation (Section 16-2). In thermodynamics we found it in the three laws described in Sections 21-2, 22-7, and 25-4.

In this chapter we seek to do the same thing for electromagnetism, proceeding in several steps. First we display, in Table 40-1, a *tentative* set of such equations. After studying this table, we will conclude from arguments of symmetry that these equations are not yet complete and that there may be (and indeed *is*) a missing term in one of them.

40-1
THE BASIC EQUATIONS OF ELECTROMAGNETISM

Table 40-1
Tentative* basic equations of electromagnetism

Symbol	Name	Equation	Text Reference
I	Gauss's law for electricity	$\epsilon_0 \oint \mathbf{E} \cdot d\mathbf{S} = q$	Eq. 28-6
II	Gauss's law for magnetism	$\oint \mathbf{B} \cdot d\mathbf{S} = 0$	Eq. 37-6
III	Faraday's law of induction	$\oint \mathbf{E} \cdot d\mathbf{l} = -\dfrac{d\Phi_B}{dt}$	Eq. 35-9
IV	Ampère's law	$\oint \mathbf{B} \cdot d\mathbf{l} = \mu_0 i$	Eq. 34-1

* "Tentative" suggests, as we shall see below, that Eq. IV is not yet complete and requires an additional term; see Table 40-2.

The missing term will prove to be no trifling correction but will round out the complete description of electromagnetism and, beyond this, will establish optics as an integral part of electromagnetism. In particular it will allow us to prove that the speed of light c in free space is related to purely electric and magnetic quantities by

$$c = \frac{1}{\sqrt{\epsilon_0 \mu_0}}. \tag{40-1}$$

It will also lead us to the concept of the electromagnetic spectrum, which lies behind the experimental discovery of radio waves.

We have seen how the principle of symmetry permeates physics and how it has often led to new insights or discoveries. For example, (a) if body A attracts body B with a force **F,** then body B attracts body A with a force **−F** (it does), and (b) if there is a negative electron, there may well be a positive electron (there is), etc.

Let us examine Table 40-1 from this point of view. First we say that when we are dealing with symmetry considerations alone (that is, not making quantitative calculations) we can ignore ϵ_0 and μ_0. These constants result from our choice of unit systems and play no role in arguments of symmetry. There are in fact unit systems in which $\epsilon_0 = \mu_0 = 1$.

With this in mind we see that the left sides of the equations in Table 40-1 are completely symmetrical, in pairs. Equations I and II are surface integrals of **E** and **B,** respectively, over closed surfaces. Equations III and IV are line integrals of **E** and **B,** respectively, around closed loops.

On the right side of these equations, things are *not* symmetrical and, in fact, there are two kinds of asymmetries, which we shall discuss separately.

1. The first asymmetry, which is not really the concern of this chapter, deals with the apparent fact that although there are isolated centers of charge (electrons and protons, say) there seem not to be isolated centers of magnetism (magnetic monopoles; see Section 37-1). Thus we account for the "q" on the right of Eq. I and for the "0" on the right of Eq. II. In the same way the term $\mu_0 i (= \mu_0 dq/dt)$ appears on the right of Eq. IV but no similar term (a current of magnetic monopoles) appears on the right of Eq. III. The resolution of *this* asymmetry depends on the not as yet discovered magnetic monopole. Considerations of symmetry have motivated physicists to search for the magnetic monopole in great earnest and in many ways. It is as though nature were hinting and guiding physicists in their explorations.

2. The second asymmetry, with which this chapter deals, sticks out like a sore thumb. On the right side of Eq. III (Faraday's law of induction; see Eq. 35-9) we find the term $-d\Phi_B/dt$ and we interpret this law loosely by saying:

If you change a magnetic field $(d\Phi_B/dt)$, you produce an electric field $(\oint \mathbf{E} \cdot d\mathbf{l})$.

We learned this in Section 35-1 where we showed that if you shove a bar magnet through a closed conducting loop, you do indeed induce an electric field, and thus a current, in that loop.

From the principle of symmetry we are entitled to suspect that the symmetrical relation holds, that is:

If you change an electric field $(d\Phi_E/dt)$, you produce a magnetic field $(\oint \mathbf{B} \cdot d\mathbf{l})$.

This symmetry principle indeed meets the test of experiment and we discuss it fully in the next section. This supposition supplies us with the important "missing" term in Eq. IV in Table 40-1.

Here we discuss in detail the evidence for the supposition of the previous section, namely:

"A changing electric field induces a magnetic field."

Although we will be guided by considerations of symmetry alone, we will also point to direct experimental verification.

Figure 40-1a shows a uniform electric field **E** filling a cylindrical region of space. It might be produced by a circular parallel-plate capacitor, as suggested in Fig. 40-1b. We assume that E is increasing at a steady rate dE/dt, which means that charge must be supplied to the capacitor plates at a steady rate; to supply this charge requires a steady current i into the positive plate and an equal steady current i out of the negative plate.

40-2
INDUCED MAGNETIC FIELDS

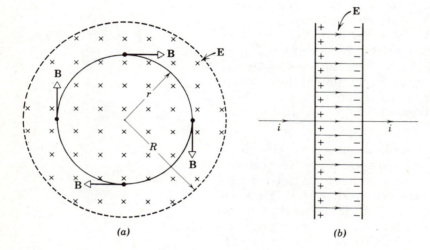

(a)   (b)

figure 40-1
(a) Showing the induced magnetic fields **B** at four points, produced by a changing electric field **E**. The electric field is increasing in magnitude. Compare Fig. 35-10.
(b) Such a changing electric field may be produced by charging a parallel-plate capacitor as shown.

Experiment shows (see Problem 4) that *a magnetic field is set up by this changing electric field.* Figure 40-1a shows **B** for four arbitrary points. Figure 40-1 suggests a beautiful example of the symmetry of nature. A changing *magnetic* field induces an *electric* field (Faraday's law); now we see that a changing *electric* field induces a *magnetic* field.

To describe this new effect quantitatively, we are guided by analogy with Faraday's law of induction,

$$\oint \mathbf{E} \cdot d\mathbf{l} = -\frac{d\Phi_B}{dt},$$ (40-2)

which asserts that an electric field (left term) is produced by a changing magnetic field (right term). For the symmetrical counterpart we might write*

$$\oint \mathbf{B} \cdot d\mathbf{l} = \mu_0 \epsilon_0 \frac{d\Phi_E}{dt}.$$ (40-3)

Equation 40-3 asserts that a magnetic field (left term) can be produced

---

* Our system of units requires that we insert the constants ϵ_0 and μ_0 in Eq. 40-3. In some unit systems they would not appear.

by a changing electric field (right term). Compare carefully Fig. 35-10, which illustrates the production of an electric field by a changing magnetic field, with Fig. 40-1a. In each case the appropriate flux Φ_B or Φ_E is *increasing.* However, experiment shows that the lines of **E** in Fig. 35-10 are *counterclockwise,* whereas those of **B** in Fig. 40-1a are *clockwise.* This difference requires that the minus sign of Eq. 40-2 be omitted from Eq. 40-3.

In Section 34-1 we saw that a magnetic field can also be set up by a current in a wire. We described this quantitatively by Ampère's law:

$$\oint \mathbf{B} \cdot d\mathbf{l} = \mu_0 i,$$

in which i is the conduction current passing through the loop around which the line integral is taken. Thus there are at least two ways of setting up a magnetic field: (a) by a changing electric field and (b) by a current. In general, both possibilities must be allowed for, or*

$$\oint \mathbf{B} \cdot d\mathbf{l} = \mu_0 \epsilon_0 \frac{d\Phi_E}{dt} + \mu_0 i. \qquad (40\text{-}4)$$

Maxwell is responsible for this important generalization of Ampère's law. It is a central and vital contribution, as we have pointed out earlier.

In Chapter 34 we assumed that no changing electric fields were present so that the term $d\Phi_E/dt$ in Eq. 40-4 was zero. In the discussion just given we assumed that there were no conduction currents in the space containing the electric field. Thus the term i in Eq. 40-4 is zero. We see now that each of these situations is a special case.

EXAMPLE 1

A parallel-plate capacitor with circular plates is being charged as in Fig. 40-1. (a) Derive an expression for the induced magnetic field at various radii r. Consider both $r < R$ and $r > R$.

From Eq. 40-3,

$$\oint \mathbf{B} \cdot d\mathbf{l} = \mu_0 \epsilon_0 \frac{d\Phi_E}{dt},$$

we can write, for $r < R$,

$$(B)(2\pi r) = \mu_0 \epsilon_0 \frac{d}{dt} [(E)(\pi r^2)] = \mu_0 \epsilon_0 \pi r^2 \frac{dE}{dt}.$$

Solving for B yields $\qquad B = \frac{1}{2}\mu_0 \epsilon_0 r \frac{dE}{dt} \qquad (r < R).$

For $r > R$, Eq. 40-3 yields

$$(B)(2\pi r) = \mu_0 \epsilon_0 \frac{d}{dt} [(E)(\pi R^2)] = \mu_0 \epsilon_0 \pi R^2 \left(\frac{dE}{dt}\right),$$

or $\qquad B = \frac{\mu_0 \epsilon_0 R^2}{2r} \frac{dE}{dt} \qquad (r > R).$

---

* Actually, there is a third way of setting up a magnetic field, by the use of magnetized bodies. In Section 37-8 we saw that this could be accounted for by inserting a *magnetizing current* term i_M on the right side of Ampère's law. This law would then read, in its full generality,

$$\oint \mathbf{B} \cdot d\mathbf{l} = \mu_0 \epsilon_0 \frac{d\Phi_E}{dt} + \mu_0 i + \mu_0 i_M.$$

In all that follows we assume that no magnetic materials are present so that $i_M = 0$.

(b) Find B at $r = R$ for $dE/dt = 10^{12}$ V/m · s and for $R = 5.0$ cm.
At $r = R$ the two equations for B reduce to the same expression, or

$$B = \tfrac{1}{2}\mu_0\epsilon_0 R \frac{dE}{dt}$$

$$= (\tfrac{1}{2})(4\pi \times 10^{-7} \text{ T} \cdot \text{m/A})(8.9 \times 10^{-12} \text{ C}^2/\text{N} \cdot \text{m}^2)(5.0 \times 10^{-2} \text{ m})(10^{12} \text{ V/m} \cdot \text{s})$$

$$= 2.8 \times 10^{-7} \text{ T} = 0.0028 \text{ gauss.}$$

This shows that the induced magnetic fields in this example are so small that they can scarcely be measured with simple apparatus, in sharp contrast to in-duced *electric* fields (Faraday's law), which can be demonstrated easily. This experimental difference is in part due to the fact that induced emfs can easily be multiplied by using a coil of many turns. No technique of comparable sim-plicity exists for magnetic fields. In experiments involving oscillations at very high frequencies dE/dt above can be very large, resulting in significantly larger values of the induced magnetic field.

40-3 DISPLACEMENT CURRENT

Equation 40-4 shows that the term $\epsilon_0 \, d\Phi_E/dt$ has the dimensions of a current. Even though no motion of charge is involved, there are advan-tages in giving this term the name *displacement* current*. Thus we can say that a magnetic field can be set up either by a conduction current i or by a displacement current i_d ($=\epsilon_0 d\Phi_E/dt$), and we can rewrite Eq. 40-4 as†

$$\oint \mathbf{B} \cdot d\mathbf{l} = \mu_0(i_d + i). \tag{40-5}$$

The concept of displacement current permits us to retain the notion that *current is continuous,* a principle established for steady conduction currents in Section 31-1. In Fig. 40-1b, for example, a current i enters the positive plate and leaves the negative plate. The *conduction* current is *not* continuous across the capacitor gap because no charge is transported across this gap. However, the displacement current i_d in the gap will prove to be exactly i, thus retaining the concept of the continuity of current.

To calculate the displacement current, recall (see Eq. 30-2) that E in the gap is given by

$$E = \frac{q}{\epsilon_0 A}.$$

Differentiating gives
$$\frac{dE}{dt} = \frac{1}{\epsilon_0 A} \frac{dq}{dt} = \frac{1}{\epsilon_0 A} \, i.$$

The displacement current i_d is by definition

$$i_d = \epsilon_0 \frac{d\Phi_E}{dt} = \epsilon_0 \frac{d(EA)}{dt} = \epsilon_0 A \frac{dE}{dt}.$$

Eliminating dE/dt between these two equations leads to

---

* The word "displacement" was introduced for historical reasons that need not concern us here.
† We may write this more generally, taking the presence of magnetic materials into ac-count, as

$$\oint \mathbf{B} \cdot d\mathbf{l} = \mu_0(i_d + i + i_M).$$

See the footnote on p. 883.

$$i_d = (\epsilon_0 A)\left(\frac{1}{\epsilon_0 A} i\right) = i,$$

which shows that the displacement current in the gap is identical with the conduction current in the lead wires.

EXAMPLE 2

What is the displacement current for the situation of Example 1? From the definition of displacement current,

$$i_d = \epsilon_0 \frac{d\Phi_E}{dt} = \epsilon_0 \frac{d}{dt}\left[(E)(\pi R^2)\right] = \epsilon_0 \pi R^2 \frac{dE}{dt}$$

$$= (8.9 \times 10^{-12} \ \text{C}^2/\text{N} \cdot \text{m}^2)(\pi)(5.0 \times 10^{-2} \ \text{m})^2(10^{12} \ \text{V/m} \cdot \text{s})$$

$$= 0.070 \ \text{A}.$$

Even though this displacement current is reasonably large, it produces only a small magnetic field (see Example 1) because it is spread out over a large area. In contrast, the *conduction* current i in the lead wires (also = 0.070 A, because $i = i_d$) can produce magnetic effects close to the (thin) wires that are easily detectable by a compass needle. The discussion of Section 34-2 shows that the magnetic effect is greatest at the surface of the wire. The capacitor of Fig. 40-1 behaves like a "conductor" of radius 5.0 cm, carrying a (displacement) current of 0.070 A. Its largest magnetic effect is at the capacitor edge, that is, at $r = 5.0$ cm, and is given by Eq. 34-4 $(B = \mu_0 i/2\pi r)$.

40-4
MAXWELL'S EQUATIONS

Equation 40-4 completes our presentation of the basic equations of electromagnetism, called *Maxwell's equations.** They are summarized in Table 40-2, which rounds out the "tentative" Table 40-1 by supplying the "missing" term in Eq. IV of that table. All equations of physics that serve, as these do, to correlate experiments in a vast area and to predict new results have a certain beauty about them and can be appreciated, by those who understand them, on an aesthetic level. This is true for Newton's laws of motion, for the laws of thermodynamics, for the theory of relativity, and for the theories of quantum physics. As for Maxwell's equations, the physicist Ludwig Boltzmann (quoting a line from Goethe) wrote, "Was it a god who wrote these lines. . . ." In more recent times J. R. Pierce, in a book chapter entitled 'Maxwell's Wonderful Equations' says, "To anyone who is motivated by anything beyond the most narrowly practical, it is worth while to understand Maxwell's equations simply for the good of his soul." The scope of these equations is remarkable, including as it does the fundamental operating principles of all large-scale electromagnetic devices such as motors, synchrotrons, television, and microwave radar.†

We suggested in Section 40-1 that Maxwell's equations (as they appear in

---

* In the preceding sections we have leaned heavily on symmetry arguments in developing Maxwell's equations as displayed in Table 40-2. Actually Maxwell did not base his work on such arguments. The British physicist Oliver Heaviside (1850–1925) seems to have been the first to point out the symmetry between **E** and **B** that these equations display. See "Maxwell, Displacement Current, and Symmetry" by Alfred M. Bork, *American Journal of Physics*, November, 1963, for an interesting historical account.

† For a recent spectacular example of an application of Maxwell's equations, see "Electromagnetic Flight" by Henry H. Kolm and Richard D. Thornton, *Scientific American*, October 1973. The article states: "The future of high-speed ground transportation may well lie not with wheeled trains but with vehicles that 'fly' a foot or so above a guideway, lifted and propelled by electromagnetic forces."

Table 40-2
The basic equations of electromagnetism (Maxwell's equations) *

Number	Name	Equation	Describes	Crucial Experiment	General Text Reference
I	Gauss's law for electricity	$\epsilon_0 \oint \mathbf{E} \cdot d\mathbf{S} = q$	Charge and the electric field	1. Like charges repel and unlike charges attract, as the inverse square of their separation. 1 . A charge on an insulated conductor moves to its outer surface.	Chapter 28
II	Gauss's law for magnetism	$\oint \mathbf{B} \cdot d\mathbf{S} = 0$	The magnetic field	2. It has thus far not been possible to verify the existence of a magnetic monopole.	Section 37-2
III	Faraday's law of induction	$\oint \mathbf{E} \cdot d\mathbf{l} = -\dfrac{d\Phi_B}{dt}$	The electrical effect of a changing magnetic field	4. A bar magnet, thrust through a closed loop of wire, will set up a current in the loop.	Chapter 35
IV	Ampère's law (as extended by Maxwell)	$\oint \mathbf{B} \cdot d\mathbf{l}$ $= \mu_0 \left(\epsilon_0 \dfrac{d\Phi_E}{dt} + i \right)$	The magnetic effect of a changing electric field or of a current	3. The speed of light can be calculated from purely electromagnetic measurements. 3'. A current in a wire sets up a magnetic field near the wire.	Section 41-8 Chapter 34

* Written on the assumption that no dielectric or magnetic material is present.

Table 40-2) bear the same relationship to electromagnetism that Newton's laws of motion do to mechanics. There is, however, an important difference. Einstein presented his special theory of relativity (see Supplementary Topic V) in 1905, roughly 200 years after Newton's laws appeared and about 40 years after Maxwell's equations. As it turns out, Newton's laws had to be drastically modified in cases in which the relative speeds approached that of light. However, *no changes whatever were required in Maxwell's equations.* In the language of Supplementary Topic V we say that "Maxwell's equations are invariant under a Lorentz transformation but Newton's laws of motion are not."

40-5 MAXWELL'S EQUATIONS AND CAVITY OSCILLATIONS

Now that we have identified Maxwell's equations as the basic equations of electromagnetism, we want to test them in several situations. In this section we address ourselves to the electromagnetic cavity oscillations of Section 38-5 (see Fig. 38-6). In Chapter 41 we will show that Maxwell's equations predict the existence of electromagnetic waves, that visible light is such a wave, and that the speed c of electromagnetic radiation (light) is indeed given by Eq. 40-1.

A completely formal treatment of the cavity oscillations of Fig. 38-6, which is beyond our scope here, would start from Maxwell's equations and would end with mathematical expressions for the variation of **B** and **E** with time and with position in the cavity for all modes of oscillation of the cavity. We confine ourselves to the fundamental mode only, shown in Fig. 38-6, for which we *postulated* the variations of **B** and **E** given in that figure; we will show that these postulated fields are completely consistent with Maxwell's equations.

Figure 40-2 presents two views of the cavity of Fig. 38-6d, in which both electric and magnetic fields are present. Study of Fig. 38-6 reveals that **B** is *decreasing* in magnitude and **E** is *increasing* at this phase of the cycle of oscillation. Let us apply Faraday's law,

$$\oint \mathbf{E} \cdot d\mathbf{l} = -\frac{d\Phi_B}{dt},$$

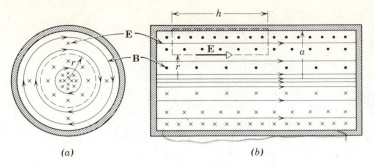

figure 40-2
Two cross sections of an electromagnetic resonant cavity at a phase of oscillation corresponding to Fig. 38-6d. (a) The dashed circle is a path suitable for applying Ampère's law. (b) The dashed rectangle is a path suitable for applying Faraday's law.

(a)   (b)

to the rectangle of dimensions h and $a - r$. There is a definite flux Φ_B through the rectangular path in question, and this flux is decreasing with time because **B** is decreasing. The line integral above is

$$\oint \mathbf{E} \cdot d\mathbf{l} = hE(r),$$

in which $E(r)$ is the value of E at a radius r from the axis of the cavity.

Note that **E** equals zero for the upper leg of the integration path (which lies in the cavity wall) and that **E** and $d\mathbf{l}$ are at right angles on the two side legs. Combining these equations yields

$$E(r) = -\frac{1}{h}\frac{d\Phi_B}{dt}. \qquad (40\text{-}6)$$

Equation 40-6 shows that $E(r)$ depends on the rate at which Φ_B through the path shown is changing with time and that it has its maximum value when $d\Phi_B/dt$ is a maximum. This occurs when **B** is zero, that is, when **B** is changing its direction; recall that a sine or cosine is changing most rapidly, that is, it has the steepest tangent, at the instant it crosses the axis between positive and negative values. Thus the electric field pattern in the cavity will have its *maximum* value when the magnetic field is *zero* everywhere, consistent with Figs. 38-6a and 38-6e and with the concept of the interchange of energy between electric and magnetic fields. You can show, by applying Lenz's law, that the electric field in Fig. 40-2b indeed points to the *right*, as shown, if the magnetic field is *decreasing*.

Figure 40-2a shows an end view of the cavity; the electric lines of force are entering the page at right angles to the page and the magnetic lines form clockwise circles. Let us apply Ampère's law in the form

$$\oint \mathbf{B} \cdot d\mathbf{l} = \mu_0\epsilon_0 \frac{d\Phi_E}{dt} + \mu_0 i, \qquad (40\text{-}4)$$

to the circular path of radius r shown in the figure. No charge is transported through the ring so that the conduction current i in Eq. 40-4 is zero. The line integral on the left is $(B)(2\pi r)$ so that the equation reduces to

$$B(r) = \frac{\mu_0\epsilon_0}{2\pi r}\frac{d\Phi_E}{dt}. \qquad (40\text{-}7)$$

Equation 40-7 shows that the magnetic field $B(r)$ is proportional to the rate at which the electric flux Φ_E through the ring is changing with time. The field $B(r)$ has its maximum value when $d\Phi_E/dt$ is at its maximum; this occurs when $\mathbf{E} = 0$, that is, when **E** is reversing its direction. Thus we see that **B** has its *maximum* value when **E** is *zero* for all points in the cavity. This is consistent with Figs. 38-6c and 38-6g and with the concept of the interchange of energy between electric and magnetic forms. A comparison with Fig. 40-1a, which like Fig. 40-2a, corresponds to an increasing electric field, shows that the lines of **B** are indeed clockwise, as viewed along the direction of the electric field.

Comparison of Eqs. 40-6 and 40-7 suggests the complete interdependence of **B** and **E** in the cavity. As the magnetic field changes with time, it induces the electric field in a way described by Faraday's law. The electric field, which also changes with time, induces the magnetic field in a way described by Maxwell's extension of Ampère's law. The oscillations, once established, sustain each other and would continue indefinitely were it not for losses due to thermal energy in the cavity walls or leakage of energy from openings that might be present in the walls. In Chapter 41 we show that a similar interplay of **B** and **E** occurs not only in standing electromagnetic waves in cavities but also in traveling electromagnetic waves, such as radio waves or visible light.

In Fig. 40-2 analyze the currents (both conduction and displacement) that occur in the cavity (both in its conducting walls and within its volume). Show the relationship between these currents and the electric and magnetic fields and also show that, considering both conduction and displacement currents together, it is reasonable to conclude that current is continuous around closed loops.

EXAMPLE 3

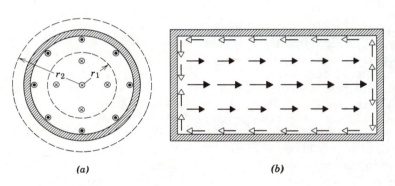

(a)   (b)

figure 40-3
The cavity of Fig. 40-2, showing (a) the conduction current coming up the walls and displacement current going down the cavity volume and (b) the displacement current (black arrowheads) in the volume of the cavity and the conduction currents (white arrowheads) at the walls. The arrows in each case represent current densities. Note the continuity of current (conduction + displacement), that is, it is possible to form closed current loops.

Figure 40-3 shows two views of the cavity, at an instant corresponding to that of Fig. 40-2. For simplicity, we do not show the **E** and **B** fields; the arrows represent currents. Because E is increasing in Figs. 40-2 and 40-3, the positive charge on the left end cap must be increasing. Thus there must be conduction currents in the walls pointing from right to left in Fig. 40-3b. These currents are also shown by the dots (representing the tips of arrows) near the cavity walls in Fig. 40-3a.

Bearing in mind that $\epsilon_0 \, d\Phi_E/dt$ is a displacement current, we can write Eq. 40-7 as

$$B(r) = \frac{\mu_0}{2\pi r}\left(\epsilon_0 \frac{d\Phi_E}{dt}\right) = \frac{\mu_0}{2\pi r} i_d.$$

This equation stresses that **B** in the cavity is associated with a displacement current; compare Eq. 34-4. Applying the right-hand rule in Fig. 40-2a shows that the displacement current i_d must be directed into the plane of the figure if it is to be associated with the clockwise lines of **B** that are present.

The displacement current is represented in Fig. 40-3b by the arrows that point to the right and in Fig. 40-3a by the crosses that represent arrows entering the page. Study of Fig. 40-3 shows that the current is continuous, being directed up the walls as a conduction current and then back down through the volume of the cavity as a displacement current. If we apply Ampère's law as extended by Maxwell,

$$\oint \mathbf{B} \cdot d\mathbf{l} = \mu_0(i_d + i), \qquad (40\text{-}8)$$

to the circular path of radius r_1 in Fig. 40-3a, we see that **B** at that path is due entirely to the displacement current, the conduction current i *within the path* being zero.

For the path of radius r_2, the *net* current enclosed is zero because the conduction current in the walls is exactly equal and opposite to the displacement current in the cavity volume. Since i equals i_d in magnitude, but is oppositely directed, it follows from Eq. 40-8 that B must be zero for all points outside the cavity, in agreement with observation.

questions

1. In your own words explain why Eq. III of Table 40-1 can be interpreted by saying, "A changing magnetic field can generate an electric field."

2. If (as is true) there are unit systems in which ϵ_0 and μ_0 do not appear, how can Eq. 40-1 be true?

3. If a uniform flux Φ_E through a plane circular ring decreases with time, is the induced magnetic field (as viewed along the direction of **E**) clockwise or counterclockwise? Does it make any difference (a) whether the electric field is uniform or (b) whether the plane of the ring is perpendicular to the electric field?

4. Compare Tables 40-1 and 40-2. Is it enough to rely on the principle of symmetry alone or do we really need experimental verification for the "missing" term in Eq. IV?

5. Why is it so easy to show that "a changing magnetic field produces an electric field" but so hard to show in a simple way that "a changing electric field produces a magnetic field"?

6. In Fig. 40-1a consider a circle with $r > R$. How can a magnetic field be induced around this circle, as Example 1 shows? After all, there is no electric field at the location of this circle and $dE/dt = 0$ here.

7. In Fig. 40-1a, **E** is into the figure and is increasing in magnitude. Find the direction of **B** if (a) **E** is into the figure and decreasing, (b) **E** is out of the figure and increasing, (c) **E** is out of the figure and decreasing, and (d) **E** remains constant.

8. In Fig. 38-1c a displacement current is needed to maintain continuity of current in the capacitor. How can one exist, considering that there is no charge on the capacitor?

9. At what parts of the cycle will (a) the conduction current and (b) the displacement current in the cavity of Fig. 38-6 be zero?

10. In Figs. 40-1a and 40-1b what is the direction of the displacement current i_d? In this same figure, can you find a rule relating the directions of **B** and **E**?

11. What advantages are there in calling the term $\epsilon_0 \, d\Phi_E/dt$ in Eq. IV, Table 40-2 a displacement *current*?

12. Why are the magnetic effects of conduction currents in wires so easy to detect but the magnetic effects of displacement currents in capacitors so hard to detect?

13. Discuss the time variation during one complete cycle of the charges that appear at various points on the inner walls of the oscillating electromagnetic cavity of Fig. 38-6.

14. Would you expect that the arrangement of the magnetic and electric fields in Fig. 40-2 is the only possible arrangement? If there are other arrangements, would you expect them to have higher or lower frequencies than those shown in Fig. 40-2?

15. In connection with Fig. 40-3, in what sense can the end caps be considered as capacitor plates? In what sense can the cylindrical walls be considered as an inductor? (*Note:* Figure 40-3 is clearly a case of distributed elements but there must be a smooth transition between distributed and lumped elements.)

16. (a) In Fig. 40-2a is it possible to apply Faraday's law usefully to the dashed circle? (b) Is it possible to apply Ampère's law usefully to the dashed rectangle of Fig. 40-2b? Discuss.

problems

SECTION 40-2

1. (a) Convince yourself that, for the conditions of Fig. 40-1, the induced magnetic field is greatest at the edge of the capacitor. (b) Make a rough, non-numerical, plot of $B(r)$ from $r = 0$ to $r \gg R$.

2. You are given a 1.0-μF parallel-plate capacitor. How would you establish an (instantaneous) displacement current of 1.0 A in the space between its plates?

3. Prove that the displacement current in a parallel-plate capacitor can be written as

$$i_d = C \frac{dV}{dt}.$$

4. In 1929 M. R. Van Cauwenerghe succeeded in measuring directly, for the first time, the displacement current i_d between the plates of a parallel-plate capacitor to which an alternating potential difference was applied, as suggested by Fig. 40-1. He used circular plates whose effective radius was 40 cm and whose capacitance was 1.0×10^{-10} F. The applied potential difference had a maximum value V_m of 174 kV at a frequency of 50 Hz. (a) What maximum displacement current is present between the plates? (See Problem 3.) (b) Why is the applied potential difference chosen to be as high as it is? [The delicacy of these measurements is such that they were only performed in a direct manner more than 60 years after Maxwell enunciated the concept of displacement current! The reference is *Journal de Physique*, No. 8, p. 303, (1929)].

SECTION 40-3

5. Figure 40-4 shows the plates P_1 and P_2 of a circular parallel-plate capacitor of radius R. They are connected as shown to long straight wires in which a constant conduction current i exists. A_1, A_2, and A_3 are hypothetical circles of radius r, two of them outside the capacitor and one between the plates.

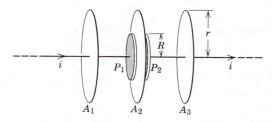

figure 40-4
Problem 5

Write expressions for the magnetic field B at the circumference of each of these circles.

Answer: $\frac{\mu_0 i}{2\pi r}$, in all cases.

6. In Example 1 show that the *displacement current density* j_d is given, for $r < R$, by

$$j_d = \epsilon_0 \frac{dE}{dt}.$$

7. A parallel-plate capacitor has square plates 1.00 m on a side as in Fig. 40-5. There is a charging current of 2.00 A flowing into (and out of) the capacitor. (a) What is the displacement current through the region between the plates?

Edge view

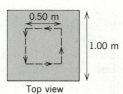

Top view

figure 40-5
Problem 7

(b) What is dE/dt in this region? (c) What is the displacement current through the square dashed path between the plates? (d) What is $\int \mathbf{B} \cdot d\mathbf{l}$ around this square dashed path?
Answer: (a) 2.00 A. (b) 2.3×10^{11} V/m·s. (c) 0.50 A. (d) 6.3×10^{-7} T·m.

8. The capacitor in Fig. 40-6 consisting of two circular plates with area $A = 0.10$ m² is connected to a source of potential $\varepsilon = \varepsilon_m \sin \omega t$, where $\varepsilon_m = 200$ V and $\omega = 100$ rad/s. The maximum value of the displacement current is $i_d = 8.9 \times 10^{-6}$ A. Neglect fringing of the electric field at the edges of the plates. (a) What is the maximum value of the current i? (b) What is the maximum value of $d\Phi_E/dt$, where Φ_E is the electric flux through the region between the plates? (c) What is the separation d between the plates? (d) Find the maximum value of the magnitude of $\mathbf{B}$ between the plates at a distance $R = 0.10$ m from the center.

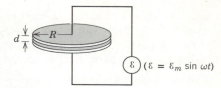

figure 40-6
Problem 8

9. In Example 1 how does the displacement current through a hypothetical concentric circular loop of radius r vary with r? Consider both (a) $r < R$ and (b) $r > R$. *Answer:* (a) $\pi\epsilon_0 r^2(dE/dt)$. (b) $\pi\epsilon_0 R^2(dE/dt)$.

10. A parallel-plate capacitor with circular plates 20.0 cm in diameter is being charged as in Fig. 40-1a. The displacement current density throughout the region is uniform, into the paper in the diagram, and has a value of 20.0 A/m². (a) Calculate the magnetic field B at a distance $R = 5.0$ cm from the axis of symmetry of the region. (b) Calculate dE/dt in this region.

11. Identify the Maxwell equation that is equivalent to or includes: (a) Electric lines of force end only on electric charges. (b) The displacement current. (c) Under static conditions, there cannot be any charge inside a conductor. (d) A changing electric field must be accompanied by a magnetic field. (e) The net magnetic flux through a closed surface is always zero. (f) A changing magnetic field must be accompanied by an electric field. (g) Magnetic flux lines have no ends. (h) The net electric flux through a closed surface is proportional to the total charge inside. (i) An electric charge is always accompanied by an electric field. (j) There are no magnetic monopoles. (k) An electric current is always accompanied by a magnetic field. (l) Coulomb's law. (m) The electrostatic field is conservative.
Answer: See Table 40-2. (a) I. (b) IV. (c) I. (d) IV. (e) II. (f) III. (g) II. (h) I. (i) I. (j) II. (k) IV. (l) I. (m) III.

12. Collect and tabulate expressions for the following four quantities, considering both $r < R$ and $r > R$. Copy down the derivations side by side and study them as interesting applications of Maxwell's equations to problems having cylindrical symmetry. (a) $B(r)$ for a current i in a long wire of radius R (see Section 34-2). (b) $E(r)$ for a long uniform cylinder of charge of radius R (see Section 28-8; also Problem 27, Chapter 28). (c) $B(r)$ for a parallel-plate capacitor, with circular plates of radius R, in which E is changing at a constant rate (see Section 40-2). (d) $E(r)$ for a cylindrical region of radius R in which a uniform magnetic field B is changing at a constant rate (see Section 35-5).

13. A long cylindrical conducting rod with radius a is centered on the x-axis as shown in Fig. 40-7. A narrow saw cut is made in the rod at $x = b$. A conduction current i, increasing with time and given by $i = \alpha t$, flows toward the right in the rod; α is a (positive) proportionality constant. At $t = 0$ there is no charge on the cut faces near $x = b$. (a) Find the magnitude of the charge on these faces, as a function of time. (b) Use Eq. I in Table 40-2 to find E in the gap as a function of time. (c) Sketch the lines of $\mathbf{B}$ for $r < a$, where r is the distance from the x-axis. (d) Use Eq. IV in Table 40-2 to find $B(r)$ in the gap for $r < a$. (e) Compare the above answer with $B(r)$ in the *rod* for $r < a$.
Answer: (a) $\frac{1}{2}\alpha t^2$. (b) $\alpha t^2/2\pi\epsilon_0 a^2$. (c) As you sight along the axis of the rod from left to right in Fig. 40-7 the lines of $\mathbf{B}$ form clockwise circles, both in the rod and in the gap. (d) $\mu_0\alpha t r/2\pi a^2$. (e) Same as (d).

figure 40-7
Problem 13

14. Using the definitions of flux Φ, volume charge density ρ, and current density $\mathbf{j}$, write the four Maxwell equations of Table 40-2 in such a manner that all the fluxes, currents, and charges appear as volume or surface integrals.

SECTION 40-4

15. Assume that the existence of magnetic monopoles is firmly established by experiment. (a) How would you modify the equations of Table 40-2? Let q_m be the expression for the strength of the presumed magnetic monopole, analogous to the basic electric charge e. (b) What SI units would q_m have?

    Answer: (a) $\oint \mathbf{B} \cdot d\mathbf{S} = q_m$, $\oint \mathbf{E} \cdot d\mathbf{l} = -\dfrac{d\Phi_B}{dt} - \dfrac{dqm}{dt}$; other equations are unchanged. (b) $V \cdot s$.

16. *A self-consistency property of two of the Maxwell equations* (numbers III and IV in Table 40-2). Two adjacent closed paths *abcda* and *efcbe* share the common edge *bc* as shown in Fig. 40-8. (a) We may apply $\oint \mathbf{E} \cdot d\mathbf{l} = -d\Phi_B/dt$ to each of these closed paths separately. Show that from this alone, $\oint \mathbf{E} \cdot d\mathbf{l} = -d\Phi_B/dt$ is *automatically* satisfied for the composite closed path *abefcda*. (b) Repeat using $\dfrac{1}{\mu_0} \oint \mathbf{B} \cdot d\mathbf{l} = i + \epsilon_0 \dfrac{d\Phi_E}{dt}$.

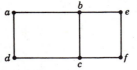

figure 40-8
Problem 16

17. *A self-consistency property of two of the Maxwell equations* (numbers I and II in Table 40-2). Two adjacent parallelepipeds share a common face as shown in Fig. 40-9. (a) We may apply $\epsilon_0 \oint \mathbf{E} \cdot d\mathbf{S} = q$ to each of the two closed surfaces separately. Show that, from this alone, it follows that $\epsilon_0 \oint \mathbf{E} \cdot d\mathbf{S} = q$ is *automatically* satisfied for the composite closed surface. (b) Repeat using $\oint \mathbf{B} \cdot d\mathbf{S} = 0$.

figure 40-9
Problem 17

SECTION 40-5

18. A cylindrical electromagnetic cavity 5.0 cm in diameter and 7.0 cm long is oscillating in the mode shown in Fig. 38-6. (a) Assume that, for points on the axis of the cavity, $E_m = 10^4$ V/m. For such axial points what is the maximum rate $(dE/dt)_m$ at which E changes? (b) Assume that the average value of $(dE/dt)_m$, for all points over a cross section of a cavity, is about one-half the value found above for axial points. On this assumption, what is the maximum value of B at the cylindrical surface of the cavity?

19. in microscopic terms the principle of continuity of current may be expressed as

$$\oint (\mathbf{j} + \mathbf{j}_d) \cdot d\mathbf{S} = 0,$$

in which $\mathbf{j}$ is the conduction current density and $\mathbf{j}_d$ is the displacement current density. The integral is to be taken over any closed surface; the equation essentially says that whatever current flows into the enclosed volume must also flow out. (a) Apply this equation to the surface shown by the dashed lines in Fig. 40-10 shortly after switch S is closed. (b) Apply it to various surfaces that may be drawn in the cavity of Fig. 40-3, including some that cut the cavity walls.

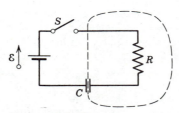

figure 40-10
Problem 19

41
electromagnetic
waves

In Chapter 40 we stated without proof that Maxwell's equations predict the existence of electromagnetic waves whose speed in a vacuum is given by Eq. 40-1, or*

$$c = \frac{1}{\sqrt{\mu_0 \epsilon_0}}.$$ (40-1)

We have called c the "speed of light" but, because neither λ nor ν (recall that $c = \lambda\nu$) appears in Eq. 40-1, c is the speed of electromagnetic waves in general, no matter what their wavelengths (λ) or their frequencies (ν).

We can classify waves as standing waves or as traveling waves. In Section 40-5 (see Figs. 40-2 and 40-3) we studied *standing* electromagnetic waves in an electromagnetic resonant cavity. In this chapter we study *traveling* electromagnetic waves, either confined within the conducting walls of a transmission line (Sections 41-4, 5, 6) or in free space (Sections 41-3, 7, 8).

Figure 41-1 suggests the range of the electromagnetic spectrum as we now know it.† From Maxwell's equations we conclude that all these waves have the same nature and speed and that they differ only in frequency, and thus in wavelength. The names attached to the various regions of the spectrum are identified with the experimental techniques

41-1
INTRODUCTION

41-2
THE ELECTROMAGNETIC SPECTRUM

---

* See Section 41-8 and Supplementary Topic VI for a proof of this relationship.

† *Spectrum* is a Latin word that means "ghost" or "apparition." It was first used in this way by Isaac Newton in 1671 to describe the wavering, rainbowlike image formed on the wall of a darkened room when he held a prism in the path of a beam of sunlight entering through a small hole in his window shade.

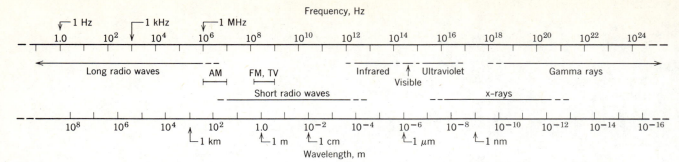

figure 41-1
The electromagnetic spectrum. Note that both scales are logarithmic. The terms μm and nm are *micrometer* and *nanometer,* respectively; both are accented on the first syllable.

for producing and detecting the waves in question. For the AM and the FM-TV bands the frequency ranges are also a matter of legal definition and are sharply defined.

There are no gaps in the spectrum. For example, we can produce electromagnetic waves with $\nu \cong 3 \times 10^{11}$ Hz, either by microwave techniques (microwave oscillators) or by infrared techniques (heated sources). Also, there are no realized upper or lower limits to the frequency or wavelength scales. As an example at one end of the spectrum, electromagnetic waves with $\nu \cong 10^{-2}$ Hz, which corresponds to a period of about 100 s and to a wavelength of about 5000 earth radii, have been detected at the earth's surface.*

It is hard to realize the extent to which we are bathed in electromagnetic waves. The sun is our predominant source, in the sense that its radiations define the environment to which we as a species have adapted.

Let us also consider earth-originated sources of electromagnetic radiation. We are criss-crossed by radio and TV signals. Microwaves from radar systems, from telephone relay systems, and so forth, may reach us. There are electromagnetic waves from light bulbs, from heated engine blocks in automobiles, from X-ray machines, from fireflies, from lightning flashes, from γ-radiation from the radioactive materials in the earth, and so on.

Many electromagnetic waves reach us from extraterrestrial sources and in fact essentially all that we know about the universe comes to us in this way. Electromagnetic waves from outside the earth in the visual range have, of course, been observed since the dawn of the human race. Three technological developments have greatly extended our horizons as far as the study of electromagnetic waves from space is concerned. They are:

The Telescope (see Section 44-6)

This instrument was first used for astronomical observations by Galileo in 1610. With it he (a) discovered the mountains and craters on the moon, which had previously been assumed to be a perfectly spherical body; (b) discovered that the Milky Way was composed of multitudes of individual stars; (c) discovered the four innermost satellites of Jupiter, revealing a model of the Copernican solar system; (d) observed the phases of the planet Venus, an important support for the Copernican theory of the solar system; (e) shed some light on the distinctive character of Saturn (his telescope could not resolve the rings of Saturn but at least he claimed it to differ from other planets) and (f) made a step forward in understanding sunspots. It would be hard to claim, for a single person and in so short a span of years, such major discoveries with astro-

* See "The Longest Electromagnetic Waves" by James R. Heirtzler, *Scientific American,* March 1962.

nomical telescopes. Ground-based optical astronomy reaches its present potential with the construction of the 6.0m (= 236 in.) reflecting telescope in the Soviet Union.*

In October, 1957 the USSR satellite Sputnik I (mass = 83 kg) orbited the earth. From this beginning has developed the major space effort with which we are all familiar. The UHURU satellite, launched by the United States in 1970, initiated the study of X-ray emmision from extraterrestrial objects.† Since that time many other satellites have studied electromagnetic waves from space in this and other regions of the electromagnetic spectrum.

Orbiting Satellites and Other Spacecraft

A Bell Laboratories engineer, Karl G. Jansky,‡ was investigating electromagnetic disturbances to transoceanic telephone traffic when he realized, in 1931, that a source of extraterrestrial signals at radio frequencies was present. This was the beginning of the science of radioastronomy,

The Development of Radioastronomy

figure 41-2
One of eight similar components of the 5-km radio telescope array at the Mullard Radio Astronomy Observatory near Cambridge, England. At a wavelength of 6 cm this array has a resolving power of 2 seconds of arc, comparable with that of large optical telescopes. See "Radio Astronomy and Cosmology" by John B. Irwin, *Sky and Telescope,* December 1976.

* See "The Soviet 6-meter Altazimuth Reflector" by Bazart K. Ioannisiani, *Sky and Telescope,* November 1977.
† See "Some Recent Advances in X-Ray Astronomy" by Alan P. Lightman, *Sky and Telescope,* October 1976. Satellites made it possible to study electromagnetic radiations from space which otherwise would be absorbed by the earth's atmosphere.
‡ The unit of energy flux in radioastronomy is called the *jansky* in his honor.

based on high performance radiotelescopes such as the 1000-ft diameter parabolic reflector at Arecibo, Puerto Rico. The radiotelescope array at the Mullard Radio Astronomy Observatory, near Cambridge, England, is another example. This 5-km array, which became operational in 1972, employs four fixed and four movable parabolic reflectors of the type shown in Fig. 41-2.

41-3
ELECTROMAGNETIC WAVES FROM SPACE

In this section we consider just four of the important discoveries about the nature of our universe made by studying electromagnetic waves from space. We restrict ourselves to the radio frequency region of the spectrum.

1. Electromagnetic radiation with $\lambda = 21.1$ cm ($\nu = 1420$ MHz) is emitted by neutral hydrogen atoms* that populate the spaces between the stars that comprise our galaxy.† The density of hydrogen atoms in interstellar space is only about 1 atom/cm³ and the mean time spent by a hydrogen atom capable of radiating before it does so is about 10^7 y. Nevertheless our galaxy is so huge that this electromagnetic radiation is readily detectable.

For hydrogen atoms at rest with respect to the terrestrial detector of their radiations the precise wavelength is 21.1061 cm ($\nu = 1420.406$ MHz). Our galaxy, however, is not a rigid structure, its outer regions move with respect to the inner core. Just as for sound, we can measure the radial velocities of light-emitting sources by the Doppler effect; see Sections 20-7 and 42-5. By observing these frequency shifts we can learn a lot about the structure and the internal motions of our galaxy. In recent years electromagnetic radiation from molecules in the interstellar regions of our galaxy has been detected. We may list formaldehyde ($HCHO$), ammonia (NH_3), and carbon monoxide (CO).

2. Quasi-stellar radio sources (*quasars*) were discovered in 1962. These are optical objects associated normally with large radioemissions. What makes them so interesting is that they have very large "red shifts" in their spectra which, with our present understanding of astrophysics, would lead us to believe that they are at enormous distances from us, well outside our own galaxy. On the other hand, if they are really so far away, they must have enormous energy outputs for which we cannot account. Their nature is under active study today.

3. The primeval fireball radiation.‡ Most astrophysicists today believe that the universe originated about 7×10^9 years ago in the explosion of a very highly concentrated matter-radiation complex. This is the so-called "Big Bang" theory of the origin of the universe. The debris from this explosion expanded outward uniformly in all directions (hence the expanding universe concept) and the matter gradually condensed to form galaxies, stars, and planets as we know them today.

In 1965 the Princeton University physicist Robert Dicke recognized that, if the remnants of the electromagnetic radiation from this primeval fireball, expected to be detectable by radiotelescopes, could be found, it would provide impressive evidence in support of the Big-Bang theory.

* See "Radio Emission from Interstellar Neutral Hydrogen," by R. D. Davies, *Contemporary Physics*, August 1961.
† Our galaxy is a disc-shaped cluster of stars about 10^5 light years in diameter. The Milky Way defines the plane of the disc. Our sun is located about halfway from the galactic center to the galactic edge. There are many other galaxies in our universe; every new high-performance telescope reveals more.
‡ See "The Primeval Fireball" by P. J. E. Peebles and D. T. Wilkinson, *Scientific American*, June 1976.

Unknown to Dicke two Bell Laboratory scientists, A. Penzias and R. Wilson, had actually discovered this electromagnetic radiation from space without realizing its cosmological significance. They made their discovery during engineering studies of communication satellite problems. Within a few years the reality of the primeval fireball radiation was well established.

This episode reminds us that Jansky, also of the Bell Laboratories, founded the science of radio astronomy in 1931. In each case a strictly engineering endeavor led to unexpected insights into our understanding of the universe.*

4. In 1968 a group of British radio astronomers at Cambridge University, headed by Antony Hewish, discovered the first of a series of objects that emit strong short bursts of electromagnetic radiation separated by the astronomically incredibly short time period of about one second. These objects were called *pulsars*.†

It is now generally agreed that pulsars are ordinary stars in which, during the process of shrinking under their own attractive gravitational forces, the electrons of the star combine with the protons, leaving a compact ball of neutrons. Typical *neutron stars* (see Problems 4-23 and 16-5) have these properties: (a) radii of about 10 km (the solar radius is about 7×10^5 km), (b) densities of about 10^{16} kg/m³ (the mean solar density is about 10^3 kg/m³), and (c) periods of rotation of about 10^{-2} s (the solar rotation period is about 10^6 s). The observed pulses seem to arise from a radio-emitting spot on the surface of the neutron star. The emitted electromagnetic rotation sweeps by a terrestrial observer as the star rotates; the analogy to the revolving beam from a lighthouse beacon is clear.

41-4
TRANSMISSION LINES

In this section we turn our attention from the passive observation of electromagnetic waves that reach us to the deliberate generation of such waves and their transmission from point to point by a *transmission line*. Figure 41-3 shows one type of line, a *coaxial cable*, its input end being connected to a switch S. For the time being we assume that the cable is infinitely long and that the cable elements have zero resistance. (The Atlantic cable, first laid successfully in 1866, is a coaxial cable, although it is far from resistanceless.)

When switch S is closed on b, the central and the outer conductors are at the same potential. If we then throw the switch to a, a potential difference V suddenly appears between these elements. This potential

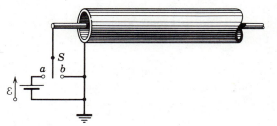

figure 41-3
An electromagnetic signal can be sent along the transmission line (coaxial cable) by throwing switch S from b to a.

---

* See "The Roots of Solid-State Research at Bell Labs" by Lillian Hartmann Hoddeson, *Physics Today*, March 1977, for a fascinating historical account of the interaction of science and technology in a particular industrial laboratory. In 1937 and 1956 Bell Laboratory scientists were awarded Nobel Prizes.
† See "Pulsars" by A. Hewish, *Scientific American*, October 1968, and "The Nature of Pulsars" by J. P. Ostriker, *Scientific American*, January 1971. Hewish received a shared Nobel Prize in 1974 for this discovery.

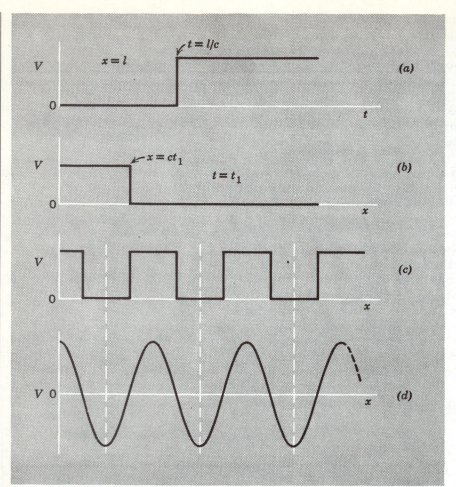

figure 41-4
(*a*) The variation with time of the potential difference between the conductors of a coaxial cable at a distance *l* from the input end. (*b*) An instantaneous "snapshot" of the pulse in the cable at a certain time t_1. (*c*) The waveform if switch *S* in Fig. 41-3 is periodically thrown between *a* and *b*. (*d*) The waveform if we replace switch *S* by an electromagnetic oscillator with a sinusoidal output.

difference does not appear instantaneously all along the line but is propagated with a finite speed *c* that will turn out to be exactly that of light, assuming a resistanceless line. Figure 41-4*a* shows that the potential difference between the conductors at a distance *l* along the line suddenly rises, at a time given by $t = l/c$, from zero to a value determined by the battery emf. We can also consider the variation of *V* with position *x* along the line at a given time t_1 after closing the switch. Figure 41-4*b* shows such an instantaneous "snapshot." It, too, suggests a traveling "wavefront" moving along the line at speed *c*. At $t = t_1$ the signal has not yet reached points where $x > ct_1$.

If we throw switch *S* periodically from *b* to *a* and back again, a wave disturbance like that of Fig. 41-4*c* is propagated. This suggests that if we replace the battery and switch arrangement by an electromagnetic oscillator with a sinusoidal output of frequency ν a wave like that of Fig. 41-4*d* will be propagated.

A traveling wave in a resistanceless transmission line will exhibit a wavelength λ given by

$$\lambda = \frac{c}{\nu}.$$

If the oscillator frequency is 60 Hz, the common commercial power frequency, the wavelength is 5×10^6 m, which is about 3000 miles. At this low frequency traveling waves are not apparent in any line of normal length. By the time the polarity of the oscillator has changed appreciably, the energy fed into the line at the oscillator end has been delivered to the load.

Frequencies in the radio or the microwave range are much higher and the wavelengths correspondingly smaller. Commercial television frequencies as established by the Federal Communications Commission range from 54 to 980 MHz. In terms of wavelength this is a range of 5.6 to 0.31 m. At these wavelengths the patterns of potential difference in the transmission lines used to send television signals across the country can be described aptly as traveling waves. Microwaves, used in radar systems and for communication purposes, have even smaller wavelengths, in the range of about 20 cm to about 0.5 mm.

These considerations suggest another way of viewing the difference between lumped and distributed circuit elements. A system is "distributed" if the wavelength is about the same size as, or less than, the dimensions of the system. If the wavelength is much larger than the dimensions of the system, we are dealing with lumped components. A transmission line 50 m long would be a lumped system for electromagnetic radiation at 60 Hz $(\lambda = 5 \times 10^6$ m$)$ but a distributed system at 100 MHz $(\lambda = 3$ m$)$. In a lumped system the circuit analysis is normally carried out in terms of lumped system parameters such as L, C, and R; in a distributed system the analysis is often carried out in terms of the fields that are set up and the charges and currents that are related to them.

EXAMPLE 1

A potential difference given by

$$V_0 = V_m \sin \omega t$$

is applied between the terminals of a long resistanceless transmission line; the frequency $\nu(=\omega/2\pi)$ is 3×10^9 Hz $(=3$ GHz; see Table 1-2$)$. Write an equation for $V(t)$ at a point P which is 1.5 wavelengths down the line from the oscillator.

The general equation for a wave traveling in the x direction (see Eq. 19-10a) is

$$V = V_m \sin (\omega t - kx),$$

where $k(=2\pi/\lambda)$ is the wave number. At $x = 0$ this gives correctly the time variation of the input terminal potential difference. At $x = 1.5\lambda$ we have

$$V_P = V_m \sin \left[\omega t - \left(\frac{2\pi}{\lambda} \right)(1.5\lambda) \right] = V_m \sin (\omega t - 3\pi)$$

$$= -V_m \sin \omega t.$$

Thus V_P is always equal in magnitude to V_0 but is opposite in sign. What is the wavelength in this example?

41-5
COAXIAL CABLE— FIELDS AND CURRENTS

Figures 41-5a and 41-5b are "snapshots" of the electric and magnetic field configurations in a coaxial cable. The electric field is radial and the magnetic field forms concentric lines about the central conductor. The entire pattern moves along the line, assumed resistanceless, at speed c.

The field patterns in this figure obey the *boundary condition* required for a line that is assumed to be resistanceless, namely, that $\mathbf{E}$ for all points on either conducting surface has no tangential component (see Example 7, Chapter 28). We can find the field patterns mathematically from Maxwell's equations by imposing this requirement. The configuration

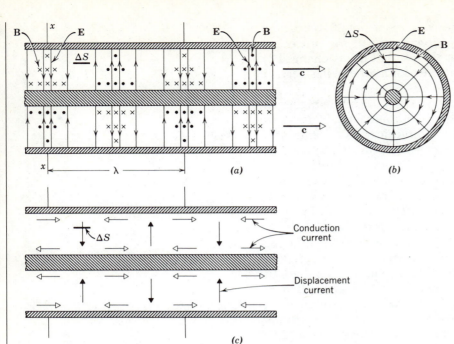

figure 41-5
(a) The electric and magnetic fields in a coaxial cable, showing a wave traveling to the right at speed c.
(b) A cross-sectional view at a plane through xx in (a); the wave is emerging from the page. (c) Conduction currents in the walls (open arrows) and displacement currents in the space between the conductors (filled arrows) associated with the wave in (a); the arrows in each case represent current density vectors.

shown is the simplest of many different wave patterns that can travel along the line. The coaxial cable, unlike the electromagnetic cavity of Fig. 38-6, is not a resonant device. The angular frequency ω of waves that travel along it can be varied continuously, as is the case for all traveling waves, such as transverse waves in a long stretched cord.

Figure 41-5c shows the currents in the cable at the instant corresponding to Figs. 41-5a and 41-5b. The arrows parallel to the cable axis represent conduction currents in the central and the outer conductors. The vertical arrows with filled heads represent displacement currents that exist in the space between the conductors. Note that the conduction current and the displacement current arrows form closed loops, preserving the concept of the continuity of current.

EXAMPLE 2

Verify that the displacement current represented in Fig. 41-5c is consistent with the pattern of **B** and **E** shown in Fig. 41-5a.

Consider a small surface element $\Delta \mathbf{S}$ shown edge-on in Fig. 41-5; it is shown as viewed from above in Fig. 41-6. This hypothetical element is stationary with respect to the cable while the field configuration moves through it at speed c. Figure 41-6a shows the electric lines of force in and near this element. It is clear from symmetry that, at the instant shown, the net flux Φ_E through this area is zero. However, even though Φ_E is zero in magnitude, it is, at this instant, *changing at its most rapid rate*, since **E** at the element ΔS is at the very moment of reversing its direction as the wave moves through. Thus the displacement current, which is given by

$$i_d = \epsilon_0 \frac{d\Phi_E}{dt},$$

also has its maximum value.

Figure 41-6b shows the magnetic field in the vicinity of the element of area. Let us apply the generalized form of Ampère's law,

$$\oint \mathbf{B} \cdot d\mathbf{l} = \mu_0(i + i_d),$$

to the element. The conduction current i is zero since no charge is transported

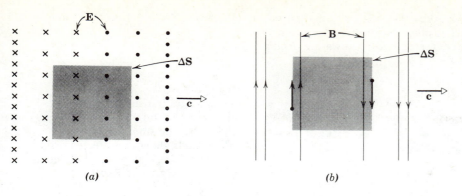

(a)   (b)

figure 41-6
The area element S in Fig. 41-5 enlarged and viewed from above, showing the adjacent (a) electric and (b) magnetic fields.

through ΔS. The displacement current i_d is not zero, having, in fact, its maximum value. Thus, since the right side of this equation $(=\mu_0 i_d)$ does not vanish, the left side must not vanish. Study of Fig. 41-6b shows that $\oint \mathbf{B} \cdot d\mathbf{l}$ around the boundary of this square has, indeed, a nonzero value. Thus the field and displacement current configurations of Fig. 41-5a–c are consistent. We have not discussed the direction of i_d; that is, does it point into the plane of Fig. 41-6, as Fig. 41-5c asserts, or out of it? We leave this as a question. You may be guided by considering the direction of the displacement current in Figs. 40-1a and 40-1b.

EXAMPLE 3

Show that the conduction currents in Fig. 41-5c are appropriately related to the magnetic field pattern.

Figure 41-7 shows a cross section of the cable at a plane through xx in Fig. 41-5a. Let us apply Ampère's law,

$$\oint \mathbf{B} \cdot d\mathbf{l} = \mu_0(i + i_d),$$

to the ring of radius r. The displacement current through this ring is zero, this current being at right angles to the central conductor. The conduction current i in the central conductor, shown by the $\times$'s in the figure, does pass through this ring so that the equation becomes

$$(B)(2\pi r) = \mu_0 i,$$

or

$$B = \frac{\mu_0 i}{2\pi r}.$$

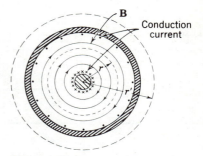

figure 41-7
Example 3. The coaxial cable of Fig. 41-5b, showing the conduction currents in the central and outer conductors. The wave is emerging from the page.

Note that $\mathbf{B}$ is related to the current in the central conductor by the usual right-hand rule and that the expression for B is that found earlier (Eq. 34-4) for a long straight wire carrying a steady current.

If we apply Ampère's law to the large ring of radius r', we must put the *net* conduction current equal to zero because the current in the outer conductor, shown by the dots, is equal and opposite to that in the inner conductor. This means that $\mathbf{B}$ must be zero for points outside the cable, in agreement with experiment.

It is interesting to compare the electromagnetic oscillations in a typical *traveling* wave, such as that of Fig. 41-5, with those in a cavity resonator, such as that of Fig. 38-6. The latter oscillations are an electromagnetic *standing* wave. In a traveling wave $\mathbf{E}$ and $\mathbf{B}$ are in phase, which means that at a given position along the transmission line they reach their maxima at the same time. However, Fig. 38-6 shows that at a given position in a cavity resonator $\mathbf{E}$ and $\mathbf{B}$ reach their maxima one-fourth of a cycle apart; they are 90° out of phase.

A complete analogy exists in mechanical systems. In the acoustic resonator of Fig. 38-5 the time variations of pressure and velocity for the standing acoustic wave are also 90° out of phase, in exact correspondence to the electromagnetic cavity oscillations of Fig. 38-6. The acoustic analogy to a transmission line (Fig. 41-8) would be an infinitely long gas-filled tube, one end being connected to an acoustic oscillator such as a loud-speaker. The entire configuration of Fig. 41-8 moves to the right with speed v. The pressure variations, suggested by the dots, and the instantaneous velocities, suggested by the arrows, are *in phase*, just as are **E** and **B** in the coaxial cable of Fig. 41-5.

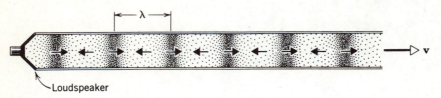

figure 41-8
An acoustic transmission line, showing a sound wave traveling to the right. The small arrows with filled heads show the directed drift velocities for small volume elements of the gas. Compare Fig. 41-5.

It is possible to send electromagnetic waves through a hollow metal pipe that has no central conductor. We assume that the inner walls of such a pipe, or *waveguide* as it is called, are resistanceless and that the cross section is rectangular.

Figure 41-9 shows a typical electric and magnetic field pattern. We imagine that a microwave oscillator is connected to the left end and sends electromagnetic energy down the guide. Figure 41-9(*i*) shows a side view of the guide and Fig. 41-9(*ii*), a top view; Fig. 41-9(*iii*) shows the cross section. As for the coaxial cable, the field patterns are such that **E** has no tangential component for any point on the inner surface of the guide. The fields **E** and **B** are in phase, again like the coaxial cable.

41-6
WAVEGUIDE

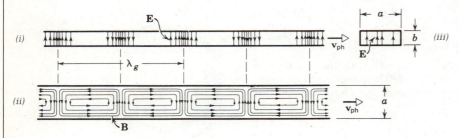

figure 41-9
A waveguide, showing (*i*) a side view of the lines of **E,** (*ii*) a top view of the lines of **B,** and (*iii*) a cross-sectional view of the lines of **E.** In (*iii*) the wave is emerging from the page. For simplicity the lines of **B** are not shown in (*i*) and (*iii*), nor are the lines of **E** shown in (*ii*).

As for all traveling waves, the angular frequency ω of electromagnetic waves traveling down a guide can be varied continuously. In a waveguide of given dimensions, however, there exists, for every mode of transmission, that is, for every pattern of **E** and **B**, a so-called *cutoff frequency* ω_0. A given guide will not transmit waves in a given mode if their frequency is below the cutoff value for that mode in that guide. The field patterns of Fig. 41-9 show the *dominant mode* for a rectangular guide; this is the mode with the lowest cutoff frequency. Given the frequency ω of electromagnetic waves to be transmitted, it is common practice to select a guide whose dimensions are such that ω is larger than the cutoff frequency ω_0 for the dominant mode but smaller than the cutoff frequencies of all other modes. Under these conditions the dominant mode of propagation is the only one possible.

In a (resistanceless) coaxial cable the wave patterns travel at speed c. In the acoustic transmission line of Fig. 41-8 (assumed "resistanceless") the waves also travel at a speed v, which is the same as the propagation speed in an infinite medium. In a waveguide, however, the speed is *not c*. In waveguides we must distinguish between (*a*) the *phase speed* v_{ph}, which is the speed at which the wave patterns of Fig. 41-9 travel, and (*b*) the *group speed* v_{gr}, which is the speed at which electromagnetic energy or information-carrying "signals" travel along the guide. These speeds, which are identical for electromagnetic waves in a coaxial cable and for acoustic waves in a tube, are different for waves in a waveguide.

The phase speed is not directly measurable. The wave pattern is a repetitive structure, and there is no way to distinguish one wave maximum from another. We can observe the waves entering one end of the guide and leaving at the other, but there is no way to identify a particular wave maximum so that we can time its passage down the guide. We can put a "signal" on the wave by increasing the power level of the oscillator for a short time. We can time this power pulse as it passes through the guide, but there is no guarantee that it travels at the same speed as the wave pattern and, indeed, it does not. The speed of such signals or markers is the speed at which *energy* is propagated, that is, the group speed.

From Maxwell's equations we can show that the phase speed and the group speed for the mode of Fig. 41-9 are

$$v_{ph} = \frac{c}{\sqrt{1 - \left(\frac{\lambda}{2a}\right)^2}} \tag{41-1}$$

and

$$v_{gr} = c\sqrt{1 - \left(\frac{\lambda}{2a}\right)^2}, \tag{41-2}$$

in which a is the width of the guide and λ the free-space wavelength. Note that as $a \to \infty$, which corresponds to free-space conditions, $v_{ph} = v_{gr} = c$.

The phase speed v_{ph} is *greater* than the velocity of light, the group speed v_{gr} being correspondingly less. In relativity theory we learn that no speed at which signals or energy travel can be faster than that of light. However, signals or energy *cannot* be transmitted down a guide at the phase speed v_{ph}; they travel with speed v_{gr} which is *always* less than c so there is no conflict with the theory of relativity.

The wavelength λ in Eqs. 41-1 and 41-2 is the wavelength that would be measured for the oscillations in free space, that is,

$$\lambda = \frac{c}{\nu}, \tag{41-3}$$

where c is the speed in free space and ν is the frequency. For waves of a given frequency, the wavelength exhibited in a guide (λ_g) must differ from the free-space wavelength λ because the speed v_{ph} has changed. The so-called *guide wavelength* λ_g is given by

$$\lambda_g = \frac{v_{ph}}{\nu} = \frac{v_{ph}}{c/\lambda} = \lambda \frac{v_{ph}}{c}.$$

From Eq. 41-1 this yields

$$\lambda_g = \frac{\lambda}{\sqrt{1 - \left(\frac{\lambda}{2a}\right)^2}}. \tag{41-4}$$

Thus the guide wavelength, which is the wavelength exhibited by the field patterns in Fig. 41-9, is larger than the free-space wavelength.

EXAMPLE 4

What must be the width a of a rectangular guide such that the energy of electromagnetic radiation whose free-space wavelength is 3.0 cm travels down the guide (a) at 95% of the speed of light? (b) At 50% of the speed of light?

From Eq. 41-2 we have

$$v_{gr} = 0.95c = c\sqrt{1 - \left(\frac{\lambda}{2a}\right)^2}.$$

Solving for a yields $a = 4.8$ cm; repeating for $v_{gr} = 0.50c$ yields $a = 1.7$ cm.

This illustrates the cutoff phenomenon described above. If $\lambda = 2a$, then $v_{gr} = 0$ and energy cannot travel down the guide. For the radiation considered in this example $\lambda = 3.0$ cm, so that the guide must have a width a of *at least* $\frac{1}{2} \times 3.0$ cm $= 1.5$ cm if it is to transmit this wave. The guide whose width we calculated in (a) above can transmit radiations whose free-space wavelength is 2×4.8 cm $= 9.6$ cm *or less*.

The acoustic transmission line of Fig. 41-8 cannot be infinitely long. Its far end may be sealed by a solid cap or left open, or it may have a flange, a horn, or some similar device mounted on it. If the far end is not sealed, energy will escape into the medium beyond. We call this *acoustic radiation*. In general, some energy will also be reflected back down the transmission line. If acoustic radiation is desirable, the designer's task is to fashion a termination (that is, an "acoustic antenna") for the transmission line such that the smallest possible fraction of the incident energy will be reflected back down the line. Such a termination might take the form of a flared horn. Acoustic radiation, of course, requires a medium such as air in order to be propagated.

An electromagnetic transmission line such as a coaxial cable or a waveguide can also be terminated in many ways, and energy can escape from the end of the line into the space beyond. In contrast to sound waves, a physical medium is not required. Thus electromagnetic energy can be radiated from the end of the transmission line, to form a traveling electromagnetic wave in free space.

Figure 41-10 shows an effective termination for a coaxial cable; it consists of two wires arranged as shown and is called an *electric dipole antenna*. The potential difference between the two conductors alternates sinusoidally as the wave reaches them, the effect being that of an electric dipole whose dipole moment **p** varies with time, both in magnitude and direction.

41-7
RADIATION

figure 41-10
An electric dipole antenna on the end of a coaxial cable.

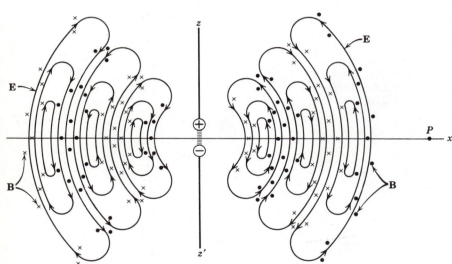

figure 41-11
Showing the fields of **E** and **B** radiated away from an oscillating electric dipole. We show only the fields at distances from the dipole that are large compared with its dimensions. The electromagnetic wave sweeping through distant point *P* is a plane wave, moving in the *x*-direction.

Figure 41-11 shows an electromagnetic wave generated by such an oscillating electric dipole. The figure is a section through a figure of revolution about the dipole axis *zz'* and the wave moves out in any direction from the dipole with speed *c*. The fields shown are those that exist at distances from the dipole which are large compared with the dipole dimensions. We call this the *radiation field* and it is the focus of our concern here. The fields close to the dipole (the *near field*) are more complex but they have no interest for us here because they die out rapidly and we want to study only the fields at large distances.

Figure 41-12 shows eight cyclical "snapshots" of the fields of **E** and **B** sweeping past an observer at point *P* in Fig. 41-11. We assume that *P* is so far distant from the oscillating dipole that the wave fronts passing

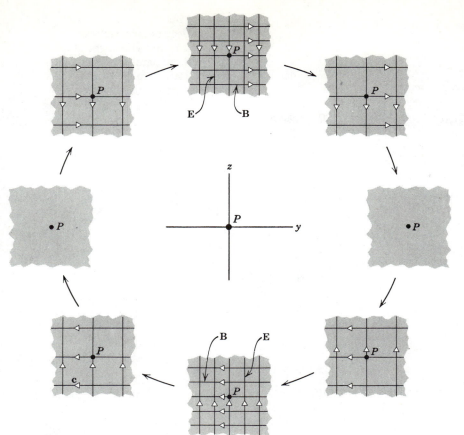

figure 41-12
Showing eight cyclical "snapshots" of the plane electromagnetic wave radiated from the oscillating dipole of Fig. 41-11 through point P. The direction of the wave (the x-direction) is out of the plane of the figure.

through and near P describe a plane wave; see Section 19-2. The speed c of the wave in free space is given by $c = \nu\lambda$, which we can write as

$$c = \frac{\omega}{k}, \qquad (41\text{-}5)$$

where ω, the angular frequency, and k, the wave number, are related to the frequency ν and the wavelength λ by

$$\omega = 2\pi\nu \qquad \text{and} \qquad k = 2\pi/\lambda.$$

41-8
TRAVELING WAVES AND MAXWELL'S EQUATIONS

In earlier sections we have postulated the existence of certain magnetic and electric field distributions, in resonant cavities, coaxial cables, and waveguides, and we have shown that these postulated distributions are consistent with Maxwell's equations, as are the distributions of conduction and displacement currents associated with the fields. If you pursue your studies of electromagnetism, you will learn how to derive mathematical expressions for **E** and **B** by subjecting Maxwell's equations to the boundary conditions appropriate to the problem at hand. In this section we continue our program by showing that the postulated patterns of **E** and **B** for a traveling electromagnetic wave are completely consistent with Maxwell's equations. In doing so, we will be able to show that the speed of such waves in free space is that of visible light and thus that visible light is itself an electromagnetic wave.

If the observer at P in Fig. 41-11 is, as we have assumed, at a considerable distance from the source, the *wavefronts* described by the electric and magnetic fields that reach him (see Fig. 41-12) will be

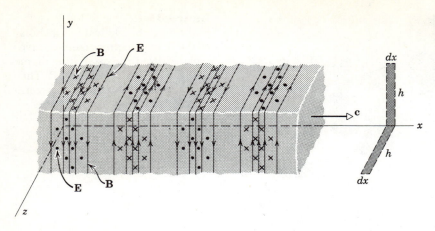

figure 41-13
A plane electromagnetic wave traveling to the right at speed c. Lines of **B** are parallel to the y axis; those of **E** are parallel to the z axis. The shaded rectangles on the right refer to Fig. 41-14.

planes and the wave that moves past him will be a *plane wave*. Figure 41-13 shows a three-dimensional "snapshot" of a plane wave traveling in the x direction. The lines of **E** are parallel to the z axis and those of **B** are parallel to the y axis. The values of **B** and **E** for this wave depend only on x and t (not on y or z). We postulate that they are given in magnitude by

$$B = B_m \sin (kx - \omega t) \tag{41-6}$$

and

$$E = E_m \sin (kx - \omega t). \tag{41-7}$$

Figure 41-14 shows two sections through the three-dimensional diagram of Fig. 41-13. In Fig. 41-14a the plane of the page is the xz plane and in Fig. 41-14b it is the xy plane. Note that, as for the traveling waves in a coaxial cable (Fig. 41-5a) and in a waveguide (Fig. 41-9), **E** and **B** are in phase, that is, at any point through which the wave is moving they reach their maximum values at the same time.

The shaded rectangle of dimensions dx and h in Fig. 41-14a is fixed in space. As the wave passes over it, the magnetic flux Φ_B through the rectangle will change, which will give rise to induced electric fields around

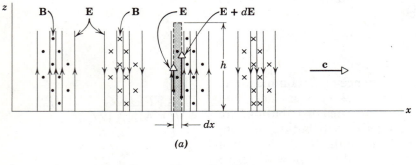

(a)

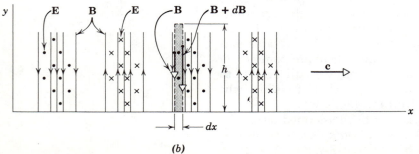

(b)

figure 41-14
The wave of Fig. 41-13 viewed (a) in the xz plane and (b) in the xy plane.

the rectangle, according to Faraday's law of induction. These induced electric fields are, in fact, simply the electric component of the traveling wave.

Let us apply Lenz's law to this induction process. The flux Φ_B for the shaded rectangle of Fig. 41-14a is *decreasing* with time because the wave is moving through the rectangle to the right and a region of weaker magnetic fields is moving into the rectangle. The induced field will act to oppose this change, which means that if we imagine that the boundary of the rectangle is a conducting loop, a *counterclockwise* induced current would appear in it. This current would produce a field of **B** that, within the rectangle, would point out of the page, thus opposing the decrease in Φ_B. There is, of course, no conducting loop, but the net induced electric field **E** does indeed act counterclockwise around the rectangle because $E + dE$, the magnitude of **E** at the right edge of the rectangle, is greater than E, the magnitude of **E** at the left edge. Thus the electric field configuration is entirely consistent with the concept that it is induced by the changing magnetic field.

For a more detailed analysis let us apply Faraday's law of induction, or

$$\oint \mathbf{E} \cdot d\mathbf{l} = -\frac{d\Phi_B}{dt}, \tag{41-8}$$

going counterclockwise around the shaded rectangle of Fig. 41-14a. There is no contribution to the integral from the top or bottom of the rectangle because **E** and $d\mathbf{l}$ are at right angles here. The integral then becomes

$$\oint \mathbf{E} \cdot d\mathbf{l} = [(E + dE)(h)] - [(E)(h)] = dE\ h.$$

The flux Φ_B for the rectangle is

$$\Phi_B = (B)(dx\ h),$$

where B is the magnitude of **B** at the rectangular strip and $dx\ h$ is the area of the strip. Differentiating gives

$$\frac{d\Phi_B}{dt} = h\ dx\ \frac{dB}{dt}.$$

From Eq. 41-8 we then have

$$dE\ h = -h\ dx\ \frac{dB}{dt},$$

or

$$\frac{dE}{dx} = -\frac{dB}{dt}. \tag{41-9}$$

Actually, both B and E are functions of x and t; see Eqs. 41-6 and 41-7. In evaluating dE/dx, we assume that t is constant because Fig. 41-14a is an "instantaneous snapshot." Also, in evaluating dB/dt we assume that x is constant since what is required is the time rate of change of B at a particular place, the strip in Fig. 41-14a. The derivatives under these circumstances are called *partial derivatives*, and a somewhat different notation is used for them; see Example 2, p. 414 and p. 636. In this notation Eq. 41-9 becomes

$$\frac{\partial E}{\partial x} = -\frac{\partial B}{\partial t}. \tag{41-10}$$

The minus sign in this equation is appropriate and necessary, for, although E is increasing with x at the site of the shaded rectangle in Fig. 41-14a, B is decreasing with t. Since $E(x,t)$ and $B(x,t)$ are known (see

Eqs. 41-6 and 41-7), Eq. 41-10 reduces to

$$kE_m \cos (kx - \omega t) = \omega B_m \cos (kx - \omega t),$$

or (see Eq. 41-5)

$$\frac{\omega}{k} = \frac{E_m}{B_m} = c. \qquad (41\text{-}11a)$$

Thus the speed of the wave c is the ratio of the amplitudes of the electric and the magnetic components of the wave. From Eqs. 41-6 and 41-7 we see that the ratio of amplitudes is the same as the ratio of instantaneous values, or

$$E = cB. \qquad (41\text{-}11b)$$

This important result will be useful in later sections.

We now turn our attention to Fig. 41-14b, in which the flux Φ_E for the shaded rectangle is decreasing with time as the wave moves through it. According to Equation IV, Table 40-2 (with $i = 0$, because there are no conduction currents in a traveling electromagnetic wave),

$$\oint \mathbf{B} \cdot d\mathbf{l} = \mu_0 \epsilon_0 \frac{d\Phi_E}{dt}, \qquad (41\text{-}12)$$

this changing flux will induce a magnetic field at points around the periphery of the rectangle. This induced magnetic field is simply the magnetic component of the electromagnetic wave. Thus, as in the cavity resonator of Section 38-5, the electric and the magnetic components of the wave are intimately connected with each other, each depending on the time rate of change of the other.

Comparison of the shaded rectangles in Fig. 41-14 shows that for each the appropriate flux, Φ_B or Φ_E, is *decreasing* with time. However, if we proceed counterclockwise around the upper and lower shaded rectangles, we see that $\oint \mathbf{E} \cdot d\mathbf{l}$ is *positive*, whereas $\oint \mathbf{B} \cdot d\mathbf{l}$ is *negative*. This is as it should be. If you compare Figs. 35-10 and 40-1a, you will be reminded that although the fluxes Φ_B and Φ_E in those figures are changing with time in the same way (both are increasing) the lines of the induced fields, $\mathbf{E}$ and $\mathbf{B}$, respectively, circulate in opposite directions.

The integral in Eq. 41-12, evaluated by proceeding counterclockwise around the shaded rectangle of Fig. 41-14b, is

$$\oint \mathbf{B} \cdot d\mathbf{l} = [-(B + dB)(h)] + [(B)(h)] = -h\, dB,$$

where B is the magnitude of $\mathbf{B}$ at the left edge of the strip and $B + dB$ is its magnitude at the right edge.

The flux Φ_E through the rectangle of Fig. 41-14b is

$$\Phi_E = (E)(h\, dx).$$

Differentiating gives

$$\frac{d\Phi_E}{dt} = h\, dx\, \frac{dE}{dt}.$$

Thus we can write Eq. 41-12 as

$$-h\, dB = \mu_0 \epsilon_0 \left(h\, dx\, \frac{dE}{dt} \right)$$

or, substituting partial derivatives,

$$-\frac{\partial B}{\partial x} = \mu_0 \epsilon_0 \frac{\partial E}{\partial t}. \qquad (41\text{-}13)$$

Again, the minus sign in this equation is appropriate and necessary, for, although B is increasing with x at the site of the shaded rectangle in Fig. 41-14b, E is decreasing with t.

Combining this equation with Eqs. 41-6 and 41-7 yields

$$-kB_m \cos(kx - \omega t) = -\mu_0 \epsilon_0 \omega E_m \cos(kx - \omega t),$$

or (see Eq. 41-5)
$$\frac{E_m}{B_m} = \frac{k}{\mu_0 \epsilon_0 \omega} = \frac{1}{\mu_0 \epsilon_0 c}. \qquad (41\text{-}14)$$

Eliminating E_m/B_m between Eqs. 41-11a and 41-14 yields

$$c = \frac{1}{\sqrt{\mu_0 \epsilon_0}}. \qquad (41\text{-}15)$$

Substituting numerical values yields

$$c = \frac{1}{\sqrt{(4\pi \times 10^{-7}\ \text{T} \cdot \text{m/A})(8.9 \times 10^{-12}\ \text{C}^2/\text{N} \cdot \text{m}^2)}}$$
$$= 3.0 \times 10^8\ \text{m/s}, \qquad (41\text{-}16)$$

which is the speed of light in free space! This emergence of the speed of light from purely electromagnetic considerations is the crowning achievement of Maxwell's electromagnetic theory. Maxwell made this prediction before radio waves were known and before it was realized that light was electromagnetic in nature. His prediction led to the concept of the electromagnetic spectrum and to the discovery of radio waves by Heinrich Hertz in 1890. It made it possible to discuss optics as a branch of electromagnetism and to derive its fundamental laws from Maxwell's equations.

A conclusion as basic as Eq. 41-15 must be subject to rigorous experimental testing and indeed this has been done. Of the three quantities in this equation, one, μ_0, has an exact assigned value of $4\pi \times 10^{-7}\ \text{T} \cdot \text{m/A}$. The other two, c and ϵ_0, are measurable quantities. Our confidence in Maxwell's equations, bolstered by many successful predictions and agreements with experiment, is such that we now use the measured speed of light (2.99792458×10^8 m/s; see Appendix B) to determine ϵ_0, by way of Eq. 41-15. This gives a much better value than we could measure by the methods of Section 30-2.

Curiously enough Maxwell himself did not view the propagation of electromagnetic waves and electromagnetic phenomena in general, in anything like the terms suggested by, say, Fig. 41-13. Like all physicists of his day he believed firmly that space was permeated by a subtle substance called the *ether* and that electromagnetic phenomena could be accounted for in terms of rotating vortices in this aether.

It is a tribute to Maxwell's genius that, with such mechanical models in his mind, he was able to deduce the laws of electromagnetism that bear his name. These laws, as we have pointed out, not only required no change when Einstein's special theory of relativity came on the scene almost half-century later but, indeed, were strongly confirmed by that theory. Today we no longer believe in the aether concept; see Supplementary Topic V, "Special Relativity—A Summary of Conclusions."

One of the important characteristics of an electromagnetic wave is that it can transport energy from point to point. As we show below, we can describe the rate of energy flow per unit area in a plane electromagnetic wave by a vector **S**, called the *Poynting vector* after John Henry Poynting (1852–1914), who first pointed out its properties. We define **S** from

41-9
THE POYNTING VECTOR

$$S = \frac{1}{\mu_0} \mathbf{E} \times \mathbf{B}. \qquad (41\text{-}17)$$

In SI units **S** is expressed in watts/meter²; the direction of **S** gives the direction in which the energy moves. The vectors **E** and **B** refer to their instantaneous values at the point in question. If we apply Eq. 41-17 to the traveling plane electromagnetic wave of Fig. 41-13, it is clear that **E** × **B**, hence **S**, point in the direction of propagation. Note, too, that **S** points parallel to the axis for all points in the coaxial cable of Fig. 41-5.

We get meaningful results if we extend the Poynting vector concept to other electromagnetic situations involving either traveling or standing electromagnetic waves, as we will see in Examples 5 and 6. If we extend it to circuit situations involving steady or almost steady currents and lumped circuit elements, we are led to some interesting conclusions, which we explore in Problems 28, 31, and 32.

EXAMPLE 5

Analyze energy flow in the cavity of Fig. 38-6, using the Poynting vector.

Study of Fig. 41-15 shows that when the energy is all electric (Figs. 41-15*a* and 41-15*e*) it is concentrated along the axis, because this is the region in which **E** has its maximum value. When the energy is all magnetic (Figs. 41-15*c* and 41-15*g*), it is concentrated near the walls. Thus the energy surges back and forth periodically between the central region and the region near the walls. The figure

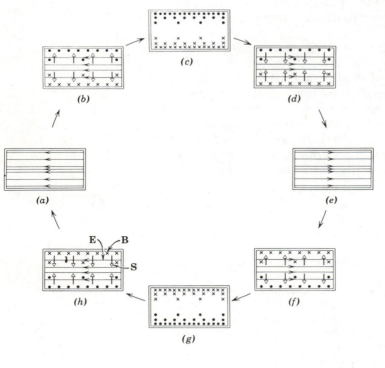

figure 41-15
Example 5. Energy surges back and forth periodically between the central region of the cavity and the region near the walls, as indicated by the Poynting vector **S** (vertical arrows).

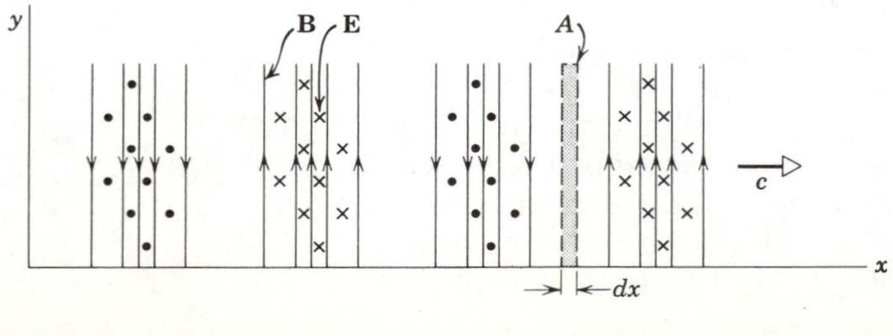

figure 41-16
A plane wave is traveling to the right at speed *c*; compare Fig. 41-14. The dashed rectangle in this figure represents a three-dimensional box, of area *A* and thickness *dx*, that extends at right angles to the plane of the figure.

shows, by the open arrows, the direction of **S** at various points in the cavity and at various times in the cycle. Note that **S** equals zero for Figs. 41-15a, c, e, and g, which is appropriate because at these instants of time the field configurations are momentarily stationary and energy is not flowing. A pendulum bob at the end of its swing forms a mechanical analogy. Verify from Eq. 41-17 that these arrows point in the correct directions.

We can use Figure 41-16 to derive the Poynting relation for the special case of a traveling plane electromagnetic wave. It shows a cross section of a traveling plane wave, along with a thin "box" of thickness dx and area A. The box, a mathematical construction, is fixed with respect to the axes while the wave moves through it. At any instant the energy stored in the box, from Eqs. 30-9 and 36-19, is

$$dU = dU_E + dU_B = (u_E + u_B)(A\ dx)$$

$$= \left(\tfrac{1}{2}\epsilon_0 E^2 + \frac{1}{2\mu_0} B^2\right)A\ dx, \qquad (41\text{-}18)$$

where $A\ dx$ is the volume of the box and E and B are the instantaneous values of the field vectors in the box.

Using Eq. 41-11b ($E = cB$) to eliminate *one* of the E's in the first term in Eq. 41-18 and *one* of the B's in the second term leads to

$$dU = \left[\tfrac{1}{2}\epsilon_0 E(cB) + \frac{1}{2\mu_0} B\left(\frac{E}{c}\right)\right]A\ dx$$

$$= \frac{(\mu_0\epsilon_0 c^2 + 1)(EBA\ dx)}{2\mu_0 c}.$$

From Eq. 41-15, however, $\mu_0\epsilon_0 c^2 = 1$, so that

$$dU = \frac{EBA\ dx}{\mu_0 c}.$$

This energy dU will pass through the right face of the box in a time dt equal to dx/c. Thus the energy per unit area per unit time, which is S, is given by

$$S = \frac{dU}{dt\ A} = \frac{EBA\ dx}{\mu_0 c\ (dx/c)\ A} = \frac{1}{\mu_0} EB.$$

This is exactly the prediction of the more general relation Eq. 41-17 for a traveling plane wave.

This relation refers to values of S, E, and B at any instant of time. We are usually more interested in the *average* value of S, taken over one or more cycles of the wave. An observer making intensity measurements on a wave moving past him would measure this average value $\overline{S}$. We can easily show (see Example 6) that $\overline{S}$ is related to the *maximum* values of E and B by

$$\overline{S} = \frac{1}{2\mu_0} E_m B_m.$$

EXAMPLE 6

An observer is at a distance r from a point light source whose power output is P_0. Calculate the magnitudes of the electric and the magnetic fields. Assume that the source is monochromatic, that it radiates uniformly in all directions, and that at distant points it behaves like the traveling plane wave of Fig. 41-13.

The power that passes through a sphere of radius r is $(\overline{S})(4\pi r^2)$, where $\overline{S}$ is the *average* value of the Poynting vector at the surface of the sphere. This power must equal P_0, or

$$P_0 = \overline{S}4\pi r^2.$$

From the definition of **S** (Eq. 41-17), we have

$$\overline{S} = \overline{\left(\frac{1}{\mu_0} EB\right)}.$$

Using the relation $E = cB$ (Eq. 41-11b) to eliminate B leads to

$$\overline{S} = \frac{1}{\mu_0 c}\,\overline{E^2}.$$

The average value of E^2 over one cycle is $\frac{1}{2}E_m^2$, since E varies sinusoidally (see Eq. 41-7). This leads to

$$P_0 = \left(\frac{E_m^2}{2\mu_0 c}\right)(4\pi r^2),$$

or

$$E_m = \frac{1}{r}\sqrt{\frac{P_0\mu_0 c}{2\pi}}.$$

For $P_0 = 10^3$ W and $r = 1.0$ m this yields

$$E_m = \frac{1}{(1.0\text{ m})}\sqrt{\frac{(10^3\text{ W})(4\pi\times 10^{-7}\text{ Wb/A·m})(3\times 10^8\text{ m/s})}{2\pi}}$$

$$= 240\text{ V/m}.$$

The relationship $E_m = cB_m$ (Eq. 41-11a) leads to

$$B_m = \frac{E_m}{c} = \frac{240\text{ V/m}}{3\times 10^8\text{ m/s}} = 8\times 10^{-7}\text{ T}.$$

Note that E_m is appreciable as judged by ordinary laboratory standards but that B_m $(= 0.008$ gauss$)$ is quite small.

---

questions

1. If you are asked what fraction of the electromagnetic spectrum lies in the visible range, how would you respond?

2. Project Seafarer was an ambitious program to construct an enormous antenna, buried underground on a site about 4000 square miles in area. Its purpose was to transmit signals to submarines while they were deeply submerged. If the effective wavelength was, say, about 10^4 earth radii, what would be the frequency and the period of the radiations emitted? Ordinarily electromagnetic radiations do not penetrate very far into conductors such as sea water. Can you think of any reason why such ELF (extremely low frequency) radiations should penetrate more effectively? Think of the limiting case of zero frequency. (Why not transmit signals at zero frequency?)

3. How would you characterize electromagnetic radiation that has a frequency of 10 kHz? 10^{20} Hz? or a wavelength of 500 nm? 10 km? 0.50 nm?

4. Electromagnetic waves reach us from the farthest depths of the universe. Do they tell us what the universe is like now, what it was like at some time in the past, or something in between? Discuss.

5. *The Dark Night Sky Paradox:** Perhaps the simplest astronomical observation that you can make is this: When the sun sets, the sky becomes dark. This is true and seems obvious but an argument can be made that it should not be so. Consider:

    "Assuming an infinite universe, uniformly populated by stars more or less like our sun, we can say that a straight line projected from the observer in any direction will eventually hit a star. The *distances R* of most of these stars will be very great indeed so that the stars illuminate him only weakly, the illumination varying as $1/R^2$. On the other hand, the *number* of distant stars located within a spherical shell whose radii are R and $R + \Delta R$ increases as R^2 (assuming that ΔR is constant). Can you prove this last statement? These two effects seem to cancel precisely. Thus the night sky should be virtually infinitely bright, the observer being illuminated by an infinity of suns."

    Can you see any flaw in this argument? Think of the finite speed of light, the large scale of the universe, the expanding universe concept, the finite

* Usually called *Olbers' paradox*.

lifetime of stars, and so on. See "The Dark Night Sky Paradox" by E. R. Harrison, *American Journal of Physics*, February 1977, for an excellent historical review and a lucid explanation.

6. When ordinary stars condense to form neutron stars (pulsars), why does the angular rotation rate become so large?

7. In the coaxial cable of Fig. 41-3, what are the directions of the conduction current (a) in the central conductor and (b) in the outer conductor, shortly after the switch is thrown to position a? Consider points that have been reached by the wavefront of Fig. 49-4b and those that have not.

8. What is the relation between the wavelength in the cable of Fig. 41-5 and that in free space for a coaxial cable?

9. What is the direction of the displacement current in Fig. 41-6? Give an argument to support your answer.

10. Compare a coaxial cable and a waveguide, used as a transmission line. Point out both similarities and differences.

11. Can traveling waves with a continuous range of wavelengths be sent down (a) a coaxial cable and (b) a waveguide? Can standing waves with a continuous range of wavelengths be set up in a resonant cavity? Develop mechanical or acoustical analogies to support your answers.

12. If a certain wavelength is larger than the cutoff wavelength for a guide in its dominant mode, can energy be sent down it in any other mode?

13. Explain why the term $\epsilon_0 \, d\Phi_E/dt$ is needed in Ampère's equation to understand the propagation of electromagnetic waves.

14. In the equation $c = 1/\sqrt{\mu_0\epsilon_0}$ (Eq. 40-1), how can c always have the same value if μ_0 is *arbitrarily* assigned and ϵ_0 is measured?

15. Is it conceivable that electromagnetic theory might some day be able to predict the value of c (3×10^8 m/s), not in terms of μ_0 and ϵ_0, but directly and numerically without recourse to any measurements?

16. Figure 41-17 shows a magnetic dipole activated by an oscillating LC circuit. Discuss the nature of the traveling wave at a distant point P.

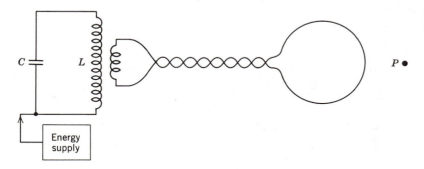

figure 41-17
Question 16

17. In a coaxial cable is the energy transported in the conductors, through the agency of the currents, or in the space between them, through the agency of the fields?

SECTION 41-2

1. In Fig. 41-1, showing the electromagnetic spectrum, the frequency ν and the wavelength λ are related by $c = \lambda\nu$. If the frequency intervals in the figure are evenly spaced, show that the wavelength intervals are also necessarily equally spaced. Bear in mind that both scales are logarithmic.

SECTION 41-3

2. The earth's mean radius is 6.4×10^6 m and the mean earth-sun distance is 1.5×10^8 km. What fraction of the electromagnetic radiation emitted by the

problems

sun is intercepted by the disc of the earth? Assume an inverse square law for the decrease of intensity, measured perhaps in watts/meter², with distance.

3. The intensity of direct solar radiation that was unabsorbed by the atmosphere on a lazy summer day is 100 W/m². How close would you have to stand to a 1.0-kW electric heater to feel the same intensity? Assume that the heater is 100% efficient and radiates equally in all directions.
Answer: 0.89 m.

4. Our closest stellar neighbor, α-Centauri, is 4.3 light years away. It has been suggested that TV programs from our planet (The Tonight Show, for example) have reached this star and may have been viewed by the hypothetical inhabitants of a hypothetical planet orbiting this star. The moon is 3.8×10^5 km from earth and perhaps its hypothetical inhabitants (who always hide when a space ship approaches) also watch these programs. From the inverse square law, what would be the ratio of the intensities of such signals, measured perhaps in W/m², for the moon and for α-Centauri?

SECTION 41-5

5. A coaxial cable is made of a center wire of radius a surrounded by a thin metal tube of radius b. The space between the conductors is evacuated. (a) Find the capacitance per unit length of this coaxial cable (*Hint:* Imagine equal but opposite charges to be on the wire and the tube). (b) Find the inductance per unit length of this coaxial cable (*Hint:* Imagine a current i flowing down the center wire and back along the tube.)
Answer: (a) $2\pi\epsilon_0/\ln b/a$. (b) $(\mu_0/2\pi) \ln b/a$.

6. Using Gauss's law, sketch the instantaneous charges that appear on the conductors of the coaxial cable of Fig. 41-5 and show that this pattern of charges is appropriately related to the conduction currents shown in Fig. 41-5c.

7. A resonant cavity is constructed by closing each end of the coaxial cable of Fig. 41-5 with a metal cap. The cavity contains three half-waves. Describe the patterns of **E** and **B** that occur, assuming the same mode of oscillation as that shown in Fig. 41-5. (Hint: Remember that **E** can have no tangential component at a conducting surface and that **B** and **E** must be 90° out of phase.)

SECTION 41-6

8. (a) For a rectangular guide of width 3.0 cm, what must the free-space wavelength of radiation be if it is to require 1.0 μs $(=10^{-6}$ s$)$ for energy to traverse a 100-m length of guide? (b) What is the phase speed under these circumstances?

9. For a rectangular guide of width 3.0 cm, plot the phase speed, the group speed, and the guide wavelength as a function of the free-space wavelength. Assume the dominant mode.

10. Under what conditions will the guide wavelength in the guide of Fig. 41-9 be double the free-space wavelength?

11. (a) What guide wavelength does 10-cm radiation (free-space wavelength) exhibit in a rectangular guide whose width is 6.0 cm? Assume the dominant mode. (b) What is the cutoff wavelength for this guide?
Answer: (a) 18 cm. (b) 12 cm.

SECTION 41-7

12. (a) At a distance of 80 miles from a radio transmitter, how much later would you observe a wave emitted from the antenna? (b) If the radiation were a radio wave emitted from the sun? (c) If it were emitted from a stellar radio source 380 light years distant?

SECTION 41-8

13. An electromagnetic wave is traveling in the negative y-direction. At a particular position and time, the electric field is along the positive z-axis and has a magnitude of 100 V/m. What are the direction and magnitude of the magnetic field?   *Answer:* 3.3×10^{-7} T, in the negative x-direction.

14. The magnetic field equations for an electromagnetic wave in free space are $B_x = B \sin{(ky + \omega t)}$, $B_y = B_z = 0$. (a) What is the direction of propagation? (b) Write the electric field equations.

15. How does the displacement current density vary with space and time in a traveling plane electromagnetic wave?

16. (a) Starting with Eqs. 41-10 and 41-13, show that $E(x, t)$ satisfies the "wave equation"

$$\frac{\partial^2 E}{\partial x^2} = \mu_0 \epsilon_0 \frac{\partial^2 E}{\partial t^2}.$$

(b) What "wave equation" does $B(x, t)$ satisfy?

17. Show that (a) through (d) below satisfy Eqs. 41-10 and 41-13. In each of these, A is a constant. (a) $E = Ac(x - ct)$, $B = A(x - ct)$. (b) $E = Ac(x + ct)^{15}$, $B = -A(x + ct)^{15}$. (c) $E = Ace^{(x-ct)}$, $B = Ae^{(x-ct)}$. (d) $E = Ac \ln{(x + ct)}$, $B = -A \ln{(x + ct)}$. (e) Generalize these examples to show that $E = Acf(x - ct)$, $B = Af(x - ct)$ is a solution where f is *any* function of $(x - ct)$. What is the corresponding situation for functions of $(x + ct)$?

18. Show that the directions of the fields $\mathbf{E}$ and $\mathbf{B}$ in Fig. 41-11 are consistent with the direction of propagation of the radiation.

19. Show that the directions of the fields E and B in the various "patches" in Fig. 41-12 are consistent with the direction of propagation of the radiation.

SECTION 41-9

20. Sunlight strikes the earth, outside its atmosphere, with an intensity of 2.0 cal/cm²-min. Calculate E_m and B_m for sunlight, assuming it to be a wave like that of Fig. 41-12.

21. A #10 copper wire (diameter, 0.10 in.; resistance per 1000 ft, 1.00 Ω) carries a current of 25 A. Calculate (a) $\mathbf{E}$, (b) $\mathbf{B}$, and (c) $\mathbf{S}$ for a point on the surface of the wire.
   Answer: (a) $\mathbf{E} = 8.2 \times 10^{-2}$ V/m, parallel to the wire. (b) $\mathbf{B} = 3.9 \times 10^{-3}$ T, tangent to the wire and perpendicular to its axis. (c) $\mathbf{S} = 260$ W/m², radially inward.

22. Prove that for any point in an electromagnetic wave such as that of Fig. 41-13 the density of energy stored in the electric field equals that stored in the magnetic field.

23. A plane radio wave has $E_m \cong 10^{-4}$ V/m. Calculate (a) B_m and (b) the intensity of the wave, as measured by $\overline{\mathbf{S}}$.
   Answer: (a) 3.3×10^{-13} T. (b) 1.3×10^{-11} W/m².

24. If a coaxial cable has resistance, energy must flow from the fields into the conducting surfaces to provide Joule heating. How must the electric lines of force of Fig. 41-5a be modified in this case? (*Hint:* The Poynting vector near the surface must have a component pointing toward the surface.)

25. Analyze the flow of energy in the waveguide of Fig. 41-9, using the Poynting vector.

26. A coaxial cable (inner radius a, outer radius b) is used as a transmission line between a battery ε and a resistor R, as shown in Fig. 41-18. (a) Calculate $\mathbf{E}$, $\mathbf{B}$ for $a < r < b$. (b) Calculate the Poynting vector $\mathbf{S}$ for $a < r < b$. (c) By suitably integrating the Poynting vector, show that the total power flowing across the annular cross section $a < r < b$ is ε^2/R. Is this reasonable?

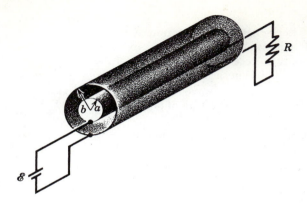

figure 41-18
Problem 26

(d) Show that the direction of **S** is always from the battery to the resistor, no matter which way the battery is connected.

27. An airplane flying at a distance of 10 km from a radio transmitter receives a signal of intensity 10 μW/m². Calculate (a) the (average) electric field at the airplane due to this signal; (b) the (average) magnetic field at the airplane; (c) the total power radiated by the transmitter, assuming the transmitter to radiate isotropically and the earth to be a perfect absorber.
Answer: (a) 6.1×10^{-2} V/m. (b) 2.1×10^{-10} T. (c) 13 kW.

28. A cube of edge a has its edges parallel to the x-, y-, and z-axes of a rectangular coordinate system. A uniform electric field **E** is parallel to the y-axis and a uniform magnetic field **B** is parallel to the x-axis. Calculate (a) the rate at which, according to the Poynting vector point of view, energy may be said to pass through each face of the cube and (b) the net rate at which the energy stored in the cube may be said to change.

29. Consider the possibility of standing waves:

$$E = E_m(\sin \omega t)(\sin kx)$$
$$B = B_m(\cos \omega t)(\cos kx).$$

(a) Show that these satisfy Eqs. 41-10 and 41-13 if E_m is suitably related to B_m and ω suitably related to k. What are these relationships? (b) Find the (instantaneous) Poynting vector. (c) Show that the time average power flow across any area is zero. (d) Describe the flow of energy in this problem.
Answer: (a) Eq. 41-11a must hold for standing waves as well as for traveling waves. (b) $S = (E_m^2/4\mu_0 c) \sin 2\omega t \sin 2kx$.

30. Figure 41-19 shows a long resistanceless transmission line, delivering power from a battery to a resistive load. A steady current i exists as shown. (a) Sketch qualitatively the electric and magnetic fields around the line, and (b) show that, according to the Poynting vector point of view, energy travels from the battery to the resistor through the space around the line and not through the line itself. (*Hint:* Each conductor in the line is an equipotential surface, since the line has been assumed to have no resistance.)

figure 41-19
Problem 30

31. Figure 41-20 shows a cylindrical resistor of length l, radius a, and resistivity ρ, carrying a current i. (a) Show that the Poynting vector **S** at the surface of the resistor is everywhere directed normal to the surface, as shown. (b) Show that the rate P at which energy flows into the resistor through its cylindrical surface, calculated by integrating the Poynting vector over this surface, is equal to the rate at which thermal energy is produced; that is,

$$\int \mathbf{S} \cdot d\mathbf{A} = i^2 R,$$

where $d\mathbf{A}$ is an element of area of the cylindrical surface. This shows that, according to the Poynting vector point of view, the energy that appears in a

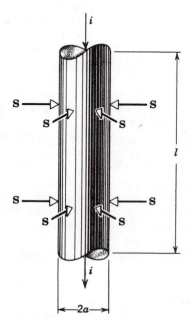

figure 41-20
Problem 31

resistor as thermal energy does not enter it through the connecting wires but through the space around the wires and the resistor. (*Hint:* **E** is parallel to the axis of the cylinder, in the direction of the current; **B** forms concentric circles around the cylinder, in a direction given by the right-hand rule.)

32. Figure 41-21 shows a parallel-plate capacitor being charged. (*a*) Show that the Poynting vector **S** points everywhere radially into the cylindrical volume. (*b*) Show that the rate P at which energy flows into this volume, calculated by integrating the Poynting vector over the cylindrical boundary of this volume, is equal to the rate at which the stored electrostatic energy increases; that is, that

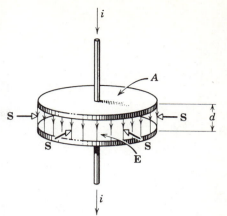

figure 41-21
Problem 32

$$\int \mathbf{S} \cdot d\mathbf{A} = Ad\,\frac{d}{dt}\left(\tfrac{1}{2}\epsilon_0 E^2\right),$$

where Ad is the volume of the capacitor and $\tfrac{1}{2}\epsilon_0 E^2$ is the energy density for all points within that volume. This analysis shows that, according to the Poynting vector point of view, the energy stored in a capacitor does not enter it through the wires but through the space around the wires and the plates. (*Hint:* To find **S** we must first find **B**, which is the magnetic field set up by the displacement current during the charging process; see Fig. 40-1. Ignore fringing of the lines of **E**.)

33. A long hollow nonconducting cylinder (radius R, length l) carries a uniform charge per unit area of σ on its surface. An externally applied torque causes the cylinder to rotate at constant acceleration $\omega(t) = \alpha t$ about the cylinder axis. (*a*) Find **B** within the cylinder (treat it as a solenoid). (*b*) Find **E** at the inner surface of the cylinder. (*c*) Find **S** at the inner surface of the cylinder. (*d*) Show that the flux of **S** entering the interior volume of the cylinder is equal to $\dfrac{d}{dt}\left(\dfrac{\pi R^2 l}{2\mu_0}\,B^2\right)$.

Answer: (*a*) $B = \sigma R\omega\mu_0$. (*b*) $E = \tfrac{1}{2}\mu_0\sigma R^2\alpha$. (*c*) $S = \tfrac{1}{2}\mu_0\sigma^2 R^3\alpha^2 t$.

42
the nature and propagation of light

This chapter connects what is true about the entire electromagnetic spectrum (see Fig. 41-1) to visible light, which is, of course, a part of that spectrum. In chapters that immediately follow we will focus on visible light. This is important to do because almost all of us, having adapted by genetic selection to the sun, have two electromagnetic receptors that function in this range and, indeed, define the range. These, of course, are our eyes. We must always remember that whatever we say about the visible part of the spectrum holds, at its foundations, for all other parts of the spectrum as well. The differences are largely in the nature of the means of production and detection in the various ranges.

Here we define ''light'' as radiation that can affect the eye. Figure 42-1, which shows the relative eye sensitivity of an assumed *standard*

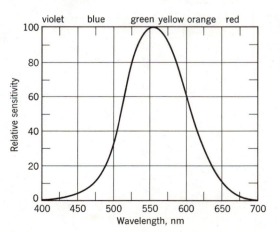

figure 42-1
The relative eye sensitivity of an assumed *standard observer* at different wavelengths for a specified level of illumination.

observer to radiations of various wavelengths, shows that the center of the visible region is about 5.55×10^{-7} m or 555 nm. Light of this wavelength produces the sensation of yellow-green.*

In optics we often use the micrometer (abbr. μm) the nanometer (abbr. nm), and the Ångstrom (abbr. Å) as units of wavelength. They are defined from

$$1 \text{ micrometer} = 10^{-6} \text{ meter}$$

$$1 \text{ nanometer} = 10^{-9} \text{ meter}$$

$$1 \text{ Ångstrom} = 10^{-10} \text{ meter}.$$

Thus we can say that the center of the visible region is 0.555 μm, 555 nm, or 5550 Å.

The limits of the visible spectrum are not well defined because the eye sensitivity curve approaches the axis asymptotically at both long and short wavelengths. If we take the limits, arbitrarily, as the wavelengths at which the eye sensitivity has dropped to 1% of its maximum value, these limits are about 430 and 690 nm, less than a factor of two in wavelength. The eye can detect radiation beyond these limits if it is intense enough. In many experiments in physics one can use photographic plates or light-sensitive electronic detectors in place of the human eye.

42-2
ENERGY AND MOMENTUM

Energy is carried by electromagnetic waves from the sun to the earth or from an open fire to a hand placed nearby. We described the transport of energy by such a wave in free space in Section 41-9 by the Poynting vector **S**, or

$$\mathbf{S} = \frac{1}{\mu_0} \mathbf{E} \times \mathbf{B}, \tag{42-1}$$

where **E** and **B** are the instantaneous values of the electric and magnetic field vectors.

Less familiar is the fact that electromagnetic waves may also transport linear momentum. In other words, it is possible to exert a pressure (a *radiation pressure*†) on an object by shining a light on it. Such forces must be small in relation to forces of our daily experience because we do not ordinarily notice them. We do not, after all, fall over backward when we raise a window shade in a dark room and let sunlight shine on us. Radiation pressure effects are, however, important in the life cycles of stars because of the incredibly high temperatures (2×10^7 K for our sun) that we associate with stellar interiors.

The first measurement of radiation pressure was made in 1901–1903 by Nichols and Hull at Dartmouth College and by Lebedev in Russia, about thirty years after the existence of such effects had been predicted theoretically by Maxwell.

---

* See "The Retinex Theory of Color Vision" by Edwin H. Land, *Scientific American*, December 1977, and "Color and Perception: the Work of Edwin Land in the Light of Current Concepts" by M. H. Wilson and R. W. Brocklebank, *Contemporary Physics*, December 1961, for a fascinating discussion of the problems of perception and the distinction between color as a characteristic of light and color as a perceived property of objects.

† See "Radiation Pressure" by G. E. Henry, *Scientific American*, June 1957; see also "The Pressure of Laser Light" by Arthur Ashkin, *Scientific American*, February 1972.

Let a parallel beam of light fall on an object for a time t, the incident light being *entirely absorbed* by the object. If energy U is absorbed during this time, the momentum p delivered to the object is given, according to Maxwell's prediction, by

$$p = \frac{U}{c} \qquad \text{(total absorption)}, \qquad (42\text{-}2a)$$

where c is the speed of light. The direction of **p** is the direction of the incident beam. If the light energy U is *entirely reflected*, the momentum delivered will be twice that given above, or

$$p = \frac{2U}{c} \qquad \text{(total reflection)}. \qquad (42\text{-}2b)$$

In the same way, twice as much momentum is delivered to an object when a perfectly elastic tennis ball is bounced from it as when it is struck by a perfectly inelastic ball (a lump of putty, say) of the same mass and speed. If the light energy U is partly reflected and partly absorbed, the delivered momentum will lie between U/c and $2U/c$.

EXAMPLE 1

A parallel beam of light with an energy flux S of 10 W/cm² falls for 1 hr on a perfectly reflecting plane mirror of 1.0-cm² area. (*a*) What momentum is delivered to the mirror in this time and (*b*) what force acts on the mirror?

(*a*) The energy that is reflected from the mirror is

$$U = (10 \text{ W/cm}^2)(1.0 \text{ cm}^2)(3600 \text{ s}) = 3.6 \times 10^4 \text{ J}.$$

The momentum delivered after 1 hour's illumination is

$$p = \frac{2U}{c} = \frac{(2)(3.6 \times 10^4 \text{ J})}{3 \times 10^8 \text{ m/s}} = 2.4 \times 10^{-4} \text{ kg m/s}.$$

(*b*) From Newton's second law, the average force on the mirror is equal to the average rate at which momentum is delivered to the mirror, or

$$F = \frac{p}{t} = \frac{2.4 \times 10^{-4} \text{ kg m/s}}{3600 \text{ s}} = 6.7 \times 10^{-8} \text{ N}.$$

This is a small force.

Nichols and Hull, in 1903, measured radiation pressures and verified Eq. 42-2, using a torsion balance technique. They allowed light to fall on mirror M in Fig. 42-2; the radiation pressure caused the balance arm to turn through a measured angle θ, twisting the torsion fiber F. Assuming a suitable calibration for their torsion fiber, the experimenters could arrive at a numerical value for this pressure. Nichols and Hull measured the intensity of their light beam by allowing it to fall on a blackened metal disk of known absorptivity and by measuring the temperature rise of this disk. In a particular run these experimenters measured a radiation pressure of 7.01×10^{-6} N/m²; for their light beam, the value predicted, using Eq. 42-2, was 7.05×10^{-6} N/m², in excellent agreement. Assuming a mirror area of 1 cm², this represents a force on the mirror of only 7×10^{-10} N, about 100 times smaller than the force calculated in Example 1.

The success of the experiment of Nichols and Hull was the result in large part of the care they took to eliminate spurious deflecting effects caused by changes in the speed distribution of the molecules in the gas

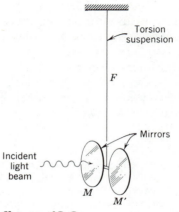

figure 42-2
Suggesting the experiment of Nichols and Hull, used to measure radiation pressure.

surrounding the mirror. These changes were brought about by the small rise in the temperature of the mirror as it absorbed light energy from the incident beam. This "radiometer effect" is responsible for the spinning action of the familiar toy radiometers when placed in a beam of sunlight. In a perfect vacuum such effects would not occur, but in the best vacuums available in 1903 radiometer effects were present and had to be taken specifically into account in the design of the experiment.

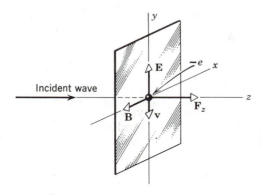

figure 42-3
An incident plane light wave falls on an electron in a thin resistive sheet. Instantaneous values of **E**, **B**, the electron velocity **v**, and the radiation force **F**$_z$ are shown.

To demonstrate the transport of momentum from Maxwell's equations in a particular case, let a plane electromagnetic wave traveling in the z-direction fall on a large thin sheet of a material of high resistivity as in Fig. 42-3. A small part of the incident energy will be absorbed within the sheet, but most of it will be transmitted if the sheet is thin enough.*

The incident wave vectors **E** and **B** vary with time at the sheet as

$$\mathbf{E} = \mathbf{E}_m \sin \omega t \qquad (42\text{-}3)$$

and

$$\mathbf{B} = \mathbf{B}_m \sin \omega t \qquad (42\text{-}4)$$

where **E** is parallel to the $\pm y$ axis and **B** is parallel to the $\pm x$ axis.

In Section 31-4 we saw that the effect of a (constant) electric force ($= -eE$) on a conduction electron in a metal was to make it move with a (constant) drift speed v_d. The electron behaves as if it is immersed in a viscous fluid, the electric force acting on it being counterbalanced by a "viscous" force, which may be taken as proportional to the electron speed. Thus for a constant field E, after equilibrium is established,

$$eE = bv_d, \qquad (42\text{-}5)$$

where b is a resistive damping coefficient. The electron equilibrium speed, dropping the subscript d, is thus

$$v = \frac{eE}{b}. \qquad (42\text{-}6)$$

If the applied electric field varies with time and if the variation is slow enough, the electron speed can continually readjust itself to the changing value of E so that its speed continues to be given essentially by its equilibrium value (Eq. 42-6) at all times. These readjustments are more rapidly made the more viscous the medium, just as a stone falling in air reaches a constant equilibrium rate of descent only relatively slowly but one falling in a viscous oil does so quite rapidly. We assume that the sheet in Fig. 42-3 is so viscous, that is, that its resistivity is so high, that Eq. 42-6 remains true even for the rapid oscillations of E in the incident light beam.

As the electron vibrates parallel to the y-axis, it experiences a *second* force

---

* Some of the incident energy will also be reflected, but the reflected wave is of such low intensity that we can ignore it in the derivation that follows.

due to the *magnetic* component of the wave. This force $\mathbf{F}_z$ $(= -e\mathbf{v} \times \mathbf{B})$ points in the z-direction, being at right angles to the plane formed by $\mathbf{v}$ and $\mathbf{B}$, that is, the xy-plane. The instantaneous magnitude of $\mathbf{F}_z$ is given by

$$F_z = evB = \frac{e^2 EB}{b}. \qquad (42\text{-}7)$$

$\mathbf{F}_z$ always points in the positive z-direction because $\mathbf{v}$ and $\mathbf{B}$ reverse their directions simultaneously; this force is, in fact, the mechanism by which the radiation pressure acts on the sheet of Fig. 42-3.

From Newton's second law, F_z is the rate dp_e/dt at which the incident wave delivers momentum to each electron in the sheet, or

$$\frac{dp_e}{dt} = \frac{e^2 EB}{b}. \qquad (42\text{-}8)$$

Momentum is delivered at this rate to every electron in the sheet and thus to the sheet itself. It remains to relate the momentum transfer to the sheet to the absorption of energy within the sheet.

The electric field component of the incident wave does work on each oscillating electron at an instantaneous rate (see Eq. 42-6) given by

$$\frac{dU_e}{dt} = F_E v = (eE)\left(\frac{eE}{b}\right) = \frac{e^2 E^2}{b}.$$

Note that the magnetic force $\mathbf{F}_z$, always being at right angles to the velocity $\mathbf{v}$, does no work on the oscillating electron. Equation 41-11b shows that for a plane wave in free space B and E are related by

$$E = Bc.$$

Substituting above for *one* of the E's leads to

$$\frac{dU_e}{dt} = \frac{e^2 EBc}{b}. \qquad (42\text{-}9)$$

This equation represents the rate, per electron, at which energy is absorbed from the incident wave.

Comparing Eqs. 42-8 and 42-9 shows that

$$\frac{dp_e}{dt} = \frac{1}{c}\frac{dU_e}{dt}.$$

Integrating yields

$$\int_0^t \frac{dp_e}{dt}\, dt = \frac{1}{c}\int_0^t \frac{dU_e}{dt}\, dt,$$

or

$$p_e = \frac{U_e}{c}, \qquad (42\text{-}10)$$

where p_e is the momentum delivered to a single electron in any given time t and U_e is the energy absorbed by that electron in the same time interval. Multiplying each side by the number of free electrons in the sheet leads to Eq. 42-2a.

Although we derived this relation (Eq. 42-10) for a particular kind of absorber, no characteristics of the absorber—for example, the resistive damping coefficient b—remain in the final expression. This is as it should be because Eq. 42-10 is a general property of radiation absorbed by *any* material.

42-3
*THE SPEED OF LIGHT**

Light travels so fast that there is nothing in our daily experience to suggest that its speed is not infinite. It calls for considerable insight even to ask, "How fast does light travel?" Galileo asked himself this question and actually tried to answer it experimentally. His book, *Two New Sciences*, published in 1638, is written in the form of a conversation

* See "The Speed of Light" by J. H. Rush, *Scientific American*, August 1955.

among three persons called Salviati, Sagredo, and Simplicio. Here is part of what they say about the speed of light.

Simplicio: Everyday experience shows that the propagation of light is instantaneous; for when we see a piece of artillery fired, at a great distance, the flash reaches our eyes without lapse of time; but the sound reaches the ear only after a noticeable interval.

Sagredo: Well, Simplicio, the only thing I am able to infer from this familiar bit of experience is that sound, in reaching our ear, travels more slowly than light; it does not inform me whether the coming of the light is instantaneous or whether, although extremely rapid, it still occupies time. . . .

Salviati, who speaks with Galileo's voice, then describes a possible method (actually carried out) for measuring the speed of light. He and an assistant stand facing each other some distance apart, at night. Each carries a lantern which can be covered or uncovered at will. Galileo started the experiment by uncovering his lantern. When the light reached the assistant he uncovered his own lantern, whose light was then seen by Galileo. Galileo tried to measure the time between the instant at which he uncovered his own lantern and the instant at which the light from his assistant's lantern reached him. For a one-mile separation we now know that the round trip travel time would be only 11 μs. This is much less than human reaction times, so the method fails.

To measure a large velocity directly, we must either measure a small time interval or use a long base line. This situation suggests that astronomy, which deals with great distances, might be able to provide an experimental value for the speed of light; this proved to be true. Although it would be desirable to time the light from the sun as it travels to the earth, there is no way of knowing when the light that reaches us at any instant left the sun; we must use subtler astronomical methods.

Note, however, that microwave pulses are quite regularly reflected from the moon; this gives a 7.68×10^8-m base line (there and back) for timing purposes. The speed of light (and of microwaves) is so well known now from other experiments that we use these measurements to measure the lunar distance accurately. Microwave signals have also been reflected from Venus.

In 1675 Ole Roemer, a Danish astronomer working in Paris, made some observations of the moons of Jupiter (see Problem 20) from which a speed of light of 2×10^8 m/s may be deduced. About fifty years later James Bradley, an English astronomer, made some astronomical observations of an entirely different kind from which a value of 3.0×10^8 m/s may be deduced.

In 1849 Hippolyte Louis Fizeau (1819–1896), a French physicist, first measured the speed of light by a nonastronomical method, obtaining a value of 3.13×10^8 m/s. Figure 42-4 shows Fizeau's apparatus. Let us first ignore the toothed wheel. Light from source S is made to converge

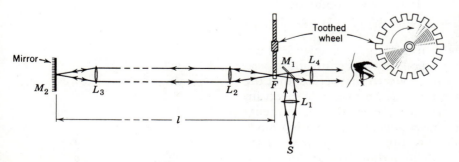

figure 42-4
Fizeau's apparatus for measuring the speed of light.

by lens L_1, is reflected from mirror M_1, and forms in space at F an image of the source. Mirror M_1 is a so-called "half-silvered mirror"; its reflecting coating is so thin that only half the light that falls on it is reflected, the other half being transmitted.

Light from the image at F enters lens L_2 and emerges as a parallel beam; after passing through lens L_3 it is reflected back along its original direction by mirror M_2. In Fizeau's experiment the distance l between M_2 and F was 8630 meters or 5.36 miles. When the light strikes mirror M_1 again, some will be transmitted, entering the eye of the observer through lens L_4.

The observer will see an image of the source formed by light that has traveled a distance $2l$ between the wheel and mirror M_2 and back again. To time the light beam a marker of some sort must be put on it. This is done by "chopping" it with a rapidly rotating toothed wheel. Suppose that during the round-trip travel time of $2l/c$ the wheel has turned just enough so that, when the light from a given "burst" returns to the wheel, point F is covered by a tooth. The light will hit the face of the tooth that is toward M_2 and will not reach the observer's eye.

If the speed of the wheel is exactly right, the observer will not see any of the bursts because each will be screened by a tooth. The observer measures c by increasing the angular speed ω of the wheel from zero until the image of source S disappears. Let θ be the angular distance from the center of a gap to the center of a tooth. The time needed for the wheel to rotate a distance θ is the round-trip travel time $2l/c$. In equation form,

$$\frac{\theta}{\omega} = \frac{2l}{c} \quad \text{or} \quad c = \frac{2\omega l}{\theta}. \tag{42-11}$$

This "chopped beam" technique, suitably modified, is used today to measure the speeds of neutrons and other particles.

EXAMPLE 2

The wheel used by Fizeau had 720 teeth. What is the smallest angular speed at which the image of the source will vanish?

The angle θ is $1/1440$ rev; solving Eq. 42-11 for ω gives

$$\omega = \frac{c\theta}{2l} = \frac{(3.00 \times 10^8 \text{ m/s})(1/1440 \text{ rev})}{(2)(8630 \text{ m})} = 12.1 \text{ rev/s.}$$

The French physicist Foucault (1819–1868) greatly improved Fizeau's method by substituting a rotating mirror for the toothed wheel. The American physicist Albert A. Michelson (1852–1931) conducted an extensive series of measurements of c, extending over a fifty-year period, using this technique.

We must view the speed of light within the larger framework of the speed of electromagnetic radiation in general. It is a significant experimental confirmation of Maxwell's theory of electromagnetism that the speed in free space of waves in all parts of the electromagnetic spectrum has the same value c. Table 42-1 shows some selected measurements that have been made of the speed of electromagnetic radiation since Galileo's day. It stands as a monument to man's persistence and ingenuity. Note in the last column how the uncertainty in the measurement has improved through the years. Note also the international character of the effort and the variety of methods.

Table 42-1
The speed of electromagnetic radiation in free space (some selected measurements)*

Date	Experimenter	Country	Method	Speed (km/s)	Uncertainty (km/s)
1600(?)	Galileo	Italy	Lanterns and shutters	"If not instantaneous, it is extraordinarily rapid"	
1675	Roemer	France	Moons of Jupiter	200,000	
1729	Bradley	England	Aberration of starlight	304,000	
1849	Fizeau	France	Toothed wheel	313,300	
1862	Foucault	France	Rotating mirror	298,000	500
1876	Cornu	France	Toothed wheel	299,990	200
1880	Michelson	U.S.A.	Rotating mirror	299,910	50
1883	Newcomb	England	Rotating mirror	299,860	30
1906	Rosa and Dorsey	U.S.A.	Electromagnetic theory	299,781	10
1923	Mercier	France	Standing waves on wires	299,782	15
1926	Michelson	U.S.A.	Rotating mirror	299,796	4
1928	Karolus and Mittelstaedt	Germany	Kerr cell	299,778	10
1932	Michelson, Pease, and Pearson	U.S.A.	Rotating mirror	299,774	11
1941	Anderson	U.S.A.	Kerr cell	299,776	14
1950	Bergstrand	Sweden	Geodimeter	299,792.7	0.25
1950	Essen	England	Microwave cavity	299,792.5	3
1950	Bol and Hansen	U.S.A.	Microwave cavity	299,789.3	0.4
1951	Aslakson	U.S.A.	Shoran radar	299,794.2	1.9
1952	Rank, Ruth, and Ven der Sluis	U.S.A.	Molecular spectra	299,776	7
1952	Froome	England	Microwave interferometer	299,792.6	0.7
1954	Florman	U.S.A.	Microwave interferometer	299,795.1	1.9
1957	Bergstrand	Sweden	Geodimeter	299,792.85	0.16
1958	Froome	England	Microwave interferometer	299,792.50	0.10
1965	Kolibayev	U.S.S.R.	Geodimeter	299,792.6	0.06
1967	Grosse	West Germany	Geodimeter	299,792.5	0.05
1973	Evenson, Wells, Peterson, Danielson, and Day	U.S.A.	Laser techniques	299,792.4574	0.0012

* See "Some Recent Determinations of the Velocity of Light, III" by Joseph F. Mulligan, *American Journal of Physics*, October 1976. Here the 1974 measurements of Blayney et al. at the National Physical Laboratory in England are described. They are in very close agreement with the results of Evenson et al. (1973).

  The task of arriving at a single "best" value for any physical quantity, such as c, from many independent measurements is usually difficult because it involves a careful evaluation of each measurement and a complex averaging process, which takes into account other physical quantities with which the quantity in question may be associated. In the case of c, however, the matter is straightforward. The latest (1973) compilers† of the "best" values of the physical constants state: "All past measurements of c have been rendered obsolete by Evenson et al.'s recent measurement . . ."; this measurement is the last entry in Table 42-1.

  These workers, at the National Bureau of Standards in Boulder, Colorado, measured the frequency ν of a certain radiation emitted by a helium-neon laser by comparing it directly with the oscillation frequency of the cesium clock, which is used to define the second (see

† See footnote to Appendix B.

Appendix A). Then, using accurate measurements of the wavelength of this radiation made by several groups of workers, they computed c from the relation $c = \lambda\nu$ and deduced the value displayed in Appendix B, namely,

$$c = (299,792.4574 \pm 0.0012) \text{ km/s}.$$

The largest source of uncertainty in this measurement is that of the definition of the meter (needed to find the wavelength) in terms of radiations from the krypton-86 atom (see Appendix A).

It is now clear that the best measurements of c are *not* made by timing the passage of light over a measured distance, as by Fizeau in 1849 and by Michelson et al. in 1932 (see Table 42-1). They are made by measuring the frequency ν and the wavelength λ of the light and computing c from $c = \lambda\nu$. This holds true for either traveling waves, as in the measurements of Evenson and his collaborators, described above, or for standing waves, an example of which we now describe. These standing wave experiments will also remind us that c in free space has the same value throughout the entire electromagnetic spectrum and is not confined to visible light.

We describe here the "microwave cavity method" used by Essen in England and by Bol and Hansen in the U.S.A. It employs standing electromagnetic waves confined to a cavity rather than traveling waves in free space.

It is possible to convert a section of waveguide such as that of Fig. 41-9 into a resonant cavity by closing it with two metal caps; see Fig. 42-5. The pattern of oscillations in the cavity is closely related to that in the guide and exhibits the same "guide wavelength" λ_g. The guide wavelength is related to the cavity length l by

$$\lambda_g = \frac{2l}{n} \qquad n = 1, 2, 3, \cdots, \qquad (42\text{-}12)$$

which is the same relationship used for acoustic waves in closed pipes; $n\ (=3$ for Fig. 42-5) gives the number of half-waves contained in the cavity.

The procedure is to measure λ_g for such a cavity, which has been tuned to resonance, and then, using Eq. 41-4,

$$\lambda_g = \frac{\lambda}{\sqrt{1 - \left(\frac{\lambda}{2a}\right)^2}}, \qquad (42\text{-}13)$$

calculate the free-space wavelength λ. From the measured resonant frequency, the speed c can be found from $c = \lambda\nu$.

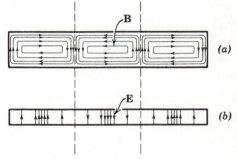

figure 42-5
A resonant cavity constructed from a section of waveguide; compare Fig. 41-9. For simplicity the lines of **E** are not shown in (*a*) and those of **B** are not shown in (*b*).

EXAMPLE 3

Essen of the National Physical Laboratory in England made a resonant cavity measurement of the speed of electromagnetic waves in 1950; see Table 42-1. His cavity was made of a circular waveguide rather than a rectangular one; it can be shown that, for the oscillation pattern used by him, the geometrical factor $2a$ in Eq. 42-13 must be replaced by $1.64062R$, where R is the guide radius. The cavity radius was 3.25876 cm; the cavity length was 15.64574 cm and it proved to resonate at 9.498300×10^9 Hz. At resonance it was determined that there were eight half-waves in the cavity. What value of c results?

From Eq. 42-12, computing only an approximate result,

$$\lambda_g = \frac{2l}{n} = \frac{(2)(15.6 \text{ cm})}{8} = 3.90 \text{ cm}.$$

Substituting into Eq. 42-13, suitably modified for a circular waveguide, yields

$$3.90 \text{ cm} = \frac{\lambda}{\sqrt{1 - \left(\frac{\lambda}{(1.64)(3.26 \text{ cm})}\right)^2}}.$$

Solving this equation for λ yields λ = 3.15 cm. Finally, we have

$$c = \lambda\nu = (3.15 \text{ cm})(9.50 \times 10^9 \text{ Hz}) = 2.99 \times 10^8 \text{ m/s.}$$

For practical reasons Essen analyzed his data in a more roundabout way than that given. His final result, based on many measurements under different conditions and carried to much greater accuracy than that illustrated in the above example, was 299,792.5 km/s with an uncertainty of 3.0 km/s.

When we say that the speed of sound in dry air at 0°C is 331.7 m/s, we imply a reference frame fixed with respect to the air mass. When we say that the speed of light in free space is 2.99792458×10^8 m/s, what reference frame is implied? It cannot be the medium through which the light wave travels because, in contrast to sound, no medium is required.

The concept of a wave requiring no medium was abhorrent to the physicists of the nineteenth century, influenced as they then were by a false analogy between light waves and sound waves or other purely mechanical disturbances. These physicists postulated the existence of an *ether*, which was a tenuous substance that filled all space and served as a medium of transmission for light. The ether was required to have a vanishingly small density to account for the fact that it could not be observed by any known means in an evacuated space.

The ether concept, although it proved useful for many years, did not survive the test of experiment. In particular, careful attempts to measure the speed of the earth through the ether always gave the result of zero.† Physicists were not willing to believe that the earth was permanently at rest in the ether and that all other bodies in the universe were in motion through it. Other hypotheses about the nature of the propagation of light also proved unsatisfactory for one reason or another.

Einstein in 1905 resolved the difficulty of understanding the propagation of light by making a bold postulate: If a number of observers are moving (at uniform velocity) with respect to each other and to a source of light and if each observer measures the speed of the light emerging from the source, *they will all obtain the same value.* This is the fundamental assumption of Einstein's special theory of relativity. It does away with the need for an ether by asserting that the speed of light is the same in *all* reference frames; none is singled out as fundamental. The theory of relativity, derived from this postulate, has been subject to many experimental tests, from which agreement with the predictions of theory has always emerged. These agreements, extending over half a century, lend strong support to Einstein's basic postulate about light propagation.

Figure 42-6 focuses specifically on the fundamental problem of light propagation. A source of light, at rest in reference frame S', emits a

42-4
MOVING SOURCES AND OBSERVERS*

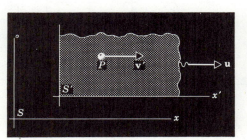

figure 42-6
Observers S and S', who are in relative motion, each observe a light pulse P. The pulse is emitted from a source, not shown, that is at rest in the S' frame of reference.

* For further information about the material of this and the next section see Supplementary Topic V.
† See Section 45-7, which describes the crucial experiment of Michelson and Morley.

light pulse P whose speed v' is measured by an observer at rest in this same frame. From the point of view of an observer in reference frame S, frame S' and its associated observer are moving in the positive x-direction at speed u. Question: What speed v would observer S measure for the light pulse P? Einstein's hypothesis asserts that *each* observer would measure the same speed c, or that

$$v = v' = c.$$

This hypothesis contradicts the classical law of addition of velocities (see Section 4-6), which asserts that

$$v = v' + u. \tag{42-14}$$

This law, which is so familiar that it seems (incorrectly) to be intuitively true, is in fact based on observations of gross moving objects in the world about us. Even the fastest of these—an earth satellite, say—is moving at a speed that is quite small compared to that of light. The body of experimental evidence that underlies Eq. 42-14 thus represents a severely restricted area of experience, namely, experiences in which $v' \ll c$ and $u \ll c$. If we assume that Eq. 42-14 holds for all particles regardless of speed, we are making a gross extrapolation. Einstein's theory of relativity predicts that this extrapolation is indeed not valid and that Eq. 42-14 is a limiting case of a more general relationship that holds for light pulses and for material particles, whatever their speed, or

$$v = \frac{v' + u}{1 + v'u/c^2}. \tag{42-15}$$

Equation 42-15 is quite indistinguishable from Eq. 42-14 at low speeds, that is, when $v' \ll c$ and $u \ll c$; see Example 4.

If we apply Eq. 42-15 to the case in which the moving object is a light pulse, and if we put $v' = c$, we obtain

$$v = \frac{c + u}{1 + cu/c^2} = c.$$

This is consistent, as it must be, with the fundamental assumption on which the derivation of Eq. 42-15 is based; it shows that *both* observers measure the same speed c for light. Equation 42-14 predicts (incorrectly) that the speed measured by S will be $c + u$. Figure 42-7 shows that the (correct) Eq. 42-15 and the (approximate) Eq. 42-14 cannot be distin-

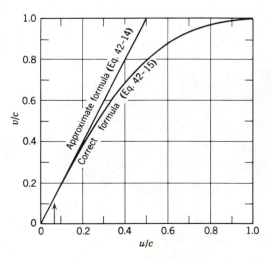

figure 42-7
The speed of a particle P, as seen by observer S in Fig. 42-6, for the special case of $v' = u$. All speeds are expressed as a ratio to c, the speed of light. The vertical arrow corresponds to about 0.25×10^8 m/s. ($= 5 \times 10^7$ mi/h.)

EXAMPLE 4

Suppose that $v' = u = 1.0 \times 10^5$ m/s $(=0.0003c = 230{,}000$ mi/h). What error is made in using Eq. 42-14 rather than Eq. 42-15 to calculate v?

Equation 42-14 gives

$$v = v' + u = 1.0 \times 10^5 \text{ m/s} + 1.0 \times 10^5 \text{ m/s} = 2.0 \times 10^5 \text{ m/s}.$$

Equation 42-15 gives

$$v = \frac{v' + u}{1 + v'u/c^2}$$

$$= \frac{1.0 \times 10^5 \text{ m/s} + 1.0 \times 10^5 \text{ m/s}}{1 + \left(\dfrac{1.0 \times 10^5 \text{ m/s}}{3.0 \times 10^8 \text{ m/s}}\right)^2}$$

$$= \frac{2.0 \times 10^5 \text{ m/s}}{1.00000011}.$$

Even at 230,000 mi/hr the error in Eq. 42-14 is immeasurably small.

EXAMPLE 5

Two electrons are ejected in opposite directions from radioactive atoms in a sample of radioactive material. Let each electron have a speed, as measured by a laboratory observer, of $0.6c$ (this corresponds to a kinetic energy of 130 keV). What is the speed of one electron as seen from the other?

Equation 42-14 gives

$$v = v' + u = 0.6c + 0.6c = 1.2c.$$

Equation 42-15 gives

$$v = \frac{v' + u}{1 + v'u/c^2} = \frac{0.6c + 0.6c}{1 + (0.6c)^2/c^2} = 0.88c.$$

This example shows that for speeds that are comparable to c, Eqs. 42-14 and 42-15 yield rather different results. A wealth of indirect experimental evidence points to the latter result as being correct.

42-5
DOPPLER EFFECT

We have seen that the same speed is measured for light no matter what the relative speeds of the light source and the observer are. The measured frequency and wavelength will change, but always in such a way that their product, which is the velocity of light, remains constant. Such frequency shifts are called *Doppler shifts*, after Johann Doppler (1803–1853), who first predicted them.

In Section 20-7 we showed that if a source of *sound* is moving away from an observer at a speed u, the frequency heard by the observer (see Eq. 20-10, which we have rearranged and in which we have substituted u for v_s) is

$$\nu' = \nu \, \frac{1}{1 + u/v} \cdot \qquad \begin{cases} 1. & \text{sound wave} \\ 2. & \text{observer fixed in medium} \\ 3. & \text{source receding from observer} \end{cases} \qquad (42\text{-}16)$$

In this equation ν is the frequency heard when the source is at rest and v is the speed of sound.

If the source is at rest in the transmitting medium but the observer is moving away from the source at speed u, the observed frequency (see Eq. 20-9b, in which u has been substituted for v_0) is

$$v' = v\left(1 - \frac{u}{v}\right) \cdot \quad \begin{cases} \text{1. sound wave} \\ \text{2. source fixed in medium} \\ \text{3. observer receding from source} \end{cases} \quad (42\text{-}17)$$

Even if the relative separation speeds u of the source and the observer are the same, the frequencies predicted by Eqs. 42-16 and 42-17 are different. This is not surprising, because a sound source moving through a medium in which the observer is at rest is physically different from an observer moving through that medium with the source at rest, as comparison of Figs. 20-10 and 20-11 shows.

We might be tempted to apply Eqs. 42-16 and 42-17 to light, substituting c, the speed of light, for v, the speed of sound. For light, as contrasted with sound, however, it has proved impossible to identify a medium of transmission relative to which the source and the observer are moving. This means that "source receding from observer" and "observer receding from source" are physically identical situations and must exhibit *exactly the same* Doppler frequency. As applied to light, either Eq. 42-16 or Eq. 42-17 or both must be incorrect. The Doppler frequency predicted by the theory of relativity is, in fact,

$$v' = v\,\frac{1 - u/c}{\sqrt{1 - (u/c)^2}} \cdot \quad \begin{cases} \text{1. light wave} \\ \text{2. source and observer} \\ \quad \text{separating} \end{cases} \quad (42\text{-}18)$$

In all three of the foregoing equations we obtain the appropriate relations for the source and the observer *approaching* each other if we replace u by $-u$.

Equations 42-16, 42-17, and 42-18 are not so different as they seem if the ratio u/c is small enough. This was made clear in Example 3, Chapter 20, for the first two of these equations. Let us expand Eqs. 42-16, 42-17, and 42-18 by the binomial theorem, as in the example referred to. The equations then become, substituting c for v,

$$v' = v\left[1 - \frac{u}{c} + \left(\frac{u}{c}\right)^2 + \cdots\right], \quad (42\text{-}16a)$$

$$v' = v\left(1 - \frac{u}{c}\right), \quad (42\text{-}17a)$$

and
$$v' = v\left[1 - \frac{u}{c} + \frac{1}{2}\left(\frac{u}{c}\right)^2 + \cdots\right] \cdot \quad (42\text{-}18a)$$

The ratio u/c for all available monochromatic light sources, even those of atomic dimensions, is small. This means that successive terms in these equations become small rapidly and, depending on the accuracy required, only a limited number of terms need be retained.

Under nearly all circumstances the differences among these three equations are not important. Nevertheless, it is of extreme interest to carry out at least one experiment precisely enough to serve as a test of Eq. 42-18a and thus, in part, of the theory of relativity.

H. E. Ives and G. R. Stilwell carried out such a precision experiment in 1938. They sent a beam of hydrogen atoms, generated in a gas discharge, down a tube at speed u, as in Fig. 42-8a. They could observe light emitted by these atoms in a direction opposite to **u** (atom 1, for example) using a mirror, and also in a direction parallel to **u** (atom 2, for example). With a precision spectrograph, they could photograph a particular characteristic spectrum line in this light, obtain-

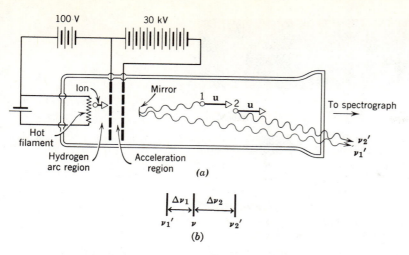

figure 42-8
The Ives-Stilwell experiment.

ing, on a frequency scale, the lines marked ν_1' and ν_2' in Fig. 42-8b. It is also possible to photograph, on the same photographic plate, a line corresponding to light emitted from *resting* atoms; such a line appears as ν in Fig. 42-9b. A fundamental measured quantity in this experiment is $\Delta\nu/\nu$, defined from

$$\frac{\Delta\nu}{\nu} = \frac{\Delta\nu_2 - \Delta\nu_1}{\nu}, \qquad (42\text{-}19)$$

(see Fig. 42-8b). It measures the extent to which the frequency of the light from resting atoms fails to lie halfway between the frequencies ν_1' and ν_2'. Table 42-2 shows that the measured results agree with the formula predicted by the theory of relativity (Eq. 42-18a) and not with the classical formula borrowed from the theory of sound propagation in a material medium (Eq. 42-16a).

Table 42-2
The Ives-Stilwell experiment*

$\dfrac{\Delta\nu}{\nu}$, 10^{-5}	Speed of moving atoms (=u), 10^6 m/s			
	0.865	1.01	1.15	1.33
Theoretical value according to classical theory (Eq. 42-16a)	1.67	2.26	2.90	3.94
Theoretical value according to the theory of relativity (Eq. 42-18a)	0.835	1.13	1.45	1.97
Experimental value	0.762	1.1	1.42	1.9

* See Eq. 42-19; the table shows only part of the data taken by Ives and Stilwell.

Ives and Stilwell did not present their experimental results as evidence for the support of Einstein's theory of relativity but rather gave them an alternative theoretical explanation. Modern observers, looking not only at their excellent experiment but at the whole range of experimental evidence, now give the Ives-Stilwell experiment the interpretation we have described for it above.

The Doppler effect for light finds many applications in astronomy, where it is used to determine the speeds at which luminous heavenly bodies are moving toward us or receding from us. Such Doppler shifts measure only the radial or line-of-sight components of the relative velocity. All galaxies† for which such measurements have been made

† See "The Red-Shift" by Allen R. Sandage, *Scientific American*, September 1956.

 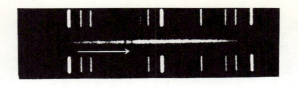

(a)   *(b)*

figure 42-9

(*a*) The central spot is a nebula in the constellation Corona Borealis; it is 130,000,000 light years distant. (*b*) The central streak shows the distribution in wavelength of the light emitted from this nebula. The two vertical dark bands show the presence of calcium. The horizontal arrow shows that these calcium lines occur at longer wavelengths than those for terrestrial light sources containing calcium, the length of the arrow representing the wavelength shift. Measurement of this shift shows that the galaxy is receding from us at 13,400 mi/s. The lines above and below the central streak represent light from a terrestrial source, used to establish a wavelength scale. (Courtesy Mount Wilson and Mount Palomar Observatories.)

(Fig. 42-9) appear to be receding from us, the recession velocity being greater for the more distant galaxies; these observations are the basis of the expanding-universe concept.

EXAMPLE 6

Certain characteristic wavelengths in the light from a galaxy in the constellation Virgo are observed to be increased in wavelength, as compared with terrestrial sources, by about 0.4%. What is the radial speed of this galaxy with respect to the earth? Is it approaching or receding?

If λ is the wavelength for a terrestrial source, then

$$\lambda' = 1.004\lambda.$$

Since we must have $\lambda'\nu' = \lambda\nu = c$, we can write this as

$$\nu' = 0.996\nu.$$

This frequency shift is so small that, in calculating the source velocity, it makes no practical difference whether we use Eq. 42-16, 42-17, or 42-18. Using Eq. 42-17 we obtain

$$\nu' = 0.996\nu = \nu\left(1 - \frac{u}{c}\right).$$

Solving yields $u/c = 0.004$, or $u = (0.004)(3 \times 10^8 \text{ m/s}) = 1.2 \times 10^6 \text{ m/s}$ or 2.7×10^6 mi/h. The galaxy is *receding*; had u turned out to be negative, the galaxy would have been moving toward us.

questions

1. How might an eye-sensitivity curve like that of Fig. 42-1 be measured?
2. Why are danger signals in red, when the eye is most sensitive to yellow-green?
3. Comment on this definition of the limits of the spectrum of visible light given by a physiologist: "The limits of the visible spectrum occur when the eye is no better adapted than any other organ of the body to serve as a detector."
4. The human body can "detect" electromagnetic radiations in parts of the electromagnetic spectrum apart from the visual range. Give examples for the infrared, ultraviolet, X-ray, and gamma ray regions. Some of the "detection processes" are not very pleasant.

5. In connection with Fig. 42-1 (*a*) do you think it possible that the wavelength of maximum sensitivity could vary if the intensity of the light is changed? (*b*) What might the curve of Fig. 42-1 look like for a group of color blind people who could not, for example, distinguish red from green? (See "The Science of Color" by S. J. Edwards, *Physics Education*, June 1975.)

6. Suppose that human eyes were insensitive to visible light but were very sensitive to infrared light. What environmental changes would be needed if you were to (*a*) walk down a long corridor and (*b*) drive a car? Would the phenomenon of color exist? How would traffic lights have to be modified?

7. How can an object absorb light energy without absorbing momentum?

8. Name two historic experiments, in addition to the radiation pressure measurements of Nichols and Hull, in which a torsion balance was used. Both are described in this book, one in Part 1 and one in Part 2.

9. A searchlight sends out a parallel beam of light. Does the searchlight experience any force associated with the emission of light?

10. In Section 42-2 we stated that the force on the mirror in the radiation pressure experiment of Nichols and Hull (see Fig. 42-2) was about 7×10^{-10} N. Identify an object whose weight at the earth's surface is about this magnitude (Table 1-4 might help).

11. Discuss this (1972) statement by Lewis M. Branscomb, Director of the National Bureau of Standards. "Scientists have looked forward to the possibility of using one gauge—one 'yardstick so to speak'—not only for the three dimensions of space but for the fourth dimension of time as well. To interchange clocks and rules scientists must know the speed with which light travels. . . . With this demonstration that both the space (wavelength) and time (frequency) dimensions of a single light source can be measured with prodigious accuracy, this goal is now within our grasp." What does he mean? (See *Science News*, February 1972.)

12. As you recline in a beach chair in the sun why are you so very conscious of the thermal energy that is delivered to you, but totally unresponsive to the linear momentum delivered from the same source?

13. Some advocate that the speed of light, now known to such high accuracy, be proclaimed as a defined constant rather than a measured one. What are the implications of this in terms of the present definitions of the meter and the second (Appendix A)?

14. When a parallel beam of light falls on an object the momentum transfers are given by Eqs. 42-2. Do these equations still hold if the light source is moving rapidly toward or away from the object at, perhaps, a speed of 0.1 c?

15. How could Galileo test experimentally that reaction times were an overwhelming source of error in his attempt to measure the speed of light, described on p. 923?

16. It has been suggested that the velocity of light may change slightly in value as time goes on. Can you find any evidence for this in Table 42-1?

17. Can you think of any "every day" observation (that is, without experimental apparatus) to show that the speed of light is not infinite? Think of lightning flashes, possible discrepancies between the predicted time of sunrise and the observed time, radio communications between earth and astronauts in orbiting space ships, and so on? Discuss.

18. A friend asserts that Einstein's postulate (that the speed of light is not affected by the uniform motion of the source or the observer) must be discarded because it violates "common sense." How would you answer him?

19. Why is the rotating mirror method for measuring the speed of light better than the toothed wheel method of Fizeau? (See Fig. 42-4.)

20. In a vacuum, does the speed of light depend on (*a*) the wavelength, (*b*) the frequency, (*c*) the intensity, (*d*) the speed of the source, or (*e*) the speed of the observer?

21. Can a galaxy be so distant that its recession speed equals c? If so, how can we see the galaxy? That is, will its light ever reach us?

22. How do the Doppler effects for light and for sound differ? In what ways are they the same?

23. Gamma rays are electromagnetic radiation emitted from radioactive nuclei. In free space, do they travel with the same speed as visible light? Does their speed depend on the speed of the nucleus that emits them?

problems

SECTION 42-1

1. (a) At what wavelengths does the eye sensitivity have half its maximum value? (b) What are the frequency and the period of the light for which the eye is most sensitive? See Fig. 42-1.
Answer: (a) 510 and 610 nm. (b) 5.5×10^{14} Hz and 1.8×10^{-15} s.

SECTION 42-2

2. Radiation from the sun striking the earth has an intensity of 1400 W/m². Assuming that the earth behaves like a flat disk at right angles to the sun's rays and that all the incident energy is absorbed, calculate the force on the earth due to radiation pressure. Compare it with the force due to the sun's gravitational attraction.

3. What is the radiation pressure 1.0 m away from a small 500-W light bulb? Assume that the surface on which the pressure is exerted faces the bulb and is perfectly absorbing and that the bulb radiates uniformly in all directions.
Answer: 1.3×10^{-7} Pa.

4. Prove, for a plane wave at normal incidence on a plane surface, that the radiation pressure on the surface is equal to the energy density in the beam outside the surface. This relation holds no matter what fraction of the incident energy is reflected.

5. Prove, for a stream of bullets striking a plane surface at right angles, that the "pressure" is *twice* the (kinetic) energy density in the stream above the surface; assume that the bullets are completely "absorbed" by the surface. Contrast this with the behavior of light. (See preceding problem.)

6. Show that for complete absorption of a parallel beam of light, the radiation pressure on the absorbing object is given by $p = S/c$, where S is the magnitude of the Poynting vector and c is the speed of light in free space.

7. It has been proposed that a spaceship might be propelled in the solar system by radiation pressure, using a large sail made of aluminum foil. How large must the sail be if the radiation force is to be equal in magnitude to the sun's gravitational attraction? Assume that the mass of the *ship + sail* is 1500 kg, that the sail is perfectly reflecting, and that the sail is oriented at right angles to the sun's rays. The sun's mass is 1.97×10^{30} kg. (Incidentally, NASA had plans to build such a "space clipper" to intercept Halley's Comet on its next swing into the inner solar system in 1986. The basic design was a sail made of polymer film coated with an aluminum film a few atoms thick. The sail area would be about 100 acres. The space clipper ship would have been well instrumented but unmanned.)
Answer: 9.2×10^5 m² (3200 ft on edge).

8. A particle in the solar system is under the combined influence of the sun's gravitational attraction and the radiation force due to the sun's rays. Assume that the particle is a sphere of density 1.0 g/cm³ and that all of the incident light is absorbed. (a) Show that all particles with radius less than some critical radius, R_0, will be blown out of the solar system. (b) Calculate R_0. (c) Does R_0 depend on the distance from the earth to the sun? (See the appendices for the necessary constants.)

9. Show that $\epsilon_0 \mathbf{E} \times \mathbf{B}$ has the dimensions of $\dfrac{\text{momentum}}{\text{area-time}}$, whereas $\dfrac{1}{\mu_0} \mathbf{E} \times \mathbf{B}$ has

the dimensions of $\dfrac{\text{energy}}{\text{area-time}}$. (The vector $\epsilon_0 \mathbf{E} \times \mathbf{B}$ may be used for computing momentum flow in the same manner that $\mathbf{S} = \dfrac{1}{\mu_0}\mathbf{E} \times \mathbf{B}$ is used to compute energy flow.)

10. A small spaceship whose mass, with occupant, is 1500 kg is drifting in outer space, where no gravitational field exists. If it shines a searchlight, which radiates 10^4 W, into space, what speed would the ship attain in one day because of the reaction force associated with the momentum carried away by the light beam?

11. A plane electromagnetic wave propagating through a vacuum is described by the following electric and magnetic fields: $E_x = E_0 \sin (kz - \omega t)$, $B_y = \dfrac{E_0}{c} \sin (kz - \omega t)$, $E_y = E_z = B_x = B_z = 0$. (a) What is the direction of propagation of the wave? (b) Find the average power per unit area. (c) What is the rate at which momentum is delivered to a perfectly absorbing surface of area A that is normal to the direction of propagation?
 Answer: (a) $+z$-direction. (b) $E_0^2/2\mu_0 c$. (c) $\epsilon_0 E_0^2 A/2$.

12. A plane electromagnetic wave, with wavelength 3.0 m, travels in free space in the $+x$-direction with its electric vector $\mathbf{E}$, of amplitude 300 V/m, directed along the y-axis. (a) What is the frequency ν of the wave? (b) What is the amplitude of the $\mathbf{B}$ field associated with the wave? (c) If $E = E_m \sin (kx - \omega t)$, what are the values of k and ω for this wave? (d) What is the time-averaged rate of energy flow per unit area associated with this wave? (e) If the wave fell upon a perfectly absorbing sheet of area A, what momentum would be delivered to the sheet per second and what is the radiation pressure exerted on the sheet?

SECTION 42-3

13. On page 926 the uncertainty in the measurement of the speed of light c is given as ± 0.0012 km/s. In Appendix B it is given as ± 0.004 parts per million. Show that these uncertainties are consistent.

14. For the value of the speed of light quoted on p. 926 (and in Appendix B) what is the uncertainty of the measurement, in feet/second? How many feet does light travel in one second?

15. Suppose that light is timed over a one-mile base line and its speed is measured to the accuracy quoted on p. 926. How large an error in the length of the base line could be tolerated, assuming other sources of error to be negligible?   *Answer:* 3×10^{-4} in.

16. *Bradley's method for determining the speed of light.* Consider a star located on a line through the sun, drawn perpendicular to the plane of the earth's orbit about the sun. The star's distance is much greater than the radius of the earth's orbit. Show that, due to the finite speed of light, a telescope through which the star is seen must be tilted at an angle of 20.5″ to the perpendicular, in the direction the earth is moving. This phenomenon, called *aberration,* is noticeable and was first explained by James Bradley in 1729. (Hint: draw an analogy to a person running through rain while holding an umbrella.)

17. Suppose that we were able to establish radio communication with the hypothetical inhabitants of a hypothetical planet orbiting our nearest star, α-Centauri, which is 4.2 light years from us. How long would it take to receive a reply to a message? Repeat for the Great Nebula in Andromeda, one of our closest extragalactic neighbors but 2×10^6 light years distant. What do these considerations lead you to conclude about the nature of our possible communication with extragalactic peoples?
 Answer: 8.4 y; 4×10^6 y.

18. In Table 42-1, which of the four determinations of c reported by Michelson and his collaborators best agrees with the currently accepted value of c as displayed in Appendix B? Take the uncertainties of measurement into ac-

count. You might make a graph showing, on a suitable scale, the three values and their uncertainties displayed as error bars.

19. The uncertainty of the distance to the moon, as measured by the reflection of radar waves from it, is about 0.8 km. Assuming that this uncertainty is associated only with the measurement of the elapsed time, what uncertainty in this time is implied?   *Answer: 5.3 μs.*

20. Roemer's method for measuring the speed of light consisted in observing the apparent times of revolution of one of the moons of Jupiter. The true period of revolution is 42.5 hr. (*a*) Taking into account the finite speed of light, how would you expect the apparent time of revolution to alter as the earth moves in its orbit from point *x* to point *y* in Fig. 42-10? (*b*) What observations would be needed to compute the speed of light? Neglect the motion of Jupiter in its orbit. Figure 42-10 is not drawn to scale.

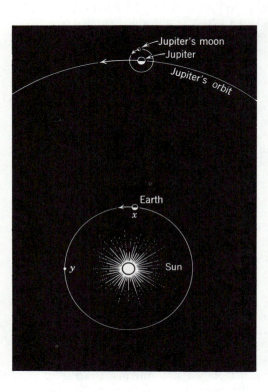

figure 42-10
Problem 20

SECTION 42-4
21. Assume that $v' = u$ in Eqs. 42-14 and 42-15. At what speed u would these expressions differ by 1.0%?   *Answer: 0.1 c.*

SECTION 42-5
22. Show that a Doppler frequency shift in kHz can be converted to a radial velocity in km/s by multiplying by -0.211.

23. A rocketship is receding from the earth at a speed of 0.2c. A light in the rocketship appears blue to passengers on the ship. What color would it appear to be to an observer on the earth? (See Fig. 42-1.)
Answer: Yellow-orange.

24. The difference in wavelength between an incident microwave beam and one reflected from an approaching or receding car is used to determine automobile speeds on the highway. (*a*) Show that if v is the speed of the car and ν the frequency of the incident beam, the change of frequency is approximately $2v\nu/c$, where c is the speed of the electromagnetic radiation. (*b*) For microwaves of frequency 2450 MHz, what is the change of frequency per mi/hr of speed?

25. The period of rotation of the sun at its equator is 24.7 days; its radius is 7.0×10^8 m. What Doppler wavelength shifts are expected for characteristic wavelengths in the vicinity of 550 nm emitted from the edge of the sun's disk?
$\qquad$ *Answer:* 3.8×10^{-3} nm.

26. An earth satellite, transmitting on a frequency of 40 MHz (exactly), passes directly over a radio receiving station at an altitude of 250 miles and at a speed of 18,000 mi/hr. Plot the change in frequency, attributable to the Doppler effect, as a function of time, counting $t = 0$ as the instant the satellite is over the station. (*Hint:* The speed u in the Doppler formula is not the actual velocity of the satellite but its component in the direction of the station. Use the nonrelativistic formula (Eq. 42-16a) and neglect the curvature of the earth and of the satellite orbit.)

27. The "red shift" of radiation from a distant nebula consists of the light (H_γ), known to have a wavelength of 4340 Å when observed in the laboratory, appearing to have a wavelength of 6562 Å. (*a*) What is the speed of the nebula in the line of sight relative to the earth? (*b*) Is it approaching or receding?
$\qquad$ *Answer:* (*a*) 1.2×10^8 m/s. (*b*) Receding.

28. Show that, for slow speeds, the Doppler shift can be written in the approximate form

$$\frac{\Delta\lambda}{\lambda} = \frac{u}{c},$$

where $\Delta\lambda$ is the change in wavelength.

29. In the experiment of Ives and Stilwell the speed u of the hydrogen atoms in a particular run was 8.61×10^5 m/s. Calculate $\Delta\nu/\nu$, on the assumptions that (*a*) Eq. 42-18a is correct and (*b*) that Eq. 42-16a is correct; compare your results with those given in Table 42-2 for this speed. Retain the first three terms only in Eqs. 42-18a and 42-16a.
Answer: (*a*) $\Delta\nu/\nu = 0.825 \times 10^{-5}$. (*b*) $\Delta\nu/\nu = 1.65 \times 10^{-5}$.

43

reflection and refraction–plane waves and plane surfaces

So far we have considered the electromagnetic spectrum, including visible light, only in unimpeded free space. Here we examine it as reflected from plane surfaces such as glass or water and especially as it behaves when it passes through such transparent materials. As we shall see in the next chapter, without the consideration of these properties, such devices as cameras, telescopes, spectacles, microscopes, etc., could not be explained. As always, what we say in this chapter for visible light has its counterpart in other regions of the electromagnetic spectrum.

In Fig. 43-1a a light beam falling on a water surface is both reflected from the surface and bent (that is, *refracted*) as it enters the water.* We represent the incident beam in Fig. 43-1b by a single line, the *incident ray*, parallel to the direction of propagation. We assume the incident beam in Fig. 43-1b to be a *plane wave*, the wavefronts being normal to the incident ray. The reflected and refracted beams are also represented

43-1
REFLECTION AND REFRACTION

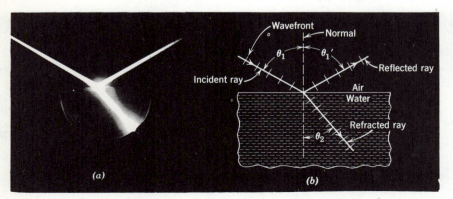

figure 43-1
(a) A photograph showing reflection and refraction at an air-water interface. (b) A representation using rays.

* A common and spectacular example of reflection and refraction is the rainbow. See "The Theory of the Rainbow" by H. Moysés Nussenzveig, *Scientific American*, April 1977.

by rays. The angles of *incidence* (θ_1), of *reflection* (θ_1'), and of *refraction* (θ_2) are measured between the normal to the surface and the appropriate ray, as shown in the figure.

By experiment, we find these laws governing reflection and refraction:

1. The reflected and the refracted rays lie in the plane formed by the incident ray and the normal to the surface at the point of incidence, that is, the plane of Fig. 43-1*b*.

2. For reflection:

$$\theta_1' = \theta_1. \qquad (43\text{-}1)$$

3. For refraction:

$$\frac{\sin \theta_1}{\sin \theta_2} = n_{21}, \qquad (43\text{-}2)$$

where n_{21} is a constant called the *index of refraction* of medium 2 with respect to medium 1. Table 43-1 shows the indices of refraction for some common substances with respect to a vacuum for a wavelength (sodium light) of 589 nm (= 5890 Å).

The index of refraction of one medium with respect to another generally varies with wavelength, as Fig. 43-2 shows. Because of this fact

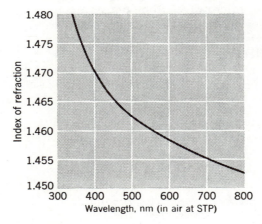

figure 43-2
The index of refraction of fused quartz with respect to a vacuum.

Table 43-1
Some indices of refraction*
[for $\lambda = 589$ nm (= 5890 Å)]

Medium	Index of Refraction
Water	1.33
Ethyl alcohol	1.36
Carbon bisulfide	1.63
Air (1 atm and 20°C)	1.0003
Methylene iodide	1.74
Fused quartz	1.46
Glass, crown	1.52
Glass, dense flint	1.66
Sodium chloride	1.53

* Measured with respect to a vacuum. The index with respect to air (except, of course, the index of air itself) will be negligibly different in most cases.

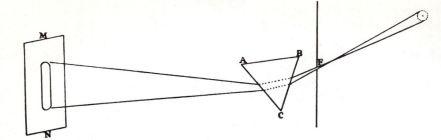

figure 43-3
Sunlight from a hole F in a screen
is refracted by prism ABC, forming
a spectrum on screen *MN*; from
Newton's Opticks (1704).

refraction, unlike reflection, can be used to analyze a beam of light into its component wavelengths. Figure 43-3, taken from Newton's *Opticks*, shows how Newton, using glass prism *ABC*, formed a spectrum of sunlight entering his window through a small hole at *F*.

The law of reflection was known to Euclid. That of refraction was discovered experimentally by Willebrod Snell (1591–1626) and deduced from the early corpuscular theory of light by René Descartes (1596–1650). The law of refraction is known as Snell's law or (in France) as Descartes' law.

We can derive the laws of reflection and refraction from Maxwell's equations, which means that these laws should hold for all regions of the electromagnetic spectrum. Figure 43-4*a* shows an experimental setup for investigating the reflection of microwaves from a large metal sheet. Figure 43-4*b* shows the reading of the detector as a function of the angular position of the mirror. The existence of a reflected beam at the proper angle confirms the law of reflection for microwaves. There is ample experimental evidence that Eqs. 43-1 and 43-2 correctly describe the behavior of reflected and refracted beams in all parts of the electromagnetic spectrum.

It is common knowledge that a polished steel surface will form a well-defined reflected beam if an incident beam falls on it, but a sheet of paper will reflect

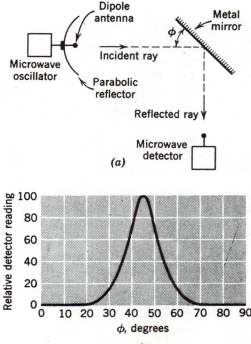

figure 43-4
(*a*) An apparatus to test the law of reflection for microwaves. (*b*) A reflected beam (for $\lambda \cong 10$ cm) appears for the expected orientation of the mirror.

light more or less in all directions (diffuse reflection). It is largely by diffuse reflection that we see nonluminous objects around us. The difference between diffuse and specular (that is, mirrorlike) reflection is a matter of surface roughness; a reflected beam will be formed only if the average depth of the surface irregularities of the reflector is substantially less than the wavelength of the incident light. This criterion of surface roughness has different implications in different regions of the electromagnetic spectrum. The bottom of a cast-iron skillet, for example, is a good reflector for microwaves of wavelength 0.5 cm but is not a good reflector for visible light (that is, one cannot shave or put on makeup by it).

A second requirement for the existence of a reflected beam is that the transverse dimensions of the reflector must be substantially larger than the wavelength of the incident beam. If a beam of visible light falls on a polished metal disk the size of a dime, a reflected beam will be formed. However, if the same disk is placed in a beam of short radio waves with, say, $\lambda = 1.0$ m, radiation will be scattered in all directions from it, and no well-defined unidirectional beam will appear. We investigate this phenomenon of *diffraction* in Chapter 46. The requirements that surfaces be "smooth" and "large" also apply to the formation of refracted beams. If these two requirements are not met the description of reflection and refraction in terms of rays, whose behavior is governed by Eqs. 43-1 and 43-2, is not valid.

EXAMPLE 1

Figure 43-5 shows an incident ray i striking a plane mirror MM' at angle of incidence θ. Trace this ray.

The reflected ray makes an angle θ with the normal at b and falls as an incident ray on mirror $M'M''$. Its angle of incidence θ' on this mirror is $\pi/2 - \theta$. A second reflected ray r' makes an angle θ' with the normal erected at b'. Rays i and r' are antiparallel for any value of θ. To see this, note that

$$\phi = \pi - 2\theta' = \pi - 2\left(\frac{\pi}{2} - \theta\right) = 2\theta.$$

Two lines are parallel if their opposite interior angles for an intersecting line (ϕ and 2θ) are equal.

Repeat the problem if the angle between the mirrors is 120° rather than 90°.

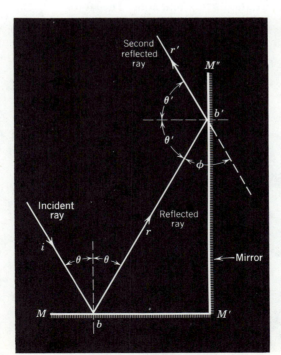

figure 43-5
Example 1.

EXAMPLE 2

An incident ray in air falls on the plane surface of a block of quartz and makes an angle of 30° with the normal. This beam contains two wavelengths, 400 and 500 nm. The indices of refraction for quartz with respect to air (n_{qa}) at these wavelengths are 1.4702 and 1.4624, respectively. What is the angle between the two refracted beams?

From Eq. 43-2 we have, for the 400-nm beam,

$$\sin \theta_1 = n_{qa} \sin \theta_2,$$

or

$$\sin 30° = (1.4702) \sin \theta_2,$$

which leads to

$$\theta_2 = 19.88°.$$

For the 500-nm beam we have

$$\sin 30° = (1.4624) \sin \theta_2',$$

or

$$\theta_2' = 19.99°.$$

The angle $\Delta\theta$ between the beams is 0.11°, the shorter wavelength component being bent through the larger angle, that is, having the smaller angle of refraction.

EXAMPLE 3

An incident ray falls on one face of a glass prism in air as in Fig. 43-6. The angle θ is so chosen that the emerging ray also makes an angle θ with the normal to the other face. Derive an expression for the index of refraction of the prism material with respect to air.

Note that $\angle\, abc = \alpha$, the two angles having their sides mutually perpendicular. Therefore,

$$\alpha = \tfrac{1}{2}\phi \text{ in which } \phi \text{ is the prism angle.} \qquad (43\text{-}3)$$

The deviation angle ψ is the sum of the two opposite interior angles in triangle aed, or

$$\psi = 2(\theta - \alpha).$$

Substituting $\tfrac{1}{2}\phi$ for α and solving for θ yields

$$\theta = \tfrac{1}{2}(\psi + \phi). \qquad (43\text{-}4)$$

At point a, θ is the angle of incidence and α the angle of refraction. The law of refraction (see Eq. 43-2) is

$$\sin \theta = n_{ga} \sin \alpha,$$

in which n_{ga} is the index of refraction of the glass with respect to air.

From Eqs. 43-3 and 43-4 this yields

$$\sin \frac{\psi + \phi}{2} = n_{ga} \sin \frac{\phi}{2}$$

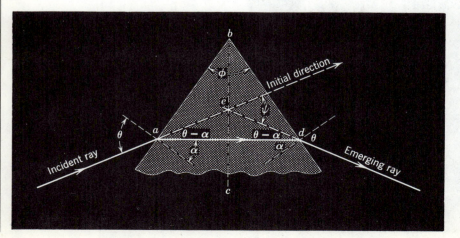

figure 43-6
Example 3.

or

$$n_{ga} = \frac{\sin \frac{1}{2}(\psi + \phi)}{\sin (\phi/2)}, \qquad (43\text{-}5)$$

which is the desired relation. This equation holds only for θ so chosen that the light ray passes symmetrically through the prism. For this condition the deviation angle ψ is a minimum; if θ is either increased or decreased, a larger deviation will be produced. ψ is called the *angle of minimum deviation*.

A theory of light would not be accepted if it were not able to predict the well-established laws of reflection and refraction. We could derive these laws from Maxwell's equations, but mathematical complexity prevents us from doing so here. Fortunately, we can derive these and several other laws of optics on the basis of a simpler but less comprehensive theory of light, put forward by the Dutch physicist Christian Huygens in 1678. This theory simply assumes that light is a wave rather than, say, a stream of particles. It says nothing about the nature of the wave and, in particular—since Maxwell's theory of electromagnetism appeared only after the lapse of a century—gives no hint of the electromagnetic character of light. Huygens did not know whether light was a transverse wave or a longitudinal one; he did not know the wavelengths of visible light; he had little knowledge of the speed of light. Nevertheless, his theory was a useful guide to experiment for many years and remains useful today for pedagogic and certain other practical purposes. We must not expect it to yield the same wealth of detailed information that Maxwell's more complete electromagnetic theory does.

Huygens' theory is based on a geometrical construction, called *Huygens' principle,* that allows us to tell where a given wavefront will be at any time in the future if we know its present position; it is: *All points on a wavefront can be considered as point sources for the production of spherical secondary wavelets. After a time t the new position of the wavefront will be the surface of tangency to these secondary wavelets.*

We illustrate this by a trivial example: Given a wavefront (ab in Fig. 43-7) in a plane wave in free space, where will the wavefront be a

43-2
HUYGENS' PRINCIPLE

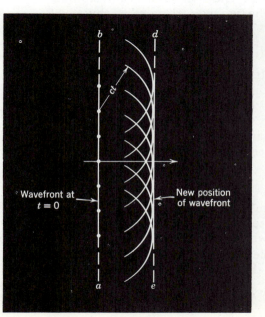

figure 43-7
The propagation of a plane wave in free space is described by the Huygens construction. Note that the ray (horizontal arrow) representing the wave is perpendicular to the wavefronts.

time t later? Following Huygens' principle, we let several points on this plane (see dots) serve as centers for secondary spherical wavelets. In a time t the radius of these spherical waves is ct, where c is the speed of light in free space. We represent the plane of tangency to these spheres at time t by de. As we expect, it is parallel to plane ab and a perpendicular distance ct from it. Thus plane wavefronts are propagated as planes and with speed c. Note that the Huygens method involves a three-dimensional construction and that Fig. 43-7 is the intersection of this construction with the plane of the page.

We might expect that, contrary to observation, a wave should be radiated backward as well as forward from the dots in Fig. 43-7. This result is avoided by assuming that the intensity of the spherical wavelets is not uniform in all directions but varies continuously from a maximum in the forward direction to a minimum of zero in the back direction. This is suggested by the shading of the circular arcs in Fig. 43-7. Huygens' method can be applied quantitatively to *all* wave phenomena; see Problem 14. The method was put on a firm mathematical footing by Augustin Fresnel (1788–1827).

Figure 43-8a shows three wavefronts in a plane wave falling on a plane mirror MM'. For convenience they are chosen to be one wavelength apart. Note that θ_1, the angle between the wavefronts and the mirror, is the same as the angle between the incident ray and the normal to the mirror. In other words, θ_1 is the *angle of incidence*. The three wavefronts are related to each other by the Huygens construction, as in Fig. 43-7.

In Fig. 43-8b a Huygens wavelet centered on point a will expand to include point l after a time λ/c. Light from point p in this same wavefront cannot move beyond the mirror but must expand upward as a spherical Huygens wavelet. Setting a compass to radius λ and swinging an arc about p provides a semicircle to which the reflected wavefront must be tangent. Since point l must lie on the new wavefront, this tangent must pass through l. Note that the angle θ_1' between the wavefront and the mirror is the same as the angle between the reflected ray and the normal to the mirror. In other words, θ_1' is the *angle of reflection*.

Consider right triangles alp and $a'lp$. They have side lp in common and side al $(=\lambda)$ is equal to side $a'p$. The two right triangles are thus congruent and we may conclude that

$$\theta_1 = \theta_1',$$

as required by the law of reflection. If you recall that the Huygens construction is three-dimensional and that the arcs shown represent segments of spherical surfaces, you will be able to convince yourself that the reflected ray lies in the plane formed by the incident ray and the normal to the mirror, that is, the plane of Fig. 43-8. This is also a requirement of the law of reflection; see p. 939.

Figure 43-9 shows four stages in the refraction of three successive wavefronts in a plane wave falling on an interface between air (medium 1) and glass (medium 2). For convenience, we assume that the incident wavefronts are separated by λ_1, the wavelength as measured in medium 1. Let the speed of light in air be v_1 and that in glass be v_2. We assume that

$$v_2 < v_1. \tag{43-6}$$

43-3
HUYGENS' PRINCIPLE AND THE LAW OF REFLECTION

43-4
HUYGENS' PRINCIPLE AND THE LAW OF REFRACTION

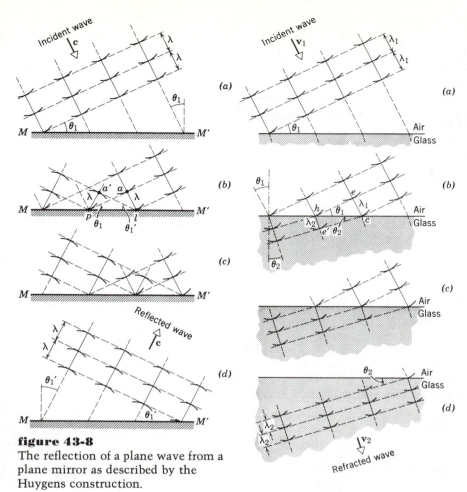

figure 43-9
The refraction of a plane wave at a plane interface as described by Huygens construction; we do not show the reflected wave, for simplicity. Note the change in wavelength on refraction.

figure 43-8
The reflection of a plane wave from a plane mirror as described by the Huygens construction.

This assumption about the speeds is vital to the derivation that follows. It was not possible to test it experimentally because of technical difficulties until 1850, when the assumption was shown by Foucault to be correct.

The wavefronts in Fig. 43-9a are related to each other by the Huygens construction of Fig. 43-7. As in Fig. 43-8, θ_1 is the angle of incidence. In Fig. 43-9b consider the time $(=\lambda_1/v_1)$ during which a Huygens wavelet from point e moves to include point c. Light from point h, traveling through glass at a reduced speed (recall the assumption of Eq. 43-6) will move a shorter distance

$$\lambda_2 = \lambda_1 \frac{v_2}{v_1} \qquad (43\text{-}7)$$

during this time. The refracted wavefront must be tangent to an arc of this radius centered on h. Since c lies on the new wavefront, the tangent must pass through this point, as shown. Note that θ_2, the angle between the refracted wavefront and the air-glass interface, is the same as the angle between the refracted ray and the normal to this interface. In other words, θ_2 is the *angle of refraction*. Note, too, that the wavelength in glass (λ_2) is less than the wavelength in air (λ_1).

For the right triangles hce and hce' we may write

$$\sin \theta_1 = \frac{\lambda_1}{hc} \qquad \text{(for } hce\text{)}$$

and

$$\sin \theta_2 = \frac{\lambda_2}{hc} \qquad \text{(for } hce'\text{)}.$$

Dividing and using Eq. 43-7 yields

$$\frac{\sin \theta_1}{\sin \theta_2} = \frac{\lambda_1}{\lambda_2} = \frac{v_1}{v_2} = \text{a constant.} \qquad (43\text{-}8)$$

The law of refraction, as stated in Eq. 43-2, is

$$\frac{\sin \theta_1}{\sin \theta_2} = n_{21}, \qquad (43\text{-}2)$$

so that n_{21} is now revealed as the ratio of the speeds of light in the two media, or

$$n_{21} = \frac{v_1}{v_2}. \qquad (43\text{-}9)$$

We may rewrite Eq. 43-8 as

$$\left(\frac{c}{v_1}\right) \sin \theta_1 = \left(\frac{c}{v_2}\right) \sin \theta_2, \qquad (43\text{-}10)$$

in which c is the speed of light in free space. The quantities (c/v_1) and (c/v_2) (see Eq. 43-9) are the indices of refraction of medium 1 and of medium 2, respectively, with respect to a vacuum. Introducing the symbols n_1 and n_2 for these quantities allows us to write the law of refraction as

$$n_1 \sin \theta_1 = n_2 \sin \theta_2. \qquad (43\text{-}11)$$

If we assume that the medium above the glass in Fig. 43-9 is a vacuum rather than air, the speed v_1 becomes c and the wavelength, called λ_1 in Fig. 43-9, assumes a value λ that is characteristic of the wave in free space. Equation 43-7 may thus be written

$$\lambda_2 = \lambda \frac{v_2}{c} = \frac{\lambda}{n_2} \qquad (43\text{-}12a)$$

or

$$\lambda_n = \frac{\lambda}{n}. \qquad (43\text{-}12b)$$

This shows specifically that the wavelength of light in a material medium is less than the wavelength of the same wave in a vacuum. Figure 43-9 shows clearly the difference in wavelength in the two media.

The application of Huygens' principle to refraction requires that if a light ray is bent toward the normal in passing from air to an optically dense medium then the speed of light in that optically dense medium (glass, say) must be *less* than that in air; see Eq. 43-6. This requirement holds for all wave theories of light. For the early particle theory of light put forward by Newton, refraction can be explained only if the speed of light in the medium in which light is bent toward the normal (the optically dense medium) is *greater* than that in air. The dense medium was thought to exert attractive forces on the light "corpuscles" as they neared the surface, speeding them up and changing their direction to cause them to make a smaller angle with the normal. Figure 43-10 shows a figure from a 1637 work of René Descartes, in which he makes an analogy between the refraction of light and the motion of a tennis ball on entering a medium in which it moves more slowly.

An experimental comparison of the speed of light in water and in air is decisive, therefore, between the wave and corpuscular theories of light. Such a measurement was first carried out by Foucault in 1850; he showed conclusively

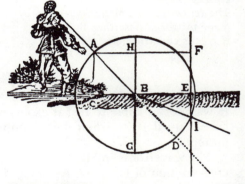

figure 43-10
According to the early (incorrect) corpuscular theory of light, ABI is the path of a ray on entering a medium in which its speed is less. The (correct) wave theory predicts that for the ray shown the speed in the lower medium (below BE) must be greater; from Descartes' La Dioptrique (1637).

that light travels more slowly in water than in air, thus ruling out the corpuscular theory of Newton.

Let light rays in an optically dense medium (glass, say) fall on a surface on the other side of which is a less optically dense medium (air, say); see Fig. 43-11. As the angle of incidence θ is increased, a situation is reached (see ray e) at which the refracted ray points along the surface, the angle of refraction being 90°. For angles of incidence larger than this *critical angle* θ_c no refracted ray exists, giving rise to a phenomenon called *total internal reflection*.

We find the critical angle by putting $\theta_2 = 90°$ in the law of refraction (see Eq. 43-11):

$$n_1 \sin \theta_c = n_2 \sin 90°,$$

or

$$\sin \theta_c = \frac{n_2}{n_1}. \qquad (43\text{-}13)$$

For glass and air $\sin \theta_c = (1.00/1.50) = 0.667$, which yields $\theta_c = 41.8°$. Total internal reflection does not occur when light originates in the medium of lower index of refraction.

43-5
TOTAL INTERNAL REFLECTION

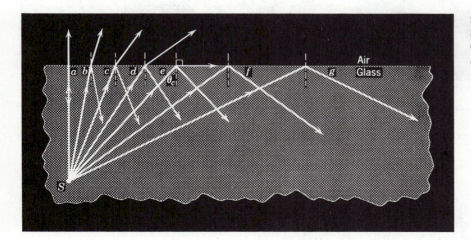

figure 43-11
Showing the total internal reflection of light from a source S; the *critical angle* is θ_e.

Light can be "piped" from one point to another with little loss by allowing it to enter one end of a rod of transparent plastic. The light will undergo total internal reflection at the boundary of the rod and will follow its contour, emerging at its far end. Images may be transferred from one location to another, using a bundle of fine glass fibers, each fiber transmitting a small fraction of the image.*

Fiber optics techniques make possible many useful optical devices for transmitting and transforming luminous images. Figure 43-12 shows a short fiber bundle constructed so that the fibers taper in diameter along its length. The wide end is shown placed over the letter S in the printed word "OPTICS." We see, with the aid of a mirror placed above the bundle, that a letter S, reduced in size, has been transmitted by total internal reflection through the bundle to its narrow end.

* See "Fiber Optics" by N. S. Kapany, *Scientific American*, November 1960.

figure 43-12

A bundle of tapered fibers (below) is placed over the letter *S*. Above, with the aid of a mirror, we see that the image, reduced in size, is transmitted to the top of the bundle by total internal reflection in the individual fibers. (Courtesy Dr. N. S. Kapany, Optics Technology, Inc.)

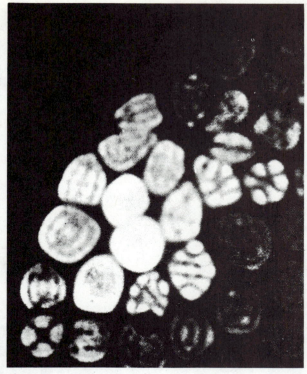

figure 43-13

A photomicrograph of light emerging from the end of a bundle of fibers. The fiber diameters approach the wavelength of light so that each fiber acts like an "optical waveguide." We have here convincing visual evidence that light is an electromagnetic wave. (Courtesy of Dr. N. S. Kapany, Optics Technology, Inc.)

Figure 43-13 is an enlarged view of a cross section of such a fiber bundle in which the diameters of the individual fibers are made so small that they are of the order of magnitude of the wavelength of light. This condition violates the spirit of our assumption of p. 941, namely, that the transverse dimensions of reflecting and refracting surfaces would be large compared to the wavelength of light. Consequently, a description of the reflection and refraction of light in terms of rays, as in Figs. 43-1*b* and 43-11, is not possible. Figure 43-13 is readily interpreted, however, on the basis of the electromagnetic wave theory of light and provides convincing pictorial supporting evidence for that theory. The fibers behave like waveguides (see Section 41-6), and the patterns of darkness and light represent the distribution of the **E** and **B** vectors for various modes of oscillation of the electromagnetic waves traveling down the "guides."

As of 1977 several experimental, field condition, tests of telephone communication* by optical fiber are underway. In a test in Chicago 24 hairlike fibers formed, with their protective sheath, a cable 0.50 in. in diameter and 1.5 miles long. The incoming voice signal is transformed into coded light pulses which are sent along the fiber at 4.47 pulses/second; the pulses are reconstituted at the far end to form a replica of the input signal. The design capability of the cable is 8064 simultaneous two-way conversations. The chemical purity requirements of the fiber materials and the design requirements of the electrical $\rightleftharpoons$ optical input and output devices at the ends of the cable push solid-state and materials research technologies close to their present-day limits.

* See "Communication by Optical Fiber" by J. S. Cook, *Scientific American*, November 1973; "Telephoning by Light, I: The Breakthroughs" by John H. Douglas, *Science News*, July 1975; "Light-Wave Communications" by W. S. Boyle, *Scientific American*, August 1977.

EXAMPLE 4

Figure 43-14a shows a triangular prism of glass, a ray incident normal to one face being totally reflected. If θ_1 is 45°, what can you conclude about the index of refraction n of the glass?

The angle θ_1 must be equal to or greater than the critical angle θ_c where θ_c is given by Eq. 43-13:

$$\sin \theta_c = \frac{n_2}{n_1} = \frac{1}{n},$$

in which, for all practical purposes, the index of refraction of air $(=n_2)$ is set equal to unity. Suppose that the index of refraction of the glass is such that total internal reflection just occurs, that is, that $\theta_c = 45°$. This would mean

$$n = \frac{1}{\sin 45°} = 1.41.$$

Thus the index of refraction of the glass must be *equal to or larger than* 1.41. If it were less, total internal reflection would not occur.

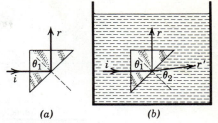

figure 43-14
Examples 4, 5

EXAMPLE 5

What happens if the prism in Example 4 (assume that $n = 1.50$) is immersed in water $(n = 1.33)$? See Fig. 43-14b.

The new critical angle, given by Eq. 43-13, is

$$\sin \theta_c = \frac{n_2}{n_1} = \frac{1.33}{1.50} = 0.887,$$

which corresponds to $\theta_c = 62.5°$. The actual angle of incidence $(=45°)$ is less than this so that we do *not* have total internal reflection.

There is a reflected ray, with an angle of reflection of 45°, as Fig. 43-14b shows. There is also a refracted ray, with an angle of refraction given by

$$n_1 \sin \theta_1 = n_2 \sin \theta_2$$

$$(1.50)(\sin 45°) = (1.33) \sin \theta_2,$$

which yields $\theta_2 = 52.9°$. Show that as $n_2 \to n_1$, $\theta_c \to 90°$.

Maxwell's equations permit us to calculate how the incident energy is divided between the reflected and the refracted beams. Figure 43-15 shows the theoretical prediction for (a) a light beam in air falling on a glass-air interface and (b) a light beam in glass falling on such an interface. Figure 43-15a shows that for angles of incidence up to about 50°, less than 10% of the light energy is reflected. At grazing incidence, however (that is, angles of incidence near 90°), the surface becomes an excellent reflector. We are all familiar with the high reflecting power of a wet road for light from automobile headlights that strikes near grazing incidence.

Figure 43-15b shows clearly that at a certain critical angle (41.8° in this case; see Eq. 43-13) *all* the light is reflected. For angles of incidence appreciably below this value, about 4% of the energy is reflected.

43-6
FERMAT'S PRINCIPLE

In 1650 Pierre Fermat discovered a remarkable principle which we can express in these terms: *A light ray traveling from one point to another will follow a path such that, compared with nearby paths, the time required is either a minimum or a maximum or will remain unchanged (that is, it will be stationary).*

We can readily derive the laws of reflection and refraction from this

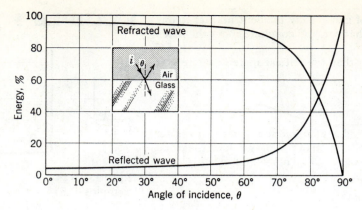

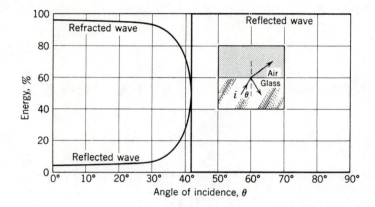

figure 43-15
(a) The percent of the energy
reflected and refracted when an
incident wave in air falls on glass
(n = 1.50). The same for the
incident wave in glass, showing
total internal reflection.

principle. Figure 43-16 shows two fixed points A and B and a ray APB connecting them.* The total length l of this ray is

$$l = \sqrt{a^2 + x^2} + \sqrt{b^2 + (d - x)^2},$$

where x locates the point P at which the ray touches the mirror.

According to Fermat's principle, P will have a position such that the time of travel of the light must be a minimum (or a maximum or must remain unchanged). Expressed in another way, the total length l of the ray must be a minimum (or a maximum or must remain unchanged). In either case, the methods of the calculus require that dl/dx be zero. Taking this derivative yields

$$\frac{dl}{dx} = (\tfrac{1}{2})(a^2 + x^2)^{-1/2}(2x) + \tfrac{1}{2}[b^2 + (d - x)^2]^{-1/2}(2)(d - x)(-1) = 0,$$

which we can rewrite as

$$\frac{x}{\sqrt{a^2 + x^2}} = \frac{d - x}{\sqrt{b^2 + (d - x)^2}}.$$

Comparison with Fig. 43-16 shows that we can rewrite this as

$$\sin \theta_1 = \sin \theta_1',$$

or

$$\theta_1 = \theta_1',$$

which is the law of reflection.

To prove the law of refraction from Fermat's principle, consider Fig.

* We assume that ray APB lies in the plane of the figure; see Problem 26.

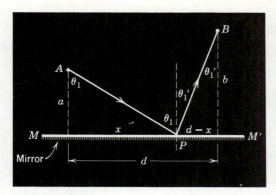

figure 43-16
A ray from A passes through B after reflection at P.

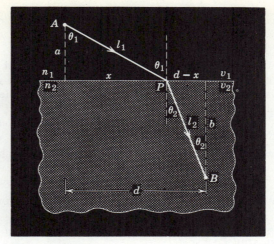

figure 43-17
A ray from A passes through B after refraction at P.

43-17, which shows two points A and B in two different media and a ray APB connecting them. The time t is given by

$$t = \frac{l_1}{v_1} + \frac{l_2}{v_2}.$$

Using the relation $n = c/v$ we can write this as

$$t = \frac{n_1 l_1 + n_2 l_2}{c} = \frac{l}{c}.$$

The quantity $l(= n_1 l_1 + n_2 l_2)$ is called the *optical path length* of the ray. Equation 43-12b $(\lambda_n = \lambda/n)$ shows that the optical path length is equal to the length that this same number of waves would have if the medium were a vacuum. We must not confuse the optical path length with the geometrical path length, which is $l_1 + l_2$.

Fermat's principle requires that l be a minimum (or a maximum or must remain unchanged) which, in turn, requires that x be so chosen that $dl/dx = 0$. The optical path length is

$$l = n_1 l_1 + n_2 l_2 = n_1 \sqrt{a^2 + x^2} + n_2 \sqrt{b^2 + (d - x)^2}.$$

Differentiating yields

$$\frac{dl}{dx} = n_1(\tfrac{1}{2})(a^2 + x^2)^{-1/2}(2x) + n_2(\tfrac{1}{2})[b^2 + (d - x)^2]^{-1/2}(2)(d - x)(-1) = 0,$$

which we can recast as

$$n_1 \frac{x}{\sqrt{a^2 + x^2}} = n_2 \frac{d - x}{\sqrt{b^2 + (d - x)^2}}.$$

Comparison with Fig. 43-17 shows that this, in turn, we can write as

$$n_1 \sin \theta_1 = n_2 \sin \theta_2,$$

which is the law of refraction.

In each of the examples of this section the time required, or, what is equivalent, the optical path length, proves to be a *minimum*.

questions

1. Describe what your immediate environment would be like if all objects were totally absorbing. Sitting in a chair in a room, could you see anything? If a person entered the room could you see her?

2. Would you expect sound waves to obey the laws of reflection and of refraction obeyed by light waves? Discuss the propagation of spherical and of cylindrical waves, using Huygens' principle. Does Huygens' principle apply to sound waves in air? If Huygens' principle predicts the laws of reflection and refraction, why is it necessary or desirable to view light as an electromagnetic wave, with all its attendant complexity?

3. A street light, viewed by reflection across a body of water in which there are ripples, appears very elongated. Explain.

4. The light beam in Fig. 43-1a is broadened on entering the water. Explain.

5. By what percent does the speed of blue light in fused quartz differ from that of red light?

6. Can (a) reflection phenomena or (b) refraction phenomena be used to determine the wavelength of light?

7. How can one determine the indices of refraction of the media in Table 43-1 relative to water, given the data in that table?

8. You are given a cube of glass. How can you find the speed of light (from a sodium light source) in this cube?

9. Describe and explain what a fish sees as he looks in various directions above his "horizon."

10. How did Foucault's measurement of the speed of light in water decide between the wave and the particle theories of light?

11. Why does a diamond "sparkle" more than a glass imitation cut to the same shape?

12. Is it plausible that the wavelength of light should change in passing from air into glass but that its frequency should not? Explain.

13. Light has (a) a wavelength, (b) a frequency, and (c) a speed. Which, if any, of these quantities remains unchanged when light passes from a vacuum into a slab of glass?

14. In reflection and refraction why do the reflected and refracted rays lie in the plane defined by the incident ray and the normal to the surface? Can you think of any exceptions?

15. What causes mirages? Does it have anything to do with the fact that the index of refraction of air is not constant but varies with its density? See "Mirages" by Alistair B. Fraser and William B. Mach, *Scientific American*, January, 1976.

16. Design a periscope, taking advantage of total internal reflection. What are the advantages compared with silvered mirrors?

17. What characteristics must a material have in order to serve as an efficient "light pipe"?

18. A certain toothbrush has a red plastic handle into which rows of nylon bristles are set. The tops of the bristles (but not their sides) appear red. Explain.

19. Discuss the formation of rainbows. See "The Theory of the Rainbow" by H. Moysés Nussenzveig, *Scientific American*, April 1977.

20. Why are optical fibers potentially more effective carriers of information than, say, microwaves or cables? Think of the frequencies involved.

21. What does "optical path length" mean? Can the optical path length ever be less than the geometrical path length? Ever greater?

22. A solution of copper sulphate appears blue when we view it through transmitted light. Does this mean that a copper sulphate solution absorbs blue light selectively? Discuss.

23. If you are an English literature major and interested in James Joyce, what would the letters *alp* and *hce* in Figures 43-8 and 43-9 mean to you?

1. The wavelength of yellow sodium light in air is 5890 Å. (a) What is its frequency? (b) What is its wavelength in glass whose index of refraction is 1.52? (c) From the results of (a) and (b) find its speed in this glass. *Answer:* (a) 5.1×10^{14} Hz. (b) 3880 Å. (c) 1.98×10^8 m/s.

2. Prove that if a mirror is rotated through an angle α, the reflected beam is rotated through an angle 2α. Is this result reasonable for $\alpha = 45°$?

3. Ptolemy, who founded the city of Alexandria in Egypt toward the end of the first century A.D., gave the following measured values for the angle of incidence θ_1 and the angle of refraction θ_2 for a light beam passing from air to water:

θ_1	θ_2	θ_1	θ_2
10°	7°45′	50°	35°0′
20°	15°30′	60°	40°30′
30°	22°30′	70°	45°30′
40°	29°0′	80°	50°0′

Are these data consistent with Snell's law? If so, what index of refraction results? These data are interesting as the oldest recorded physical measurements. *Answer:* Yes; $n = 1.34$ vs. today's value of 1.33.

4. What is the speed in fused quartz of light of wavelength 550 nm (= 5500 Å)? (See Fig. 43-2.)

5. The speed of yellow sodium light in a certain liquid is measured to be 1.92 $\times 10^8$ m/s. What is the index of refraction of this liquid, with respect to air, for sodium light? *Answer:* 1.56.

6. Suppose that the speed of light in air has been measured with an uncertainty of, say, 1 km/s. In calculating the speed in vacuum, suppose that it is not certain whether n for air is 1.00029 or 1.00030. (a) How much extra uncertainty is introduced into the calculated value for c? (b) Estimate how accurately n should be known for this purpose.

7. A bottom-weighted vertical pole extends 2.0 m above the bottom of a swimming pool and 0.5 m above the water. Sunlight is incident at 45°. What is the length of the shadow of the pole on the bottom of the pool? *Answer:* 1.4 m.

8. A 60° prism is made of fused quartz. A ray of light falls on one face, making an angle of 45° with the normal. Trace the ray through the prism graphically with some care, showing the paths traversed by rays representing (a) blue light, (b) yellow-green light, and (c) red light. (See Figs. 43-2 and 43-6.)

9. Prove that a ray of light incident on the surface of a sheet of plate glass of thickness t emerges from the opposite face parallel to its initial direction but displaced sideways, as in Fig. 43-18. Show that, for small angles of incidence θ, this displacement is given by

$$x = t\theta \frac{n-1}{n}$$

where n is the index of refraction and θ is measured in radians.

10. Assume that the index of refraction of the earth's atmosphere varies, with altitude only, from the value one at the "edge" of the atmosphere to some larger value at the surface of the earth. (a) Neglecting the earth's curvature, show that the apparent angle of a star from the zenith direction is independent of how the refractive index of the atmosphere varies with altitude and depends only on the value of n at the earth's surface. (*Hint:* Compare a uniform atmosphere with one consisting of layers of increasing refractive index.) (b) How does the earth's curvature affect the analysis?

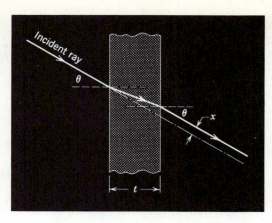

figure 43-18
Problem 9

11. Two perpendicular mirrors form the sides of a vessel filled with water, as shown in Fig. 43-19. A light ray is incident from above, normal to the water surface. (*a*) Show that the emerging ray is parallel to the incident ray. Assume that there are two reflections at the mirror surfaces. (*b*) Repeat the analysis for the case of oblique incidence, the ray lying in the plane of the figure. (*c*) Using three mirrors, state and prove the three-dimensional analog to this problem.

figure 43-19
Problem 11

12. Show that for a thin prism (ϕ small) and light not far from normal incidence (θ_1 small) the deviation angle is independent of the angle of incidence and is equal to $(n-1)\phi$ (see Fig. 43-6).

13. In Fig. 43-6 show by graphical ray tracing, using a protractor, that if θ for the incident ray is *either* increased *or* decreased, the deviation angle ψ is increased. The symmetrical situation shown in this figure is called the *position of minimum deviation*.

SECTION 43-2

14. One end of a stick is dragged through water at a speed v which is greater than the speed u of water waves. Applying Huygens' construction to the water waves, show that a conical wavefront is set up and that its half-angle α is given by

$$\sin \alpha = u/v.$$

This is familiar as the bow wave of a ship or the shock wave caused by an object moving through air with a speed exceeding that of sound, as in Fig. 20-12.

15. When an electron moves through a medium at a speed exceeding the speed of light in that medium, it radiates electromagnetic energy (the Cerenkov effect, see Section 20-7). What minimum speed must an electron have in a liquid of refractive index 1.54 in order to radiate?
Answer: 1.9×10^8 m/s.

SECTION 43-5

16. A ray of light is incident normally on the face *ab* of a glass prism ($n = 1.52$), as shown in Fig. 43-20. (*a*) Assuming that the prism is immersed in air, find the largest value for the angle ϕ so that the ray is totally reflected at face *ac*. (*b*) Find ϕ if the prism is immersed in water.

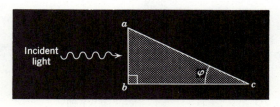

figure 43-20
Problem 16

17. A light ray falls on a square glass slab as in Fig. 43-21. What must the index of refraction of the glass be if total internal reflection occurs at the vertical face? *Answer: n > 1.22.*

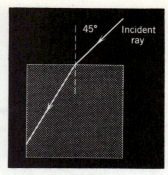

figure 43-21
Problem 17

18. A given monochromatic light ray, initially in air, strikes the 90° prism at P (see Fig. 43-22) and is refracted there and at Q to such an extent that it just grazes the right-hand prism surface after it emerges into air at Q. (a) Determine the index of refraction, relative to air, of the prism for this wavelength in terms of the angle of incidence θ_1 which gives rise to this situation. (b) Give a numerical upper bound for the index of refraction of the prism. (c) Show, by a ray diagram, what happens if the angle of incidence at P is slightly greater than θ_1, is slightly less than θ_1.

19. A glass prism with an apex angle of 60° has $n = 1.60$. (a) What is the smallest angle of incidence for which a ray can enter one face of the prism and emerge from the other? (b) What angle of incidence would be required for the ray to pass through the prism symmetrically, as in Fig. 43-6? *Answer: (a) 36°. (b) 53°.*

20. A point source is 80 cm below the surface of a body of water. Find the diameter of the largest circle at the surface through which light can emerge from the water.

21. A drop of liquid may be placed on a semicircular slab of glass as in Fig. 43-23. (a) Show how to determine the index of refraction of the liquid by observing total internal reflection. The index of refraction of the glass is unknown and must also be determined. Is the range of indices of refraction that can be measured in this way restricted in any sense? (b) In reality, how practical is this method?

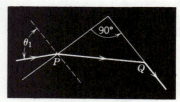

figure 43-22
Problem 18

figure 43-23
Problem 21

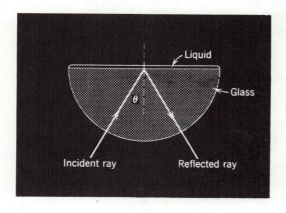

22. A point source of light is placed a distance h below the surface of a large deep lake. (a) Show that the fraction f of the light energy that escapes directly from the water surface is independent of h and is given by

$$f = \tfrac{1}{2}(1 - \sqrt{1 - 1/n^2})$$

where n is the index of refraction of water. (*Note:* Absorption within the water and reflection at the surface — except where it is total — have been neglected.) (b) Evaluate this fraction for $n = 1.33$.

23. Figure 43-24 shows a *constant-deviation prism*. Although made of one piece of glass, it is equivalent to two 30° 60° 90° prisms and one 45° 45° 90° prism. White light is incident in the direction i. θ_1 is changed by rotating the prism so that, in turn, light of any desired wavelength may be made to follow the path shown, emerging at r. Show that, if $\sin \theta_1 = \tfrac{1}{2}n$, then $\theta_2 = \theta_1$ and beams i and r are at right angles.

figure 43-24
Problem 23

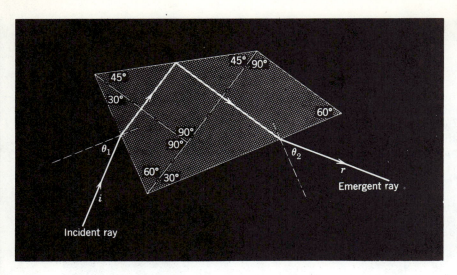

24. A plane wave of white light traveling in fused quartz strikes a plane surface of the quartz, making an angle of incidence θ. Is it possible for the internally reflected beam to appear (a) bluish or (b) reddish? Roughly what value of θ must be used? (*Hint:* White light will appear bluish if wavelengths corresponding to red are removed from the spectrum.)

25. A glass cube has a small spot at its center. (a) What parts of the cube face must be covered to prevent the spot from being seen, no matter what the direction of viewing? (b) What fraction of the cube surface must be so covered? Assume a cube edge of 1.0 cm and an index of refraction of 1.50. (Neglect the subsequent behavior of an internally reflected ray.)
Answer: (a) Cover the center of each face with an opaque disk of radius 0.45 cm. (b) About 0.63.

SECTION 43-6

26. Using Fermat's principle, prove that the reflected ray, the incident ray, and the normal lie in one plane.

27. Prove that the optical path lengths for reflection and refraction in Figs. 43-16 and 43-17 are minimal when compared with other nearby paths connecting the same two points.

28. Light of free space wavelength 600 nm (= 6000 Å) travels 1.6×10^{-4} cm in a medium of index of refraction 1.5. Find (a) the optical path length, (b) the wavelength in the medium, and (c) the phase difference after moving that distance, with respect to light traveling the same distance in free space.

44

reflection and refraction-spherical waves and spherical surfaces

In Chapter 43 we described the reflection and refraction of plane waves at plane surfaces. In this chapter we consider the more general case of spherical waves falling on spherical reflecting and refracting surfaces. All of the results of Chapter 43 will emerge as special cases of the results of this chapter, because we can view a plane as a spherical surface with an infinite radius of curvature.

Both in Chapter 43 and in this chapter we make extensive use of *rays*. Although a ray is a convenient construction, it proves impossible to isolate one physically. Figure 44-1a shows schematically a plane wave of wavelength λ falling on a slit of width $a = 5\lambda$. We find that the light flares out into the geometrical shadow of the slit, a phenomenon called *diffraction*, which we will study in Chapter 46. Figures 44-1b $(a = 3\lambda)$ and 44-1c $(a = \lambda)$ show that diffraction becomes more pronounced as $a/\lambda \to 0$ and that attempts to isolate a single ray from the incident plane wave are futile.

Figure 44-2 shows water waves in a shallow ripple tank, produced by tapping the water surface periodically with the edge of a stick. We see that the plane wave so generated also flares out by diffraction when it encounters a gap in a barrier placed across it. Diffraction is characteristic of waves of all types. We can hear around corners, for example, because of the diffraction of sound waves.

The diffraction of waves at a slit (or at an obstacle such as a wire) is expected from Huygens' principle. Consider the portion of the wavefront that arrives at the position of the slit in Fig. 44-1. We can view every point on it as the site of an expanding spherical Huygens' wavelet. The "bending" of light into the region of the geometrical shadow is as-

44-1
GEOMETRICAL OPTICS AND WAVE OPTICS

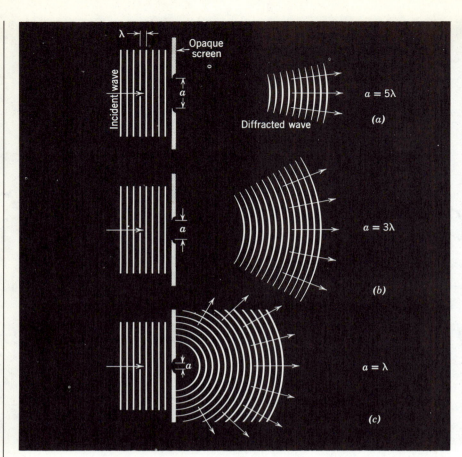

figure 44-1
An attempt to isolate a ray by reducing the slit width a fails because of diffraction, which becomes more pronounced as a/λ approaches zero. Certain details of the diffracted waves are omitted for simplicity (we will discuss them fully in Chapter 46; see also Fig. 44-3). The main point is clear, however: for a fixed wavelength λ the diffracted wave broadens as we reduce the slit width a.

sociated with the blocking off of Huygens' wavelets from those parts of the incident wavefront that lie behind the slit edges.

Figure 44-3 was made by allowing parallel light to fall on a slit placed 50 cm in front of a photographic plate. In Fig. 44-3a the slit width was about 6 μm. The central band of light is much wider than this, showing that light has "flared out" into the geometric shadow of the slit. In addition, many secondary maxima, omitted from Fig. 44-1 for simplicity, appear. Figure 44-3b shows what happens when the slit width is reduced by a factor of two. The central maximum becomes wider, in agreement with Fig. 44-1. Figure 44-3c shows the effect of reducing the slit width by an additional factor of 7, to 0.4 μm. The central maximum is now much wider and the secondary maxima, whose intensities relative to the central maximum have been deliberately overemphasized by long exposure, are very evident.

Diffraction can be ignored if the ratio a/λ is large enough, a being a measure of the smallest sideways dimension of the slit or obstacle. If $a \gg \lambda$, light appears to travel in straight lines which we can represent by rays that obey the laws of reflection (Eq. 43-1) and refraction (Eq. 43-2). In Chapter 43 this condition, called *geometrical optics*, prevailed, the lateral dimensions of all mirrors, prisms, etc., being much greater than the wavelength. We also assume in this chapter that the conditions for geometrical optics are *also* satisfied.

If the requirement for geometrical optics is not met, we cannot describe the behavior of light by rays but must take its wave nature specifically into account. This subject is called *wave optics*; it includes geometrical optics as an important limiting case in much the same

figure 44-2
Diffraction of water waves at a slit in a ripple tank. Note that the slit width is about the same size as the wavelength. Compare with Fig. 44-1c. (Courtesy of Educational Services Incorporated.)

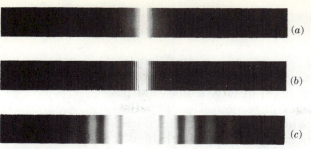

figure 44-3

(*a*) The intensity of light diffracted from a slit of width $a \cong 6\ \mu\text{m}$ and falling on a screen 50 cm beyond. (*b*) The slit width is reduced by a factor of two. (*c*) The slit width is further reduced by an additional factor of seven. Note that secondary maxima, made prominent in this case by deliberate overexposure, appear on either side of the central maximum. These secondary maxima have been omitted from Fig. 44-1 for simplicity.

sense that we can view an ideal gas as a limiting case of a real gas; both abstractions are exceedingly useful. We will treat wave optics in later chapters.

Figure 44-4 shows a point source of light *O*, the *object,* placed a distance *o* in front of a plane mirror. The light falls on the mirror as a spherical wave represented in the figure by rays emanating from *O*.* At the point at which each ray strikes the mirror we construct a reflected ray. If we

44-2
SPHERICAL WAVES— PLANE MIRROR

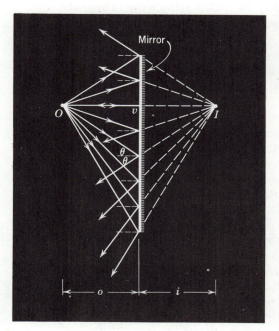

figure 44-4

A point object *O* forms a virtual image *I* in a plane mirror. The rays *appear* to emanate from *I*, but actually light does not pass through this point.

---

* In our discussion of reflection from mirrors in Chapter 43 (see Fig. 43-8) we assumed an incident *plane* wave; the incident rays are parallel to each other in that case. Here we have a *point* source and the rays striking the mirror are *diverging* from that point source.

extend the reflected rays backward, they intersect at a point I which is the same distance behind the mirror that the object O is in front of it; I is called the *image* of O.

Images may be *real* or *virtual*. In a real image light actually passes through the image point; in a virtual image the light *behaves* as though it diverges from the image point, although, in fact, it does not pass through this point; see Fig. 44-4. Images in plane mirrors are always virtual. We know from daily experience how "real" such a virtual image appears to be and how definite is its location in the space behind the mirror, even though this space may, in fact, be occupied by a brick wall.

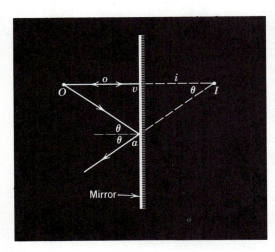

figure 44-5
Two rays from Fig. 44-4; ray Oa makes an arbitrary angle θ with the normal.

Figure 44-5 shows two rays from Fig. 44-4. One strikes the mirror at v, along a perpendicular line. The other strikes it at an arbitrary point a, making an angle of incidence θ with the normal at that point. Elementary geometry shows that the angles aOv and aIv are also equal to θ. Thus the right triangles $aOva$ and $aIva$ are congruent and

$$o = -i, \tag{44-1}$$

in which we arbitrarily introduce the minus sign to show that I and O are on opposite sides of the mirror. Equation 44-1 does not involve θ, which means that *all* rays striking the mirror pass through I when extended backward, as we have seen above. Beyond assuming that the mirror is truly plane and that the conditions for geometrical optics hold, we have made no approximations in deriving Eq. 44-1. A point object produces a point image in a plane mirror, with $o = -i$, no matter how large the angle θ in Fig. 44-5.

Because of the finite diameter of the pupil of the eye, only rays that lie fairly close together can enter the eye after reflection at a mirror. For the eye position shown in Fig. 44-6 only a small patch of the mirror near point a is effective in forming the image; the rest of the mirror may be covered up or removed. If we move our eye to another location, a different patch of the mirror will be effective; the location of the virtual image I will remain unchanged, however, as long as the object remains fixed.

If the object is an extended source such as the head of a person, a virtual image is also formed. From Eq. 44-1, every point of the source has an image point that lies an equal distance directly behind the plane of

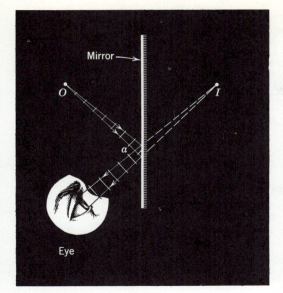

figure 44-6
A "pencil" of rays from O enters the eye after reflection at the plane mirror. Only a small portion of the mirror near a is effective. The small arcs represent portions of spherical wavefronts.

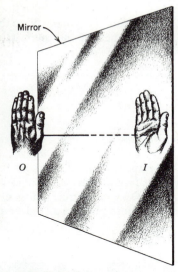

figure 44-7
A plane mirror reverses left and right. The object O is a left hand; the image I is a right hand. Try it in a mirror.

the mirror. Thus the image reproduces the object point by point. Most of us prove this every day by looking into a mirror.

Images in plane mirrors differ from objects in that left and right are interchanged. The image of a printed page is different from the page itself. Similarly, if a top is made to spin clockwise, the image, viewed in a vertical mirror, will seem to spin counterclockwise. Figure 44-7 shows an image of a left hand, constructed by using point-by-point application of Eq. 44-1; the image has the symmetry of a right hand.*

EXAMPLE 1

How tall must a vertical mirror be if a person 6 ft high is to be able to see her entire length? Assume that her eyes are 4 in. below the top of her head.

Figure 44-8 shows the paths followed by light rays leaving the top of the

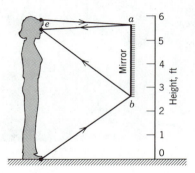

figure 44-8
Example 1. A person can view her full-length image in a mirror that is only half her height.

---

* See "The Overthrow of Parity" by Philip Morrison, *Scientific American*, April 1957, for a discussion of the distinction in nature between right and left.

woman's head and the tips of her toes. These rays, chosen so that they will enter the eye e after reflection, strike the vertical mirror at points a and b, respectively. The mirror need occupy only the region between these two points. Calculation shows that b is 2 ft, 10 in. and a is 5 ft, 10 in. above the floor. The length of the mirror is thus 3.0 ft, or half the height of the person. Note that this height is independent of the distance between the person and the mirror. Mirrors that extend below point b only show reflections of the floor between the person and the mirror.

Two plane mirrors are placed at right angles, and a point object O is located on the perpendicular bisector, as shown in Fig. 44-9a. Locate the images. Images I_1 and I_2 are formed in mirrors ab and cd, respectively. There is also a third image; it may be considered to be the image of I_1 in mirror cd or the image of I_2 in mirror ab. The three images and the object O lie on a circle whose center is on the line of intersection of the mirrors and whose plane is at right angles to that line.

EXAMPLE 2

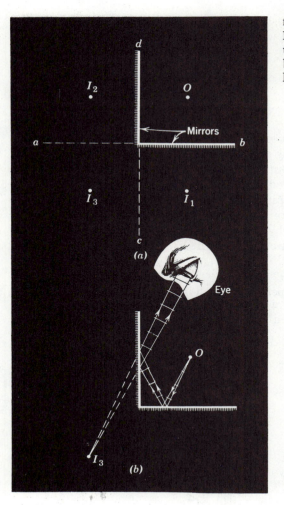

figure 44-9
Example 2. (a) Real object O has three virtual images. (b) A typical bundle of rays used to view I_3. The light originates from O.

In viewing I_3 the light entering the observer's eye is reflected *twice* after leaving the source. Figure 44-9b shows a typical bundle of rays. In viewing I_1 or I_2, the light is reflected only once, as in Fig. 44-6.

In Fig. 44-10 a spherical light wave from a point object O falls on a concave spherical mirror whose radius of curvature is r.* A line through O and the center of curvature C makes a convenient reference axis.

A ray from O that makes an arbitrary angle α with this axis intersects the axis at I after reflection from the mirror at a. A ray that leaves O along the axis will be reflected back along itself at v and will also pass through I. Thus, for these two rays at least, I is the image of O; it is a *real* image because light actually passes through I. Let us find the location of I.

44-3
SPHERICAL WAVES— SPHERICAL MIRROR

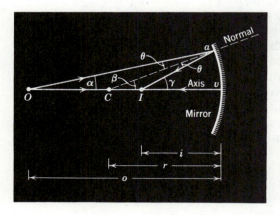

figure 44-10
Two rays from O converge after reflection in a spherical concave mirror, forming a real image at I.

A useful theorem is that the exterior angle of a triangle is equal to the sum of the two opposite interior angles. Applying this to triangles $OaCO$ and $OaIO$ in Fig. 44-10 yields

$$\beta = \alpha + \theta$$

and

$$\gamma = \alpha + 2\theta.$$

Eliminating θ between these equations leads to

$$\alpha + \gamma = 2\beta. \tag{44-2}$$

In radian measure we can write angles α, β, and γ as

$$\alpha \cong \frac{av}{vO} = \frac{av}{o}$$

$$\beta = \frac{av}{vC} = \frac{av}{r} \tag{44-3}$$

$$\gamma \cong \frac{av}{vI} = \frac{av}{i}.$$

Note that only the equation for β is exact, for the reason that the center of curvature of arc av is at C and not at O or I. However, the equations for α and for γ are approximately correct if these angles are sufficiently small. *In all that follows we assume that the rays diverging from the object make only a small angle α with the axis of the mirror.* We call such rays, which lie close to the mirror axis, *paraxial rays*. We did not find it necessary to make such an assumption for plane mirrors. Sub-

* A spherical shell, viewed from inside, is everywhere *concave*; viewed from outside it is everywhere *convex*. In this chapter we will always judge concave and convex from the point of view of an observer sighting along the direction of the incident light.

stituting these equations into Eq. 44-2 and canceling av yields

$$\frac{1}{o} + \frac{1}{i} = \frac{2}{r}, \qquad (44\text{-}4)$$

in which o is the *object distance* and i is the *image distance.* Both these distances are measured from the *vertex* of the mirror, which is the point v at which the axis intercepts the mirror.

Significantly, Eq. 44-4 does not contain α (or β, γ, or θ), so that it holds for all rays that strike the mirror provided that they are sufficiently paraxial. In an actual case the rays can be made as paraxial as one likes by putting a circular diaphragm in front of the mirror, centered about the vertex v; this will impose a certain maximum value of α.

As α in Fig. 44-10 is permitted to become larger, it will become less true that a point object will form a point image; the image will become extended and fuzzy. No sharp criterion for deciding whether a given ray is paraxial can be laid down. If the maximum permitted value of α is reduced, the rays will become more paraxial and the image will become sharper. Unfortunately, the image will also become fainter because less total light energy will be reflected from the mirror. A compromise must often be made between image brightness and image quality.

As for plane mirrors, the image (real or virtual) in a spherical mirror can be seen only if the eye is located so that light rays from the object can enter it after reflection. In Fig. 44-11 a bundle of light rays is shown entering the eye in position x; only the small patch of mirror near a is effective for this eye position. If the observer moves his eye to position y, the image will vanish for him because the mirror does not exist near point a'.

Although Eq. 44-4 was derived for the special case in which the object is located beyond the center of curvature, it is generally true, no matter where the object is located. It is also true for convex mirrors, as in Fig. 44-12.

In applying Eq. 44-4, we must be careful to follow a consistent convention of signs for o, i, and r. As the basis for the sign conventions to be used in this book, we start from this statement:

> In Fig. 44-10, in which light *diverges* from a *real* object, falls on a *concave* mirror, and *converges* after reflection to form a *real* image, the quantities o, i, and r in Eq. 44-4 and f in Eq. 44-5 are given positive numerical values.

We used Fig. 44-10 to derive Eq. 44-4 and you should associate them in your mind as an aid in getting the signs correct.

Let us fix our minds on the side of the mirror from which incident light comes. Because mirrors are opaque, the light, after reflection, must remain on this side, and if an image is formed here it will be a *real* image. Therefore, we call the side of the mirror from which the light comes the *R-side* (for *real image*). We call the back of the mirror the *V-side* (for *virtual image*), because images formed on this side of the mirror must be virtual, no light being actually present on this side.

In the indented statement above we associated real images with positive image distances. This suggests our first sign convention:

1. The image distance i is positive if the image (real) lies on the R-side of the mirror, as in Fig. 44-10; i is negative if the image (virtual) lies on the V-side, as in Fig. 44-12.

If the mirror in Fig. 44-10, which is concave as viewed from the direction of the incident light, is made convex, the rays will *diverge* after re-

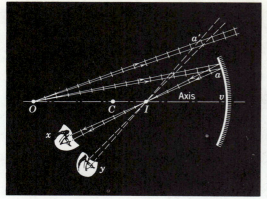

figure 44-11
The eye must be properly located to see (real) image *I*. An observer at *x* can see the image. One at *y* cannot.

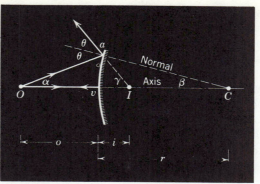

figure 44-12
Two rays from *O* diverge after reflection in a spherical convex mirror, forming a virtual image at *I*, the point from which they appear to originate. Compare Fig. 44-10.

flection and will form a *virtual* image, as Fig. 44-12 shows. Thus the indented statement above suggests our second sign convention:

2. The radius of curvature *r* is positive if the center of curvature of the mirror lies on the R-side, as in Fig. 44-10; *r* is negative if the center of curvature lies on the V-side, as in Fig. 44-12.

You should not commit these sign conventions to memory but should deduce them in each case from the basic statement on p. 964, using Figs. 44-10, 12 as mnemonic aids.

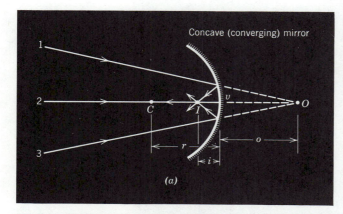

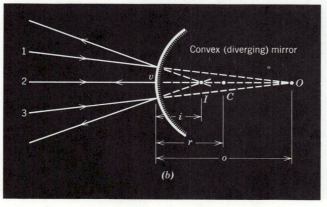

figure 44-13
(*a*) *Converging* rays (see 1, 2, and 3) fall on a *concave* mirror. A *virtual* object *O* produces a *real* image *I*; note that no light passes through *O* but it does pass through *I*. Also note that *o* is negative but *i, r* (and thus *f*), are positive. (*b*) Again *converging* rays (see 1, 2, and 3) fall on a *convex* mirror. A *virtual* object *O* produces a *virtual* image *I*. Note again that no light passes through *O* or *I*. Here *o, i, r* (and thus *f*), are all negative. Compare these figures with Fig. 44-10, in which a *real* object (*diverging* rays) produces a *real* image on reflection from a *concave* mirror. Here *o, i, r* (and thus *f*), are all positive. See also Fig. 44-12, in which a *real* object (*diverging* rays) produces a *virtual* image in a *convex* mirror. Here *o* is positive but *i, r,* (and thus *f*), are negative. These four figures show all possibilities for arrangements of virtual/real objects and virtual/real images in mirrors. Compare them carefully. Figure 44-23 shows a similar comparison for thin lenses.

For all cases that we have considered so far, including plane mirrors, we have assumed that the incident light striking the mirror was *diverging* from a point source (the *object*) when it struck the mirror. In such cases, as we have seen, we have taken the object distance o in Eq. 44-4 as *positive*.

It is possible, by arrangements of mirrors and/or lenses, to make *converging* light fall on the mirror in question. In such cases we declare the *object* to be *virtual* and we assign a *negative* sign to the object distance o in Eq. 44-4. Figure 44-13 is an example. No matter what the source of light, O, which lies on the V-side, is a virtual object and o, the object distance, is taken as negative.

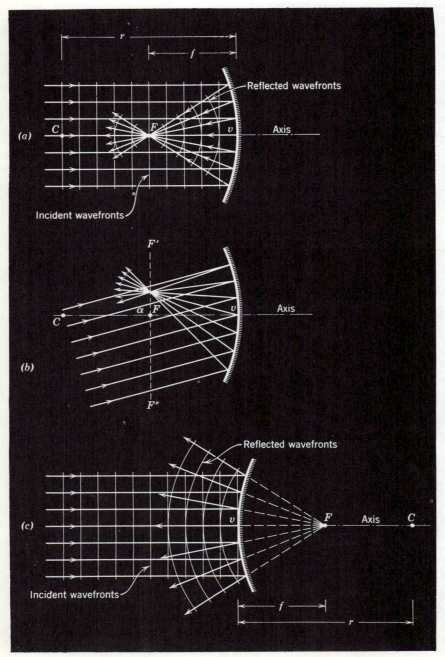

figure 44-14
(a) The focal point for a concave spherical mirror, showing both the rays and the wavefronts. F and C lie on the R-side, the focal point is real, and the focal length f of the mirror is positive (as is r). (b) The same, except that the incident light makes an angle α with the mirror axis; the rays are focused at a point in the *focal plane F' F"*. (c) Same as (a) except that the mirror is convex; F and C lie on the V-side of the mirror. The focal point is virtual and the focal length f is negative (as is r).

EXAMPLE 3

A convex mirror has a radius of curvature of 20 cm. If a point source is placed 14 cm away from the mirror, as in Fig. 44-12, where is the image?

A rough graphical construction, applying the law of reflection at a in the figure, shows that the image will be on the V-side of the mirror and thus will be virtual. We may verify this quantitatively and analytically from Eq. 44-4, noting that r is negative here because the center of curvature of the mirror is on its V-side. We have

$$\frac{1}{o} + \frac{1}{i} = \frac{2}{r}$$

or

$$\frac{1}{+14 \text{ cm}} + \frac{1}{i} = \frac{2}{-20 \text{ cm}},$$

which yields $i = -5.8$ cm, in agreement with the graphical prediction. The negative sign for i reminds us that the image is on the V-side of the mirror and thus is virtual.

When *parallel* light falls on a mirror (Fig. 44-14), we call the image point (real or virtual) the *focal point F* of the mirror. The focal length f is the distance between F and the vertex. If we put $o \to \infty$ in Eq. 44-4, thus insuring parallel incident light, we have

$$i = \tfrac{1}{2}r = f.$$

Equation 44-4 can then be rewritten

$$\frac{1}{o} + \frac{1}{i} = \frac{1}{f}, \qquad (44\text{-}5)$$

where f, like r, is taken as positive for mirrors whose centers of curvature are on the R-side (that is, for *concave*, or *converging* mirrors; see Fig. 44-14a) and negative for those whose centers of curvature are on the V-side (that is, for *convex*, or *diverging* mirrors; see Fig. 44-13c). Figure 44-14b shows an incident plane wave that makes a small angle α with the mirror axis. The rays are focused at a point in the *focal plane* of the mirror. This is a plane at right angles to the mirror axis at the focal point.

We now consider objects that are not points. Figure 44-15 shows a candle in front of (a) a concave mirror and (b) a convex mirror. We choose

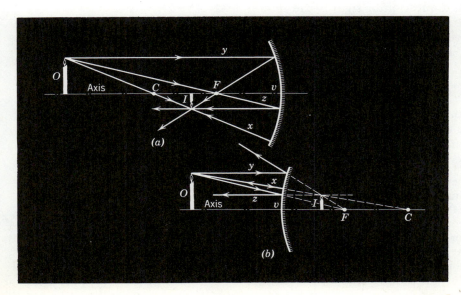

figure 44-15
The image of an extended object in (a) a concave mirror and (b) a convex mirror is located graphically. Any two of the three special rays shown are sufficient.

to draw the mirror axis through the foot of the candle and, of course, through the center of curvature. We can find the image of any off-axis point, such as the tip of the candle, graphically, using the following facts:

1. A ray that strikes the mirror after passing (either directly or upon being extended) through the center of curvature C returns along itself (ray x in Fig. 44-15). Such rays strike the mirror at right angles.
2. A ray that strikes the mirror parallel to its axis passes (or will pass when extended) through the focal point (ray y).
3. A ray that strikes the mirror after passing (either directly or upon being extended) through the focal point emerges parallel to the axis (ray z).

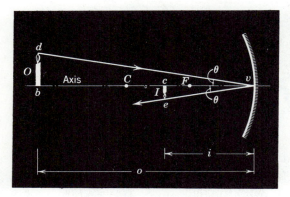

figure 44-16
A particular ray for the arrangement of Fig. 44-15, used to show that the *lateral magnification m* is given by $-i/o$.

Figure 44-16 shows a ray (dve) that originates on the tip of the object candle of Fig. 44-15a, is reflected from the mirror at point v, and passes through the tip of the image candle. The law of reflection demands that this ray make equal angles θ with the mirror axis as shown. For the two similar right triangles in the figure we can write

$$\frac{ce}{bd} = \frac{vc}{vb}.$$

The quantity on the left (apart from a question of sign) is the *lateral magnification m* of the mirror. Since we want to represent an *inverted* image by a *negative* magnification, we arbitrarily define m for this case as $-(ce/bd)$. Since $vc = i$ and $vb = o$, we have at once

$$m = -\frac{i}{o}. \tag{44-6}$$

This equation gives the magnification for spherical and plane mirrors under all circumstances. For a plane mirror, $o = -i$ and the predicted magnification is $+1$ which, in agreement with experience, indicates an *erect* image the same size as the object.

Images in spherical mirrors suffer from several "defects" that arise because the assumption of paraxial rays is never completely justified. In general, a point source will not produce a point image; see Problem 14. Apart from this, distortion arises because the magnification varies somewhat with distance from the mirror axis, Eq. 44-6 being strictly correct only for paraxial rays. Superimposed on these defects are *diffraction effects* which come about because the basic assumption of geometrical optics, that light travels in straight lines, must always be considered an approximation.

Figure 44-17 shows a point source O near a convex spherical *refracting* surface of radius of curvature r. The surface separates two media whose indices of refraction differ, that of the medium in which the incident light falls on the surface being n_1 and that on the other side of the surface being n_2.

figure 44-17
Two rays from O converge after refraction at a spherical surface, forming a real image at I. Here o (diverging rays), i, and r are all positive; see Example 4.

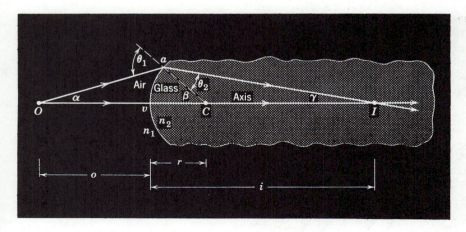

From O we draw a line through the center of curvature C of the refracting surface, thus establishing a convenient axis which intercepts the surface at vertex v. From O we draw a ray that makes a small but arbitrary angle α with the axis and strikes the refracting surface at a, being refracted according to

$$n_1 \sin \theta_1 = n_2 \sin \theta_2.$$

The refracted ray intersects the axis at I. A ray from O that travels along the axis will not be bent on entering the surface and will also pass through I. Thus, for these two rays at least, I is the image of O.

As in the derivation of the mirror equation, we use the theorem that the exterior angle of a triangle is equal to the sum of the two opposite interior angles. Applying this to triangles $COaC$ and $ICaI$ yields

$$\theta_1 = \alpha + \beta \tag{44-7}$$

and

$$\beta = \theta_2 + \gamma. \tag{44-8}$$

As α is made small, angles β, γ, θ_1, and θ_2 in Fig. 44-17 also become small. We assume that α, hence all these angles, *are arbitrarily small.* We made this same paraxial ray assumption for spherical mirrors. Replacing the sines of the angles by the angles themselves—since the angles are required to be small—permits us to write the law of refraction as

$$n_1 \theta_1 \cong n_2 \theta_2. \tag{44-9}$$

Combining Eqs. 44-8 and 44-9 leads to

$$\beta = \frac{n_1}{n_2} \theta_1 + \gamma.$$

Eliminating θ_1 between this equation and Eq. 44-7 leads, after rearrangement, to

$$n_1 \alpha + n_2 \gamma = (n_2 - n_1)\beta. \tag{44-10}$$

In radian measure the angles α, β, and γ in Fig. 44-17 are

$$\alpha \cong \frac{av}{o}$$

$$\beta = \frac{av}{r} \qquad\qquad (44\text{-}11)$$

$$\gamma \cong \frac{av}{i}.$$

Only the second of these equations is exact. The other two are approximate because I and O are *not* the centers of circles of which av is an arc. However, for paraxial rays (α small enough) the inaccuracies in Eq. 44-11 can be made as small as desired.

Substituting Eqs. 44-11 into Eq. 44-10 leads readily to

$$\frac{n_1}{o} + \frac{n_2}{i} = \frac{n_2 - n_1}{r}. \qquad\qquad (44\text{-}12)$$

This equation holds whenever light is refracted from point objects at spherical surfaces, assuming only that the rays are paraxial. As with the mirror formula, care must be taken to use Eq. 44-12 with consistent signs for o, i, and r. Once again we establish our sign conventions by physical reasoning from a particular case, that of Fig. 44-17:

> In Fig. 44-17, in which light *diverges* from a *real* object, falls on a *convex* refracting surface, and *converges* after refraction to form a *real* image, the quantities o, i, and r in Eq. 44-12 have positive numerical values.

We used Fig. 44-17 to derive Eq. 44-12, and you should associate it in your mind as an aid in getting the signs correct. This basic statement is quite similar to that which we made for mirrors on p. 964.

We fix our attention on the side of the refracting surface from which the incident light falls on the surface. In contrast to mirrors, the light energy *passes through* a refracting surface to the other side, and if a real image is formed it must appear on the far side, which we call the R-side. The side from which the incident light comes is called the V-side because virtual images must appear here. Figure 44-18 suggests this important distinction between reflection and refraction.

In the indented statement above we associated real images with positive-image distances. Thus we are led to the sign convention:

1. The image distance i is positive if the image (real) is on the R-side of the refracting surface, as in Fig. 44-17; i is negative if the image (virtual) lies on the V-side, as in Fig. 44-19.

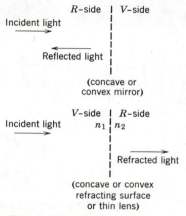

figure 44-18

Real images are formed on the same side as the incident light for mirrors but on the opposite side for refracting surfaces and thin lenses. This is so because the incident light is reflected back by mirrors but is transmitted through by refracting surfaces. Note that o is positive if rays *diverge* from a *real* object O.

figure 44-19

Two rays from O diverge after refraction at a spherical surface, forming a virtual image at I. Here o is positive (the rays from O are diverging) but i and r are negative.

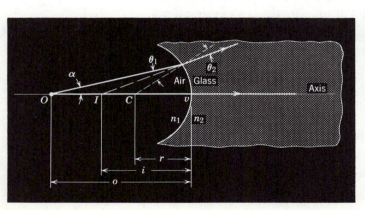

The refracting surface in Fig. 44-17 is convex. If it is made concave (still assuming that $n_2 > n_1$), the rays will *diverge* after refraction and form a *virtual* image, as Fig. 44-19 shows. Thus we are led to our second sign convention:

2. The radius of curvature r is positive if the center of curvature of the refracting surface lies on the R-side, as in Fig. 44-17; r is negative if the center of curvature is on the V-side, as in Fig. 44-19.

The sign conventions for refracting surfaces are the same as for mirrors (p. 964), the fundamental difference between the two situations being absorbed in the definitions of R-side and V-side in Fig. 44-18. This difference is easily remembered on physical grounds.

EXAMPLE 4

Locate the image for the geometry shown in Fig. 44-17, assuming the radius of curvature to be 10 cm, n_2 to be 2.0, and n_1 to be 1.0. Let the object be 20 cm to the left of v.

From Eq. 44-12,

$$\frac{n_1}{o} + \frac{n_2}{i} = \frac{n_2 - n_1}{r},$$

we have

$$\frac{1.0}{+20 \text{ cm}} + \frac{2.0}{i} = \frac{2.0 - 1.0}{+10 \text{ cm}}.$$

Note that r is positive because the center of curvature of the surface lies on the R-side. This relation yields $i = +40$ cm in agreement with the graphical construction. The light energy actually passes through I so that the image is real, as indicated by the positive sign for i.

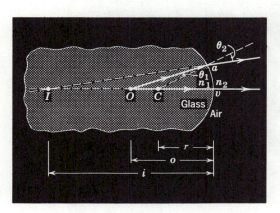

figure 44-20
Two rays from O appear to originate from I (virtual image) after refraction at a spherical surface.

EXAMPLE 5

An object is immersed in a medium with $n_1 = 2.0$, being 15 cm from the spherical surface whose radius of curvature is -10 cm, as in Fig. 44-20, r is negative because C lies on the V-side. Locate the image.

Figure 44-20 shows a ray traced through the surface by applying the law of refraction at point a. A second ray from O along the axis emerges undeflected at v. We find the image I by extending these two rays backward; it is virtual.

From Eq. 44-12,

$$\frac{n_1}{o} + \frac{n_2}{i} = \frac{n_2 - n_1}{r},$$

we have

$$\frac{2.0}{+15 \text{ cm}} + \frac{1.0}{i} = \frac{1.0 - 2.0}{-10 \text{ cm}},$$

which yields $i = -30$ cm, in agreement with Fig. 44-20 and with the sign conventions. Note that n_1 always refers to the medium on the side of the surface from which the light comes.

What is the relationship between i and o if the refracting surface is plane?

A plane surface has an infinite radius of curvature. Putting $r \to \infty$ in Eq. 44-12 leads to

$$i = -o\,\frac{n_2}{n_1}.$$

Figure 44-21 illustrates the situation graphically (a) for an object in air as seen from below water and (b) for an object in water with air above. This shows that a diver, looking upward at, say, an overhanging tree branch, will think it higher than it is by the factor 1.33/1.00. Similarly, an observer in air will think that objects on the bottom of a water tank are closer to the surface than they actually are, in the ratio 1.00/1.33. These considerations, being based on Eq. 44-12, hold

EXAMPLE 6

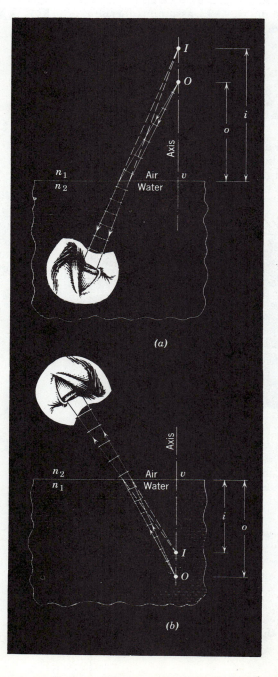

figure 44-21
Refraction at a plane surface at near-normal incidence, showing a pencil of rays and the corresponding wavefronts entering the pupil.
(a) Source in air and (b) source in water.

only for paraxial rays, which means that the incident rays can make only a small angle with the normal; this angle has been exaggerated in the figure for clarity. Note again that n_1 is always identified with the medium that lies on the side of the surface containing the incident light.

In most refraction situations there is more than one refracting surface. This is true even for a spectacle lens, the light passing from air into glass and then from glass into air. In microscopes, telescopes, cameras,* etc., there are often many more than two surfaces.

Figure 44-22a shows a thick glass "lens" of length l whose surfaces are ground to radii r' and r''. A point object O' is placed near the left surface as shown. A ray leaving O' along the axis is not deflected on entering or leaving the lens because it falls on each surface along a normal.

A second ray leaving O', at an arbitrary angle α with the axis, strikes the surface at point a', is refracted, and strikes the second surface at point a''. The ray is again refracted and crosses the axis at I'', which, being the intersection of two rays from O'', is the image of point O', formed after refraction at two surfaces.

44-5
THIN LENSES

* For insight into the difficult task of designing real high-performance lenses see "The Photographic Lens" by William H. Price, *Scientific American,* August 1976.

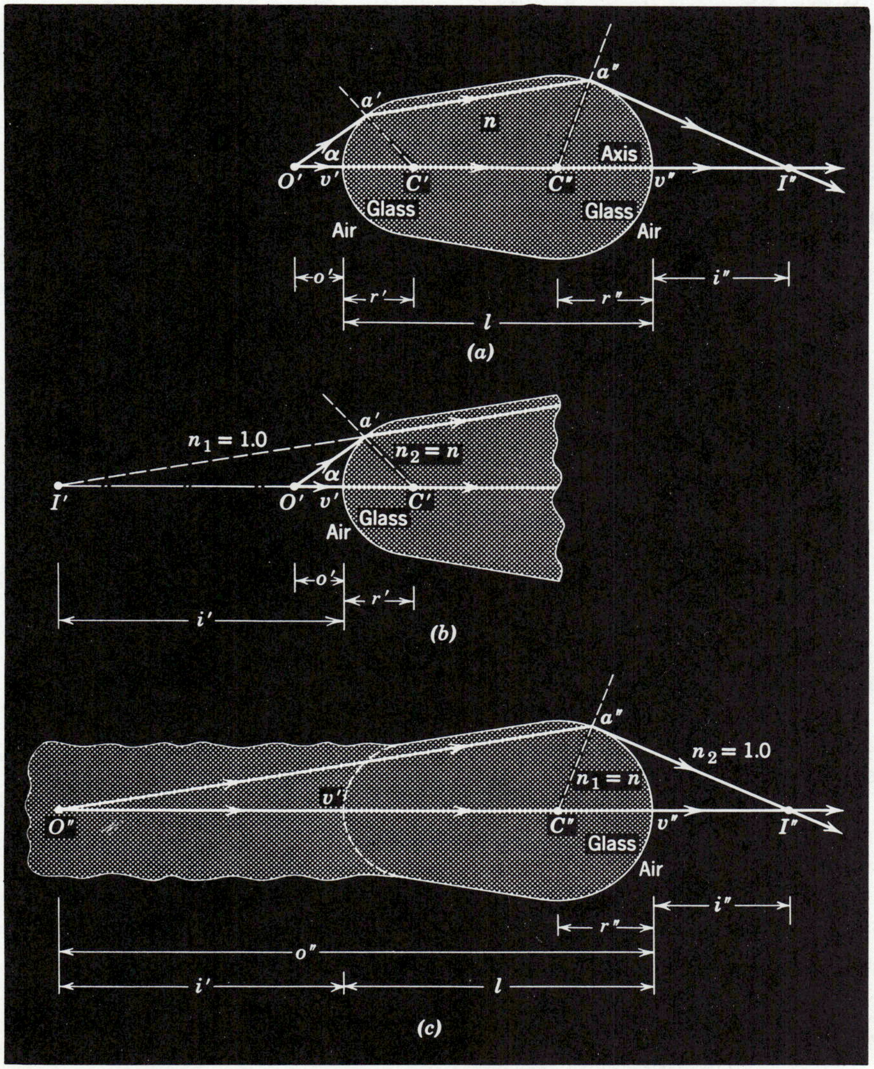

figure 44-22
(a) Two rays from O' intersect at I'' (real image) after refraction at two spherical surfaces. (b) The first surface and (c) the second surface shown separately. The quantities α and n have been exaggerated for clarity.

Figure 44-22b shows the first surface, which forms a virtual image of O' at I'. To locate I', we use Eq. 44-12,

$$\frac{n_1}{o} + \frac{n_2}{i} = \frac{n_2 - n_1}{r}.$$

Putting $n_1 = 1.0$ and $n_2 = n$ and bearing in mind that the image distance is negative (that is, $i = -i'$ in Fig. 44-22b), we obtain

$$\frac{1}{o'} - \frac{n}{i'} = \frac{n-1}{r'}. \tag{44-13}$$

In this equation i' will be a positive number because we have arbitrarily introduced the minus sign appropriate to a virtual image.

Figure 44-22c shows the second surface. Unless an observer at point a'' were aware of the existence of the first surface, he would think that the light striking that point originated at point I' in Fig. 44-22b and that the region to the left of the surface was filled with glass. Thus the (virtual) image I' formed by the first surface serves as a real object O'' for the second surface. The distance of this object from the second surface is

$$o'' = i' + l. \tag{44-14}$$

In applying Eq. 44-12 to the second surface, we insert $n_1 = n$ and $n_2 = 1.0$ because the object behaves as if it were imbedded in glass. If we use Eq. 44-14, Eq. 44-12 becomes

$$\frac{n}{i' + l} + \frac{1}{i''} = \frac{1-n}{r''}. \tag{44-15}$$

Let us now assume that the thickness l of the "lens" in Fig. 44-22 is so small that we can neglect it in comparison with other linear quantities in this figure (such as o', i', o'', i'', r', and r''). In all that follows we make this *thin-lens approximation*. Putting $l = 0$ in Eq. 44-15 leads to

$$\frac{n}{i'} + \frac{1}{i''} = -\frac{n-1}{r''}. \tag{44-16}$$

Adding Eqs. 44-13 and 44-16 leads to

$$\frac{1}{o'} + \frac{1}{i''} = (n-1)\left(\frac{1}{r'} - \frac{1}{r''}\right).$$

Finally, calling the original object distance simply o and the final image distance simply i leads to

$$\frac{1}{o} + \frac{1}{i} = (n-1)\left(\frac{1}{r'} - \frac{1}{r''}\right). \tag{44-17}$$

This equation holds only for paraxial rays and only if the lens is so thin that it essentially makes no difference from which surface of the lens the quantities o and i are measured. In Eq. 44-17 r' refers to the first surface struck by the light as it traverses the lens and r'' to the second surface.

The sign conventions for Eq. 44-17 are the same as those for mirrors and for single refracting surfaces. Because the lens is assumed to be thin, we refer to the R-side and the V-side of the lens itself (see Fig. 44-18) rather than those of its separate surfaces. The sign conventions are suggested by Fig. 44-23.

1. The image distance i is positive if the image (real) lies on the R-side

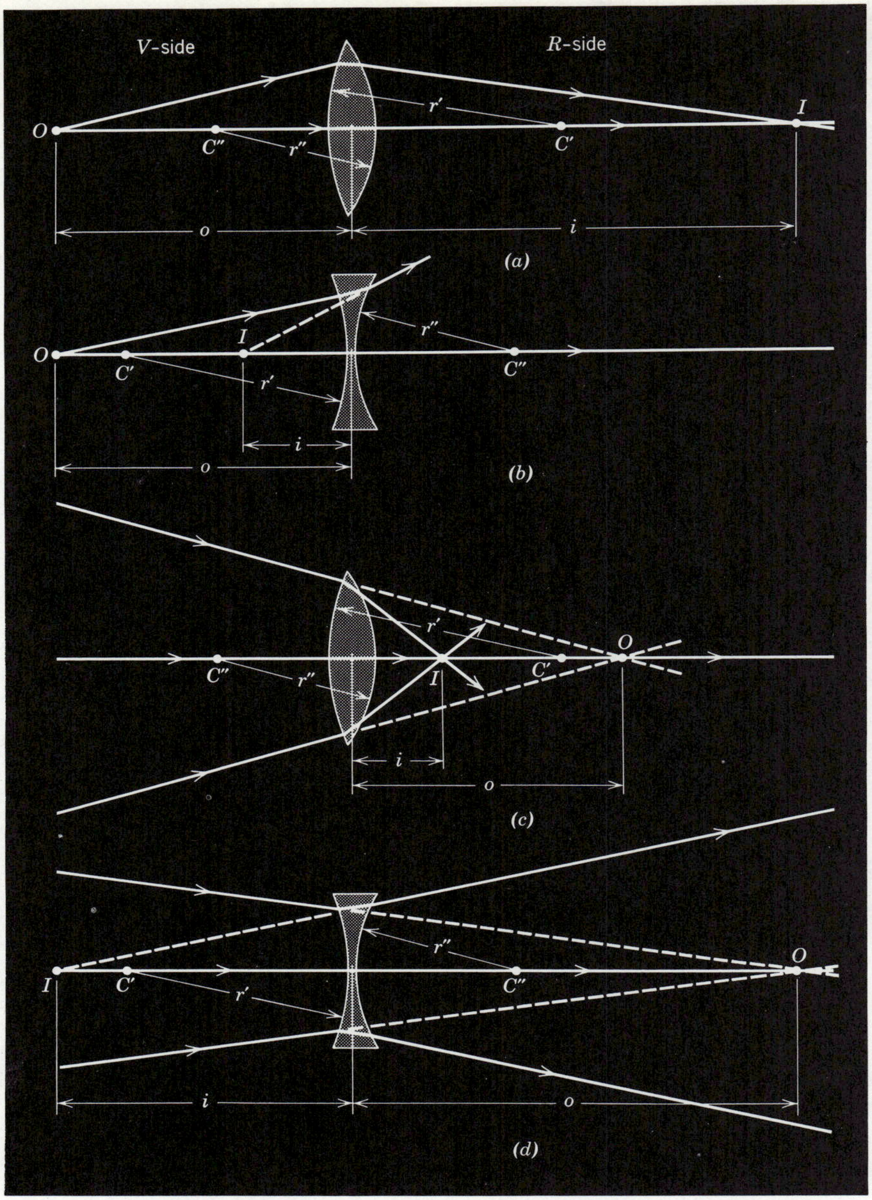

figure 44-23

(a) A real image and a real object. Both o and i are positive. (b) A virtual image and a real object. Note that o is positive but i is negative. (c) A real image and a virtual object. Note that o is negative but i is positive. (d) A virtual image and a virtual object. Both o and i are negative. Note that the sign of o is determined by whether the incident rays are diverging (first two cases) or converging (second two cases) when they strike the thin lens. For diverging incident rays the object is real; for converging rays it is virtual. The sign of i is determined by whether I is on the R-side of the lens (in which case I is real and i is positive, as in the first and third figures) or by whether I is on the V-side of the lens (in which case I is virtual and i is negative, as in the second and fourth figures).

of the lens, as in Figs. 44-23a and 44-23c. It is negative if the image (virtual) lies on the V-side of the lens, as in Figs. 44-23b and 44-23d.

2. The object distance o is positive if *diverging* rays fall on the lens, as in Figs. 44-23a and 44-23b; the object is real in such cases. The object distance o is negative if *converging* rays fall on the lens, as in Figs. 44-23c and 44-23d; the object is virtual in such cases.

3. The radii of curvature r' and r'' refer, respectively, to the first and to the second surfaces to be struck by the incident light. Each of these quantities is positive if the corresponding centers of curvature C' and C'' lie on the R-side of the lens; they are negative otherwise.

Figure 44-24a and 44-24c shows parallel light from a distant object falling on a thin lens. The image location is called the *second focal point*

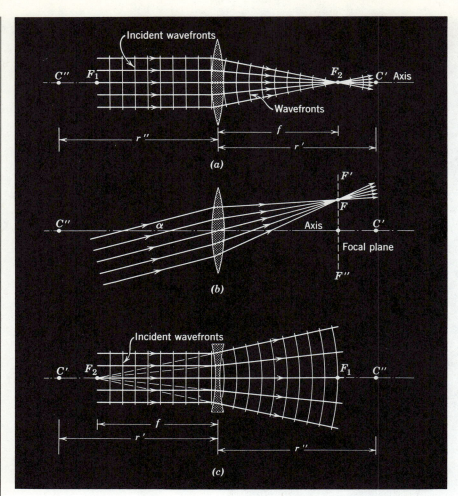

(a)

(b)

(c)

figure 44-24

(a) Parallel light passes through the second focal point F_2 of a converging lens. (b) The incident light makes an angle α with the lens axis, the rays being focused in the focal plane $F'F''$. (c) Parallel light, passing through a diverging lens, seems to originate at the second focal point F_2. C' and C'' are centers of curvature for the lens surfaces; F_1 is the first focal point.

F_2 of the lens. The distance from F_2 to the lens is called the *focal length* f. The *first focal point* for a thin lens (F_1 in figure) is the object position for which the image is at infinity. For thin lenses the first and second focal points are on opposite sides of the lens and are equidistant from it.

We can find the focal length from Eq. 44-17 by inserting $o \rightarrow \infty$ and $i = f$. This yields

$$\frac{1}{f} = (n-1)\left(\frac{1}{r'} - \frac{1}{r''}\right). \tag{44-18}$$

This relation is called the *lens maker's equation* because it allows us to compute the focal length of a lens in terms of the radii of curvature and the index of refraction of the material. Combining Eqs. 44-17 and 44-18 allows us to write the thin-lens equation as

$$\frac{1}{o} + \frac{1}{i} = \frac{1}{f}. \tag{44-19}$$

Figure 44-24b shows parallel incident rays that make a small angle α with the lens axis; they are brought to a focus in the *focal plane* $F'F''$, as shown. This is a plane normal to the lens axis at the focal point.

In Fig. 44-24a we note that all rays in the figure contain the same number of wavelengths; in other words, they have the same *optical path*

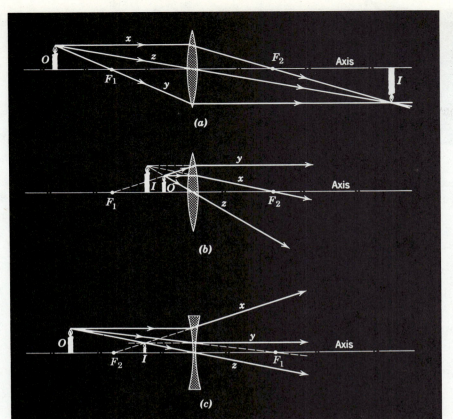

lengths; see Section 43-6. The optical path lengths are the same because the wavefronts are surfaces over which the wave disturbance has the same constant value and because all the rays shown pass through the same number of wavefronts.

EXAMPLE 7

The lenses of Fig. 44-24 have radii of curvature of magnitude 40 cm and are made of glass with $n = 1.65$. Compute their focal lengths.

Since C' lies on the R-side of the lens in Fig. 44-24a, r' is positive (=+40 cm). Since C'' lies on the V-side, r'' is negative (=−40 cm). Substituting in Eq. 44-18 yields

$$\frac{1}{f} = (n - 1)\left(\frac{1}{r'} - \frac{1}{r''}\right) = (1.65 - 1)\left(\frac{1}{+40 \text{ cm}} - \frac{1}{-40 \text{ cm}}\right),$$

or
$$f = +31 \text{ cm}.$$

A positive focal length indicates that in agreement with Fig. 44-24a the focal point F_2 is on the R-side of the lens and parallel incident light converges after refraction to form a real image.

In Fig. 44-24c C' lies on the V-side of the lens so that r' is negative (=−40 cm). Since r'' is positive (=+40 cm), Eq. 44-18 yields

$$f = -31 \text{ cm}.$$

A negative focal length indicates that in agreement with Fig. 44-24c the focal point F_2 is on the V-side of the lens and incident light diverges after refraction to form a virtual image.

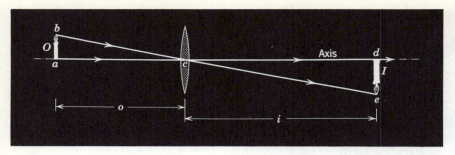

figure 44-26
Two rays for the situation of Fig.
44-25a.

We can locate the image of an extended object such as a candle
(Fig. 44-25) graphically by using the following three facts:

1. A ray parallel to the axis and falling on the lens passes, either directly
   or when extended, through the second focal point F_2 (ray x in Fig.
   44-25).
2. A ray falling on a lens after passing, either directly or when extended,
   through the first focal point F_1 will emerge from the lens parallel to
   the axis (ray y).
3. A ray falling on the lens at its center will pass through undeflected.
   There is no deflection because the lens, near its center, behaves like a
   thin piece of glass with parallel sides. The direction of the light rays
   is not changed and the sideways displacement can be neglected because the lens thickness has been assumed to be negligible (ray z).

Figure 44-26, which represents part of Fig. 44-25a, shows a ray passing
from the tip of the object through the center of curvature to the tip of the
image. For the similar triangles abc and dec we may write

$$\frac{de}{ab} = \frac{dc}{ac}.$$

The right side of this equation is i/o and the left side is $-m$, where m is
the *lateral magnification*. The minus sign is required because we wish
m to be negative for an inverted image. This yields

$$m = -\frac{i}{o}, \tag{44-20}$$

which holds for all types of thin lenses and for all object distances.

A converging thin lens has a focal length of +24 cm. An object is placed 9.0 cm   **EXAMPLE 8**
from the lens as in Fig. 44-25b. Describe the image.
   From Eq. 44-19,

$$\frac{1}{o} + \frac{1}{i} = \frac{1}{f},$$

we have

$$\frac{1}{+9.0 \text{ cm}} + \frac{1}{i} = \frac{1}{+24 \text{ cm}},$$

which yields $i = -14.4$ cm, in agreement with the figure. The minus sign means
that the image is on the V-side of the lens and is thus virtual.
   The lateral magnification is given by

$$m = -\frac{i}{o} = -\frac{-14.4 \text{ cm}}{+9.0 \text{ cm}} = +1.6,$$

again in agreement with the figure. The plus signifies an erect image.

Images formed by lenses suffer from defects similar to those discussed for mirrors on p. 968. There are effects connected with the failure of a point object to form a point image, with the variation of magnification with distance from the lens axis, and with diffraction. For lenses, but not for mirrors, there are also *chromatic aberrations* associated with the fact that the refracting properties of the lens vary with wavelength because the index of refraction of the lens material does. If a point object on the lens axis emits white light, the image, neglecting other lens defects, will be a series of colored points spread out along the axis. We have all seen the colored images produced by inexpensive lenses. A great deal of ingenious optical engineering goes into the design of lenses (more commonly lens systems) in which the various lens defects are minimized. The lens surfaces are normally not spherical nor are the lenses "thin."

Note that we can write the mirror formula (Eq. 44-5) and the (identical) thin lens formula (Eq. 44-19) in this form:

$$\frac{1}{o/|f|} + \frac{1}{i/|f|} = \pm 1 \qquad (44\text{-}21)$$

in which $|f|$, the absolute value of the focal length f, is always positive. On the right side of the equation we choose $+1$ for a converging lens or a concave mirror and -1 for a diverging lens or a convex mirror. See Problem 41.

Figure 44-27 is a graphical representation of Eq. 44-21, for both mirrors and lenses, Fig. 44-27*a* being for concave mirrors and converging lenses and Fig. 44-27*b* for convex mirrors and diverging lenses.

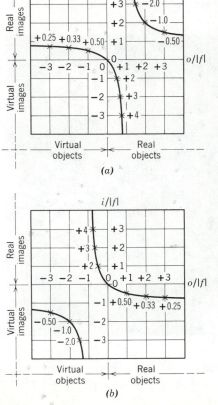

(a)

(b)

figure 44-27
(*a*) A representation of $i/|f|$ and $o/|f|$ for concave mirrors and converging lenses. Note that (lower left quadrant) a virtual object cannot produce a virtual image. The numbers near the x's are the magnifications (see Eq. 44-20). A positive sign means an erect image, a negative sign an inverted image. Compare previous figures.

(*b*) A representation of $i/|f|$ and $o/|f|$ for convex mirrors and diverging lenses. Note that (upper right quadrant) a real object cannot produce a real image. Can you verify this? Once again, the numbers near the x's are the magnifications (Eq. 44-20). See "Image Formation in Lenses & Mirrors, a Complete Representation" by Albert A. Bartlett, *The Physics Teacher*, May 1976.

The human eye is a remarkably effective organ* but its range can be extended in many ways by a host of optical instruments such as spectacles, simple magnifiers, motion picture projectors, cameras (including TV cameras), microscopes, telescopes, etc. In many cases these devices extend the scope of our vision beyond the visible range; satellite-borne infrared cameras and X-ray microscopes are examples.

In almost all cases of modern sophisticated optical instruments the mirror and thin lens formulas (Eqs. 44-5 and 44-19) hold only as an approximation. The rays may not be paraxial, as anyone who has used a camera knows; in astronomical telescopes, however, the rays are indeed paraxial. In typical laboratory microscopes the lens cannot be considered "thin" in the sense in which that term was defined in Section 44-5. In most optical instruments lenses are compound, that is, they are made of several components, cemented together, the interfaces rarely being exactly spherical. This is done to improve image quality and brightness and to relax greatly the dependence on paraxial rays.

In what follows we describe three optical instruments, assuming, for simplicity of illustration, that the thin lens formula applies.

a. The Simple Magnifier. The normal human eye can focus a sharp image of an object on the retina if the object O is located anywhere from infinity, the stars, say, to a certain point called the near point P_n, which we take to be about 25 cm from the eye. If you move the object within the near point the perceived retinal image becomes fuzzy. The location of the near point normally varies with age. We have all heard stories about people who claim not to need glasses but who read their newspapers at arm's length; their near points are receding! Find your own near point by moving this page closer to your eyes, considered separately, until you reach a position at which the image begins to become indistinct.

Figure 44-28a shows an object O placed at the near point P_n. The size

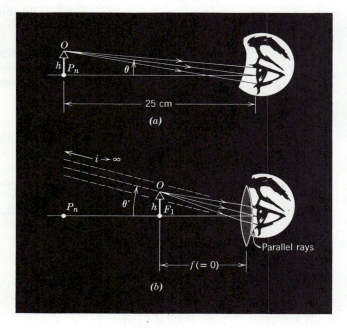

figure 44-28
(a) An object O of height h is placed at the near point P_n of the human eye. If moved any closer, it would fail to form a clear image on the retina. (b) A converging lens (that is, a simple magnifier) is placed close to the eye and the object O is moved from P_n to F_1. Not drawn to scale.

---

* Insect's eyes are even more versatile than human eyes. (Have you ever tried to catch a fly?) See "The Compound Eye of Insects" by G. Adrian Horridge, *Scientific American*, July 1977.

of the perceived image on the retina is measured by the angle θ. In Fig. 44-28b we insert a converging lens of focal length f just in front of the eye and move the object O to the first focal point F_1 of the lens. The eye now perceives an image at infinity, the angle of the image rays being θ', where $\theta' > \theta$. The *angular magnification* m_θ, not to be confused with the *lateral magnification* m given by Eq. 44-20, can be found from

$$m_\theta = \theta'/\theta$$

where $\qquad \theta \cong h/25 \text{ cm} \quad \text{and} \quad \theta' \cong h/f,$

or

$$m_\theta \cong 25 \text{ cm}/f. \qquad (44\text{-}22)$$

Note that, as expected, if $f = 25$ cm, then $m_\theta = 1$, that is, $\theta' = \theta$. Lens aberrations limit the angular magnifications for a simple converging lens to a few orders of magnitude. This is enough, however, for stamp collectors and for actors portraying Sherlock Holmes. More sophisticated magnifier designs have appreciably greater angular magnifications.

b. A Compound Microscope. Figure 44-29 shows a thin lens version of a compound microscope, used for viewing small objects that are very close to the instrument. The object O, of height h, is placed just outside the first focal point F_1 of the objective lens, whose focal length is f_{ob}. A real, inverted image I of height h' is formed by the objective, the lateral magnification being given by Eq. 44-20, or

$$m = \frac{h'}{h} = -\frac{s \tan \theta}{f_{ob} \tan \theta} = -\frac{s}{f_{ob}}. \qquad (44\text{-}23)$$

As usual, the minus sign indicates an inverted image.

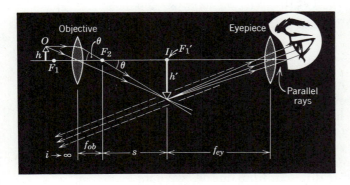

figure 44-29
A simplified version of a compound microscope, using "thin" lenses.

The distance s (sometimes called the tube length) is so chosen that the image I falls on the first focal point F_1' of the eyepiece, which then acts as a simple magnifier as in subsection a above. Parallel rays enter the eye and a final image I' forms at infinity. The final magnification M is given by the product of the linear magnification m for the objective lens, given by Eq. 44-23, and the angular magnification m_θ for the eyepiece, given by Eq. 44-22, or

$$M = m \times m_\theta = -\left(\frac{s}{f_{ob}}\right)\left(\frac{25 \text{ cm}}{f_{ey}}\right). \qquad (44\text{-}24)$$

c. An Astronomical Telescope. Like microscopes, telescopes come in a large variety of forms. The form we describe here is the simple re-

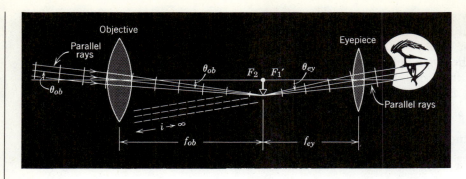

figure 44-30
A simplified version of an astronomical telescope, using "thin" lenses.

fracting telescope* that consists of an objective lens and an eyepiece, both represented in Fig. 44-30 by thin lenses, although in practice, as for microscopes, they will each be compound lens systems.

At first glance it may seem that the lens arrangements for telescopes (Fig. 44-30) and for microscopes (Fig. 44-29) are similar. However, telescopes are designed to view large objects, such as galaxies, stars, and planets, at large distances, whereas microscopes are designed for just the opposite purpose. Note also that in Fig. 44-30 the second focal point of the objective F_2 coincides with the first focal point of the eyepiece F_1', but in Fig. 44-29 these points are separated by a distance s (sometimes called the *tube length*).

In Fig. 44-30 parallel rays from a distant object strike the objective lens, making an angle θ_{ob} with the telescope axis and forming a real, inverted image at the common focal point F_2, F_1'. This image acts as an object for the eyepiece and a (still inverted) virtual image is formed at infinity. The rays defining the image make an angle θ_{ey} with the telescope axis.

The angular magnification m_θ of the telescope is given by θ_{ey}/θ_{ob}. For paraxial rays we can write $\theta_{ob} = h'/f_{ob}$ and $\theta_{ey} = -h'/f_{ey}$ or

$$m_\theta = -\frac{f_{ob}}{f_{ey}}. \tag{44-25}$$

Magnification is only one of the design factors of an astronomical telescope and is indeed easily achieved (How?). There is also *light gathering power* which determines how bright the image is. This is important when viewing faint objects such as distant galaxies and is accomplished by making the objective lens diameter as large as possible. There is *field of view*. An instrument designed for galactic observation (narrow field of view) must be quite different from one designed for the observation of meteors (wide field of view). Also, there are lens and mirror aberrations including *spherical aberration* (that is, lenses and mirrors with truly spherical surfaces do not form sharp images) and *chromatic aberration* (that is, for simple lenses the focal length varies with wavelength so that fuzzy images are formed, displaying unnatural colors). There is also *resolving power* which describes the ability of any optical instrument to distinguish between two objects (stars, say) whose

---

* For an example of the diversity of astronomical telescopes, see "The MMT Observatory on Mount Hopkins" by Nathaniel P. Carleton and Thomas E. Hoffman, *Sky and Telescope*, July 1976. "MMT" stands for Multiple Mirror Telescope.
* See also *Physics Today*, p. 18, April 1977 for a preview of a 2.4-meter (aperture) Space Telescope, planned to orbit the earth in 1983. The device is 46 ft long and will weigh 17,000 lb. It will be launched into orbit by a space shuttle, which will return every two or three years to make repairs and change or add instruments.

angular separation is small. We will discuss this fully in Section 46-5. This by no means exhausts the design parameters of astronomical telescopes. We could also make a similar listing for compound microscopes and, indeed, for any high performance optical instrument.

EXAMPLE 9

Figure 44-31 shows a simple astronomical telescope like that of Fig. 44-30 with the exceptions that (a) the incident parallel rays are parallel to the telescope axis, and (b) the incident rays fill the objective lens, whose diameter is d_{ob}, as, they normally would. Find the diameter d of the *emergent pencil*, as it is called, which contains all the information that is available for entering the eye.

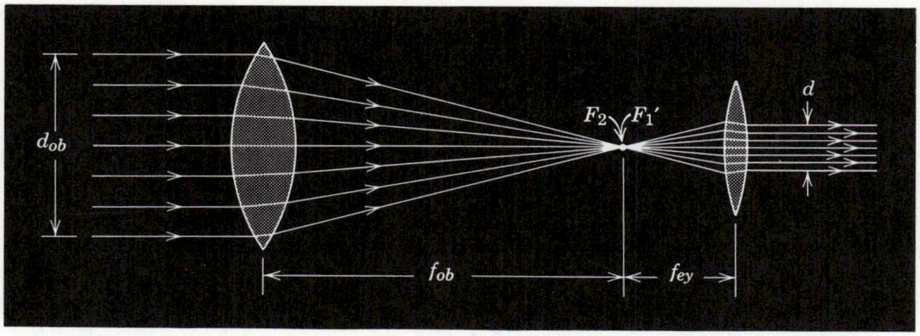

figure 44-31
The formula for d (Eq. 44-26) holds even if the incident rays are not parallel to the telescope axis, as in Fig. 44-30. The rays must be close to the axis, however, or most of the quantitative considerations of this section do not hold true.

In Fig. 44-31 we have, from similar triangles,

$$\frac{d_{ob}/2}{f_{ob}} = \frac{d/2}{f_{ey}},$$

or (see Eq. 44-25)

$$d = d_{ob}\left(\frac{f_{ey}}{f_{ob}}\right) = -\frac{d_{ob}}{m_\theta}. \qquad (44\text{-}26)$$

Note that m_θ is inherently negative so that d is positive, as it must be.

We must compare d with d_p, the diameter of the pupil of the eye. The pupil diameter is not constant but varies with illumination, between the limits of about 2 mm (sunlight) to 9 mm (darkness). The ideal condition is $d = d_p$. For example, if $d > d_p$ some light does not enter the eye and is wasted.

Actually these considerations are modified in practice. For most major optical instruments we use photographic recording. Also, telescopes (and cameras) in orbiting satellites and space probes often send data to earth electronically, where it is processed by computor techniques into photographic form. The emergent pencil concept remains important however, not only for visual observation but for many lens systems design considerations.

questions

1. Can you think of a simple test or observation to prove that the law of reflection is the same for all wavelengths, under conditions in which geometrical optics prevails?

2. We all know that when we look into a mirror right and left are reversed. Our right hand will seem to be a left hand; if we part our hair on the left it will seem to be parted on the right, etc. Can you think of a system of mirrors that would let us see ourselves as others see us? If so draw it and prove your point by sketching some typical rays.

3. We have seen that a single reflection in a plane mirror reverses right and left. When we drive down a highway, for example, the letters on the highway signs are reversed as seen through the rear view mirror. And yet, as

seen through this same mirror, you still seem to be driving down the right lane. Why does the mirror reverse the signs and not the lanes? Or does it? Discuss.

4. A young women peers closely at her face in a high-quality plane mirror. She sees an image of her face in perfect focus. (Ignore the right-left reversal). Why the "perfect focus"? After all her nose is closer to the mirror than her earlobes are.

5. Note (Fig. 44-8) that any portion of the mirror that extends below *b* or above *a* is totally wasted if you wish to see yourself at full length in a mirror. If you have a "full length" mirror available convince yourself that this is true by taping a sheet of newspaper across the mirror at points *a* and *b*, leaving only the region between *a* and *b* exposed. Then, by stepping backward and forward, convince yourself that this theorem is true no matter what your distance from the mirror. Finally tape a sheet of paper *between* *a* and *b* and note what you see in the mirror.

6. If a mirror reverses right and left, why doesn't it reverse up and down?

7. Devise a system of plane mirrors that will let you see the back of your head. Trace the rays to prove your point.

8. If converging rays fall on a plane mirror, is the image virtual?

9. In many city buses a convex mirror is suspended over the door, in full view of the driver. Why not a plane or a concave mirror?

10. Dentists and dental hygienists use a small mirror with a long handle attached to examine your teeth. Is the mirror concave, convex, or plane, and why?

11. What approximations were made in deriving the mirror equation (Eq. 44-4):

$$\frac{1}{o} + \frac{1}{i} = \frac{2}{r}?$$

12. Under what conditions will a spherical mirror, which may be concave or convex, form (*a*) a real image, (*b*) an inverted image, and (*c*) an image smaller than the object?

13. Is it possible to see a perfect reflector? That is, if a perfectly reflecting concave spherical mirror is placed in a dark room and illuminated by, say, a luminous point object placed on its optical axis, could you see the surface of the mirror? Discuss.

14. Can a virtual image be projected onto a screen? Photographed? If you put a piece of paper at the site of a virtual object (assuming a high-intensity light beam) will it ignite after sufficient exposure? Consider these two questions for real images and real objects and note the differences, if any. Discuss.

15. The human eye has often been likened to a camera and, indeed, this comparison is valid. There is a difference however; the eye has no shutter (the eyelids do not serve this purpose). If you leave the shutter of a camera open and scan it across the horizon, the developed image on the film will be a blur. On the other hand, if you scan the horizon with your eye you will see every object distinctly. Have you any ideas to explain the difference between these two situations? See "Visual Motion Perception" by Gunnar Johansson, *Scientific American,* June 1975.

16. In connection with Question 15 consider this further difference between the eye and the camera. If you photograph, say, an elm tree, the image on the film can easily be overexposed beyond the point of recognition. On the other hand, if you simply look at an elm tree your perceived image remains constant for hours on end. Why this difference?

17. For the human eye we assert that the main focusing device is the cornea, that is, the curved outer surface of the eye and that the "lens" of the eye serves to make minor focussing adjustments. True or false? What function does the iris have? Can you think of advantages that the camera has over the human eye? There are some.

18. Does the apparent depth of an object below water depend on the angle of view of the observer in air? Explain and illustrate with ray diagrams.

19. An unsymmetrical thin lens forms an image of a point object on its axis. Is the image location changed if the lens is reversed?

20. Why has a lens two focal points and a mirror only one?

21. Under what conditions will a thin lens, which may be converging or diverging, form (a) a real image, (b) an inverted image, and (c) an image smaller than the object?

22. A skin diver wants to use an air-filled plastic bag as a converging lens for underwater use. Sketch a suitable cross section for the bag.

23. What approximations were made in deriving the thin lens equation (Eq. 44-19):

$$\frac{1}{o} + \frac{1}{i} = \frac{1}{f}?$$

24. Under what conditions will a thin lens have a lateral magnification (a) of -1 and (b) of $+1$?

25. How does the focal length of a glass lens for blue light compare with that for red light, assuming the lens is (a) diverging and (b) converging?

26. Does the focal length of a lens depend on the medium in which the lens is immersed? Is it possible for a given lens to act as a converging lens in one medium and a diverging lens in another medium?

27. Are the following statements true for a glass lens in air? (a) A lens that is thicker at the center than at the edges is a converging lens for parallel light. (b) A lens that is thicker at the edges than at the center is a diverging lens for parallel light. Explain and illustrate, using wavefronts.

28. Under what conditions would the lateral magnification $(m = -i/o)$ for lenses and mirrors become infinite? Is there any practical significance to such a condition?

29. Light rays are reversible. Discuss the situation in terms of objects and images if all rays in Figs. 44-10, 44-14, 44-17, 44-19, 44-23, and 44-25 are reversed in direction.

30. In connection with Fig. 44-24a, we pointed out that all rays originating on the same wavefront in the incident wave have the same optical path length to the image point. Discuss this in connection with Fermat's principle (Section 43-6).

31. Is the focal length of a spherical mirror affected by the medium in which it is immersed? . . . of a thin lens . . . ? What's the difference?

32. What is the significance of the origin of coordinates in Figs. 44-27a and 44-27b?

33. In Fig. 44-23 what are the signs of o, i, r', and r'' in the four cases shown?

34. In Fig. 44-27a how do you interpret $o/|f| = +1$ and $i/|f| = +1$? Draw a ray diagram to illustrate these two situations, for thin converging lenses. In Fig. 44-27b answer the same question for $o/|f| = -1$ and $i/|f| = -1$, for thin diverging lenses.

35. Why is the magnification of a simple magnifier (see the derivation leading to Eq. 44-22) defined in terms of angles rather than image/object size?

36. Ordinary spectacles do not magnify but a simple magnifier does. What then, is the function of spectacles?

37. The "f-number" of a camera lens (see Problem 47) is its focal length divided by its aperture, that is, its effective diameter. Why is this useful to know in photography? How can the f-number of the lens by changed? How is exposure time related to f-number?

38. Does it matter whether (a) an astronomical telescope, (b) a compound microscope, (c) a simple magnifier, (d) a camera, including a TV camera, or (e) a projector, including a slide projector and a motion picture projector produce erect or inverted images? What about real or virtual images? Discuss each case.

39. Why does chromatic aberration occur in simple lenses but not in mirrors?

40. The unaided human eye produces a real but *inverted* image on the retina. (*a*) Why then don't we perceive objects such as people and trees as upside down? (*b*) We don't, of course, but suppose that we wore special glasses so that we did. If you then turned this book upside down, could you read this question with the same facility that you do now? Discuss.

41. In the movies, when the director wishes to portray a scene as viewed through a pair of binoculars, a mask with a horizontal figure-eight opening usually appears on the screen. What is wrong with this?

42. Why are all recent large astronomical telescopes of the reflecting rather than the refracting variety? Think of mechanical mounting problems for lenses and mirrors, the difficulties of shaping (that is, "figuring") the various optical surfaces involved, problems with small flaws in the optical glass blanks used to make lenses and mirrors, etc.

problems

SECTION 44-2

1. A small object is 10 cm in front of a plane mirror. If you stand behind the object, 30 cm from the mirror, and look at its image, for what distance must you focus your eyes?   *Answer:* 40 cm.

2. Suppose you wished to photograph an object seen in a plane mirror. If the object is 5.0 m to your right and 1.0 m closer to the plane of the mirror than you, for what distance must you focus the lens of your camera?

3. A point object is 10 cm away from a mirror while the eye of an observer (pupil diameter 5.0 mm) is 20 cm away. Assuming both the eye and the point to be on the same line perpendicular to the surface, find the area of the mirror used in observing the reflection of the point.   *Answer:* 2.2 mm².

4. Two plane mirrors make an angle of 90° with each other. What is the largest number of images of an object placed between them that can be seen by a properly placed eye? The object need *not* lie on the mirror bisector.

5. Solve Example 2 if the angle between the mirrors is (*a*) 45°, (*b*) 60°, (*c*) 120°, the object always being placed on the bisector of the mirrors.
   Answer: (*a*) 7, (*b*) 5, (*c*) 2.

6. How many images of himself can an observer see in a room whose ceiling and two adjacent walls are mirrors? Explain.

7. A small object *O* is placed one-third of the way between two parallel plane mirrors as in Fig. 44-32. Trace appropriate bundles of rays for viewing the four images that lie closest to the object.

8. In Fig. 44-6 you rotate the mirror 30° counterclockwise, leaving the point object *O* in place. Is the (virtual) image point displaced? If so, where is it? Can the eye still see the image without being moved? Sketch a figure showing the new situation, with *O* and the eye remaining unchanged in position.

9. You are peering into a peephole into an illuminated box. A plane mirror is on the interior wall facing you. A square wire framework is rotating slowly in a counterclockwise direction (as seen from above) about a vertical axis.

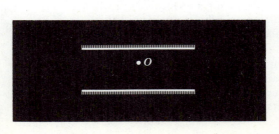

figure 44-32
Problem 7

The distance from the peephole to the mirror is 15 cm. (a) How far from the mirror should the vertical axis of rotation of the square be located so that its image in the mirror would appear at half size? (b) Compare the directions of rotation of the object and image. (c) Describe in your own words what you would see, assuming that you have not been told that the box contains a mirror.

10. Extend Fig. 44-9 to three dimensions by adding a mirror perpendicular to the common axis of the two mirrors shown. This forms a *corner reflector*, much used in optical, microwave, and other applications. It has the property that an incident ray is returned, after three reflections, along the same direction. Can you prove this?

SECTION 44-3

11. For clarity, the rays in figures like Fig. 44-10 are not drawn paraxial enough for Eq. 44-4 to hold with great accuracy. With a ruler, measure r and o in this figure and calculate, from Eq. 44-4, the predicted value of i. Compare this with the measured value of i.

12. Fill in this table, each column of which refers to a spherical mirror and a real object. Check your results by graphical analysis. Distances are in centimeters; if a number has no plus or minus sign in front of it, find the correct sign.

	a	b	c	d	e	f	g	h
Type	Concave						Convex	
f	20		+20			20		
r				−40			40	
i				−10			4	
o	+10	+10	+30	+60				+24
m		+1		−0.5		+0.10		0.50
Real image?		no						
Erect image?								no

Answer: For alternate vertical columns: (a) +, +40, −20, +2, no, yes. (c) Concave, +40, +60, −2, yes, no. (e) Convex, −20, +20, +0.5, no, yes. (g) −20, −, −, +5, no, yes.

13. A short linear object of length l lies on the axis of a spherical mirror, a distance o from the mirror. (a) Show that its image will have a length l' where

$$l' = l\left(\frac{f}{o-f}\right)^2.$$

(b) Show that the *longitudinal magnification* $m'(=l'/l)$ is equal to m^2 where m is the lateral magnification discussed in Section 44-3. (c) Is there any condition such that, neglecting all mirror defects, the image of a small cube would also be a cube?   *Answer:* (c) Yes; object at center of curvature.

14. Redraw Fig. 44-33 on a large sheet of paper and trace carefully the reflected rays, using the law of reflection. Is a point focus formed? Discuss.

15. A thin flat plate of partially reflecting glass is a distance b from a convex mirror. A point source of light S is placed a distance a in front of the plate (see Fig. 44-34) so that its image in the partially reflecting plate coincides with its image in the mirror. If $b = 7.5$ cm and the focal length of the mirror is $f = -30$ cm, find a and draw the ray diagram.   *Answer:* 23 cm.

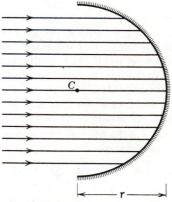

figure 44-33
Problem 14

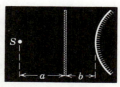

figure 44-34
Problem 15

16. Verify that Eq. 44-4 is consistent with the situations of Figs. 44-5, 10, 12 (see Example 3), 13, 14, and 15. In some cases only a qualitative answer is possible.

17. Modify Fig. 44-12 in such a way (possibly by adding a second mirror to the left of the one shown) that the object for Fig. 44-12 is virtual. Trace the rays.

SECTION 44-4

18. A penny lies on the bottom of a swimming pool 10 ft. deep. What is its apparent depth as viewed from above the water? The index of refraction of water is 1.33.

19. Fill in the following table, each column of which refers to a spherical surface separating two media with different indices of refraction. Distances are measured in centimeters. The object is real in all cases.

	a	b	c	d	e	f	g	h
n_1	1.0	1.0	1.0	1.0	1.5	1.5	1.5	1.5
n_2	1.5	1.5	1.5		1.0	1.0	1.0	
o	+10	+10		+20	+10		+70	+100
i		−13	+600	−20	−6	−7.5		+600
r	+30		+30	−20		−30	+30	−30
Real image?								

Draw a figure for each situation and construct the appropriate rays graphically. Assume a point object.
Answer: For alternate vertical columns: (a) −18, no. (c) +71, yes. (e) +30, no. (g) −26, no.

20. A layer of water ($n = 1.33$) 2.0 cm thick floats on carbon tetrachloride ($n = 1.46$) 4.0 cm thick. How far below the water surface, viewed at normal incidence, does the bottom of the tank seem to be?

21. As an example of the importance of the paraxial ray assumption, consider this problem. You place a coin at the bottom of a swimming pool filled with water ($n = 1.33$) to a depth of 8.0 ft. What is the apparent depth of the coin below the surface when viewed (a) at near normal incidence (that is, by paraxial rays) and (b) by rays that leave the coin making an angle of 30° with the normal (that is, definitely not paraxial rays)? What do you conclude?
Answer: (a) 6.0 ft. (b) 5.6 ft.

22. Define and locate the first and second focal points (see p. 975) for a single spherical refracting surface such as that of Fig. 44-17.

23. A parallel incident beam falls on a solid glass sphere at normal incidence. Locate the image in terms of the index of refraction n and the sphere radius r.

Answer: Assuming the light incident from the left, $i = \dfrac{2-n}{2(n-1)} r$, to the right

of the right edge of the sphere if n < 2, as it is for glass.

SECTION 44-5

24. A double-convex lens is to be made of glass with an index of refraction of 1.50. One surface is to have twice the radius of curvature of the other and the focal length is to be 6.0 cm. What are the radii?

25. A lens is made of glass having an index of refraction of 1.5. One side of the lens is flat and the other convex with a radius of curvature of 20 cm. (a) Find the focal length of the lens. (b) If an object is placed 40 cm to the left of the

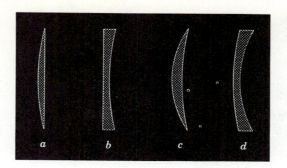

figure 44-35
Problem 26

989 PROBLEMS CHAP. 44

lens, where will the image be located? (c) Would either of these answers change if we rotated the lens through 360°?

Answer: (a) +40 cm. (b) Image to the right at +∞. (c) No.

26. Using the lensmaker's equation (Eq. 44-18), decide which of the thin lenses in Fig. 44-35 are converging and which diverging for parallel incident light.

27. An object is placed at a center of curvature of a double-concave lens, both of whose radii of curvature have the same magnitude. (a) What are the signs of the two radii of curvature? (b) Find the location of the image in terms of the radius of curvature r and the index of refraction, n, of the glass. (c) Describe the nature of the image. (d) Verify your result with a ray diagram.

Answer: (a) r' is negative and r'' is positive; see Fig. 44-23b and 44-23c.

(b) $i = -\dfrac{r}{2n-1}$. (c) Virtual and erect.

28. Show that the focal length f' for a thin lens whose index of refraction is n and which is immersed in a medium, water, say, whose index of refraction is n' is given by

$$\frac{1}{f'} = \frac{n-n'}{n'}\left(\frac{1}{r'} - \frac{1}{r''}\right).$$

29. Reproduce Fig. 44-27a from first principles, that is, from Eq. 44-19. How do you know (a) that the lens is diverging or converging? (b) That the image is real or virtual? (c) That the object is real or virtual? (d) That the lateral magnification is > 1 or < 1?

30. Fill in this table, each column of which refers to a thin lens, to the extent possible. Check your results by graphical analysis. Distances are in centimeters; if a number (except in row n) has no plus sign or minus sign in front of it, find the correct sign.

	a	b	c	d	e	f	g	h	i
Type	converging								
f	10	+10	10	10					
r'					+30	−30	−30		
r''					−30	+30	−60		
i									
o	+20	+5	+5	+5	+10	+10	+10	+10	+10
n					1.5	1.5	1.5		
m			>1	<1				0.5	0.5
Real image?									yes
Erect image?								yes	

Draw a figure for each situation and construct the appropriate rays graphically. The object is real in all cases.

Answer: Alternate vertical columns (an X means that the quantity cannot be found from the data given): (*a*) +, X, X, +20, X, −1, yes, no. (*c*) Converging, +, X, X, −10, X, no, yes. (*e*) Converging, +30, −15, +1.5, no, yes. (*g*) Diverging, −120, −9.2, +0.92, no, yes. (*i*) Converging, +3.3, X, X, +5, X, −, no.

31. A luminous object and a screen are a fixed distance D apart. (*a*) Show that a converging lens of focal length f will form a real image on the screen for two positions that are separated by

$$d = \sqrt{D(D - 4f)}.$$

(*b*) Show that the ratio of the two image sizes for these two positions is

$$\left(\frac{D - d}{D + d}\right)^2.$$

32. A converging lens with a focal length of +20 cm is located 10 cm to the left of a diverging lens having a focal length of −15 cm. If a real object is located 40 cm to the left of the first lens, locate and describe completely the image formed.

33. Two thin lenses of focal length f_1 and f_2 are in contact. Show that they are equivalent to a single thin lens with a focal length given by

$$f = \frac{f_1 f_2}{f_1 + f_2}.$$

34. Show that the distance between a real object and its real image formed by a thin converging lens is always greater than or equal to four times the focal length of the lens.

35. The formula

$$\frac{1}{o} + \frac{1}{i} = \frac{1}{f}$$

is called the *Gaussian* form of the thin lens formula. Another form of this formula, the *Newtonian* form, is obtained by considering the distance x from the object to the first focal point and the distance x' from the second focal point to the image. Show that

$$xx' = f^2.$$

36. An erect object is placed a distance in front of a converging lens equal to twice the focal length f_1 of the lens. On the other side of the lens is a converging mirror of focal length f_2 separated from the lens by a distance $2(f_1 + f_2)$. (*a*) Find the location, nature and relative size of the final image. (*b*) Draw the appropriate ray diagram. See Fig. 44-36.

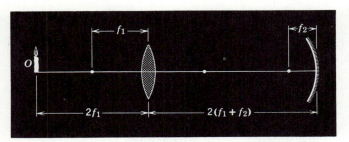

figure 44-36
Problem 36

37. Two thin lenses, one having $f = +12.0$ cm and the other having $f = −10.0$ cm, are separated by 7.0 cm. A small object is placed 43.5 cm from the center of

the lens system on the principal axis first on one side and next on the other side. Find the location of the final image in each case.

Answer: For object nearer to the converging lens the image is 8.5 cm to the opposite side of the center of the lens system. For an object nearer the diverging lens, the image is 63.5 cm to the opposite side of the center of the lens system.

38. (a) Show that a thin converging lens of focal length f followed by a thin diverging lens of focal length $-f$ will bring parallel light to a focus beyond the second lens provided that the separation of the lenses L satisfies $0 < L < f$. (b) Does this property change if the lenses are interchanged? (c) What happens when $L = 0$?

39. An object is placed 1.0 m in front of a converging lens, of focal length 0.5 m, which is 2.0 m in front of a plane mirror. (a) Where is the final image, measured from the lens, that would be seen by an eye looking toward the mirror through the lens? (b) Is the final image real or virtual? (c) Is the final image erect or inverted? (d) What is the lateral magnification?

Answer: (a) 0.60 m on the side of the lens away from the mirror. (b) Real. (c) Erect. (d) +0.20.

40. An object is 20 cm to the left of a lens with a focal length of +10 cm. A second lens of focal length +12.5 cm is 30 cm to the right of the first lens. (a) Using the image formed by the first lens as the object for the second, find the location and relative size of the final image. (b) Verify your conclusions by drawing the lens system to scale and constructing a ray diagram. (c) Describe the final image.

41. Show that Eq. 44-21 is correct.

42. An object is placed 1.0 m in front of a converging lens, of focal length 0.50 m, which is 2.0 m in front of a plane mirror. (a) Where is the final image, measured from the lens, that would be seen by an eye looking toward the mirror through the lens? (b) Is the final image real or virtual? (c) Is the final image erect or inverted? (d) What is the lateral magnification?

SECTION 44-6

43. In connection with Fig. 44-28b, (a) show that if the object O is moved from the first focal point F_1 toward the eye, the image moves in from infinity and the angle θ' (and thus the angular magnification m_θ) is increased. (b) If you continue this process, at what image location will m_θ have its maximum useable value? (c) Show that the maximum useable value of m_θ is $1 + (25\text{ cm})/f$. (d) Show that in this situation the angular magnification is equal to the linear magnification.

44. A microscope of the type shown in Fig. 44-29 has a focal length for the objective lens of 4.0 cm and for the eyepiece lens of 8.0 cm. The distance between the lenses is 25 cm. (a) What is the distance s in Fig. 44-29? (b) To reproduce the conditions of Fig. 44-29 how far beyond F_1 in that figure should the object be placed? (c) What is the lateral magnification m of the objective? (d) What is the angular magnification m_θ of the eyepiece? (e) What is the overall magnification M of the microscope?

45. *The eye—the basic optical instrument:* Figure 44-37a suggests a normal human eye. Parallel rays, entering a relaxed eye gazing at infinity, produce a real, inverted image on the retina. The eye thus acts as a converging lens. Most of the refraction occurs at the outer surface of the eye, the *cornea.* Assume a focal length f for the eye of 2.50 cm.

In Fig. 44-37b the object is moved in to a distance o (=40.0 cm) from the

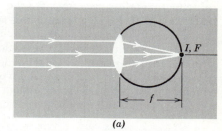

(a)

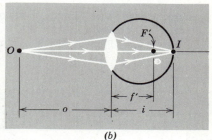

(b)

figure 44-37
Problem 45

eye. To form an image on the retina the effective focal length of the eye must be reduced to f'. This is done by the action of the ciliary muscles that change the shape of the lens and thus the effective focal length of the eye.

(a) Find f' from the above data. (b) Would the effective radii of curvature of the lens become larger or smaller in the transition from a to b in Fig. 44-37? (In the figure the structure of the eye is only roughly suggested and Fig. 44-37b is not to scale.)   *Answer:* (a) 2.35 cm. (b) Smaller.

46. In an eye that is *farsighted* the eye focuses parallel rays so that the image would form behind the retina, as in Fig. 44-38a. In an eye that is *nearsighted* the image is formed in front of the retina, as in Fig. 44-38b. (a) How would you design a corrective lens for each eye defect? Make a ray diagram for each case. (b) If you need spectacles only for reading, are you nearsighted or farsighted? (c) What is the function of bifocal spectacles, in which the upper parts and lower parts have different focal lengths? (d) Some musicians in symphony orchestras (and others as well) wear *trifocals*. What occupational optical problem is involved here?

47. *The camera:* Figure 44-39 shows an idealized camera focused on an object at infinity. A real, inverted image I is formed on the film, the image distance i being equal to the (fixed) focal length f (=5.0 cm, say) of the lens system. In Fig. 44-39b the object O is closer to the camera, the object distance o being, say, 100 cm. To focus an image I on the film, we must extend the lens away from the camera (why?). (a) Find i' in Fig. 44-39b. (b) By how much must the lens be moved? Note that the camera differs from the eye (see Problem 45) in this respect. In the camera, f remains constant and the image distance i must be adjusted by moving the lens. For the eye the image distance i remains constant and the focal length f is adjusted by distorting the lens. Compare Figs. 44-37 and 44-39 carefully.
Answer: (a) 5.3 cm. (b) 3.0 mm.

48. *The reflecting telescope:* Isaac Newton, having convinced himself (erroneously as it turned out) that chromatic aberration was an inherent property of refracting telescopes, invented the reflecting telescope, shown schematically in Fig. 44-40. He presented his second model of this telescope, which has a magnifying power of 38, to the Royal Society, which still has it.

In Fig. 44-40 incident light falls, closely parallel to the telescope axis, on the objective mirror M. After reflection from small mirror M' (the figure is not to scale), the rays form a real, inverted image in the focal plane through F. This image is then viewed through an eyepiece.

(a) Show that the angular magnification m_θ is also given by Eq. 44-25, or

$$m_\theta = -f_{ob}/f_{ey}$$

where f_{ob} is the focal length of the objective mirror and f_{ey} that of the eyepiece. (b) The 200-in. mirror of the reflecting telescope at Mt. Palomar in California has a focal length of 16.8 m. Estimate the size of the object formed

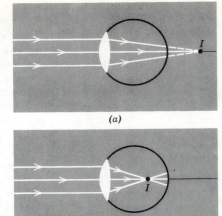

(a)

(b)

figure 44-38
Problem 46

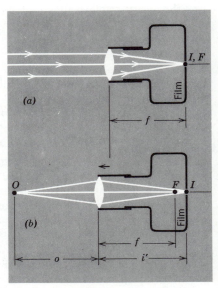

(a)

(b)

figure 44-39
Problem 47

figure 44-40
Problem 48

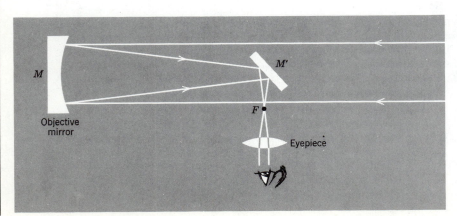

in the focal plane of this mirror when the object is a meter stick 2.0 km away. Assume parallel incident rays. (*c*) The mirror of a reflecting astronomical telescope has an effective radius of curvature ("effective" because such mirrors are ground to a parabolic rather than a spherical shape, to eliminate spherical aberration defects) of 10 m. To give an angular magnification of 200, what must be the focal length of the eyepiece?

45
interference

In Section 19-7 we saw that if two mechanical waves of the same frequency travel in approximately the same direction and have a phase difference that remains constant with time, they may combine so that their energy is not distributed uniformly in space but is a maximum at certain points and a minimum (possibly even zero) at others. The demonstration of such *interference* effects for light by Thomas Young in 1801 first established the wave theory of light on a firm experimental basis. Young was able to deduce the wavelength of light from his experiments, the first measurement of this important quantity.

Young allowed sunlight to fall on a pinhole S_0 punched in a screen A in Fig. 45-1. The emerging light spreads out by diffraction (see Section 44-1) and falls on pinholes S_1 and S_2 punched into screen B. Again diffraction occurs and two overlapping spherical waves expand into the space to the right of screen B.

The condition for geometric optics, namely, that $a \gg \lambda$ where a is the diameter of the pinholes, is definitely *not* met in this experiment. The pinholes do not cast geometrical shadows but act as sources of expanding Huygens' wavelets. We are dealing here (and in the three succeeding chapters) with *wave optics* rather than with geometrical optics.

Figure 45-2, taken from an 1803 paper of Young, shows the region between screens B and C. The blackening represents the minima of the wave disturbance; the white space between represents the maxima. If you hold the page with your eye close to the left edge and look at a grazing angle along the figure, you will see that along lines marked by x's there is cancellation of the wave; between them there is reinforcement. If we place a screen anywhere across the superimposed waves, we expect to find alternate bright and dark spots on it. Figure 45-3 shows a

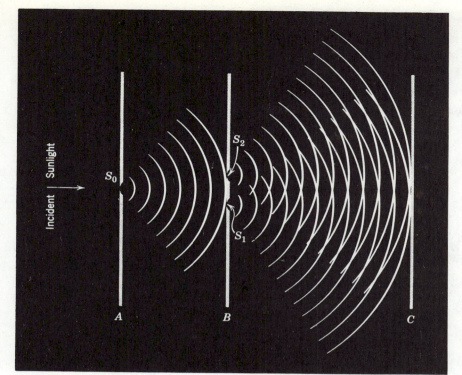

figure 45-1
Showing how Thomas Young produced an interference pattern by allowing diffracted waves from pinholes S_1 and S_2 to overlap on screen C.

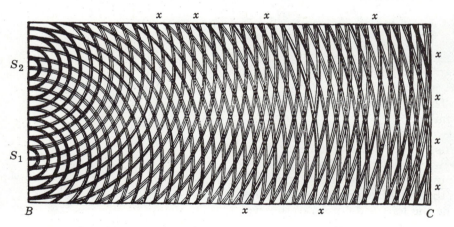

figure 45-2
Thomas Young's original drawing showing interference effects in overlapping waves. Place the eye near the left edge and sight at a grazing angle along the figure. (From Thomas Young, *Phil. Transactions*, 1803.) See also *Great Experiments in Physics*, p. 93, Morris H. Shamos, ed., Holt and Company, N.Y., 1959, for a readable annotated account of this experiment.

figure 45-3
Interference fringes for monochromatic light, made with an arrangement like that of Fig. 45-1, using long narrow slits rather than pinholes.

photograph of such *interference fringes*; in keeping with modern technique, long narrow slits rather than pinholes were used in preparing this figure.

Interference is not limited to light waves but is a characteristic of all wave phenomena. Figure 45-4, for example, shows the interference pattern of water waves in a shallow *ripple tank*. The waves are generated by two vibrators that tap the water surface in synchronism, producing two expanding spherical waves.

Let us analyze Young's experiment quantitatively, assuming that the incident light consists of a single wavelength only. In Fig. 45-5 P is an arbitrary point on the screen, a distance r_1 and r_2 from the narrow slits S_1 and S_2, respectively. Let us draw a line from S_2 to b in such a way that the lines PS_2 and Pb are equal. If d, the slit spacing, is much smaller than the distance D between the two screens (the ratio d/D in the figure has been exaggerated for clarity), S_2b is then almost perpendicular to both r_1 and r_2. This means that angle S_1S_2b is almost equal to angle PaO, both

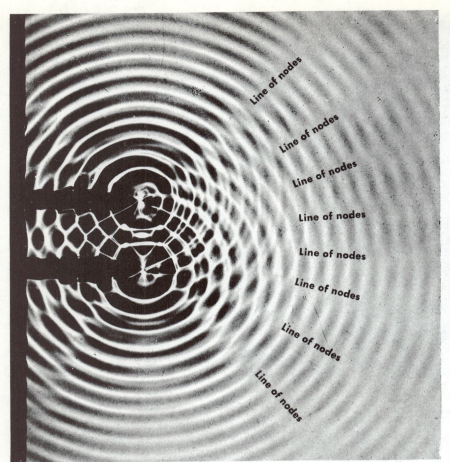

figure 45-4
The interference of water waves in a
ripple tank. There is destructive
interference along the lines marked
"Line of nodes" and constructive
interference between these lines.
(Courtesy Physical Science Study
Committee.)

angles being marked θ in the figure. This is equivalent to saying that the lines r_1 and r_2 may be taken as parallel.

We often put a lens in front of the two slits, as in Fig. 45-6, the screen C being in the focal plane of the lens. Under these conditions light focused at P must have struck the lens parallel to the line Px, drawn from P through the center of the (thin) lens. Under these conditions rays r_1 and r_2 are strictly parallel even though the requirement $D \gg d$ is not met. The lens L may in practice be the lens and cornea of the eye, screen C being the retina.

The two rays arriving at P in Figs. 45-5 or 45-6 from S_1 and S_2 are in phase at the source slits, both being derived from the same wavefront in the incident plane wave. Because the rays have different optical path lengths, they arrive at P with a phase difference. The number of wavelengths contained in S_1b, which is the path difference, determines the nature of the interference at P.

To have a *maximum* at P, S_1b $(=d \sin \theta)$ must contain an integral number of wavelengths, or

$$S_1b = m\lambda \qquad m = 0, 1, 2, \ldots,$$

which we can write as

$$d \sin \theta = m\lambda \qquad m = 0, 1, 2, \ldots \text{ (maxima).} \qquad (45\text{-}1)$$

Note that each maximum above O in Figs. 45-5 and 45-6 has a symmetrically located maximum below O. There is a central maximum described by $m = 0$.

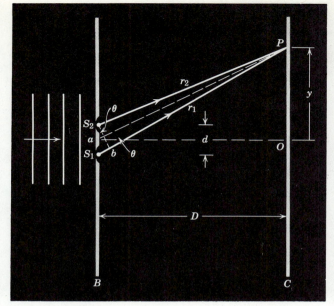

figure 45-5
Rays from S_1 and S_2 combine at P. The wave fronts of light falling on screen B have been taken as parallel. Actually, $D \gg d$, the figure being distorted for clarity. We represent the midpoint of the slits by a.

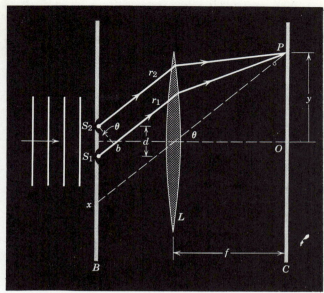

figure 45-6
A lens is normally used to produce interference fringes; compare with Fig. 45-5. The figure is again distorted for clarity in that $f \gg d$ in practice.

For a *minimum* at P, S_1b ($=d \sin \theta$) must contain a half-integral number of wavelengths, or

$$d \sin \theta = (m + \tfrac{1}{2})\lambda \qquad m = 0, 1, 2, \ldots \text{(minima)}. \qquad (45\text{-}2)$$

If a lens is used as in Fig. 45-6, it may seem that a phase difference should develop between the rays beyond the plane represented by S_2b, the path lengths between this plane and P being clearly different. In Section 44-5, however, we saw that for such parallel rays focused by a lens the *optical path lengths* are identical. Two rays with the same optical path lengths contain the same number of wavelengths, so that no phase difference will result because of the light passing through the lens.

The double-slit arrangement in Fig. 45-5 is illuminated with light from a mercury vapor lamp so filtered that only the strong green line ($\lambda = 546$ nm or 5460 Å) is effective. The slits are 0.10 mm apart, and the screen on which the interference pattern appears is 20 cm away. What is the angular position of the first minimum? Of the tenth maximum?

At the first minimum we put $m = 0$ in Eq. 45-2, or

$$\sin \theta = \frac{(m + \tfrac{1}{2})\lambda}{d} = \frac{(\tfrac{1}{2})(546 \times 10^{-9} \text{ m})}{0.10 \times 10^{-3} \text{ m}} = 0.0027.$$

This value for $\sin \theta$ is so small that we can take it to be the value of θ, expressed in radians; expressed in degrees it is 0.16°.

At the tenth maximum (not counting the central maximum) we must put $m = 10$ in Eq. 45-1. Doing so and calculating as before leads to an angular position of 3.8°. For these conditions we see that the angular spread of the first dozen or so fringes is small.

EXAMPLE 1

EXAMPLE 2

In Example 1 what is the linear distance on screen C between adjacent maxima?

If θ is small enough, we can use the approximation

$$\sin \theta \cong \tan \theta \cong \theta.$$

From Fig. 45-5 we see that

$$\tan \theta = \frac{y}{D}.$$

Substituting this into Eq. 45-1 for $\sin \theta$ leads to

$$y = m\frac{\lambda D}{d} \qquad m = 0,\ 1,\ 2,\ \ldots \ \text{(maxima).}$$

The positions of any two adjacent maxima are given by

$$y_m = m\frac{\lambda D}{d}$$

and

$$y_{m+1} = (m+1)\frac{\lambda D}{d}.$$

We find their separation Δy by subtracting

$$\Delta y = y_{m+1} - y_m = \frac{\lambda D}{d}$$

$$= \frac{(546 \times 10^{-9}\ \text{m})(20 \times 10^{-2}\ \text{m})}{0.10 \times 10^{-3}\ \text{m}} = 1.09\ \text{mm}.$$

As long as θ in Figs. 45-5 and 45-6 is small, the separation of the interference fringes is independent of m; that is, the fringes are evenly spaced. Note that if the incident light contains more than one wavelength the separate interference patterns, which will have different fringe spacings, will be superimposed.

We can use Eq. 45-1 to measure the wavelength of light; to quote Thomas Young:

From a comparison of various experiments, it appears that the breadth of the undulations [that is, the wavelength] constituting the extreme red light must be supposed to be, in air, about one 36 thousandth of an inch, and those of the extreme violet about one 60 thousandth; the mean of the whole spectrum, with respect to the intensity of light, being one 45 thousandth.

Young's value for the average effective wavelength present in sunlight (1/45,000 in.) can be written as 570 nm, which agrees rather well with the wavelength at which the eye sensitivity is a maximum, 555 nm (see Fig. 42-1). It must not be supposed that Young's work was received without criticism. One of his contemporaries, a firm believer in the corpuscular theory of light, wrote in part:

We wish to raise our feeble voice against innovations that can have no other effect than to check the progress of science, and renew all those wild phantoms of the imagination which Bacon and Newton put to flight from her temple. This paper contains nothing that deserves the name of either experiment or discovery.

Needless to say, posterity has decided in favor of Young.

45-2
COHERENCE

Analysis of the derivation of Eqs. 45-1 and 45-2 shows that a fundamental requirement for the existence of well-defined interference fringes on screen C in Figs. 45-5, 6 is that the light waves that travel from S_1 and S_2 to any point P on this screen must have a sharply defined phase difference ϕ that remains constant with time. If this condition is satisfied, a stable, well-defined fringe pattern will appear. At certain points P, ϕ

will be given, independent of time, by $n\pi$ where $n = 1, 3, 5, \ldots$ so that the resultant intensity will be strictly zero and will remain so throughout the time of observation. At other points ϕ will be given by $n\pi$ where $n = 0, 2, 4 \ldots$ and the resultant intensity will be a maximum. Under these conditions the two beams emerging from slits S_1 and S_2 are said to be completely *coherent*.

Let the source in Fig. 45-5 be removed and let slits S_1 and S_2 be replaced by two completely independent light sources, such as two fine incandescent wires placed side by side in a glass envelope. No interference fringes will appear on screen C but only a relatively uniform illumination. We can interpret this if we make the reasonable assumption that for completely independent light sources the phase difference between the two beams arriving at P will vary with time in a random way. At a certain instant conditions may be right for cancellation and a short time later (perhaps 10^{-8} s) they may be right for re-enforcement. This same random phase behavior holds for all points on screen C with the result that this screen is uniformly illuminated. The intensity at any point is equal to the sum of the intensities that each source S_1 and S_2 produces separately at that point. Under these conditions the two beams emerging from S_1 and S_2 are said to be completely *incoherent*.

Note that for completely *coherent* light beams one (1) combines the amplitudes vectorially, taking the (constant) phase difference properly into account, and then (2) squares this resultant amplitude to obtain a quantity proportional to the resultant intensity. For completely *incoherent* light beams, on the other hand, one (1) squares the individual amplitudes to obtain quantities proportional to the individual intensities, and then (2) adds the individual intensities to obtain the resultant intensity. This procedure is in agreement with the experimental fact that for completely independent light sources the resultant intensity at every point is always greater than the intensity produced at that point by either light source acting alone.

It remains to investigate under what experimental conditions coherent or incoherent beams may be produced and to give an explanation for coherence in terms of the mode of production of the radiation. Consider first a parallel beam of microwave radiation emerging from an antenna connected by a coaxial cable to an oscillator based on an electromagnetic resonant cavity. The cavity oscillations (see Section 38-5) are completely periodic with time and produce, at the antenna, a completely periodic variation of **E** and **B** with time. The radiated wave at large enough distances from the antenna is well represented by Fig. 41-13. Note that (1) the wave has essentially infinite extent in time, including both future times ($t > 0$, say) and past times ($t < 0$); see Fig. 45-7a. At any point, as the wave passes by, the wave disturbance (i.e., **E** or **B**) varies with time in a perfectly periodic way. (2) The wavefronts at points

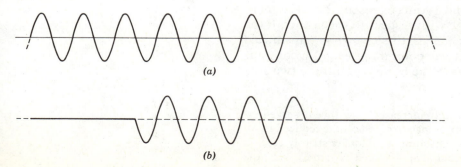

(a)

(b)

figure 45-7
(a) A section of an infinite wave and (b) a wavetrain.

far removed from the antenna are parallel planes of essentially infinite extent at right angles to the propagation direction. At any instant of time the wave disturbance varies with distance along the propagation direction in a perfectly periodic way.

Two beams generated from a single traveling wave like that of Fig. 41-13 will be completely coherent. One way to generate two such beams is to put an opaque screen containing two slits in the path of the beam. The waves emerging from the slits will always have a constant phase difference at any point in the region in which they overlap and interference fringes will be produced. Coherent radio beams can also be readily established, as can coherent elastic waves in solids, liquids and gases. The two prongs of the vibrating tapper in Fig. 45-4, for example, generate two coherent waves in the water of the ripple tank.

The technique of producing two beams from a single beam (and thus from a single source) tests specifically whether the wavefronts in the single parallel beam are truly planes, that is, whether all points in a plane at right angles to the direction of propagation have the same phase at any given instant. By dividing the beam in another way it is possible to test whether the beam is truly periodic over a large number of cycles of oscillation. This can be done, as we show in detail in Section 45-7, by inserting in the beam at 45° to it a thin sheet of material possessing the property that two beams are produced, one (which will be at right angles to the incident beam) by reflection and a second (which will be in the direction of the incident beam) by transmission. In the visible region such a sheet, called a *half-silvered mirror*, may be formed from a glass plate by depositing on it an appropriately thin film of silver. By appropriate use of mirrors (see Section 45-7) these two sub-beams can be recombined into a single beam traveling in a chosen direction. If the beams travel different distances before they are recombined, we are comparing in the combined beam a sample of the original beam with another sample an arbitrarily large number of cycles away. If the original beam is truly periodic in space and time, the two sub-beams will be completely coherent and interference fringes will be produced when they are recombined.

If we turn from microwave sources to common sources of visible light, such as incandescent wires or an electric discharge passing through a gas, we become aware of a fundamental difference. In both of these sources the fundamental light emission processes occur in individual atoms and these atoms do not act together in a cooperative (that is, *coherent*) way. The act of light emission by a single atom takes, in a typical case, about 10^{-8} s and the emitted light is properly described as a *wavetrain* (Fig. 45-7b) rather than as a wave (Fig. 45-7a). For emission times such as these the wavetrains are a few meters long.

Interference effects from ordinary light sources may be produced by putting a very narrow slit (S_0 in Fig. 45-1) directly in front of the source. This insures that the wavetrains that strike slits S_1 and S_2 in screen B in this figure originate from the same small region of the source. The diffracted beams emerging from S_1 and S_2 thus represent the same population of wavetrains and are coherent with respect to each other. If the phase of the light emitted from S_0 changes, this change is transmitted simultaneously to S_1 and S_2. Thus, at any point on screen C, a constant phase difference is maintained between the beams from these two slits and a stationary interference pattern occurs.

If the width of slit S_0 in Fig. 45-1 is gradually increased, it will be observed experimentally that the maxima of the interference fringes become reduced in intensity and that the intensity in the fringe minima is no longer strictly zero. In other words, the fringes become less distinct. If S_0 is opened extremely wide, the lowering of the maximum intensity and the raising of the minimum intensity will be such that the fringes disappear, leaving only a uniform illumination. Under these conditions we say that the beams from S_1 and S_2 pass continuously from a condition of complete coherence to one of complete incoherence. When not at either of these two limits, the beams are said to be partially coherent.

Partial coherence can also be demonstrated in two beams that are produced, as described on p. 1000, by inserting a "half-silvered mirror" in a beam at an angle of 45° to the direction of propagation. The two beams so produced, by reflection and transmission, traverse paths of different lengths before they are recombined. If the path difference is small compared with the average length of a wavetrain, the interference fringes will be sharply defined and will go essentially to zero at their minima. If the path difference is deliberately made longer, the fringes will become less distinct, and finally, when the path difference is larger than the average length of a wavetrain, the fringes will disappear altogether. Thus it is possible once again to progress smoothly in an experimental arrangement from complete coherence, through partial coherence, to complete incoherence.

The lack of coherence of the light from ordinary sources such as glowing wires is due to the fact that the emitting atoms do not act cooperatively (that is, *coherently*). Since 1960 it has proved possible to construct sources of visible light in which the atoms *do* act cooperatively and in which the emitted light is highly coherent. Such devices are called *lasers**; their light output is extremely monochromatic, intense, and highly collimated. The coherence of the emitted light can be demonstrated by placing a screen with two holes in it in the beam emerging from the laser. An interference pattern results, as Fig. 45-8 shows. These methods permit, for the first time, a degree of control of visible light approaching that possible for radio and for microwaves. The practical applications of lasers, which include the amplification of weak light signals, the use of light beams as highly efficient carriers of in-

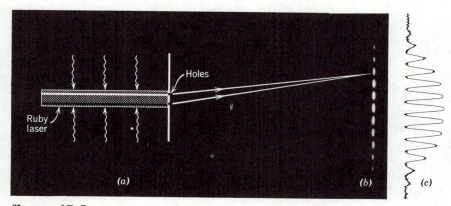

figure 45-8
A screen with two small holes is placed against the end of a laser. (*a*) The light passing through the holes forms an interference pattern on a strip of photographic film placed at (*b*). The fact that the pattern is formed shows that the light emitted from the laser is coherent across the beam cross section. At (*b*) is shown the image as formed on a photographic strip exposed in this manner. At (*c*) is an intensity plot of the film made by measuring the degree to which the film has been exposed. (Courtesy of D. F. Nelson and R. J. Collins, Bell Telephone Laboratories.)

* Laser, is a coined word, meaning "*l*ight *a*mplification through *s*timulated *e*mission of *r*adiation." For an early report, see "Optical Masers" by Arthur L. Schawlow, *Scientific American*, June 1961.

formation from point to point,* and the production of high temperatures by intense local heating, are being actively exploited.†

Let us assume that the electric field components of the two waves in Fig. 45-5 vary with time at point P as

$$E_1 = E_0 \sin \omega t \qquad (45\text{-}3)$$

and

$$E_2 = E_0 \sin (\omega t + \phi) \qquad (45\text{-}4)$$

where $\omega (= 2\pi\nu)$ is the angular frequency of the waves and ϕ is the phase difference between them. Note that ϕ depends on the location of point P, which, in turn, for a fixed geometrical arrangement, is described by the angle θ in Figs. 45-5 and 45-6. We assume that the slits are so narrow that the diffracted light from each slit illuminates the central portion of the screen uniformly. This means that near the center of the screen E_0 is independent of the position of P, that is, of the value of θ.

The resultant wave disturbance‡ at P is found from

$$E = E_1 + E_2 \qquad (45\text{-}5)$$

and (see Eq. 19-16) proves to be

$$E = E_\theta \sin (\omega t + \beta), \qquad (45\text{-}6a)$$

where

$$\beta = \tfrac{1}{2}\phi \qquad (45\text{-}6b)$$

and

$$E_\theta = 2E_0 \cos \beta = E_m \cos \beta. \qquad (45\text{-}6c)$$

E_m, the maximum possible amplitude for E_θ, is equal to twice the amplitude of the combining waves $(= 2E_0)$, corresponding to complete reinforcement. You should verify Eq. 45-6 carefully. The amplitude E_θ of the resultant wave disturbance, which determines the intensity of the interference fringes, will turn out to depend strongly on the value of θ, that is, on the location of point P in Figs. 45-5 and 45-6.

In Section 19-6 we showed that the intensity of a wave I, measured perhaps in watts/meter², is proportional to the square of its amplitude. For the resultant wave then, ignoring the proportionality constant,

$$I_\theta \propto E_\theta^2. \qquad (45\text{-}7)$$

This relationship seems reasonable if we recall (Eq. 30-9) that the energy density in an electric field is proportional to the *square* of the electric field strength. This is true for rapidly varying electric fields, such as those in a light wave, as well as for static fields.

The ratio of the intensities of two light waves is the ratio of the squares of the amplitudes of their electric fields. If I_θ is the intensity of the resultant wave at P, and I_0 is the intensity that a single wave acting alone would produce, then

$$\frac{I_\theta}{I_0} = \left(\frac{E_\theta}{E_0}\right)^2. \qquad (45\text{-}8)$$

45-3
INTENSITY IN YOUNG'S EXPERIMENT

---

* See "Light-Wave Communications" by W. S. Boyle, *Scientific American*, August 1977.
† For an example of one of the many applications of lasers, see "Processing Materials with Lasers" by Edward N Breinan, Bernard H. Kear, and Conrad M. Banas, *Physics Today*, November 1976.
‡ The electric field **E** in the light wave rather than the magnetic field **B** is normally identified with the "wave disturbance" because the effects of **B** on the human eye and on various light detectors are exceedingly small. Note too that, although Eq. 45-5 should be a vector equation, in most cases of interest the **E** vectors in the two interfering waves are closely parallel so that an algebraic equation suffices.

Combining with Eq. 45-6c leads to

$$I_\theta = 4I_0 \cos^2 \beta = I_m \cos^2 \beta. \qquad (45\text{-}9)$$

Note that the intensity of the resultant wave at any point P varies from zero [for a point at which $\phi\ (=2\beta) = \pi$, say] to I_m, which is four times the intensity I_0 of each individual wave [for a point at which $\phi\ (=2\beta) = 0$, say]. Let us compute I_θ as a function of the angle θ in Figs. 45-5 or 45-6.

The phase difference ϕ in Eq. 45-4 is associated with the path difference $S_1 b$ in Fig. 45-5 or 45-6. If $S_1 b$ is $\frac{1}{2}\lambda$, ϕ will be π; if $S_1 b$ is λ, ϕ will be 2π, etc. This suggests that

$$\frac{\text{phase difference}}{2\pi} = \frac{\text{path difference}}{\lambda},$$

$$\phi = \frac{2\pi}{\lambda}\,(d\,\sin\theta),$$

or, finally, from Eq. 45-6b,

$$\beta = \tfrac{1}{2}\phi = \frac{\pi d}{\lambda}\sin\theta. \qquad (45\text{-}10)$$

We can substitute this expression for β into Eq. 45-9 for I_θ, yielding the latter quantity as a function of θ. For convenience we collect here the expressions for the amplitude and the intensity in double-slit interference.

[Eq. 45-6c]	$E_\theta = E_m \cos\beta$	interference	$(45\text{-}11a)$
[Eq. 45-9]	$I_\theta = I_m \cos^2 \beta$	from narrow	$(45\text{-}11b)$
		slits (that is,	
[Eq. 45-10]	$\beta\,(=\tfrac{1}{2}\phi) = \dfrac{\pi d}{\lambda}\sin\theta$	$a \ll \lambda$)	$(45\text{-}11c)$

To find the positions of the intensity maxima, we put

$$\beta = m\pi \qquad m = 0, 1, 2, \ldots$$

in Eq. 45-11b. From Eq. 45-11c this reduces to

$$d\,\sin\theta = m\lambda \qquad m = 0, 1, 2, \ldots \text{ (maxima)},$$

which is the equation derived in Section 45-1 (Eq. 45-1). To find the intensity minima we write

$$\frac{\pi d\,\sin\theta}{\lambda} = (m + \tfrac{1}{2})\pi \qquad m = 0, 1, 2, \ldots \text{ (minima)},$$

which reduces to the previously derived Eq. 45-2.

Figure 45-9 shows the intensity pattern for double-slit interference.

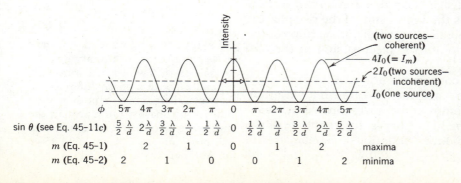

figure 45-9
The intensity pattern for double-slit interference. The heavy arrow in the central peak represents the half-width of the peak. This figure is constructed on the assumption that the two interfering waves each illuminate the central portion of the screen uniformly, that is, I_0 is independent of position as shown.

The horizontal solid line is I_0; this describes the (uniform) intensity pattern on the screen if one of the slits is covered up. If the two sources were incoherent, the intensity would be uniform over the screen and would be $2I_0$; see the horizontal dashed line in Fig. 45-9. For coherent sources we expect the energy to be merely redistributed over the screen, because energy is neither created nor destroyed by the interference process. Thus the *average* intensity in the interference pattern should be $2I_0$, as for incoherent sources. This follows at once if, in Eq. 45-11b, we substitute one-half for the cosine-squared term and if we recall that $I_m = 4I_0$. We have seen several times that the average value of the square of a sine or a cosine term over one or more half-cycles is one-half.

In Section 45-3 we combined two time-varying wave disturbances, namely,

$$E_1 = E_0 \sin \omega t \qquad (45\text{-}3)$$

and

$$E_2 = E_0 \sin (\omega t + \phi), \qquad (45\text{-}4)$$

which have the same angular frequency ω and amplitude E_0 but which have a phase difference ϕ between them. In this case the result (Eqs. 45-11a and 45-11c) is easily obtained algebraically.

In later chapters we will want to add larger numbers of wave disturbances, often an infinite number, with infinitesimal individual amplitudes. Since analytic methods become more difficult in such cases we describe a graphical method, illustrating it by rederiving Eq. 45-11a.

A sinusoidal wave disturbance such as that represented by Eq. 45-3 can be represented graphically, using a rotating vector. In Fig. 45-10a a vector of magnitude E_0 is allowed to rotate about the origin in a counter-clockwise direction with an angular frequency ω. Following electrical engineering practice we call such a rotating vector a *phasor*; see Section 39-2. The alternating wave disturbance E_1 (Eq. 45-3) is represented by the projection of this phasor on the vertical axis.

A second wave disturbance E_2, which has the same amplitude E_0 but a phase difference ϕ with respect to E_1,

$$E_2 = E_0 \sin (\omega t + \phi), \qquad (45\text{-}4)$$

can be represented graphically (Fig. 45-10b) as the projection on the vertical axis of a second phasor of magnitude E_0 which makes an angle ϕ with the first phasor. As this figure shows, the sum E of E_1 and E_2 is the sum of the projections of the two phasors on the vertical axis. This is revealed more clearly if we redraw the phasors, as in Fig. 45-10c, placing the foot of one arrow at the head of the other, maintaining the proper phase difference, and letting the whole assembly rotate counterclockwise about the origin.

In Fig. 45-10c E can also be regarded as the projection on the vertical axis of a phasor of length E_θ, which is the vector sum of the two phasors of magnitude E_0. Note that the (algebraic) sum of the projections of the two phasors is equal to the projection of the (vector) sum of the two phasors.

In most problems in optics we are concerned only with the *amplitude* E_θ of the resultant wave disturbance and not with its time variation. This is because the eye and other common measuring instruments respond to the resultant intensity of the light (that is, to the square of the amplitude) and cannot detect the rapid time variations that characterize visible light. For sodium light ($\lambda = 589$ nm $= 5890$ Å), for example, the

45-4
ADDING WAVE DISTURBANCES

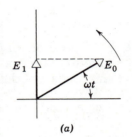

(a)

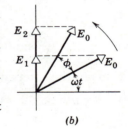

(b)

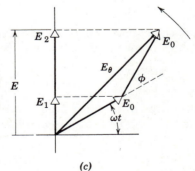

(c)

figure 45-10
(*a*) A time-varying wave disturbance E_1 is represented by a rotating vector or *phasor*. (*b*) Two wave disturbances E_1 and E_2, with a phase difference ϕ between them, are so represented. These two phasors can represent the two wave disturbances in the double-slit problem; see Eqs. 45-3 and 45-4. (*c*) Another way of drawing (*b*).

$\nu(= \omega/2\pi)$ is 5.1×10^{14} Hz. Often, then, we need not consider the rotation of the phasors but can confine our attention to finding the magnitude of the resultant phasor.

Figure 45-11a shows the phasors for double-slit interference at time $t = 0$; compare Fig. 45-10c. We see that

$$E_\theta = 2E_0 \cos \beta = E_m \cos \beta,$$

in which, from the theorem that the exterior angle of a triangle (ϕ) is equal to the sum of its two opposite interior angles $(\beta + \beta)$,

$$\beta = \tfrac{1}{2}\phi.$$

This is exactly the result arrived at earlier algebraically; compare Eqs. 45-11a and 45-11c.

In a more general case we might want to find the resultant of a number (>2) of sinusoidally varying wave disturbances. The general procedure is the following:

1. Construct a series of phasors representing the functions to be added. Draw them end to end, maintaining the proper phase relationships between adjacent phasors.

2. Construct the vector sum of this array. Its length gives the amplitude of the resultant. The angle between it and the first phasor is the phase of the resultant with respect to this first phasor. The projection of this phasor on the vertical axis gives the time variation of the resultant wave disturbance.

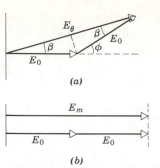

(a)

(b)

figure 45-11
(a) A construction to find the amplitude E_θ of two wave disturbances of amplitude E_0 and phase difference ϕ. (b) The maximum possible amplitude for these wave disturbances occurs for $\phi = 0$ and has the value $E_m = 2E_0$.

Find graphically the resultant $E(t)$ of the following wave disturbances:

$$E_1 = 10 \sin \omega t$$
$$E_2 = 10 \sin (\omega t + 15°)$$
$$E_3 = 10 \sin (\omega t + 30°)$$
$$E_4 = 10 \sin (\omega t + 45°).$$

Figure 45-12 in which E_0 equals 10, shows the assembly of four phasors that

EXAMPLE 3

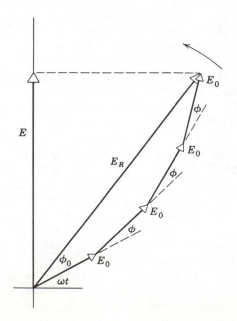

figure 45-12
Example 3. Four wave disturbances are added graphically, using the method of phasors.

represents these functions. Their vector sum, by graphical measurement, has an amplitude E_R of 38 and a phase ϕ_0 with respect to E_1 of 23°. In other words,

$$E(t) = E_1 + E_2 + E_3 + E_4 = 38 \sin{(\omega t + 23°)}.$$

Check this result by trigonometric calculation.

The colors of soap bubbles, oil slicks, and other thin films are the result of interference. Figure 45-13 shows interference effects in a thin vertical film of soapy water illuminated by monochromatic light.

Figure 45-14 shows a film of uniform thickness d and index of refraction n, the eye being focused on spot a. The film is illuminated by a broad source of monochromatic light S. There exists on this source a point P such that two rays, identified by the single and double arrows, respectively, can leave P and enter the eye as shown, after passing through point a. These two rays follow different paths in going from P to the eye, one being reflected from the upper surface of the film, the other from the lower surface. Whether point a appears bright or dark depends on the nature of the interference between the two waves that diverge from a. These waves are coherent because they both originate from the same point P on the light source.

If the eye looks at another part of the film, say a', the light that enters the eye must originate from a different point P' of the extended source, as suggested by the dashed lines in Fig. 45-14.

For near-normal incidence ($\theta \cong 0$ in Fig. 45-14) the geometrical path difference for the two rays from P will be close to $2d$. We might expect the resultant wave reflected from the film near a to be an interference maximum if the distance $2d$ is an integral number of wavelengths. This statement must be modified for two reasons.

First, the wavelength must refer to the wavelength of the light in the film λ_n and not to its wavelength in air λ; that is, we are concerned with optical path lengths rather than geometrical path lengths. The wavelengths λ and λ_n (see Eq. 43-12b) are related by

$$\lambda_n = \lambda/n. \qquad (45\text{-}12)$$

To bring out the second point, let us assume that the film is so thin that $2d$ is very much less than a wavelength. The phase difference between the two waves would be close to zero on our assumption, and we would expect such a film to appear bright on reflection. However, it appears dark. This is clear from Fig. 45-13, in which the action of gravity produces a wedge-shaped film, extremely thin at its top edge. As drainage continues, the dark area increases in size. To explain this and many similar phenomena, we assume that one or the other of the two rays of Fig. 45-14 suffers an abrupt phase change of $\pi (=180°)$ associated either with reflection at the air-film interface or transmission through it. As it turns out, the ray reflected from the upper surface suffers this phase change. The other ray is not changed abruptly in phase, either on transmission through the upper surface or on reflection at the lower surface.

In Section 19-9 we discussed phase changes on reflection for transverse waves in strings. To extend these ideas, consider the composite string of Fig. 45-15, which consists of two parts with different masses per unit length, stretched to a given tension. If a pulse moves to the right in Fig. 45-15a, approaching the junction, there will be a reflected and a transmitted pulse, the reflected pulse being *in phase* with the incident pulse. In Fig. 45-15b the situation is reversed, the incident pulse now

45-5
INTERFERENCE FROM THIN FILMS

figure 45-13
A soapy water film on a wire loop, viewed by reflected light. The black segment at the top is not a tear. It arises because the film, by drainage, is so thin here that there is destructive interference between the light reflected from its front surface and that reflected from its back surface. We shall see that these two waves differ in phase by 180°.

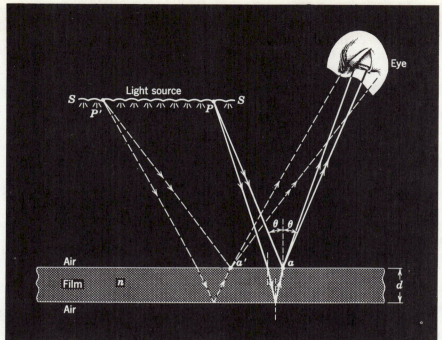

figure 45-14
Interference by reflection from a thin film, assuming an extended source S.

being in the lighter string. In this case the reflected pulse will differ in phase from the incident pulse by π (=180°). In each case the transmitted pulse will be in phase with the incident pulse.

Figure 45-15a suggests a light wave in glass, say, approaching a surface beyond which there is a less optically dense medium (one of lower index of refraction) such as air. Figure 45-15b suggests a light wave in air approaching glass. To sum up the optical situation, when reflection occurs from an interface beyond which the medium has a *lower* index of refraction, the reflected wave undergoes *no phase change*; when the medium beyond the interface has a *higher* index, there is a phase change of π.* The transmitted wave does not experience a change of phase in either case.

We are now able to take into account both factors that determine the nature of the interference, namely, differences in optical path length and phase changes on reflection. For the two rays of Fig. 45-14 to combine to give a *maximum* intensity, assuming normal incidence, we must have

$$2d = (m + \tfrac{1}{2})\lambda_n \qquad m = 0, 1, 2, \ldots.$$

The term $\tfrac{1}{2}\lambda_n$ is introduced because of the phase change on reflection, a phase change of 180° being equivalent to half a wavelength. Substituting λ/n for λ_n yields finally

$$2dn = (m + \tfrac{1}{2})\lambda \qquad m = 0, 1, 2, \ldots \text{ (maxima).} \qquad (45\text{-}13)$$

The condition for a *minimum* intensity is

$$2dn = m\lambda \qquad m = 0, 1, 2, \ldots \text{ (minima).} \qquad (45\text{-}14)$$

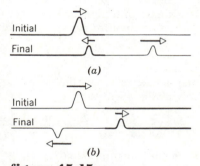

figure 45-15
Phase changes on reflection at a junction between stretched composite string. (a) Incident pulse in heavy string and (b) incident pulse in light string.

* These statements, which can be proved rigorously from Maxwell's equations (see also Section 45-6), must be modified for light falling on a less dense medium at an angle such that total internal reflection occurs. They must also be modified for reflection from metallic surfaces.

These equations hold when the index of refraction of the film is either greater or less than the indices of the media on *each* side of the film. Only in these cases will there be a relative phase change of 180° for reflections at the two surfaces. A water film in air and an air film in the space between two glass plates provide examples of cases to which Eqs. 45-13 and 45-14 apply. Example 5 provides a case in which they do not apply.

If the film thickness is not uniform, as in Fig. 45-13, where the film is wedge-shaped, constructive interference will occur in certain parts of the film and destructive interference will occur in others. Lines of maximum and of minimum intensity will appear—these are the interference fringes. They are called *fringes of constant thickness*, each fringe being the locus of points for which the film thickness d is a constant. If the film is illuminated with white light rather than monochromatic light, the light reflected from various parts of the film will be modified by the various constructive or destructive interferences that occur. This accounts for the brilliant colors of soap bubbles and oil slicks.

Only if the film is "thin," which implies that d is no more than a few wavelengths of light, will fringes of the type described, that is, fringes that appear localized on the film and associated with a variable film thickness, be possible. For very thick films (say $d \cong 1$ cm), the path difference between the two rays of Fig. 45-14 will be many wavelengths and the phase difference at a given point on the film will change rapidly as we move even a small distance away from a. For "thin" films, however, the phase difference at a also holds for reasonably nearby points; there is a characteristic "patch brightness" for any point on the film, as Fig. 45-13 shows. Interference fringes can be produced for thick films; they are not localized on the film but are at infinity. See Section 45-7.

EXAMPLE 4

A water film ($n = 1.33$) in air is 320 nm thick. If it is illuminated with white light at normal incidence, what color will it appear to be in reflected light?

By solving Eq. 45-13 for λ,

$$\lambda = \frac{2dn}{m + \frac{1}{2}} = \frac{(2)(320 \text{ nm})(1.33)}{m + \frac{1}{2}} = \frac{850 \text{ nm}}{m + \frac{1}{2}} \quad \text{(maxima)}.$$

From Eq. 45-14 the minima are given by

$$\lambda = \frac{850 \text{ nm}}{m} \quad \text{(minima)}.$$

Maxima and minima occur for the following wavelengths:

m	0 (max)	1 (min)	1 (max)	2 (min)	2 (max)
λ, nm	1700	850	570	425	340

Only the maximum corresponding to $m = 1$ lies in the visible region (see Fig. 42-1); light of this wavelength appears yellow-green. If white light is used to illuminate the film, the yellow-green component will be enhanced when viewed by reflection.

EXAMPLE 5

*Nonreflecting glass.*** Lenses are often coated with thin films of transparent substances such as MgF$_2$ ($n = 1.38$) in order to reduce the reflection from the glass surface, using interference. How thick a coating is needed to produce a minimum reflection at the center of the visible spectrum (550 nm)?

* See "Optical Interference Coatings" by Philip Baumeister and Gerald Pineus, *Scientific American*, December 1970.

We assume that the light strikes the lens at near-normal incidence (θ is exaggerated for clarity in Fig. 45-16), and we seek destructive interference between rays r and r_1. Equation 45-14 does not apply because in this case a phase change of 180° is associated with *each* ray, for at *both* the upper and lower surfaces of the MgF$_2$ film the reflection is from a medium of greater index of refraction.

There is no net change in phase produced by the two reflections, which means that the optical path difference for destructive interference is $(m + \frac{1}{2})\lambda$ (compare Eq. 45-13), leading to

$$2dn = (m + \tfrac{1}{2})\lambda \qquad m = 0, 1, 2, \ldots \text{ (minima).}$$

Solving for d and putting $m = 0$ yields

$$d = \frac{(m + \frac{1}{2})\lambda}{2n} = \frac{\lambda}{4n} = \frac{550 \text{ nm}}{(4)(1.38)} = 100 \text{ nm.}$$

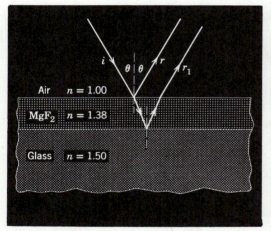

figure 45-16
Example 5. Unwanted reflections from glass can be reduced by coating the glass with a thin transparent film.

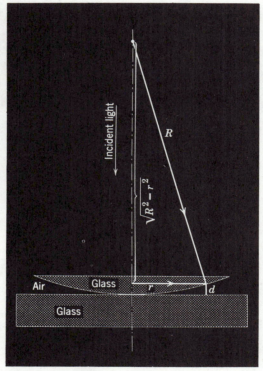

figure 45-17
Example 6. Apparatus for observing Newton's rings.

Newton's rings. Figure 45-17 shows a lens of radius of curvature R resting on an accurately plane glass plate and illuminated from above by light of wavelength λ. Figure 45-18 shows that circular interference fringes (Newton's rings) appear, associated with the variable thickness air film between the lens and the plate. Find the radii of the circular interference maxima.

Here it is the ray from the *bottom* of the (air) film rather than from the top that undergoes a phase change of 180°, for it is the one reflected from a medium of higher refractive index. The condition for a maximum remains unchanged (Eq. 45-13), however, and is

$$2d = (m + \tfrac{1}{2})\lambda \qquad m = 0, 1, 2, \ldots, \qquad (45\text{-}15)$$

the index of refraction of the air film being assumed to be unity. From Fig. 45-17

EXAMPLE 6

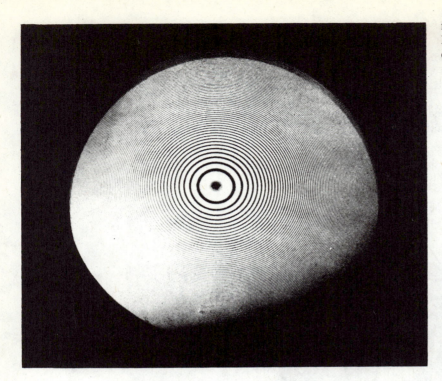

figure 45-18
Example 6. Newton's rings. (Courtesy of Bausch and Lomb Optical Co.)

we can write

$$d = R - \sqrt{R^2 - r^2} = R - R \left[1 - \left(\frac{r}{R} \right)^2 \right]^{1/2}$$

If $r/R \ll 1$, we can expand the square bracket by the binomial theorem, keeping only two terms, or

$$d = R - R \left[1 - \frac{1}{2} \left(\frac{r}{R} \right)^2 + \cdots \right] \cong \frac{r^2}{2R}.$$

Combining with Eq. 45-15 yields

$$r = \sqrt{(m + \tfrac{1}{2})\lambda R} \qquad m = 0, 1, 2, \ldots,$$

which gives the radii of the bright rings. If white light is used, each spectrum component will produce its own set of circular fringes, the sets all overlapping.

G. G. Stokes (1819–1903) used the principle of optical reversibility to investigate the reflection of light at an interface between two media. The principle states that if there is no absorption of light, a light ray that is reflected or refracted will retrace its original path if its direction is reversed. This reminds us that any mechanical system can run backward as well as forward, provided there is no absorption of energy because of friction, etc.

Figure 45-19a shows a wave of amplitude E reflected and refracted at a surface separating media 1 and 2, where $n_2 > n_1$. The amplitude of the reflected wave is $r_{12}E$, in which r_{12} is an *amplitude reflection coefficient*. The amplitude of the refracted wave is $t_{12}E$, where t_{12} is an *amplitude transmission coefficient*.

We consider only the possibility of phase changes of 0 or 180°. If $r_{12} = +0.5$, for example, we have a reduction in amplitude on reflection by one-half and no change in phase. For $r_{12} = -0.5$ we have a phase

45-6
OPTICAL REVERSIBILITY AND PHASE CHANGES ON REFLECTION

change of 180° because

$$E \sin (\omega t + 180°) = -E \sin \omega t.$$

Figure 45-19b suggests that if we reverse these two rays, they should combine to produce the original ray reversed in direction. Ray $r_{12}E$, identified by the single arrows in the figure, is reflected and refracted, producing the rays of amplitudes $r_{12}{}^2E$ and $r_{12}t_{12}E$. Ray $t_{12}E$, identified by the triple arrows, is also reflected and refracted, producing the rays of amplitudes $t_{12}t_{21}E$ and $t_{12}r_{21}E$ as shown. Note that r_{12} describes a ray in medium 1 reflected from medium 2, and r_{21} describes a ray in medium 2 reflected from medium 1. Similarly, t_{12} describes a ray that passes from medium 1 to medium 2; t_{21} describes a ray that passes from medium 2 to medium 1.

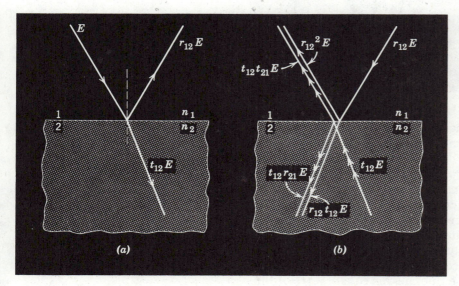

figure 45-19
(a) A ray is reflected and refracted at an air-glass interface. (b) The optically reversed situation; the two rays in the lower left must cancel.

The two rays in the upper left of Fig. 45-19b must be equivalent to the incident ray of Fig. 45-19a, reversed; the two rays in the lower left of Fig. 45-19b must cancel. This second requirement leads to

$$r_{12}t_{12}E + t_{12}r_{21}E = 0,$$

or

$$r_{12} = -r_{21}.$$

This result tells us that if we compare a wave reflected from medium 1 with one reflected from medium 2, they behave differently in that one or the other undergoes a phase change of 180°. We must rely on experiment or analysis by Maxwell's equations to show that, as we pointed out earlier, the ray reflected from the more optically dense medium is the one that experiences the phase change of 180°.

An *interferometer* is a device that can be used to measure lengths or changes in length with great accuracy by means of interference fringes. We describe the form originally built by A. A. Michelson (1852–1931) in 1881.

Consider light that leaves point P on extended source S (Fig. 45-20) and falls on half-silvered mirror M. This mirror has a silver coating just

45-7
MICHELSON'S INTERFEROMETER*

* See "Michelson: America's first Nobel Prize Winner in Science" by R. S. Shankland, *The Physics Teacher*, January 1977. See also "Michelson and his Interferometer" by R. S. Shankland, *Physics Today*, April 1974.

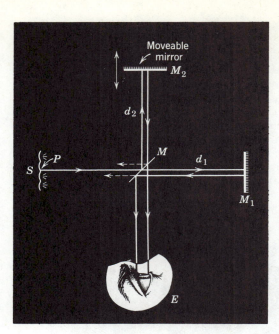

figure 45-20
Michelson's interferometer, showing the path of a particular ray originating at point P of an extended source S.

thick enough to transmit half the incident light and to reflect half; in the figure we have assumed that this mirror, for convenience, possesses negligible thickness. At M the light divides into two waves. One proceeds by transmission toward mirror M_1; the other proceeds by reflection toward M_2. The waves are reflected at each of these mirrors and are sent back along their directions of incidence, each wave eventually entering the eye E. Because the waves are coherent, being derived from the same point on the source, they will interfere.

If the mirrors M_1 and M_2 are exactly perpendicular to each other, the effect is that of light from an extended source S falling on a uniformly thick slab of air, between glass, whose thickness is equal to $d_2 - d_1$. Interference fringes appear, caused by small changes in the angle of incidence of the light from different points on the extended source as it strikes the equivalent air film. For *thick* films a path difference of one wavelength can be brought about by a very small change in the angle of incidence.

If M_2 is moved backward or forward, the effect is to change the thickness of the equivalent air film. Suppose that the center of the (circular) fringe pattern appears bright and that M_2 is moved just enough to cause the first bright circular fringe to move to the center of the pattern. The path of the light beam striking M_2 has been changed by one wavelength. This means (because the light passes twice through the equivalent air film) that the mirror must have moved one-half a wavelength.

The interferometer is used to measure changes in length by counting the number of interference fringes that pass the field of view as mirror M_2 is moved. Length measurements made in this way can be accurate if large numbers of fringes are counted.

Michelson measured the length of the standard meter, kept in Paris, in terms of the wavelength of certain monochromatic red light emitted from a light source containing cadmium. He showed that the standard meter was equivalent to 1,553,163.5 wavelengths of the red cadmium light. For this he received the Nobel prize in 1907.

Physicists have long speculated on the advantages of discarding the standard meter bar as the basic standard of length and of *defining* the meter in terms of the wavelength of some carefully chosen monochromatic radiation. This would make the primary length standard readily available in laboratories all over the world. It would improve the accuracy of length measurements, since one would no longer need to compare an unknown object with a standard object (the meter bar), using interferometer techniques, but could measure the unknown object *directly* and in an absolute sense, using these techniques. There is the additional advantage that if the standard meter bar were destroyed, it could never be replaced, whereas light sources and interferometers will (presumably) always be available.

In 1961 such an atomic standard of length was adopted by international agreement. Quoting from an article* describing the event:

The wavelength of the orange-red light of krypton-86 has replaced the platinum iridium bar as the world standard of length. Formerly the wavelength of this light was defined as a function of the length of the meter bar. Now the meter is defined as a multiple (1,650,763.73) of the wavelength of the light.

The light from krypton-86 was used in preference to that from other atomic sources because it produces sharper interference fringes in the interferometer over the long optical paths sometimes used in length measurement.

45-8
MICHELSON'S INTERFEROMETER AND LIGHT PROPAGATION†

In Section 42-4 we presented Einstein's hypothesis, now well verified, that in free space light is propagated with the same speed c no matter what the relative velocity of the source and the observer may be. We pointed out that this hypothesis contradicted the views of nineteenth-century physicists regarding wave propagation. It was difficult for these physicists, trained as they were in the classical physics of the time, to believe that a wave could be propagated without a medium. If such a medium could be established, the speed c of light would naturally be construed as the speed *with respect to that medium,* just as the speed of sound always refers to a medium such as air.

Although no medium for light propagation was obvious, the physicists postulated one, called the *ether*‡ and hypothesized that its properties were such that it was undetectable by ordinary means such as weighing.

In 1881 (24 years before Einstein's hypothesis) A. A. Michelson set himself the task of forcing the ether, assuming that it existed, to submit to direct physical verification. In particular, Michelson, later joined by E. W. Morley, tried to measure the speed u with which the earth moves through the ether. Michelson's interferometer was their instrument of choice for this now-famous Michelson-Morley experiment.

The earth together with the interferometer moving with velocity **u** through the ether is equivalent to the interferometer at rest with the ether streaming through it with velocity **−u,** as shown in Fig. 45-21. Consider a wave moving along the path MM_1M and one moving along MM_2M. The first corresponds classically to a man rowing a boat a distance d downstream and the same distance upstream; the second corresponds to rowing a boat a distance d across a stream and back.

---

* *Scientific American*, p. 75, December 1960.
† See Supplementary Topic V.
‡ More fully, the *luminiferous* (or light-carrying) *ether.*

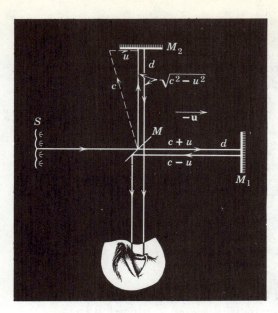

figure 45-21
The "ether" is streaming with velocity **−u** through Michelson's interferometer. The wave speeds shown are on the basis of the (incorrect) ether hypothesis.

On the ether hypothesis the speed of light on the path MM_1 is $c + u$; on the return path M_1M it is $c - u$. The time required for the complete trip is

$$t_1 = \frac{d}{c + u} + \frac{d}{c - u} = d\frac{2c}{c^2 - u^2} = \frac{2d}{c}\frac{1}{1 - (u/c)^2}.$$

The speed of light, *on the ether hypothesis*, for path MM_2 is $\sqrt{c^2 - u^2}$, as Fig. 45-21 suggests. This same speed holds for the return path M_2M, so that the time required for this complete path is

$$t_2 = \frac{2d}{\sqrt{c^2 - u^2}} = \frac{2d}{c}\frac{1}{\sqrt{1 - (u/c)^2}}.$$

The difference of time for the two paths is

$$\Delta t = t_1 - t_2$$
$$= \frac{2d}{c}\left\{\left[1 - \left(\frac{u}{c}\right)^2\right]^{-1} - \left[1 - \left(\frac{u}{c}\right)^2\right]^{-1/2}\right\}$$

Assuming $u/c \ll 1$, we can expand the quantities in the square brackets by using the binomial theorem, retaining only the first two terms. This leads to

$$\Delta t = \frac{2d}{c}\left\{\left[1 + \left(\frac{u}{c}\right)^2 + \cdots\right] - \left[1 + \frac{1}{2}\left(\frac{u}{c}\right)^2 + \cdots\right]\right\}$$
$$= \frac{2d}{c}\left\{\frac{1}{2}\left(\frac{u}{c}\right)^2\right\} = \frac{du^2}{c^3}. \qquad (45\text{-}16)$$

Now let the entire interferometer be rotated through 90°. This will interchange the roles of the two light paths, MM_1M now being the "cross-stream" path and MM_2M the "down- and up-stream" path. The time difference between the two waves entering the eye is also reversed; this changes the phase difference between the combining waves and alters the positions of the interference maxima. *The experiment consists of looking for a shift of the interference fringes as the apparatus is rotated.*

The *change* in time difference is $2\Delta t$, which corresponds to a fringe shift of $2\Delta t/T$ where $T(= \lambda/c)$ is the period of vibration of the light. The expected maximum shift in the number of fringes on a 90° rotation (see Eq. 45-16) is

$$\Delta N = \frac{2\Delta t}{T} = \frac{2\Delta tc}{\lambda} = \frac{2d}{\lambda}\left(\frac{u}{c}\right)^2 \qquad (45\text{-}17)$$

In the Michelson-Morley interferometer let $d = 11$ m (obtained by multiple reflection in the interferometer) and $\lambda = 5.9 \times 10^{-7}$ m. If u is assumed to be roughly the orbital speed of the earth, then $u/c \cong 10^{-4}$. The expected maximum fringe shift when the interferometer is rotated through 90° is then

$$\Delta N = \frac{2d}{\lambda}\left(\frac{u}{c}\right)^2 = \frac{(2)(11 \text{ m})}{5.9 \times 10^{-7} \text{ m}}(10^{-4})^2 = 0.4.$$

Even though a shift of only about 0.4 of a fringe was expected, Michelson and Morley were confident that they could observe a shift of 0.01 fringe. *They found from their experiment, however, that there was no observable fringe shift!*

The analogy between a light wave in the supposed ether and a boat moving in water, which seemed so evident in 1881, is simply incorrect. The derivation based on this analogy is incorrect for light waves. When the analysis is carried through on Einstein's hypothesis, the observed negative result is clearly predicted, the speed of light being c for all paths. The motion of the earth around the sun and the rotation of the interferometer have, in Einstein's view, no effect whatever on the speed of the light waves in the interferometer.

It should be made clear that although Einstein's hypothesis is completely consistent with the negative result of the Michelson-Morley experiment, this experiment standing alone cannot serve as a proof for Einstein's hypothesis. Einstein said that no number of experiments, however large, could prove him right but that a single experiment could prove him wrong. Our present-day belief in Einstein's hypothesis rests on consistent agreement in a large number of experiments designed to test it. The "single experiment" that might prove Einstein wrong has never been found.

questions

1. Is Young's experiment an interference experiment or a diffraction experiment, or both?

2. Do interference effects occur for sound waves? Recall that sound is a longitudinal wave and that light is a transverse wave.

3. In Young's double-slit interference experiment, using a monochromatic laboratory light source, why is screen A in Fig. 45-1 necessary? What would happen if one gradually enlarged the hole in this screen?

4. What changes occur in the pattern of interference fringes if the apparatus of Fig. 45-5 is placed under water?

5. If interference between light waves of different frequencies is possible, one should observe light beats, just as one obtains sound beats from two sources of sound with slightly different frequencies. Discuss how one might experimentally look for this possibility.

6. Why are parallel slits preferable to the pinholes that Young used in demonstrating interference?

7. Is coherence important in reflection and refraction?

8. If your source of light is a laser beam, you do not need (see Fig. 45-8) the equivalent of screen A in Fig. 45-1. Why?

9. Defend this statement: Fig. 45-7a is a sine (or cosine) wave but Fig. 45-7b is not. Indeed you cannot assign a unique frequency to the curve of Fig. 45-7b. Why not? (Hint: think of Fourier analysis.)

10. Most of us are familiar with rotating or oscillating radar antennas which produce rotating or oscillating beams of microwave radiation. It is also possible to produce an oscillating beam of microwave radiation *without* any mechanical motion of the transmitting antenna. This is done by periodically changing the phase of the radiation as it emerges from various sections of the (long) transmitting antenna. Convince yourself that, by constructive interference from various parts of the fixed antenna, an oscillating microwave beam could indeed be so produced. A very large radar installation of this type has been installed by the United States in the Aleutian Islands to monitor USSR missile experiments.

11. Describe the pattern of light intensity on screen C in Fig. 45-5 if one slit is covered with a red filter and the other with a blue filter, the incident light being white.

12. Define carefully, and distinguish between, the angles θ and ϕ that appear in Eq. 45-10.

13. If one slit in Fig. 45-5 is covered, what change occurs in the intensity of light in the center of the screen?

14. In Young's double-slit experiment suppose that screen A in Fig. 45-1 contained *two* very narrow parallel slits instead of one. (a) Show that if the spacing between these slits is properly chosen, the interference fringes can be made to disappear. (b) Under these conditions, would you call the beams emerging from slits S_1 and S_2 in screen B coherent? They do not produce interference fringes. (c) Discuss what would happen to the interference fringes in the case of a single slit in screen A if the slit width were gradually increased.

15. What are the requirements for a maximum intensity when viewing a thin film by *transmitted* light?

16. In a Newton's rings experiment, is the central spot, as seen by reflection, dark or light? Explain.

17. Why must the film of Fig. 45-14 be "thin" for us to see an interference pattern of the type described?

18. Why do coated lenses (see Example 5) look purple by reflected light?

19. A person wets his eyeglasses to clean them. As the water evaporates he notices that for a short time the glasses become markedly more nonreflecting. Explain.

20. A lens is coated to reduce reflection, as in Example 5. What happens to the energy that had previously been reflected? Is it absorbed by the coating?

21. Very small changes in the angle of incidence do not change the interference conditions much for "thin" films but they do change them for "thick" films. Why?

22. In connection with the phase change on reflection at an interface between two transparent media, do you think that phase shifts other than 0 or π are possible? Do you think that phase shifts can be calculated rigorously from Maxwell's equations?

23. Consider the following objects that produce colors when exposed to sunlight: (1) soap bubbles, (2) rose petals, (3) the inner surface of an oyster shell (irridescence), (4) thin oil slicks, (5) non-reflecting coatings on camera lenses, and (6) peacock tail feathers. All but one of these are purely interference phenomena, no pigments being involved. Which one is it? Discuss.

24. Why does a film (soap bubble, oil slick, etc.) have to be "thin" to display interference effects? Or does it? How thin is "thin"? Discuss.

25. An *optical flat* is a slab of glass that has been ground flat to within a small fraction of a wavelength. How may it be used to test the flatness of a second slab of glass?

26. The directional characteristics of a certain radar antenna as a receiver of radiation are known. What can be said about its directional characteristics as a transmitter?

27. A person in a dark room, looking through a small window, can see a second person standing outside in bright sunlight. The second person cannot see the first person. Is this a failure of the principle of optical reversibility? Assume no absorption of light.

28. Why is it necessary to rotate the interferometer in the Michelson-Morley experiment?

29. How is the negative result of the Michelson-Morley experiment interpreted according to Einstein's theory of relativity?

30. An automobile directs its headlights onto the side of a barn. Why are interference fringes not produced?

31. In principle, could you construct an acoustical version of Michelson's interferometer and use it to measure wind velocities? Discuss some of its design difficulties. Are there simpler ways to measure wind velocities?

problems

SECTION 45-1

1. Design a double-slit arrangement that will produce interference fringes 1° apart on a distant screen. Assume sodium light ($\lambda = 589$ nm $= 5890$ Å). *Answer:* Slit separation must be 0.034 mm.

2. In a double-slit arrangement the distance between slits is 5.0 mm and the slits are 1.0 m from the screen. Two interference patterns can be seen on the screen, one due to light of 480 nm and the other 600 nm. What is the separation on the screen between the third-order interference fringes of the two different patterns?

3. Sodium light ($\lambda = 589$ nm) falls on a double slit of separation $d = 2.0$ mm. D in Fig. 45-5 is 4 cm. What percent error is made in locating the tenth bright fringe if it is *not* assumed that $D \gg d$? *Answer:* 0.03%

4. In a double-slit arrangement the slits are separated by a distance equal to 100 times the wavelength of the light passing through the slits. (*a*) What is the angular separation between the first and second maxima? (*b*) What is the linear distance between the first and second maxima if the screen is at a distance of 50 cm from the slits?

5. A double-slit arrangement produces interference fringes for sodium light ($\lambda = 589$ nm) that are 0.20° apart. For what wavelength would the angular separation be 10 percent greater? *Answer:* 650 nm

6. Sodium light ($\lambda = 589$ nm) falls on a double slit of separation $d = 0.20$ mm. A thin lens ($f = +1.0$ m) is placed near the slit as in Fig. 45-6. What is the linear fringe separation on a screen placed in the focal plane of the lens?

7. In the front of a lecture hall, a coherent beam of monochromatic light from a helium-neon laser ($\lambda = 640$ nm $= 6400$ Å) illuminates a double slit. From there it travels a distance d (20 m) to a mirror at the back of the hall, and returns the same distance to a screen. (*a*) In order that the distance between interference maxima be s (10 cm) what should be the distance between the two slits? (*b*) State briefly what you will see if the lecturer slips a thin sheet of cellophane over one of the slits. The optical path length through the cellophane is 2.5 wavelengths longer than the equivalent air path length. *Answer:* (*a*) 0.26 mm. (*b*) In place of the central maximum you get a minimum.

8. In Young's interference experiment in a large ripple tank (see Fig. 45-4) the coherent vibrating sources are placed 12.0 cm apart. The distance between

maxima 2.0 m away is 18.0 cm. If the speed of ripples is 25.0 cm/s, find the frequency of the vibrators.

9. A thin flake of mica $(n = 1.58)$ is used to cover one slit of a double-slit arrangement. The central point on the screen is occupied by what used to be the seventh bright fringe. If $\lambda = 550$ nm, what is the thickness of the mica?
Answer: 6.4×10^{-3} mm

10. One slit of a double-slit arrangement is covered by a thin glass plate of refractive index 1.4, and the other by a thin glass plate of refractive index 1.7. The point on the screen where the central maximum fell before the glass plates were inserted is now occupied by what had been the fifth bright fringe before. Assume $\lambda = 480$ nm and that the plates have the same thickness t and find the value of t.

11. A double-slit arrangement produces interference fringes for sodium light $(\lambda = 589$ nm) that are 0.20° apart. What is the angular fringe separation if the entire arrangement is immersed in water? *Answer:* 0.15°.

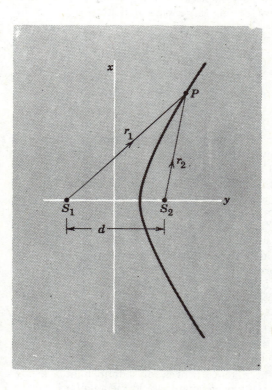

figure 45-22
Problem 12

12. Two point sources in Fig. 45-22 emit coherent waves. Show that curves, such as that given, over which the phase difference for rays r_1 and r_2 is a constant, are hyperbolas. (*Hint:* A constant phase difference implies a constant difference in length between r_1 and r_2.)

13. In Fig 45-23, the source emits monochromatic light of wavelength λ. S is a narrow slit in an otherwise opaque screen I. A plane mirror, whose surface includes the axis of the lens shown, is located a distance h below S. Screen II is at the focal plane of the lens. (a) Find the condition for maximum and minimum brightness of fringes on screen II in terms of the usual angle θ, the wavelength λ, and the distance h. (b) Do fringes appear only in region A (above the axis of the lens), only in region B (below the axis of the lens), or in both regions A and B? Explain. (*Hint:* Consider the image of S formed by the mirror.)

Answer: (a) 2 h sin θ = mλ (minimum); 2 h sin θ = (m + $\frac{1}{2}$)λ (maximum).
(b) Only in region A.

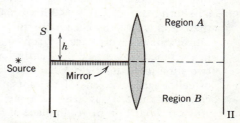

figure 45-23
Problem 13

14. Design, semi-quantitatively, a Young-type double slit interference experiment (see Fig. 45-1) for sound waves in air. Assume a frequency of 500 Hz and a speed of 330 m/s. Discuss some of the design parameters such as the nature of the source, the size of screen B, the width, height, and separation of the two slits, the "fringe" separation, the nature of the fringe detector, etc.

SECTION 45-3

15. Light of wavelength 600 nm is incident normally on a system of two parallel narrow slits separated by 0.60 mm. Sketch the intensity in the pattern observed on a distant screen as a function of angle θ, as in Fig. 45-5, for the range of values $0 \leqslant \theta \leqslant 0.0040$ radians.

16. Show that the half-width $\Delta\theta$ of the double-slit interference fringes (see arrow in Fig. 45-9) is given by

$$\Delta\theta = \frac{\lambda}{2d}$$

if θ is small enough so that $\sin\theta \cong \theta$.

17. One of the slits of a double-slit system is wider than the other, so that the amplitude of the light reaching the central part of the screen from one slit, acting alone, is twice that from the other slit, acting alone. Derive an expression for I_θ in terms of θ, corresponding to Eqs. 45-11b and 45-11c.

Answer: $I_\theta = \frac{1}{9} I_m \left[1 + 8 \cos^2\left(\frac{\pi d \sin\theta}{\lambda}\right) \right]$

18. S_1 and S_2 in Fig. 45-24 are effective point sources of radiation, excited by the same oscillator. They are coherent and in phase with each other. Placed 4.0 m apart, they emit equal amounts of power in the form of 1.0-m wavelength electromagnetic waves. (a) Find the positions of the first (that is, the nearest), the second, and the third maxima of the received signal, as the detector is moved out along Ox. (b) Is the intensity at the nearest minimum equal to zero? Justify your answer.

figure 45-24
Problem 18

SECTION 45-4

19. Find the sum of the following quantities (a) by the vector method and (b) analytically:

$$y_1 = 10 \sin \omega t$$

$$y_2 = 8 \sin (\omega t + 30°).$$

Answer: $y = 17 \sin (\omega t + 13°)$.

20. Add the following quantities graphically, using the vector method:

$$y_1 = 10 \sin \omega t$$

$$y_2 = 15 \sin (\omega t + 30°)$$

$$y_3 = 5 \sin (\omega t - 45°).$$

21. Consider the problem of determining the sum

$$A_1 \sin (\omega t + \phi_1) + A_2 \sin (\omega t + \phi_2) + \cdots + A_n \sin (\omega t + \phi_n)$$

from the phasor diagram.
(a) Show that the sum may always be written in the form

$$B \sin \omega t + C \cos \omega t.$$

(b) Show that $B^2 + C^2 \leq (A_1 + A_2 + \cdots + A_n)^2$.
(c) When does the equality sign in (b) hold?
Answer: (c) When all phase angles ϕ_i are equal.

SECTION 45-5

22. We wish to coat a flat slab of glass ($n = 1.50$) with a transparent material ($n = 1.25$) so that light of wavelength 600 nm (in vacuum) incident normally is not reflected. How can this be done?

23. A thin film 4.0×10^{-5} cm thick is illuminated by white light normal to its surface. Its index of refraction is 1.5. What wavelengths within the visible spectrum will be intensified in the reflected beam?
    Answer: 480 nm (blue).

24. White light reflected at perpendicular incidence from a soap bubble has, in the visible spectrum, a single interference maximum (at $\lambda = 600$ nm) and a single minimum at the violet end of the spectrum. If $n = 1.33$ for the film, calculate its thickness.

25. A plane wave of monochromatic light falls normally on a uniformly thin film of oil which covers a glass plate. The wavelength of the source can be varied continuously. Complete destructive interference of the reflected light is observed for wavelengths of 500 and 700 nm and for no other wavelengths in between. If the index of refraction of the oil is 1.30 and that of the glass is 1.50, find the thickness of the oil film.
    Answer: 670 nm.

26. A plane monochromatic light wave in air falls at normal incidence on a thin film of oil which covers a glass plate. The wavelength of the source may be varied continuously. Complete destructive interference in the reflected beam is observed for wavelengths of 500 and 700 nm and for no other wavelength in between. The index of refraction of glass is 1.50. Show that the index of refraction of the oil must be less than 1.50.

27. A thin film of acetone (refractive index 1.25) is floated on a thick glass plate (refractive index 1.50). Plane light waves of variable wavelengths are incident normal to the film. When one views the reflected wave it is noted that complete destructive interference occurs at 600 nm and constructive interference at 700 nm. Calculate the thickness of the acetone film.
    Answer: 840 nm.

28. White light reflected at perpendicular incidence from a soap film has, in the visible spectrum, an interference maximum at 600 nm ($= 6000$ Å) and a minimum at 450 nm ($= 4500$ Å), with no minimum in between. If $n = 1.33$ for the film, what is the film thickness, assumed uniform?

29. In Example 5 assume that there is zero reflection for light of wavelength 550 nm at normal incidence. Calculate the factor by which the reflection is diminished by the coating at 450 and at 650 nm.
    Answer: The intensity is diminished by 88% at 450 nm and by 94% at 650 nm.

30. A tanker has dumped a large quantity of kerosene ($n = 1.2$) into the Gulf of Mexico, creating a large slick on top of the water ($n = 1.3$). (a) If you are looking straight down from an airplane onto a region of the slick where its thickness is 460 nm, for which wavelength(s) of visible light is the reflection the greatest? (b) If you are scuba-diving directly under this same region of the slick, for which wavelengths of visible light is the transmitted intensity the strongest?

31. Light of wavelength 630 nm is incident normally on a thin wedge-shaped film of index of refraction 1.5. There are ten bright and nine dark fringes over the length of film. By how much does the film thickness change over this length?   *Answer:* 1.9×10^{-6} m ($= 1.9$ μm).

32. A broad source of light ($\lambda = 680$ nm) illuminates normally two glass plates 12 cm long that touch at one end and are separated by a wire 0.048 mm in diameter at the other (Fig. 45-25). How many bright fringes appear over the 12-cm distance?

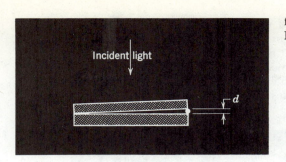

figure 45-25
Problem 32

Incident light

d

33. An oil drop $(n = 1.20)$ floats on a water $(n = 1.33)$ surface and is observed from above by reflected light (see Fig. 45-26). (a) Will the outer (thinnest) regions of the drop correspond to a bright or a dark region? (b) Approximately how thick is the oil film where one observes the third blue region from the outside of the drop? (c) Why do the colors gradually disappear as the oil thickness becomes larger?   *Answer:* (a) Bright. (b) 594 nm.

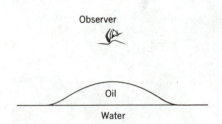

figure 45-26
Problem 33

Observer

Oil

Water

34. The diameter of the tenth bright ring in a Newton's rings apparatus changes from 1.40 to 1.27 cm as a liquid is introduced between the lens and the plate. Find the index of refraction of the liquid.

35. In a Newton's rings experiment the radius of curvature R of the lens is 5.0 m and its diameter is 2.0 cm. (a) How many rings are produced? (b) How many rings would be seen if the arrangement were immersed in water $(n = 1.33)$? Assume that $\lambda = 589$ nm.   *Answer:* (a) 34. (b) 46.

36. In the Newton's rings experiment, use the result of Example 6 to show that the difference in radius between adjacent rings is

$$\Delta r = r_{m+1} - r_m \cong \tfrac{1}{2}\sqrt{\frac{\lambda R}{m}}.$$

Assume $m \gg 1$ and use the binomial theorem. Is this result qualitatively consistent with Fig. 45-18?

37. In the Newton's rings experiment use the result of Example 6 and Problem 36 to show that the *area* between adjacent rings is given, for $m \gg 1$, by

$$A_m = \pi \lambda R \, (= \text{a constant}).$$

Is this result qualitatively consistent with Fig. 45-18?

SECTION 45-7

38. If mirror M_2 in Michelson's interferometer is moved through 0.233 mm, 792 fringes are counted. What is the wavelength of the light?

39. A thin film with $n = 1.40$ for light of wavelength 589 nm $(= 5890 \text{ Å})$ is placed in one arm of a Michelson interferometer. If a shift of 7.0 fringes occurs, what is the film thickness?   *Answer:* 5200 nm.

40. A Michelson interferometer is used with a sodium discharge tube as a light source. The yellow sodium light consists of two wavelengths, 589.0 (=5890 Å) and 589.6 nm (=5896 Å). It is observed that the interference pattern disappears and reappears periodically as one moves mirror M_2 in Fig. 45-20. (a) Explain this effect. (b) Calculate the change in path difference between two successive reappearances of the interference pattern.

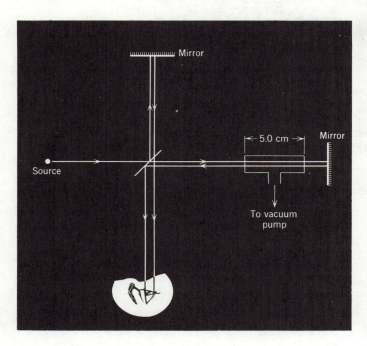

figure 45-27
Problem 41

41. An air-tight chamber 5.0 cm long with glass windows is placed in one arm of a Michelson interferometer as indicated in Fig. 45-27. Light of $\lambda = 5000$ Å is used. The air is slowly evacuated from the chamber using a vacuum pump. While the air is being removed, 60 fringes are observed to pass through the view. From these data, find the index of refraction of air at atmospheric pressure. *Answer:* 1.0003

42. Write an expression for the intensity observed in Michelson's interferometer (Fig. 45-20) as a function of the position of the moveable mirror. Measure the position of the mirror from the point at which $d_1 = d_2$.

46
diffraction

Diffraction, which is illustrated in Fig. 44-3, is the bending of light around an obstacle such as the edge of a slit. We can see the diffraction of light by looking through a crack between two fingers at a distant light source such as a tubular neon sign or by looking at a street light through a cloth umbrella. Usually diffraction effects are small and we must look for them carefully. Also, most sources of light have an extended area so that a diffraction pattern produced by one point of the source will overlap that produced by another. Finally, common sources of light are not monochromatic. The patterns for the various wavelengths overlap and again the effect is less apparent.

Diffraction was discovered by Francesco Grimaldi (1618–1663), and the phenomenon was known both to Huygens and to Newton. Newton did not see in it any justification for a wave theory for light. Huygens, although he believed in a wave theory, did not believe in diffraction! He imagined his secondary wavelets to be effective only at the point of tangency to their common envelope, thus denying the possibility of diffraction (see Section 43-3). In his words:

And thus we see the reasons why light . . . proceeds only in straight lines in such a way that it does not illuminate any object except when the path from the source to the object is open along such a line.

Jean Augustin Fresnel (1788–1827) correctly applied Huygens' principle to explain diffraction. In these early days the light waves were believed to be mechanical waves in an all-pervading ether. We have seen (Section 41-8) how Maxwell showed that light waves were not mechanical in nature but electromagnetic. Einstein rounded out our

46-1
INTRODUCTION

modern view of light waves by eliminating the need to postulate an ether (see Section 42-4).

Figure 46-1 shows a general diffraction situation. Surface *A* is a wavefront that falls on *B*, which is an opaque screen containing an aperture of arbitrary shape; *C* is a diffusing screen that receives the light that passes through this aperture. We can calculate this pattern of light intensity on *C* by subdividing the wavefront into elementary areas *d*S, each of which becomes a source of an expanding Huygens' wavelet. The light intensity at an arbitrary point *P* is found by superimposing the wave disturbances (that is, the **E** vectors) caused by the wavelets reaching *P* from all these elementary radiators.

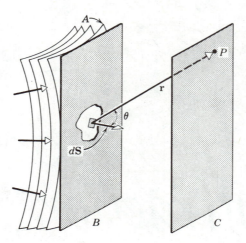

figure 46-1
Light is diffracted at the aperture in screen *B* and illuminates screen *C*. The intensity at *P* is found by dividing the wavefront at *B* into elementary radiators *d*S and combining their effects at *P*.

The wave disturbances reaching *P* differ in amplitude and in phase because (*a*) the elementary radiators are at varying distances from *P*; (*b*) the light leaves the radiators at various angles to the normal to the wavefront (see Section 43-2), and (*c*) some radiators are blocked by screen *B*; others are not. Diffraction calculations—simple in principle— may become difficult in practice. The calculation must be repeated for every point on screen *C* at which we wish to know the light intensity. We followed exactly this program in calculating the double-slit intensity pattern in Section 45-3. The calculation there was simple because we assumed only two elementary radiators, the two narrow slits.

Figure 46-2*a* shows the general case of *Fresnel diffraction,* in which the light source and/or the screen on which the diffraction pattern is displayed are a finite distance from the diffracting aperture; the wavefronts that fall on the diffracting aperture in this case and that leave it to illuminate any point *P* of the diffusing screen are not planes; the corresponding rays are not parallel.

A simplification results if source *S* and screen *C* are moved to a large distance from the diffraction aperture, as in Fig. 46-2*b*. This limiting case is called *Fraunhofer diffraction.* The wavefronts arriving at the diffracting aperture from the distant source *S* are planes, and the rays associated with these wavefronts are parallel to each other. Fraunhofer conditions can be established in the laboratory by using two converging lenses, as in Fig. 46-2*c*. The first of these converts the diverging wave from the source into a plane wave. The second lens causes plane waves leaving the diffracting aperture to converge to point *P*. All rays that

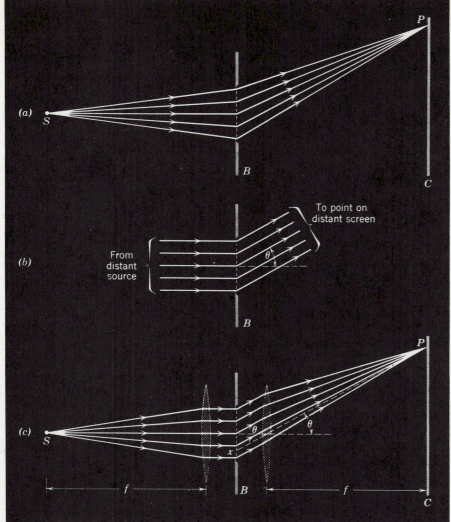

figure 46-2

(a) Fresnel diffraction. (b) Source S and screen C are moved to a large distance, resulting in Fraunhofer diffraction. (c) Fraunhofer diffraction conditions produced by lenses, leaving source S and screen C in their original positions.

illuminate P will leave the diffracting aperture parallel to the dashed line Px drawn from P through the center of this second (thin) lens. We assumed Fraunhofer conditions for Young's double-slit experiment in Section 45-1 (see Fig. 45-5).

Although Fraunhofer diffraction is a limiting case of the more general Fresnel diffraction, it is an important limiting case and is easier to handle mathematically. This book deals only with Fraunhofer diffraction.

46-2
SINGLE SLIT

Figure 46-3 shows a plane wave falling at normal incidence on a long narrow slit of width a. Let us focus our attention on the central point P_0 of screen C. The parallel rays extending from the slit to P_0 all have the same optical (though not geometrical) path lengths, as we saw in Section 44-5. Since they are in phase at the plane of the slit, they will still be in phase at P_0, and the central point of the diffraction pattern that appears on screen C has a maximum intensity.

We now consider another point on the screen. Light rays which reach

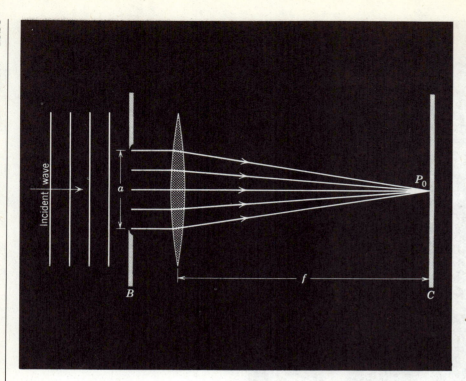

figure 46-3
Conditions at the central maximum of the diffraction pattern. The slit extends a distance above and below the figure, this distance being much greater than the slit width a.

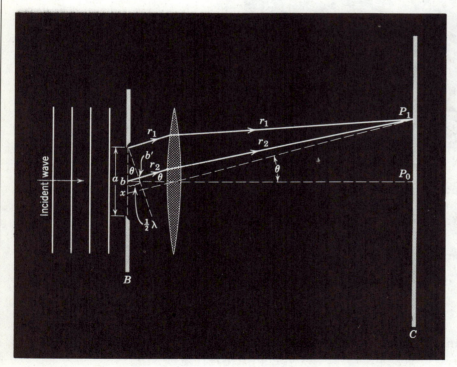

figure 46-4
Conditions at the first minimum of the diffraction pattern.

P_1 in Fig. 46-4 leave the slit at an angle θ as shown. (Note that the ray represented by the dashed line xP_1 is drawn to pass through the center of the lens and is thus undeflected; this ray determines θ.) Ray r_1 originates at the top of the slit and ray r_2 at its center. If θ is chosen so that the distance bb' in the figure is one-half a wavelength, r_1 and r_2 will be out of phase and will produce no effect at P_1.* In fact, every ray from the upper half of the slit will be canceled by a ray from the lower half, originating at a point $a/2$ below the first ray. The point P_1, the first minimum of the diffraction pattern, will have zero intensity (compare Fig. 44-3).

The condition shown in Fig. 46-4 is

$$\frac{a}{2} \sin \theta = \frac{\lambda}{2},$$

or

$$a \sin \theta = \lambda. \tag{46-1}$$

As we stated earlier (see Fig. 44-1), the central maximum becomes wider as the slit is made narrower. If the slit width is as small as one wavelength ($a = \lambda$), the first minimum occurs at $\theta = 90°$ ($\sin \theta = 1$ in Eq. 46-1), which implies that the central maximum fills the entire forward hemisphere. We assumed a condition approaching this in our discussion of Young's double-slit interference experiment in Section 45-1.

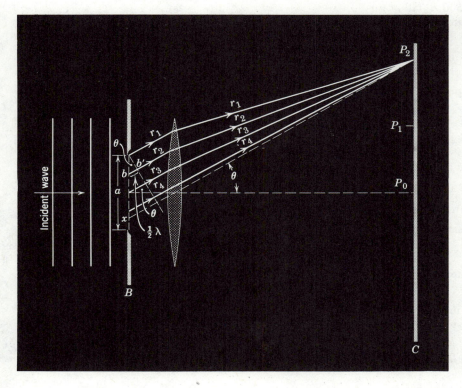

figure 46-5
Conditions at the second minimum of the diffraction pattern.

In Fig. 46-5 the slit is divided into four equal zones, with a ray leaving the top of each zone. Let θ be chosen so that the distance bb' is one-half a wavelength. Rays r_1 and r_2 will then cancel at P_2. Rays r_3 and r_4 will also be half a wavelength out of phase and will also cancel. Consider four other rays, emerging from the slit a given distance below the four rays above. The two rays below r_1 and r_2 will cancel uniquely, as will the two rays below r_3 and r_4. We can proceed across the entire slit and conclude again that no light reaches P_2; we have located a second point of zero intensity.

* Whatever phase relation exists between r_1 and r_2 at the plane represented by the sloping dashed line in Fig. 46-4 that passes through b' also exists at P_1, not being affected by the lens (see Section 44-5).

The condition described (see Fig. 46-5) requires that

$$\frac{a}{4} \sin \theta = \frac{\lambda}{2},$$

or

$$a \sin \theta = 2\lambda.$$

By extension, the general formula for the minima in the diffraction pattern on screen C is

$$a \sin \theta = m\lambda \qquad m = 1, 2, 3, \ldots \text{(minima)}. \qquad (46\text{-}2)$$

There is a maximum approximately halfway between each adjacent pair of minima. You should consider the simplification that has resulted in this analysis by examining Fraunhofer (Fig. 46-2c) rather than Fresnel conditions (Fig. 46-2a).

EXAMPLE 1

A slit of width a is illuminated by white light. For what value of a will the first minimum for red light ($\lambda = 650$ nm $= 6500$ Å) fall at $\theta = 30°$?

At the first minimum we put $m = 1$ in Eq. 46-2. Doing so and solving for a yields

$$a = \frac{m\lambda}{\sin \theta} = \frac{(1)(650 \text{ nm})}{\sin 30°} = 1300 \text{ nm}.$$

Note that the slit width must be twice the wavelength in this case.

EXAMPLE 2

In Example 1 what is the wavelength λ' of the light whose first diffraction maximum (not counting the central maximum) falls at $\theta = 30°$, thus coinciding with the first minimum for red light?

This maximum is about halfway between the first and second minima. We can find it without too much error by putting $m = 1.5$ in Eq. 46-2, or

$$a \sin \theta \cong 1.5\lambda'.$$

From Example 1, however, $\qquad a \sin \theta = \lambda.$

Dividing gives $\qquad \lambda' = \dfrac{\lambda}{1.5} = \dfrac{650 \text{ nm}}{1.5} = 430 \text{ nm}.$

Light of this color is violet. The second maximum for light of wavelength 430 nm will *always* coincide with the first minimum for light of wavelength 650 nm, no matter what the slit width. If the slit is relatively narrow, the angle θ at which this overlap occurs will be relatively large.

46-3
SINGLE SLIT—
QUALITATIVE

In Section 46-2 we located the positions of the maxima and minima for a single-slit diffraction pattern. We now wish to find an expression for the intensity of the *entire* diffraction pattern as a function of the diffraction angle θ. We do so qualitatively in this section and quantitatively in the next section.

Figure 46-6 shows a slit of width a divided into N parallel strips of width Δx. Each strip acts as a radiator of Huygens' wavelets and produces a characteristic wave disturbance at point P, whose position on the screen, for a particular arrangement of apparatus, can be described by the angle θ.

If the strips are narrow enough—which we assume—all points on a given strip have essentially the same optical path length to P, and therefore all the light from the strip will have the same phase when it ar-

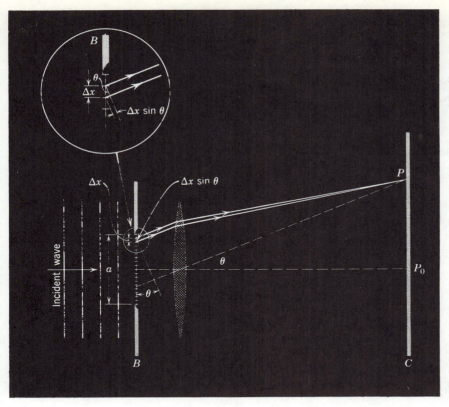

figure 46-6
We divide a slit of width a into N strips of width Δx. The insert shows conditions at the second strip more clearly. In the differential limit the slit is divided into an infinite number of strips (that is, $N \to \infty$) of differential width dx. For clarity in this and the following figure, we take $N = 18$.

rives at P. We may take the amplitudes ΔE_0 of the wave disturbances at P from the various strips may be taken as equal if θ in Fig. 46-6 is not too large.

We limit our considerations to points that lie in, or very close to, the plane of Fig. 46-6. It can be shown that this procedure is valid for a slit whose length is much greater than its width a. We made this same assumption tacitly both earlier in this chapter and in Chapter 45; see Figs. 45-5 and 46-3, for example.

The wave disturbances from adjacent strips have a constant phase difference $\Delta\phi$ between them at P given by

$$\frac{\text{phase difference}}{2\pi} = \frac{\text{path difference}}{\lambda},$$

or
$$\Delta\phi = \left(\frac{2\pi}{\lambda}\right)(\Delta x \sin \theta), \qquad (46\text{-}3)$$

where $\Delta x \sin \theta$ is, as the figure insert shows, the path difference for rays originating at the top edges of adjacent strips. Thus, at P, N vectors with the same amplitude ΔE_0, the same frequency, and the same phase difference $\Delta\phi$ between adjacent members combine to produce a resultant disturbance. We ask, for various values of $\Delta\phi$ [that is, for various points P on the screen, corresponding to various values of θ (see Eq. 46-3)], what is the amplitude E_θ of the resultant wave disturbance? We find the answer by representing the individual wave disturbances ΔE_0 by phasors and calculating the resultant phasor amplitude, as described in Section 45-4.

At the center of the diffraction pattern θ equals zero, and the phase shift between adjacent strips (see Eq. 46-3) is also zero. As Fig. 46-7a shows, the phasor arrows in this case are laid end to end and the ampli-

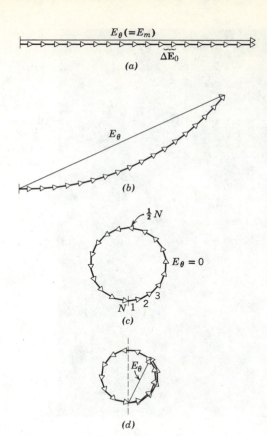

figure 46-7
Conditions at (a) the central maximum, (b) a direction slightly removed from the central maximum, (c) the first minimum, and (d) the first maximum beyond the central maximum for single-slit diffraction. This figure corresponds to $N = 18$ in Fig. 46-6.

tude of the resultant has its maximum value E_m. This corresponds to the center of the central maximum.

As we move to a value of θ other than zero, $\Delta\phi$ assumes a definite nonzero value (again see Eq. 46-3), and the array of arrows is now as shown in Fig. 46-7b. The resultant amplitude E_θ is less than before. Note that the length of the "arc" of small arrows is the same for both figures and indeed for all figures of this series. As θ increases further, we reach a situation (Fig. 46-7c) in which the chain of arrows curls around through 360°, the tip of the last arrow touching the foot of the first arrow. This corresponds to $E_\theta = 0$, that is, to the first minimum. For this condition the ray from the top of the slit (1 in Fig. 46-7c) is 180° out of phase with the ray from the center of the slit ($\frac{1}{2}N$ in Fig. 46-7c). These phase relations are consistent with Fig. 46-4, which also represents the first minimum.

As θ increases further, the phase shift continues to increase, and the chain of arrows coils around through an angular distance greater than 360°, as in Fig. 46-7d, which corresponds to the first maximum beyond the central maximum. This maximum is much smaller than the central maximum. In making this comparison, recall that the arrows marked E_θ in Fig. 46-7 correspond to the *amplitudes* of the wave disturbance and not to the *intensity*. The amplitudes must be squared to obtain the corresponding relative intensities (see Eq. 45-7).

The "arc" of small arrows in Fig. 46-8 shows the phasors representing, in amplitude and phase, the wave disturbances that reach an arbitrary point P on the screen of Fig. 46-6, corresponding to a particular angle θ. The resultant amplitude at P is E_θ. If we divide the slit of Fig. 46-6 into

46-4
SINGLE SLIT—QUANTITATIVE

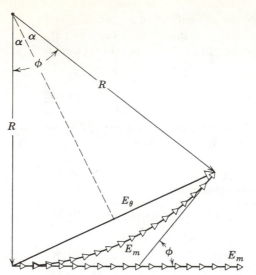

figure 46-8
A construction used to calculate the intensity in single-slit diffraction. The situation corresponds to that of Fig. 46-7b.

infinitesimal strips of width dx, the arc of arrows in Fig. 46-8 approaches the arc of a circle, its radius R being indicated in that figure. The length of the arc is E_m, the amplitude at the center of the diffraction pattern, for at the center of the pattern the wave disturbances are all in phase and this "arc" becomes a straight line as in Fig. 46-7a.

The angle ϕ in the lower part of Fig. 46-8 is revealed as the difference in phase between the infinitesimal vectors at the left and right ends of the arc E_m. This means that ϕ is the phase difference between rays from the top and the bottom of the slit of Fig. 46-6. From geometry we see that ϕ is also the angle between the two radii marked R in Fig. 46-8. From this figure we can write

$$E_\theta = 2R \sin \frac{\phi}{2}.$$

In radian measure ϕ, from the figure, is

$$\phi = \frac{E_m}{R}.$$

Combining yields

$$E_\theta = \frac{E_m}{\phi/2} \sin \frac{\phi}{2},$$

or

$$E_\theta = E_m \frac{\sin \alpha}{\alpha}, \qquad (46\text{-}4)$$

in which

$$\alpha = \frac{\phi}{2}. \qquad (46\text{-}5)$$

From Fig. 46-6, recalling that ϕ is the phase difference between rays from the top and the bottom of the slit and that the path difference for these rays is $a \sin \theta$, we have

$$\frac{\text{phase difference}}{2\pi} = \frac{\text{path difference}}{\lambda},$$

or

$$\phi = \left(\frac{2\pi}{\lambda}\right)(a \sin \theta).$$

Combining with Eq. 46-5 yields

$$\alpha = \frac{\phi}{2} = \frac{\pi a}{\lambda} \sin \theta. \qquad (46\text{-}6)$$

Equation 46-4, taken together with the definition of Eq. 46-6, gives the amplitude of the wave disturbance for a single-slit diffraction pattern at any angle θ. The intensity I_θ for the pattern is proportional to the square of the amplitude, or

$$I_\theta = I_m \left(\frac{\sin \alpha}{\alpha}\right)^2. \qquad (46\text{-}7)$$

For convenience we display together, and renumber, the formulas for the amplitude and the intensity in single-slit diffraction.

[Eq. 46-4] $\quad E_\theta = E_m \dfrac{\sin \alpha}{\alpha}$ $\left.\begin{array}{l}\\ \text{single-} \\ \\ \text{slit} \\ \\ \text{diffraction}\end{array}\right.$ $(46\text{-}8a)$

[Eq. 46-7] $\quad I_\theta = I_m \left(\dfrac{\sin \alpha}{\alpha}\right)^2$ $(46\text{-}8b)$

[Eq. 46-6] $\quad \alpha \left(= \tfrac{1}{2}\phi\right) = \dfrac{\pi a}{\lambda} \sin \theta$ $(46\text{-}8c)$

Figure 46-9 shows plots of I_θ for several values of the ratio a/λ. Note that the pattern becomes narrower as we increase a/λ; compare this figure with Figs. 44-1 and 44-3.

Minima occur in Eq. 46-8b when

$$\alpha = m\pi \qquad m = 1, 2, 3, \ldots . \qquad (46\text{-}9)$$

Combining with Eq. 46-8c leads to

$$a \sin \theta = m\lambda \qquad m = 1, 2, 3, \ldots \text{ (minima)},$$

which is the result derived in the preceding section (Eq. 46-2). In that section, however, we derived *only* this result, obtaining no quantitative information about the intensity of the diffraction pattern at places in which it was not zero. Here (Eqs. 46-8) we have complete intensity information.

Intensities of the secondary diffraction maxima. Calculate, approximately, the relative intensities of the secondary maxima in the single-slit Fraunhofer diffraction pattern.   **EXAMPLE 3**

The secondary maxima lie approximately halfway between the minima and are found (compare Eq. 46-9) from

$$\alpha \cong (m + \tfrac{1}{2})\pi \qquad m = 1, 2, 3, \ldots .$$

Substituting into Eq. 46-8b yields

$$I_\theta = I_m \left[\frac{\sin (m + \tfrac{1}{2})\pi}{(m + \tfrac{1}{2})\pi}\right]^2,$$

which reduces to $\quad \dfrac{I_\theta}{I_m} = \dfrac{1}{(m + \tfrac{1}{2})^2 \pi^2}.$

This yields, for $m = 1, 2, 3, \ldots$, $I_\theta/I_m = 0.045, 0.016, 0.0083$, etc. The successive maxima decrease rapidly in intensity.

Width of the central diffraction maximum. Derive the *half-width* $\Delta\theta$ of the central maximum in a single-slit Fraunhofer diffraction (see Fig. 46-9b). The half-width is the angle between the two points in the pattern where the intensity is one-half that at the center of the pattern.   **EXAMPLE 4**

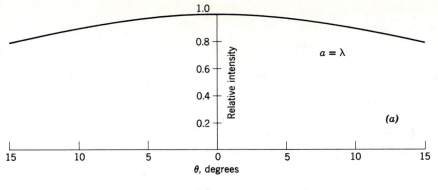

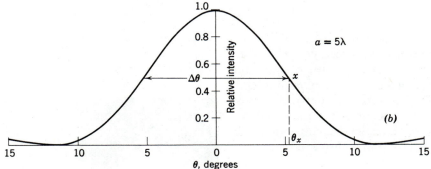

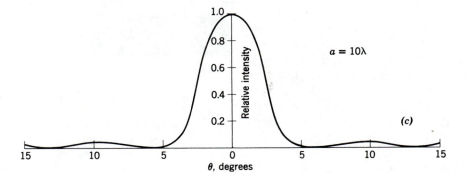

figure 46-9
The relative intensity in single-slit
diffraction for three values of the
ratio a/λ. The arrow in (b) shows the
half-width $\Delta\theta$ of the central
maximum.

Point x in Fig. 46-9b is so chosen that $I_\theta = \tfrac{1}{2}I_m$, or, from Eq. 46-8b,

$$\frac{1}{2} = \left(\frac{\sin \alpha_x}{\alpha_x}\right)^2.$$

This equation cannot be solved analytically for α_x. We can solve it graphically, as accurately as we wish, by plotting the quantity $(\sin \alpha_x/\alpha_x)^2$ as ordinate versus α_x as abscissa and noting the value of α_x at which the curve intersects the line "one-half" on the ordinate scale (see Problem 11). However, if only an approximate answer is desired, it is often quicker to use trial-and-error methods.

We know that α equals π at the first minimum; we guess that α_x is perhaps $\pi/2$ ($= 90° = 1.57$ rad). Trying this in Eq. 46-8b yields

$$\frac{I_\theta}{I_m} = \left[\frac{\sin (\pi/2)}{\pi/2}\right]^2 = 0.4.$$

This intensity ratio is *less* than 0.5, so that α_x must be *less* than 90°. After a few more trials we find easily enough that

$$\alpha_x = 1.40 \text{ rad} = 80°$$

does yield a ratio close to the correct value of 0.5.

We now use Eq. 46-8c to find the corresponding angle θ:

$$\alpha_x = \frac{\pi a}{\lambda} \sin \theta_x = 1.40,$$

or, noting that $a/\lambda = 5$ for Fig. 46-9b,

$$\sin \theta_x = \frac{1.40\lambda}{\pi a} = \frac{1.40}{5\pi} = 0.0892.$$

The half-width $\Delta\theta$ of the central maximum (see Fig. 46-9b) is given by

$$\Delta\theta = 2\theta_x = 2 \sin^{-1} 0.0892 = 2 \times 5.1° = 10.2°,$$

which is in agreement with the figure.

Diffraction will occur when a wavefront is partially blocked off by an opaque object such as a metal disk or an opaque screen containing an aperture. Here we consider diffraction at a circular aperture of diameter d, the aperture constituting the boundary of a circular converging lens.

Our previous treatment of lenses was based on geometrical optics, diffraction being specifically assumed not to occur. A rigorous analysis would be based from the beginning on wave optics, since geometrical optics is always an approximation, although often a very good one. Diffraction phenomena would emerge in a natural way from such a wave-optical analysis.

Figure 46-10 shows the image of a distant point source of light (a star) formed on a photographic film placed in the focal plane of the converging lens of a telescope.* It is not a point, as the (approximate) geometrical optics treatment suggests, but a circular disk surrounded by several progressively fainter secondary rings. Comparison with Fig. 44-3c leaves little doubt that we are dealing with a diffraction phenomenon in which, however, the aperture is a circle rather than a long narrow slit. The ratio d/λ, where d is the diameter of the lens (or of a circular aperture placed in front of the lens), determines the scale of the diffraction pattern, just as the ratio a/λ does for a slit.

Analysis shows that the first minimum for the diffraction pattern of a circular aperture of diameter d, assuming Fraunhofer conditions, is given by

$$\sin \theta = 1.22 \frac{\lambda}{d}. \qquad (46\text{-}10)$$

This is to be compared with Eq. 46-1, or

$$\sin \theta = \frac{\lambda}{a},$$

which locates the first minimum for a long narrow slit of width a. The factor 1.22 emerges from the mathematical analysis when we integrate over the elementary radiators into which the circular aperture may be divided.

In actual lenses the image of a distant point object will be somewhat larger than that shown in Fig. 46-10 and may not have radial symmetry. This is caused by the various lens "defects" mentioned in Section 44-5. However, even if all of these defects could be eliminated by suitable shaping of the lens surfaces or by intro-

46-5
DIFFRACTION AT A CIRCULAR APERTURE

figure 46-10
The image of a star formed by a converging lens is a diffraction pattern. Note the central maximum, sometimes called the Airy disk (after Sir George Airy, who first solved the problem of diffraction at a circular aperture in 1835), and the circular secondary maximum. Other secondary maxima occur at larger radii but are too faint to be seen.

---

*Diffraction phenomena also occur in microscopes and other optical instruments. They are, in fact, characteristic of the entire electromagnetic spectrum.

ducing correcting lenses, the diffraction pattern of Fig. 46-10 would remain. It is an inherent property of the lens aperture and of the wavelength of light used.

The fact that lens images are diffraction patterns is important when we wish to distinguish two distant point objects whose angular separation is small. Figure 46-11 shows the visual appearances and the corresponding intensity patterns for two distant point objects (stars, say) with small angular separations and approximately equal central intensities. In (a) the objects are not resolved; that is, they cannot be distinguished from a single point object. In (b) they are barely resolved and in (c) they are fully resolved.*

In Fig. 46-11b the angular separation of the two point sources is such that the maximum of the diffraction pattern of one source falls on the first minimum of the diffraction pattern of the other. This is called *Rayleigh's criterion*. This criterion, though useful, is arbitrary; other criteria for deciding when two objects are resolved are sometimes used. From Eq. 46-10, two objects that are barely resolvable by Rayleigh's criterion must have an angular separation θ_R of

$$\theta_R = \sin^{-1}\frac{1.22\lambda}{d}.$$

Since the angles involved are rather small, we can replace $\sin\theta_R$ by θ_R, or

$$\theta_R = 1.22\frac{\lambda}{d}, \tag{46-11}$$

in which θ_R is expressed in radians. If the angular separation θ between the objects is greater than θ_R, we can resolve the two objects; if it is less, we cannot. The angle θ_R is the smallest angular separation for which resolution is possible, using Raleigh's criterion.

EXAMPLE 5

A converging lens 3.0 cm in diameter has a focal length f of 20 cm. (a) What angular separation must two distant point objects have to satisfy Rayleigh's criterion? Assume that $\lambda = 550$ nm.

From Eq. 46-11,

$$\theta_R = 1.22\frac{\lambda}{d} = \frac{(1.22)(5.5 \times 10^{-7}\text{ m})}{3.0 \times 10^{-2}\text{ m}} = 2.2 \times 10^{-5}\text{ rad}.$$

(b) How far apart are the centers of the diffraction patterns in the focal plane of the lens? The linear separation is

$$x = f\theta = (20\text{ cm})(2.2 \times 10^{-5}\text{ rad}) = 4400\text{ nm}.$$

This is 8.0 wavelengths of the light employed.

When we wish to use a lens to resolve objects of small angular separation, it is desirable to make the central disk of the diffraction pattern as small as possible. This can be done (see Eq. 46-11) by increasing the lens diameter or by using a shorter wavelength. One reason for constructing large telescopes is to produce *sharper* images so that we can examine celestial objects in finer detail. The images are also *brighter*, not only because the energy is concentrated into a smaller diffraction

---

* Diffraction effects set the so-called "bottom line" on resolution. However, other effects combine to hamper resolution. For atmospheric effects on the observation of stars see "Toward the Diffraction Limit" by Dietrick E. Thomsen, *Science News*, August 1975.

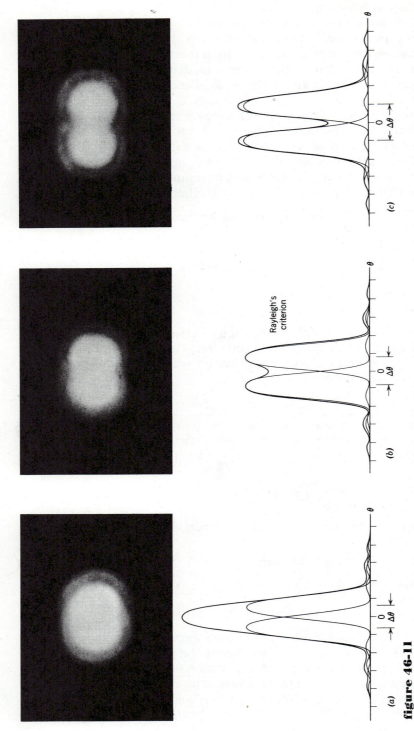

figure 46-11
The images of two distant point objects (simulated stars) are formed by a converging lens whose diameter (= 10 cm) is 200,000 times the effective wavelength (= 500 nm). Sketches of the images as they appear in the focal plane of the telescope objective lens are shown with the corresponding intensity plots below them. (a) The angular separation of the objects (see vertical ticks) is so small that the images are not resolved. (b) The objects are further apart and the images meet Rayleigh's criterion for resolution. (c) The objects are still farther apart and the images are well resolved.

disk but because the larger lens collects more light. Thus fainter objects, for example, distant galaxies, can be seen.

To reduce diffraction effects in *microscopes* we often use ultraviolet light, which, because of its shorter wavelength, permits finer detail to be examined than would be possible for the same microscope operated with visible light. We shall see in Chapter 50 that beams of electrons behave like waves under some circumstances. In the *electron microscope* such beams may have an effective wavelength of 4×10^{-3} nm, of the order of 10^5 times shorter than visible light ($\lambda \cong 500$ nm or 5000 Å). This permits the detailed examination of tiny objects like viruses. If a virus were examined with an optical microscope, its structure would be hopelessly concealed by diffraction.

46-6
DOUBLE SLIT INTERFERENCE AND DIFFRACTION COMBINED

In Young's double-slit experiment (Section 45-1) we assumed that the slits were arbitrarily narrow (that is, $a \ll \lambda$), which means that the central part of the diffusing screen was uniformly illuminated by the diffracted waves from each slit. When such waves interfere, they produce fringes of uniform intensity, as in Fig. 45-9. This idealized situation cannot occur with actual slits because the condition $a \ll \lambda$ cannot usually be met. Waves from the two actual slits combining at different points of the screen will have intensities that are *not* uniform but are governed by the diffraction pattern of a single slit. The effect of relaxing the assumption that $a \ll \lambda$ in Young's experiment is to leave the fringes relatively unchanged in location but to alter their intensities.

The interference pattern for infinitesimally narrow slits is given by Eq. 45-11b and 45-11c, or, with a small change in nomenclature,

$$I_{\theta,\text{int}} = I_{m,\text{int}} \cos^2 \beta, \tag{46-12}$$

where

$$\beta = \frac{\pi d}{\lambda} \sin \theta \tag{46-13}$$

in which d is the distance between the center-lines of the slits.

The intensity for the diffracted wave from either slit is given by Eqs. 46-8b and 46-8c, or, with a small change in nomenclature,

$$I_{\theta,\text{dif}} = I_{m,\text{dif}} \left(\frac{\sin \alpha}{\alpha}\right)^2, \tag{46-14}$$

where

$$\alpha = \frac{\pi a}{\lambda} \sin \theta. \tag{46-15}$$

We find the combined effect by regarding $I_{m,\text{int}}$ in Eq. 46-12 as a variable amplitude, given in fact by $I_{\theta,\text{dif}}$ of Eq. 46-14. This assumption, for the combined pattern, leads to

$$I_\theta = I_m \,(\cos \beta)^2 \left(\frac{\sin \alpha}{\alpha}\right)^2, \tag{46-16}$$

in which we have dropped all subscripts referring separately to interference and diffraction.

Let us express this result in words. At any point on the screen the available light intensity from each slit, considered separately, is given by the diffraction pattern of that slit (Eq. 46-14). The diffraction patterns for the two slits, again considered separately, coincide because parallel rays in Fraunhofer diffraction are focused at the same spot (see Fig. 46-5). Because the two diffracted waves are coherent, they will interfere.

The effect of interference is to redistribute the available energy over

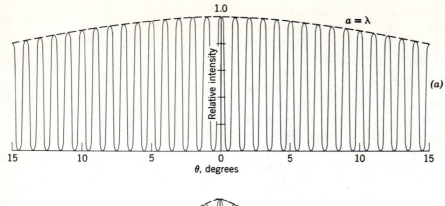

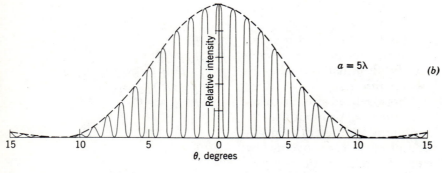

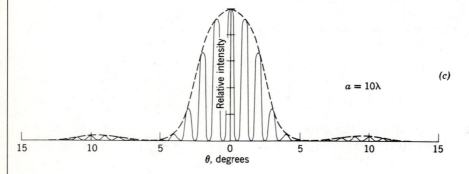

figure 46-12
Interference fringes for a double slit with slit separation $d = 50\lambda$. Three different slit widths, described by $a/\lambda = 1$, 5, and 10 are shown.

the screen, producing a set of fringes. In Section 45-1, where we assumed $a \ll \lambda$, the available energy was virtually the same at all points on the screen so that the interference fringes had virtually the same intensities (see Fig. 45-9). If we relax the assumption $a \ll \lambda$, the available energy is *not* uniform over the screen but is given by the diffraction pattern of a slit of width a. In this case the interference fringes will have intensities that are determined by the intensity of the diffraction pattern at the location of a particular fringe. Equation 46-16 is the mathematical expression of this argument.

Figure 46-12 is a plot of Eq. 46-16 for $d = 50\lambda$ and for three values of a/λ. It shows clearly that for narrow slits $(a = \lambda)$ the fringes are nearly uniform in intensity. As the slits are widened, the intensities of the fringes are markedly modulated by the "diffraction factor" in Eq. 46-16, that is, by the factor $(\sin \alpha/\alpha)^2$.

Equation 46-16 shows that the fringe envelopes of Fig. 46-12 are precisely the single-slit diffraction patterns of Fig. 46-9. This is especially clear in Fig. 46-13, which shows, for the curve of Fig. 46-12b, (a) the "interference factor" in Eq. 46-16 (that is, the factor $\cos^2 \beta$), (b) the "diffraction factor" $(\sin \alpha/\alpha)^2$, and (c) their product.

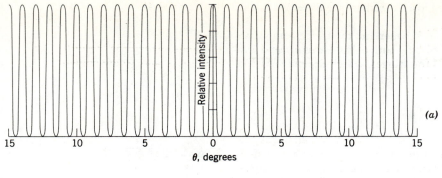

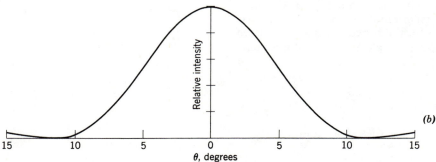

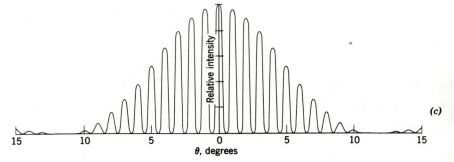

(a)

(b)

(c)

(a)

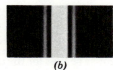

(b)

figure 46-14
(a) Interference fringes for a double-slit system in which the slit width is *not* negligible in comparison with the wavelength. The fringes are modulated in intensity by the diffraction pattern of a single slit. (b) If one of the slits is covered up, the interference fringes disappear and we see the single slit diffraction pattern. (Courtesy G. H. Carragan, Rensselaer Polytechnic Institute.)

If we put $a = 0$ in Eq. 46-16, then (see Eq. 46-15) $\alpha = 0$ and $\sin \alpha / \alpha \underset{\alpha \to 0}{=} \alpha/\alpha = 1$. Thus this equation reduces, as it must, to the intensity equation for a pair of vanishingly narrow slits (Eq. 46-12). If we put $d = 0$ in Eq. 46-16, the two slits coalesce into a single slit of width a, as Fig. 46-17 shows; $d = 0$ implies $\beta = 0$ (see Eq. 46-13) and $\cos^2 \beta = 1$. Thus Eq. 46-16 reduces, as it must, to the diffraction equation for a single slit (Eq. 46-14).

Figure 46-14 shows some actual double-slit interference photographs. The uniformly spaced interference fringes and their intensity modulation by the diffraction pattern of a single slit are clear. If one slit is covered up, as in Fig. 46-14b, the interference fringes disappear and we see the diffraction pattern of a single slit.

Starting from the curve of Fig. 46-12b, what is the effect of (a) increasing the slit width, (b) increasing the slit separation, and (c) increasing the wavelength?

(a) If we increase the slit width a, the envelope of the fringe pattern changes so that its central peak is sharper (compare Fig. 46-12c). The fringe spacing, which depends on d/λ, does not change.

EXAMPLE 6

(b) If we increase d, the fringes become closer together, the envelope of the pattern remaining unchanged.

(c) If we increase λ, the envelope becomes broader and the fringes move further apart. Increasing λ is equivalent to decreasing both of the ratios a/λ and d/λ. The general relationship of the envelope to the fringes, which depends only on d/a, does not change with wavelength.

EXAMPLE 7

In double-slit Fraunhofer diffraction what is the fringe spacing on a screen 50 cm away from the slits if they are illuminated with blue light ($\lambda = 480$ nm = 4800 Å), if $d = 0.10$ mm, and if the slit width $a = 0.02$ mm? What is the linear distance from the central maximum to the first minimum of the fringe envelope?

The intensity pattern is given by Eq. 46-16, the fringe spacing being determined by the interference factor $\cos^2 \beta$. From Example 2, Chapter 45, we have

$$\Delta y = \frac{\lambda D}{d},$$

where D is the distance of the screen from the slits. Substituting yields

$$\Delta y = \frac{(480 \times 10^{-9} \text{ m})(50 \times 10^{-2} \text{ m})}{0.10 \times 10^{-3} \text{ m}} = 2.4 \times 10^{-3} \text{ m} = 2.4 \text{ mm}.$$

The distance to the first minimum of the envelope is determined by the diffraction factor $(\sin \alpha / \alpha)^2$ in Eq. 46-16. The first minimum in this factor occurs for $\alpha = \pi$.

From Eq. 46-15,

$$\sin \theta = \frac{\alpha \lambda}{\pi a} = \frac{\lambda}{a} = \frac{480 \times 10^{-9} \text{ m}}{0.02 \times 10^{-3} \text{ m}} = 0.024.$$

This is so small that we can assume that $\theta \cong \sin \theta \cong \tan \theta$, or

$$y = D \tan \theta \cong D \sin \theta = (50 \text{ cm})(0.024) = 1.2 \text{ cm}.$$

There are about ten fringes in the central peak of the fringe envelope.

EXAMPLE 8

What requirements must be met for the central maximum of the envelope of the double-slit Fraunhofer pattern to contain exactly eleven fringes?

The required condition will be met if the sixth minimum of the interference factor $(\cos^2 \beta)$ in Eq. 46-16 coincides with the first minimum of the diffraction factor $(\sin \alpha / \alpha)^2$.

The sixth minimum of the interference factor occurs when

$$\beta = \tfrac{11}{2}\pi$$

in Eq. 46-12.

The first minimum in the diffraction term occurs for

$$\alpha = \pi.$$

Dividing (see Eqs. 46-13 and 46-15) yields

$$\frac{\beta}{\alpha} = \frac{d}{a} = \frac{11}{2}.$$

This condition depends only on the slit geometry and not on the wavelength. For long waves the pattern will be broader than for short waves, but there will always be eleven fringes in the central peak of the envelope.

The double-slit problem as illustrated in Fig. 46-12 combines interference and diffraction in an intimate way. At root both are superposition effects and depend on adding wave disturbances at a given point, taking phase differences properly into account. If the waves to be combined originate from a *finite* (and usually small) number of elementary coherent radiators, as in Young's double-slit experiment, we call the effect *interference*. If the waves to be combined originate by subdividing a wave into *infinitesimal* coherent radiators, as in our treatment of a single slit (Fig. 46-6), we call the effect *diffraction*. This distinction between interference and diffraction is convenient and useful. However, it should not cause us to lose sight of the fact that both are superposition effects and that often both are present simultaneously, as in Young's experiment.*

questions

1. Why is the diffraction of sound waves more evident in daily experience than that of light waves?

2. Why do radio waves diffract around buildings, although light waves do not?

3. A person holds a single narrow vertical slit in front of the pupil of his eye and looks at a distant light source in the form of a long heated filament. Is the diffraction pattern that he sees a Fresnel or a Fraunhofer pattern?

4. Do diffraction effects occur for virtual images as well as for real images? Explain.

5. Do diffraction effects occur for images formed by (a) plane mirrors and (b) spherical mirrors? Explain.

6. Comment on this statement: "Diffraction occurs in all regions of the electromagnetic spectrum." Consider the X-ray region and the micro-wave region for example, and give your arguments for believing it to be true or false.

7. Figure 46-1 shows the general case of diffraction from an aperture of arbitrary shape in an otherwise opaque screen. Discuss in general terms the inverse case, following the arguments of Section 46-1, in which diffraction occurs around an opaque *object*, such as a key or a paper clip, no screen B being present.

8. Distinguish between Fresnel and Fraunhofer diffraction. Do different physical principles underlie them? If so, what are they? If the same broad principle underlies them, what is it?

9. Most of us have noticed that the reception quality on a car radio deteriorates when we drive under a high-voltage transmission line. Does this have anything to do with the current in the line; with diffraction effects; with interference effects? Discuss.

10. *Fresnel's Bright Spot:* If a coin or a ball bearing (or even a bowling ball) is suspended between a point source of light and a photographic film a bright spot, called the Fresnel Bright Spot appears at the center of the geometrical shadow; see Fig. 46-15. Diffraction rings appear, both within the shadow and out of it. One might think that the center of the geometrical shadow, being most shielded from the light source, would be dark but just the reverse is true. Can you see the qualitative possibility that this experimental result can be consistent with Fresnel's diffraction theory?

*In this chapter we have not discussed *holography*, a technique in which a *three-dimensional* image of an object (rather than a two-dimensional image as in ordinary photography) can be produced. Diffraction plays a major role. To follow this up see "Holography, 1948–1971" by Dennis Gabor, *Science*, July 1972. Gabor won the Nobel Prize in 1971 for this discovery (made in 1948) and this article is an adaptation of his lecture on receiving the prize. See also "An Elementary Introduction to Practical Holography," by Alan G. Porter and S. George, *American Journal of Physics*, November 1975.

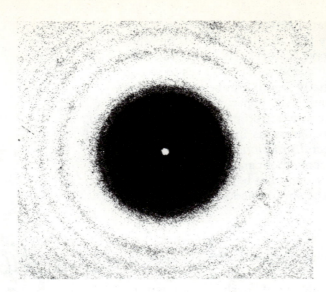

figure 46-15
Question 10
Fresnel's Bright Spot.

(Fresnel, who was trained as a civil engineer, submitted an essay on the diffraction of light to the French Academy of Sciences in response to a prize competition. He was then 30 years old. Siméon Poisson, a member of the prize committee and a distinguished mathematical physicist, was a confirmed believer in Newton's particle theory of light and totally disbelieved in the wave theory of Fresnel and others. He pointed out (as Fresnel had not) that Fresnel's theory predicted the existence of the bright spot that now bears Fresnel's name. Laboratory confirmation quickly followed. Although Fresnel won the prize, Poisson remained unconvinced of the wave theory of light and at his death, 22 years later, he still held to Newton's ideas).

11. A loud-speaker horn has a rectangular aperture 4 ft high and 1 ft wide. Will the pattern of sound intensity be broader in the horizontal plane or in the vertical?

12. We have claimed (correctly) that Maxwell's equations predict all the classical optical phenomena. Yet in Chapter 45 (Interference) and in this chapter (Diffraction) there is no mention of Maxwell's equations. Is there an inconsistency here? Where is the impact of Maxwell's equations felt? Discuss.

13. A radar antenna is designed to give accurate measurements of the height of an aircraft but only reasonably good measurements of its direction in a horizontal plane. Must the height-to-width ratio of the radar antenna be less than, equal to, or greater than unity?

14. In a single-slit Fraunhofer diffraction, what is the effect of increasing (a) the wavelength and (b) the slit width?

15. Sunlight falls on a single slit of width 10^3 nm. Describe qualitatively what the resulting diffraction pattern looks like.

16. In Fig. 46-5 rays r_1 and r_3 are in phase; so are r_2 and r_4. Why isn't there a *maximum* intensity at P_2 rather than a minimum?

17. Describe what happens to a Fraunhofer single-slit diffraction pattern if the whole apparatus is immersed in water.

18. When we speak of diffraction by a single "slit" we imply that the width of the slit must be much less than its height. Suppose that, in fact, the height was equal to twice the width. Make a rough guess at what the diffraction pattern would look like? An exact solution is of course not asked for.

19. In Fig. 46-3 we stated, correctly, that the optical path lengths from the slit to point P_0 are all the same. Why? Also, why are the optical path lengths of the rays in Fig. 46-5 from the slit to point P_2 *not* the same? What is the difference between these two figures?

20. In Fig. 46-7d why is E_θ, which represents the first maximum beyond the central maximum, not vertical? (Hint: Consider the effects of a slight winding or unwinding of the coil of vectors in this figure.)

21. Distinguish clearly between θ, α, and ϕ in Eq. 46-8c.

22. If we were to redo our analysis of the properties of lenses in Section 44-5 by the methods of geometrical optics but *without* restricting our considerations to paraxial rays and to "thin" lenses, would diffraction phenomena, such as that of Fig. 46-10, emerge from the analysis? Discuss.

23. Give at least two reasons why the usefulness of large reflecting telescopes increases as we increase the mirror diameter.

24. We have seen that diffraction limits the resolving power of optical telescopes (see Fig. 46-11). Does it also do so for radio telescopes, such as that of Fig. 41-2?

25. Distinguish carefully between interference and diffraction in Young's double-slit experiment.

26. In what way are interference and diffraction similar? In what way are they different?

27. In double-slit interference patterns such as that of Fig. 46-14a we said that the interference fringes were modulated in intensity by the diffraction pattern of a single slit. Could we reverse this statement and say that the diffraction pattern of a single slit is intensity-modulated by the interference fringes? Discuss.

SECTION 46-1

1. *Babinet's principle:* A monochromatic beam of parallel light is incident on a "collimating" hole of diameter $x \gg \lambda$. Point P lies in the geometrical shadow region on a distance screen, as shown in Fig. 46-16a. Two obstacles, shown in Fig. 46-16b, are placed in turn over the collimating hole. A is an opaque circle with a hole in it and B is the "photographic negative" of A. Using superposition concepts, show that the intensity at P is identical for each of the two diffracting objects A and B.*

problems

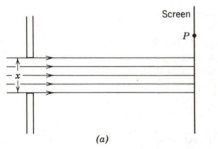

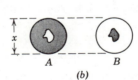

figure 46-16
Problem 1

SECTION 46-2

2. If the distance between the first and fifth minima of a single slit pattern is 0.35 mm with the screen 40 cm away from the slit and using light having a wavelength of 550 nm, what is the width of the slit?

3. In a single-slit diffraction pattern the distance between the first minimum on the right and the first minimum on the left is 5.2 mm. The screen on which the pattern is displayed is 80 cm from the slit and the wavelength is 5460 Å. Calculate the slit width.   *Answer:* 0.17 mm.

4. For a single slit the first minimum occurs at $\theta = 90°$, thus filling the forward hemisphere beyond the slit with light. What must be the ratio of the slit width to the wavelength be for this to take place?

5. A slit 1.0 mm wide is illuminated by light of wavelength 589 nm = 5890 Å. We see a diffraction pattern on a screen 3.0 m away. What is the distance

* In this connection, it can be shown that the diffraction pattern of a wire is that of a slit of equal width. See "Measuring the Diameter of a Hair by Diffraction" by S. M. Curry and A. L. Schawlow, *American Journal of Physics,* May 1974.

between the first two diffraction minima on either side of the central diffraction maximum?   *Answer: 1.8 mm.*

6. A plane wave ($\lambda = 590$ nm) falls on a slit with $a = 0.40$ mm. A converging lens ($f = +70$ cm) is placed behind the slit and focuses the light on a screen. What is the linear distance on the screen from the center of the pattern to (a) the first minimum and (b) the second minimum?

SECTION 46-4

7. What is the half-width of a diffracted beam for a slit whose width is (a) 1, (b) 5, and (c) 10 wavelengths?   *Answer: (a) 53°, (b) 10°, (c) 5.1°.*

8. A single slit is illuminated by light whose wavelengths are λ_a and λ_b, so chosen that the first diffraction minimum of λ_a coincides with the second minimum of λ_b. (a) What relationship exists between the two wavelengths? (b) Do any other minima in the two patterns coincide?

9. In Fig. 46-7d, calculate the angle E_θ makes with the vertical, assuming the slit to be divided into infinitesimal strips of width dx. See Question 20. *Answer: 12°.*

10. (a) Show that the values of α at which intensity maxima for single-slit diffraction occur can be found exactly by differentiating Eq. 46-8b with respect to α and equating to zero, obtaining the condition

$$\tan \alpha = \alpha.$$

(b) Find the values of α satisfying this relation by plotting graphically the curve $y = \tan \alpha$ and the straight line $y = \alpha$ and finding their intersections; alternatively, use a pocket calculator. (c) Find the (nonintegral) values of m corresponding to successive maxima in the single-slit pattern. Note that the secondary maxima do not lie exactly halfway between minima.

11. In Example 4 solve the transcendental equation

$$\frac{1}{2} = \left(\frac{\sin \alpha_x}{\alpha_x}\right)^2$$

graphically for α_x, to an accuracy of three significant figures. *Answer: 79.7°.*

12. If you double the width of a single slit the intensity of the central maximum of the diffraction pattern increases by a factor of four, even though the energy passing through the slit only doubles. Can you explain this quantitatively?

SECTION 46-5

13. The two headlights of an approaching automobile are 140 cm apart. At what maximum distance will the eye resolve them? Assume a pupil diameter of 5.0 mm and $\lambda = 550$ nm. Assume also that diffraction effects alone limit the resolution.   *Answer: 10 km (= 6.3 mi).*

14. An astronaut in a satellite claims he can just barely resolve two point sources on the earth, 160 km below him. What is their separation, assuming ideal conditions? Take $\lambda = 550$ nm, and the pupil diameter to be 5.0 mm.

15. How closely may two small objects be located if they are to be resolved when viewed through the telescope of a transit having a 3.0-cm objective lens if the transit is 370 m from the objects? Take the wavelength to be 550 nm.   *Answer: 8.3 mm.*

16. (a) How small is the angular separation of two stars if their images are barely resolved by the Thaw refracting telescope at the Allegheny Observatory in Pittsburgh? The lens diameter is 30 in. and its focal length is 46 ft. Assume $\lambda = 550$ nm. (b) Find the distance between these barely resolved stars if each of them is 10 light years distant from the earth. (c) For the image of a single star in this telescope, find the diameter of the first dark ring in the diffraction pattern, as measured on a photographic plate placed at the focal plane.

Tho

| | |
 Assume that the star image structure is associated entirely with diffraction at the lens aperture and not with (small) lens "errors," etc.

17. The wall of a large room is covered with acoustic tile in which small holes are drilled 5.0 mm from center to center. How far can a person be from such a tile and still distinguish the individual holes, assuming ideal conditions? Assume the diameter of the pupil to be 4.0 mm and λ to be 550 nm. *Answer: 37 m.*

18. Find the separation of two points on the moon's surface that can just be resolved by the 200-in. telescope at Mount Palomar, assuming that this distance is determined by diffraction effects. The distance from the earth to the moon is 3.8×10^5 km. Assume $\lambda = 550$ nm $(= 5500$ Å).

19. Under ideal conditions, estimate the linear separation of two objects on the planet Mars which can just be resolved by an observer on earth (a) using the naked eye, and (b) using the 200-in. Mt. Palomar telescope. Use the following data: distance to Mars $= 8.0 \times 10^7$ km; diameter of pupil $= 5.0$ mm; wavelength of light $= 550$ nm.   *Answer: (a) 1.1×10^4 km. (b) 11 km.*

20. A "spy in the sky" satellite orbiting at, say, 100 mi above the earth's surface, has a lens with a focal length of 8.0 ft. Its resolving power for objects on the ground is 1.2 ft, that is, it could easily detect an automobile. What is the effective lens diameter, determined by diffraction considerations alone? Assume $\lambda = 550$ nm. Even more effective satellites are reported to be in operation today. See "Reconnaissance and Arms Control" by Ted Greenwood, *Scientific American* February 1973.

21. (a) A circular diaphragm 0.60 m in diameter oscillates at a frequency of 25 kHz in an underwater source of sound for submarine detection. Far from the source the sound intensity is distributed as a Fraunhofer diffraction pattern for a circular hole whose diameter equals that of the diaphragm. Take the speed of sound in water to be 1450 m/s and find the angle between the normal to the diaphragm and the direction of the first minimum. (b) Repeat for a source having an (audible) frequency of 1.0 kHz. *Answer: (a) 6.7°. (b) Because $1.22\lambda > d$, there is no answer for 1.0 kHz; explain.*

22. It can be shown that, except for $\theta = 0$, a circular obstacle produces the same diffraction pattern as a circular hole of the same diameter. Furthermore, if there are many such obstacles, water droplets, say, located randomly, then the interference effects vanish leaving only the diffraction associated with a single obstacle. (a) Explain why one sees a "ring" around the moon on a foggy night. Usually the ring is reddish in color: explain. (b) Calculate the size of the water droplet in the air if the ring around the moon appears to have a diameter 1.5 times that of the moon. (c) At what distance from the moon might a bluish ring be seen? Sometimes the rings are white: explain. (d) The color arrangement is opposite to that in a rainbow: why should this be so?

23. Suppose that, as in Example 8, the envelope of the central peak contains eleven fringes. How many fringes lie between the first and second minima of the envelope?   *Answer: 5.*

24. For $d = 2a$ in Fig. 46-17, how many interference fringes lie in the central diffraction envelope?

25. If we put $d = a$ in Fig. 46-17, the two slits coalesce into a single slit of width $2a$. Show that Eq. 46-16 reduces to the diffraction pattern for such a slit.

26. Construct qualitative vector diagrams like those of Fig. 46-7 for the double slit interference pattern. For simplicity consider $d = 2a$ (see Fig. 46-17). Can you interpret the main features of the intensity pattern this way?

27. (a) How many complete fringes appear between the first minima of the fringe envelope to either side of the central maximum for a double slit pattern if $\lambda = 550$ nm, $d = 0.15$ mm and $a = 0.030$ mm? (b) What is the ratio of the intensity of the third fringe to the side of the center to that of the central fringe.   *Answer: (a) 9. (b) 0.25.*

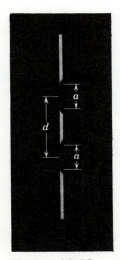

figure 46-17
Problems 24, 25, and 26

28. (*a*) Design a double-slit system in which the fourth fringe, not counting the central maximum, is missing. (*b*) What other fringes, if any, are also missing?

figure 46-18
Problem 29

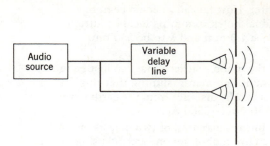

29. An acoustic double-slit system (slit separation *d*, slit width *a*) is driven by two loudspeakers as shown in Fig. 46-18. By use of a variable delay line, the phase of one of the speakers may be varied. Describe in detail what changes occur in the intensity pattern at large distances as this phase difference is varied from zero to 2π. Take both the interference and diffraction effects into account.

47
gratings
and spectra

In connection with Young's experiment (Sections 45-1 and 45-3) we discussed the interference of two coherent waves formed by diffraction at two elementary radiators (pinholes or slits). In our first treatment we assumed that the slit width was much less than the wavelength, so that light diffracted from each slit illuminated the observation screen essentially uniformly. Later, in Section 46-6, we took the slit width into account and showed that the intensity pattern of the interference fringes is modulated by a "diffraction factor" $(\sin \alpha/\alpha)^2$ (see Eq. 46-16).

Here we extend our treatment to cases in which the number N of radiators or diffracting centers is larger — and usually much larger — than two. We consider two situations:

1. An array of N parallel equidistant slits, called a *diffraction grating.*
2. A three-dimensional array of periodically arranged radiators — the atoms in a crystalline solid such as NaCl, for example. In this case the average spacing between the elementary radiators is so small that interference effects must be sought at wavelengths much smaller than those of visible light. We speak of *X-ray diffraction.*

In each case we distinguish carefully between the diffracting properties of a single radiator (slit or atom) and the interference of the waves diffracted, coherently, from the assembly of radiators.

A logical extension of Young's double-slit interference experiment is to increase the number of slits from two to a larger number N. An arrangement like that of Fig. 47-1, usually involving many more slits (as many as 10^4 slits/cm is not uncommon), is called a *diffraction grating.* As for

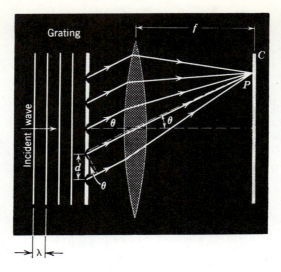

figure 47-1
An idealized diffraction grating containing five slits. The slit width a is shown for convenience to be considerably smaller than λ, although this condition is not realized in practice. The figure is also distorted in that f is much greater than d in practice.

a double slit, the intensity pattern that results when monochromatic light of wavelength λ falls on a grating consists of a series of interference fringes. The *angular separations* of these fringes are determined by the ratio λ/d, where d is the spacing between the centers of adjacent slits. The relative *intensities* of these fringes are determined by the diffraction pattern of a single grating slit, which depends on the ratio λ/a, where a is the slit width.

Figure 47-2, which compares the intensity patterns for $N = 2$ and $N = 5$, shows clearly that the "interference" fringes are modulated in intensity by a "diffraction" envelope, as in Fig. 46-14. Figure 47-3 presents a theoretical calculation of the intensity patterns for three fringes near the centers of the patterns of Fig. 47-2. These two figures show that increasing N (a) does not change the spacing between the (principal) interference fringe maxima, provided d and λ remain unchanged, (b) sharpens the (principal) maxima, and (c) introduces small secondary maxima between the principal maxima. Three such secondaries are present (but not readily visible) between each pair of adjacent principal maxima in Fig. 47-2b.

A principal maximum in Fig. 47-1 will occur when the path difference between rays from adjacent slits ($= d \sin \theta$) is given by

$$d \sin \theta = m\lambda \qquad m = 0, 1, 2, \ldots \qquad \text{(principal maxima)}, \quad (47\text{-}1)$$

where m is called the *order number.* This equation is identical with Eq. 45-1, which locates the intensity maxima for a double slit. The *loca-*

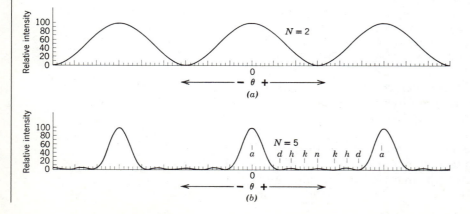

figure 47-2
Intensity patterns for "gratings" with (a) $N = 2$ and (b) $N = 5$ for the same value of d and λ. Note how the intensities of the fringes are modulated by a diffraction envelope as in Fig. 46-14; thus the assumption $a \ll \lambda$ is not realized in these actual "gratings." For $N = 5$ three very faint secondary maxima, not visible in this photograph, appear between each pair of adjacent primary maxima.

figure 47-3
Calculated intensity patterns for (a) a two-slit and (b) a five-slit grating for the same value of d and λ. This figure shows the sharpening of the principal maxima and the appearance of faint secondary maxima for $N > 2$. The letters on the five-slit pattern refer to Fig. 47-5. Three fringes only, centered around $\theta = 0$ in Fig. 47-1, are shown; compare Fig. 47-2. We may thus assume essentially equal intensities for each of the principal maxima shown.

tions of the (principal) maxima are thus determined only by the ratio λ/d and are independent of N. As for the double slit, the ratio a/λ determines the relative *intensities* of the principal maxima but does not alter their locations appreciably.

The sharpening of the principal maxima as N is increased can be understood by a graphical argument, using phasors. Figures 47-4a and 47-4b show conditions at any of the principal maxima for a two-slit and a nine-slit grating. The small arrows represent the amplitudes of the wave disturbances arriving at the screen at the position of each principal maximum. For simplicity we consider the central principal maximum only, for which $m = 0$, and thus $\theta = 0$, in Eq. 47-1.

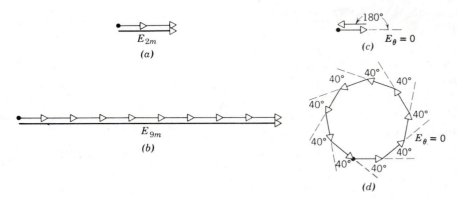

(a)

(b)

(c)

(d)

figure 47-4
Drawings (a) and (b) show conditions at the central principal maximum for a two-slit and a nine-slit grating, respectively. Drawings (c) and (d) show conditions at the minimum of zero intensity that lies on either side of this central principal maximum. In going from (a) to (c) the phase shift between waves from adjacent slits changes by 180° ($\Delta\phi = 2\pi/2$); in going from (b) to (d) it changes by 40° ($\Delta\phi = 2\pi/9$).

Consider the angle $\Delta\theta_0$ corresponding to the position of zero intensity that lies on either side of the central principal maximum. Figures 47-4c and 47-4d show the phasors at this point. The phase difference between waves from adjacent slits, which is zero at the central principal maximum, must increase by an amount $\Delta\phi$ chosen so that the array of phasors just closes on itself, yielding zero resultant intensity. For $N = 2$, $\Delta\phi = 2\pi/2$ (=180°); for $N = 9$, $\Delta\phi = 2\pi/9$ (=40°). In the general case it is given by

$$\Delta\phi = \frac{2\pi}{N}.$$

This increase in phase difference for adjacent waves corresponds to an increase in the path difference Δl given by

$$\frac{\text{phase difference}}{2\pi} = \frac{\text{path difference}}{\lambda},$$

or

$$\Delta l = \left(\frac{\lambda}{2\pi}\right)\Delta\phi = \left(\frac{\lambda}{2\pi}\right)\left(\frac{2\pi}{N}\right) = \frac{\lambda}{N}.$$

From Fig. 47-1, however, the path difference Δl at the first minimum is also given by $d \sin \Delta\theta_0$, so that we can write

$$d \sin \Delta\theta_0 = \frac{\lambda}{N},$$

or

$$\sin \Delta\theta_0 = \frac{\lambda}{Nd}.$$

Since $N \gg 1$ for actual gratings, $\sin \Delta\theta_0$, will ordinarily be quite small (that is, the lines will be sharp), and we may replace it by $\Delta\theta_0$, expressed in radians, to good approximation, or

$$\Delta\theta_0 = \frac{\lambda}{Nd} \qquad \text{(central principal maximum)}. \qquad (47\text{-}2)$$

This equation shows specifically that if we increase N for a given λ and d, $\Delta\theta_0$ will decrease, which means that the central principal maximum becomes sharper.

We state without proof,* and for later use, that for principal maxima other than the central one (that is, for $m \neq 0$) the angular distance between the position θ_m of the principal maximum of order m and the minimum that lies on either side is given by

$$\Delta\theta_m = \frac{\lambda}{Nd \cos\theta_m} \qquad \text{(any principal maximum)}. \qquad (47\text{-}3)$$

For the central principal maximum we have $m = 0$, $\theta_m = 0$, and $\Delta\theta_m = \Delta\theta_0$, so that Eq. 47-3 reduces, as it must, to Eq. 47-2.

The origin of the secondary maxima that appear for $N > 2$ can also be understood using the phasor method. Figure 47-5a shows conditions at the central principal maximum for a five-slit grating. The vectors are in phase. As we depart from the central maximum, θ in Fig. 47-1 increases from zero and the phase difference between adjacent vectors increases from zero to $\Delta\phi = \frac{2\pi}{\lambda} d \sin\theta$. Successive figures show how the resultant wave amplitude E_θ varies with $\Delta\phi$. Verify by graphical construction that a given figure represents conditions for both $\Delta\phi$ and $2\pi - \Delta\phi$. Thus we start at $\Delta\phi = 0$, proceed to $\Delta\phi = 180°$, and then trace backward through the sequence, following the phase differences shown in parentheses, until we reach $\Delta\phi = 360°$. This sequence corresponds to traversing the

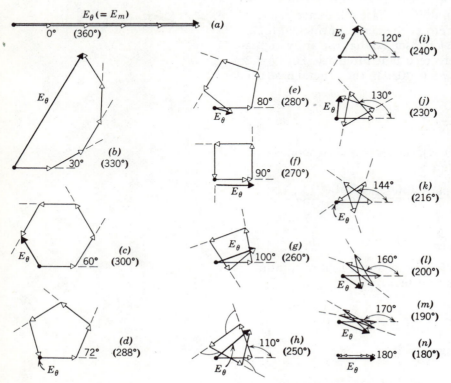

figure 47-5
The figures taken in sequence from (a) to (n) and then from (n) to (a) show conditions as the intensity pattern of a five-slit grating is traversed from the central principal maximum to an adjacent principal maximum. Phase differences between waves from adjacent slits are shown directly or, when going from (n) to (a), in parentheses. Principal maxima occur at (a), secondary maxima at, or near, (h) and (n), and points of zero intensity at (d) and (k). Compare Fig. 47-3b.

* See Problem 10.

intensity pattern from the central principal maximum to an adjacent one. Figure 47-5, which should be compared with Fig. 47-3b, shows that for $N = 5$ there are three secondary maxima, corresponding to $\Delta\phi = 110°$, 180°, and 250°. Make a similar analysis for $N = 3$ and show that only one secondary maximum occurs. In actual gratings, which commonly contain 10,000 to 50,000 "slits," the secondary maxima lie so close to the principal maxima or are so reduced in intensity that they cannot be distinguished from them experimentally.

47-3
DIFFRACTION GRATINGS

The *grating spacing d* for a typical grating that contains 12,000 "slits" distributed over a 1-in. width is 2.54 cm/12,000, or 2100 nm. Gratings are often used to measure wavelengths and to study the structure and intensity of spectrum lines. Few devices have contributed more to our knowledge of modern physics.

Gratings are made by ruling equally spaced parallel grooves on a glass or a metal* plate, using a diamond cutting point whose motion is automatically controlled by an elaborate ruling engine. Once such a master grating has been prepared, replicas can be formed by pouring a collodion solution on the grating, allowing it to harden, and stripping it off. The stripped collodion, fastened to a flat piece of glass or other backing, forms a good grating.

Figure 47-6 shows a cross section of a common type of grating ruled on glass. In the rudimentary grating of Fig. 47-1 open slits were separated by opaque strips; the *amplitude* of the wave disturbance varied in a periodic way as light passed through the grating, dropping to zero on the opaque strips. The grating of Fig. 47-6 is transparent everywhere, so that there is little periodic change in amplitude as the grating is crossed. The effect of the rulings is to change the *optical thickness* of the grating in a periodic way, rays traversing the grating between the rulings (b in Fig. 47-6) containing more wavelengths than rays traversing the grating in the center of the rulings (a in Fig. 47-6). This results in a periodic change of *phase* as light crosses the grating at right angles to the rulings. Reflection gratings also depend for their operation on a periodic change in phase of the reflected wave as one crosses the grating, the change in amplitude under these conditions being negligible. The principal maxima for *phase gratings*, assuming that the incident light falls on the grating at right angles, is given by the same formula derived earlier for idealized amplitude or slit gratings, namely,

$$d \sin\theta = m\lambda \qquad m = 0, 1, 2\ldots,$$

where d is the distance between the rulings and the integer m is called the *order* of the particular principal maximum. Essentially all gratings used in the visible spectrum, whether of the transmission type, as in Fig. 47-6, or the reflection type, are phase gratings.

Figure 47-7 shows a simple grating spectroscope, used for viewing the spectrum of a light source, assumed to emit a number of discrete wavelengths, or *spectrum lines*. The light from source S is focused by lens L_1 on a slit S_1 placed in the focal plane of lens L_2. The parallel light emerging from collimator C falls on grating G. Parallel rays associated with a particular interference maximum occurring at angle θ fall on lens L_3, being brought to a focus in plane F-F'. The image formed in this

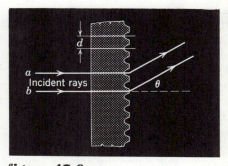

figure 47-6
An enlarged cross section of a diffraction grating ruled on glass. Such gratings, in which the phase of the emerging wave changes as one crosses the grating, are called *phase gratings.*

---

* Gratings ruled on metal are called *reflection gratings* because the interference effects are viewed in reflected rather than in transmitted light. For an insight into the highly complex technique of preparing gratings, see "Ruling Engines," by A. G. Ingolls, *Scientific American*, June, 1952, and "Diffraction Gratings," by E. W. Palmer and J. F. Verrill, *Contemporary Physics*, May, 1968.

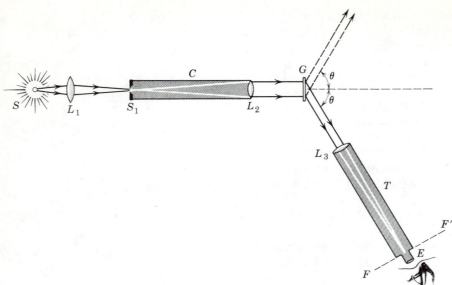

plane is examined, using a magnifying lens arrangement *E*, called an eyepiece. A symmetrical interference pattern is formed on the other side of the central position, as shown by the dotted lines. The entire spectrum can be viewed by rotating telescope *T* through various angles. Instruments used for scientific research or in industry are more complex than the simple arrangement of Fig. 47-7. They invariably employ photographic or photoelectric recording and are called *spectrographs.* Figure 49-12 shows a small portion of the spectrum of iron, produced by examining the light produced in an arc struck between iron electrodes, using a research type spectrograph with photographic recording. Each line in the figure represents a different wavelength that is emitted from the source.

Grating instruments can be used to make absolute measurements of wavelength, since the grating spacing *d* in Eq. 47-1 can be measured accurately with a traveling microscope. Several spectra are normally produced in such instruments, corresponding to $m = \pm1, \pm2$, etc., in Eq. 47-1 (see Fig. 47-8). This may cause some confusion if the spectra overlap. Further, this multiplicity of spectra reduces the recorded intensity of any given spectrum line because the available energy is divided among a number of spectra.

This disadvantage of the grating instrument can be overcome by shaping the profile of the grating grooves so that a large fraction of the light is thrown into a particular order on a particular side (for a given wavelength). This technique, called *blazing*, so alters the diffracting properties of the individual grooves (by controlling their profiles) that the light of wavelength λ diffracted by a single groove has a sharp peak of maximum intensity at a selected angle θ ($\neq 0$).

Gratings can, of course, separate wavelengths that are distributed continuously and not as sharp spectral lines. The emission from a red hot poker is an example. At the Flandrau Planetarium at the University of Arizona a "live" (m = 1) spectrum* of the visible radiation

---

* See "Eight Feet of the Solar Spectrum," O. Richard Norton, *Sky and Telescope*, September 1977. Even here sharp spectral lines are present. They appear as *absorption* lines (rather than *emission* lines) as the sun's radiation passes through its atmosphere. These are the so-called (dark) Fraunhofer lines. The element helium (from a Greek word meaning the sun) was discovered from an analysis of these lines. This is the only one of the elements not first discovered on earth.

from the sun is displayed to visitors; it is 8 ft long and about 5.5 in. high.

Light can also be analyzed into its component wavelengths if the grating in Fig. 47-7 is replaced by a prism. In a *prism spectrograph* each wavelength in the incident beam is deflected through a definite angle θ, determined by the index of refraction of the prism material for that wavelength. Curves such as Fig. 43-2, which gives the index of refraction of fused quartz as a function of wavelength, show that the shorter the wavelength, the larger the angle of deflection θ. Such curves vary from substance to substance and must be found by measurement. Prism instruments are not adequate for accurate *absolute* measurements of wavelength because the index of refraction of the prism material at the wavelength in question is usually not known precisely enough. Both prism and grating instruments make accurate *comparisons* of wavelength, using a suitable comparison spectrum such as that shown in Fig. 49-12, in which careful absolute determinations have been made of the wavelengths of the spectrum lines. The prism instrument has an advantage over a grating instrument in that its light energy is concentrated into a single spectrum so that brighter lines may be produced.

EXAMPLE 1

A grating with 8000 rulings/in. is illuminated with white light at perpendicular incidence. Describe the diffraction pattern. Assume that the wavelength of the light extends from 4000 to 7000 Å (= 400 to 700 nm).

The grating spacing d is 2.54 cm/8000, or 3170 nm. The central or zero-order maximum corresponds to $m = 0$ in Eq. 47-1. All wavelengths present in the incident light are superimposed at $\theta = 0$, as Fig. 47-8 shows.

The first-order diffraction pattern corresponds to $m = 1$ in Eq. 47-1. The 400 nm line occurs at an angle given by

$$\theta = \sin^{-1}\frac{m\lambda}{d} = \sin^{-1}\frac{(1)(400 \text{ nm})}{3170 \text{ nm}} = \sin^{-1} 0.126 = 7.3°.$$

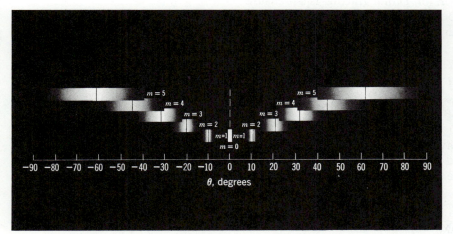

figure 47-8
Example 1. The spectrum of white light as viewed in a grating instrument like that of Fig. 47-7. The different orders, identified by the order number m, are shown separated vertically for clarity. As actually viewed, they would not be so displaced. The central line in each order corresponds to $\lambda = 550$ nm (= 5500 Å).

In the same way the angle for the 7000-A line is found to be 12.8°, and the entire pattern of Fig. 47-8 can be calculated. Note that the *first-order spectrum* ($m = 1$) is isolated but that the second-, third-, and fourth-order spectra overlap.

EXAMPLE 2

A diffraction grating has 10^4 rulings uniformly spaced over 1 in. It is illuminated at normal incidence by yellow light from a sodium vapor lamp. This light contains two closely spaced lines (the well-known *sodium doublet*) of wavelengths 5890.0 and 5895.9 Å (= 589.00 and 589.59 nm). (*a*) At what angle will the first-order maximum occur for the first of these wavelengths?

The grating spacing d is 10^{-4} in., or 2540 nm. The first-order maximum corresponds to $m = 1$ in Eq. 47-1. We thus have

$$\theta = \sin^{-1}\frac{m\lambda}{d} = \sin^{-1}\frac{(1)(589 \text{ nm})}{2540 \text{ nm}} = \sin^{-1} 0.232 = 13.3°.$$

(*b*) What is the angular separation between the first-order maxima for these lines?

The straightforward way to find this separation is to repeat this calculation for $\lambda = 589.59$ nm and to subtract the two angles. A difficulty, which can best be appreciated by carrying out the calculation, is that we must carry a large number of significant figures to obtain a meaningful value for the difference between the angles. To calculate the difference in angular positions *directly*, let us write down Eq. 47-1, solved for $\sin \theta$, and differentiate it, treating θ and λ as variables:

$$\sin \theta = \frac{m\lambda}{d}$$

$$\cos \theta \, d\theta = \frac{m}{d} \, d\lambda.$$

If the wavelengths are close enough together, as in this case, $d\lambda$ can be replaced by $\Delta\lambda$, the actual wavelength difference; $d\theta$ then becomes $\Delta\theta$, the quantity we seek. This gives

$$\Delta\theta = \frac{m \, \Delta\lambda}{d \cos \theta} = \frac{(1)(0.59 \text{ nm})}{(2540 \text{ nm})(\cos 13.3°)} = 2.4 \times 10^{-4} \text{ rad} = 0.014°.$$

Note that although the wavelengths involve five significant figures, our calculation, done this way, involves only two or three, with consequent reduction in numerical manipulation.

The quantity $d\theta/d\lambda$, called the *dispersion D* of a grating, is a measure of the angular separation produced between two incident monochromatic waves whose wavelengths differ by a small wavelength interval. From this example we see that

$$D = \frac{d\theta}{d\lambda} = \frac{m}{d \cos \theta}. \tag{47-4}$$

47-4
RESOLVING POWER OF A GRATING

To distinguish light waves whose wavelengths are close together, the maxima of these wavelengths formed by the grating should be as narrow as possible. Expressed otherwise, the grating should have a high *resolving power R*, defined from

$$R = \frac{\lambda}{\Delta\lambda}. \tag{47-5}$$

Here λ is the mean wavelength of two spectrum lines that can barely be recognized as separate and $\Delta\lambda$ is the wavelength difference between them. The smaller $\Delta\lambda$ is, the closer the lines can be and still be resolved; hence the greater the resolving power R of the grating. It is to achieve a high resolving power that gratings with many rulings are constructed.

The resolving power of a grating is usually determined by the same consideration (that is, the Rayleigh criterion) that we used in Section 46-5 to determine the resolving power of a lens. If two principal maxima are to be barely resolved, they must, according to this criterion, have an angular separation $\Delta\theta$ such that the maximum of one line coincides with the first minimum of the other; see Fig. 46-11. If we apply this criterion, we can show that

$$R = Nm, \tag{47-6}$$

where N is the total number of rulings in the grating and m is the order. As expected, the resolving power is zero for the central principal maximum ($m = 0$), all wavelengths being undeflected in this order.

Let us derive Eq. 47-6. The angular separation between two principal maxima whose wavelengths differ by $\Delta\lambda$ is found from Eq. 47-4, which we recast as

$$\Delta\theta = \frac{m \, \Delta\lambda}{d \cos\theta}. \tag{47-4}$$

The Rayleigh criterion (Section 46-5) requires that this be equal to the angular separation between a principal maximum and its adjacent minimum. This is given from Eq. 47-3, dropping the subscript m in $\cos\theta_m$, as

$$\Delta\theta_m = \frac{\lambda}{Nd \cos\theta}. \tag{47-3}$$

Equating Eqs. 47-4 and 47-3 leads to

$$R \,(=\lambda/\Delta\lambda) = Nm,$$

which is the desired relation.

EXAMPLE 3

In Example 2 how many rulings must a grating have if it is barely to resolve the sodium doublet in the third order?

From Eq. 47-5 the required resolving power is

$$R = \frac{\lambda}{\Delta\lambda} = \frac{589 \text{ nm}}{(589.59 - 589.00)\text{nm}} = 1000.$$

From Eq. 47-6 the number of rulings needed is

$$N = \frac{R}{m} = \frac{1000}{3} = 330.$$

This is a modest requirement.

The resolving power of a grating must not be confused with its dispersion. Table 47-1 shows the characteristics of three gratings, each illuminated with light of $\lambda = 589$ nm (= 5890 Å), the diffracted light being viewed in the first order ($m = 1$ in Eq. 47-1).

Table 47-1
Some characteristics of three gratings
($\lambda = 589$ nm, $m = 1$)

Grating	N	$d,$ nm	θ	R	D 10^{-2} degrees/nm
A	10,000	2540	13.3°	10,000	2.32
B	20,000	2540	13.3°	20,000	2.32
C	10,000	1370	25.5°	10,000	4.64

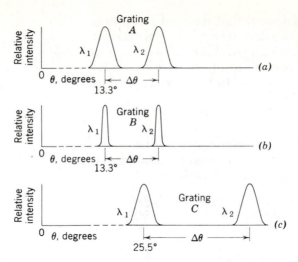

figure 47-9
The intensity patterns for light of wavelengths λ_1 and λ_2 near 589 nm (= 5890 Å), incident on the gratings of Table 47-1. Grating B has the highest resolving power and grating C the highest dispersion.

You should verify that the values of D and R given in the table can be calculated from Eqs. 47-4 and 47-6, respectively.

For the conditions of use noted in Table 47-1, gratings A and B have the same *dispersion* and A and C have the same *resolving power*. Figure 47-9 shows the intensity patterns that would be produced by these gratings for two incident waves of wavelengths λ_1 and λ_2, in the vicinity of $\lambda = 589$ nm. Grating B, which has high resolving power, has narrow intensity maxima and is inherently capable of distinguishing lines that are much closer together in wavelength than those of Fig. 47-9. Grating C, which has high dispersion, produces twice the angular separation between rays λ_1 and λ_2 that grating B does.

EXAMPLE 4

The grating of Example 1 has 8000 lines illuminated by light from a mercury vapor discharge. (*a*) What is the expected dispersion, in the third order, in the vicinity of the intense green line ($\lambda = 546$ nm = 5460 Å)?

Noting that $d = 3170$ nm, we have, from Eq. 47-1,

$$\theta = \sin^{-1}\frac{m\lambda}{d} = \sin^{-1}\frac{(3)(546 \text{ nm})}{3170 \text{ nm}} = \sin^{-1}0.517 = 31.1°.$$

From Eq. 47-4 we have

$$D = \frac{m}{d\cos\theta} = \frac{3}{(3170 \text{ nm})(\cos 31.1°)} = 1.1 \times 10^{-3} \text{ rad/nm} = 6.3 \times 10^{-2} \text{ deg/nm}.$$

(*b*) What is the expected resolving power of this grating in the fifth order? Equation 47-6 gives

$$R = Nm = (8000)(5) = 40,000.$$

Thus near $\lambda = 546$ nm a wavelength difference $\Delta\lambda$ given by Eq. 47-5, or

$$\Delta\lambda = \frac{\lambda}{R} = \frac{546 \text{ nm}}{40,000} = 0.014 \text{ nm},$$

can be distinguished.

47-5
X-RAY DIFFRACTION

Figure 47-10 shows how X-rays are produced when electrons from a heated filament F are accelerated by a potential difference V and strike a metal target T. X-rays are electromagnetic radiation with wavelengths of the order of 0.1 nm (or 1 Å). This value is to be compared with 550 nm

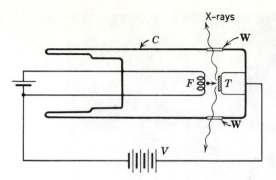

figure 47-10
X-rays are generated when electrons from heated filament F, accelerated by potential difference V, are brought to rest on striking metallic target T. W is a "window"—transparent to X-rays—in the evacuated metal container C.

(or 5500 Å) for the center of the visible spectrum. For such small wavelengths a standard optical diffraction grating, as normally employed, cannot be used. For $\lambda = 0.10$ nm and $d = 3000$ nm, for example, Eq. 47-1 shows that the first-order maximum occurs at

$$\theta = \sin^{-1}\frac{m\lambda}{d} = \sin^{-1}\frac{(1)(0.10 \text{ nm})}{3 \times 10^3 \text{ nm}} = \sin^{-1} 0.33 \times 10^{-4} = 0.002°.$$

This is too close to the central maximum to be practical. A grating with $d \cong \lambda$ is desirable, but, because X-ray wavelengths are about equal to atomic diameters, such gratings cannot be constructed mechanically.

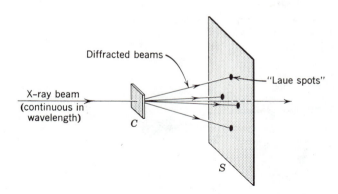

Diffracted beams

X-ray beam
(continuous in
wavelength)

"Laue spots"

C

S

figure 47-11
A nonmonochromatic beam of X-rays falls on a crystal C, which may be NaCl. Strong diffracted beams appear in certain directions, forming a so-called Laue pattern on a photographic film S.

In 1912 it occurred to physicist Max von Laue that a crystalline solid, consisting as it does of a regular array of atoms, might form a natural three-dimensional "diffraction grating" for X-rays. Figure 47-11 shows that if a collimated beam of X-rays, continuously distributed in wavelength, is allowed to fall on a crystal, such as sodium chloride, intense beams corresponding to constructive interference from the many diffracting centers of which the crystal is made up appear in certain sharply defined directions. If these beams fall on a photographic film, they form an assembly of "Laue spots." Figure 47-12, which is an actual example of these spots, shows that the hypothesis of Laue is indeed correct. The atomic arrangements in the crystal can be deduced from a careful study of the positions and intensities of the Laue spots* in much the same way that we might deduce the structure of an optical grating (that is, the detailed profile of its slits) by a study of the positions and intensities of the lines in the interference pattern.

* Other experimental arrangements have supplanted the Laue technique to a considerable extent today; the principle remains unchanged, however (see Question 21).

figure 47-12
Laue X-ray diffraction pattern from sodium chloride. A crystal of ordinary table salt was used in making this plate. (Courtesy of W. Arrington and J. L. Katz, X-ray Laboratory, Rensselaer Polytechnic Institute.)

Figure 47-13 shows how sodium and chlorine atoms (strictly, Na^+ and Cl^- ions) are stacked to form a crystal of sodium chloride. This pattern, which has *cubic* symmetry, is one of the many atomic arrangements exhibited by solids. The model represents the *unit cell* for sodium chloride. This is the smallest unit from which the crystal may be built up by repetition in three dimensions. You should verify that no smaller assembly of atoms possesses this property. For sodium chloride the length of the cube edge of the unit cell is 0.562737 nm.

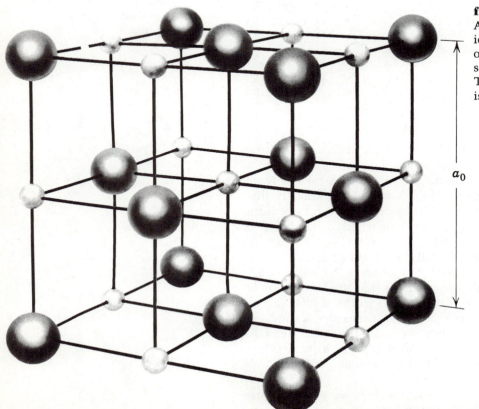

figure 47-13
A model showing how Na^+ and Cl^- ions are stacked to form a unit cell of NaCl. The small spheres represent sodium ions, the large ones chlorine. The edge a_0 of the (cubical) unit cell is 0.562737 nm.

a_0

Each unit cell in sodium chloride has four sodium ions and four chlorine ions associated with it. In Fig. 47-13 the sodium ion in the center belongs entirely to the cell shown. Each of the other twelve sodium ions shown is shared with three adjacent unit cells so that each contributes one-fourth of an ion to the cell under consideration. The total number of sodium ions is then $1 + \frac{1}{4}(12) = 4$. By similar reasoning you can show that although there are fourteen chlorine ions in Fig. 47-13, only four are associated with the unit cell shown.

The unit cell is the fundamental repetitive diffracting unit in the crystal, corresponding to the slit (and its adjacent opaque strip) in the optical diffraction grating of Fig. 47-1. Figure 47-14a shows a particular plane in a sodium chloride crystal. If each unit cell intersected by this plane is represented by a small cube, Fig. 47-14b results. You may imagine each of these figures extended indefinitely in three dimensions.

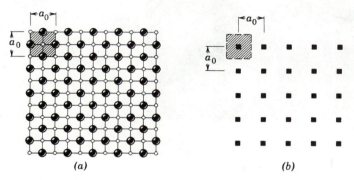

(a)   (b)

figure 47-14
(a) A section through a crystal of sodium chloride, showing the sodium and chlorine ions. (b) The corresponding unit cells in this section, each cell being represented by a small black square.

Let us treat each small cube in Fig. 47-14b as an elementary diffracting center, corresponding to a slit in an optical grating. The *directions* (but not the intensities) of all the diffracted X-ray beams that can emerge from a sodium chloride crystal (for a given X-ray wavelength and a given orientation of the incident beam) are determined by the geometry of this three-dimensional lattice of diffracting centers. In exactly the same way the *directions* (but not the intensities) of all the diffracted beams that can emerge from a particular optical grating (for a given wavelength and orientation of the incident beam) are determined only by the geometry of the grating, that is, by the grating spacing d. Representing the unit cell by what is essentially a point, as in Fig. 47-14b, corresponds to representing the slits in a diffraction grating by lines, as we did in discussing Young's experiment in Section 45-1.

The *intensities* of the lines from an optical diffraction grating are determined by the diffracting characteristics of a single slit, as Fig. 46-14 shows. In the idealized case of Fig. 47-1 these characteristics depend on the slit width a. In practical optical gratings these characteristics depend on the detailed shape of the profile of the grating rulings.

In exactly the same way the *intensities* of the diffracted beams emerging from a crystal depend on the diffracting characteristics of the unit cell.* Fundamentally the X-rays are diffracted by electrons, diffraction by nuclei being negligible in most cases. Thus the diffracting characteristics of a unit cell depend on how the electrons are distributed

---

* For some directions in which a beam might be expected to emerge, from interference considerations, no beam will be found because the diffracting characteristics of the unit cell are such that no energy is diffracted in that direction. Similarly, in optical gratings some lines, permitted by interference considerations, may not appear if their predicted positions coincide with a null in the single-slit diffraction pattern (see Fig. 46-12).

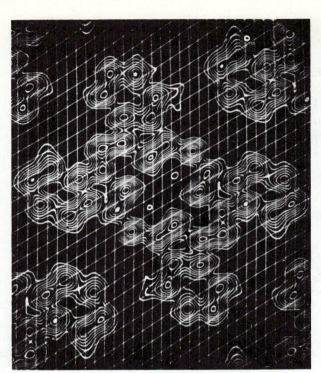

(a)

(b)

figure 47-15
(*a*) A photograph of an oscilloscope screen arranged to display projected electron density contours for phthalocyanine ($C_{32}H_{18}N_8$). Such plots, constructed electronically from X-ray diffraction data by an analog computer, provide a vivid picture of the structure of molecules. (Courtesy of Ray Pepinsky.) (*b*) A structural representation of the molecule phthalocyanine. The student should make a detailed comparison with (*a*), locating the various atoms identified in (*b*). Note that hydrogen atoms, which contain only a single electron, are not prominent in (*a*).

throughout the volume of the cell. By studying the *directions* of diffracted X-ray beams, we can learn the basic symmetry of the crystal. By also studying the *intensities* we can learn how electrons are distributed in the unit cell. Figure 47-15 shows the average density of electrons projected onto a particular plane through a unit cell of a crystal of phthalocyanine. This remarkable figure suggests something of the full power of X-ray methods for studying the structure of solids.

Bragg's law predicts the conditions under which diffracted X-ray beams from a crystal are possible. In deriving it, we ignore the structure of the unit cell, which is related only to the intensities of these beams. The dashed sloping lines in Fig. 47-16*a* represent the intersection with the plane of the figure of an arbitrary set of planes passing through the ele-

47-6
BRAGG'S LAW

mentary diffracting centers. The perpendicular distance between adjacent planes is d. Many other such families of planes, with different *interplanar spacings*, can be defined.

Figure 47-16b shows a plane wave that lies in the plane of the figure falling on one member of the family of planes defined in Fig. 47-16a, the incident rays making an angle θ with the plane.* Consider a family of diffracted rays lying in the plane of Fig. 47-16b and making an angle β with the plane containing the elementary diffracting centers. The diffracted rays will combine to produce maximum intensity if the path difference between adjacent rays is an integral number of wavelengths or

$$ae - bd = h(\cos \beta - \cos \theta) = l\lambda \qquad l = 0, 1, 2, \ldots \qquad (47\text{-}7)$$

For $l = 0$ this leads to $\qquad\qquad \beta = \theta,$

and the plane of atoms acts like a mirror for the incident wave, no matter what the value of θ.

For other values of l, β does not equal θ, but the diffracted beam can always be regarded as being "reflected" from a *different* set of planes than that shown in Fig. 47-16a with a different interplanar spacing d. Since we wish to describe each diffracted beam as a "reflection" from a particular set of planes and since we are dealing in the present argument only with the particular set of planes shown in Fig. 47-16a, we ignore all values of l other than $l = 0$ in Eq. 47-7. It can also be shown that a plane of diffracting centers acts like a mirror (that is, $\beta = \theta$) whether or not the incident wave lies in the plane of Fig. 47-16b.

Figure 47-16c shows an incident wave striking the *family* of planes, a single member of which was considered in Fig. 47-16b. For a single plane, mirrorlike "reflection" occurs for *any* value of θ, as we have seen. To have a constructive interference in the beam diffracted from the entire family of planes in the direction θ, the rays from the separate planes must reinforce each other. This means that the path difference for rays from adjacent planes (*abc* in Fig. 47-16c) must be an integral number of wavelengths or

$$2d \sin \theta = m\lambda \qquad m = 1, 2, 3, \ldots \qquad (47\text{-}8)$$

This relation is called *Bragg's law* after W. L. Bragg who first derived it. The quantity d in this equation (the interplanar spacing) is the perpendicular distance between the planes. For the planes of Fig. 47-16a analysis shows that d is related to the unit cell dimension a_0 by

$$d = \frac{a_0}{\sqrt{5}}. \qquad (47\text{-}9)$$

If an incident *monochromatic* X-ray beam falls at an *arbitrary* angle θ on a particular set of atomic planes, a diffracted beam will *not* result because Eq. 47-8 will not, in general, be satisfied. If the incident X-rays are *continuous* in wavelength, diffracted beams will result when wavelengths given by

$$\lambda = \frac{2d \sin \theta}{m} \qquad m = 1, 2, 3, \ldots$$

are present in the incident beam (see Eq. 47-8).

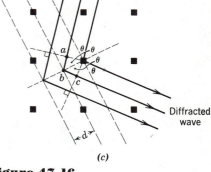

figure 47-16
(a) A section through the NaCl unit cell lattice of Fig. 45-14b. The dashed sloping lines represent an arbitrary family of planes, with interplanar spacing d. (b) An incident wave falls, at grazing angle θ, on one of the planes, xx', shown in (a). (c) An incident wave falls on the entire family of planes shown in (a). A strong diffracted wave is formed.

---

* In X-ray diffraction it is customary to specify the direction of a wave by giving the angle between the ray and the plane (the *glancing angle*) rather than the angle between the ray and the normal.

At what angles must an X-ray beam with $\lambda = 0.110$ nm fall on the family of planes represented in Fig. 47-16c if a diffracted beam is to exist? Assume the material to be sodium chloride.

EXAMPLE 5

The interplanar spacing d for these planes is given by Eq. 47-9 or

$$d = \frac{a_0}{\sqrt{5}} = \frac{0.563 \text{ nm}}{2.24} = 0.252 \text{ nm}.$$

Equation 47-8 gives

$$\sin \theta = \frac{m\lambda}{2d} = \frac{(m)(0.110 \text{ nm})}{(2)(0.252 \text{ nm})} = 0.218 \, m.$$

Diffracted beams are possible at $\theta = 12.6°$ ($m = 1$), $\theta = 25.9°$ ($m = 2$), $\theta = 40.9°$ ($m = 3$), and $\theta = 60.7°$ ($m = 4$). Higher-order beams cannot exist because they require $\sin \theta$ to exceed unity. Actually, the odd-order beams ($m = 1, 3$) prove to have zero intensity because the unit cell in cubic crystals such as NaCl has diffracting properties such that the intensity of the light scattered in these orders is zero.

X-ray diffraction is a powerful tool for studying the arrangements of atoms in crystals.[*] To do so quantitatively requires that the wavelength of the X-rays be known. In one of several approaches to this problem the unit cell dimension for NaCl[†] is determined by a method that does not involve X-rays. X-ray diffraction measurements on NaCl can then be used to determine the wavelength of the X-ray beam which in turn can be used to determine the structures of solids other than NaCl.

If ρ is the measured density of NaCl, we have, for the unit cell of Fig. 47-13,[‡] recalling that each unit cell contains four NaCl "molecules,"

$$\rho = \frac{m}{V} = \frac{4m_{NaCl}}{a_0^3}.$$

Here m_{NaCl}, the mass of a NaCl molecule, is given by

$$m_{NaCl} = \frac{M}{N_0},$$

where M is the molecular weight of NaCl and N_0 is Avogadro's number. Combining these two equations and solving for a_0 yields

$$a_0 = \left(\frac{4M}{N_0\rho}\right)^{1/3}$$

which permits us to calculate a_0. Once a_0 is known, the wavelengths of monochromatic X-ray beams can be found, using Bragg's law (Eq. 47-8).

questions

1. Discuss this statement: "A diffraction grating can just as well be called an interference grating."
2. How would the spectrum of an enclosed source that is formed by a diffraction grating on a screen change (if at all) when the source, grating and screen are all submerged in water?

[*] See "X-ray Crystallography" by Sir Lawrence Bragg, *Scientific American*, July 1968.
[†] In practice, calcite ($CaCO_3$) proves to be more useful as a standard crystal for a number of technical reasons.
[‡] This relation cannot be written down unless it is known that the structure of NaCl is cubic. This can be determined, however, by inspection of the symmetry of the spots in Fig. 47-12; the wavelength of the X-rays need not be known.

3. Assume that the limits of the visible spectrum are 4300 and 6800 Å. Is it possible to design a grating, assuming that the incident light falls normally on it, such that the first-order spectrum *barely overlaps* the second-order spectrum?

4. In a grating spectrograph, several lines having different wavelengths and formed in different orders might appear near a certain angle. How could you distinguish between their orders?

5. For the simple spectroscope of Fig. 47-7, show (a) that θ increases with λ for a grating and (b) that θ decreases with λ for a prism.

6. You are given a photograph of a spectrum on which the angular positions and the wavelengths of the spectrum lines are marked. (a) How can you tell whether the spectrum was taken with a prism or a grating instrument? (b) What information could you gather about either the prism or the grating from studying such a spectrum?

7. Explain in your own words why increasing the number of slits, N, in a diffraction grating sharpens the maxima. Why does decreasing the wavelength do so? Why does increasing the slit spacing, d, do so?

8. (a) What is a "phase grating"? (b) Can you make a diffraction grating out of parallel rows of fine wires strung closely together?

9. How much information can you discover about the structure of a diffraction grating by analyzing carefully the spectrum it forms of a monochromatic light source. Let $= 589$ nm, for an example.

10. (a) Why does a diffraction grating have closely spaced rulings? (b) Why does it have a large number of rulings?

11. Two nearly equal wavelengths are incident on a grating of N slits and are just not resolvable. However, they become resolved if the number of slits is increased. Formulas aside, is the explanation of this that: (a) more light can get through the grating? (b) the principal maxima become more intense and hence resolvable? (c) the diffraction pattern is spread more and hence the wavelengths become resolved? (d) there are a larger number of orders? or (e) the principal maxima become narrower and hence resolvable?

12. The relation $R = Nm$ suggests that the resolving power of a given grating can be made as large as desired by choosing an arbitrarily high order of diffraction. Discuss.

13. Show that at a given wavelength and a given angle of diffraction the resolving power of a grating depends only on its width W $(= Nd)$.

14. According to Eq. 47-3 the principal maxima become wider (that is, $\Delta\theta_m$ increases) the higher the order m (that is, the larger θ_m becomes). According to Eq. 47-6 the resolving power becomes greater the higher the order m. Explain this apparent paradox.

15. How would you *measure* (a) the dispersion D and (b) the resolving power R for either a prism or a grating spectrograph.

16. Is the pattern of Fig. 47-12 more properly described as a diffraction pattern or as an interference pattern?

17. For a given family of planes in a crystal, can the wavelength of incident X-rays be (a) too large or (b) too small to form a diffracted beam?

18. Why cannot a simple cube of edge $a_0/2$ in Fig. 47-13 be used as a unit cell for sodium chloride?

19. In some respects Bragg reflection *differs from* plane grating diffraction. Of the following statements which one is *true for Bragg reflection but not true for grating diffraction*? (a) two different wavelengths may be superposed. (b) radiation of a given wavelength may be sent in more than one direction. (c) long waves are deviated more than short waves. (d) there is only one grating spacing. (e) diffraction maxima of a given wavelength occur only for particular angles of incidence.

20. If a parallel beam of X-rays of wavelength λ is allowed to fall on a randomly oriented crystal of any material, generally no intense diffracted beams will

occur. Such beams appear if (*a*) the X-ray beam consists of a continuous distribution of wavelengths rather than a single wavelength or (*b*) the specimen is not a single crystal but a finely divided powder. Explain.

21. In Fig. 47-17(*a*) we show schematically the Debye-Scherrer experimental arrangement and in Fig. 47-17(*b*) a corresponding x-ray diffraction pattern. (*a*) Keeping in mind that the Laue method uses a large single crystal and an x-ray beam continuously distributed in wavelength, explain the origin of the spots. (Hint: each spot corresponds to the direction of scattering from a family of planes.)

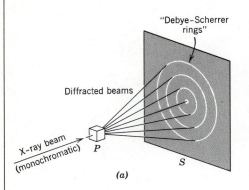

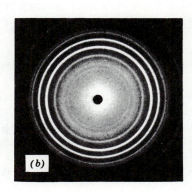

figure 47-17
Question 21 (a) and (b)
(a) A monochromatic beam of X-rays falls on an aggregate P of very small crystals oriented at random (called a microcyrstalline powder). Strong diffracted beams form rings surrounding a central spot, the so-called Debye-Scherrer rings, on a photographic film S.
(b) Debye-Scherrer X-ray diffraction pattern from zirconium oxide crystals. (From U. Fano and L. Fano, *Basic Physics of Atoms and Molecules*, John Wiley's Sons, 1959)

(*b*) Keeping in mind that the Debye-Scherrer method uses a large number of small single crystals randomly oriented and a monochromatic beam of x-rays, explain the origin of the rings. (Hint: because the small crystals are randomly oriented, all possible angles of incidence are obtained.)

problems

SECTION 47-2

1. Show that in a grating with alternately transparent and opaque strips of *equal* width, all the even orders (except $m = 0$) are absent.

2. The central intensity maximum formed by a grating, along with its subsidiary secondary maxima, can be viewed as the diffraction pattern of a single "slit" whose width is that of the entire grating. Treating the grating as a single wide slit, assuming that $m = 0$, and using the methods of Section 46-4, show that Eq. 47-2 can be derived.

3. A diffraction grating is made up of slits of width 300 nm with a 900 nm separation between their centers. The grating is illuminated by monochromatic plane waves, $\lambda = 600$ nm, the angle of incidence being zero. (*a*) How many diffracted lines are there? (*b*) What is the angular width of the spectral lines observed, if the grating has 1000 slits? *Angular width* is defined to be the angle between the two positions of zero intensity on either side of the maximum.   *Answer:* (*a*) 3, corresponding to $m = 0, \pm 1$. (*b*) 1.8×10^{-3} rad.

4. Assume that light is incident on a grating at an angle ψ as shown in Fig. 47-18. Show that the condition for a diffraction maximum is

$$d(\sin \psi + \sin \theta) = m\lambda \qquad m = 0, 1, 2, \ldots.$$

Only the special case $\psi = 0$ has been treated in this chapter (compare Eq. 47-1).

5. A transmission grating with $d = 1.50 \times 10^{-4}$ cm is illuminated at various angles of incidence by light of wavelength 600 nm. Plot as a function of angle of incidence (0 to 90°) the angular deviation of the first-order diffracted beam from the incident direction.

6. Derive this expression for the intensity pattern for a three-slit "grating":

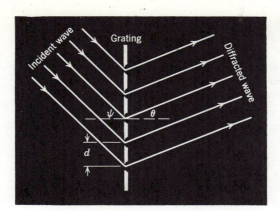

figure 47-18
Problem 4

$$I_\theta = \tfrac{1}{9} I_m(1 + 4\cos\phi + 4\cos^2\phi),$$

where

$$\phi = \frac{2\pi d \sin\theta}{\lambda}.$$

Assume that $a \ll \lambda$ and be guided by the derivation of the corresponding double-slit formula (Eq. 45-9).

7. (a) Using the result of Problem 6, show that the half-width of the fringes for a three-slit diffraction pattern, assuming θ small enough so that $\sin\theta \cong \theta$, is

$$\Delta\theta \cong \frac{\lambda}{3.2d}.$$

(b) Compare this with the expression derived for the two-slit pattern in Problem 16, Chapter 45. (c) Do these results support the conclusion that for a fixed slit spacing the interference maxima become sharper as the number of slits is increased?

8. (a) Using the result of Problem 6, show that a three-slit "grating" has only one secondary maximum. Find (b) its location and (c) its relative intensity.

9. A three-slit grating has separation d between adjacent slits. If the middle slit is covered up, will the half-width of the intensity maxima become broader or narrower? See Problem 7.
   Answer: Narrower by the factor 0.78.

10. Derive Eq. 47-3, that is, the expression for $\Delta\theta_m$, the angular distance between a principal maximum of order m and either adjacent minimum.

11. A diffraction grating has a large number N of slits, each of width d. Let I_{max} denote the intensity at some principal maximum, and let I_k denote the intensity of the kth adjacent secondary maximum. (a) If $k \ll N$, show from the phasor diagram that, approximately, $I_k/I_{max} = 1/(k + \tfrac{1}{2})^2\pi^2$. (Compare this with the single-slit formula.) (b) For those secondary maxima which lie roughly midway between two adjacent principal maxima, show that roughly $I_k/I_{max} = 1/N^2$. (c) Consider the central principal maximum and those adjacent secondary maxima for which $k \ll N$. Show that this part of the diffraction pattern quantitatively resembles that for one single slit of width Nd.

SECTION 47-3

12. A diffraction grating 2.0 cm wide has 6000 rulings. At what angles will maximum-intensity beams occur if the incident radiation has a wavelength of 589 nm?

13. With light from a gaseous discharge tube incident normally on a grating with a distance 1.73×10^{-4} cm between adjacent slit centers, a green line appears with sharp maxima at measured transmission angles $\theta = \pm 17.6°$, $37.3°$, $-37.1°$, $65.2°$, and $-65.0°$. Compute the wavelength of the green line that best fits the data. *Answer:* 524 nm.

14. A diffraction grating has 5000 rulings/in., and a strong diffracted beam is noted at $\theta = 30°$. (a) What are the possible wavelengths of the incident light? (b) What colors are they (see Figure 42-1)?

15. A grating has 3150 rulings/cm. For what wavelengths in the visible spectrum can fifth-order diffraction be observed?
    Answer: All wavelengths shorter than 635 nm.

16. Given a grating with 4000 lines/cm, how many orders of the entire visible spectrum (400–700 nm) can be produced?

17. A diffraction grating 3.00 cm wide produces a deviation of 30.0° in the second order with light of wavelength 600 nm. What is the total number of lines on the grating?   Answer: 12,500.

18. Light of wavelength 600 nm is incident normally on a diffraction grating. Two *adjacent* principal maxima occur at $\sin \theta = 0.2$ and $\sin \theta = 0.3$, respectively. The fourth order is a missing order. (a) What is the separation between adjacent slits? (b) What is the smallest possible individual slit width? (c) Name *all* orders actually appearing on the screen with the values chosen in (a) and (b).

19. Assume that the limits of the visible spectrum are arbitrarily chosen as 430 and 680 nm (= 4300 and 6800 Å). Design a grating that will spread the first-order spectrum through an angular range of 20°.
    Answer: 11,000 lines/cm.

20. Light containing a mixture of two wavelengths, 500 nm and 600 nm, is incident normally on a diffraction grating. It is desired (1) that the first and second principal maxima for each wavelength appear at $\theta \leq 30°$, (2) that the dispersion be as high as possible, *and* (3) that the third order for 600 nm be a missing order. (a) What is the separation between adjacent slits? (b) What is the smallest possible individual slit width? (c) Name *all* orders for 600 nm that actually appear on the screen with the values chosen in (a) and (b).

21. A narrow beam of monochromatic light strikes a grating at normal incidence and produces sharp maxima at the following angles from the normal: 6°40', 13°30', 20°20', 35°40'. No other maxima appear at any angle between 0° and 35°40'. The separation between adjacent slit centers in the grating is 5.04×10^{-4} cm. (a) Compute the wavelength of the light used. (b) Make the most complete *quantitative* statement that can be inferred from the above data concerning the width of each slit.
    Answer: (a) 586 nm. (b) The slit width lies between 1.01 and 1.68 μm.

22. A grating designed for use in the infrared region of the electromagnetic spectrum is "blazed" to concentrate all its intensity in the first order $(m = 1)$ for $\lambda = 8,000$ nm. If visible light (400 nm $< \lambda <$ 700 nm) were allowed to fall on this grating, what visual appearance would the diffracted beams present?

23. Show that the dispersion of a grating can be written as

$$D = \frac{\tan \theta}{\lambda}.$$

24. A grating has 3500 rulings/cm and is illuminated at normal incidence by white light. A spectrum is formed on a screen 30 cm from the grating. If a 1.0-cm square hole is cut in the screen, its inner edge being 5.0 cm from the central maximum, what range of wavelengths passes through the hole?

25. Two spectral lines have wavelengths λ and $\lambda + \Delta\lambda$, respectively, where $\Delta\lambda \ll \lambda$. Show that their angular separation $\Delta\theta$ in a grating spectrometer is given approximately by $\Delta\theta = \Delta\lambda/\sqrt{(d/m)^2 - \lambda^2}$, where d is the separation of adjacent slit centers and m is the order at which lines are observed. Notice that the angular separation is greater in the higher orders.

26. An optical grating with a spacing $d = 15,000$ Å is used to analyze soft X-rays of wavelength $\lambda = 5.0$ Å. The angle of incidence θ is $90° - \gamma$, where γ is a *small* angle. The first-order maximum is found at an angle $\theta = 90° - 2\beta$. Find the value of β.

SECTION 47-4

27. A grating has 6000 rulings/cm and is 6.0 cm wide. (a) What is the smallest wavelength interval that can be resolved in the third order at $\lambda = 500$ nm? (b) How many higher orders can be seen? Assume normal incidence of light on the grating throughout.   *Answer:* (a) 4.6×10^{-3} nm. (b) None.

28. A grating has 40,000 rulings spread over 3.0 in. (a) What is its expected dispersion D for sodium light ($\lambda = 589$ nm $= 5890$ Å) in the first three orders? (b) What is its resolving power in these orders?

29. A source containing a mixture of hydrogen and deuterium atoms emits a red doublet at $\lambda = 656.3$ nm ($= 6563$ Å) whose separation is 0.18 nm ($= 1.8$ Å). Find the minimum number of lines needed in a diffraction grating which can resolve these lines in the first order.   *Answer:* 3650.

30. In a particular grating the sodium doublet (see Example 2) is viewed in third order at 10° to the normal and is barely resolved. Find (a) the grating spacing and (b) the total width of the rulings.

31. A diffraction grating has a resolving power $R = \lambda/\Delta\lambda = Nm$. (a) Show that the corresponding frequency range $\Delta\nu$ that can just be resolved is given by $\Delta\nu = c/Nm\lambda$. (b) From Fig. 47-1, show that the "times of flight" of the two extreme rays differ by an amount $\Delta t = Nd \sin\theta/c$. (c) Show that $(\Delta\nu)(\Delta t) = 1$, this relation being independent of the various grating parameters. Assume $N \gg 1$.

32. A diffraction grating is made up of slits of width 300 nm with a 900 nm separation between centers. The grating is illuminated by monochromatic plane waves, $\lambda = 600$ nm, the angle of incidence being zero. (a) How many diffracted lines are there? (b) What is the angular width of the spectral lines observed, if the grating has 1000 slits? (c) How is the angular width of the spectral lines related to the resolving power of the grating?

SECTION 47-5

33. Consider an infinite two-dimensional square lattice as in Fig. 47-14b. One interplanar spacing is obviously a_o itself. (a) Calculate the next five smaller interplanar spacings by sketching figures similar to Figure 47-16a. (b) Show that your answers obey the general formula

$$d = a_o/\sqrt{h^2 + k^2}$$

where h, k are both relatively prime integers (no common factor other than unity).

SECTION 47-6

34. Monochromatic high energy x-rays are incident on a crystal. If first order reflection is observed at Bragg angle 3.4°, at what angle would second order reflection be expected?

35. Monochromatic X-rays are incident on a NaCl crystal whose lattice spacing is 0.3 nm. When the beam is rotated 60° from the normal, first-order Bragg reflection is observed. What is the wavelength of the x-rays? *Answer:* 0.30 nm.

36. In comparing the wavelengths of two monochromatic X-ray lines, it is noted that line A gives a first-order reflection maximum at a glancing angle of 30° to the smooth face of a crystal. Line B, known to have a wavelength of 0.097 nm, gives a third-order reflection maximum at an angle of 60° from the same face of the same crystal. Find the wavelength of line A.

37. Monochromatic X-rays ($\lambda = 1.25$ Å) fall on a crystal of sodium chloride, making an angle of 45° with the reference line shown in Fig. 47-19. The planes shown are those of Fig. 47-16a, for which $d = 2.52$ Å. Through what angles must the crystal be turned to give a diffracted beam associated with the planes shown? Assume that the crystal is turned about an axis that is perpendicular to the plane of the page. Ignore the possibility (see Problem 40) that some of these beams may be of zero intensity.
Answer: 30.6°, 15.3°, (clockwise); 3.1°, 37.8°, (counterclockwise).

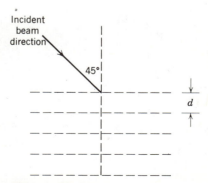

figure 47-19
Problems 37, 38

38. Assume that the incident X-ray beam in Fig. 47-19 is not monochromatic but contains wavelengths in a band from 0.095 to 0.130 nm. Will diffracted beams, associated with the planes shown, occur? Assume $d_0 = 0.275$ nm.

39. Prove that it is impossible to determine both wavelength of radiation and spacing of Bragg reflecting planes in a crystal by measuring the angles for Bragg reflection in several orders.

40. *Missing orders in X-ray diffraction.* In Example 5 the $m = 1$ beam, permitted by interference considerations, has zero intensity because of the diffracting properties of the unit cell for this geometry of beams and crystal. Prove this. (*Hint:* Show that the "reflection" from an atomic plane through the top of a layer of unit cells is canceled by a "reflection" from a plane through the middle of this layer of cells. All odd-order beams prove to have zero intensity.)

48
polarization

Light, like all electromagnetic radiation, is predicted by electromagnetic theory to be a *transverse wave,* the directions of the vibrating electric and magnetic vectors being at right angles to the direction of propagation instead of parallel to it, as in a longitudinal wave. The transverse waves of Figs. 48-1 and 41-13 have the additional characteristic that they are *plane-polarized.* This means that the vibrations of the **E** vector are parallel to each other for all points in the wave. At any such point the vibrating **E** vector and the direction of propagation form a plane, called the *plane of vibration;* in a plane-polarized wave all such planes are parallel.

The transverse nature of light waves cannot be deduced from the interference or diffraction experiments so far described because longitudinal waves such as sound waves also show these effects. Thomas Young in 1817 provided an experimental basis for believing that light waves are transverse. Two of his contemporaries, Dominique-François Arago (1786–1853) and Augustin Jean Fresnel (1788–1827), were able, by allowing a light beam to fall on a crystal of calcite, to produce two separate

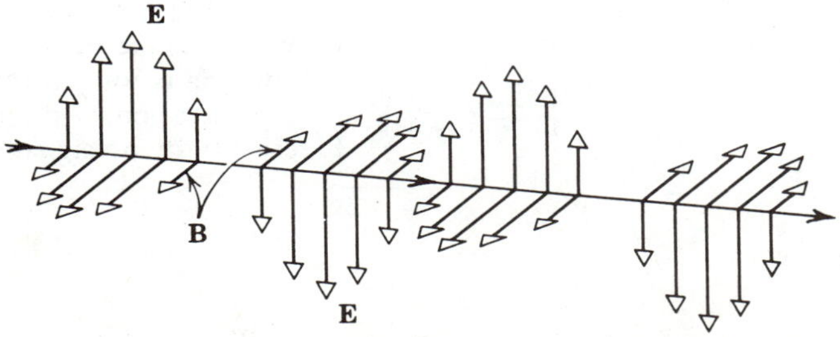

48-1
POLARIZATION

figure 48-1
An instantaneous "snapshot" of a plane-polarized wave showing the vectors **E** and **B** along a particular ray. The wave is moving to the right with speed *c*. The plane containing the vibrating **E** vector and the direction of propagation is a *plane of vibration.*

beams (see Section 48-4). Astonishingly, these beams, although co-herent, produced no interference fringes but only a uniform illumination. Young deduced from this that light must be a transverse wave and that the planes of vibration in the two beams must be at right angles to each other. Wave disturbances that act at right angles to each other cannot show interference effects; you are asked to prove this in Problem 1. Young's words to Arago were these:

I have been reflecting on the possibility of giving an imperfect explanation of the affection of light which constitutes polarization without departing from the genuine doctrine of undulations. It is a principle in this theory that all undulations are simply propagated through homogeneous mediums in concentric spherical surfaces like the undulations of sound, consisting simply in the direct and retrograde motions of the particles in the direction of the radius with their concomitant condensation and rarefactions [that is, longitudinal waves]. And yet, it is possible to explain in this theory a transverse vibration, propagated also in the direction of the radius, and with equal velocity, the motions of the particles being in a certain constant direction with respect to that radius; and this is a *polarization*.

Note how Young presents the possibility of a transverse vibration as a novel idea, light having been generally—but incorrectly—assumed to be a longitudinal vibration.

In a plane-polarized transverse wave it is necessary to specify two directions, that of the wave disturbance (**E**, say) and that of propagation. In a longitudinal wave these directions are identical. In plane-polarized transverse waves, but not in longitudinal waves, we may thus expect a lack of symmetry about the direction of propagation. Electromagnetic waves in the radio and microwave range exhibit this lack of symmetry readily. Such a wave, generated by the surging of charge up and down in the dipole that forms the transmitting antenna of Fig. 48-2, has (at large distances from the dipole and at right angles to it) an electric field vector parallel to the dipole axis. When this plane-polarized wave falls on a second dipole connected to a microwave detector, the alternating electric component of the wave will cause electrons to surge back and forth in the receiving antenna, producing a reading on the detector. If we turn the receiving antenna through 90° about the direction of propagation, the detector reading drops to zero. In this orientation the electric field vector is not able to cause charge to move along the dipole axis because it points at right angles to this axis. We can reproduce the experiment of Fig. 48-2 by turning the receiving antenna of a television set (assumed an electric dipole type) through 90° about an axis that points toward the transmitting station.

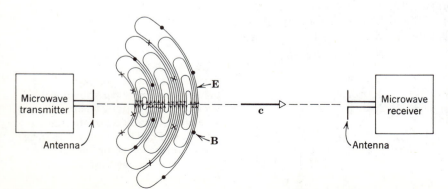

figure 48-2
The vectors **E** in the transmitted wave are parallel to the axis of the receiving antenna so that the wave will be detected. If the receiving antenna is rotated through 90° about the direction of propagation, essentially no signal will be detected.

Common sources of visible light differ from radio and microwave sources in that the elementary radiators, that is, the atoms and molecules, act independently. The light propagated in a given direction consists of independent wavetrains whose planes of vibration are randomly oriented about the direction of propagation, as in Fig. 48-3b. Such light, though still transverse, is *unpolarized*. The random orientation of the planes of vibration produces symmetry about the propagation direction, which, on casual study, conceals the true transverse nature of the waves. To study this transverse nature, we must find a way to unsort the different planes of vibration.

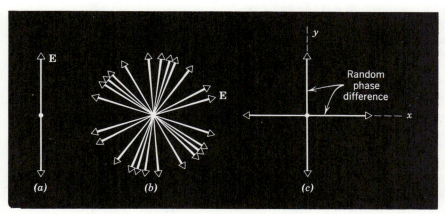

figure 48-3
(a) A plane-polarized transverse wave moving toward the reader, showing only the electric vector. (b) An unpolarized transverse wave viewed as a random superposition of many plane-polarized wavetrains. (c) A second, completely equivalent, description of an unpolarized transverse wave; here the unpolarized wave is viewed as two plane-polarized waves with a random phase difference. The orientation of the x and y axes about the propagation direction is completely arbitrary.

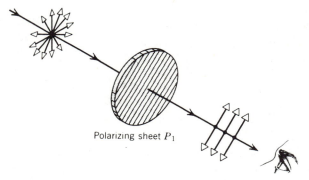

Polarizing sheet P_1

figure 48-4
A polarizing sheet produces plane-polarized light from unpolarized light. The parallel lines, which are not actually visible on the sheet, suggest the characteristic polarizing direction of the sheet.

48-2
POLARIZING SHEETS

Figure 48-4 shows unpolarized light falling on a sheet of commercial polarizing material called *Polaroid.** There exists in the sheet a certain characteristic polarizing direction, shown by the parallel lines. The sheet will transmit only those wave-train components whose electric vectors vibrate parallel to this direction and will absorb those that vibrate at right angles to this direction. The emerging light will be plane-polarized. This polarizing direction is established during the manufacturing process by embedding certain long-chain molecules in a flexible plastic sheet and then stretching the sheet so that the molecules are aligned parallel to each other. Polarizing sheets more than 2 ft wide and 100 ft long may be produced.

In Fig. 48-5 the polarizing sheet or *polarizer* lies in the plane of the

---

* There are other ways of producing polarized light without using this well-known commerical product. We mention some of them below. Also see "The Amateur Scientist" by Jearl Walker, *Scientific American*, December 1977 for ways of making polarizing sheets, quarter-wave and half-wave plates and doing various experiments with them.

page and the direction of propagation is into the page. The arrow **E** shows the plane of vibration of a randomly selected wavetrain falling on the sheet. Two vector components, $\mathbf{E}_x$ (of magnitude $E \sin \theta$) and $\mathbf{E}_y$ (of magnitude $E \cos \theta$), can replace **E,** one parallel to the polarizing direction and one at right angles to it. Only the former will be transmitted; the latter is absorbed within the sheet.

Let us place a second polarizing sheet P_2 (usually called, when so used, an *analyzer*) as in Fig. 48-6. If P_2 is rotated about the direction of propagation, there are two positions, 180° apart, at which the transmitted light intensity is almost zero; these are the positions in which the polarizing directions of P_1 and P_2 are at right angles.

If the amplitude of the plane-polarized light falling on P_2 is E_m, the amplitude of the light that emerges is $E_m \cos \theta$, where θ is the angle between the polarizing directions of P_1 and P_2. Recalling that the intensity of the light beam is proportional to the square of the amplitude, we see that the transmitted intensity I varies with θ according to

$$I = I_m \cos^2 \theta, \tag{48-1}$$

in which I_m is the maximum value of the transmitted intensity. It occurs when the polarizing directions of P_1 and P_2 are parallel, that is, when $\theta = 0$ or 180°. Figure 48-7a, in which two overlapping polarizing sheets are in the parallel position ($\theta = 0$ or 180° in Eq. 48-1) shows that the light transmitted through the region of overlap has its maximum value. In Fig. 48-7b one or the other of the sheets has been rotated through 90° so that θ in Eq. 48-1 has the value 90 or 270°; the light transmitted through the region of overlap is now a minimum.

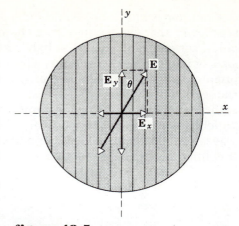

figure 48-5
A wavetrain **E** is equivalent to two component wavetrains $\mathbf{E}_y$ and $\mathbf{E}_x$. Only $\mathbf{E}_y$ is transmitted by the polarizer.

figure 48-6
Unpolarized light is not transmitted by crossed polarizing sheets.

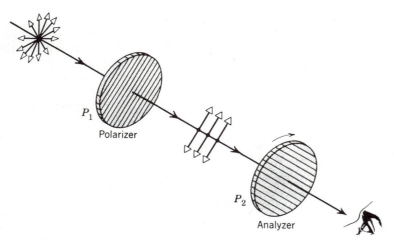

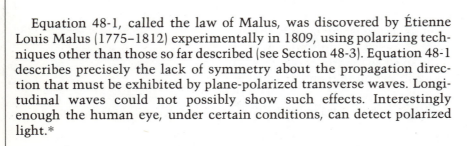

Equation 48-1, called the law of Malus, was discovered by Étienne Louis Malus (1775–1812) experimentally in 1809, using polarizing techniques other than those so far described (see Section 48-3). Equation 48-1 describes precisely the lack of symmetry about the propagation direction that must be exhibited by plane-polarized transverse waves. Longitudinal waves could not possibly show such effects. Interestingly enough the human eye, under certain conditions, can detect polarized light.*

* The so-called *Haidinger's brushes;* the interested student is referred to *Concepts of Classical Optics* by John Strong, W. H. Freeman & Co., 1958.

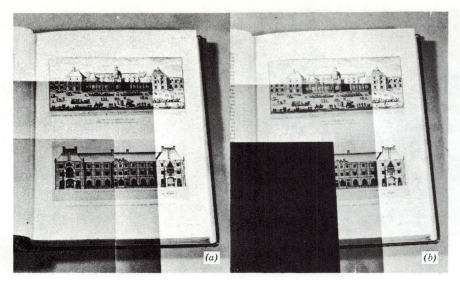

figure 48-7
Two square sheets of Polaroid are laid over a book. In (a) the axes of polarization of the two sheets are parallel and light passes through both sheets. In (b) one sheet has been rotated 90° and no light passes through. The book is opened to an illustration of the Luxembourg Palace in Paris. Malus discovered the phenomenon of polarization by reflection while looking at sunlight reflected off the palace windows through a calcite crystal.

EXAMPLE 1

Two polarizing sheets have their polarizing directions parallel so that the intensity I_m of the transmitted light is a maximum. Through what angle must either sheet be turned if the intensity is to drop by one-half?

From Eq. 48-1, since $I = \frac{1}{2}I_m$, we have

$$\tfrac{1}{2}I_m = I_m \cos^2 \theta$$

or

$$\theta = \cos^{-1} \pm \frac{1}{\sqrt{2}} = \pm 45°, \pm 135°.$$

The same effect is obtained no matter which sheet is rotated or in which direction.

Historically polarization studies were made to investigate the nature of light. Today we reverse the procedure and deduce something about the nature of an object from the polarization state of the light emitted by, or scattered from, that object. It has been possible to deduce, from studies of the polarization of light reflected from them, that the grains of cosmic dust present in our galaxy have been oriented in the weak galactic magnetic field ($\sim 5 \times 10^{-10}$ T) so that their long dimension is parallel to this field. Polarization studies have shown that Saturn's rings consist of ice crystals. The size and shape of virus particles can be determined by the polarization of ultraviolet light scattered from them. Much useful information about the structure of atoms and nuclei is gained from polarization studies of their emitted radiations in all parts of the electromagnetic spectrum. Thus we have a useful research technique for structures ranging in size from a galaxy ($\sim 10^{+22}$ m) to a nucleus ($\sim 10^{-15}$ m). Polarized light also has many practical applications in industry and in engineering science.

48-3
POLARIZATION BY REFLECTION

Malus discovered in 1809 that light can be partially or completely polarized by reflection. Anyone who has watched the sun's reflection in water, while wearing a pair of sunglasses made of polarizing sheet, has probably noticed the effect. It is necessary only to tilt the head from side to side, thus rotating the polarizing sheets, to observe that the intensity of the reflected sunlight passes through a minimum.

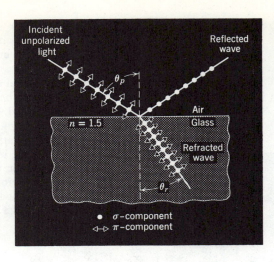

figure 48-8
For a particular angle of incidence θ_p, the reflected light is completely polarized, as shown. The transmitted light is partially polarized.

Figure 48-8 shows an unpolarized beam falling on a glass surface. The **E** vector for each wavetrain in the beam can be resolved into two components, one perpendicular to the plane of incidence—which is the plane of Fig. 48-8—and one lying in this plane. The first component, represented by the dots, is called the *σ-component*, from the German *senkrecht*, meaning perpendicular. The second component, represented by the arrows, is called the *π-component* (for *parallel*). On the average, for completely unpolarized incident light, these two components are of equal amplitude.

Experimentally, for glass or other dielectric materials, there is a particular angle of incidence, called the *polarizing angle* θ_p, at which the reflection coefficient for the π-component is zero. This means that the beam reflected from the glass, although of low intensity, is plane-polarized, with its plane of vibration at right angles to the plane of incidence. This polarization of the reflected beam can easily be verified by analyzing it with a polarizing sheet.

The π-component at the polarizing angle is entirely refracted; the

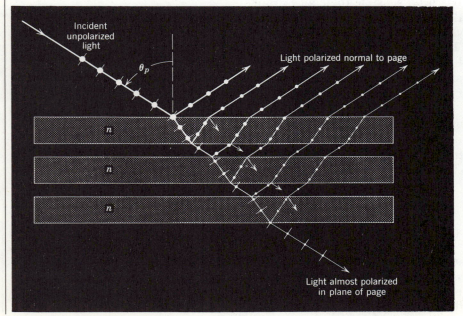

figure 48-9
Polarization of light by a stack of glass plates. Unpolarized light is incident on a stack of glass plates at Brewster's angle θ_p. (Polarization in the plane of the page is shown by the short lines and polarization normal to the page by the dots.) All light reflected out of the original ray is polarized normal to the page. After passing through several reflecting interfaces, the transmitted light no longer contains any appreciable component polarized normal to the page.

σ-component is only partially refracted. Thus the transmitted beam, which is of high intensity, is only partially polarized. By using a stack of glass plates rather than a single plate, we obtain reflections from successive surfaces and we can increase the intensity of the emerging reflected (σ-component) beam (see Fig. 48-9). By the same token, the σ-components are progressively removed from the transmitted beam, making it more completely π-polarized.

At the polarizing angle it is found experimentally that the reflected and the refracted beams are at right angles, or (Fig. 48-8)

$$\theta_p + \theta_r = 90°.$$

From Snell's law, $\qquad n_1 \sin \theta_p = n_2 \sin \theta_r.$

Combining these equations leads to

$$n_1 \sin \theta_p = n_2 \sin (90° - \theta_p) = n_2 \cos \theta_p$$

or $\qquad\qquad \tan \theta_p = \dfrac{n_2}{n_1}, \qquad\qquad\qquad (48\text{-}2)$

where the incident ray is in medium 1 and the refracted ray in medium 2. This can be written as

$$\tan \theta_p = n, \qquad\qquad\qquad (48\text{-}3)$$

where $n\ (= n_2/n_1)$ is the index of refraction of medium 2 with respect to medium 1. Equation 48-3 is known as *Brewster's law* after Sir David Brewster (1781–1868), who deduced it empirically in 1812. It is possible to prove this law rigorously from Maxwell's equations (see also Question 6).

EXAMPLE 2

We wish to use a plate of glass ($n = 1.50$) as a polarizer. What is the polarizing angle? What is the angle of refraction?

From Eq. 48-3,

$$\theta_p = \tan^{-1} 1.50 = 56.3°.$$

The angle of refraction follows from Snell's law:

$$(1)\ \sin \theta_p = n \sin \theta_r$$

or $\qquad\qquad \sin \theta_r = \dfrac{\sin 56.3°}{1.50} = 0.555 \qquad \theta_r = 33.7°.$

48-4
DOUBLE REFRACTION

In earlier chapters we assumed that the speed of light, and thus the index of refraction, is independent of the direction of propagation in the medium and of the state of polarization of the light. Liquids, amorphous solids such as glass, and crystalline solids having cubic symmetry normally show this behavior and are said to be *optically isotropic.* Many other crystalline solids are optically *anisotropic* (that is, not isotropic).*

Solids may be anisotropic in many properties. Mica cleaves readily in

---

* Many transparent amorphous solids such as glasses and plastics become optically anisotropic when they are mechanically stressed. This fact is useful in engineering design studies in that strains in gears, bridge structures, etc., can be studied quantitatively by building plastic models, stressing them appropriately, and examining the optical anisotropy that results, using polarization techniques. The interested reader should consult "Photoelasticity," a chapter by H. T. Jessop in Vol. 6 of the *Encyclopedia of Physics,* edited by H. Flugge (1958), Springer Verlag, Berlin.

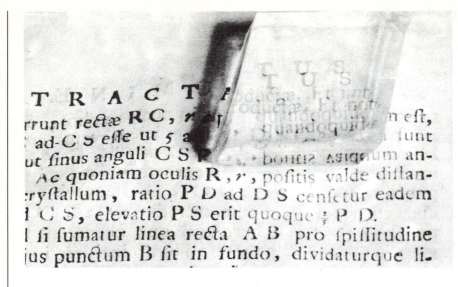

figure 48-10
Two images, one polarized 90°
relative to the other, are formed by a
calcite crystal. The book on which
the crystal is lying is Huygens' *Opera
Reliqua,* wherein the phenomenon
of double refraction is discussed.

one plane only; a cube of crystalline graphite does not have the same electric resistance between all pairs of opposite faces; a cube of crystalline nickel magnetizes more readily in certain directions than in others, etc. If a solid is a mixture of a large number of tiny crystallites, it may appear to be isotropic because of the random orientations of the crystallites. Powdered mica, for example, compacted to a solid mass with a binder, does not exhibit the cleavage properties that characterize the crystallites making it up.

Figure 48-10, in which a polished crystal of calcite ($CaCO_3$) is laid over some printed letters, shows the optical anisotropy of this material; *the image appears double.* Figure 48-11 shows a beam of unpolarized light falling on a calcite crystal at right angles to one of its faces. The

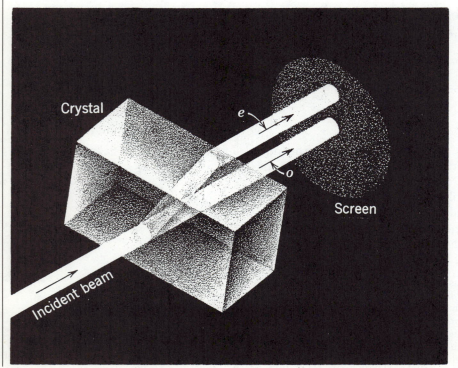

Crystal

e

o

Screen

Incident beam

figure 48-11
A beam of unpolarized light falling
on a calcite crystal is split into two
beams which are polarized at right
angles to each other.

single beam splits into two at the crystal surface. The "double-bending" of a beam transmitted through calcite, exhibited in Figs. 48-10 and 48-11, is called *double refraction*.

If the two emerging beams in Fig. 48-11 are analyzed with a polarizing sheet, they are found to be plane-polarized with their planes of vibration at right angles to each other, a fact discovered by Huygens in 1678. Huygens used a second calcite crystal to investigate the polarization states of the beams labeled *o* and *e* in the figure.

If experiments are carried out at various angles of incidence, one of the beams in Fig. 48-11 (represented by the *ordinary ray*, or *o*-ray) will be found to obey Snell's law of refraction at the crystal surface, just like a ray passing from one isotropic medium into another. The second beam (represented by the *extraordinary ray*, or *e*-ray) will not. In Fig. 48-11, for example, the angle of incidence for the incident light is zero but the angle of refraction of the *e*-ray, contrary to the prediction of Snell's law, is not. In general, the *e*-ray does not even lie in the plane of incidence.

This difference between the waves represented by the *o*- and *e*-rays with respect to Snell's law can be explained in these terms:

1. The *o*-wave travels in the crystal with the same speed v_o in all directions. In other words, the crystal has, for this wave, a single index of refraction n_o, just like an isotropic solid.

2. The *e*-wave travels in the crystal with a speed that varies with direction from v_o to a larger value (for calcite) v_e. In other words, the index of refraction, defined as c/v, varies with direction from n_o to a smaller value (for calcite) n_e.

The quantities n_o and n_e are called the *principal indices of refraction* for the crystal. Problem 12 suggests how to measure them. Table 48-1 shows these indices for six doubly refracting crystals. For three of them the *e*-wave is slower; for the other three it is faster than the *o*-wave. Some doubly refracting crystals (mica, topaz, etc.) are more complex optically than calcite and require *three* principal indices of refraction for a complete description of their optical properties. Crystals whose basic crystal structure is cubic (see Fig. 47-13) are optically isotropic, requiring only *one* index of refraction.

Table 48-1
Principal indices of refraction of several doubly refracting crystals (for sodium light, $\lambda = 589$ nm $= 5890$ Å)

Crystal	Formula	n_o	n_e	$n_e - n_o$
Ice	H_2O	1.309	1.313	+0.004
Quartz	SiO_2	1.541	1.553	+0.012
Wurzite	ZnS	2.356	2.378	+0.022
Calcite	$CaCO_3$	1.658	1.486	−0.172
Dolomite	$CaO \cdot MgO \cdot 2CO_2$	1.681	1.500	−0.181
Siderite	$FeO \cdot CO_2$	1.875	1.635	−0.240

The behavior for the speeds of the two waves traveling in calcite is summarized by Fig. 48-12, which shows two wave surfaces spreading out from an imaginary point light source *S* imbedded in the crystal. The *o*-wave surface is a sphere, as we would expect if the medium were isotropic. The *e*-wave surface is an ellipsoid of revolution about a characteristic direction in the crystal called the *optic axis*. The two wave

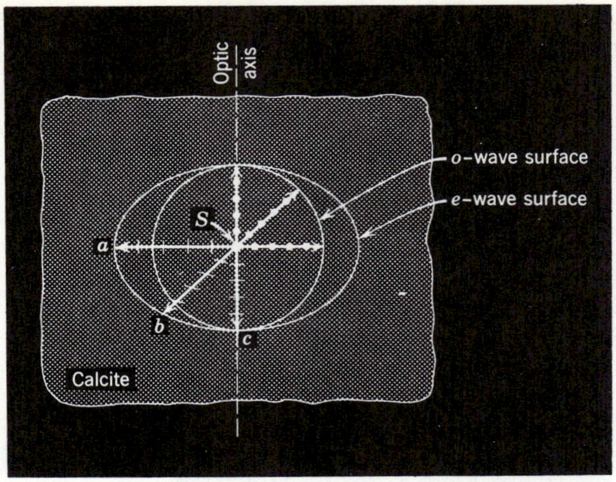

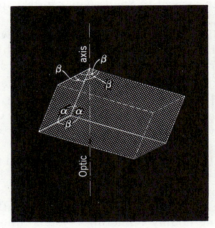

figure 48-12
Huygens' wave surfaces generated by a point source S imbedded in calcite. The polarization states for three o-rays and three e-rays are shown by the dots and lines, respectively. Note that in general (ray Sb) the bars representing the polarization direction are not perpendicular to the e-rays.

surfaces represent light having two different polarization states. If we consider for the present only rays lying in the plane of Fig. 48-12, then (a) the plane of polarization for the o-rays is perpendicular to the figure, as suggested by the dots, and (b) that for the e-rays coincides with the plane of the figure, as suggested by the short lines. We describe the polarization states more fully at the end of this section.

Figure 48-13, which shows a typical calcite crystal that may be obtained by cleavage from a naturally occurring crystal, shows how to locate the optic axis. The edges of calcite crystals may have any lengths but the angles at which the edges intersect always have one or another of two values, 78° 05′ or 101° 55′. The optic axis is found by erecting a line at either of the two corners where three obtuse angles meet (the "blunt" corners), making equal angles with the crystal edges. *Any line in the crystal parallel to this line is also an optic axis.*

We can use Huygens' principle to study the propagation of light waves in doubly refracting crystals. Figure 48-14a shows the special case in which unpolarized light falls at normal incidence on a calcite slab cut from a crystal in such a way that the optic axis is normal to the surface. Consider a wavefront that, at time $t = 0$, coincides with the crystal surface. Following Huygens, we may let any point on this surface serve as a radiating center for a double set of Huygens' wavelets, such as those in Fig. 48-12. The plane of tangency to these wavelets represents the new position of this wavefront at a later time t. The incident beam in Fig. 48-14a is propagated through the crystal without deviation at speed v_o. The beam emerging from the slab will have the same polarization character as the incident beam. The calcite slab, in these special circumstances only, behaves like an isotropic material, and no distinction can be made between the o- and the e-waves.

Figure 48-14b shows two views of another special case, namely, unpolarized incident light falling at right angles on a slab cut so that the optic axis is parallel to its surface. In this case too the incident beam is propagated without deviation. However, we can now identify o- and e-waves that travel through the crystal with different speeds, v_o and v_e, respectively. These waves are polarized at right angles to each other.

Some doubly refracting crystals have the interesting property, called *dichroism*, in which one of the polarization components is strongly absorbed within the crystal, the other being transmitted with little loss. Dichroism, illustrated in Fig. 48-15, is the basic operating principle of the commercial Polaroid sheet. The

figure 48-13
A calcite crystal; α is 78° 05′; β is 101° 55′.

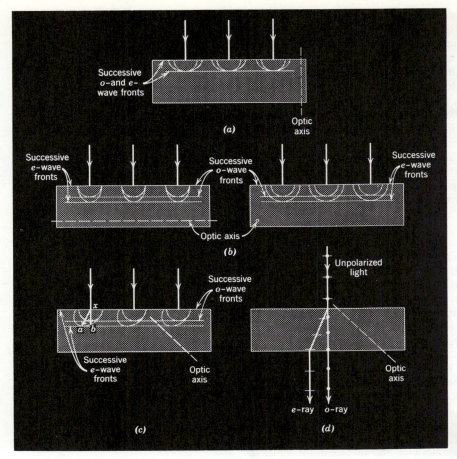

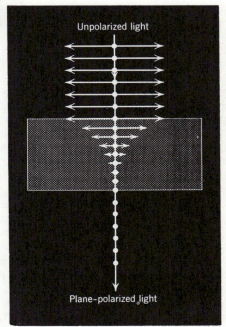

figure 48-14
Unpolarized light falls at normal incidence on slabs cut from a calcite crystal. The Huygens' wavelets are appropriate sections of the figure of revolution about the optic axis represented by Fig. 48-13. (a) No double refraction or speed difference occurs. (b) No double refraction occurs but there is a speed difference. (c) Both double refraction and a speed difference occur. (d) Same as (c) but showing the polarization states and the emerging rays.

figure 48-15
Showing the absorption of one polarization component inside a dichroic crystal of the type used in Polaroid sheets.

many small crystallites, imbedded in a plastic sheet with their optic axes parallel, have a polarizing action equivalent to that of a single large crystal slab.

Figure 48-14c shows unpolarized light falling at normal incidence on a calcite slab cut so that its optic axis makes an arbitrary angle with the crystal surface. Two spatially separated beams are produced, as in Fig. 48-11. They travel through the crystal at different speeds, that for the o-wave being v_o and that for the e-wave being intermediate between v_o and v_e. Note that ray xa represents the shortest *optical* path for the transfer of light energy from point x to the e-wavefront. Energy transferred along any other ray, in particular along ray xb, would have a longer transit time, a consequence of the fact that the speed of e-waves varies with direction.* Figure 48-14d represents the same case as Fig. 48-14c. It shows the rays emerging from the slab, as in Fig. 48-11, and makes clear that the emerging beams are polarized at right angles to each other, that is, they are *cross-polarized*.

We now seek to understand, in terms of the atomic structure of optically anisotropic crystals, how cross-polarized light waves with different speeds can exist. Light is propagated through a crystal by the action of the vibrating **E** vectors of the wave on the electrons in the crystal. These electrons, which experience electrostatic restoring forces if they are moved from their equilibrium positions, are set into forced periodic oscillation about these positions and pass along the transverse wave disturbance that constitutes the light wave. The strength of the

* The reader who has not previously read Section 43-6 on Fermat's principle may wish to do so now.

restoring forces may be measured by a force constant k, as for the simple harmonic oscillator discussed in Chapter 15 (see Eq. 15-4).

In optically isotropic materials the force constant k is the same for all directions of displacement of the electrons from their equilibrium positions. In doubly refracting crystals, however, k varies with direction. For electron displacements that lie in a plane at right angles to the optic axis k has the constant value k_o, no matter how the displacement is oriented in this plane. For displacements parallel to the optic axis, k has the larger value (for calcite) k_e.* Note carefully that the speed of a wave in a crystal is determined by the direction in which the E vectors vibrate and *not* by the direction of propagation. It is the transverse E-vector vibrations that call the restoring forces into play and thus determine the wave speed. Note too that the stronger the restoring force, that is, the larger k, the faster the wave. For waves traveling along a stretched cord, for example, the restoring force for the transverse displacements is determined by the tension F in the cord. Equation 19-12 shows that an increase in F means an increase in the wave speed v.

Figure 48-16, a long weighted "tire chain" supported at its upper end, provides a one-dimensional mechanical analogy for double refraction. It applies specifically to o- and e-waves traveling at right angles to the optic axis, as in Fig. 48-14b. If the supporting block is oscillated, as in Fig. 48-16a, a transverse wave travels along the chain with a certain speed. If the block is oscillated lengthwise, as in Fig. 48-16b, another transverse wave is also propagated. The restoring force for the second wave is greater than for the first, the chain being more rigid in the plane of Fig. 48-16b than in the plane of Fig. 48-16a. Thus the second wave travels along the chain with a greater speed.

In the language of optics we would say that the speed of a transverse wave in the chain depends on the orientation of the plane of vibration of the wave. If we oscillate the top of the chain in a random way, the wave disturbance at a point along the chain can be described as the sum of two waves, polarized at right angles and traveling with different speeds. This corresponds exactly to the optical situation of Fig. 48-14b.

For waves traveling parallel to the optic axis, as in Fig. 48-14a, or for waves in optically isotropic materials, the appropriate mechanical analogy is a single weighted hanging chain. Here there is only one speed of propagation, no matter how the upper end is oscillated. The restoring forces are the same for all orientations of the plane of polarization of waves traveling along such a chain.

These considerations allow us to understand more clearly the polarization states of the light represented by the double-wave surface of Fig. 48-12. For the (spherical) o-wave surface, the E-vector vibrations must be everywhere at right angles to the optic axis. If this is so, the same force constant k_o will always hold, and the o-waves will travel with the same speed in all directions. More specifically, if we draw a ray in Fig. 48-12 from S to the o-wave surface, considered three-dimensionally (that is, as a sphere), the E-vector vibrations will always be at right angles to the plane defined by this ray and the optic axis. Thus these vibrations will always be at right angles to the optic axis.

For the (ellipsoidal) e-wave surface, the E-vector vibrations in general have a component parallel to the optic axis. For rays such as Sa in Fig. 48-12 or for the e-rays of Fig. 48-14b, the vibrations are completely parallel to this axis. Thus a relatively strong force constant (in calcite) k_e is operative, and the wave speed v_e will be relatively high. For e-rays, such as Sb in Fig. 48-12, the parallel component of the E-vector vibrations is less than 100%, so that the corresponding wave speed will be less than v_e. For ray Sc, in Fig. 48-12, the parallel component is zero, and the distinction between o- and e-rays disappears.

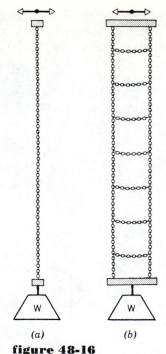

(a)   *(b)*

figure 48-16
Two views of a one-dimensional mechanical model for double refraction.

---

* For doubly refracting crystals with $n_e > n_o$ (see Table 48-1) k for displacements parallel to the optic axis is *smaller* than for those at right angles to it. Also, for crystals with three principal indices of refraction, there will be three principal force constants. Such crystals have two optic axes and are called *biaxial*. The crystals listed in Table 48-1 have only a single optic axis and are called *uniaxial*.

Let plane-polarized light of angular frequency ω (=$2\pi\nu$) fall at normal incidence on a slab of calcite cut so that the optic axis is parallel to the face of the slab, as in Fig. 48-17. The two waves that emerge will be plane-polarized at right angles to each other, and, if the incident plane of vibration is at 45° to the optic axis, they will have equal amplitudes. Since the waves travel through the crystal at different speeds, there will be a phase difference ϕ between them when they emerge from the crystal. If the crystal thickness is chosen so that (for a given frequency of light) $\phi = 90°$, the slab is called a *quarter-wave plate*. The emerging light is said to be *circularly polarized*.

In Section 15-7 we saw that the two emerging plane-polarized waves just described (vibrating at right angles with a 90° phase difference) can be represented as the projections on two perpendicular axes of a vector rotating with angular frequency ω about the propagation direction. These two descriptions of circularly polarized light are completely equivalent. Figure 48-18 clarifies the relationship between these two descriptions.

48-5
CIRCULAR POLARIZATION

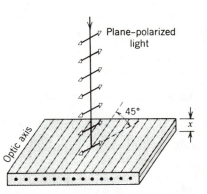

figure 48-17
Plane-polarized light falls on a doubly refracting slab of thickness x cut with its optic axis parallel to the surface. The plane of vibration of the incident light is oriented to make an angle of 45° with the optic axis.

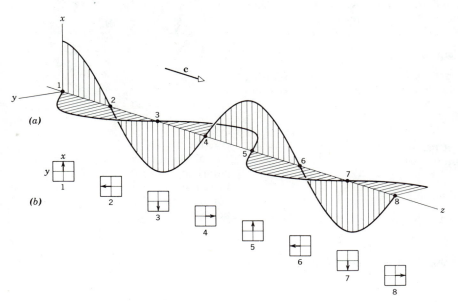

figure 48-18
(*a*) Two plane-polarized waves of equal amplitude and at right angles to each other are moving in the z direction. They differ in phase by 90°; where one wave has maximum values, the other is zero. (*b*) Views of the resultant amplitude of the approaching wave as seen by observers located at the positions shown on the z axis. Note that the resultant vector rotates clockwise with time.

A quartz quarter-wave plate is to be used with sodium light ($\lambda = 589$ nm = 5890 Å). What must its thickness be?

Two waves travel through the slab at speeds corresponding to the two principal indices of refraction given in Table 48-1 ($n_e = 1.553$ and $n_o = 1.541$). If the

EXAMPLE 3

crystal thickness is x, the number of wavelengths of the first wave contained in the crystal is

$$N_e = \frac{x}{\lambda_e} = \frac{x n_e}{\lambda},$$

where λ_e is the wavelength of the e-wave in the crystal and λ is the wavelength in air. For the second wave the number of wavelengths is

$$N_o = \frac{x}{\lambda_o} = \frac{x n_o}{\lambda},$$

where λ_o is the wavelength of the o-wave in the crystal. The difference $N_e - N_o$ must be one-fourth, or

$$\frac{1}{4} = \frac{x}{\lambda}(n_e - n_o).$$

This equation yields

$$x = \frac{\lambda}{4(n_e - n_o)} = \frac{589 \text{ nm}}{(4)(1.553 - 1.541)} = 0.012 \text{ mm}.$$

This plate is rather thin; most quarter-wave plates are made from mica, splitting the sheet to the correct thickness by trial and error.

EXAMPLE 4

A beam of circularly polarized light falls on a polarizing sheet. Describe the emerging beam.

The circularly polarized light, as it enters the sheet, can be represented by

$$E_x = E_m \sin \omega t$$

and

$$E_y = E_m \cos \omega t,$$

where x and y represent arbitrary perpendicular axes. These equations correctly represent the fact that a circularly polarized wave is equivalent to two plane-polarized waves with equal amplitude and a 90° phase difference.

The resultant amplitude in the incident circularly polarized wave is

$$E_{\text{cp}} = \sqrt{E_x^2 + E_y^2} = \sqrt{E_m^2(\sin^2 \omega t + \cos^2 \omega t)} = E_m,$$

an expected result if the circularly polarized wave is represented as a rotating vector. The resultant intensity in the incident circularly polarized wave is proportional to E_m^2, or

$$I_{\text{cp}} \propto E_m^2. \tag{48-4}$$

Let the polarizing direction of the sheet make an arbitrary angle θ with the x axis as shown in Fig. 48-19. The instantaneous value of the plane-polarized wave transmitted by the sheet is

$$E = E_y \sin \theta + E_x \cos \theta$$

$$= E_m \cos \omega t \sin \theta + E_m \sin \omega t \cos \theta$$

$$= E_m \sin(\omega t + \theta).$$

The intensity of the wave transmitted by the sheet is proportional to E^2, or

$$I \propto E_m^2 \sin^2(\omega t + \theta).$$

The eye and other measuring instruments respond only to the average intensity $\bar{I}$, which is found by replacing $\sin^2(\omega t + \theta)$ by its average value over one or more cycles $(= \frac{1}{2})$, or

$$\bar{I} \propto \tfrac{1}{2}E_m^2.$$

Comparison with Eq. 48-4 shows that inserting the polarizing sheet reduces the

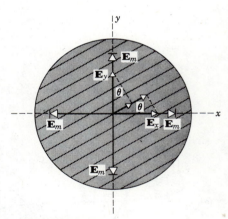

figure 48-19
Circularly polarized light falls on a polarizing sheet. E_x and E_y are instantaneous values of the two components, their maximum values being E_m.

intensity by one-half. The orientation of the sheet makes no difference, since θ does not appear in this equation; this is to be expected if circularly polarized light is represented by a rotating vector, all azimuths about the propagation direction being equivalent. Inserting a polarizing sheet in an *unpolarized* beam has just the same effect, so that a simple polarizing sheet cannot be used to distinguish between unpolarized and circularly polarized light.

A beam of light is thought to be circularly polarized. How may this be verified?

EXAMPLE 5

Insert a quarter-wave plate. If the beam is circularly polarized, the two components will have a phase difference of 90° between them. The quarter-wave plate will introduce a further phase difference of ±90° so that the emerging light will have a phase difference of either zero or 180°. In either case the light will now be *plane-polarized* and can be made to suffer complete extinction by rotating a polarizer in its path.

Does the quarter-wave plate have to be oriented in any particular way to carry out this test?

A plane-polarized light wave of amplitude E_0 falls on a calcite quarter-wave plate with its plane of vibration at 45° to the optic axis of the plate, which is taken as the y axis; see Fig. 48-20. The emerging light will be circularly polarized. In what direction will the rotating electric vector appear to rotate? The direction of propagation is out of the page.

EXAMPLE 6

The wave component whose vibrations are parallel to the optic axis (the e-wave) can be represented as it emerges from the plate as

$$E_y = (E_0 \cos 45°) \sin \omega t = \frac{1}{\sqrt{2}} E_0 \sin \omega t = E_m \sin \omega t.$$

The wave component whose vibrations are at right angles to the optic axis (the o-wave) can be represented as

$$E_x = (E_0 \sin 45°) \sin (\omega t - 90°) = -\frac{1}{\sqrt{2}} E_0 \cos \omega t = -E_m \cos \omega t,$$

the 90° phase shift representing the action of the quarter-wave plate. Note that E_x reaches its maximum value one-fourth of a cycle *later* than E_y does, for, in calcite, wave E_x (the o-wave) travels *more slowly* than wave E_y (the e-wave).

To decide the direction of rotation, let us locate the tip of the rotating electric vector at two instants of time, (a) $t = 0$ and (b) a short time t_1 later chosen so that ωt_1 is a small angle. At $t = 0$ the coordinates of the tip of the rotating vector (see Fig. 48-20a) are

$$E_y = 0 \quad \text{and} \quad E_x = -E_m.$$

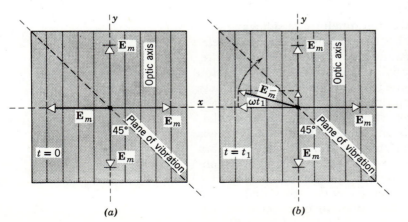

(a)   (b)

figure 48-20
Plane-polarized light falls from behind on a quarter-wave plate oriented so that the light emerging from the page is circularly polarized. In this case the electric vector E_m rotates clockwise as seen by an observer facing the light source.

At $t = t_1$ these coordinates become, approximately,

$$E_y = E_m \sin \omega t_1 \cong E_m(\omega t_1)$$

$$E_x = -E_m \cos \omega t_1 \cong -E_m.$$

Figure 48-20b shows that the vector which represents the emerging circular polarized light is rotating clockwise; by convention such light is called *right-circularly polarized*, the observer always being considered to face the light source.

If the plane of vibration of the incident light in Fig. 48-20 is rotated through $\pm 90°$, the emerging light will be *left-circularly polarized*.

That light waves can deliver *linear momentum* to an absorbing screen or to a mirror is in accord with classical electromagnetism, with quantum physics, and with experiment. The facts of circular polarization suggest that light so polarized might also have *angular* momentum associated with it. This is indeed the case; once again the prediction is in accord with classical electromagnetism and with quantum physics. Experimental proof was provided in 1936 by Richard A. Beth, who showed that when circularly polarized light is produced in a doubly refracting slab, the slab experiences a reaction torque.

The angular momentum carried by light plays a vital role in understanding the emission of light from atoms and of γ-rays from nuclei. If light carries away angular momentum as it leaves the atom, the angular momentum of the residual atom must change by exactly the amount carried away; otherwise the angular momentum of the isolated system *atom plus light* will not be conserved.

Classical and quantum theory both predict that if a beam of circularly polarized light is completely absorbed by an object on which it falls, an angular momentum given by

$$L = \frac{U}{\omega} \tag{48-5}$$

is transferred to the object, where U is the amount of absorbed energy and ω the angular frequency of the light. You should verify that the dimensions in Eq. 48-5 are consistent.

48-6
ANGULAR MOMENTUM OF LIGHT

A light wave, falling on a transparent solid, causes the electrons in the solid to oscillate periodically in response to the time-varying electric vector of the incident wave. The wave that travels through the medium is the resultant of the incident wave and of the radiations from the oscillating electrons. The resultant wave has a maximum intensity in the direction of the incident beam, falling off rapidly on either side. The lack of sideways scattering, which would be essentially complete in a large "perfect" crystal, comes about because the oscillating charges in the medium act cooperatively or coherently.

When light passes through a gas, we find much more sideways scattering. The oscillating electrons in this case, being separated by relatively large distances and not being bound together in a rigid structure, act independently rather than cooperatively. Thus the rigid cancellation of wave disturbances that are not in the forward direction is less likely to occur; there is more sideways scattering.

Light scattered sideways from a gas can be wholly or partially polarized, even though the incident light is unpolarized. Figure 48-21 shows an unpolarized beam moving upward on the page and striking a gas atom at *a*. The electrons at *a* will oscillate in response to the electric components of the incident wave, their motion being equivalent to two oscillating dipoles whose axes are represented by the arrow and the dot at *a*. An oscillating dipole does not radiate along its own line of action.

48-7
SCATTERING OF LIGHT

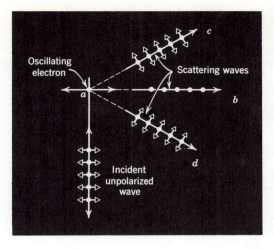

figure 48-21
Light is polarized either partially (*c* and *d*) or completely (*b*) by scattering from a gas molecule at *a*.

Thus an observer at *b* would receive no radiation from the dipole represented by the arrow at *a*. The radiation reaching him would come entirely from the dipole represented by the dot at *a*; thus this radiation would be plane-polarized, the plane of vibration passing through the line *ab* and being normal to the page.

Observers at *c* and *d* would detect partially polarized light, since the dipole represented by the arrow at *a* would radiate somewhat in these directions. Observers viewing the transmitted or the back-scattered light would not detect any polarization effects because both dipoles at *a* would radiate equally in these two directions.

A familiar example is the scattering of sunlight by the molecules of the earth's atmosphere. If the atmosphere were not present, the sky would appear black except when we looked directly at the sun. This has been verified by measurements made above the atmosphere in manned and unmanned space vehicles. We can easily check with a polarizer that the light from the cloudless sky is at least partially polarized. This fact is used in polar exploration in the so-called *solar compass.* In this device we establish direction by noting the nature of the polarization of the scattered sunlight. As is well known, magnetic compasses are not useful in these regions. It has been learned* that bees orient themselves in their flights between their hive and the pollen sources by means of polarization of the light from the sky; bees' eyes contain built-in polarization-sensing devices.

It still remains to be explained why the light scattered from the sky is predominantly blue and why the light received directly from the sun — particularly at sunset when the length of the atmosphere that it must traverse is greatest — is red. The cross section of an atom or molecule for light scattering depends on the wavelength, blue light being scattered more effectively than red light. Since the blue light is largely scattered, the transmitted light will have the color of normal sunlight with the blues largely removed; it is therefore more reddish in appearance.

The fact that the scattering cross section for blue light is higher than that for red light can be made reasonable. An electron in an atom or molecule is bound there by strong restoring forces. It has a definite natural frequency, like a small mass suspended in space by an assembly of springs. The natural frequency for

* See "Polarized Light and Animal Navigation" by Talbot H. Watermann in *Scientific American,* July 1955; *Bees: Their Vision, Chemical Sense, and Language* by K. von Frisch, Cornell University Press, 1950; and "Polarized-Light Navigation by Insects" by Rüdiger Wehner, *Scientific American,* July 1976.

electrons in atoms and molecules is usually in a region corresponding to violet or ultraviolet light.

When light is allowed to fall on such bound electrons, it sets up forced oscillations at the frequency of the incident light beam. In mechanical resonant systems it is possible to "drive" the system most effectively if we impress on it an external force whose frequency is as close as possible to that of the natural resonant frequency. In the case of light the blue is closer to the natural resonant frequency of the bound electron than is the red light. Therefore, we would expect the blue light to be more effective in causing the electron to oscillate, and thus it will be more effectively scattered.

When X-rays were discovered in 1898, there was much speculation whether they were waves or particles. In 1906 they were established as transverse waves by Charles Glover Barkla (1877–1944) by means of a polarization experiment.

When the unpolarized X-rays strike scattering block S_1 in Fig. 48-22, they set the electrons into oscillatory motion. The considerations of the preceding section require that the X-rays scattered toward the second block be plane-polarized as shown in the figure. Let this wave be scattered from the second scattering block, and let us examine the radiation scattered from it by rotating a detector D in a plane at right angles to the line joining the blocks. The electrons will oscillate parallel to each other, and the positions of maximum and zero intensity will be as shown. A plot of detector reading as a function of the angle ϕ supports the hypothesis that X-rays are transverse waves. If the X-rays were a stream of particles or a longitudinal wave, these effects could by no means be so readily understood. Thus Barkla's important experiment established that X-rays are a part of the electromagnetic spectrum.

48-8
DOUBLE SCATTERING

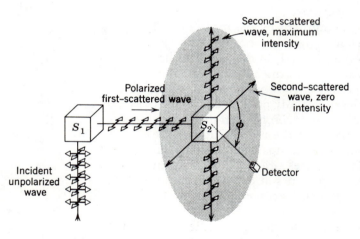

figure 48-22
A double-scattering experiment used by Barkla to show that X-rays are transverse waves.

In later studies you will learn that beams of particles such as electrons, protons, and pions can be viewed as waves. Scattering (including double scattering) techniques are often used to investigate the polarization characteristics of such beams.*

questions

1. It is said that light from ordinary sources is unpolarized. Can you think of any common sources that emit polarized light?
2. Why do sunglasses made of polarizing materials have a marked advantage over those that simply depend on absorption effects?
3. Unpolarized light falls on two polarizing sheets so oriented that no light is

*See "Polarized Accelerator Targets" by Gilbert Shapiro, *Scientific American,* July, 1976.

transmitted. If a third polarizing sheet is placed between them, can light be transmitted?

4. 3-D movies were very popular in the early 1950's. Viewers wore polarizing glasses and a polarizing sheet was placed in front of each of the *two* projectors needed. Explain how the system worked. Can you suggest any problems that may have led to the early abandonment of the system?

5. A wire grid, consisting of an array of wires arranged parallel to one another, can polarize an incident unpolarized beam of electromagnetic waves that pass through it. Explain the facts that (a) the diameter of the wires and the spacing between them must be much less than the incident wavelength to obtain effective polarization and (b) the transmitted component is the one whose electric vector oscillates in a direction perpendicular to the wires.

6. Brewster's law determines the polarizing angle on reflection from a dielectric. A plausible interpretation for zero reflection of the π-component at that angle is that the charges in the dielectric are caused to oscillate parallel to the reflected ray by this component and produce no radiation in this direction. Explain this and comment on the plausibility.

7. Can polarization by reflection occur if the light is incident on the interface from the side with the higher index of refraction (glass to air, for example)?

8. Devise a way to identify the polarizing direction of a sheet of Polaroid.

9. Is the optic axis of a doubly refracting crystal simply a line or a direction in space? Has it a direction sense, like an arrow? What about the characteristic direction of a polarizing sheet?

10. If ice is doubly refracting (see Table 48-1), why don't we see two images of objects viewed through an ice cube?

11. Is it possible to produce interference effects between the o-beam and the e-beam, which are separated by the calcite crystal from the incident unpolarized beam in Fig. 48-11, by recombining them? Explain your answer.

12. From Table 48-1, would you expect a quarter-wave plate made from calcite to be thicker than one made from quartz?

13. Does the *e*-wave in doubly refracting crystals always travel at a speed given by c/n_e?

14. In Fig. 48-14a and 48-14b describe qualitatively what happens if the incident beam falls on the crystal with an angle of incidence that is not zero. Assume in each case that the incident beam remains in the plane of the figure.

15. Devise a way to identify the direction of the optic axis in a quarter-wave plate.

16. If plane-polarized light falls on a quarter-wave plate with its plane of vibration making an angle of (a) 0° or (b) 90° with the axis of the plate, describe the transmitted light. (c) If this angle is arbitrarily chosen, the transmitted light is called *elliptically polarized*; describe such light.

17. You are given an object which may be (a) a disk of grey glass, (b) a polarizing sheet, (c) a quarter-wave plate, or (d) a half-wave plate (see Problem 14). How could you identify it?

18. Can a plane-polarized light beam be represented as a sum of two circularly polarized light beams of opposite rotation? What effect has changing the phase of one of the circular components on the resultant beam?

19. How can a right-circularly polarized light beam be transformed into a left-circularly polarized beam?

20. Could (a) a radar beam and (b) a sound wave in air be circularly polarized?

21. A beam of light is said to be unpolarized, plane-polarized, or circularly polarized. How could you choose among them experimentally?

22. A parallel beam of light is absorbed by an object placed in its path. Under what circumstances will (a) linear momentum and (b) angular momentum be transferred to the object?

23. When observing a clear sky through a polarizing sheet, one finds that the intensity varies by a factor of two on rotating the sheet. This does not happen when one views a cloud through the sheet. Can you devise an explanation?

24. In 1949 it was discovered that light from distant stars in our galaxy is slightly plane-polarized, with the preferred plane of vibration being parallel to the plane of our galaxy. This is probably due to non-isotropic scattering of the starlight by elongated and slightly aligned interstellar grains. If the grains are oriented with their long axis parallel to the interstellar magnetic field lines, as discussed in the text, and they absorb and radiate electromagnetic waves like the oscillating electrons in a radio antenna, how must the magnetic field be oriented with respect to the galactic plane?

problems

SECTION 48-1

1. Prove that two plane-polarized light waves of equal amplitude, their planes of vibration being at right angles to each other, cannot produce interference effects. (*Hint:* Prove that the intensity of the resultant light wave, averaged over one or more cycles of oscillation, is the same no matter what phase difference exists between the two waves.)

SECTION 48-2

2. Unpolarized light falls on two polarizing sheets placed one on top of the other. What must be the angle between the characteristic directions of the sheets if the intensity of the transmitted light is (a) one-third the maximum intensity of the transmitted beam or (b) one-third the intensity of the incident beam? Assume that the polarizing sheet is ideal, that is, that it reduces the intensity of unpolarized light by exactly 50%.

3. A beam of plane polarized light strikes two polarizing sheets. The characteristic direction of the first sheet is at an angle θ with respect to the incident beam, the characteristic direction of the second sheet being at 90° with respect to the incident beam. Find the angle θ for a transmitted beam intensity that is 1/10 the incident beam intensity.   *Answer:* 20° or 70°.

4. An unpolarized beam of light is incident on a group of four polarizing sheets which are lined up so that the characteristic direction of each is rotated by 30° clockwise with respect to the preceding sheet. What fraction of the incident intensity is transmitted?

5. A beam of light is a mixture of plane polarized light and randomly polarized light. When it is sent through a Polaroid sheet, it is found that the transmitted intensity can be varied by a factor of five depending on the orientation of the Polaroid. Find the relative intensities of these two components of the incident beam.
Answer: Plane polarized $\frac{2}{3}$, randomly polarized $\frac{1}{3}$.

6. Partially polarized light (a mixture of unpolarized and plane-polarized beams) can be represented as two plane-polarized beams of unequal intensities, I along the x-axis and i along the y-axis say, with a random phase difference. The degree of polarization is defined as $p = (I - i)/(I + i)$.
   (a) Suppose a beam of partially polarized light passes through a Polaroid sheet with its characteristic direction at an angle θ with the x-axis; show that the transmitted intensity I_t is

$$I_t = I \frac{1 + p \cos 2\theta}{1 + p}.$$

   (b) Does this reduce to expected results for $p = 1$ and $p = 0$?

7. It is desired to rotate the plane of polarization of a beam of plane-polarized light by 90°. (a) How might this be done using only Polaroid sheets? (b) How many sheets are required in order that the total intensity loss is less than 5%? Assume that each Polaroid sheet is ideal.   *Answer:* 48.

SECTION 48-3

8. (a) At what angle of incidence will the light reflected from water be completely polarized? (b) Does this angle depend on the wavelength of the light?

9. Calculate the range of polarizing angles for white light incident on fused quartz. Assume that the wavelength limits are 400 and 700 nm (4000 and 7000 Å) and use the dispersion curve of Fig. 43-2.
    Answer: 55°30′ to 55°46′

SECTION 48-4

10. Plane-polarized light of wavelength 525 nm strikes, at normal incidence, a wurzite crystal, cut with its faces parallel to the optic axis. What is the smallest possible thickness of the crystal if the emergent o and e-rays combine to form plane-polarized light? See Table 48-1.

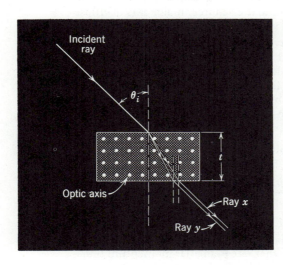

figure 48-23
Problem 11

11. A narrow beam of unpolarized light falls on a calcite crystal cut with its optic axis as shown in Fig. 48-23. (a) For $t = 1.0$ cm and for $\theta_i = 45°$, calculate the perpendicular distance between the two emerging rays x and y. (b) Which is the o-ray and which the e-ray? (c) What are the states of polarization of the emerging rays? (d) Describe what happens if a polarizer is placed in the incident beam and rotated. (*Hint:* Inside the crystal the E-vector vibrations for one ray are always perpendicular to the optic axis and for the other ray they are always parallel. The two rays are described by the indices n_o and n_e; *in this plane* each ray obeys Snell's law.)
    Answer: (a) 0.55 mm. (b) x is the e-ray, y is the o-ray. (c) E in ray y lies in the plane of the figure; E in ray x lies at right angles to the plane of the figure. (d) Every 90° one or the other beam, alternately, will be extinguished.

12. A prism is cut from calcite so that the optic axis is parallel to the prism edge as shown in Fig. 48-24. Describe how such a prism might be used to measure the two principal indices of refraction for calcite. (*Hint:* See hint in Problem 11; see also Example 3, Chapter 43.)

13. How thick must a sheet of mica be if it is to form a quarter-wave plate for yellow light ($\lambda = 589$ nm $= 5890$ Å)? Mica cleaves in such a way that the appropriate indices of refraction, for transmission at right angles to the cleavage plane, are 1.6049 and 1.6117.   *Answer:* 0.022 mm.

SECTION 48-5

14. What would be the action of a *half-wave plate* (that is, a plate twice as thick as a quarter-wave plate) on (a) plane-polarized light (assume the plane of vibration to be at 45° to the optic axis of the plate), (b) circularly polarized light, and (c) unpolarized light.

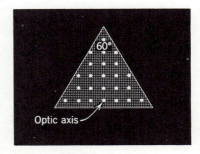

figure 48-24
Problem 12

15. Describe the state of polarization represented by these sets of equations:

(a) $E_x = E \sin(kz - \omega t)$

$E_y = E \cos(kz - \omega t)$,

(b) $E_x = E \cos(kz - \omega t)$

$E_y = E \cos\left(kz - \omega t + \dfrac{\pi}{4}\right)$,

(c) $E_x = E \sin(kz - \omega t)$

$E_y = -E \sin(kz - \omega t)$.

Answer: Assuming that a right-handed coordinate system is used, (a) circular, counterclockwise as seen facing the scource; (b) elliptical, counterclockwise as seen facing the scource, major axis of ellipse along $y = x$; (c) plane, along $y = -x$.

16. A sheet of Polaroid and a quarter-wave plate are glued together in such a way that, if the combination is placed with face A against a shiny coin, the face of the coin can be seen when illuminated with light of appropriate wavelength. When the combination is placed with face A away from the coin, the coin cannot be seen. (a) Which component is on face A and (b) what is the relative orientation of the components?

SECTION 48-6

17. Show that in a parallel beam of circularly polarized light the angular momentum per unit volume L_r is given by

$$L_r = \frac{P}{\omega c},$$

where P is the power per unit area of the beam. Start from Eq. 48-5.

18. Assume that a parallel beam of circularly polarized light whose intensity is 100 W is absorbed by an object, (a) At what rate is angular momentum transferred to the object? (b) If the object is a flat disk of diameter 5.0 mm and mass 1.0×10^{-2} g, after how long a time (assuming it is free to rotate about its axis) would it attain an angular speed of 1.0 rev/s? Assume a wavelength of 500 nm.

49
light and
quantum physics

We have studied many properties of light, including its propagation, reflection, refraction, diffraction, and interference. This chapter deals in part with the production of light and with the way that such studies led, in 1900, to the birth of quantum physics.

The most common light sources are heated solids and gases through which an electric discharge is passing. The tungsten filament of an incandescent lamp and the familiar neon sign are examples in each category. By analyzing the light from a source with a spectrometer, we can learn how strongly it radiates at various wavelengths. Figure 49-1, which is typical of spectra for heated solids, shows the results of such measurements for a heated tungsten ribbon at 2000 K.

The ordinate $\mathscr{R}_\lambda$ in Fig. 49-1 is called the *spectral radiancy*, defined so that the quantity $\mathscr{R}_\lambda \, d\lambda$ is the rate at which energy is radiated per unit area of surface for wavelengths lying in the interval λ to $\lambda + d\lambda$. Typical units for $\mathscr{R}_\lambda$ are W/cm$^2 \cdot \mu$m; the corresponding units of $\mathscr{R}_\lambda \, d\lambda$ are W/cm^2. In measuring $\mathscr{R}_\lambda$, all radiation emerging into the forward hemisphere is included.

Sometimes we wish to discuss the total radiated energy without regard to its wavelength. An appropriate quantity here is the *radiancy $\mathscr{R}$*, defined as the rate per unit surface area at which energy is radiated into the forward hemisphere, appropriate units being W/cm^2. We can find it by integrating the radiation present in all wavelength intervals:

$$\mathscr{R} = \int_0^\infty \mathscr{R}_\lambda \, d\lambda. \qquad (49\text{-}1)$$

The radiancy $\mathscr{R}$ can be interpreted as the area under the plot of $\mathscr{R}_\lambda$ against λ. In Fig. 49-1 this area—and thus $\mathscr{R}$—is 23.5 W/cm^2. Note the

49-1
SOURCES OF LIGHT

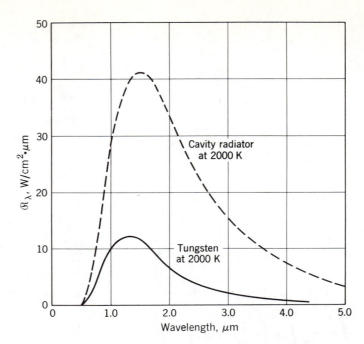

figure 49-1
The spectral radiancy of tungsten at 2000 K. The dashed curve refers to a cavity radiator at the same temperature. One micrometer $(=1\ \mu m) = 10^{-6}$ m $= 10^4$ Å $= 10^3$ nm.

formal similarity between such curves and the Maxwell speed distribution curve of Section 24-2.

For every material there exists a family of spectral radiancy curves like that of Fig. 49-1, one curve for every temperature. If such families of curves are compared, no obvious regularities stand out. A quantitative understanding in terms of a basic theory presents serious difficulties. Fortunately, it is possible to work with an idealized heated solid, called a *cavity radiator.* Its light-emitting properties prove to be independent of any particular material and to vary in a simple way with temperature. In much the same way it proved convenient earlier to deal with an ideal gas rather than to analyze the properties of the infinite variety of real gases. The cavity radiator is the *ideal solid* as far as its light-emitting characteristics are concerned. We shall describe in the next two sections how the theoretical study of cavity radiation in 1900 by the German physicist Max Planck (1858–1947) laid the foundations of modern quantum physics.

Let us construct a cavity in each of three metal blocks through the walls of which a small hole is drilled. Let the blocks be made of any suitable materials; for example, tungsten, tantalum, and molybdenum. Let each block be raised to the same uniform temperature (say 2000 K) as determined by a suitable thermometer. Finally, let us observe the blocks by their emitted light in a dark room. Measurements of $\mathscr{R}$ and $\mathscr{R}_\lambda$ show the following:

1. The radiation from the cavity interior is always more intense than the radiation from the outside wall. Comparison of the two curves in Fig. 49-1 makes this clear for tungsten. For the three materials given,

49-2
CAVITY RADIATORS

at 2000 K the ratio of the radiancy for the outside surface to that for the cavity is 0.259 (tungsten), 0.212 (molybdenum), and 0.232 (tantalum).

2. At a given temperature the radiancy of the hole is *identical for all three radiators,* in spite of the fact that the radiancies of the outer surfaces are different. At 2000 K the cavity radiancy (that is, the hole radiancy) is 90.0 W/cm².

3. In contrast to the radiancy of the outer surfaces, the cavity radiancy $\mathcal{R}_c$ varies with temperature in a simple way, namely, as

$$\mathcal{R}_c = \sigma T^4, \tag{49-2}$$

where σ is a universal constant (the Stefan-Boltzmann constant) whose measured value is 5.67×10^{-8} W/m² · K⁴. The radiancy of the outer surfaces varies with temperature in a more complicated way and is different for different materials. We often write it as

$$\mathcal{R} = e\mathcal{R}_c = e\sigma T^4, \tag{49-3}$$

where e, the *emissivity,* depends on the material and the temperature.

4. $\mathcal{R}_\lambda$ for the cavity radiation varies with temperature in the way shown in Fig. 49-2. These curves depend only on the temperature and are quite independent of the material and of the shape and size of the cavity.

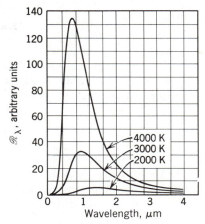

figure 49-2
The spectral radiancy for cavity radiation at three different temperatures.

Figure 49-3 shows an actual cavity, consisting of a hollow thin-walled cylinder of tungsten heated by sending an electric current through it. The cylinder is mounted in an evacuated glass bulb, and a tiny hole is drilled through the cylinder wall. You can see from the photograph that the radiancy of the cavity interior is greater than that of the cavity walls.

We can deduce many of the facts just given about cavity radiation from Fig. 49-4, which shows two cavities made of different materials, of arbitrary shapes, but with the same wall temperature T. Radiation, described by $\mathcal{R}_A$, goes from cavity A to cavity B and radiation described by $\mathcal{R}_B$ moves in the opposite direction. If these two rates of energy transfer are not equal, one end of the composite block will start to heat up and the other end will start to cool down, which is a violation of the second law of thermodynamics. (Why?) Thus we must have

$$\mathcal{R}_A = \mathcal{R}_B = \mathcal{R}_c, \tag{49-4}$$

where $\mathcal{R}_c$ describes the total radiation for *all* cavities.

Not only the total radiation but also the distribution of radiant energy with wavelength must be the same for each cavity in Fig. 49-4. We can show this by placing a filter between the two cavity openings, so chosen that it permits only a selected narrow band of wavelengths to pass. Applying the same argument, we can show that we must have

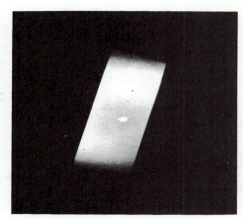

figure 49-3
An incandescent tungsten tube with a small hole drilled in its wall. The radiation emerging from the hole is cavity radiation.

figure 49-4
Two radiant cavities initially at the same temperature are placed together as shown.

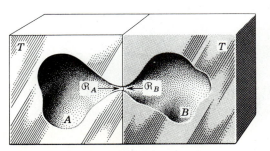

$$\mathscr{R}_{\lambda A} = \mathscr{R}_{\lambda B} = \mathscr{R}_{\lambda c}, \qquad (49\text{-}5)$$

where $\mathscr{R}_{\lambda c}$ is a spectral radiancy characteristic of all cavities.

A theoretical explanation for the cavity radiation was the outstanding unsolved problem in physics during the years before the turn of the present century. A number of capable physicists advanced theories based on classical physics, which, however, had only limited success. Figure 49-5, for example, shows the theory of Wien; the fit to the experimental points is reasonably good, but definitely not within the experimental error of the data. Wien's formula is

$$\mathscr{R}_\lambda = \frac{c_1}{\lambda^5} \frac{1}{e^{c_2/\lambda T}},$$

where c_1 and c_2 are constants that must be determined empirically by fitting the theoretical formula to the experimental data.

In 1900 Max Planck pointed out that if Wien's formula were modified in a simple way, it would prove to fit the data precisely. Planck's formula, announced to the Berlin Physical Society on October 19, 1900, was

$$\mathscr{R}_\lambda = \frac{c_1}{\lambda^5} \frac{1}{e^{c_2/\lambda T} - 1}. \qquad (49\text{-}6)$$

This formula, though interesting and important, was still empirical at that stage and did not constitute a theory.

Planck sought such a theory in terms of a detailed model of the atomic processes taking place at the cavity walls. He assumed that the atoms that make up these walls behave like tiny electromagnetic oscillators, each with a characteristic frequency of oscillation. The oscillators emit electromagnetic energy into the cavity and absorb electromagnetic energy from it. Thus it should be possible to deduce the characteristics of the cavity radiation from those of the oscillators with which it is in equilibrium.

Planck was led to make two radical assumptions about the atomic oscillators. As eventually formulated, these assumptions are the following:

1. An oscillator cannot have *any* energy but only energies given by

$$E = nh\nu, \qquad (49\text{-}7)$$

49-3
PLANCK'S RADIATION FORMULA

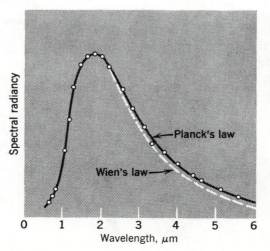

figure 49-5
The circles show the experimental spectral radiancy data of Coblentz for cavity radiation. The theoretical formulas of Wien and Planck are also shown, Planck's providing an excellent fit to the data.

where ν is the oscillator frequency, h is a constant (now called *Planck's constant*), and n is a number (now called a *quantum number*) that can take on only integral values. Equation 49-7 asserts that the oscillator energy is *quantized*. Later developments show that the correct formula for a harmonic oscillator is $E = (n + \frac{1}{2})h\nu$. This change makes no difference to Planck's conclusions, however.

2. The oscillators do not radiate energy continuously, but only in "jumps," or *quanta*. These quanta of energy are emitted when an oscillator changes from one to another of its quantized energy states. Thus, if n changes by one unit, Eq. 49-7 shows that an amount of energy given by

$$\Delta E = \Delta n h\nu = h\nu \qquad (49\text{-}8)$$

is radiated. As long as an oscillator remains in one of its quantized states (or *stationary states* as they are called), it neither emits nor absorbs energy.

These assumptions were radical ones and, indeed, Planck himself resisted accepting them for many years. In his words, "My futile attempts to fit the elementary quantum of action [that is, the quantity h] somehow into the classical theory continued for a number of years, and they cost me a great deal of effort."

Consider the application of Planck's hypotheses to a large-scale oscillator such as a mass-spring system or an *LC* circuit. It would be a stoutly defended common belief that oscillations in such systems could take place with *any* value of total energy and not with only certain discrete values. In the decay of such oscillations (by friction in the mass-spring system or by resistance and radiation in the *LC* circuit), it would seem that the mechanical or electromagnetic energy would decrease in a perfectly continuous way and not by "jumps." There is no basis in everyday experience, however, to dismiss Planck's assumptions as violations of "common sense," for Planck's constant proves to have a very small value, namely,

$$h = 6.626 \times 10^{-34} \text{ joule} \cdot \text{second}.$$

The following example makes this clear.

EXAMPLE 1

A mass-spring system has a mass $m = 1.0$ kg and a spring constant $k = 20$ N/m and is oscillating with an amplitude of 1.0 cm. (a) If its energy is quantized according to Eq. 49-7, what is the quantum number n?

From Eq. 15-11 the frequency is

$$\nu = \frac{1}{2\pi} \sqrt{\frac{k}{m}} = \frac{1}{2\pi} \sqrt{\frac{20 \text{ N/m}}{1.0 \text{ kg}}} = 0.71 \text{ Hz}.$$

From Eq. 7-8 the mechanical energy is

$$E = \tfrac{1}{2}kx_{max}^2 = \tfrac{1}{2}(20 \text{ N/m})(1.0 \times 10^{-2} \text{ m})^2 = 1.0 \times 10^{-3} \text{ J}.$$

From Eq. 49-7 the quantum number is

$$n = \frac{E}{h\nu} = \frac{1.0 \times 10^{-3} \text{ J}}{(6.6 \times 10^{-34} \text{ J} \cdot \text{s})(0.71 \text{ Hz})} = 2.1 \times 10^{30}.$$

(b) If n changes by unity, what fractional change in energy occurs?

The fractional change in energy is given by dividing Eq. 49-8 by Eq. 49-7, or

$$\frac{\Delta E}{E} = \frac{h\nu}{nh\nu} = \frac{1}{n} = \sim 10^{-30}.$$

Thus for large-scale oscillators the quantum numbers are enormous and the quantized nature of the energy of the oscillations will not be apparent. Similarly, in large-scale experiments we are not aware of the discrete nature of mass and the quantized nature of charge, that is, of the existence of atoms and electrons.

On the basis of his two assumptions, Planck was able to derive his radiation law (Eq. 49-6) entirely from theory. His theoretical expressions for the hitherto empirical constants c_1 and c_2 were

$$c_1 = 2\pi c^2 h \qquad \text{and} \qquad c_2 = \frac{hc}{k},$$

where k is Boltzmann's constant (see Section 23-5) and c is the speed of light. By inserting the experimental values for c_1 and c_2, Planck was able to derive the values of both h and k. Planck described his theory to the Berlin Physical Society on December 14, 1900. Quantum physics started on that date. Planck's ideas soon received re-enforcement from Einstein, who, in 1905, applied the concepts of energy quantization to a new area of physics, the photoelectric effect.

Before discussing this effect, it is important to realize that although Planck had quantized the energies of the oscillators in the cavity walls he still treated the radiation within the cavity as an electromagnetic wave. Einstein's analysis of the photoelectric effect first pointed out the inadequacy of the wave picture of light in certain situations.

49-4 PHOTOELECTRIC EFFECT

Figure 49-6 shows an apparatus used to study the photoelectric effect. Monochromatic light, falling on metal plate A, will liberate *photoelectrons*, which can be detected as a current if they are attracted to metal cup B by means of a potential difference V applied between A and B. Galvanometer G serves to measure this *photoelectric current*.

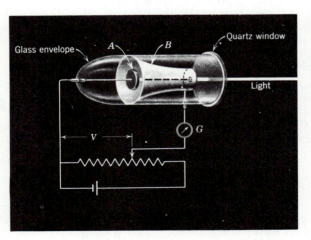

figure 49-6
An apparatus used to study the photoelectric effect. V can be varied continuously and can also be reversed in sign by a switching arrangement not shown.

Figure 49-7 (curve a) is a plot of the photoelectric current in an apparatus like that of Fig. 49-6, as a function of the potential difference V. If V is large enough, the photoelectric current reaches a certain limiting value at which all photoelectrons ejected from plate A are collected by cup B.

If we reverse V in sign, the photoelectric current does not immediately drop to zero, which proves that the electrons are emitted from A

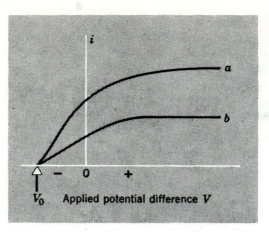

figure 49-7
Some data taken with the apparatus of Fig. 49-6. The applied potential difference V is called positive when the cup B in Fig. 49-6 is positive with respect to the photoelectric surface A. In curve b the incident light intensity has been reduced to one-half that of curve a.

with a finite velocity. Some will reach cup B in spite of the fact that the electric field opposes their motion. However, if this reversed potential difference is made large enough, a value V_0 (the *stopping potential*) is reached at which the photoelectric current does drop to zero. This potential difference V_0, multiplied by the electron charge, measures the kinetic energy K_{max} of the fastest ejected photoelectron. In other words,

$$K_{max} = eV_0. \qquad (49\text{-}9)$$

Here K_{max} turns out to be independent of the intensity of the light as shown by curve b in Fig. 49-7, in which the light intensity has been reduced to one-half.

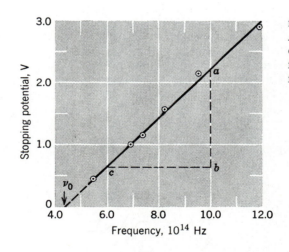

figure 49-8
A plot of Millikan's measurements of the stopping potential at various frequencies for sodium. The cutoff frequency ν_0 is 4.39×10^{14} Hz.

Figure 49-8 shows the stopping potential V_0 as a function of the frequency of the incident light for sodium. Note that there is a definite cutoff frequency ν_0, below which no photoelectric effect occurs. These data were taken by R. A. Millikan (1868–1953), whose painstaking work on the photoelectric effect won him the Nobel prize in 1923. Because the photoelectric effect is largely a surface phenomenon, it is necessary to avoid oxide films, grease, or other surface contaminants. Millikan devised a technique to cut shavings from the metal surface under vacuum conditions, a "machine shop *in vacuo*" as he called it.

Three major features of the photoelectric effect cannot be explained in terms of the wave theory of light:

1. Wave theory suggests that the kinetic energy of the photoelectrons should increase as the light beam is made more intense. However, Fig. 49-7 shows that K_{max} $(=eV_0)$ is independent of the light intensity; this has been tested over a range of intensities of 10^7.

2. According to the wave theory, the photoelectric effect should occur for any frequency of the light, provided only that the light is intense enough. However, Fig. 49-8 shows that there exists, for each surface, a characteristic *cutoff frequency* ν_0. For frequencies less than this, the photoelectric effect disappears, no matter how intense the illumination.

3. If the energy of the photoelectrons is "soaked up" from the incident wave by the metal plate, it is not likely that the "effective target area" for an electron in the metal is much more than a few atomic diameters. Thus, if the light is feeble enough, there should be a measurable time lag (see Example 2) between the impinging of the light on the surface and the ejection of the photoelectron. During this interval the electron should be "soaking up" energy from the wave until it has accumulated enough energy to escape. However, no detectable time lag has ever been measured. This disagreement is particularly striking when the photoelectric substance is a gas; under these circumstances the energy of the emitted photoelectron must certainly be "soaked out of the wave" by a single atom.

EXAMPLE 2

A metal plate is placed 5 m from a monochromatic light source whose power output is 10^{-3} W. Consider that a given ejected photoelectron may collect its energy from a circular area of the plate as large as ten atomic diameters (10^{-9} m) in radius. The energy required to remove an electron through the metal surface is about 5.0 eV. Assuming light to be a wave, how long would it take for such a "target" to soak up this much energy from such a light source?

The target area is π $(10^{-9}$ m$)^2$ or 3×10^{-18} m²; the area of a 5-m sphere centered on the light source is 4π $(5$ m$)^2 \cong 300$ m². Thus, if the light source radiates uniformly in all directions, the rate P at which energy falls on the target is given by

$$P = (10^{-3} \text{ W})\left(\frac{3 \times 10^{-18} \text{ m}^2}{300 \text{ m}^2}\right) = 10^{-23} \text{ J/s}.$$

Assuming that all this power is absorbed, we may calculate the time required from

$$t = \left(\frac{5 \text{ eV}}{10^{-23} \text{ J/s}}\right)\left(\frac{1.6 \times 10^{-19} \text{ J}}{1 \text{ eV}}\right) \cong 20 \text{ hours}.$$

However, no detectable time lag has been measured.

49-5
EINSTEIN'S PHOTON THEORY

Einstein succeeded in explaining the photoelectric effect by making a remarkable assumption, namely, that the energy in a light beam travels through space in concentrated bundles, later called *photons*.* The

---

* The word *photon* was coined by an American physical chemist, G. N. Lewis, in 1926. He wrote, "I therefore take the liberty of proposing for this hypothetical new atom . . . the name *photon*." When Max Planck proposed Einstein for membership in the Berlin Physical Society, he deprecated Einstein's 1905 paper on the photon theory, saying that such a great man should be excused a little fault. Planck, you will recall, was the (almost reluctant) discoverer of quantization. Einstein received the Nobel prize in 1921, not for the theory of relativity but for his concept of photons.

energy E of a single photon (see Eq. 49-8) is given by

$$E = h\nu. \tag{49-10}$$

Recall that Planck believed that light, although emitted from its source discontinuously, travels through space as an electromagnetic wave. Einstein's hypothesis suggests that light traveling through space behaves not like a wave at all but like a particle. Millikan, whose experiments verified Einstein's ideas in every detail, spoke of Einstein's "bold, not to say reckless, hypothesis."

If we apply Einstein's photon concept to the photoelectric effect, we obtain

$$h\nu = E_0 + K_{max} \tag{49-11}$$

where $h\nu$ is the energy of the photon. Equation 49-11 says that a photon carries an energy $h\nu$ into the surface. Part of this energy (E_0) is used in causing the electron to pass through the metal surface. The excess energy $(h\nu - E_0)$ is given to the electron in the form of kinetic energy; if the electron does not lose energy by internal collisions as it escapes from the metal, it will exhibit it all as kinetic energy after it emerges. Thus K_{max} represents the maximum kinetic energy that the photoelectron can have outside the surface; in nearly all cases it will have less energy than this because of internal losses.

Consider how Einstein's photon hypothesis meets the three objections raised against the wave-theory interpretation of the photoelectric effect. As for objection 1 (the lack of dependence of K_{max} on the intensity of illumination), there is complete agreement of the photon theory with experiment. If we double the light intensity, we double the number of photons and thus double the photoelectric current; we do not change the energy $(=h\nu)$ of the individual photons or the nature of the individual photoelectric processes described by Eq. 49-11.

Objection 2 (the existence of a cutoff frequency) follows from Eq. 49-11. If K_{max} equals zero, we have

$$h\nu_0 = E_0,$$

which asserts that the photon has just enough energy to eject the photoelectrons and none extra to appear as kinetic energy. This quantity E_0 is called the *work function* of the substance. If ν is reduced below ν_0, the individual photons, no matter how many of them there are (that is, no matter how intense the illumination), will not have enough energy to eject photoelectrons.

Objection 3 (the absence of a time lag) follows from the photon theory because the required energy is supplied in a concentrated bundle. It is not spread uniformly over a large area, as in the wave theory.

Although the photon hypothesis certainly fits the facts of photoelectricity, it seems to be in direct conflict with the wave theory of light which, as we have seen in earlier chapters, has been verified in many experiments. Our modern view of the nature of light is that it has a dual character, behaving like a wave under some circumstances and like a particle, or photon, under others. We discuss the wave-particle duality at length in Chapter 50. Meanwhile, let us continue our studies of the firm experimental foundation on which the photon concept rests.

Let us rewrite Einstein's photoelectric equation (Eq. 49-11) by substituting eV_0 for K_{max} (see Eq. 49-9). This yields, after rearrangement,

$$V_0 = \frac{h}{e}\nu - \frac{E_0}{e}. \tag{49-12}$$

Thus Einstein's theory predicts a linear relationship between V_0 and ν, in complete agreement with experiment; see Fig. 49-8. The slope of the experimental curve in this figure should be h/e, or

$$\frac{h}{e} = \frac{ab}{bc} = \frac{2.20 \text{ V} - 0.65 \text{ V}}{(10 \times 10^{14} - 6.0 \times 10^{14}) \text{ Hz}} = 3.9 \times 10^{-15} \text{ V} \cdot \text{s}.$$

We can find h by multiplying this ratio by the electron charge e,

$$h = (3.9 \times 10^{-15} \text{ V} \cdot \text{s})(1.6 \times 10^{-19} \text{ C}) = 6.2 \times 10^{-34} \text{ J} \cdot \text{s}.$$

From a more careful analysis of these and other data, including data taken with lithium surfaces, Millikan found the value $h = 6.57 \times 10^{-34}$ J·s with an accuracy of about 0.5%. This agreement with the value of h derived from Planck's radiation formula is a striking confirmation of Einstein's photon concept.*

EXAMPLE 3

Deduce the work function that follows from Fig. 49-8.

The intersection of the straight line in Fig. 49-8 with the horizontal axis is the cutoff frequency ν_0. Substituting these values yields

$$E_0 = h\nu_0 = (6.63 \times 10^{-34} \text{ J} \cdot \text{s})(4.39 \times 10^{14} \text{ Hz})$$

$$= 2.92 \times 10^{-19} \text{ J} = 1.82 \text{ eV}.$$

Compelling confirmation of the concept of the photon as a concentrated bundle of energy was provided in 1923 by A. H. Compton (1892–1962) who earned a Nobel prize for this work in 1927.† Compton allowed a beam of X-rays of sharply defined wavelength λ to fall on a graphite block, as in Fig. 49-9, and he measured, for various angles of scattering, the intensity of the scattered X-rays as a function of their wavelength. Figure 49-10 shows his experimental results. We see that although the incident beam consists essentially of a single wavelength λ the scattered X-rays have intensity peaks at *two* wavelengths; one of them is the same as the incident wavelength, the other, λ', being larger by an amount $\Delta\lambda$. This so-called *Compton shift* $\Delta\lambda$ varies with the angle at which the scattered X-rays are observed.

The presence of a scattered wave of wavelength λ' cannot be understood if the incident X-rays are regarded as an electromagnetic wave like that of Fig. 41-13. On this picture the incident wave of frequency ν causes electrons in the scattering block to oscillate at that same frequency. These oscillating electrons, like charges surging back and forth in a small radio transmitting antenna, radiate electromagnetic waves that again have this same frequency ν. Thus, on the wave picture

49-6
THE COMPTON EFFECT

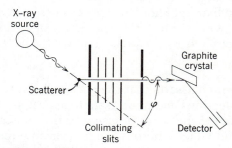

figure 49-9
Compton's experimental arrangement. Monochromatic X-rays of wavelength λ fall on a graphite scatterer. The distribution of intensity with wavelength is measured for X-rays scattered at any selected angle φ. The scattered wavelengths are measured by observing Bragg reflections from a crystal; see Eq. 47-8. Their intensities are measured by a detector, such as an ionization chamber.

---

* In an actual experiment, we would have to take into account any difference in the contact potential that might exist between the metal of plate A and the (different) metal of cup B in Fig. 49-6. See, for example, *Experiments in Modern Physics* by A. C. Melissinos, Academic Press, New York, 1966 for an exact treatment of such a situation.
† For an historical account of Compton's researches read "The Scattering of X Rays as Particles," by A. H. Compton, *American Journal of Physics,* December 1961. In this article Compton recalled that the distinguished Indian physicist, Sir C. V. Raman, who won a Nobel prize in 1930, said to him: "Compton, you are a good debator but the truth is not in you." This, and the earlier comment by Planck on Einstein, suggests the reluctance with which new ideas are accepted even by brilliant scientists, such as Planck and Raman. Compton (and Einstein) turned out to be right.

the scattered wave should have the same frequency ν and the same wavelength λ as the incident wave.

Compton* was able to explain his experimental results by postulating that the incoming X-ray beam was not a wave but an assembly of photons of energy $E\ (=h\nu)$ and that these photons experienced billiard-ball-like collisions with the free electrons in the scattering block. The "recoil" photons emerging from the block constitute, on this view, the scattered radiation. Since the incident photon transfers some of its energy to the electron with which it collides, the scattered photon must have a lower energy E'; it must therefore have a lower frequency $\nu'\ (=E'/h)$, which implies a larger wavelength $\lambda'\ (=c/\nu')$. This point of view accounts, at least qualitatively, for the wavelength shift $\Delta\lambda$. Notice how different this particle model of X-ray scattering is from that based on the wave picture. Now let us analyze a single photon-electron collision quantitatively.

Figure 49-11 shows a collision between a photon and an electron, the electron assumed to be initially at rest and essentially free, that is, not bound to the atoms of the scatterer. Let us apply the law of conservation of energy to this collision. Since the recoil electrons may have a speed v that is comparable with that of light, we must use the relativistic expression for the kinetic energy of the electron. From Eqs. 49-10 and 8-21 we may write

$$h\nu = h\nu' + (m - m_0)c^2,$$

in which the second term on the right is the relativistic expression for the kinetic energy of the recoiling electron, m being the relativistic mass and m_0 the rest mass of that particle. Substituting c/λ for ν (and c/λ' for ν') and using Eq. 8-20 to eliminate the relativistic mass m leads us to

$$\frac{hc}{\lambda} = \frac{hc}{\lambda'} + m_0c^2\left(\frac{1}{\sqrt{1-(v/c)^2}} - 1\right). \qquad (49\text{-}13)$$

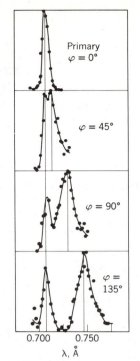

figure 49-10

Compton's experimental results. The solid vertical line on the left corresponds to the wavelength λ, that on the right to λ'. Results are shown for four different angles of scattering φ. Note that the Compton shift $\Delta\lambda$ for $\varphi = 90°$ is $h/m_0c = 0.0243$ Å.

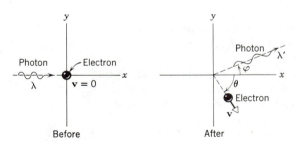

figure 49-11

A photon of wavelength λ is incident on an electron at rest. On collision, the photon is scattered at an angle φ with increased wavelength λ', whereas the electron moves off with speed v in direction θ.

Now let us apply the (vector) law of conservation of linear momentum to the collision of Fig. 49-11. We first need an expression for the momentum of a photon. In Section 42-2 we saw that if an object completely absorbs an energy U from a parallel light beam that falls on it, the light beam, according to the wave theory of light, will simultaneously transfer to the object a linear momentum given by U/c. On the photon picture we imagine this momentum to be carried along by the individual photons, each photon transporting linear momentum in amount $p = h\nu/c$, where $h\nu$ is the photon energy. Thus, if we substitute

*P. W. Debye simultaneously and independently offered the same interpretation.

λ for c/ν, we can write

$$p = \frac{E}{c} = \frac{h\nu}{c} = \frac{h}{\lambda}. \qquad (49\text{-}14)$$

This conclusion, that the momentum of a photon is given by h/λ, may also be deduced from the theory of relativity.

For the electron, the relativistic expression for the linear momentum is given by Eq. 9-13, or

$$\mathbf{p}_e = \frac{m_0 \mathbf{v}}{\sqrt{1 - (v/c)^2}}.$$

We can then write for the conservation of the x-component of linear momentum

$$\frac{h}{\lambda} = \frac{h}{\lambda'} \cos \varphi + \frac{m_0 v}{\sqrt{1 - (v/c)^2}} \cos \theta \qquad (49\text{-}15)$$

and for the y-component

$$0 = \frac{h}{\lambda'} \sin \varphi - \frac{m_0 v}{\sqrt{1 - (v/c)^2}} \sin \theta. \qquad (49\text{-}16)$$

Our immediate aim is to find $\Delta\lambda\ (=\lambda' - \lambda)$, the wavelength shift of the scattered photons, so that we may compare it with the experimental results of Fig. 49-10. Compton's experiment did not involve observations of the recoil electron in the scattering block. Of the five collision variables (λ, λ', v, φ, and θ) that appear in the three equations (49-13, 49-15, and 49-16) we may eliminate two. We chose to eliminate v and θ, which deal only with the electron, thereby reducing the three equations to a single relation among the variables.

Carrying out the necessary algebraic steps (see Problem 31) leads to this simple result:

$$\Delta\lambda\ (=\lambda' - \lambda) = \frac{h}{m_0 c} (1 - \cos \varphi). \qquad (49\text{-}17)$$

Thus the Compton shift $\Delta\lambda$ depends only on the scattering angle φ and *not* on the initial wavelength λ. Equation 49-17 predicts within experimental error the experimentally observed Compton shifts of Fig. 49-10. Note from the equation that $\Delta\lambda$ varies from zero (for $\varphi = 0$, corresponding to a "grazing" collision in Fig. 49-11, the incident photon being scarcely deflected) to $2h/m_0 c$ (for $\varphi = 180°$, corresponding to a "head-on" collision, the incident photon being reversed in direction).

It remains to explain the presence of the peak in Fig. 49-10 for which the wavelength does *not* change on scattering. This peak can be understood as resulting from a collision between a photon and electrons bound in an ionic core in the scattering block. During photon collisions the bound electrons behave like the free electrons that we considered in Fig. 49-11, with the exception that their effective mass is much greater. This is because the ionic core as a whole recoils during the collision. The effective mass M for a carbon scatterer is approximately the mass of a carbon nucleus. Since this nucleus contains six protons and six neutrons, we have approximately, $M = 12 \times 1840 m_0 = 22{,}000 m_0$. If we replace m_0 by M in Eq. 49-17, we see that the Compton shift for collisions with tightly bound electrons is immeasurably small.

As in the cavity radiation problem (see Eq. 49-7) and the photoelectric effect (see Eq. 49-11), Planck's constant h is centrally involved in the

Compton effect. The quantity h is the central constant of quantum physics. In a universe in which $h = 0$ there would be no quantum physics and classical physics would be valid in the subatomic domain. In particular, as Eq. 49-17 shows, there would be no Compton effect (that is, $\Delta\lambda = 0$) in such a universe.

EXAMPLE 4

X-rays with $\lambda = 1.00$ Å $(=0.10$ nm) are scattered from a carbon block. The scattered radiation is viewed at 90° to the incident beam. (a) What is the Compton shift $\Delta\lambda$? (b) What kinetic energy is imparted to the recoiling electron?

(a) Putting $\varphi = 90°$ in Eq. 49-17, we have, for the Compton shift,

$$\Delta\lambda = \frac{h}{m_0 c}(1 - \cos\varphi)$$

$$= \frac{6.63 \times 10^{-34} \text{ J} \cdot \text{s}}{(9.11 \times 10^{-31} \text{ kg})(3.00 \times 10^8 \text{ m/s})}(1 - \cos 90°)$$

$$= 2.43 \times 10^{-12} \text{ m} = 0.0243 \text{ Å} = 2.43 \times 10^{-3} \text{ nm}.$$

(b) If we put K for the kinetic energy of the electron, we can write Eq. 49-13 as

$$\frac{hc}{\lambda} = \frac{hc}{\lambda'} + K.$$

Since $\lambda' = \lambda + \Delta\lambda$, we obtain

$$\frac{hc}{\lambda} = \frac{hc}{\lambda + \Delta\lambda} + K,$$

which reduces to

$$K = \frac{hc\,\Delta\lambda}{\lambda(\lambda + \Delta\lambda)}$$

$$= \frac{(6.63 \times 10^{-34} \text{ J} \cdot \text{s})(3.00 \times 10^8 \text{ m/s})(2.43 \times 10^{-12} \text{ m})}{(1.00 \times 10^{-10} \text{ m})(1.00 + 0.024) \times 10^{-10} \text{ m}}$$

$$= 4.73 \times 10^{-17} \text{ J} = 295 \text{ eV}.$$

You may show that the initial photon energy E in this case $(=h\nu = hc/\lambda)$ is 12,400 eV so that the photon lost about 2.3% of its energy in this collision. A photon whose energy was ten times as large $(= 124,000$ eV) can be shown to lose 23% of its energy in a similar collision. This follows from the fact that $\Delta\lambda$ does not depend on the initial wavelength. Hence more energetic X-rays, which have smaller wavelengths, will experience a larger percent increase in wavelength and thus a larger percent loss in energy.

49-7
LINE SPECTRA

We have seen how Planck successfully explained the nature of the radiation from heated solid objects of which the cavity radiator formed the prototype. Such radiations form *continuous spectra* and are contrasted with *line spectra* such as those of Fig. 49-12, which show the radiation emitted from iron ions and atoms in an electric arc struck between iron electrodes. We shall see that Planck's quantization ideas, suitably ex-

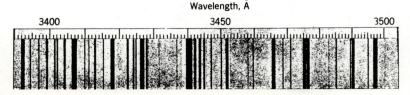

Wavelength, Å

3400   3450   3500

figure 49-12
A small portion of the spectrum of iron, in the region of 3400 to 3500 Å $(= 340$ to 350 nm).

tended, lead to an understanding of line spectra also. The prototype for the study of line spectra is that of atomic hydrogen; being the simplest atom it has the simplest spectrum.

Line spectra are common in all parts of the electromagnetic spectrum. Figure 49-13 shows a spectrum of the γ-rays $(\lambda \cong 10^{-12}$ m) emitted from a particular radioactive nucleus, an isotope of mercury. Figure 49-14 shows a spectrum of X-rays $(\lambda \cong 10^{-10}$ m) emitted from a molybdenum target when struck by a 35-KeV electron beam. The sharp emission lines are superimposed on a continuous background.

Figure 49-15 shows a spectrum associated with the molecule HCl. It occurs in the infrared, with $\lambda \cong 10^{-6}$ m. This is an *absorption* spectrum rather than an emission spectrum, as in Fig. 49-12. Experiment shows that isolated atoms and molecules absorb radiation, as well as emit it, at discrete wavelengths.

Figure 49-16 shows a portion of the absorption spectrum of ammonia (NH_3) in the microwave region $(\lambda \cong 10^{-2}$ m). Finally, Fig. 49-17 shows how radiation in the radio-frequency region $(\lambda \cong 43$ m) is absorbed by hydrogen molecules placed in a magnetic field.

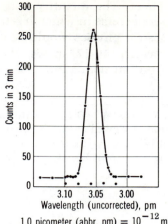

1.0 picometer (abbr. pm) = 10^{-12} m

figure 49-13
A wavelength plot for a gamma ray emitted by the nucleus Hg¹⁹⁸. (From data by DuMond and co-workers.)

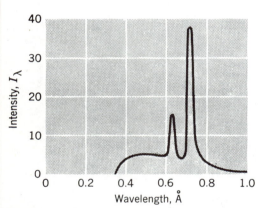

figure 49-14
X-rays from a molybdenum target struck by 35-keV electrons. Note the two sharp lines rising above a broad continuous base. The wavelength of the most intense line is 7.1×10^{-2} nm or 0.71 Å. (From data by Ulrey.)

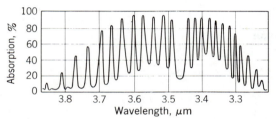

figure 49-15
An absorption spectrum of the HCl molecule near $\lambda = 3.5 \times 10^{-6}$ m = 3.5 μm. (From data by E. S. Imes.)

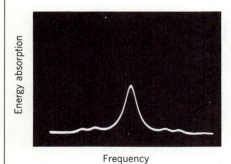

Frequency

figure 49-16
An oscilloscope trace showing one strong line and four weak lines in the absorption spectrum of ammonia at microwave frequencies.

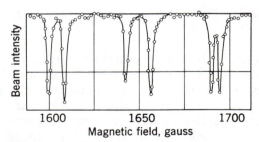

figure 49-17
A portion of the absorption spectrum of the protons in molecular hydrogen at $\lambda \cong 43$ m. In this technique the frequency is left fixed and the sample is placed in a magnetic field, which is varied to scan the spectrum. (From data by Kellogg, Rabi, and Zacharias.)

The attempts of physicists to explain observable phenomena in terms of theoretical models of the physical world which can be given mathematical expression is nowhere better illustrated than in the development of models of the atom. In this case, key evidence leading finally to the wave-mechanical atom was the line spectrum of hydrogen.

In 1815 Prout (1785–1850) suggested that the elements were made up of hydrogen, using as evidence the fact that the atomic weights of many elements are nearly integer multiples of that of hydrogen. With the discovery of the electron by J. J. Thomson (1856–1940) in 1897 the level of sophistication increased greatly. Thomson proposed the "plum pudding" model where the positive charge of the atom was thought to be spread out through the whole atom (a sphere of radius about 10^{-10} m) with the electrons located here and there like plums in the pudding. Then in 1911 Ernest Rutherford (1871–1937) showed the inconsistency between the α-particle scattering experiments of Geiger and Marsden and Thomson's atom and proposed the nuclear model of the atom which is the basis of present theories. Here the positive charge is confined to the nucleus, a very small sphere of radius about 10^{-14} m. The electrons circulate about the nucleus in a volume of the same order of magnitude as Thomson's sphere.

Investigation of the hydrogen spectrum led Niels Bohr (1885–1962) to postulate that the circular orbits of the electrons were quantized, that is, that their angular momentum could have only integral multiples of a basic value. We shall present this Bohr atom in some detail here. The Bohr atom, while deficient in several details, illustrates the ideas of quantization within the simpler mathematical framework of classical physics. Before proceeding, however, we should point out that the wavemechanical atom subsequently replaced the Bohr atom, as we will point out in Sections 50-3 and 50-4. Furthermore, models of the nucleus, while retaining the basic assumptions of Rutherford, have been highly refined and now assume the presence of subnuclear particles which themselves move within and make up the nucleus.

We might wish to associate the frequency of an emitted spectrum line, such as from hydrogen (Fig. 49-18), with the frequency of an electron revolving in an orbit inside the atom. Classical electromagnetism predicts that charges will radiate energy when they are accelerated. In this way electromagnetic waves are emitted from a radio transmitting antenna in which electrons are caused to surge back and forth. This radiation represents a loss of energy for the moving electrons which, in a radio antenna, is compensated for by supplying energy from an oscillator. In an isolated atom, however, no energy is supplied from external sources. We would expect the frequency of the electron, and thus that of the emitted radiation, to change continuously as the energy drains away. This prediction of classical theory cannot be reconciled with the existence of sharp spectrum lines. Thus classical physics cannot explain the hydrogen, or any other, spectrum.

Bohr circumvented this difficulty by assuming that, like Planck's oscillators, the hydrogen atom exists in certain stationary states in which it does not radiate. Radiation occurs only when the atom makes a transition from one state, with energy E_k, to a state with lower energy E_j. In equation form

$$h\nu = E_k - E_j, \qquad (49\text{-}18)$$

where $h\nu$ is the quantum of energy carried away by the photon that is emitted from the atom during the transition.

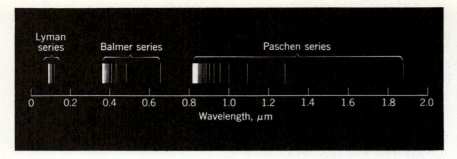

figure 49-18
The spectrum of hydrogen. It consists of a number of series of lines, three of which are shown. Within each series the spectrum lines follow a regular pattern, approaching a so-called *series limit* at the short-wave end of the series.

In order to learn the allowed frequencies predicted by Eq. 49-18, we need to know the energies of the various stationary states in which a hydrogen atom can exist. This calculation was first carried out by Bohr on the basis of a specific model for the hydrogen atom put forward by him. Bohr's model was highly successful for hydrogen and had a tremendous influence on the further development of the subject; it is now regarded as an important preliminary stage in the development of a more complete theory of quantum physics.

Let us assume that the electron in the hydrogen atom moves in a circular orbit of radius r centered on its nucleus. We assume that the nucleus, which is a single proton, is so massive that the center of mass of the system is essentially at the position of the proton. Let us calculate the energy E of such an atom.

Writing Newton's second law for the motion of the electron, we have (using Coulomb's law)

$$F = ma,$$

or

$$\frac{e^2}{4\pi\epsilon_0 r^2} = m\frac{v^2}{r}.$$

This allows us to calculate the kinetic energy of the electron, which is

$$K = \tfrac{1}{2}mv^2 = \frac{e^2}{8\pi\epsilon_0 r}. \tag{49-19}$$

The potential energy U of the proton-electron system is given by

$$U = V(-e) = -\frac{e^2}{4\pi\epsilon_0 r}, \tag{49-20}$$

where V ($=e/4\pi\epsilon_0 r$) is the potential of the proton at the radius of the electron.

The total energy E of the system is

$$E = K + U = -\frac{e^2}{8\pi\epsilon_0 r}. \tag{49-21}$$

Since the orbit radius can apparently take on any value, so can the energy E. The problem of quantizing E reduces to that of quantizing r.

Every property of the orbit is fixed if the radius is given. Equations 49-19, 49-20, and 49-21 show this specifically for the energies K, U, and E. From Eq. 49-19 we can show that the linear speed v for the electron is also given in terms of r by

$$v = \sqrt{\frac{e^2}{4\pi\epsilon_0 mr}}. \tag{49-22}$$

The rotational frequency ν_0 follows at once from

$$\nu_0 = \frac{v}{2\pi r} = \sqrt{\frac{e^2}{16\pi^3\epsilon_0 mr^3}}. \tag{49-23}$$

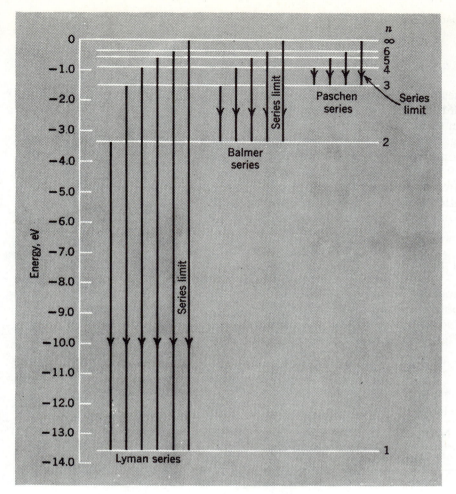

figure 49-19
An energy level diagram for hydrogen showing the quantum number *n* for each level and some of the transitions that appear in the spectrum. An infinite number of levels is crowded in between the levels marked $n = 6$ and $n = \infty$. Compare this figure carefully with Fig. 49-18.

The linear momentum p follows from Eq. 49-22:

$$p = mv = \sqrt{\frac{me^2}{4\pi\epsilon_0 r}}. \tag{49-24}$$

The angular momentum L is given by

$$L = pr = \sqrt{\frac{me^2 r}{4\pi\epsilon_0}}. \tag{49-25}$$

Thus if r is known, the orbit parameters K, U, E, v, ν_o, p, and L are also known. If any one of these quantities is quantized, all of them must be.

At this stage Bohr had no rules to guide him and so made (after some indirect reasoning which we do not reproduce) a bold hypothesis, namely, that the necessary quantization of the orbit parameters shows up most simply when applied to the angular momentum and that, specifically, L can take on only values given by

$$L = n\frac{h}{2\pi} \qquad n = 1, 2, 3, \ldots . \tag{49-26}$$

Planck's constant appears again in a fundamental way; the integer n is a *quantum number*.

Combining Eqs. 49-25 and 49-26 leads to

$$r = n^2 \frac{h^2\epsilon_0}{\pi me^2} \qquad n = 1, 2, 3, \ldots , \tag{49-27}$$

which tells how r is quantized. Substituting Eq. 49-27 into Eq. 49-21 produces

$$E = -\frac{me^4}{8\epsilon_0^2 h^2}\frac{1}{n^2} \qquad n = 1, 2, 3, \ldots, \qquad (49\text{-}28)$$

which gives directly the energy values of the allowed stationary states.

Figure 49-19 shows the energies of the stationary states and their associated quantum numbers. Equation 49-27 shows that the orbit radius increases as n^2. The upper level in Fig. 49-19, marked $n = \infty$, corresponds to a state in which the electron is completely removed from the atom (that is, $E = 0$ and $r = \infty$). Figure 49-19 also shows some of the quantum jumps that take place between the different stationary states.

Combining Eqs. 49-18 and 49-28 allows us to write a completely theoretical formula for the frequencies of the lines in the hydrogen spectrum. It is

$$\nu = \frac{me^4}{8\epsilon_0^2 h^3}\left(\frac{1}{j^2} - \frac{1}{k^2}\right) \qquad (49\text{-}29)$$

in which j and k are integers describing, respectively, the lower and the upper stationary states. The corresponding wavelengths can easily be found from $\lambda = c/\nu$. Table 49-1 shows some wavelengths so calculated; it should be compared carefully with Figs. 49-18 and 49-19.

Table 49-1
The hydrogen spectrum
(Some selected lines)

Name of series	Quantum Number		Wavelength, Å
	j (lower state)	k (upper state)	
Lyman	1	2	1216
	1	3	1026
	1	4	970
	1	5	949
	1	6	940
	1	∞	912
Balmer	2	3	6563
	2	4	4861
	2	5	4341
	2	6	4102
	2	7	3970
	2	∞	3650
Paschen	3	4	18751
	3	5	12818
	3	6	10938
	3	7	10050
	3	8	9546
	3	∞	8220

EXAMPLE 5

Calculate the binding energy of the hydrogen atom (the energy binding the electron to the nucleus) from Eq. 49-28.

The binding energy is numerically equal to the energy of the lowest state in Fig. 49-19. The largest negative value of E in Eq. 49-28 is found for $n = 1$. This yields

$$E = -\frac{me^4}{8\epsilon_0^2 h^2}$$

$$= -\frac{(9.11 \times 10^{-31} \text{ kg})(1.60 \times 10^{-19} \text{ C})^4}{(8)(8.85 \times 10^{-12} \text{ C}^2/\text{N} \cdot \text{m}^2)^2 (6.63 \times 10^{-34} \text{ J} \cdot \text{s})^2}$$

$$= -2.17 \times 10^{-18} \text{ J} = -13.6 \text{ eV},$$

which agrees with the experimentally observed binding energy for hydrogen.

In principle any negatively charged particle could replace the electron in the hydrogen (or other) atom, still be bound to the nucleus, and orbit the nucleus in a manner similar to the electron. Such a newly formed atom is called an *exotic atom*.* In practice, it is difficult to build exotic atoms, for many of the substitute negative particles are unstable or, when captured by a nucleus, form unstable atoms. Nevertheless, such atoms have been made and studied.

The Bohr model also applies to an exotic hydrogenlike atom, and Eq. 49-27 shows that the radii are smaller the larger the mass of the orbiting particle. A muon (see Appendix F), for example, will orbit much closer to the nucleus than an electron and its behavior will be strongly affected by the distribution of charge within the nucleus. Studies of muonic atoms (see Problem 49-46) give us information, therefore, about how the positive charges (protons) are distributed within nuclei.

In general, the motivation to investigate exotic atoms is either to learn about the size and detailed structure of the nucleus or to learn about properties (such as mass) of the negative particles themselves, which are more easily determined when they are in a bound system than when they are moving about freely.

Although all theories in physics have limitations, they usually do not break down abruptly but in a continuous way, yielding results that agree less and less well with experiment. Thus the predictions of Newtonian mechanics become less and less accurate as the speed is made to approach that of light. A similar relationship must exist between quantum physics and classical physics; it remains to find the circumstances under which the latter theory is revealed as a special case of the former.

The radius of the lowest energy state in hydrogen (the so-called *ground state*) is found by putting $n = 1$ in Eq. 49-27; it turns out to be 5.3×10^{-11} m. If $n = 10,000$, however, the radius is $(10,000)^2$ times as large or 5.3 mm. This "atom" is so large that we suspect that its behavior should be accurately described by classical physics. Let us test this by computing the frequency of the emitted light on the basis of both classical and quantum assumptions. These calculations should differ at small quantum numbers but should agree at large quantum numbers. The fact that quantum physics reduces to classical physics at large quantum numbers is called the *correspondence principle*. This principle, credited to Niels Bohr, was very useful during the years in which quantum physics was being developed. Bohr, in fact, based his theory of the hydrogen atom on correspondence principle arguments.

Classically, the frequency of the light emitted from an atom is equal to ν_0, its frequency of revolution† in its orbit. We can express this in terms of a quantum number n by combining Eqs. 49-23 and 49-27 to

49-9
THE CORRESPONDENCE PRINCIPLE

---

* See "Exotic Atoms" by Clyde E. Wiegand in *Scientific American*, November 1972.
† Integral multiples of the frequency also exist but may be ignored without affecting the present argument.

obtain

$$\nu_0 = \frac{me^4}{8\epsilon_0^2 h^3} \frac{2}{n^3}. \qquad (49\text{-}30)$$

Quantum physics predicts that the frequency ν of the emitted light is given by Eq. 49-29. Considering a transition between an orbit with quantum number $k = n$ and one with $j = n - 1$ leads to

$$\nu = \frac{me^4}{8\epsilon_0^2 h^3} \left[\frac{1}{(n-1)^2} - \frac{1}{n^2} \right]$$

$$= \frac{me^4}{8\epsilon_0^2 h^3} \left[\frac{2n-1}{(n-1)^2 n^2} \right]. \qquad (49\text{-}31)$$

As $n \to \infty$ the expression in the square brackets above approaches $2/n^3$ so that $\nu \to \nu_0$ as $n \to \infty$. Table 49-2 illustrates this example of the correspondence principle.

Table 49-2
The correspondence principle as applied to the hydrogen atom

Quantum Number, n	Frequency of Revolution in Orbit (Eq. 49-30) Hz	Frequency of Transition to Next Lowest State (Eq. 49-31) Hz	Difference, %
2	8.20×10^{14}	24.6×10^{14}	67
5	5.26×10^{13}	7.38×10^{13}	29
10	6.57×10^{12}	7.72×10^{12}	14
50	5.25×10^{10}	5.42×10^{10}	3
100	6.578×10^{9}	6.677×10^{9}	1.5
1,000	6.5779×10^{6}	6.5878×10^{6}	0.15
10,000	6.5779×10^{3}	6.5789×10^{3}	0.015
25,000	4.2099×10^{2}	4.2102×10^{2}	0.007

questions

1. "Pockets" formed by the coals in a coal fire seem brighter than the coals themselves. Is the temperature in such pockets appreciably higher than the surface temperature of an exposed glowing coal?

2. The relation $\mathcal{R} = \sigma T^4$ (Eq. 49-2) is exact for true cavities and holds for all temperatures. Why don't we use this relation as the basis of a definition of temperature at, say, 100°C?

3. Do all incandescent solids obey the fourth-power law of temperature, as Eq. 49-3 seems to suggest?

4. A hole in the wall of a cavity radiator is sometimes called a *black body*. Why?

5. It is stated that if we look into a cavity whose walls are maintained at a constant temperature, no details of the interior are visible. Does this seem reasonable?

6. A piece of metal glows with a bright red color at 1100 K. At this same temperature, however, a piece of quartz does not glow at all. Explain. (*Hint:* Quartz is transparent to visible light.)

7. Are there quantized quantities in classical physics? If so, give examples. Is energy quantized in classical physics?

8. Does it make sense to speak of charge quantization in physics? How, if at all, is this different from energy quantization?

9. Show that Planck's constant has the dimensions of angular momentum.

Does this necessarily suggest that angular momentum is a quantized quantity?

10. For quantum effects to be "everyday" phenomena in our lives, what order of magnitude value would h need to have? (See G. Gamow, *Mr. Tompkins in Wonderland* (Cambridge University Press, Cambridge, 1957), for a delightful popularization of a world in which the physical constants c, G, and h make themselves obvious.)

11. In Fig. 49-7, why doesn't the photoelectric current rise vertically to its maximum (saturation) value when the applied potential difference is slightly more positive than V_0?

12. In the photoelectric effect, why does the existence of a cutoff frequency speak in favor of the photon theory and against the wave theory?

13. Why are photoelectric measurements so sensitive to the nature of the photoelectric surface?

14. Why is it that even for incident radiation that is monochromatic the photoelectrons are emitted with a spread of velocities?

15. How can a *photon* energy be given by $E = h\nu$ (Eq. 49-10) when the very presence of the frequency ν in the formula implies that light is a *wave*?

16. Does Einstein's theory of photoelectricity, in which light is postulated to be a photon, invalidate Young's interference experiment?

17. Explain the statement that one's eyes could not detect faint starlight if light were not corpuscular.

18. Assume that the emission of photons from a source of radiation is random in direction. Would you expect the intensity (or energy density) to vary inversely as the square of the distance from the source in the photon theory as it does in the wave theory?

19. What is the direction of a Compton scattered electron with maximum kinetic energy compared with the direction of the incident monochromatic photon beam?

20. Why, in the Compton scattering picture (Fig. 49-11), would you expect $\Delta\lambda$ to be independent of the materials of which the scatterer is composed?

21. Why don't we observe a Compton effect with visible light?

22. Light from distant stars is Compton scattered many times by free electrons in outer space before reaching us. This shifts the light toward the red. How can this shift be distinguished from the Doppler red shift due to the motion of receding stars?

23. Why was the Balmer series, rather than the Lyman or Paschen series, the first to be detected and analyzed in the hydrogen spectrum?

24. Only a relatively small number of Balmer lines can be observed from laboratory discharge tubes, whereas a large number are observed in stellar spectra. Explain this in terms of the small density, high temperature, and large volume of gases in stellar atmospheres.

25. In Bohr's theory for the hydrogen atom orbits, what is the implication of the fact that the potential energy is negative and is greater in magnitude than the kinetic energy?

26. (a) Can a hydrogen atom absorb a photon whose energy exceeds its binding energy (13.6 eV)? (b) What minimum energy must a photon have to initiate the photoelectric effect in hydrogen gas?

27. On emitting a photon, the hydrogen atom recoils to conserve momentum. Explain the fact that the energy of the emitted photon is less than the energy difference between the energy levels involved in the emission process.

28. Would you expect to observe all the lines of atomic hydrogen if such a gas were excited by electrons of energy 13.6 eV? Explain.

29. How would you estimate the temperature of hydrogen gas at which atomic collisions cause significant ionization of the atoms?

30. List and discuss the assumptions made by Planck in connection with the cavity radiation problem, by Einstein in connection with the photoelectric effect, and by Bohr in connection with the hydrogen atom problem.

31. Describe several methods that can be used to determine experimentally the value of Planck's constant h.

32. Discuss Example 1 in terms of the correspondence principle.

33. According to classical mechanics, an electron moving in an orbit should be able to do so with any angular momentum whatever. According to Bohr's theory of the hydrogen atom, however, the angular momentum is quantized according to $L = nh/2\pi$. Reconcile these two statements, using the correspondence principle.

34. The correspondence principle is expressed by taking n to infinity or by taking h to zero. Explain the relation between these two techniques for finding the classical limit.

35. Can you use the device of letting $h \rightarrow 0$ to obtain classical results from quantum results in the case of the photoelectric effect? Explain.

36. (a) Newton's light corpuscles were assumed to behave according to the laws of Newtonian mechanics. Is the photon concept a return to this idea of a light corpuscle? (b) The ether was invented as a medium in which light waves undulate. Does the photon concept eliminate the need for an ether?

problems

SECTION 49-2

1. Figure 49-1 compares the spectral radiancy of tungsten at 2000 K with that of a cavity radiator. If the radiancy $\mathscr{R}$—the area under the curve for tungsten—is 23.3 W/cm², verify that the emissivity of tungsten at 2000 K is 0.259.

2. What is the power radiated from a nichrome wire 1.0 m long and having a diameter of 0.060 in. at a temperature of 800°C if the emissivity of nichrome is 0.92?

3. The power rating of a light bulb tells how much electrical power is supplied. (a) An incandescent bulb of power rating $P = 100$ W has a tungsten filament of diameter $d = 0.40$ mm and length $l = 30$ cm. The emissivity of tungsten is 0.26. At what temperature does the filament operate? (b) A fluorescent tube rated at 40 W gives as much visible light as a 100-W incandescent bulb. Explain why.   *Answer:* (a) 1470 K.

4. The average rate of solar radiation incident per unit area on the earth is 0.485 cal/cm² · min (or 335 W/m²). (a) Explain the consistency of this number with the solar constant (the solar energy falling per unit time at normal incidence on unit area of the earth's surface) whose value is 1.94 cal/cm² · min (or 1340 W/m²). (b) Consider the earth to behave like a cavity radiating energy into space at this same rate. What surface temperature would the earth have under these circumstances?

5. (a) Assuming the surface temperature of the sun to be 5700 K, use the Stefan-Boltzmann law to determine the rest mass lost per second to radiation by the sun. Take the sun's diameter to be 1.4×10^9 m. (b) What fraction of the sun's rest mass is lost each year by electromagnetic radiation? Take the sun's rest mass to be 2.0×10^{30} kg.
Answer: (a) 4.1×10^9 kg. (b) 6.5×10^{-14}.

6. An oven with an inside temperature $T = 227$°C is in a room having a temperature $T_r = 27$°C. There is a small opening of area 5.0 cm² in one side of the oven. How much net power is transferred from the oven to the room? (*Hint:* Consider both oven and room as cavities.)

SECTION 49-3

7. The wavelength λ_{max} at which the spectral radiancy has its maximum value per unit wavelength for a particular temperature T is given by the

Wien displacement law, $\lambda_{max} T = $ constant. At what wavelength does a cavity radiator at 6000 K radiate most per unit wavelength? The experimentally determined value of Wien's constant is 2.898×10^{-3} m·K.
Answer: 4800 Å = 480 nm.

8. Show that Wien's law is a special case of Planck's law (Eq. 49-6) for short wavelengths or low temperatures.

9. Prove the Stefan-Boltzmann law by showing directly from Planck's formula that

$$\int_0^\infty \mathcal{R}_\lambda \, d\lambda = AT^4$$

where A is a constant.

10. A cavity whose walls are held at 4000 K has a circular aperture 5.0 mm in diameter. (*a*) At what rate does energy in the visible range (defined to extend from 0.40 to 0.70 μm) escape from this hole? (*b*) What fraction of the total radiation escaping from the cavity does this represent? Solve either analytically or graphically.

SECTION 49-4

11. In Example 2 suppose that the "target" is a single gas atom of 1.0Å (= 0.10 nm) radius and that the intensity of the light source is reduced to 1.0×10^{-5} W. If the binding energy of the most loosely bound electron in the atom is 2.0 eV, what time lag for the photoelectric effect is expected on the basis of the wave theory of light? *Answer:* 10 years.

SECTION 49-5

12. Show that the energy E of a photon (in eV) is related to the wavelength λ (in nm) by

$$E = 1.24 \times 10^3 / \lambda.$$

13. Solar radiation falls on the earth at a rate of 2.0 cal/cm²·min. How many photons/cm²·min is this, assuming an average wavelength of 550 nm? *Answer:* 2.3×10^{19}.

14. A 100-W sodium vapor lamp radiates uniformly in all directions. (*a*) At what distance from the lamp will the average density of photons be 10/cm³? (*b*) What is the average density of photons 2.0 m from the lamp? Assume the light to be monochromatic, with $\lambda = 589$ nm (= 5890 Å).

15. An atom absorbs a photon having a wavelength of 375 nm and immediately emits another photon having a wavelength of 580 nm. How much energy was absorbed by the atom in this process? Ease the computation by using the result of Problem 12. *Answer:* 1.17 eV.

16. What are (*a*) the frequency, (*b*) the wavelength, and (*c*) the momentum of a photon whose energy equals the rest energy of the electron?

17. The energy required to remove an electron from sodium is 2.3 eV. Does sodium show a photoelectric effect for orange light, with $\lambda = 680$ nm (= 6800 Å)? *Answer:* No.

18. You wish to pick a substance for a photocell operable with visible light. Which of the following will do (work function in parentheses): tantalum (4.2 eV); tungsten (4.5 eV); aluminum (4.2 eV); barium (2.5 eV); lithium (2.3 eV)?

19. Incident photons strike a sodium surface having a work function $E_0 = 2.2$ eV causing photoelectric emission. When a stopping potential $V_0 = 5.0$ V is imposed, there is no photocurrent. What is the wavelength of the incident photons? *Answer:* 172 nm.

20. Find the maximum kinetic energy of photoelectrons if the work function of the material is 2.0×10^{-19} J and the frequency of the radiation is 3.0×10^{15} Hz.

21. (*a*) If the work function for a metal is 1.8 eV, what would be the stopping

potential for light having a wavelength of 4000 Å? (b) What would be the maximum velocity of the emitted photoelectrons at the metal's surface? *Answer:* (a) 1.3 V. (b) 6.8×10^5 m/s.

22. Light of a wavelength 200 nm falls on an aluminum surface. In aluminum 4.2 eV are required to remove an electron. What is the kinetic energy of (a) the fastest and (b) the slowest emitted photoelectrons? (c) What is the stopping potential? (d) What is the cutoff wavelength for aluminum?

23. The work function for a clean lithium surface is 2.3 eV. Make a rough plot of the stopping potential V_0 versus the frequency of incident light for such a surface.

24. Show, by analyzing a collision between a photon and a free electron (using relativistic mechanics), that it is impossible for a photon to give all of its energy to the free electron. In other words, the photoelectric effect cannot occur for completely free electrons; the electrons must be bound in a solid or in an atom.

SECTION 49-6

25. Photons of wavelength 0.024 Å are incident on free electrons. (a) Find the wavelength of a photon which is scattered 30° from the incident direction. (b) Do the same if the scattering angle is 120°.
Answer: (a) 0.027 Å. (b) 0.060 Å.

26. What is the maximum kinetic energy of the Compton-scattered electrons from a sheet of copper struck by a monochromatic photon beam in which the incident photons each have a wavelength of 1.4×10^{-3} nm? (See Example 4.)

27. An X-ray photon of wavelength $\lambda = 0.10$ Å strikes an electron head on $(\varphi = 180°)$. Determine (a) the change in wavelength of the photon, (b) the change in energy of the photon, and (c) the final kinetic energy of the electron.   *Answer:* (a) 0.049 Å. (b) −41 keV. (c) 41 keV.

28. A 0.20-nm photon falling on a carbon block is scattered by a Compton collision and its frequency is shifted by 0.010%. (a) Through what angle is the photon scattered? (b) How much energy does the electron which scattered the photon gain? [Note that for any wave motion, $\Delta\nu = -\left(\dfrac{c}{\lambda^2}\right)\Delta\lambda$.]

29. Calculate the percent change in photon energy for a Compton collision with φ in Fig. 49-11 equal to 90° for radiation in (a) the microwave range, with $\lambda = 3.0$ cm, (b) the visible range, with $\lambda = 5000$ Å = 500 nm, (c) the X-ray range, with $\lambda = 1.0$ Å = 0.10 nm, and (d) the gamma-ray range, the energy of the gamma-ray photons being 1.0 MeV. What are your conclusions about the importance of the Compton effect in these various regions of the electromagnetic spectrum, judged solely by the criterion of energy loss in a single Compton encounter?
Answer: (a) 8.1×10^{-9}%. (b) 4.9×10^{-4}%. (c) 2.4%. (d) 68%.

30. A photon "hits" an electron "head-on" and recoils backward directly along the line of incidence. If the electron moves off at a speed βc, where $\beta \ll 1$ ($=10^{-3}$, for example), show that the ratio of the final electron kinetic energy to the initial photon energy is just β. (*Hint:* Set up the problem as a non-relativistic Compton "collision.")

31. Carry out the necessary algebra to eliminate v and θ from Eqs. 49-13, 49-15, and 49-16 to obtain the Compton shift relation (Eq. 49-17).

SECTION 49-8

32. A line in the X-ray spectrum of gold consists of photons all having nearly the same wavelength 0.185 Å. If the energy in these photons comes from the atoms jumping from one specific energy level, −13.7 KeV, to another lower one, what is the second energy?

33. Light of wavelength 4863 Å = 486.3 nm is emitted by a hydrogen atom. (a) What transition of the hydrogen atom is responsible for this radiation? (b) To what series does this radiation belong?
Answer: (a) $n = 4$ to $n = 2$. (b) The Balmer series.

34. Show on an energy-level diagram for hydrogen, the quantum numbers corresponding to a transition in which the wavelength of the emitted photon is 1216 Å = 121.6 nm.

35. What are (a) the energy, (b) momentum, and (c) wavelength of the photon that is emitted when a hydrogen atom undergoes a transition from the state $n = 3$ to $n = 1$?   *Answer:* (a) 12 eV. (b) 6.5×10^{-27} kg·m/s. (c) 1030 Å.

36. (a) Using Bohr's formula for the frequencies of the lines of the hydrogen spectrum, calculate the three longest wavelengths of the Balmer series. (b) Between what wavelength limits does the Balmer series lie?

37. In the ground state of the hydrogen atom, according to Bohr's theory, what are (a) the quantum number, (b) the orbit radius, (c) the angular momentum, (d) the linear momentum, (e) the angular velocity, (f) the linear speed, (g) the force on the electron, (h) the acceleration of the electron, (i) the kinetic energy, (j) the potential energy, and (k) the total energy?
Answer: (a) 1. (b) 5.3×10^{-11} m. (c) 1.1×10^{-34} J·s. (d) 2.0×10^{-24} kg·m/s. (e) 4.1×10^{16} rad/s. (f) 2.2×10^6 m/s. (g) 8.2×10^{-8} N. (h) 9.0×10^{22} m/s^2. (i) +13.6 eV. (j) −27.2 eV. (k) −13.6 eV.

38. How do the quantities (b) to (k) in Problem 37 vary with the quantum number?

39. How much energy is required to remove an electron from a hydrogen atom in the state with $n = 8$?   *Answer:* 0.21 eV.

40. A hydrogen atom is excited from a state with $n = 1$ to one with $n = 4$. (a) Calculate the energy that must be absorbed by the atom. (b) Calculate and display on an energy-level diagram the different photon energies that may be emitted if the atom returns to the $n = 1$ state. (c) Calculate the recoil speed of the hydrogen atom, assumed initially at rest, if it makes the transition from $n = 4$ to $n = 1$ in a single quantum jump.

41. A hydrogen atom in a state having a *binding energy* (the energy required to remove an electron) of 0.85 eV makes a transition to a state with an *excitation energy* (the difference in energy between the state and the ground state) of 10.2 eV. (a) Find the energy of the emitted photon. (b) Show this transition on an energy-level diagram for hydrogen, labeling the appropriate quantum numbers.
Answer: (a) 2.6 eV. (b) $n = 4$ to $n = 2$.

42. A neutron, with kinetic energy of 6.0 eV, collides with a resting hydrogen atom in its ground state. Show that this collision must be elastic (that is, kinetic energy must be conserved).

43. From the energy-level diagram for hydrogen, explain the observation that the frequency of the second Lyman-series line is the sum of the frequencies of the first Lyman-series line and the first Balmer-series line. This is an example of the empirically discovered *Ritz combination principle.* Use the diagram to find some other valid combination.

44. Using Bohr's theory, calculate the energy required to remove the electron from the ground state of singly ionized helium, that is, helium with one electron removed.

45. Use Bohr's theory to compare the spectrum of singly ionized helium with the spectrum of hydrogen.
Answer: $\lambda_{\text{He}} = \tfrac{1}{4}\lambda_{\text{H}}$, for corresponding spectral lines.

46. *Muonic atoms.* Apply Bohr's theory to a muonic atom, which consists of a nucleus of charge Ze with a negative muon (an elementary particle with a charge of $-e$ and a mass m that is 207 times as large as the electron mass) circulating about it. Calculate (a) the radius of the first Bohr orbit, (b) the ionization energy, and (c) the wavelength of the most energetic photon that can be emitted. Assume that the muon is circulating about a hydrogen nucleus ($Z = 1$).

47. *Positronium.* Apply Bohr's theory to the positronium atom. This consists of a positive and a negative electron revolving around their center of mass, which lies halfway between them. (a) What relationship exists between this

spectrum and the hydrogen spectrum? (b) What is the radius of the ground state orbit? (*Hint:* It will be necessary to analyze this problem from first principles because this "atom" has no nucleus; both particles revolve about a point halfway between them.)

Answer: (a) Corresponding positronium wavelengths are longer by a factor of two. (b) Radius to center of mass is equal to the corresponding radius for hydrogen.

48. Perhaps an atom could be formed by an electron and a neutron binding together by gravitational forces. Calculate the ground state radius of an electron in such an atom by using a Bohr-type model in which the Coulomb attractive electrical force is replaced by the attractive gravitational force.

49. A diatomic gas molecule consists of two atoms of mass m separated by a fixed distance d rotating about an axis as indicated in Fig. 49-20. Assuming that its angular momentum is quantized as in the Bohr atom, determine (a) the possible angular velocities, and (b) the possible quantized rotational energies. (c) Show these on an energy-level diagram.

Answer: (a) $nh/\pi md^2$. (b) $n^2h^2/4\pi^2md^2$.

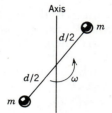

figure 49-20
Problem 49

SECTION 49-9

50. (a) If the angular momentum of the earth due to its motion around the sun were quantized according to Bohr's relation $L = nh/2\pi$, what would the quantum number be? (b) Could such quantization be detected if it existed?

51. In Table 49-2 show that the quantity in the last column is given by

$$\frac{100(\nu - \nu_0)}{\nu} \cong \frac{150}{n}$$

for large quantum numbers.

52. If an electron is revolving in an orbit at frequency ν_0, classical electromagnetism predicts that it will radiate energy not only at this frequency but also at $2\nu_0$, $3\nu_0$, $4\nu_0$, and so on. Show that this is also predicted by Bohr's theory of the hydrogen atom in the limiting case of large quantum numbers.

50
waves
and particles

In 1924 Louis de Broglie of France reasoned that (*a*) nature is strikingly symmetrical in many ways; (*b*) our observable universe is composed entirely of light and matter; (*c*) if light has a dual, wave-particle nature, perhaps matter has also. Since matter was then regarded as being composed of particles, de Broglie's reasoning suggested that one should search for a wavelike behavior for matter.

De Broglie's suggestion might not have received serious attention had he not predicted what the expected wavelength of the so-called matter waves would be. We recall that about 1680 Huygens put forward a wave theory of light that did not receive general acceptance, not only because it seemed to contradict Newton's particle — or corpuscular — theory but also because Huygens was not able to state what the wavelength of the light was. When Thomas Young rectified this defect in 1800, the wave theory of light started on its way to acceptance.

De Broglie assumed that the wavelength of the predicted matter waves was given by the same relationship that held for light, namely, Eq. 49-14, or

$$\lambda = \frac{h}{p},\qquad(50\text{-}1)$$

which connects the wavelength of a light wave with the momentum of the associated photons. The dual nature of light shows up strikingly in this equation and also in Eq. 49-10 ($E = h\nu$). Each equation contains within its structure both a wave concept (ν and λ) and a particle concept

50-1
MATTER WAVES

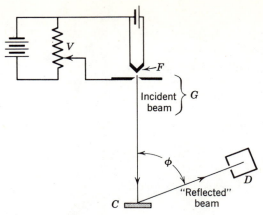

figure 50-1
The apparatus of Davisson and Germer. Electrons from the hot filament F are accelerated by an adjustable potential difference V. After "reflection" from crystal C they are collected by detector D.

(E and p). De Broglie predicted that the wavelength of matter waves would also be given by Eq. 50-1, where p would now be the momentum of the particle of matter.

EXAMPLE 1

What wavelength is predicted by Eq. 50-1 for a beam of electrons whose kinetic energy is 100 eV?

We can find the speed of the electrons from $K = \frac{1}{2}mv^2$, or

$$v = \sqrt{\frac{2K}{m}} = \sqrt{\frac{(2)(100 \text{ eV})(1.6 \times 10^{-19} \text{ J/eV})}{9.1 \times 10^{-31} \text{ kg}}}$$

$$= 5.9 \times 10^6 \text{ m/s}.$$

The momentum follows from

$$p = mv = (9.1 \times 10^{-31} \text{ kg})(5.9 \times 10^6 \text{ m/s}) = 5.4 \times 10^{-24} \text{ kg} \cdot \text{m/s}.$$

The wavelength (called the *de Broglie wavelength*) is found from Eq. 50-1 or

$$\lambda = \frac{h}{p} = \frac{6.6 \times 10^{-34} \text{ J} \cdot \text{s}}{5.4 \times 10^{-24} \text{ kg} \cdot \text{m/s}} = 1.2 \text{ Å } (= 0.12 \text{ nm}).$$

This is the same order of magnitude as the size of an atom or the spacing between adjacent planes of atoms in a solid.

In 1926 Elsasser pointed out that the wave nature of matter might be tested in the same way that the wave nature of X-rays was first tested, namely, by allowing a beam of electrons of the appropriate energy to fall on a crystalline solid. The atoms of the crystal serve as a three-dimensional array of diffracting centers for the electron "wave"; we should look for strong diffracted peaks in certain characteristic directions, just as for X-ray diffraction.

This idea was tested by C. J. Davisson and L. H. Germer in this country and by G. P. Thomson in Scotland.* Figure 50-1 shows the apparatus of Davisson and Germer. Electrons from a heated filament are accelerated by an adjustable potential difference V and emerge from the "electron gun" G with kinetic energy eV. This electron beam is allowed to

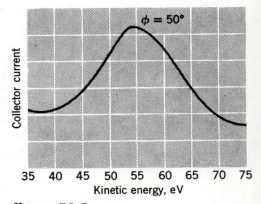

figure 50-2
The collector current in detector D in Fig. 50-1 as a function of the kinetic energy of the incident electrons, showing a diffraction maximum. The angle ϕ in Fig. 50-1 is adjusted to 50°. If an appreciably smaller or larger value is used, the diffraction maximum disappears.

---

* For an historical account of Thomson's researches see "Early Work in Electron Diffraction" by Sir George Thomson, *Scientific American,* December 1961. Interestingly, G. P. Thomson, who demonstrated that electrons are wave-like, is the son of J. J. Thomson who demonstrated (see Section 33-8) that they are particle-like. Both father and son received Nobel prizes, 31 years apart.

fall at normal incidence on a single crystal of nickel at C. Detector D is set at a particular angle ϕ and readings of the intensity of the "reflected" beam are taken at various values of the accelerating potential V. Figure 50-2 shows that a strong beam occurs at $\phi = 50°$ for $V = 54$ volts.

All such strong "reflected" beams can be accounted for by assuming that the electron beam has a wavelength, given by $\lambda = h/p$, and that "Bragg reflections" occur from certain families of atomic planes precisely as described for X-rays in Section 47-6.

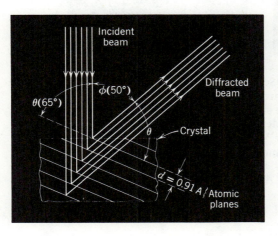

figure 50-3
The strong diffracted beam at $\phi = 50°$ and $V = 54$ volts arises from wavelike "reflection" from the family of atomic planes shown, for $d = 0.91$ Å. The Bragg angle θ is 65° For simplicity, refraction of the wave as it enters and leaves the surface is ignored.

Figure 50-3 shows such a Bragg reflection, obeying the Bragg relationship

$$m\lambda = 2d \sin \theta \qquad m = 1, 2, 3, \ldots \qquad (50\text{-}2)$$

For the conditions of Fig. 50-3 the effective interplanar spacing d can be shown by X-ray analysis to be 0.91 Å. Since ϕ equals 50°, it follows that θ equals $90° - \frac{1}{2} \times 50°$ or 65°. The wavelength to be calculated from Eq. 50-2, if we assume $m = 1$, is

$$\lambda = 2d \sin \theta = 2(0.91 \text{ Å})(\sin 65°) = 1.65 \text{ Å} = 0.165 \text{ nm}.$$

The wavelength calculated from the de Broglie relationship $\lambda = h/p$ is, for 54-eV electrons (see Example 1), 1.64 Å = 0.164 nm. This excellent agreement, combined with much similar evidence*, is a convincing argument for believing that electrons are wavelike in some circumstances.

Not only electrons but all other particles, charged or uncharged, show wavelike characteristics. Beams of slow neutrons from nuclear reactors are routinely used to investigate the atomic structure of solids.† See Fig. 50-4, which suggests the wave-like character of (a) electrons and (b) neutrons.

The evidence for the existence of matter waves with wavelengths given by Eq. 50-1 is strong indeed. Nevertheless, the evidence that matter is composed of particles remains equally strong; see Fig. 10-12. Thus, for matter as for light, we must face up to the existence of a dual character; matter behaves in some circumstances like a particle and in others like a wave.

* An interferometer operating with electron beams has been designed, and diffraction patterns have been obtained from electron beams passing through a multiple slit system. See "An Experiment on Electron Interference" by O. Donati, G. F. Missiroli, and G. Possi, *American Journal of Physics*, May 1973, and "Electron Diffraction at Multiple Slits" by Clauss Jönsson, *American Journal of Physics*, January 1974.
† An apparatus that exhibits the interference of neutron waves has been constructed, as well. See "Neutrons as Waves" by Dietrick E. Thomsen in *Science News*, April 24, 1976.

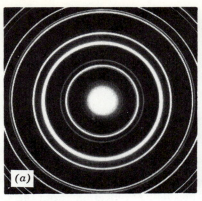

figure 50-4

In Figure 50-4a we show a Debye-Scherrer pattern of *electron* diffraction by gold crystals (compare with the X-ray diffraction pattern of Fig. 47-17) and in Fig. 50-4b we show a Laue pattern of diffraction of *neutrons* from a nuclear reactor by a single sodium chloride crystal (compare with the X-ray diffraction pattern of Fig. 47-12). (Figure 50-4a is from U. Fano and L. Fano, *Basic Physics of Atoms and Molecules*, John Wiley and Sons, 1959, and Fig. 50-4b is from Blackwood et al., *Outline of Atomic Physics*, John Wiley and Sons, 1955.)

The motion of electrons in beams is not bounded or limited in the beam direction. We can make an analogy to a sound wave in a long gas-filled tube, a wave traveling down a long string, or an electromagnetic wave in a long waveguide. All four cases can be described by appropriate traveling waves and, significantly, waves of *any* wavelength (within a certain range) can be propagated.

Let these last three waves be bounded by imposing physical restrictions. For the sound wave this corresponds to inserting end walls on a section of the long gas-filled pipe, thus forming an acoustic resonant cavity (Fig. 38-5). For the waves in the string it corresponds to removing a finite section of string and clamping it at each end, as a violin string (Fig. 20-4). For the electromagnetic wave it corresponds to inserting end caps on a finite length of waveguide, thus forming an electromagnetic resonant cavity (Fig. 38-6).

Two important changes occur: (a) the motions are now represented by standing rather than traveling waves and (b) only certain wavelengths (or frequencies) can now exist. This *quantization* of the wavelength is a direct result of confining the wave to a finite region. We expect that if electrons are limited in their motions by being localized in an atom, that (a) the electron motion can be represented by a standing matter wave, and (b) the electron motion will become quantized, that is, its energy can take on only certain discrete values.

De Broglie was able to derive the Bohr quantization condition for angular momentum by applying proper boundary conditions to matter waves in the hydrogen atom. Figure 50-5 suggests an instantaneous "snapshot" of a standing matter wave associated with an orbit of radius r. The de Broglie wavelength $(\lambda = h/p)$ has been chosen so that the orbit of radius r contains an integral number n of the matter waves, or

50-2
ATOMIC STRUCTURE AND STANDING WAVES

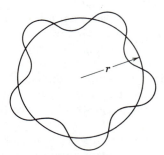

figure 50-5

Showing how an electron wave can be adjusted in wavelength to fit an integral number of times around the circumference of a Bohr orbit of radius r. This concept, like the Bohr orbit concept, is now regarded as oversimplified.

$$\frac{2\pi r}{\lambda} = \frac{2\pi r}{(h/p)} = n \qquad n = 1, 2, 3, \ldots.$$

This leads at once to

$$L = pr = n\frac{h}{2\pi} \qquad n = 1, 2, 3, \ldots$$

which (see Section 49-8) is the Bohr quantization condition for L.

The idea that the stationary states in atoms correspond to standing matter waves was taken up by Erwin Schrödinger in 1926 and used by him as the foundation of *wave mechanics,* one of several equivalent formulations of quantum physics.

An important quantity in wave mechanics is the *wave function* Ψ, which measures the "wave disturbance" of matter waves. For waves on strings the "wave disturbance" may be measured by a transverse displacement y; for sound waves it may be measured by a pressure variation p; for electromagnetic waves it may be measured by the electric field vector **E**.

We make the physical meaning of the wave disturbance Ψ clear in Section 50-4. Meanwhile, let us study the wave function $\Psi(x)$ for a simple, one-dimensional problem, that of the possible motions of a particle of mass m confined between rigid walls of separation l as in Fig. 50-6b. The wave function can be obtained by analogy with a known mechanical problem, that of the natural modes of vibration of a string of length l, clamped at each end as in Fig. 50-6a.

In the vibrating string the boundary conditions require that nodes exist at each end. This means that the wavelength λ must be chosen so that

$$l = n\frac{\lambda}{2} \qquad n = 1, 2, 3, \ldots,$$

or that the wavelength λ is "quantized" by the requirement that

$$\lambda = \frac{2l}{n} \qquad n = 1, 2, 3, \ldots, \tag{50-3}$$

The wave disturbance for the string is represented by a standing wave whose spatial dependence is $A \sin kx$ (see Section 19-9), where A is a constant and $k(= 2\pi/\lambda)$ is the wave number. Since λ is quantized, k must be also, or

$$k = \frac{2\pi}{\lambda} = \frac{n\pi}{l} \qquad n = 1, 2, 3, \ldots,$$

which leads to

$$y = A \sin\frac{n\pi x}{l} \qquad n = 1, 2, 3, \ldots. \tag{50-4}$$

Inspection of Eq. 50-4 shows that no matter what value of n we select, nodes exist at $x = 0$ and at $x = l$, as required by the boundary conditions. Figure 50-7 shows plots of this equation (the spatial dependence of the standing wave) for the modes of vibration of the string corresponding to $n = 1, 2,$ and 3.

Consider now the particle confined between rigid walls. Since we assume the walls to be perfectly rigid, the particle cannot penetrate them so that Ψ, which represents the particle motion in some way not

50-3
WAVE MECHANICS

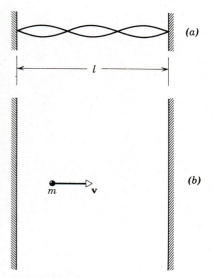

(a)

(b)

figure 50-6
(a) A stretched string of length l clamped between rigid supports. (b) A particle of mass m and velocity v confined to move between rigid walls a distance l apart.

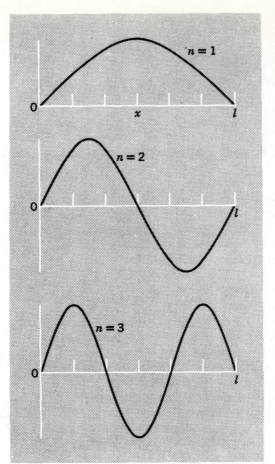

figure 50-7
Three quantized modes of vibration for the string of Fig. 50-6a. The figure also represents three of the quantized wave functions for the particle of Fig. 50-6b. The ordinate is a displacement amplitude in the first case and a wave-function amplitude in the second.

yet clearly specified (see Section 50-4), must vanish at $x = 0$ and $x = l$. The allowed wavelengths of the matter waves must be given by Eq. 50-3, or

$$\lambda = \frac{2l}{n}.$$

Replacing λ by h/p (see Eq. 50-1) leads to

$$p = \frac{nh}{2l}, \tag{50-5}$$

which shows that the linear momentum of the particle is quantized. The momentum $p(= mv)$ is related to the energy E (which is entirely kinetic and is equal to $\frac{1}{2}mv^2$) by

$$p = \sqrt{2mE}. \tag{50-6}$$

Combining Eqs. 50-5 and 50-6 leads to the quantization condition for E, or

$$E = n^2 \frac{h^2}{8ml^2} \qquad n = 1, 2, 3, \ldots. \tag{50-7}$$

The particle cannot have any energy, as we would expect classically, but only energies given by Eq. 50-7.

We describe the matter wave in strict analogy with Eq. 50-4, by

$$\Psi = A \sin \frac{n \pi x}{l} \qquad n = 1, 2, 3, \ldots. \qquad (50\text{-}8)$$

Figure 50-7 can serve equally well to show how the amplitude of the standing matter waves for the states of motion corresponding to $n = 1, 2,$ and 3 varies throughout the box. We see clearly in this problem how the act of localizing or bounding a particle leads to energy quantization.

EXAMPLE 2

Consider an electron $(m = 9.1 \times 10^{-31}$ kg) confined by electrical forces to move between two rigid "walls" separated by 1.0×10^{-9} m, which is about five atomic diameters. Find the quantized energy values for the three lowest stationary states.

From Eq. 50-7, for $n = 1$, we have

$$E = n^2 \frac{h^2}{8ml^2} = (1)^2 \frac{(6.6 \times 10^{-34} \text{ J} \cdot \text{s})^2}{(8)(9.1 \times 10^{-31} \text{ kg})(1.0 \times 10^{-9} \text{ m})^2}$$

$$= 6.0 \times 10^{-20} \text{ J} = 0.38 \text{ eV}.$$

The energies for the next two states ($n = 2$ and $n = 3$) are $2^2 \times 0.38$ eV $= 1.5$ eV and $3^2 \times 0.38$ eV $= 3.4$ eV.

EXAMPLE 3

Consider a grain of dust $(m = 1.0 \ \mu\text{g} = 1.0 \times 10^{-9}$ kg) confined to move between two rigid walls separated by 0.1 mm $(= 10^{-4}$ m). Its speed is only 10^{-6} m/s, so that it requires 100 s to cross the gap. What quantum number describes this motion?

The energy is

$$E \ (= K) = \tfrac{1}{2} mv^2 = \tfrac{1}{2} (10^{-9} \text{ kg})(10^{-6} \text{ m/s})^2$$

$$= 5 \times 10^{-22} \text{ J}.$$

Solving Eq. 50-7 for n yields

$$n = \sqrt{8mE} \frac{l}{h} = \sqrt{(8)(10^{-9} \text{ kg})(5 \times 10^{-22} \text{ J})} \left(\frac{10^{-4} \text{ m}}{6.6 \times 10^{-34} \text{ J} \cdot \text{s}} \right)$$

$$= 3 \times 10^{14}.$$

Even in these extreme conditions the quantized nature of the motion would never be apparent; we cannot distinguish experimentally between $n = 3 \times 10^{14}$ and $n = 3 \times 10^{14} + 1$. Classical physics, which fails completely for the problem of Example 2, works extremely well for this problem.

50-4
MEANING OF Ψ

Max Born first suggested that the quantity Ψ^2 at any particular point is a measure of the probability that the particle will be near that point. More exactly, if a volume element dV is constructed at that point, the probability that the particle will be found in the volume element at a given instant is $\Psi^2 \, dV$. This interpretation of Ψ provides a statistical connection between the wave and the associated particle; it tells us where the particle is likely to be, not where it is.

For the particle confined between rigid walls, the probability that the particle will lie between two planes that are distance x and $x + dx$ from one wall (see Fig. 50-8) is given by

$$\Psi^2 \, dx = A^2 \sin^2 \frac{n \pi x}{l} \, dx.$$

Thus the *probability density* is

$$\Psi^2 = A^2 \sin^2 \frac{n\pi x}{l} \qquad n = 1, 2, 3, \dots . \qquad (50\text{-}9)$$

Figure 50-8 shows Ψ^2 for the three stationary states corresponding to $n = 1$, 2, and 3. Note that for $n = 1$ the particle is more likely to be near the center than the ends. This is in sharp contradiction to the results of classical physics, according to which the particle has the same probability of being located anywhere between the walls, as shown by the horizontal line in Fig. 50-8.

The problem of a particle confined between rigid walls has little real application in physics. We would prefer to illustrate the wave mechanics of Schrödinger by applying it to a more experimentally realizable situation, such as the hydrogen atom. Only mathematical complexity prevents us from doing this. We state without proof that when this problem is solved by wave mechanics, the motion of the electron in the ground state of the atom, defined by putting $n = 1$ in Eq. 49-28, is described by the following wave function:

$$\Psi = \sqrt{\frac{1}{\pi a^3}}\, e^{-r/a} \qquad (50\text{-}10)$$

where
$$a = \frac{h^2 \epsilon_0}{\pi m e^2}.$$

Putting $n = 1$ in Eq. 49-27 shows that a is the radius of the ground-state orbit in Bohr's theory. This special interpretation has little meaning in wave mechanics; a is taken here merely as a convenient unit of length when dealing with atomic problems, having the value 0.529 Å = 0.0529 nm.

Consider two hypothetical spherical shells centered on the nucleus of a hydrogen atom with radii r and $r + dr$. What is the probability $P(r)$ that the electron will lie between these shells, as a function of r?

This probability is $\Psi^2\, dV$, where dV is the volume between the shells, or $4\pi r^2\, dr$. Thus

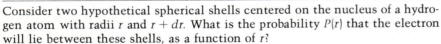

$$\Psi^2\, dV = \left(\sqrt{\frac{1}{\pi a^3}}\, e^{-r/a}\right)^2 (4\pi r^2\, dr) = P(r)\, dr$$

or
$$P(r) = \frac{4r^2}{a^3}\, e^{-2r/a}.$$

Figure 50-9 shows a plot of this function. Note that the most probable location for the electron corresponds to the first Bohr radius. Thus in wave mechanics we do not say that the electron in the $n = 1$ state in hydrogen goes around the nucleus in a circular orbit of 0.529 Å radius, but only that the electron is more likely to be found at this distance from the nucleus than at any other distance, either larger or smaller.

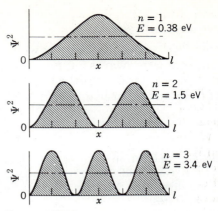

figure 50-8
The "probability density" for three states of motion of the particle of Fig. 50-6b, along with the corresponding quantized energies for the conditions of Example 2. The horizontal lines show the predictions of classical mechanics, in which the probability function is constant for all positions of the particle.

EXAMPLE 4

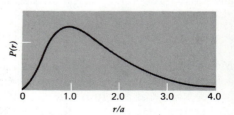

figure 50-9
The probability function for the ground state of the hydrogen atom, as calculated from wave mechanics. The separation between the nucleus and the electron is r; a is the radius of the first Bohr orbit (0.529×10^{-10} m), used here merely as a convenient unit of distance.

Some physicists believe that only those quantities that can be measured have any real meaning in physics. If we could focus a "super" microscope on an electron in an atom and see it moving around in an orbit, we would declare that such orbits have meaning. However, we shall show that it is fundamentally impossible to make such an observation — even with the most ideal instruments that could conceivably be constructed. Therefore, one might say that such orbits have no physical meaning.

We observe the moon traveling around the earth by means of the sunlight that it reflects in our direction. Now light transfers linear momentum to an object from which it is reflected. In principle, this reflected light would disturb the course of the moon in its orbit, although a little thought shows that this disturbing effect is negligible in this case.

For electrons the situation is quite different. Here, too, we can hope to "see" the electron only if we reflect light, or another particle, from it. In this case the recoil that the electron experiences when the light (photon) bounces from it completely alters the electron's motion in a way that cannot be avoided or even corrected for.

It is not surprising that the probability curve of Fig. 50-9 is the most detailed information that we can hope to obtain, by measurement, about the distribution of negative charge in the hydrogen atom. If orbits such as those envisaged by Bohr existed, they would be broken up completely in our attempts to verify their existence. Under these circumstances, we prefer to say that it is the probability function, and not the orbits, that represents physical reality.

The fact that we can't describe the motions of electrons in a classical way finds expression in the *uncertainty principle,* enunciated by Werner Heisenberg in 1927.* To formulate this principle, consider a beam of monoenergetic electrons of speed v_0 moving from left to right in Fig. 50-10. Let us set ourselves the task of measuring the position of a particular electron in the vertical (y) direction and also its velocity component v_y in this direction. If we succeed in carrying out these measurements with unlimited accuracy, we can then claim to have established the position and motion of the electron (or one component of it at least) with precision. However, we shall see that it is impossible to make these two measurements simultaneously with unlimited accuracy.

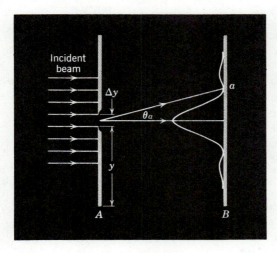

figure 50-10
An incident beam of electrons is diffracted at the slit in screen A, forming a typical diffraction pattern on screen B. If the slit is made narrower, the pattern becomes wider.

* See "The Uncertainty Principle" by G. Gamow in *Scientific American,* January 1958.

To measure y we block the beam with an absorbing screen A in which we put a slit of width Δy. If an electron gets through the slit, its vertical position must be known to this accuracy. By making the slit narrower, we can improve the accuracy of this vertical position measurement as much as we wish.

Since the electron behaves like a wave, it will undergo diffraction at the slit, and a photographic plate placed at B in Fig. 50-10 will reveal a typical diffraction pattern. The existence of this diffraction pattern means that there is an uncertainty in the values of v_y possessed by the electrons emerging from the slit. Let v_{ya} be the value of v_y that corresponds to an electron landing at the first minimum on the screen, marked by point a and described by a characteristic angle θ_a. We take v_{ya} as a rough measure of the uncertainty Δv_y in v_y for electrons emerging from the slit.

The first minimum in the diffraction pattern is given by Eq. 46-2, or

$$\sin \theta_a = \frac{\lambda}{\Delta y}.$$

If we assume that θ_a is small enough, we can write this equation as

$$\theta_a \cong \frac{\lambda}{\Delta y}. \qquad (50\text{-}11)$$

To reach point a, v_{ya} $(= \Delta v_y)$ must be such that

$$\theta_a \cong \frac{\Delta v_y}{v_0}. \qquad (50\text{-}12)$$

Combining Eqs. 50-11 and 50-12 leads to

$$\frac{\Delta v_y}{v_0} = \frac{\lambda}{\Delta y},$$

which we rewrite as $\qquad \Delta v_y \, \Delta y \cong \lambda v_0. \qquad (50\text{-}13)$

Now λ, the wavelength of the electron beam, is given by h/p or h/mv_0; substituting this into Eq. 50-13 yields

$$\Delta v_y \, \Delta y \cong \frac{hv_0}{mv_0}.$$

We rewrite this as $\qquad \Delta p_y \, \Delta y \cong h. \qquad (50\text{-}14)$

In Eq. 50-14 Δp_y $(=m\Delta v_y)$ is the uncertainty in our knowledge of the vertical momentum of the electrons; Δy is the uncertainty in our knowledge of their vertical position. The equation tells us that, since the product of these uncertainties is a constant, we cannot measure p_y and y simultaneously with unlimited accuracy.

If we want to improve our measurement of y (that is, if we want to reduce Δy), we use a finer slit. However, this will produce a wider diffraction pattern (see Eq. 50-11). A wider pattern means that our knowledge of the vertical momentum component of the electron has deteriorated, or, in other words, Δp_y has increased; this is exactly what Eq. 50-14 predicts.

The limits on measurement imposed by Eq. 50-14 have nothing to do with the crudity of our measuring instruments. We are permitted to postulate the existence of the finest conceivable measuring equipment. Equation 50-14 represents a fundamental limitation, imposed by nature.

Equation 50-14 is a derivation, for a special case, of a general principle known as the *uncertainty principle*. As applied to position and momentum measurements, it asserts that

$$\Delta p_x \, \Delta x \gtrsim h$$

$$\Delta p_y \, \Delta y \gtrsim h$$

$$\Delta p_z \, \Delta z \gtrsim h. \qquad (50\text{-}15)$$

Thus no component of the motion of an electron, free or bound, can be described with unlimited precision.

Planck's constant h probably appears nowhere that has more deep-seated significance than in Eq. 50-15. If this product had been zero instead of h, the classical ideas about particles and orbits would be correct; it would then be possible to measure both momentum and position with unlimited precision. The fact that h appears means that the classical ideas are wrong; the magnitude of h tells us under what circumstances these classical ideas must be replaced by quantum ideas. Gamow* has speculated, in an interesting and readable fantasy, what our world would be like if the constant h were much larger than it is, so that nonclassical ideas would be apparent to our sense perceptions.

EXAMPLE 5

An electron has a speed of 300 m/s, accurate to 0.010%. With what fundamental accuracy can we locate the position of this electron?

The electron momentum is

$$p = mv = (9.1 \times 10^{-31} \text{ kg})(300 \text{ m/s}) = 2.7 \times 10^{-28} \text{ kg} \cdot \text{m/s}.$$

The uncertainty in momentum is given to be 0.010% of this, or

$$\Delta p = (0.00010)(2.7 \times 10^{-28} \text{ kg} \cdot \text{m/s}) = 2.7 \times 10^{-32} \text{ kg} \cdot \text{m/s}.$$

The minimum uncertainty in position, from Eq. 50-15, is

$$\Delta x = \frac{h}{\Delta p} = \frac{6.6 \times 10^{-34} \text{ J} \cdot \text{s}}{2.7 \times 10^{-32} \text{ kg} \cdot \text{m/s}}$$

$$= 2.4 \text{ cm.}$$

If the electron momentum has really been determined by measurement to have the accuracy stated, there is no hope whatever that its position can be known to any better accuracy than that stated, namely, about 1 in. The concept of the electron as a tiny dot is not very valid under these circumstances.

EXAMPLE 6

A bullet has a speed of 300 m/s, accurate to 0.010%. With what fundamental accuracy can we locate its position? Its mass is 50 g (=0.050 kg).

This example is the same as Example 5 in every respect save the mass of the particle involved. The momentum is

$$p = mv = (0.05 \text{ kg})(300 \text{ m/s}) = 15 \text{ kg} \cdot \text{m/s}$$

and

$$\Delta p = (0.00010)(15 \text{ kg} \cdot \text{m/s}) = 1.5 \times 10^{-3} \text{ kg} \cdot \text{m/s}.$$

Equation 50-15 yields

$$\Delta x = \frac{6.6 \times 10^{-34} \text{ J} \cdot \text{s}}{1.5 \times 10^{-3} \text{ kg} \cdot \text{m/s}} = 4.4 \times 10^{-13} \text{ m.}$$

This is so far beyond the possibility of measurement (a nucleus is only about 10^{-15} m in diameter) that we can assert that for heavy objects like bullets the uncertainty principle sets no limits whatever on our measuring procedures. Once again the correspondence principle shows us how quantum physics reduces to classical physics under the appropriate circumstances.

* Mr. Tompkins in Wonderland, New York: Macmillan, 1940.

The uncertainty relation shows us why it is possible for both light and matter to have a dual, wave-particle, nature. It is because these two views, so obviously opposite to each other, can never be brought face to face in the same experimental situation. If we invent an experiment that forces the electron to reveal its wave character strongly, its particle character will always be inherently fuzzy. If we change the experiment to bring out the particle character more strongly, the wave character necessarily becomes fuzzy. Matter and light are like coins that can be made to display either face at will but not both simultaneously. Niels Bohr first pointed out in his principle of complementarity how the ideas of wave and of particle complement rather than contradict each other.

questions

1. How can the wavelength of an electron be given by $\lambda = h/p$ when the very presence of the momentum p in this formula implies that the electron is a particle?

2. How could Davisson and Germer be sure that the "54-eV" peak of Fig. 50-2 was a first-order diffraction peak, that is, that $m = 1$ in Eq. 50-2?

3. In a repetition of Thomson's experiment for measuring e/m for the electron (see Section 33-8), a beam of electrons is collimated by passage through a slit. Why is the beamlike character of the emerging electrons not destroyed by diffraction of the electron wave at this slit? (See Problem 9.)

4. Why is the wave nature of matter not more apparent to our daily observations?

5. Considering the wave behavior of electrons, we should expect to be able to construct an electron microscope. This, indeed, has been done. (a) Compare the lens system of a light microscope with that of an electron microscope. (b) What advantages might an electron microscope have over a light microscope? (For recent developments, see "The Scanning Electron Microscope" by T. E. Everhart and T. L. Hayes, *Scientific American*, January 1972; "A High-Resolution Scanning Electron Microscope" by Albert V. Crewe, *Scientific American*, April 1971.)

6. How can electron diffraction be used to study properties of the surface of a solid? (See, in this connection, "Electron Microscopy of Atoms in Crystals" by John M. Cowley and Sumio Iijima, *Physics Today*, March 1977.)

7. How would an electron interferometer work? . . . a neutron interferometer? (See footnotes, p. 1119).

8. Is an electron a particle? Is it a wave? Explain.

9. Considering electrons and photons to be particles, how are they different from each other?

10. Can the de Broglie wavelength of a particle be smaller than a linear dimension of the particle? . . . ; Larger? Is there any relation necessarily between such quantities?

11. If, in the de Broglie formula $\lambda = h/mv$, we let $m \to \infty$, do we get the classical result for particles of matter?

12. Apply the correspondence principle to the problem of a particle confined between rigid walls, showing that those features which seem "strange" (that is, the quantization of energy and the nonuniformity of the probability functions of Fig. 50-8) become undetectable experimentally at large quantum numbers.

13. In the $n = 1$ mode, for a particle confined between rigid walls, what is the probability that the particle will be found in a small volume element at the surface of either wall?

14. A standing wave can be viewed as the superposition of two traveling waves.

Can you apply this to the problem of a particle confined between rigid walls, giving an interpretation in terms of the motion of the electron?

15. Discuss similarities and differences between a matter wave and an electromagnetic wave.

16. What is the physical significance of the wave function Ψ?

17. How can the predictions of wave mechanics be so exact if the only information we have about the positions of the electrons is statistical?

18. (a) Give examples of how the process of measurement disturbs the system being measured. (b) Can the disturbances be taken into account ahead of time by suitable calculations?

19. Why does the concept of Bohr orbits violate the uncertainty principle?

20. Make up some numerical examples to show the difficulty of getting the uncertainty principle to reveal itself during experiments with an object whose mass is about 1 g.

21. Figure 50-8 shows that for $n = 3$ the probability function for a particle confined between rigid walls is zero at two points between the walls. How can the particle ever move across these positions? (*Hint:* Consider the implications of the uncertainty principle.)

22. We can state the uncertainty principle in terms of angular quantities (compare Eq. 50-15) as

$$\Delta L \, \Delta \phi \gtrsim h$$

where ΔL is the uncertainty in the *angular* momentum and $\Delta \phi$ the uncertainty in the *angular* position. For electrons in atoms the angular momentum has definite quantized values, with no uncertainty whatever. What can we conclude about the uncertainty in the angular position and about the validity of the orbit concept?

problems

SECTION 50-1

1. A bullet of mass 40 g travels at 1000 m/s. (a) What wavelength can we associate with it? (b) Why does the wave nature of the bullet not reveal itself through diffraction effects?   *Answer:* (a) $\lambda = 1.7 \times 10^{-35}$ m.

2. If the de Broglie wavelength of a proton is 1.0×10^{-13} m, (a) what is the speed of the proton and (b) through what electric potential would the proton have to be accelerated to acquire this speed?

3. Sodium ions are accelerated through a potential difference of 300 V. (a) What is the momentum acquired by the ions? (b) What is their de Broglie wavelength?   *Answer:* (a) 1.9×10^{-21} kg·m/s. (b) 3.5×10^{-13} m.

4. What wavelength do we associate with a beam of neutrons whose energy is 0.025 eV?

5. An electron and a photon each have a wavelength of 2.0 Å = 0.20 nm. What are their (a) momenta and (b) energies?
   Answer: (a) 3.3×10^{-24} kg·m/s for each. (b) 38 eV for the electron and 6200 eV for the photon.

6. The 50-GeV electron accelerator at Stanford (1 GeV = 10^9 eV) provides an electron beam of small wavelength, suitable for probing the fine details of nuclear structure by scattering experiments. What will this wavelength be and how does it compare with the size of an average nucleus? (At these energies it is sufficient to use the extreme relativistic relationship between momentum and energy, namely, $p = E/c$. This is the same relationship used for light (Eq. 42-2a) and is justified whenever the kinetic energy of a particle is much greater than its rest energy $m_0 c^2$, as in this case.)

7. The highest achievable resolving power of a microscope is limited only by the wavelength used; that is, the smallest detail that can be separated is about equal to the wavelength. Suppose one wishes to "see" inside an atom. Assuming the atom to have a diameter of 1.0 Å = 0.10 nm, this means that

we wish to resolve detail of separation about 0.10 Å = 0.010 nm. (a) If an electron microscope is used, what minimum energy of electrons is needed? (b) If a light microscope is used, what minimum energy of photons is needed? (c) Which microscope seems more practical for this purpose?
Answer: (a) 1.5×10^4 eV. (b) 1.2×10^5 eV.

8. What accelerating voltage would be required for electrons in an electron microscope to obtain the same ultimate resolving power as that which could be obtained from a gamma-ray microscope using 0.20 MeV-gamma rays?

9. In a repetition of Thomson's experiment for measuring e/m for the electron, a beam of 1.0×10^4-eV electrons is collimated by passage through a slit of width 0.50 mm. Why is the beamlike character of the emergent electrons not destroyed by diffraction of the electron wave at this slit?
Answer: The de Broglie wavelength (0.12 Å) is much smaller than the slit width.

10. In the experiment of Davisson and Germer (a) at what angles would the second- and third-order diffracted beams corresponding to a strong maximum of Fig. 50-2 occur, provided they are present, and (b) at what angle would the first-order diffracted beam occur if the accelerating potential were changed from 54 to 60 V? The experimenter is free to rotate the crystal.

11. A neutron crystal spectrometer utilizes crystal planes of spacing $d = 0.7323$ Å in a beryllium crystal. What must be the Bragg angle θ so that only neutrons of energy $K = 4.0$ eV are reflected? Consider only first-order reflections.   *Answer:* 5.6°.

12. Make a plot of de Broglie wavelength against kinetic energy for (a) electrons and (b) protons. Restrict the range of energy values to those in which classical mechanics applies reasonably well. A convenient criterion is that the maximum kinetic energy on each plot only be about 5% of the rest energy $m_0 c^2$ for the particular particle.

13. What is the wavelength of a hydrogen atom moving with a velocity corresponding to the mean kinetic energy for thermal equilibrium at 20°C?
Answer: 1.5 Å = 0.15 nm.

SECTION 50-2
14. According to the correspondence principle, as $n \to \infty$ we expect classical results in the Bohr atom. Hence, the de Broglie wavelength associated with the electron (a quantum result) should get smaller compared with the radius of the orbit as n increases. Indeed, we expect that $\lambda/r \to 0$ as $n \to \infty$. Is this the case?

SECTION 50-3
15. (a) What is the separation in energy between the lowest two energy levels for a container 20 cm on a side containing argon atoms? (b) How does this compare with the thermal energy of the argon atoms at 300 K? (c) At what temperature does the thermal energy equal the spacing between these two energy levels?
Answer: (a) 3.9×10^{-22} eV. (b) The thermal energy is about 10^{20} times as great. (c) 3.0×10^{-18} K.

16. (a) Find, approximately, the smallest allowed energy of an electron confined to an atomic nucleus (diameter about 1.4×10^{-14} m)? (b) Compare this with the several MeV of energy binding protons and neutrons inside the nucleus; on this basis should we expect to find electrons inside nuclei?

SECTION 50-4
17. If an electron moves from a state represented by $n = 3$ in Fig. 50-8 to one represented by $n = 2$ in that figure, emitting electromagnetic radiation in the process, what are (a) the energy of the single emitted photon and (b) the corresponding wavelength?   *Answer:* (a) 1.9 eV. (b) 6600 Å.

18. A particle is confined between rigid walls separated by a distance l. What is the probability that it will be found within a distance $l/3$ from one wall

(a) for $n = 1$, (b) for $n = 2$, (c) for $n = 3$, and (d) under the assumption of classical physics?

19. In the ground state of the hydrogen atom show that the probability P_r that the electron lies within a sphere of radius r is given by

$$P_r = 1 - e^{-2r/a}\left(\frac{2r^2}{a^2} + \frac{2r}{a} + 1\right).$$

Does this yield expected values for (a) $r = 0$ and (b) $r = \infty$? (c) State clearly the difference in meaning between this expression and that given in Section 50-4.   *Answer:* (a) Zero. (b) 1; both as expected.

20. In the ground state of the hydrogen atom, what is the probability that the electron will lie within a sphere whose radius is that of the first Bohr orbit?

SECTION 50-5

21. A microscope using photons is employed to locate an electron in an atom to within a distance of 0.10 Å = 0.010 nm. What is the uncertainty in the momentum of the electron located in this way?
Answer: 6.6×10^{-23} kg·m/s.

22. The uncertainty in the position of an electron is given as about 0.050 nm, which is the radius of the first Bohr orbit in hydrogen. What is the uncertainty in the linear momentum of the electron?

23. Show that if the uncertainty in the location of a particle is equal to its de Broglie wavelength, the uncertainty in its velocity is equal to its velocity.

24. In Example 2, the electron's energy was determined exactly by the size of the box. How do you reconcile this with the fact that the uncertainty in the location of the electron cannot exceed 1.0×10^{-9} m if the uncertainty principle is to be obeyed?

25. From the uncertainty principle $\Delta p_x \, \Delta x \gtrsim h$, show that, if L is the angular momentum component along a line perpendicular to the x-direction and φ is the azimuthal angle about this line (see Fig. 50-11), then

$$\Delta L \Delta \varphi \gtrsim h;$$

(see Question 22).

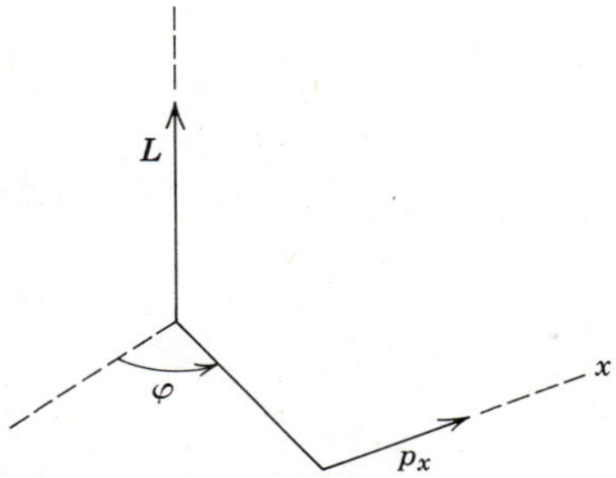

figure 50-11
Problem 25

supplementary topics

In Section 11-6 we discussed the relations between the linear and angular kinematic variables for a particle moving in a plane but confined to move in a circle about an axis at right angles to the plane. Such a particle might be any particle in a rigid body rotating about a fixed axis. Here we relax the restriction and allow the particle to move freely in the plane. A planet moving in an elliptical orbit about the sun is an example.

We start from Eq. 11-11, $\mathbf{r} = \mathbf{u}_r r$, in which, however, we now take *both* $\mathbf{u}_r$ and r to be variables; the particle is no longer confined to a circle of constant radius. We find the velocity by differentiation, or

$$\mathbf{v} = \frac{d\mathbf{r}}{dt} = \mathbf{u}_r \frac{dr}{dt} + r \frac{d\mathbf{u}_r}{dt}.$$

Equation 11-13 shows us that $d\mathbf{u}_r/dt = \mathbf{u}_\theta \omega$. Thus we can write

$$\mathbf{v} = \mathbf{u}_r \frac{dr}{dt} + \mathbf{u}_\theta \omega r, \qquad (\text{I-1})$$

which shows that $\mathbf{v}$ has two components, a radial component $v_r = dr/dt$ and a component at right angles, $v_\theta = \omega r$. If we hold r constant, then $dr/dt = 0$ and Eq. I-1 reduces to Eq. 11.14a, as it must.

To find the acceleration we differentiate Eq. I-1, remembering that *all five* quantities on the right are variables. We obtain

$$\mathbf{a} = \frac{d\mathbf{v}}{dt} = \mathbf{u}_r \frac{d^2r}{dt^2} + \frac{dr}{dt}\frac{d\mathbf{u}_r}{dt} + (\mathbf{u}_\theta)\left(\omega \frac{dr}{dt} + r \frac{d\omega}{dt}\right) + (\omega r)\left(\frac{d\mathbf{u}_\theta}{dt}\right).$$

Now $d\mathbf{u}_r/dt = \mathbf{u}_\theta \omega$, $d\mathbf{u}_\theta/dt = -\mathbf{u}_r \omega$ (see Eq. 11-16), and $d\omega/dt = \alpha$. Substituting and rearranging leads us finally to

$$\mathbf{a} = \mathbf{u}_r\left(\frac{d^2r}{dt^2} - \omega^2 r\right) + \mathbf{u}_\theta\left(\alpha r + 2\omega \frac{dr}{dt}\right). \qquad (\text{I-2})$$

$$= \mathbf{u}_r(a_r - \omega^2 r) + \mathbf{u}_\theta(\alpha r + 2\omega v_r).$$

SUPPLEMENTARY TOPIC I
RELATION BETWEEN LINEAR AND ANGULAR KINEMATICS FOR A PARTICLE MOVING IN A PLANE

Once again, if $r =$ a constant, then $dr/dt = d^2r/dt^2 = 0$ and Eq. I-2 reduces to Eq. 11-17, which we derived especially for this case.

The two new terms in Eq. I-2, $\mathbf{u}_r d^2r/dt^2$ and $\mathbf{u}_\theta 2\omega\, dr/dt$, need a little explanation. We can understand the first of these terms by imagining that the particle moving in the plane is *not* rotating about the axis. If we put $\omega = \alpha = 0$ in Eq. I-2 this equation reduces to

$$\mathbf{a} = \mathbf{u}_r \frac{d^2r}{dt^2} = \mathbf{u}_r a_r,$$

which is just the familiar acceleration of a particle moving along a straight line. Hence this term in Eq. I-2 gives the radial acceleration due to the change in the *magnitude* of $\mathbf{r}$, the other radial acceleration term arising from the changing *direction* of $\mathbf{r}$ as the particle rotates.

There are also two θ-directed acceleration terms. The first one, $\mathbf{u}_\theta \alpha r$, arises simply from the angular acceleration α of a particle in circular motion ($r =$ constant) and is the tangential acceleration of Section 11-5. To understand the second term, $\mathbf{u}_\theta 2\omega\, dr/dt$, consider a man walking outward along a radial line painted on the floor of a merry-go-round. The merry-go-round is rotating with constant angular velocity ω so that its angular acceleration α is now zero. If the man were simply to stand still on the merry-go-round, ($d^2r/dt^2 = dr/dt = 0$, and $r =$ constant) his acceleration, as seen by an observer in a reference frame on the ground (see Eq. I-2), would be simply the familiar centripetal acceleration $-\mathbf{u}_r \omega^2 r$, directed radially inward. If he walks outward, however, $dr/dt \neq 0$ and then Eq. I-2 predicts that the ground observer would also measure a θ-directed acceleration given by $\mathbf{u}_\theta 2\omega v_r$, where $v_r = dr/dt$. This is called a *Coriolis acceleration*. It arises from the fact that even though the angular velocity of the man is constant his speed increases as r increases. Let us convince ourselves that this effect really exists.*

Figure I-1a shows the walking man (point P) as he appears to the ground observer at times t and $t + \Delta t$. We show at time t his radially directed velocity $\mathbf{v}_r (= \mathbf{u}_r\, dr/dt)$ and also a θ-directed velocity caused by the rotation of the merry-go-round and given by $\mathbf{v}_\theta (= \mathbf{u}_\theta \omega r)$. At a time Δt later each of these velocities has changed. The radial velocity has changed in direction, although its magnitude remains dr/dt. The θ-directed velocity has not only changed direction (we have learned to account for this as a centripetal acceleration), but, because the man has moved outward to a point at which the floor is moving faster, its *magnitude* has also changed, from ωr to $\omega(r + \Delta r)$.

Figure I-1b shows the change in velocity caused by the change in direction of the radial line along which the man is walking. If $\Delta\theta$ in the triangle shown is small enough, we have

$$\Delta v_r = v_r\, \Delta\theta.$$

Dividing by Δt and letting Δt approach zero yields

$$a' = \frac{dv_r}{dt} = v_r \frac{d\theta}{dt} = v_r \omega.$$

This is just half the term $2\omega v_r$ in Eq. I-2. However, we have considered only the change in the *radial* velocity; there is also a change in the *tangential* velocity.

The change in tangential velocity, caused by the fact that the man is moving radially outward, is

$$\Delta v_\theta = \omega(r + \Delta r) - \omega r = \omega \Delta r.$$

Dividing by Δt and letting Δt approach zero yields

$$a'' = \frac{dv_\theta}{dt} = \omega \frac{dr}{dt} = \omega v_r.$$

Now both a' and a'' are magnitudes of vectors that point in the same direction,

---

* See "The Coriolis Effect," James E. McDonald, *Scientific American*, May 1952 and also "The Case of the Coriolis Force," Malcolm Correll, *The Physics Teacher*, January 1976.

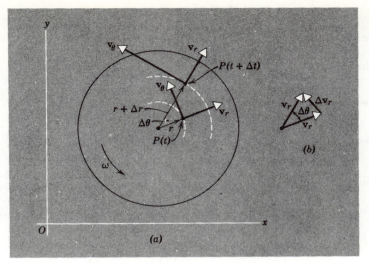

figure 1-1

(a) A merry-go-round, rotating about a fixed axis, is observed by an observer in inertial reference frame x, y. A man walks along a radial line at constant speed v. In a time interval Δt this line, as seen by the ground observer, sweeps through an angle $\Delta\theta$ and the man moves between the positions shown. His r- and θ-directed velocities are shown for each position. *(b)* Showing the change Δv_r in the walking man's r-directed velocity. Note that, as $\Delta t \to 0$, Δv_r points in the θ-direction at P.

namely the direction of increasing θ at point $P(t)$. The total acceleration in this direction is then

$$a' + a'' = v_r\omega + \omega v_r = 2\omega v_r,$$

which is just what we set out to prove.

If there is indeed an acceleration in the θ-direction in Fig. I-1, there must be a force in this direction. For a man walking outward along a radial line on a rotating merry-go-round this force can only be provided by the friction between his feet and the floor.

We remember that we can interpret classical mechanics most simply if we always view events from an inertial frame. If we do so, we can always associate accelerations with forces exerted by bodies that we can point to in the environment. We can still apply classical mechanics, however, if we select a noninertial reference frame, such as a rotating frame. The small penalty that we must pay is that we must introduce inertial forces, that is, forces that we cannot associate with objects in the environment and which cannot be detected by an observer in an inertial frame. In Section 6-4 we saw that centrifugal force is such an inertial force.

Consider an observer on the rotating merry-go-round watching a man walk along a radial line at a constant speed $v_r = dr/dt$. He would say that the man is in equilibrium because he has no acceleration. Yet the floor is exerting a (very real) frictional force on the soles of the man's feet. This force has one component $(-\mathbf{u}_r F_r)$ that points radially inward and one $(\mathbf{u}_\theta F_\theta)$ that points in the θ-direction, that is, in the direction of rotation.

From the point of view of the ground observer these forces are understandable and, indeed, quite necessary. F_r is associated with the centripetal acceleration $\omega^2 r$ and F_θ with the Coriolis acceleration $2\omega v_r$. The observer on the merry-go-round does not see either of these accelerations however; to him the walking man is in equilibrium. How can this be, in view of the frictional forces that act on the soles of the walking man's shoes? The man himself is well aware of these forces; if he did not lean to compensate for their turning effect, they would knock him off his feet!

The observer on the merry-go-round saves the situation by declaring that two inertial forces act on the walking man, just canceling the (real) frictional forces. One of these inertial forces, called the *centrifugal force*, has magnitude F_r and acts radially *outward*. The other, called the *Coriolis force*, has magnitude F_θ and acts in the negative θ-direction, that is, *opposite* to the direction of rotation. By introducing these forces, which seem quite "real" to him although he cannot point to any body in the environment that is causing them, the observer in the rotating (noninertial) reference frame can apply classical mechanics in the usual way. The ground observer, who is in an inertial frame, cannot detect these inertial forces. Indeed there is no need for them—and no room for them—in his applications of classical mechanics.

Equations I-1 and I-2 are general kinematical descriptions for the motion of a particle in two dimensions. An obvious extension, which we will not attempt here, is to derive corresponding descriptions for motion in three dimensions; this will require us to introduce a third unit vector to define the third dimension.*

Some vectors called *axial vectors*, such as ω, α, τ, and l, differ in a rather important way from other vectors called *polar vectors*, of which r, v, a, F, and p are examples. Although we shall not need to take this difference into account in this book, it may prove to be instructive and interesting to examine briefly what the difference is.

Consider a typical polar vector such as r. If a student leaves his dormitory and goes to a classroom, his displacement vector r points *from* the dormitory *to* the classroom; there is no question as to our choice of direction. This direction is both "physical" and "natural." Similar remarks apply to the other typical polar vectors listed, namely, v, a, F, and p.

If a student sees a wheel rotating about a fixed axis, he can assign an angular velocity ω to the wheel and can give direction to ω by the right-hand rule (see Section 11-4). This direction, however, *is a convention only*, based on this arbitrary rule. A left-hand rule would have given the opposite direction. The things that are "physical" and "natural" about the wheel are the axis of rotation and the sense of rotation, that is, is it going clockwise or counterclockwise as the student looks at it from a particular end of the axis? Whether ω is chosen to point in one way or the other along the axis does not really matter as long as we are consistent. The same remarks apply to the angular acceleration α and to the other axial vectors listed, namely, $\tau (= r \times F)$ and $l (= r \times p)$. It is for this reason that we sometimes find it more comfortable to say "torque *around* an axis" than "torque *along* an axis" although they mean the same thing. All vectors defined as the vector product of two *polar* vectors are axial vectors because they all depend for their direction assignment on the (arbitrary) right-hand rule.

We have stressed that the laws of physics remain the same no matter how we change the inertial reference frame in which they are expressed. In Section 2-5 we discussed this for translations and rotations of the reference frame and noted that laws expressed in vector form remained unchanged (that is, *invariant*) under such transformations. We also noted that something special may occur when we change the reference frame in another way, namely, by substituting a left-handed frame for a right-handed one. There is an easy way to make such a transformation: Build a right-handed frame and look at its image in a mirror; it will be converted to a left-handed frame (see Fig. II-1) because of the well-known property of a mirror to reverse right and left.

Figure II-1*a* shows the vector displacement of a student from his dormitory to each of three classrooms. In the mirror each displacement is *still* from the dormitory D to a classroom C. In Fig. II-1*b*, however, we show a rotating wheel in three orientations. If we establish the directions of ω for both the wheels and their mirror images by the right-hand rule, we see that the image vectors are reversed in comparison to the corresponding image vectors in Fig. II-1*a* (toward the origin rather than away from the origin). Polar vectors and axial vectors behave differently when we transform reference frames by mirror reflection! This behavior of axial vectors under mirror reflection is not hard to understand. If we imagine ourselves physically applying the right-hand rule to a real rotating wheel, in the mirror, we shall *seem* to be applying a left-hand rule because the image of our right hand is our left hand. A left-hand rule, of course, will give us the opposite direction for ω.

Hence an axial vector is a vector whose sense of direction depends on the

---

* See, for example, *Mechanics*, Section 3-5, by Keith R. Symon, Addison-Wesley Publishing Co., 3d ed., 1971.

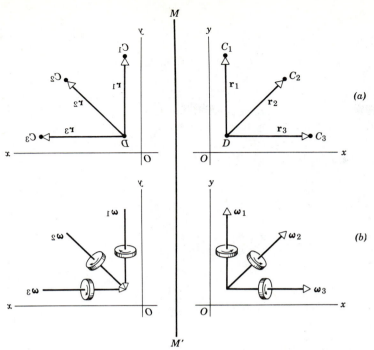

(a)

(b)

figure II-1
(a) Polar vectors, showing, on the right, the displacements $\mathbf{r}_1$, $\mathbf{r}_2$, and $\mathbf{r}_3$ between a dormitory D and three classrooms C_1, C_2, and C_3. On the left we have the mirror images of D, C_1, C_2, and C_3, along with the corresponding displacements. *(b)* *Axial vectors*, showing, on the right, the angular velocities $\boldsymbol{\omega}_1$, $\boldsymbol{\omega}_2$, and $\boldsymbol{\omega}_3$ of three wheels rotating as shown. On the left we have the mirror images of these wheels, along with the angular velocities assigned using the usual right hand rule.

handedness of the reference frame. It is sometimes called a *pseudovector*. A polar vector is a vector that has a direction independent of the reference frame. We mention these facts (1) to stress the arbitrary character of the direction assigned to axial vectors and (2) to stress the importance of testing experiments and physical laws for invariance under translation, rotation, and mirror reflection of the inertial reference frame. In Section 2-5 we referred briefly to some experiments that were *not* invariant under a reflection transformation. This fact, which constituted a violation under certain circumstances of a law of physics previously thought to be well founded (the law of *conservation of parity*), has posed some challenging problems and is leading us to an understanding of the physical world at a deeper level.*

Figure III-1 shows a section of a long string which is under tension F. The string has been pulled transversely in the y-direction so that a displacement wave travels along the string in the x-direction. We consider a differential element of the string dx and apply Newton's second law of motion to it in order to find how the wave moves along the string.

Let μ be the mass per unit length of the string, so that the mass of element dx is $\mu\,dx$. The net force in the y-direction acting on this element is

$$F \sin \theta_{x+dx} - F \sin \theta_x.$$

We consider only small transverse displacements of the string, so that the restoring force will vary linearly with displacement and the principle of superposition will hold (see Section 19-4). This means that θ in Fig. III-1 will be small, so that we may replace $\sin \theta$ by $\tan \theta$. Now $\tan \theta$ is simply the slope of the string, that is, it equals $\partial y/\partial x$. We must use partial derivatives because the transverse displacement y depends not only on x but also on t. The net force in the y-direction is then

$$F\left(\frac{\partial y}{\partial x}\right)_{x+dx} - F\left(\frac{\partial y}{\partial x}\right)_x,$$

* See "The Overthrow of Parity," by Philip Morrison, *Scientific American*, April 1957.

SUPPLEMENTARY TOPIC III
THE WAVE EQUATION FOR A STRETCHED STRING

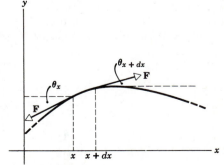

figure III-1

which we may write as

$$F \frac{\partial}{\partial x}\left(\frac{\partial y}{\partial x}\right) dx$$

or

$$F \frac{\partial^2 y}{\partial x^2} dx.$$

The mass of the element of the string is $\mu\, dx$ and its transverse acceleration is simply $\partial^2 y/\partial t^2$. Hence, Newton's second law, applied to the transverse motion of the string, is

$$F \frac{\partial^2 y}{\partial x^2} dx = (\mu\, dx) \frac{\partial^2 y}{\partial t^2}$$

or

$$\frac{\partial^2 y}{\partial x^2} = \frac{\mu}{F} \frac{\partial^2 y}{\partial t^2}. \tag{III-1}$$

Equation III-1, called the *wave equation*, is the differential equation that describes wave propagation in a string of mass per unit length μ and tension F.

To prove this we show that Eqs. 19-2 and 19-3

$$y = f(x \pm vt), \tag{III-2}$$

which is the general equation representing a wave of any shape traveling along x, is a solution of Eq. III-1. Recall that v in Eq. III-2 is the speed of the wave disturbance and f is any reasonable function of $(x \pm vt)$.

Let us see whether Eq. III-2 is indeed a solution of Eq. III-1 by substituting the former equation into the latter. To do so we note that the two second partial derivatives of y are

$$\frac{\partial^2 y}{\partial x^2} = f'' \qquad \text{and} \qquad \frac{\partial^2 y}{\partial t^2} = v^2 f''$$

in which f'' is the second derivative of the function f of Eq. III-2 with respect to $(x \pm vt)$. Substitution of these derivatives into Eq. III-1 yields

$$f'' = \frac{\mu}{F} v^2 f'',$$

which we may write as (see Eq. 19-12)

$$v = \sqrt{\frac{F}{\mu}}. \tag{III-3}$$

Thus we conclude that Eq. III-2 is indeed a solution of the partial differential equation Eq. III-1 if the speed of the wave disturbance described by this equation is given by Eq. III-3.

In particular, let us check that Eq. 19-10

$$y = y_m \sin (kx \pm \omega t) \tag{19-10}$$

is a solution of Eq. III-1. We know that it must be because Eq. 19-10 is simply a special case of the general relation Eq. III-2, which we have just shown to be a solution. Even so it is instructive to test this important specific function of $(x \pm vt)$ by substitution into Eq. III-1.

The second derivatives of Eq. 19-10 are

$$\frac{\partial^2 y}{\partial x^2} = -k^2 y_m \sin (kx \pm \omega t)$$

and

$$\frac{\partial^2 y}{\partial t^2} = -\omega^2 y_m \sin (kx \pm \omega t).$$

Substitution into Eq. III-1 yields

$$-k^2 y_m \sin (kx \pm \omega t) = \left(\frac{\mu}{F}\right) [-\omega^2 y_m \sin (kx \pm \omega t)]$$

or
$$\frac{\omega}{k} = \sqrt{\frac{F}{\mu}}.$$

Since $\omega/k = v$ (see Eq. 19-11), this relation is identical with Eq. III-3, and Eq. 19-10, as we expect, is indeed a solution of Eq. III-1.

SUPPLEMENTARY TOPIC IV
DERIVATION OF MAXWELL'S SPEED DISTRIBUTION LAW

Boltzmann, in 1876, derived the Maxwell speed distribution law (see Eq. 24-2) from this line of argument: Let a uniform gravitational field **g** act on an ideal gas maintained at a fixed temperature T. The number of molecules per unit volume n_v will then decrease with altitude z according to the law of atmospheres (see Example 1, Chapter 17). From what we know about the statistical-mechanical interpretation of temperature, however, the speed distribution law — whose form we assume that we do not yet know — must remain the same at all altitudes because it depends only on the temperature. However, this law determines the rate at which molecules move vertically in the atmosphere at any altitude and must thus be intimately related to the decrease of n_v with z. By exploring this relationship in detail we can, in fact, deduce the speed distribution law.

The weight of gas per unit area between the levels z and $z + dz$ in Fig. IV-1 is $n_v mg \, dz$ in which m is the mass of a single molecule. For equilibrium, this weight per unit area must equal the difference in pressure between z and $z + dz$, or

$$n_v mg \, dz = -dp \tag{IV-1}$$

in which we have inserted a minus sign because p decreases as z increases.

We can write the equation of state of an ideal gas, $pV = nRT$, as

$$p = n_v kT \tag{IV-2}$$

because $n = n_v V/N_0$, where N_0 ($= R/k$) is Avogadro's number, the number of molecules per mole, and k is Boltzmann's constant. Combining Eqs. IV-1 and IV-2 yields

$$\frac{dp}{p} = \frac{dn_v}{n_v} = -\frac{mg}{kT} \, dz.$$

For a constant temperature, we can integrate this relation to yield

$$n_v = \text{constant } e^{-mgz/kT} \tag{IV-3}$$

which, in view of Eq. IV-2, agrees with the result of Example 1, Chapter 17.

We can find the change in n_v as we go from z to $z + dz$ by differentiating Eq. IV-3, or

$$dn_v = -\text{constant } e^{-mgz/kT} \, dz. \tag{IV-4}$$

We associate this decrease in n_v over the interval dz with the fact that, at $z = 0$ (which can be any level we choose) there are some upward-directed molecules — we call them "special molecules" temporarily for convenience — whose vertical velocity components lie in a particular range v_z to $v_z + dv_z$ such that (neglecting collisions; see below) they can rise as high as z but not as high as $z + dz$. Such molecules pass upward through the level z, reverse their direction and pass downward again, as Fig. IV-1 shows. At this point we see more clearly the relationship between Eq. IV-3 and the speed distribution law. Molecules that pass through the interval dz (from above or below) or molecules that never reach the interval cannot contribute to the decrease dn_v of Eq. IV-4.

The rate per unit area at which "special molecules" leave level $z = 0$ (and arrive at level z) is $v_z n_v(v_z) \, dv_z$. Here $n_v(v_z) \, dv_z$ is the number of molecules per unit volume whose vertical velocity components lie between v_z and $v_z + dv_z$.

Now the rate per unit area at which the "special molecules" arrive at level z, but not as high as level $z + dz$, is proportional to the magnitude of the density difference dn_v between z and $z + dz$, or, from Eq. IV-4,

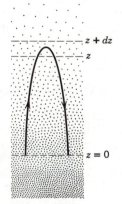

figure IV-1

$$v_z n_r(v_z)\, dv_z = \text{constant } e^{-mgz/kT}\, dz, \qquad\qquad \text{(IV-5)}$$

in which the constant is independent of z. Equation IV-5, which requires that the change dn_v be accounted for by the "special molecules" is, in fact, the defining equation for $n_v(v_z)$.

From conservation of energy the special molecules have the property that*

$$\tfrac{1}{2}mv_z{}^2 = mgz$$

or
$$mv_z\, dv_z = mg\, dz.$$

We use these two relations to eliminate z and dz from Eq. IV-5, obtaining, as you should verify,

$$n_v(v_z)\, dv_z = \text{constant } e^{-mv_z{}^2/2kT}\, dv_z \qquad\qquad \text{(IV-6a)}$$

in which $n_v(v_z)\, dv_z$ is the number of molecules per unit volume whose vertical velocity components lie between v_z and $v_z + dv_z$. Note that Eq. IV-6a does not contain g or z. The gravitational field of Fig. IV-1, introduced to allow us to calculate the speed distribution, has served its purpose. We may apply Eq. IV-6a to a gas for which $g = 0$ or in which gravitational effects are negligible. In such a case the vertical direction, which we have identified as the z-direction, no longer has any special meaning. That is, the speed distribution for one component of velocity should be the same for another component of velocity since there is no special or preferred direction in a gas in equilibrium free of external forces. Thus we can write

$$n_v(v_x)\, dv_x = \text{constant } e^{-mv_x{}^2/2kT}\, dv_x \qquad\qquad \text{(IV-6b)}$$

and
$$n_v(v_y)\, dv_y = \text{constant } e^{-mv_y{}^2/2kT}\, dv_y, \qquad\qquad \text{(IV-6c)}$$

for the other two velocity components.

We now seek to find Maxwell's speed distribution (Eq. 24-2); it is expressed in terms of the speed v, rather than in terms of the separate components v_x, v_y, and v_z. We are not concerned here with the direction of $\mathbf{v}$, because we assume it to be completely random. We can represent any velocity $\mathbf{v}$ as a vector extending from the origin in Fig. IV-2; the projections of the vector in the x-y- and z-directions are v_x, v_y, and v_z, respectively. We commonly say that the axes of Fig. IV-2 define a "velocity space," which has many formal similarities to ordinary (or coordinate) space, in which the axes are x, y, and z.

We also show in Fig. IV-2 a small "volume" element, whose sides are dv_x, dv_y, and dv_z; we say that this element has a volume $dv_x\, dv_y\, dv_z$ in velocity space. A point in this element corresponds to a particle whose velocity components lie between v_x and $v_x + dv_x$; v_y and $v_y + dv_y$; and v_z and $v_z + dv_z$. We can regard $n_v(v_z)$ in Eq. IV-6a as giving the *probability* that a given molecule will have a velocity component in the specified range v_z to $v_z + dv_z$, with similar interpretations for $n_v(v_x)$ and $n_v(v_y)$. The probability that a given molecule will have *all three* of its velocity components in the specified ranges, that is, the probability that the tip of the velocity vector $\mathbf{v}$ will lie inside the volume element of Fig. IV-2, is the *product* of the three (independent) probabilities given in Eq. IV-6, or

$$\text{constant } e^{-mv_x{}^2/2kT}e^{-mv_y{}^2/2kT}e^{-mv_z{}^2/2kT}\, dv_x\, dv_y\, dv_z$$

which, since
$$v^2 = v_x{}^2 + v_y{}^2 + v_z{}^2,$$

we may write as

$$\text{constant } e^{-mv^2/2kT}(dv_x\, dv_y\, dv_z). \qquad\qquad \text{(IV-7)}$$

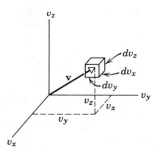

v_z

$\mathbf{v}$

dv_z

dv_x

dv_y

v_z

v_x

v_y

v_y

v_x

figure IV-2

---

* If we consider collisions, this result is still true *on the average* for the many molecules that start at $z = 0$ with a given value of v_z and move to the interval z to $z + dz$, having $v_z = 0$ there, even though such molecules would follow very erratic paths because of the collisions.

The quantity in parentheses above is a volume element in velocity space. Since in Maxwell's speed distribution law we are not concerned with the direction of molecular velocities but only with their speeds, it is more convenient to substitute a different volume element for the above, namely, one corresponding to all molecules whose speeds lie between v and $v + dv$, *regardless of direction*. This volume element is not a "cube" but is the space between two concentric spheres, one of radius v and one of radius $v + dv$. The volume of this element in velocity space is $(4\pi v^2)(dv)$. Substituting this for the quantity enclosed in parentheses in Eq. IV-7 yields for the number of molecules per unit volume whose speeds lie between v and $v + dv$,

$$n_v(v)\ dv = \text{constant } e^{-mv^2/2kT}(4\pi v^2\ dv)$$

or

$$n_v(v) = Cv^2 e^{-mv^2/2kT}$$

in which C is a constant. If we sum up over all possible speeds, we simply obtain the total number of molecules per unit volume, regardless of speed. Hence, we can find C by requiring that

$$\int_0^\infty n_v(v)\ dv = n_v,$$

where n_v is the total number of particles per unit volume, regardless of speed. Guided by the methods of Example 3 (Chapter 24), you should show that

$$C = 4\pi n_v(m/2\pi kT)^{3/2}$$

so that

$$n_v(v) = 4\pi n_v(m/2\pi kT)^{3/2}v^2 e^{-mv^2/2kT}. \qquad \text{(IV-8)}$$

Let us consider a finite number N of molecules contained in a box of volume V. If we multiply each side of the above equation by V, we can replace $n_v V$ on the right by N and $n_v(v)V$ on the left by $N(v)$, which gives us Eq. 24-2.

Here we simply display in one place some conclusions drawn from the special theory of relativity (hereafter, SR) proposed by Einstein in 1905. We omit all proofs and make only a modest attempt to make the conclusions acceptable in terms of "common sense."

SUPPLEMENTARY TOPIC V
SPECIAL RELATIVITY— A SUMMARY OF CONCLUSIONS*

V-1
Introduction

V-2
The Postulates (RR, Section 1.9)

Einstein based his theory on two postulates and *all* of the conclusions of SR derive from them.

a. The First Postulate. From the time of Galileo it was known that the laws of mechanics were the same in all inertial frames (see Fig. V-1 and p. 66). This means that all inertial observers having relative motion, even though they may measure different values for the velocities, momenta, etc., of the particles involved in a given experiment (a game of pool, perhaps), would nevertheless agree on the laws of mechanics involved (conservation of linear momentum, etc.) and on the outcome of the experiment (who won).

Einstein took the bold step of extending this invariance principle to *all* of physics, not only mechanics, including especially electromagnetism. Einstein's first postulate is:

---

* For a fuller treatment, geared to the level of this book, see *Introduction to Special Relativity*, Robert Resnick, John Wiley and Sons, Inc., New York, 1968. References to this work will be in the style RR, p. 187; RR, Section 1.9, etc.

The laws of physics are the same in all inertial frames. No preferred inertial frame exists.

b. The Second Postulate. In pre-SR days a vexing question was this: Given that the speed of light c is 2.988×10^8 m/s, with respect to what is this speed measured? For sound waves in air the answer is simple—it is with respect to the medium (air) through which the sound wave travels. Light, however, travels through a vacuum. Even so, is there a tenuous medium (the luminiferous, or light carrying, ether) that plays the same role for light that air does for sound? If such an ether exists, can we detect it? Alternatively, should c be measured with respect to the source that emits the light?

All attempts to make experimental verifications along these lines failed completely (see Section 45-8* and RR, Sections 1.5 through 1.8). Einstein made a second bold postulate.

The speed of light is the same in all inertial frames.

Note that no ether is needed or involved. This second postulate means, for example, that if you consider three light sources (a) one at rest with respect to you, (b) one moving toward you at speed 0.9 c, say, and (c) one moving away from you at speed 0.9 c, you would measure the *same* speed of light from all three sources.

This second postulate has been tested directly (see RR, p. 34) using as a moving "light" source π^0 mesons generated in a proton synchrotron at speeds of 0.99975 c. These mesons disintegrate by emitting γ-rays which, like light, are electromagnetic in character and travel with the same speed. The speed measured for the radiation emitted by these fast moving sources was, within experimental error, just c, as Einstein's second postulate predicts.

V-3

Special Relativity and Newtonian Mechanics (RR, Section 2.8)

Many of the conclusions of SR simply don't seem reasonable on the basis of everyday experience. Even Einstein's second postulate seems to violate common sense. If you catch a pitched baseball thrown by a pitcher (a) at rest with respect to you, (b) moving toward you (in an automobile, say) at 30 mi/h and, (c) moving away from you at this same speed, you expect a different baseball speed in each case with respect to you. But if you extend this experience to a source (the pitcher) emitting light (photons), you would contradict Einstein's second postulate. And yet experiment shows that light does have the same speed in each case, in support of Einstein's postulate.

The solution to this dilemma comes about when we realize that the basis of our "common sense" experience is very limited indeed. It is restricted to speeds v such that $v \ll c$, where c is the speed of light. For example, the speed of a satellite in earth orbit may be about 8000 m/s, which seems fast to us, but in terms of the speed of light (3.0×10^8 m/s) it is only 0.000027 c. We simply have no personal experience in regions of high relative velocity.

As an example, to accelerate an average person (to say nothing of a spaceship) to 0.90 c would require no less than 13 percent of this country's 1971 total energy consumption. However, the particles of physics (electrons, mesons, protons, etc.) can readily be accelerated to high speeds. Electrons emerging from the two-mile long linear accelerator at Stanford University have speeds of 0.999 c, for example. In the arena of particle physics SR is absolutely necessary for the solution of mechanical problems.

It turns out that in nature there is a certain finite speed that cannot be exceeded and which we call the limiting speed. This limiting speed is the speed

* This book is published in a combined volume (Chapters 1–50) and separate volumes (Part I, Chapters 1–25; Part II, Chapters 26–50). Whether cited references are accessible depends on which of these three volumes you are reading.

of light, c, the greatest speed with which signals can be transmitted. Classical physics assumes that signals can be transmitted with infinite speed, but nature contradicts that assumption, and it really does seem fanciful that such a signal could exist. Experiment confirms c as the limiting speed, so that in a sense the speed of light plays the role in relativity that infinity does in classical physics. It is then not difficult to understand — in fact, it becomes very plausible — that the finite speed of the source of light cannot affect the measured value of the speed of an emitted signal already having the limiting value.

The world in which we live and develop our sense perceptions is a world of Newtonian mechanics, in which $v \ll c$. Newtonian mechanics is revealed as a special case of SR for the limit of low speeds. Indeed, a test of SR is to allow $c \to \infty$ (in which case $v \ll c$ always holds true) and see that the corresponding formulas of Newtonian mechanics emerge.

Newtonian mechanics, although a special case, is an all-important one. It describes the essential motions of our solar system, the tides, our space ventures, the behavior of baseballs and pinball machines, etc. It works beautifully in the vastly important realm $v \ll c$. But it breaks down at speeds approaching that of light.

Few theories have been subject to more rigorous experimental tests than SR. Not the least among them is the fact that particle accelerators work. They are designed using SR at the level of engineering and technology. An accelerator designed on the basis of Newtonian mechanics simply would not work. Nuclear reactors and, alas, nuclear bombs, are further proof that SR really works.

Einstein once said that no number of experiments could prove him right but a single experiment could prove him wrong. To date this single experiment has not been found.

The basic observation made in SR (or in Newtonian mechanics for that matter) is this. Consider observers to be in different inertial frames, S and S' (Fig. V-1). The corresponding axes of S and S' are parallel, the x-x' axes being common, and S' moves to the right with speed v as seen by S; the two origins coincide at $t = t' = 0$. Each observer, S and S', records the same event, which might be the detonation of a tiny flashbulb, and assigns space and time coordinates to the event, namely, x, y, z, t and x', y', z', t'. What are the relations between these two sets of numbers written down in the observers' notebooks?

Before SR the accepted relations were

$$x' = x - vt \qquad y' = y$$
$$t' = t \qquad z' = z, \tag{V-1}$$

called the *Galilean transformation equations* (RR, Section 1.2). Though impressively correct in the important region $v \ll c$ they nevertheless fail as $v \to c$.

The corresponding equations used in SR, called the *Lorentz transformation equations*, are (RR, Table 2-1)

$$x' = \frac{x - vt}{\sqrt{1 - (v/c)^2}} \qquad y' = y$$
$$t' = \frac{t - (v/c^2)x}{\sqrt{1 - (v/c)^2}} \qquad z' = z \tag{V-2}$$

Note certain things about these equations. (a) Space and time coordinates are thoroughly intertwined. In particular, time is not the same for each observer; t' depends on x as well as on t. (b) If we let $c \to \infty$, the Lorentz equations reduce to the Galilean equations, as promised! Finally, (c) We must have $v < c$ or else the quantities x' and t' become indeterminate $(v = c)$ or imaginary $(v > c)$. The speed of light is an upper limit for the speeds of material objects.

The Lorentz equations, like everything else in SR, can be derived from Einstein's two postulates (RR, Section 2.2).

V-4

The Transformation Equations
(RR, Section 2.2)

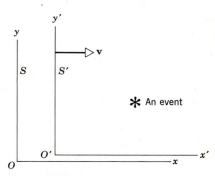

figure V-1
Two inertial frames with parallel axes, the $x - x'$ axis being common. S' moves to the right with speed v as seen by S. At $t = t' = O$ the two origins, O and O', coincide.

Let S' observe two events that occur at the same place in his reference frame. They might be two successive positions of the hand of a clock located at a fixed position, x'. Let S' measure the time interval $\Delta t'$ between these events. S, for whom the clock appears to be moving, observes the same two events and measures a different time interval Δt, which is given by

$$\Delta t = \frac{\Delta t'}{\sqrt{1 - (v/c)^2}} . \tag{V-3}$$

This fact, that $\Delta t > \Delta t'$, is called *time dilation*, and we often verbalize it as "moving clocks run slow." Observer S records a longer time interval than shown to have transpired on the moving clock.

Equation V-3 has been tested experimentally and found to be correct. In one test the "moving clocks" were fast particles called pions $(\pi \pm)$. Pions are radioactive, and their rate of radioactive decay is a measure of their time-keeping ability. See RR, Example 3, p. 75.

Now let us consider a rod, parallel to the $x - x'$ axes, to be at rest in the S' frame. S' will measure a length $\Delta x'$ for it. S, however, measuring the same rod, which is moving with respect to him, would find a length Δx, which is given by

$$\Delta x = \sqrt{1 - (v/c)^2}\, \Delta x' \tag{V-4}$$

This fact, that $\Delta x < \Delta x'$, is called *length contraction*.

The length contraction has been verified in the design of, say, the linear electron accelerator at Stanford University. At an exit speed of $v = 0.999975c$, each meter of the accelerating tube seems like 7.1 mm to an observer moving with the electron. If these length contraction considerations had not been taken into account the machine simply would not work.

The simplest way to understand these results—the time dilation and the length contraction—is to note that one observer, S', is at rest with respect to what he is measuring (clock or rod) whereas for the other observer, S, the objects are in motion. Relativity therefore asserts that *motion affects measurement*. If we had interchanged frames, letting the clock and rod be at rest in S, for example, we would have found the observers again disagreeing on the measured values, but now we would have $\Delta x' < \Delta x$ and $\Delta t' > \Delta t$. So the results are reciprocal, neither observer being "absolutely" right or wrong.

What both observers *will* agree on however, is the *rest length* of a given rod (they will both measure the rod to have the same length when the rod is at rest with respect to their measuring instruments) and the *proper time interval* of a given clock (they will both measure the successive positions of the hand of the clock to have taken the same elapsed time when the clock is at rest with respect to their measuring instruments).

That motion should affect measurement is not a strange idea, even in classical physics. For example, the measured frequency of sound or of light depends on the motion of the source with respect to the observer; we call this the Doppler effect and everyone is familiar with it. And, in mechanics the measured values of the speed, the momentum, or kinetic energy of moving particles are different for observers on the ground than those on a moving train. However, in classical physics measurements of space intervals and time intervals *are* absolute whereas in SR such measurements are relative to the observer. Not only does experiment contradict classical physics but only by adopting the relativity of space and time do we arrive at the invariance (the absoluteness) of all of the laws of physics for all observers. Surely, giving up the absoluteness of the laws of physics (would they then be laws?), as classical notions of time and length require, would leave us with an arbitrary and complex world. By comparison, relativity is absolute and simple.

Let S observe a particle moving with speed u' parallel to the x'-axis. What speed u would S measure? From the Galilean transformation equations (Eq. V-1) we can easily show that

$$u = u' + v \tag{V-5}$$

This relation, which seems to most of us to be intuitively obvious, is, alas, not

V-5
Time Dilation and Length Contraction (RR, Sections 2.3 and 2.4)

V-6
Relativistic Addition of Velocities and the Doppler Effect (Sections 4.6, 6.5, 42.4, and 42.5; RR, Sections 2.6 and 2.7)

true (except for the very important special case of $v \ll c$). The Lorentz transformation equations lead us to

$$u = \frac{u' + v}{1 + (u'v/c^2)}.\qquad(V\text{-}6)$$

As we expect, for $c \to \infty$, Eq. V-6 reduces to Eq. V-5. Prove that if $u' < c$ and $v < c$, then it must always be true that $u < c$. There is no way to generate speeds $\geqslant c$ by compounding velocities.

Using the relativistic velocity addition result (Eq. V-6), we can deduce the Doppler effect for light. In relativity theory there is no difference between the two cases, which are different in classical theory (namely, source at rest — observer moving and observer at rest — source moving); only the relative motion v of source and observer counts. This fact and the result

$$\nu = \nu' \sqrt{\frac{c \pm v}{c \mp v}}\qquad(V\text{-}7)$$

are in agreement with experiment. Here, ν' is the frequency of the source at rest in S' and ν is the frequency observed in frame S with respect to which the source moves at speed v; the upper signs refer to source and observer moving *toward* one another and the lower signs refer to source and observer moving *away* from one another. Equation V-7 is called the *longitudinal* Doppler effect, and v refers to the relative velocity of source and observer along the line connecting them.

There is in relativity, however, an effect not predicted by classical physics, namely a *transverse* Doppler effect; that is, if the relative velocity v is at right angles to the line connecting source and observer, we find

$$\nu = \nu' \sqrt{1 - v^2/c^2}.\qquad(V\text{-}8)$$

This result, confirmed by experiment, can be interpreted simply as a time dilation, moving clocks appearing to run slow.

We have seen that time and length measurements are functions of velocity v. Should mass be any different? SR tells us that the *relativistic mass m* of a particle moving at speed v with respect to the observer is

$$m = \frac{m_0}{\sqrt{1 - (v/c)^2}}\qquad(V\text{-}9)$$

in which m_0 is the rest mass, that is, the mass measured when the particle is at rest $(v = 0)$ with respect to the observer.

It is m and not m_0 that must be taken into account when designing magnets to bend charged particles in arcs of circles. By these techniques, Eq. V-9 has been thoroughly tested. Incidentally, the ratio m/m_0 for electrons emerging from the Stanford University linear accelerator at $K = 30$ GeV is the order of 60,000.

To preserve the law of conservation of linear momentum in SR, we redefine the momentum of a particle of rest mass m_0 and speed v as,

$$p = mv = \frac{m_0 v}{\sqrt{1 - (v/c)^2}}.$$

As a result of the considerations above, in SR the kinetic energy of a particle is no longer given by $\frac{1}{2} m_0 v^2$ but by

$$K = mc^2 - m_0 c^2$$

$$= m_0 c^2 \left(\frac{1}{\sqrt{1 - (v/c)^2}} - 1 \right).\qquad(V\text{-}10)$$

Can you show that $K \to \frac{1}{2} m_0 v^2$ as $c \to \infty$?

V-7

Mass, Momentum, and Kinetic Energy (Sections 8.9 and 9.3; RR, Sections 3.3 and 3.5)

The best known result of SR is the so-called mass-energy equivalence. That is, the conservation of total energy is equivalent to the conservation of relativistic mass. Mass and energy are equivalent; they form a single invariant that we can call mass-energy. The relation

$$E = mc^2 \qquad\qquad (V\text{-}11)$$

expresses the fact that mass-energy can be expressed in energy (E) units or equivalently in mass ($m = E/c^2$) units. In fact, it has become common practice to refer to masses in terms of electron volts, such as saying that the rest mass of an electron is 0.51 MeV, for convenience in energy calculations. Likewise entities of zero rest mass, such as photons, may be assigned an effective mass equivalent to their energy. We associate mass with each of the various forms of energy.

In classical physics we had two separate conservation principles: (1) the conservation of (classical) mass, as in chemical reactions, and (2) the conservation of energy. In relativity, these merge into one conservation principle, that of the conservation of mass-energy. The two classical laws may be viewed as special cases that would be expected to agree with experiment only if energy transfers into or out of the system are so small compared with the system's rest mass that the corresponding fractional change in rest mass of the system is too small to be measured.

For example, the rest mass of a hydrogen atom is 1.00797 u (= 938.8 MeV). If enough energy (13.58 eV) is added to ionize hydrogen, that is, to break it up into its constituent parts, a proton and an electron, the fractional change in rest mass of the system is

$$\frac{13.58 \text{ eV}}{938.8 \times 10^6 \text{ eV}} = 1.45 \times 10^{-8}$$

or 1.45×10^{-6} percent, too small to measure. However, for a nucleus such as the deuteron, whose rest mass is 2.01360 u (= 1876.4 MeV), one must add an energy of 2.22 MeV to break it up into its constituent parts, a proton and a neutron. The fractional change in rest mass of the system is

$$\frac{2.22 \text{ MeV}}{1876.4 \text{ MeV}} = 1.18 \times 10^{-3}$$

or 0.12 percent, which is readily measureable. This is characteristic of the fractional rest-mass changes in nuclear reactions, so that one must use the relativistic law of conservation of mass-energy to get agreement between theory and experiment in nuclear reactions. The classical (rest) mass is *not* conserved, but total energy (mass-energy) is.

In Chapter 41 we sought to make the existence of electromagnetic waves plausible by showing that such waves are consistent with Maxwell's equations as expressed in Table 40-2. Here we seek to start from Maxwell's equations and derive from them a differential equation whose solutions will describe electromagnetic waves. We will show directly that the speed c of such waves is given by Eq. 40-1, or $c = 1/\sqrt{\epsilon_0\mu_0}$.

We followed a similar program in Supplementary Topic III for mechanical waves on a stretched string. Starting from Newton's laws of motion we derived a differential equation (Eq. III-1) whose solutions (Eq. III-2) described such waves. We showed further that the speed v of these waves is given by Eq. III-3, or $v = \sqrt{F/\mu}$.

In Table 40-2 we wrote Maxwell's equations as

$$\epsilon_0 \oint \mathbf{E}\cdot d\mathbf{S} = q, \qquad\qquad (VI\text{-}1)$$

* Supplementary Topics I to V appear in Part I.

V-8
The Equivalence of Mass and Energy (Section 8.9; RR, Section 3.6)

SUPPLEMENTARY TOPIC VI*
THE DIFFERENTIAL FORM OF MAXWELL'S EQUATIONS AND THE ELECTROMAGNETIC WAVE EQUATION

VI-1
Introduction

$$\oint \mathbf{B} \cdot d\mathbf{S} = 0, \tag{VI-2}$$

$$\oint \mathbf{B} \cdot d\mathbf{l} = \mu_0(i + \epsilon_0 \, d\Phi_E/dt), \tag{VI-3}$$

and

$$\oint \mathbf{E} \cdot d\mathbf{l} = -d\Phi_B/dt. \tag{VI-4}$$

These equations are said to be written in integral form. The field variables $\mathbf{E}$ and $\mathbf{B}$, which are usually the unknown quantities, appear in the integrands. Only in a few symmetric cases (see Sections 28-8 and 34-2, for example) can we "factor them out." In more general problems we cannot do so.

The situation is somewhat analogous to computing the density ρ of a body if we know its mass m and volume τ. In general these are related by the integral equation

$$m = \int \rho \, d\tau.$$

Only if ρ is a constant over all parts of the volume can we factor it out and write $\rho = m/\tau$.

To carry out our program it is desirable to recast Maxwell's equations in the form of equalities that apply at each *point* in space rather than as integrals that apply to various *regions* of space. In other words, we wish to convert Maxwell's equations from the integral form of Eqs. VI-1 to 4 into *differential form*. We will then be able to relate $\mathbf{E}$ and $\mathbf{B}$ at a point to the charge density and current density at that point.

VI-2
The Operator ∇

To transform Maxwell's equations into differential form we must deepen our understanding of vector methods and, in particular, become familiar with the vector operator ∇.

In Section 29-7 we saw how to obtain the components of the (vector) electrostatic field $\mathbf{E}$ at any point from the (scalar) potential function $V(x,y,z)$ by partial differentiation. Thus,

$$E_x = -\frac{\partial V}{\partial x}, \quad E_y = -\frac{\partial V}{\partial y}, \quad \text{and} \quad E_z = -\frac{\partial V}{\partial z}$$

so that the electrostatic field

$$\mathbf{E} = \mathbf{i}E_x + \mathbf{j}E_y + \mathbf{k}E_z$$

can be written as

$$\mathbf{E} = -\left(\mathbf{i}\frac{\partial V}{\partial x} + \mathbf{j}\frac{\partial V}{\partial y} + \mathbf{k}\frac{\partial V}{\partial z} \right). \tag{VI-5}$$

We can write Eq. VI-5 in compact vector notation as

$$\mathbf{E} = -\nabla V,$$

where ∇ ("del") is a vector operator defined as

$$\nabla = \mathbf{i}\frac{\partial}{\partial x} + \mathbf{j}\frac{\partial}{\partial y} + \mathbf{k}\frac{\partial}{\partial z}. \tag{VI-6}$$

This operator is useful in dealing with scalar and vector fields (see Sections 16-8 and 18-7 for examples of such fields). Given any scalar field ψ we may form a vector field, called the gradient of ψ and written as grad ψ or $\nabla\psi$, simply by applying the operator ∇ to ψ. Given a vector field $\mathbf{U} = U_x\mathbf{i} + U_y\mathbf{j} + U_z\mathbf{k}$ we may apply the operator ∇ to it in two different ways. One way is to take the dot product of ∇ and $\mathbf{U}$, yielding the scalar field called the divergence of $\mathbf{U}$ and written as div $\mathbf{U}$ or $\nabla \cdot \mathbf{U}$. The other way is to take the cross product of ∇ and $\mathbf{U}$, yielding the vector field called the curl of $\mathbf{U}$ and written curl $\mathbf{U}$ or $\nabla \times \mathbf{U}$. These operations may be summarized as

$$\text{grad } \psi \equiv \nabla\psi = \mathbf{i}\frac{\partial\psi}{\partial x} + \mathbf{j}\frac{\partial\psi}{\partial y} + \mathbf{k}\frac{\partial\psi}{\partial z}, \tag{VI-7}$$

$$\text{div } \mathbf{U} \equiv \nabla\cdot\mathbf{U} = \frac{\partial U_x}{\partial x} + \frac{\partial U_y}{\partial y} + \frac{\partial U_z}{\partial z}, \tag{VI-8}$$

$$\text{curl } \mathbf{U} \equiv \nabla\times\mathbf{U} = \mathbf{i}\left(\frac{\partial U_z}{\partial y} - \frac{\partial U_y}{\partial z}\right)$$

$$+ \mathbf{j}\left(\frac{\partial U_x}{\partial z} - \frac{\partial U_z}{\partial x}\right)$$

$$+ \mathbf{k}\left(\frac{\partial U_y}{\partial x} - \frac{\partial U_x}{\partial y}\right). \tag{VI-9}$$

Note that grad ψ and curl $\mathbf{U}$ are vectors, whereas div $\mathbf{U}$ is a scalar. You can gain some familiarity with these operations by the following exercises: (1) prove that curl (grad ψ) = 0 and (2) prove that div (curl $\mathbf{U}$) = 0.

Another frequently occurring operator is ∇^2 ("del squared"). It is simply $\nabla\cdot\nabla$, or, as the student can show from Eq. VI-6,

$$\nabla^2 = \nabla\cdot\nabla = \frac{\partial^2}{\partial x^2} + \frac{\partial^2}{\partial y^2} + \frac{\partial^2}{\partial z^2}.$$

When we apply ∇^2 to a scalar field ψ, we obtain

$$\nabla^2\psi = \frac{\partial^2\psi}{\partial x^2} + \frac{\partial^2\psi}{\partial y^2} + \frac{\partial^2\psi}{\partial z^2}. \tag{VI-10}$$

For a vector field $\mathbf{U}$, the operation $\nabla^2\mathbf{U}$ is

$$\nabla^2\mathbf{U} = \mathbf{i}\left(\frac{\partial^2}{\partial x^2} + \frac{\partial^2}{\partial y^2} + \frac{\partial^2}{\partial z^2}\right)U_x + \mathbf{j}\left(\frac{\partial^2}{\partial x^2} + \frac{\partial^2}{\partial y^2} + \frac{\partial^2}{\partial z^2}\right)U_y + \mathbf{k}\left(\frac{\partial^2}{\partial x^2} + \frac{\partial^2}{\partial y^2} + \frac{\partial^2}{\partial z^2}\right)U_z \tag{VI-11}$$

As an exercise, show that curl (curl $\mathbf{U}$) = $-\nabla^2\mathbf{U}$ + grad (div $\mathbf{U}$).

In this section we show how to cast the first two of Maxwell's equations (Eqs. VI-1, 2) into differential form. Let us apply Eq. VI-1 to a differential volume element shaped like a rectangular parallelepiped and containing a point P at (and near) which an electric field exists (see Fig. VI-1a). Point P is located at x, y, z in the reference frame of Fig. VI-1b and the edges of the parallelepiped have lengths dx, dy, and dz.

We can write the surface area vector for the rear face of the parallelepiped as $d\mathbf{S} = -\mathbf{i}\ dy\ dz$. The minus sign enters because $d\mathbf{S}$ is defined to point in the direction of the *outward* normal, which is defined by $-\mathbf{i}$. For the front face we have $d\mathbf{S} = +\mathbf{i}\ dy\ dz$.

If the electric field at the rear face is $\mathbf{E}$, that at the front face, which is a distance dx away from the rear face, is $\mathbf{E} + (\partial\mathbf{E}/\partial x)\ dx$, the latter term representing the change in $\mathbf{E}$ associated with the change dx in x.

The flux through the entire surface of the parallelepiped is $\oint \mathbf{E}\cdot d\mathbf{S}$ and the contribution to this flux due to these two faces alone is

$$(\mathbf{E})\cdot(-\mathbf{i}\ dy\ dz) + \left(\mathbf{E} + \frac{\partial\mathbf{E}}{\partial x}\ dx\right)\cdot(+\mathbf{i}\ dy\ dz)$$

$$= dx\ dy\ dz\left(\frac{\partial\mathbf{E}}{\partial x}\cdot\mathbf{i}\right) = dx\ dy\ dz\ \frac{\partial}{\partial x}(\mathbf{E}\cdot\mathbf{i})$$

$$= dx\ dy\ dz\ \frac{\partial E_x}{\partial x}.$$

With similar contributions from the other four faces the total electric flux becomes

$$\oint \mathbf{E}\cdot d\mathbf{S} = dx\ dy\ dz\left(\frac{\partial E_x}{\partial x} + \frac{\partial E_y}{\partial y} + \frac{\partial E_z}{\partial z}\right).$$

VI-3
Maxwell's Equations in Differential Form—I

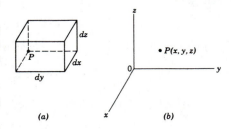

(a)   *(b)*

From Eq. VI-8 we may write this as

$$\oint \mathbf{E} \cdot d\mathbf{S} = dx\,dy\,dz\,\text{div }\mathbf{E}. \qquad \text{(VI-12)}$$

Now the right-hand side of Eq. VI-1, which gives the charge enclosed by the surface, may be written in general as $q = \int \rho\,d\tau$ and, in particular, for the differential volume element at P, as

$$q = \rho\,dx\,dy\,dz, \qquad \text{(VI-13)}$$

where ρ is the charge per unit volume at P. Substituting Eqs. VI-12 and 13 into Eq. VI-1 and canceling the common factor $dx\,dy\,dz$, we have finally

$$\epsilon_0\,\text{div }\mathbf{E} = \rho, \qquad \text{(VI-14)}$$

which is Maxwell's first equation (Eq. VI-1) in differential form.

Using the same technique, we can express Maxwell's second equation (Eq. VI-2) in differential form as

$$\text{div }\mathbf{B} = 0. \qquad \text{(VI-15)}$$

We now seek to transform Maxwell's third and fourth equations (Eqs. VI-3, 4) into differential form. We start by applying Eq. VI-3 to a differential surface element of rectangular shape at a point P in some region of a magnetic field, as shown in Fig. VI-2a. The point P is located at x, y, z in the reference frame of Fig. VI-2b and the sides of the rectangle, which is parallel to the x-y plane, have lengths dx and dy. Going around the path, as shown by the arrows, we have

$$\oint \mathbf{B} \cdot d\mathbf{l} = \mathbf{B} \cdot (-\mathbf{j}\,dy) \qquad \text{(rear side)}$$

$$+ \mathbf{B} \cdot (+\mathbf{i}\,dx) \qquad \text{(left side)}$$

$$+ \left(\mathbf{B} + \frac{\partial \mathbf{B}}{\partial x}\,dx\right) \cdot (+\mathbf{j}\,dy) \qquad \text{(front side)}$$

$$+ \left(\mathbf{B} + \frac{\partial \mathbf{B}}{\partial y}\,dy\right) \cdot (-\mathbf{i}\,dx), \qquad \text{(right side)}$$

where $\mathbf{B}$ is the magnetic field at P.

Collecting terms we obtain

$$\oint \mathbf{B} \cdot d\mathbf{l} = dx\,dy\left(\frac{\partial \mathbf{B}}{\partial x} \cdot \mathbf{j} - \frac{\partial \mathbf{B}}{\partial y} \cdot \mathbf{i}\right)$$

$$= dx\,dy\left[\frac{\partial}{\partial x}(\mathbf{B} \cdot \mathbf{j}) - \frac{\partial}{\partial y}(\mathbf{B} \cdot \mathbf{i})\right]$$

$$= dx\,dy\left(\frac{\partial B_y}{\partial x} - \frac{\partial B_x}{\partial y}\right). \qquad \text{(VI-16)}$$

Now in the right-hand side of Eq. VI-3 i is the current enclosed by the path and $d\Phi_E/dt$ is the change in electric flux through the enclosed surface. Hence, if $\mathbf{J}$ is taken to represent the current density and $d\mathbf{S}(= \mathbf{k}\,dx\,dy)$ the surface area vector, we can write

$$i = \mathbf{J} \cdot d\mathbf{S} = \mathbf{J} \cdot (\mathbf{k}\,dx\,dy) = dx\,dy\,J_z \qquad \text{(VI-17)}$$

and

$$\frac{d\Phi_E}{dt} = \frac{\partial \mathbf{E}}{\partial t} \cdot d\mathbf{S} = \frac{\partial \mathbf{E}}{\partial t} \cdot (\mathbf{k}\,dx\,dy)$$

or

$$\frac{d\Phi_E}{dt} = \frac{\partial E_z}{\partial t}\,dx\,dy. \qquad \text{(VI-18)}$$

Substituting Eqs. VI-16, 17, and 18 into Eq. VI-3 and canceling the common factor $dx\,dy$, we get

$$\frac{1}{\mu_0}\left(\frac{\partial B_y}{\partial x} - \frac{\partial B_x}{\partial y}\right) = J_z + \epsilon_0\frac{\partial E_z}{\partial t}. \qquad \text{(VI-19)}$$

VI-4

Maxwell's Equations in Differential Form — II

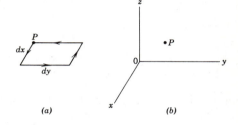

(a)

(b)

We could have proceeded exactly as above if we had started with a rectangle parallel to the y-z plane or one parallel to the z-x plane. Each rectangle would have given us a different component of an arbitrarily oriented differential surface at P. Equation VI-19 is obviously the z-component equation corresponding to Eq. VI-3. If we multiply it by **k** and add to it the two similar vector equations, which may be obtained by cyclically permuting x, y, z and **i, j, k**, corresponding to the x-component and y-component equations, we obtain

$$\text{curl } \mathbf{B} = \mu_0(\mathbf{J} + \epsilon_0 \, \partial \mathbf{E}/\partial t), \qquad (\text{VI-20})$$

which is the third Maxwell equation in differential form.

Similarly, starting with Eq. VI-4, we may show that

$$\text{curl } \mathbf{E} = -\partial \mathbf{B}/\partial t, \qquad (\text{VI-21})$$

which is the fourth Maxwell equation in differential form.

We have derived four differential equations (see VI-22 to VI-25 below) from the four integral equations (VI-1 to VI-4). It can be shown that the integral equations can be derived from the differential equations, that is, the two sets of equations are *equivalent*.

We have now obtained from their integral form the four basic equations of electromagnetism, Maxwell's equations, in differential form. Corresponding to the integral equations, Eqs. VI-1, 2, 3, and 4 respectively, we have

VI-5
The Wave Equation

$$\epsilon_0 \text{ div } \mathbf{E} = \rho, \qquad (\text{VI-22})$$

$$\text{div } \mathbf{B} = 0, \qquad (\text{VI-23})$$

$$\text{curl } \mathbf{B} = \mu_0(\mathbf{J} + \epsilon_0 \, \partial \mathbf{E}/\partial t), \qquad (\text{VI-24})$$

$$\text{curl } \mathbf{E} = -\partial \mathbf{B}/\partial t, \qquad (\text{VI-25})$$

which are four coupled partial differential equations. They apply at each point of space in an electromagnetic field.

We will now derive the wave equation for electromagnetic waves in free space. In free space, the charge density ρ and the current density **J** are zero, so that the Maxwell equations there become

$$\text{div } \mathbf{E} = 0,$$

$$\text{div } \mathbf{B} = 0,$$

$$\text{curl } \mathbf{B} = \mu_0\epsilon_0 \, \partial \mathbf{E}/\partial t,$$

and

$$\text{curl } \mathbf{E} = -\partial \mathbf{B}/\partial t.$$

Let us take the curl of the equation for curl **E**; we obtain

$$\text{curl curl } \mathbf{E} = -\text{curl } \frac{\partial \mathbf{B}}{\partial t} = -\frac{\partial}{\partial t} \text{ curl } \mathbf{B}.$$

But, from above, curl $\mathbf{B} = \mu_0\epsilon_0(\partial \mathbf{E}/\partial t)$, so that

$$\text{curl curl } \mathbf{E} = -\mu_0\epsilon_0 \frac{\partial^2 \mathbf{E}}{\partial t^2}. \qquad (\text{VI-26})$$

From the exercise in VI-2, we know that curl curl $\mathbf{E} = -\nabla^2 E + \text{grad div } \mathbf{E}$ and from above that div $\mathbf{E} = 0$. Thus,

$$\text{curl curl } \mathbf{E} = -\nabla^2 \mathbf{E}. \qquad (\text{VI-27})$$

Combining Eqs. VI-26 and VI-27, we obtain finally

$$\nabla^2 \mathbf{E} = \mu_0\epsilon_0 \frac{\partial^2 \mathbf{E}}{\partial t^2}. \qquad (\text{VI-28}a)$$

Proceeding as above, show that **B** satisfies the same equation, or

$$\nabla^2 \mathbf{B} = \mu_0\epsilon_0 \frac{\partial^2 \mathbf{B}}{\partial t^2}. \qquad (\text{VI-28}b)$$

Equations VI-28 are the equations of electromagnetic wave motion. Being vector equations, they are equivalent to six scalar equations, one for each component of **E** and of **B**.

There are many solutions of Eqs. VI-28, corresponding to different kinds of electromagnetic waves—plane, spherical, and cylindrical waves being three examples. Let us consider a solution in which two components of **E** and two of **B** vanish, that is, in which

$$E_x = E_z = 0 \qquad \text{and} \quad B_x = B_y = 0.$$

Equations VI-28 are satisfied for these assumptions. For the nonvanishing components, E_y and B_z, Eqs. VI-28 reduce to (see Eq. VI-11)

$$\frac{\partial^2 E_y}{\partial x^2} + \frac{\partial^2 E_y}{\partial y^2} + \frac{\partial^2 E_y}{\partial z^2} = \mu_0 \epsilon_0 \frac{\partial^2 E_y}{\partial t^2} \qquad \text{(VI-29}a)$$

and

$$\frac{\partial^2 B_z}{\partial x^2} + \frac{\partial^2 B_z}{\partial y^2} + \frac{\partial^2 B_z}{\partial z^2} = \mu_0 \epsilon_0 \frac{\partial^2 B_z}{\partial t^2}. \qquad \text{(VI-29}b)$$

If we make the additional assumption that E_y and B_z are functions of x and t only, the simplified wave equation that results, namely

$$\frac{\partial^2 E_y}{\partial x^2} = \mu_0 \epsilon_0 \frac{\partial^2 E_y}{\partial t^2} \qquad \text{and} \quad \frac{\partial^2 B_z}{\partial x^2} = \mu_0 \epsilon_0 \frac{\partial^2 B_z}{\partial t^2}, \qquad \text{(VI-29}c)$$

is similar to Eq. III-1 for the vibrating string.

A solution to these equations, as you may verify by substitution, is

$$E_y = E_m \sin (kx - \omega t) \qquad \text{(VI-30}a)$$

and

$$B_z = B_m \sin (kx - \omega t). \qquad \text{(VI-30}b)$$

We interpret Eqs. VI-30 as an electromagnetic wave traveling in the positive x-direction, as in Fig. 41-13, with a speed $c = \omega/k$. Show that substituting Eq-VI-30a into Eq. VI-29a (or Eq. VI-30b into Eq. VI-29b) yields

$$c = \omega/k = 1/\sqrt{e_0 \mu_0},$$

which (see Eq. 40-1) gives the speed of electromagnetic waves in free space.

appendices

SI base units[a]

Quantity	Name	Symbol	Definition
length	meter	m	" . . . the length equal to 1,650,763.73 wavelengths in vacuum of the radiation corresponding to the transition between the levels $2p_{10}$ and $5d_5$ of the krypton-86 atom." (1960)
mass	kilogram	kg	" . . . this prototype [a certain platinum-iridium cylinder] shall henceforth be considered to be the unit of mass." (1889)
time	second	s	" . . . the duration of 9,192,-631,770 periods of the radiation corresponding to the transition between the two hyperfine levels of the ground state of the cesium-133 atom." (1967)

* Adapted from "The International System of Units (SI)," National Bureau of Standards Special Publication 330, 1972 edition.

[a] The definitions of these base units were adopted by the General Conference of Weights and Measures, an international body, on the dates shown. In this book we will not use the candela.

SI base units (Continued)

Quantity	Name	Symbol	Definition
electric current	ampere	A	" . . . that constant current which, if maintained in two straight parallel conductors of infinite length, of negligible circular cross section, and placed 1 meter apart in vacuum, would produce between these conductors a force equal to 2×10^{-7} newton per meter of length." (1946)
thermodynamic temperature	kelvin	K	" . . . the fraction 1/273.16 of the thermodynamic temperature of the triple point of water." (1967)
amount of substance	mole	mol	" . . . the amount of substance of a system which contains as many elementary entities as there are atoms in 0.012 kilogram of carbon-12." (1971)
luminous intensity	candela	cd	" . . . the luminous intensity, in the perpendicular direction, of a surface of 1/600,000 square meter of a blackbody at the temperature of freezing platinum under a pressure of 101,325 newton per square meter." (1967)

Some SI derived units with special names

Quantity	SI unit			
	Name	Symbol	Expression in terms of other units	Expression in terms of SI base units
frequency	hertz	Hz		s^{-1}
force	newton	N		$m \cdot kg/s^2$
pressure	pascal	Pa	N/m^2	$kg/m \cdot s^2$
energy, work, quantity of heat	joule	J	$N \cdot m$	$kg \cdot m^2/s^2$
power, radiant flux	watt	W	J/s	$kg \cdot m^2/s^3$
quantity of electricity, electric charge	coulomb	C		$A \cdot s$
electric potential, potential difference, electromotive force	volt	V	W/A	$kg \cdot m^2/A \cdot s^3$
capacitance	farad	F	C/V	$A^2 \cdot s^4/kg \cdot m^2$
electric resistance	ohm	Ω	V/A	$kg \cdot m^2/A^2 \cdot s^3$
conductance	siemens	S	A/V	$A^2 \cdot s^3/kg \cdot m^2$
magnetic flux	weber	Wb	$V \cdot s$	$kg \cdot m^2/A \cdot s^2$
magnetic field	tesla	T	Wb/m^2	$kg/A \cdot s^2$
inductance	henry	H	Wb/A	$kg \cdot m^2/A^2 \cdot s^2$

Some symbols for units of physical quantities

SI Symbols		Symbols other than SI that are Commonly Used	
Name	Abbreviation	Name	Abbreviation
ampere	A	angstrom	Å
candela	cd	British thermal unit	Btu
coulomb	C	calorie	cal
farad	F	day	d
henry	H	degree	°
hertz	Hz	dyne	dyn
joule	J	electron volt	eV
kelvin	K	foot	ft
kilogram	kg	gauss	G
meter	m	gram	g
mole	mol	horsepower	hp
newton	N	hour	h
ohm	Ω	inch	in.
pascal	Pa	mile	mi
radian	rad	minute (of arc)	'
second	s	minute (of time)	min
siemens	S	pound	lb
steradian	sr	revolution	rev
tesla	T	second (of arc)	"
volt	V	standard atmosphere	atm
watt	W	unified atomic mass unit	u
weber	Wb	year	yr

Over the years many hundreds of measurements of fundamental physical quantities, alone and in combination, have been made by hundreds of scientists in many countries. These measurements have different precisions and, most important, they are interdependent. For example, the direct measurements of e, e/m, h/e, etc., are obviously interrelated. Sorting out the best values of e, m, h, etc., from a large mass of overlapping data is not simple.†

For most problems in this book three significant figures will do, and the computational (rounded) values may be used.

APPENDIX B
SOME FUNDAMENTAL CONSTANTS OF PHYSICS*

Constant	Symbol	Computational value	Best (1973) Value	
			Value[a]	Uncertainty[b]
Speed of light in a vacuum	c	3.00×10^8 m/s	2.99792458	0.004
Elementary charge	e	1.60×10^{-19} C	1.6021892	2.9
Electron rest mass	m_e	9.11×10^{-31} kg	9.109534	5.1
Permittivity constant	ϵ_0	8.85×10^{-12} F/m	8.854187818	0.008
Permeability constant	μ_0	1.26×10^{-6} H/m	4π (exactly)	—
Electron charge to mass ratio	e/m_e	1.76×10^{11} C/kg	1.7588047	2.8
Proton rest mass	m_p	1.67×10^{-27} kg	1.6726485	5.1
Ratio of proton mass to electron mass	m_p/m_e	1840	1836.15152	0.38
Neutron rest mass	m_n	1.68×10^{-27} kg	1.6749543	5.1
Muon rest mass	m_μ	1.88×10^{-28} kg	1.883566	5.6
Planck constant	h	6.63×10^{-34} J·s	6.626176	5.4
Electron Compton wavelength	λ_C	2.43×10^{-12} m	2.4263089	1.6
Molar gas constant	R	8.31 J/mol·K	8.31441	31
Avogadro constant	N_A	6.02×10^{23}/mol	6.022045	5.1
Boltzmann constant	k	1.38×10^{-23} J/K	1.380662	32
Molar volume of ideal gas at STP[c]	V_m	2.24×10^{-2} m³/mol	2.241383	31
Faraday constant	F	9.65×10^4 C/mol	9.648456	2.8
Stefan–Boltzmann constant	σ	5.67×10^{-8} W/m²·K⁴	5.67032	125
Rydberg constant	R	1.10×10^7/m	1.097373177	0.075
Gravitational constant	G	6.67×10^{-11} m³/s²·kg	6.6720	615
Bohr radius	a_0	5.29×10^{-11} m	5.2917706	0.82
Electron magnetic moment	μ_e	9.28×10^{-24} J/T	9.284832	3.9
Proton magnetic moment	μ_p	1.41×10^{-26} J/T	1.4106171	3.9
Bohr magneton	μ_B	9.27×10^{-24} J/T	9.274078	3.9
Nuclear magneton	μ_N	5.05×10^{-27} J/T	5.050824	3.9

[a] Same unit and power of ten as the computational value.
[b] Parts per million.
[c] STP-standard temperature and pressure $=0°$ C and 1.0 atm.

* The values in this table were selected from a longer list developed by E. Richard Cohen and B. N. Taylor, *Journal of Physical and Chemical Reference Data*, vol. 2, no. 4 (1973).
† See "A Pilgrim's Progress in Search of the Fundamental Constants," by J. W. M. Du Mond, *Physics Today*, October 1965, and "The Fundamental Physical Constants" by Taylor, Langenberg, and Parker, *Scientific American*, October, 1970.

APPENDIX C
SOLAR, TERRESTRIAL, AND LUNAR DATA

The sun

Mass	1.99×10^{30} kg
Radius	6.96×10^5 km
Mean density	1,410 kg/m^3
Surface gravity	274 m/s^2
Surface temperature	6000 K
Total radiation rate	3.92×10^{26} W

The earth

Mass	5.98×10^{24} kg
Equatorial radius	6.378×10^6 m
	3963 mi
Polar radius	6.357×10^6 m
	3950 mi
Radius of a sphere of the same volume	6.37×10^6 m
	3600 mi
Mean density	5522 kg/m^3
Acceleration of gravity[a]	9.80665 m/s^2
	32.1740 ft/s^2
Mean orbital speed	29,770 m/s
	18.50 mi/s
Angular speed	7.29×10^{-5} rad/s
Solar constant[b]	1340 W/m^2
Magnetic field (at Washington, D.C.)	5.7×10^{-5} T
Magnetic dipole moment	8.1×10^{22} A·m^2
Standard atmosphere	1.013×10^5 Pa
	14.70 lb/in.^{2}
	760.0 mm-Hg
Density of dry air at STP[c]	1.29 kg/m^3
Speed of sound in dry air at STP	331.4 m/s
	1089 ft/s
	742.5 mi/h

[a] This value, adopted by the General Committee on Weights and Measures in 1901, approximates the value at 45° latitude at sea level.
[b] This is the rate per unit area at which solar energy falls, at normal incidence, just outside the earth's atmosphere.
[c] STP = standard temperature and pressure = 0° C and 1 atm.

The moon

Mass	7.36×10^{22} kg
Radius	1738 km
Mean density	3340 kg/m^3
Surface gravity	1.67 m/s^2
Mean earth-moon distance	3.80×10^5 km

	MERCURY	VENUS	EARTH	MARS	JUPITER	SATURN	URANUS	NEPTUNE	PLUTO
Maximum distance from sun (10^6 km)	69.7	109	152.1	249.1	815.7	1,507	3,004	4,537	7,375
Minimum distance from sun (10^6 km)	45.9	107.4	147.1	206.7	740.9	1,347	2,735	4,456	4,425
Mean distance from sun (10^6 km)	57.9	108.2	149.6	227.9	778.3	1,427	2,869.6	4,496.6	5,900
Mean distance from sun (astronomical units)	.387	.723	1	1.524	5.203	9.539	19.18	30.06	39.44
Period of revolution	88 d	224.7 d	365.26 d	687 d	11.86 y	29.46 y	84.01 y	164.8 y	247.7 y
Rotation period	59 d	−243 d retrograde	23 h 56 min 4 s	24 h 37 min 23 s	9 h 50 min 30 s	10 h 14 min	−11 h retrograde	16 h	6 d 9 h
Orbital velocity (km/s)	47.9	35	29.8	24.1	13.1	9.6	6.8	5.4	4.7
Inclination of axis	<28°	3°	23°27'	23°59'	3°05'	26°44'	82°5'	28°48'	?
Inclination of orbit to ecliptic	7°	3.4°	0°	1.9°	1.3°	2.5°	.8°	1.8°	17.2°
Eccentricity of orbit	.206	.007	.017	.093	.048	.056	.047	.009	.25
Equatorial diameter (km)	4,880	12,104	12,756	6,787	142,800	120,000	51,800	49,500	6,000 (?)
Mass (earth = 1)	.055	.815	1	.108	317.9	95.2	14.6	17.2	.1 (?)
Volume (earth = 1)	.06	.88	1	.15	1,316	755	67	57	.1 (?)
Density (water = 1)	5.4	5.2	5.5	4.0	1.3	.7	1.2	1.7	?
Oblateness	0	0	.003	.009	.06	.1	.06	.02	?
Atmosphere (main components)	none	carbon dioxide	nitrogen, oxygen	carbon dioxide, argon	hydrogen, helium	hydrogen, helium	hydrogen, helium, methane	hydrogen, helium, methane	none detected
Mean temperature at visible surface (degrees Celsius) S = solid, C = clouds	350(S) d −170(S) night	−33 (C) 480 (S)	22 (S)	−23 (S)	−150 (C)	−180 (C)	−210 (C)	−220 (C)	−230(?)
Atmospheric pressure at surface (millibars)	10^{-9}	90,000	1,000	6	?	?	?	?	?
Surface gravity (earth = 1)	.37	.88	1	.38	2.64	1.15	1.17	1.18	?
Mean apparent diameter of sun as seen from planet	1°22'40"	44'15"	31'59"	21'	6'09"	3'22"	1'41"	1'04"	49"
Known satellites	0	0	1	2	13	10	5	2	0

* Reprinted by permission from "The Solar System," Carl Sagan, *Scientific American*, September 1975.

APPENDIX E
PERIODIC TABLE OF THE ELEMENTS

NOBLE GASES

atomic number → 1
atomic mass → H 1.0079

PERIODS

Period	IA	IIA	IIIB	IVB	VB	VIB	VIIB	VIII	VIII	VIII	IB	IIB	IIIA	IVA	VA	VIA	VIIA	VIIIA / O
1	1 H 1.00797																	2 He 4.00260
2	3 Li 6.941	4 Be 9.01218											5 B 10.81	6 C 12.01115	7 N 14.0067	8 O 15.9994	9 F 18.99840	10 Ne 20.179
3	11 Na 22.98977	12 Mg 24.305											13 Al 26.98154	14 Si 28.086†	15 P 30.97376	16 S 32.06	17 Cl 35.453	18 Ar 39.948
4	19 K 39.098	20 Ca 40.08	21 Sc 44.9559	22 Ti 47.90	23 V 50.9414	24 Cr 51.996	25 Mn 54.9380	26 Fe 55.847	27 Co 58.9332	28 Ni 58.71	29 Cu 63.546	30 Zn 65.38	31 Ga 69.72	32 Ge 72.59	33 As 74.9216	34 Se 78.96	35 Br 79.904	36 Kr 83.80
5	37 Rb 85.4678	38 Sr 87.62	39 Y 88.9059	40 Zr 91.22	41 Nb 92.9064	42 Mo 95.94	43 Tc 98.9062	44 Ru 101.07	45 Rh 102.9055	46 Pd 106.4	47 Ag 107.868	48 Cd 112.40	49 In 114.82	50 Sn 118.69	51 Sb 121.75	52 Te 127.60	53 I 126.9045	54 Xe 131.30
6	55 Cs 132.9054	56 Ba 137.34	57 *La 138.9055	72 Hf 178.49	73 Ta 180.9479	74 W 183.85	75 Re 186.2	76 Os 190.2	77 Ir 192.22	78 Pt 195.09	79 Au 196.9665	80 Hg 200.59	81 Tl 204.37	82 Pb 207.19	83 Bi 208.9804	84 Po (210)	85 At (210)	86 Rn (222)
7	87 Fr (223)	88 Ra 226.0254	89 †Ac (227)	104 (261)	105 (260)	106 (263)												

* Lanthanides

58 Ce 140.12	59 Pr 140.9077	60 Nd 144.24	61 Pm (147)	62 Sm 150.4	63 Eu 151.96	64 Gd 157.25	65 Tb 158.9254	66 Dy 162.50	67 Ho 164.9304	68 Er 167.26	69 Tm 168.9342	70 Yb 173.04	71 Lu 174.97

† Actinides

90 Th 232.0381	91 Pa 231.0359	92 U 238.029	93 Np 237.0482	94 Pu (244)	95 Am (243)	96 Cm (247)	97 Bk (247)	98 Cf (251)	99 Es (254)	100 Fm (257)	101 Md (258)	102 No (255)	103 Lr (256)

APPENDIX F
THE PARTICLES OF PHYSICS*

Family name	Particle name	Symbol — Particle	Symbol — Antiparticle	Spin	Charge, e	Strangeness	Rest mass, MeV	Mean life, seconds	Typical decay mode
—	Photon	γ	γ	1	0	0	0	Stable	—
L E P T O N S	Electron	e^-	$\overline{e^-}$	$\frac{1}{2}$	∓ 1	0	0.5110	Stable	—
	Muon	μ^+	$\overline{\mu^+}$	$\frac{1}{2}$	± 1	0	105.7	2.197×10^{-6}	$e + \nu + \bar{\nu}$
	Electron's neutrino	ν_e	$\overline{\nu_e}$	$\frac{1}{2}$	0	0	0	Stable	—
	Muon's neutrino	ν_μ	$\overline{\nu_\mu}$	$\frac{1}{2}$	0	0	0	Stable	—
H A D R O N S — M E S O N S	Pion	π^+ π^0	$\overline{\pi^+}$ π^0	0 0	± 1 0	0 0	139.6 135.0	2.603×10^{-8} 8.28×10^{-17}	$\mu + \nu$ $\gamma + \gamma$
	K-meson	K^+	$\overline{K^+}$	0	± 1	± 1	493.7	1.237×10^{-8}	$\mu + \nu$
		K^0	$\overline{K^0}$	0	0	± 1	497.7	$\begin{cases} 8.930 \times 10^{-11} \\ 5.181 \times 10^{-8} \end{cases}$	$\pi^+ + \pi^-$ $\pi^0 + \pi^0 + \pi^0$
	Eta-meson	η^0	η^0	0	0	0	548.8	?	$\gamma + \gamma$
BARYONS — NUCLEON	Proton	p	$\bar{p}$	$\frac{1}{2}$	± 1	0	938.3	Stable	—
	Neutron	n	$\bar{n}$	$\frac{1}{2}$	0	0	939.6	918	$p + e^- + \nu$
	Lambda particle	Λ^0	$\overline{\Lambda^0}$	$\frac{1}{2}$	0	∓ 1	1116	2.578×10^{-10}	$p + \pi^-$
	Sigma particle	Σ^+	$\overline{\Sigma^+}$	$\frac{1}{2}$	$+1$	∓ 1	1189	8.00×10^{-11}	$p + \pi^0$
		Σ^0	$\overline{\Sigma^0}$	$\frac{1}{2}$	0	∓ 1	1192	$< 1.0 \times 10^{-14}$	$\Lambda^0 + \gamma$
		Σ^-	$\overline{\Sigma^-}$	$\frac{1}{2}$	-1	∓ 1	1197	1.482×10^{-10}	$n + \pi^-$
	Xi particle	Ξ^0	$\overline{\Xi^0}$	$\frac{1}{2}$	0	∓ 2	1315	2.96×10^{-10}	$\Lambda^0 + \pi^0$
		Ξ^-	$\overline{\Xi^-}$	$\frac{1}{2}$	∓ 1	∓ 2	1321	1.652×10^{-10}	$\Lambda^0 + \pi^-$
	Omega particle	Ω^-	$\overline{\Omega^-}$	$\frac{3}{2}$	∓ 1	∓ 3	1672	1.3×10^{-10}	$\Xi^0 + \pi^-$

* See (1) "Review of Particle Properties," *Reviews of Modern Physics*, vol. 48, no. 2, Part II, April (1976).
(2) "Quarks with Color and Flavor," by Sheldon Lee Glashow, *Scientific American*, October (1975).
(3) "The New Elementary Particles and Charm," by Lewis Ryder, *Physics Education*, January (1976) for fuller information. Particle physics is one of the sharp cutting edges of contemporary physics.

Conversion factors may be read off directly from the tables. For example, 1 degree = 2.778×10^{-3} revolutions, so $16.7° = 16.7 \times 2.778 \times 10^{-3}$ rev. The SI quantities are capitalized. The prefix "ab" refers to electromagnetic units (emu); "stat" refers to electrostatic units (esu). Adapted in part from G. Shortley and D. Williams, *Elements of Physics*, Prentice-Hall, Englewood Cliffs, N.J., 1965.

Plane angle

	°	′	″	RADIAN	rev
1 degree =	1	60	3600	1.745×10^{-2}	2.778×10^{-3}
1 minute =	1.667×10^{-2}	1	60	2.909×10^{-4}	4.630×10^{-5}
1 second =	2.778×10^{-4}	1.667×10^{-2}	1	4.848×10^{-6}	7.716×10^{-7}
1 RADIAN =	57.30	3438	2.063×10^{5}	1	0.1592
1 revolution =	360	2.16×10^{4}	1.296×10^{6}	6.283	1

Solid angle

1 sphere = 4π steradians = 12.57 steradians

Length

	cm	METER	km	in.	ft	mi
1 centimeter =	1	10^{-2}	10^{-5}	0.3937	3.281×10^{-2}	6.214×10^{-6}
1 METER =	100	1	10^{-3}	39.3	3.281	6.214×10^{-4}
1 kilometer =	10^{5}	1000	1	3.937×10^{4}	3281	0.6214
1 inch =	2.540	2.540×10^{-2}	2.540×10^{-5}	1	8.333×10^{-2}	1.578×10^{-5}
1 foot =	30.48	0.3048	3.048×10^{-4}	12	1	1.894×10^{-4}
1 mile =	1.609×10^{5}	1609	1.609	6.336×10^{4}	5280	1

1 angstrom = 10^{-10} m   1 light-year = 9.4600×10^{12} km   1 yard = 3 ft
1 nautical mile = 1852 m   1 parsec = 3.084×10^{13} km   1 rod = 16.5 ft
   = 1.151 miles = 6076 ft   1 fathom = 6 ft   1 mil = 10^{-3} in.

Area

	METER²	cm²	ft²	in.²	circ mil
1 SQUARE METER =	1	10^{4}	10.76	1550	1.974×10^{9}
1 square centimeter =	10^{-4}	1	1.076×10^{-3}	0.1550	1.974×10^{5}
1 square foot =	9.290×10^{-2}	929.0	1	144	1.833×10^{8}
1 square inch =	6.452×10^{-4}	6.452	6.944×10^{-3}	1	1.273×10^{6}
1 circular mil =	5.067×10^{-10}	5.067×10^{-6}	5.454×10^{-9}	7.854×10^{-7}	1

1 square mile = 2.788×10^{8} ft² = 640 acres   1 acre = 43,600 ft²
1 barn = 10^{-28} m²

Volume

	METER³	cm³	li	ft³	in.³
1 CUBIC METER =	1	10^6	1000	35.31	6.102×10^4
1 cubic centimeter =	10^{-6}	1	1.000×10^{-3}	3.531×10^{-5}	6.102×10^{-2}
1 liter =	1.000×10^{-3}	1000	1	3.531×10^{-2}	61.02
1 cubic foot =	2.832×10^{-2}	2.832×10^4	28.32	1	1728
1 cubic inch =	1.639×10^{-5}	16.39	1.639×10^{-2}	5.787×10^{-4}	1

1 U. S. fluid gallon = 4 U. S. fluid quarts = 8 U. S. pints = 128 U. S. fluid ounces = 231 in.³
1 British imperial gallon = 277.4 in.³   1 liter = 10^{-3} m³.

Mass

Quantities in the shaded areas are not mass units but are often used as such. When we write, for example, 1 kg "=" 2.205 lb this means that a kilogram is a *mass* that *weighs* 2.205 pounds under standard condition of gravity (g = 9.80665 m/s²).

	gm	KG	slug	u	oz	lb	ton
1 gram =	1	0.001	6.852×10^{-5}	6.024×10^{23}	3.527×10^{-2}	2.205×10^{-3}	1.102×10^{-6}
1 KILOGRAM =	1000	1	6.852×10^{-2}	6.024×10^{26}	35.27	2.205	1.102×10^{-3}
1 slug =	1.459×10^4	14.59	1	8.789×10^{27}	514.8	32.17	1.609×10^{-2}
1 u =	1.660×10^{-24}	1.660×10^{-27}	1.137×10^{-28}	1	5.855×10^{-26}	3.660×10^{-27}	1.829×10^{-30}
1 ounce =	28.35	2.835×10^{-2}	1.943×10^{-3}	1.708×10^{25}	1	6.250×10^{-2}	3.125×10^{-5}
1 pound =	453.6	0.4536	3.108×10^{-2}	2.732×10^{26}	16	1	0.0005
1 ton =	9.072×10^5	907.2	62.16	5.465×10^{29}	3.2×10^4	2000	1

Density

Quantities in the shaded areas are weight densities and, as such, are dimensionally different from mass densities. See note for mass table.

	slug/ft³	KG/METER³	g/cm³	lb/ft³	lb/in.³
1 slug per ft³ =	1	515.4	0.5154	32.17	1.862×10^{-2}
1 KILOGRAM per METER³ =	1.940×10^{-3}	1	0.001	6.243×10^{-2}	3.613×10^{-5}
1 gram per cm³ =	1.940	1000	1	62.43	3.613×10^{-2}
1 pound per ft³ =	3.108×10^{-2}	16.02	1.602×10^{-2}	1	5.787×10^{-4}
1 pound per in.³ =	53.71	2.768×10^4	27.68	1728	1

Time

	yr	d	h	min	SECOND
1 year =	1	365.2	8.766×10^3	5.259×10^5	3.156×10^7
1 day =	2.738×10^{-3}	1	24	1440	8.640×10^4
1 hour =	1.141×10^{-4}	4.167×10^{-2}	1	60	3600
1 minute =	1.901×10^{-6}	6.944×10^{-4}	1.667×10^{-2}	1	60
1 SECOND =	3.169×10^{-8}	1.157×10^{-5}	2.778×10^{-4}	1.667×10^{-2}	1

Speed

	ft/s	km/h	METER/ SECOND	mi/h	cm/s	knot
1 foot per second =	1	1.097	0.3048	0.6818	30.48	0.5925
1 kilometer per hour =	0.9113	1	0.2778	0.6214	27.78	0.5400
1 METER per SECOND =	3.281	3.6	1	2.237	100	1.944
1 mile per hour =	1.467	1.609	0.4470	1	44.70	0.8689
1 centimeter per second =	3.281×10^{-2}	3.6×10^{-2}	0.01	2.237×10^{-2}	1	1.944×10^{-2}
1 knot =	1.688	1.852	0.5144	1.151	51.44	1

1 knot = 1 nautical mi/h   1 mi/min = 88.00 ft/s = 60.00 mi/h

Force

Quantities in the shaded areas are not force units but are often used as such. For instance, if we write 1 gram-force "=" 980.7 dynes, we mean that a gram-*mass* experiences a *force* of 980.7 dynes under standard conditions of gravity ($g = 9.80665$ m/s²).

	dyne	NEWTON	lb	pdl	gf	kgf
1 dyne =	1	10^{-5}	2.248×10^{-6}	7.233×10^{-5}	1.020×10^{-3}	1.020×10^{-6}
1 NEWTON =	10^5	1	0.2248	7.233	102.0	0.1020
1 pound =	4.448×10^5	4.448	1	32.17	453.6	0.4536
1 poundal =	1.383×10^4	0.1383	3.108×10^{-2}	1	14.10	1.410×10^{-2}
1 gram-force =	980.7	9.807×10^{-3}	2.205×10^{-3}	7.093×10^{-2}	1	0.001
1 kilogram-force =	9.807×10^5	9.807	2.205	70.93	1000	1

Pressure

	atm	dyne/cm²	inch of water	cm-Hg	PASCAL	lb/in.²	lb/ft²
1 atmosphere =	1	1.013×10^6	406.8	76	1.013×10^5	14.70	2116
1 dyne per cm² =	9.869×10^{-7}	1	4.015×10^{-4}	7.501×10^{-5}	0.1	1.450×10^{-5}	2.089×10^{-3}
1 inch of water[a] at 4° C =	2.458×10^{-3}	2491	1	0.1868	249.1	3.613×10^{-2}	5.202
1 centimeter of mercury[a] at 0° C =	1.316×10^{-2}	1.333×10^4	5.353	1	1333	0.1934	27.85
1 PASCAL =	9.869×10^{-6}	10	4.015×10^{-3}	7.501×10^{-4}	1	1.450×10^{-4}	2.089×10^{-2}
1 pound per in.² =	6.805×10^{-2}	6.895×10^4	27.68	5.171	6.895×10^3	1	144
1 pound per ft² =	4.725×10^{-4}	478.8	0.1922	3.591×10^{-2}	47.88	6.944×10^{-3}	1

[a] Where the acceleration of gravity has the standard value 9.80665 m/s².

1 bar = 10^6 dyne/cm² = 0.1 MPa   1 millibar = 10^3 dyne/cm² = 10^2 Pa

Energy, work, heat

Quantities in the shaded areas are not properly energy units but are included for convenience. They arise from the relativistic mass-energy equivalence formula $E = mc^2$ and represent the energy released if a kilogram or unified atomic mass unit (u) is completely converted to energy.

	Btu	erg	ft·lb	hp·h	JOULE	cal	kW·h	eV	MeV	kg	u
1 British thermal unit =	1	1.055×10^{10}	777.9	3.929×10^{-4}	1055	252.0	2.930×10^{-4}	6.585×10^{21}	6.585×10^{15}	1.174×10^{-14}	7.074×10^{12}
1 erg =	9.481×10^{-11}	1	7.376×10^{-8}	3.725×10^{-14}	10^{-7}	2.389×10^{-8}	2.778×10^{-14}	6.242×10^{11}	6.242×10^{5}	1.113×10^{-24}	670.5
1 foot-pound =	1.285×10^{-3}	1.356×10^{7}	1	5.051×10^{-7}	1.356	0.3239	3.766×10^{-7}	8.464×10^{18}	8.464×10^{12}	1.509×10^{-17}	9.092×10^{9}
1 horsepower-hour =	2545	2.685×10^{13}	1.980×10^{6}	1	2.685×10^{6}	6.414×10^{5}	0.7457	1.676×10^{25}	1.676×10^{19}	2.988×10^{-11}	1.800×10^{16}
1 JOULE =	9.481×10^{-4}	10^{7}	0.7376	3.725×10^{-7}	1	0.2389	2.778×10^{-7}	6.242×10^{18}	6.242×10^{12}	1.113×10^{-17}	6.705×10^{9}
1 calorie =	3.968×10^{-3}	4.186×10^{7}	3.087	1.559×10^{-6}	4.186	1	1.163×10^{-6}	2.613×10^{19}	2.613×10^{13}	4.659×10^{-17}	2.807×10^{10}
1 kilowatt-hour =	3413	3.6×10^{13}	2.655×10^{6}	1.341	3.6×10^{6}	8.601×10^{5}	1	2.247×10^{25}	2.247×10^{19}	4.007×10^{-11}	2.414×10^{16}
1 electron volt =	1.519×10^{-22}	1.602×10^{-12}	1.182×10^{-19}	5.967×10^{-26}	1.602×10^{-19}	3.827×10^{-20}	4.450×10^{-26}	1	10^{-6}	1.783×10^{-36}	1.074×10^{-9}
1 million electron volts =	1.519×10^{-16}	1.602×10^{-6}	1.182×10^{-13}	5.967×10^{-20}	1.602×10^{-13}	3.827×10^{-14}	4.450×10^{-20}	10^{6}	1	1.783×10^{-30}	1.074×10^{-3}
1 kilogram =	8.521×10^{13}	8.987×10^{23}	6.629×10^{16}	3.348×10^{10}	8.987×10^{16}	2.147×10^{16}	2.497×10^{10}	5.610×10^{35}	5.610×10^{29}	1	6.025×10^{26}
1 unified atomic mass unit =	1.415×10^{-13}	1.492×10^{-3}	1.100×10^{-10}	5.558×10^{-17}	1.492×10^{-10}	3.564×10^{-11}	4.145×10^{-17}	9.31×10^{8}	931.0	1.660×10^{-27}	1

Power

	Btu/h	ft·lb/s	hp	cal/s	kW	WATT
1 British thermal unit per hour =	1	0.2161	3.929×10^{-4}	7.000×10^{-2}	2.930×10^{-4}	0.2930
1 foot-pound per second =	4.628	1	1.818×10^{-3}	0.3239	1.356×10^{-3}	1.356
1 horsepower =	2545	550	1	178.2	0.7457	745.7
1 calorie per second =	14.29	3.087	5.613×10^{-3}	1	4.186×10^{-3}	4.186
1 kilowatt =	3413	737.6	1.341	238.9	1	1000
1 WATT =	3.413	0.7376	1.341×10^{-3}	0.2389	0.001	1

Charge

	abcoul	A·h	COULOMB	statcoul
1 abcoulomb =	1	2.778×10^{-3}	10	2.998×10^{10}
1 ampere-hour =	360	1	3600	1.079×10^{13}
1 COULOMB =	0.1	2.778×10^{-4}	1	2.998×10^{9}
1 statcoulomb =	3.336×10^{-11}	9.266×10^{-14}	3.336×10^{-10}	1

1 electronic charge = 1.602×10^{-19} coulomb

Current

	abamp	AMPERE	statamp
1 abampere =	1	10	2.998×10^{10}
1 AMPERE =	0.1	1	2.998×10^{9}
1 statampere =	3.336×10^{-11}	3.336×10^{-10}	1

Potential, electromotive force

	abvolt	VOLT	statvolt
1 abvolt =	1	10^{-8}	3.336×10^{-11}
1 VOLT =	10^{8}	1	3.336×10^{-3}
1 statvolt =	2.998×10^{10}	299.8	1

Resistance

	abohm	OHM	statohm
1 abohm =	1	10^{-9}	1.113×10^{-21}
1 OHM =	10^{9}	1	1.113×10^{-12}
1 statohm =	8.987×10^{20}	8.987×10^{11}	1

Capacitance

	abf	FARAD	μF	statf
1 abfarad =	1	10^{9}	10^{15}	8.987×10^{20}
1 FARAD =	10^{-9}	1	10^{6}	8.987×10^{11}
1 microfarad =	10^{-15}	10^{-6}	1	8.987×10^{5}
1 statfarad =	1.113×10^{-21}	1.113×10^{-12}	1.113×10^{-6}	1

Inductance

	abhenry	HENRY	μH	mH	stathenry
1 abhenry =	1	10^{-9}	0.001	10^{-6}	1.113×10^{-21}
1 HENRY =	10^{9}	1	10^{6}	1000	1.113×10^{-12}
1 microhenry =	1000	10^{-6}	1	0.001	1.113×10^{-18}
1 millihenry =	10^{6}	0.001	1000	1	1.113×10^{-15}
1 stathenry =	8.987×10^{20}	8.987×10^{11}	8.987×10^{17}	8.987×10^{14}	1

Magnetic flux

	maxwell	WEBER
1 maxwell =	1	10^{-8}
1 WEBER =	10^{8}	1

Magnetic field

	gauss	TESLA	milligauss
1 gauss =	1	10^{-4}	1000
1 TESLA =	10^{4}	1	10^{7}
1 milligauss =	0.001	10^{-7}	1

1 tesla = 1 weber/meter^2

Mathematical Signs and Symbols

= equals
≅ equals approximately
≠ is not equal to
≡ is identical to, is defined as
> is greater than (≫ is much greater than)
< is less than (≪ is much less than)
≧ is more than or equal to (or, is no less than)
≦ is less than or equal to (or, is no more than)
± plus or minus ($\sqrt{4} = \pm 2$)
∝ is proportional to (Hooke's law: $F \propto x$, or $F = -kx$)
Σ the sum of
$\bar{x}$ the average value of x

The Greek Alphabet

Alpha	A	α	Nu	N	ν
Beta	B	β	Xi	Ξ	ξ
Gamma	Γ	γ	Omicron	O	o
Delta	Δ	δ	Pi	Π	π
Epsilon	E	ϵ	Rho	P	ρ
Zeta	Z	ζ	Sigma	Σ	σ
Eta	H	η	Tau	T	τ
Theta	Θ	θ	Upsilon	Υ	υ
Iota	I	ι	Phi	Φ	ϕ, φ
Kappa	K	κ	Chi	X	χ
Lambda	Λ	λ	Psi	Ψ	ψ
Mu	M	μ	Omega	Ω	ω

APPENDIX I
MATHEMATICAL FORMULAS

Geometry

Circle of radius r: circumference $= 2\pi r$; area $= \pi r^2$.
Sphere of radius r: area $= 4\pi r^2$; volume $= \frac{4}{3}\pi r^3$.
Right circular cylinder of radius r and height h: area $= 2\pi r^2 + 2\pi rh$; volume $= \pi r^2 h$.

Quadratic Formula

If $ax^2 + bx + c = 0$, then $x = \dfrac{-b \pm \sqrt{b^2 - 4ac}}{2a}$.

Trigonometric Functions of Angle θ

$$\sin \theta = \frac{y}{r} \qquad \cos \theta = \frac{x}{r}$$

$$\tan \theta = \frac{y}{x} \qquad \cot \theta = \frac{x}{y}$$

$$\sec \theta = \frac{r}{x} \qquad \csc \theta = \frac{r}{y}$$

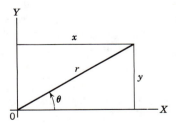

Pythagorean Theorem

$$x^2 + y^2 = r^2$$

Trigonometric Identities

$$\sin^2 \theta + \cos^2 \theta = 1 \qquad \sec^2 \theta - \tan^2 \theta = 1 \qquad \csc^2 \theta - \cot^2 \theta = 1$$

$$\sin 2\theta = 2 \sin \theta \cos \theta$$

$$\cos 2\theta = \cos^2 \theta - \sin^2 \theta = 2 \cos^2 \theta - 1 = 1 - 2 \sin^2 \theta$$

$$\sin \theta = \frac{e^{i\theta} - e^{-i\theta}}{2i} \qquad \cos \theta = \frac{e^{i\theta} + e^{-i\theta}}{2}$$

$$e^{\pm i\theta} = \cos \theta \pm i \sin \theta$$

$$\sin (\alpha \pm \beta) = \sin \alpha \cos \beta \pm \cos \alpha \sin \beta$$

$$\cos (\alpha \pm \beta) = \cos \alpha \cos \beta \mp \sin \alpha \sin \beta$$

$$\tan (\alpha \pm \beta) = \frac{\tan \alpha \pm \tan \beta}{1 \mp \tan \alpha \tan \beta}$$

$$\sin \alpha \pm \sin \beta = 2 \sin \tfrac{1}{2}(\alpha \pm \beta) \cos \tfrac{1}{2}(\alpha \mp \beta)$$

Taylor's Series

$$f(x_0 + x) = f(x_0) + f'(x_0)x + f''(x_0) \frac{x^2}{2!} + f'''(x_0) \frac{x^3}{3!} + \cdots$$

Binomial Expansion

$$(1 + x)^n = 1 + \frac{nx}{1!} + \frac{n(n - 1)}{2!} x^2 + \cdots$$

Exponential Expansion

$$e^x = 1 + x + \frac{x^2}{2!} + \frac{x^3}{3!} + \cdots$$

Logarithmic Expansion

$$\ln(1 + x) = x - \tfrac{1}{2}x^2 + \tfrac{1}{3}x^3 - \cdots$$

Trigonometric Expansions (θ in radians)

$$\sin \theta = \theta - \frac{\theta^3}{3!} + \frac{\theta^5}{5!} - \cdots$$

$$\cos \theta = 1 - \frac{\theta^2}{2!} + \frac{\theta^4}{4!} - \cdots$$

Derivatives and Indefinite Integrals

In what follows, the letters u and v stand for any functions of x, and a and m are constants. To each of the integrals should be added an arbitrary constant of integration. The *Handbook of Chemistry and Physics* (Chemical Rubber Publishing Co.) gives a more extensive tabulation.

1. $\dfrac{dx}{dx} = 1$

2. $\dfrac{d}{dx} (au) = a \dfrac{du}{dx}$

3. $\dfrac{d}{dx} (u + v) = \dfrac{du}{dx} + \dfrac{dv}{dx}$

4. $\dfrac{d}{dx} x^m = mx^{m-1}$

1. $\int dx = x$

2. $\int au \, dx = a \int u \, dx$

3. $\int (u + v) \, dx = \int u \, dx + \int v \, dx$

4. $\int x^m \, dx = \dfrac{x^{m+1}}{m + 1} \qquad (m \neq -1)$

5. $\dfrac{d}{dx} \ln x = \dfrac{1}{x}$

5. $\displaystyle\int \dfrac{dx}{x} = \ln |x|$

6. $\dfrac{d}{dx} (uv) = u \dfrac{dv}{dx} + v \dfrac{du}{dx}$

6. $\displaystyle\int u \dfrac{dv}{dx} dx = uv - \int v \dfrac{du}{dx} dx$

7. $\dfrac{d}{dx} e^x = e^x$

7. $\displaystyle\int e^x dx = e^x$

8. $\dfrac{d}{dx} \sin x = \cos x$

8. $\int \sin x \, dx = - \cos x$

9. $\dfrac{d}{dx} \cos x = - \sin x$

9. $\int \cos x \, dx = \sin x$

10. $\dfrac{d}{dx} \tan x = \sec^2 x$

10. $\int \tan x \, dx = \ln |\sec x|$

11. $\dfrac{d}{dx} \cot x = - \csc^2 x$

11. $\int \cot x \, dx = \ln |\sin x|$

12. $\dfrac{d}{dx} \sec x = \tan x \sec x$

12. $\int \sec x \, dx = \ln |\sec x + \tan x|$

13. $\dfrac{d}{dx} \csc x = - \cot x \csc x$

13. $\int \csc x \, dx = \ln |\csc x - \cot x|$

14. $\dfrac{d}{dx} \arctan x = \dfrac{1}{1 + x^2}$

14. $\displaystyle\int \dfrac{dx}{1 + x^2} = \arctan x$

15. $\dfrac{d}{dx} \arcsin x = \dfrac{1}{\sqrt{1 - x^2}}$

15. $\displaystyle\int \dfrac{dx}{\sqrt{1 - x^2}} = \arcsin x$

16. $\dfrac{d}{dx} \text{arcsec } x = \dfrac{1}{x\sqrt{x^2 - 1}}$

16. $\displaystyle\int \dfrac{dx}{x\sqrt{x^2 - 1}} = \text{arcsec } x$

Vector Products

Let $\mathbf{i}, \mathbf{j}, \mathbf{k}$ be unit vectors in the x, y, z directions. Then

$$\mathbf{i} \cdot \mathbf{i} = \mathbf{j} \cdot \mathbf{j} = \mathbf{k} \cdot \mathbf{k} = 1, \qquad \mathbf{i} \cdot \mathbf{j} = \mathbf{j} \cdot \mathbf{k} = \mathbf{k} \cdot \mathbf{i} = 0,$$

$$\mathbf{i} \times \mathbf{i} = \mathbf{j} \times \mathbf{j} = \mathbf{k} \times \mathbf{k} = 0,$$

$$\mathbf{i} \times \mathbf{j} = \mathbf{k}, \qquad \mathbf{j} \times \mathbf{k} = \mathbf{i}, \qquad \mathbf{k} \times \mathbf{i} = \mathbf{j}.$$

Any vector $\mathbf{a}$ with components a_x, a_y, a_z along the x, y, z axes can be written

$$\mathbf{a} = a_x\mathbf{i} + a_y\mathbf{j} + a_z\mathbf{k}.$$

Let $\mathbf{a}, \mathbf{b}, \mathbf{c}$ be arbitrary vectors with magnitudes a, b, c. Then

$$\mathbf{a} \times (\mathbf{b} + \mathbf{c}) = \mathbf{a} \times \mathbf{b} + \mathbf{a} \times \mathbf{c}$$

$$(s\mathbf{a}) \times \mathbf{b} = \mathbf{a} \times (s\mathbf{b}) = s(\mathbf{a} \times \mathbf{b}) \qquad (s = \text{a scalar}).$$

Let θ be the smaller of the two angles between $\mathbf{a}$ and $\mathbf{b}$. Then

$$\mathbf{a} \cdot \mathbf{b} = \mathbf{b} \cdot \mathbf{a} = a_x b_x + a_y b_y + a_z b_z = ab \cos \theta$$

$$\mathbf{a} \times \mathbf{b} = -\mathbf{b} \times \mathbf{a} = \begin{vmatrix} \mathbf{i} & \mathbf{j} & \mathbf{k} \\ a_x & a_y & a_z \\ b_x & b_y & b_z \end{vmatrix} = (a_y b_z - b_y a_z)\mathbf{i} + (a_z b_x - b_z a_x)\mathbf{j}$$
$$+ (a_x b_y - b_x a_y)\mathbf{k}$$

$$|\mathbf{a} \times \mathbf{b}| = ab \sin \theta$$

$$\mathbf{a} \cdot (\mathbf{b} \times \mathbf{c}) = \mathbf{b} \cdot (\mathbf{c} \times \mathbf{a}) = \mathbf{c} \cdot (\mathbf{a} \times \mathbf{b})$$

$$\mathbf{a} \times (\mathbf{b} \times \mathbf{c}) = (\mathbf{a} \cdot \mathbf{c})\mathbf{b} - (\mathbf{a} \cdot \mathbf{b})\mathbf{c}$$

Degrees	Radians	Sine	Tangent	Cotangent	Cosine		
0	0	0	0	∞	1.0000	1.5708	**90**
1	.0175	.0175	.0175	57.290	.9998	1.5533	89
2	.0349	.0349	.0349	28.636	.9994	1.5359	88
3	.0524	.0523	.0524	19.081	.9986	1.5184	87
4	.0698	.0698	.0699	14.301	.9976	1.5010	86
5	.0873	.0872	.0875	11.430	.9962	1.4835	**85**
6	.1047	.1045	.1051	9.5144	.9945	1.4661	84
7	.1222	.1219	.1228	8.1443	.9925	1.4486	83
8	.1396	.1392	.1405	7.1154	.9903	1.4312	82
9	.1571	.1564	.1584	6.3138	.9877	1.4137	81
10	.1745	.1736	.1763	5.6713	.9848	1.3963	**80**
11	.1920	.1908	.1944	5.1446	.9816	1.3788	79
12	.2094	.2079	.2126	4.7046	.9781	1.3614	78
13	.2269	.2250	.2309	4.3315	.9744	1.3439	77
14	.2443	.2419	.2493	4.0108	.9703	1.3265	76
15	.2618	.2588	.2679	3.7321	.9659	1.3090	**75**
16	.2793	.2756	.2867	3.4874	.9613	1.2915	74
17	.2967	.2924	.3057	3.2709	.9563	1.2741	73
18	.3142	.3090	.3249	3.0777	.9511	1.2566	72
19	.3316	.3256	.3443	2.9042	.9455	1.2392	71
20	.3491	.3420	.3640	2.7475	.9397	1.2217	**70**
21	.3665	.3584	.3839	2.6051	.9336	1.2043	69
22	.3840	.3746	.4040	2.4751	.9272	1.1868	68
23	.4014	.3907	.4245	2.3559	.9205	1.1694	67
24	.4189	.4067	.4452	2.2460	.9135	1.1519	66
25	.4363	.4226	.4663	2.1445	.9063	1.1345	**65**
26	.4538	.4384	.4877	2.0503	.8988	1.1170	64
27	.4712	.4540	.5095	1.9626	.8910	1.0996	63
28	.4887	.4695	.5317	1.8807	.8829	1.0821	62
29	.5061	.4848	.5543	1.8040	.8746	1.0647	61
30	.5236	.5000	.5774	1.7321	.8660	1.0472	**60**
31	.5411	.5150	.6009	1.6643	.8572	1.0297	59
32	.5585	.5299	.6249	1.6003	.8480	1.0123	58
33	.5760	.5446	.6494	1.5399	.8387	.9948	57
34	.5934	.5592	.6745	1.4826	.8290	.9774	56
35	.6109	.5736	.7002	1.4281	.8192	.9599	**55**
36	.6283	.5878	.7265	1.3764	.8090	.9425	54
37	.6458	.6018	.7536	1.3270	.7986	.9250	53
38	.6632	.6157	.7813	1.2799	.7880	.9076	52
39	.6807	.6293	.8098	1.2349	.7771	.8901	51
40	.6981	.6428	.8391	1.1918	.7660	.8727	**50**
41	.7156	.6561	.8693	1.1504	.7547	.8552	49
42	.7330	.6691	.9004	1.1106	.7431	.8378	48
43	.7505	.6820	.9325	1.0724	.7314	.8203	47
44	.7679	.6947	.9657	1.0355	.7193	.8029	46
45	.7854	.7071	1.0000	1.0000	.7071	.7854	**45**
		Cosine	Cotangent	Tangent	Sine	Radians	Degrees

1901	Wilhelm Konrad Röntgen	1845–1923	for the discovery of the remarkable rays subsequently named after him
1902	Hendrik Antoon Lorentz Pieter Zeeman	1853–1928 1865–1943	for their researches into the influence of magnetism upon radiation phenomena
1903	Antoine Henri Becquerel	1852–1908	for his discovery of spontaneous radioactivity

* See *Nobel Lectures, Physics,* 1901–1970, Elsevier Publishing Company, for the Nobel presentations, lectures and biographies. The attributions are, almost without exception, quotations from the Nobel Prize citations.

	Pierre Curie	1859–1906	for their joint researches on the radiation phenomena discovered by Professor Henri Becquerel
	Marie Sklowdowska-Curie	1867–1934	
1904	Lord Rayleigh (John William Strutt)	1842–1919	for his investigations of the densities of the most important gases and for his discovery of argon
1905	Philipp Eduard Anton von Lenard	1862–1947	for his work on cathode rays
1906	Joseph John Thomson	1856–1940	for his theoretical and experimental investigations on the conduction of electricity by gases
1907	Albert Abraham Michelson	1852–1931	for his optical precision instruments and metrological investigations carried out with their aid
1908	Gabriel Lippmann	1845–1921	for his method of reproducing colors photographically based on the phenomena of interference
1909	Guglielmo Marconi	1874–1937	for their contributions to the development of wireless telegraphy
	Carl Ferdinand Braun	1850–1918	
1910	Johannes Diderik van der Waals	1837–1923	for his work on the equation of state for gases and liquids
1911	Wilhelm Wien	1864–1928	for his discoveries regarding the laws governing the radiation of heat
1912	Nils Gustaf Dalén	1869–1937	for his invention of automatic regulators for use in conjunction with gas accumulators for illuminating lighthouses and buoys
1913	Heike Kamerlingh Onnes	1853–1926	for his investigations of the properties of matter at low temperatures which led, *inter alia*, to the production of liquid helium
1914	Max von Laue	1879–1960	for his discovery of the diffraction of Röntgen rays by crystals
1915	William Henry Bragg	1862–1942	for their services in the analysis of crystal structure by means of Röntgen rays
	William Lawrence Bragg	1890–1971	
1917	Charles Glover Barkla	1877–1944	for his discovery of the characteristic Röntgen radiation of the elements
1918	Max Planck	1858–1947	for his discovery of energy quanta
1919	Johannes Stark	1874–1957	for his discovery of the Doppler effect in canal rays and the splitting of spectral lines in electric fields
1920	Charles-Édouard Guillaume	1861–1938	for the service he has rendered to precision measurements in Physics by his discovery of anomalies in nickel steel alloys
1921	Albert Einstein	1879–1955	for his services to Theoretical Physics, and especially for his discovery of the law of the photoelectric effect
1922	Niels Bohr	1885–1962	for the investigation of the structure of atoms, and of the radiation emanating from them
1923	Robert Andrews Millikan	1868–1953	for his work on the elementary charge of electricity and on the photoelectric effect
1924	Karl Manne Georg Siegbahn	1886–1954	for his discoveries and research in the field of x-ray spectroscopy
1925	James Franck	1882–1964	for their discovery of the laws governing the impact of an electron upon an atom
	Gustav Hertz	1887–1975	

1926	Jean Baptiste Perrin	1870–1942	for his work on the discontinuous structure of matter, and especially for his discovery of sedimentation equillibrium
1927	Arthur Holly Compton	1892–1962	for his discovery of the effect named after him
	Charles Thomson Rees Wilson	1869–1959	for his method of making the paths of electrically charged particles visible by condensation of vapor
1928	Owen Willans Richardson	1879–1959	for his work on the thermionic phenomenon and especially for the discovery of the law named after him
1929	Prince Louis-Victor de Broglie	1892–	for his discovery of the wave nature of electrons
1930	Sir Chandrasekhara Venkata Raman	1888–1970	for his work on the scattering of light and for the discovery of the effect named after him
1932	Werner Heisenberg	1901–1976	for the creation of quantum mechanics, the application of which has, among other things, led to the discovery of the allotropic forms of hydrogen
1933	Erwin Schrödinger	1887–1961	for the discovery of new productive forms of atomic theory
	Paul Adrien Maurice Dirac	1902–	
1935	James Chadwick	1891–1974	for his discovery of the neutron
1936	Victor Franz Hess	1883–1964	for his discovery of cosmic radiation
	Carl David Anderson	1905–	for his discovery of the positron
1937	Clinton Joseph Davisson	1881–1958	for their experimental discovery of the diffraction of electrons by crystals
	George Paget Thomson	1892–1975	
1938	Enrico Fermi	1901–1954	for his demonstrations of the existence of new radioactive elements produced by neutron irradiation, and for his related discovery of nuclear reactions brought about by slow neutrons
1939	Ernest Orlando Lawrence	1901–1958	for the invention and development of the cyclotron and for results obtained with it, especially with regard to artificial radioactive elements
1943	Otto Stern	1888–1969	for his contribution to the development of the molecular ray method and his discovery of the magnetic moment of the proton
1944	Isidor Isaac Rabi	1898–	for his resonance method for recording the magnetic properties of atomic nuclei
1945	Wolfgang Pauli	1900–1958	for the discovery of the Exclusion Principle, also called the Pauli Principle
1946	Percy Williams Bridgman	1882–1961	for the invention of an apparatus to produce extremely high pressures, and for the discoveries he made therewith in the field of high-pressure physics
1947	Sir Edward Victor Appleton	1892–1965	for his investigations of the physics of the upper atmosphere, especially for the discovery of the so-called Appleton layer
1948	Patrick Maynard Stuart Blackett	1897–1974	for his development of the Wilson cloud chamber method, and his discoveries therewith in the fields of nuclear physics and cosmic radiation

1949	Hideki Yukawa	1907–	for his prediction of the existence of mesons on the basis of theoretical work on nuclear forces
1950	Cecil Frank Powell	1903–1969	for his development of the photographic method of studying nuclear processes and his discoveries regarding mesons made with this method
1951	Sir John Douglas Cockcroft	1897–1967	for their pioneer work on the transmutation of atomic nuclei by artificially accelerated atomic particles
	Ernest Thomas Sinton Walton	1903–	
1952	Felix Bloch	1905–	for their development of new methods for nuclear magnetic precision methods and discoveries in connection therewith
	Edward Mills Purcell	1912–	
1953	Frits Zernike	1888–1966	for his demonstration of the phase-contrast method, especially for his invention of the phase-contrast microscope
1954	Max Born	1882–1970	for his fundamental research in quantum mechanics, especially for his statistical interpretation of the wave function
	Walther Bothe	1891–1957	for the coincidence method and his discoveries made therewith
1955	Willis Eugene Lamb	1913–	for his discoveries concerning the fine structure of the hydrogen spectrum
	Polykarp Kusch	1911–	for his precision determination of the magnetic moment of the electron
1956	William Shockley	1910–	for their researches on semiconductors and their discovery of the transistor effect
	John Bardeen	1908–	
	Walter Houser Brattain	1902–	
1957	Chen Ning Yang	1922–	for their penetrating investigation of the so-called parity laws which has led important discoveries regarding the elementary particles
	Tsung Dao Lee	1926–	
1958	Pavel Aleksejevič Čerenkov	1904–	for the discovery and the interpretation of the Čerenkov effect
	Il' ja Michajlovič Frank	1908–	
	Igor' Evgen'evič Tamm	1895–1971	
1959	Emilio Gino Segrè	1905–	for their discovery of the antiproton
	Owen Chamberlain	1920–	
1960	Donald Arthur Glaser	1926–	for the invention of the bubble chamber
1961	Robert Hofstadter	1915–	for his pioneering studies of electron scattering in atomic nuclei and for his thereby achieved discoveries concerning the structure of the nucleons
	Rudolf Ludwig Mössbauer	1929–	for his researches concerning the resonances absorption of γ-radiation and his discovery in this connection of the effect which bears his name
1962	Lev Davidovič Landau	1908–	for his pioneering theories of condensed matter, especially liquid helium
1963	Eugene P. Wigner	1902–	for his contributions to the theory of the atomic nucleus and the elementary particles, particularly through the discovery and application of fundamental symmetry principles
	Maria Goeppert Mayer	1906–1972	for their discoveries concerning nuclear shell structure
	J. Hans D. Jensen	1907–1973	

1964	Charles H. Townes	1915–	for fundamental work in the
	Nikolai G. Basov	1922–	field of quantum electronics
	Alexander M. Prochorov	1916–	which has led to the construction of oscillators and amplifiers based on the maser-laser principle
1965	Sin-Itiro Tomonaga	1906–	for their fundamental work in
	Julian Schwinger	1918–	quantum electrodynamics, with
	Richard P. Feynman	1918–	deep-ploughing consequences for the physics of elementary particles
1966	Alfred Kastler	1902–	for the discovery and development of optical methods for studying Hertzian resonance in atoms
1967	Hans Albrecht Bethe	1906–	for his contributions to the theory of nuclear reactions, especially his discoveries concerning the energy production in stars
1968	Luis W. Alvarez	1911–	for his decisive contribution to elementary particle physics, in particular the discovery of a large number of resonance states, made possible through his development of the technique of using hydrogen bubble chamber and data analysis
1969	Murray Gell-Mann	1929–	for his contributions and discoveries concerning the classification of elementary particles and their interactions
1970	Hannes Alvén	1908–	for fundamental work and discoveries in magneto-hydrodynamics with fruitful applications in different parts of plasma physics
	Louis Néel	1904–	for fundamental work and discoveries concerning antiferromagnetism and ferrimagnetism which have led to important applications in solid state physics
1971	Dennis Gabor	1900–	for his discovery of the principles of holography
1972	John Bardeen	1908–	for their development of a theory
	Leon N. Cooper	1930–	of superconductivity
	J. Robert Schrieffer	1931–	
1973	Leo Esaki	1925–	for his discovery of tunneling in semiconductors
	Ivar Giaever	1929–	for his discovery of tunneling in superconductors
	Brian D. Josephson	1940–	for his theoretical prediction of the properties of a super-current through a tunnel barrier
1974	Antony Hewish	1924–	for the discovery of pulsars
	Sir Martin Ryle	1918–	for his pioneering work in radioastronomy
1975	Aage Bohr	1922–	for the discovery of the connec-
	Ben Mottelson	1926–	tion between collective motion
	James Rainwater	1917–	and particle motion and the development of the theory of the structure of the atomic nucleus based on this connection
1976	Burton Richter	1931–	for their (independent) discovery
	Samuel Chao Chung Ting	1936–	of an important fundamental particle.
1977	Philip Warren Anderson	1923–	for their fundamental theoretical
	Nevill Francis Mott	1905–	investigations of the electronic
	John Hasbrouck Van Vleck	1899–	structure of magnetic and disordered solids

Much of the literature of physics is written, and continues to be written, in the Gaussian system of units. In electromagnetism many equations have slightly different forms depending on whether it is intended, as in this book, that mks variables be used or that Gaussian variables be used. Equations in this book can be cast in Gaussian form by replacing the symbols listed below under "SI" by those listed under "Gaussian." For example, Eq. 37-26,

$$\mathbf{B} = \mu_0(\mathbf{H} + \mathbf{M})$$

becomes

$$\frac{\mathbf{B}}{c} = \left(\frac{4\pi}{c^2}\right)\left(\frac{c}{4\pi}\mathbf{H} + c\mathbf{M}\right)$$

or

$$\mathbf{B} = \mathbf{H} + 4\pi\mathbf{M}$$

in Gaussian form. Symbols used in this book that are not listed below remain unchanged. The quantity c is the speed of light.

Quantity	SI	Gaussian
Permittivity constant	ϵ_0	$1/4\pi$
Permeability constant	μ_0	$4\pi/c^2$
Electric displacement	$\mathbf{D}$	$\mathbf{D}/4\pi$
Magnetic induction	$\mathbf{B}$	$\mathbf{B}/c$
Magnetic flux	Φ_B	Φ_B/c
Magnetic field strength	$\mathbf{H}$	$c\mathbf{H}/4\pi$
Magnetization	$\mathbf{M}$	$c\mathbf{M}$
Magnetic dipole moment	$\boldsymbol{\mu}$	$c\boldsymbol{\mu}$

In addition to casting the equations in the proper form it is of course necessary to use a consistent set of units in those equations. Below we list some equivalent quantities in SI and Gaussian units. This table can be used to transform units from one system to the other.

Quantity	Symbol	SI	Gaussian system
Length	l	1 m	10^2 cm
Mass	m	1 kg	10^3 g
Time	t	1 s	1 s
Force	$\mathbf{F}$	1 N	10^5 dynes
Work or Energy	W, E	1 J	10^7 ergs
Power	P	1 W	10^7 ergs/s
Charge	q	1 C	3×10^9 statcoul
Current	i	1 A	3×10^9 statamp
Electric field strength	$\mathbf{E}$	1 V/m	$\frac{1}{3} \times 10^{-4}$ statvolt/cm
Electric potential	V	1 V	$\frac{1}{300}$ statvolt
Electric polarization	$\mathbf{P}$	1 C/m²	3×10^5 statcoul/cm²
Electric displacement	$\mathbf{D}$	1 C/m²	$12\pi \times 10^5$ statvolt/cm
Resistance	R	1 Ω	$\frac{1}{9} \times 10^{-11}$ s·cm⁻¹
Capacitance	C	1 F	9×10^{11} cm
Magnetic flux	Φ_B	1 Wb	10^8 maxwells
Magnetic induction	$\mathbf{B}$	1 T	10^4 gauss
Magnetic field strength	$\mathbf{H}$	1 A-turn/m	$4\pi \times 10^{-3}$ oersted
Magnetization	$\mathbf{M}$	1 Wb/m²	$1/4\pi \times 10^4$ gauss
Inductance	I	1 H	$\frac{1}{9} \times 10^{-11}$

All factors of 3 in the above table, apart from exponents, should be replaced by 2.99792458 for accurate work; this arises from the numerical value of the speed of light. For example the SI unit of capacitance

(= 1 farad) is actually $8.987551787 \times 10^{11}$ cm rather than 9 (= 3^2) $\times 10^{11}$ cm as listed above. This example also shows that not only units but also the dimensions of physical quantities may differ between the two systems.

Consult *Classical Electromagnetism*, second edition, by J. D. Jackson (John Wiley and Sons, 1975) for a fuller treatment of units and dimensions.

index to part two

SOME USEFUL NUMBERS

$\sqrt{2} = 1.414$ $\sqrt{3} = 1.732$ $\sqrt{10} = 3.162$ $\pi = 3.142$

$\pi^2 = 9.870$ $\sqrt{\pi} = 1.773$ $\log \pi = 0.4971$ $4\pi = 12.57$

$e = 2.718$ $1/e = 0.3679$ $\log e = 0.4343$ $\ln 2 = 0.6932$

$\sin 30° = \cos 60° = 0.5000$ $\cot 30° = \tan 60° = 1.7321$

$\cos 30° = \sin 60° = 0.8660$ $\sin 45° = \cos 45° = 0.7071$

$\tan 30° = \cot 60° = 0.5774$ $\tan 45° = \cot 45° = 1.0000$

Change of Base

$\log x = \ln x / \ln 10 = 0.4343 \ln x$

$\ln x = \log x / \log e = 2.303 \log x$